Routledge's
Decimal Ready Reckoner

With all computations

from $\frac{1}{2}$p to £2·00

and 40 other values of special importance

for metrication conversions

London Routledge & Kegan Paul

First published in 1969
by Routledge & Kegan Paul Limited
Broadway House, 68–74 Carter Lane
London, EC4V 5EL
Reprinted 1969, 1971
Printed in Great Britain
by C. Tinling & Co. Ltd
Prescot

ISBN 0 7100 6660 0

Introduction

This Ready Reckoner has several features which are novel, and will be found of a great help to the user. It was computed and set in type entirely by a computer, thereby eliminating the possibility of human error in computation, typesetting and proofreading.

Notes

1 At the top left corner of each page is the reciprocal of the number being multiplied. This will be found of particular use in foreign currency transactions (e.g. $2·39=£1. Therefore turn to the nearest reciprocal, and you will find $1·00=£0·42.)

Reciprocals are useful in division: e.g. to divide 47 by 25, turn to the page 25·0=£1 (£0·04), look up 47:
Answer: 1·88

2 Fractions that are commonly required are given at the bottom of the page. The superior figure is rounded off, as tenths of a penny cannot be used.

3 To make the reckoner more useful for metric conversion, common conversions have been indicated at the foot of each page, and forty numbers above 200, selected as numbers which are commonly referred to are given to help with these conversions (e.g. 353: 1 gramme=0·3527 ounces). A list of these pages may be found on page viii, together with a list of useful conversion factors.

Suggestions for changes and improvements will be welcomed by the publishers.

Useful conversion factors

1 Acre=0·4047 hectares (2·471)
1 British Thermal Unit=0·252 Calories (3·968)
1 British Thermal Unit=1·055 kilojoules (0·947)
1 British Thermal Unit=0·293 watt hours (3·41)
1 British Thermal Unit per second=1·415 horse power (0·707)
1 calorie=3·968 British Thermal Units (0·252)
1 calorie=4·186 joules (0·239)
1 Centimetre=0·3937 inches (2·54)
Centimetre: see also metre or millimetre
1 Cubic centimetre=0·0610 cubic inches (16·387)
1 Cubic foot=0·0283 cubic metres (35·3)
1 Cubic foot=28·316 litres (0·0351)
1 Cubic foot=6·23 gallons (0·1605)
1 Cubic inch=16·387 cubic centimetres (0·0610)
1 Cubic metre=35·3 cubic feet (0·0283)
1 Cubic metre=1·308 cubic yards (0·764)
1 Cubic yard=0·7645 cubic metres (1·308)
1 Cwt=50·8 kilogrammes (0·01968)
1 Cwt per acre=125·5 kilogrammes per hectare (0·007968)
1 Foot=0·3048 metres (3·28)
1 Gallon (imp.)=0·1605 cubic feet (6·23)
1 Gallon (imp)=1·201 gallons U.S. (0·8327)
1 Gallon (U.S.)=3·785 litres (0·264)
1 Gallon (imp.)=4·546 litres (0·220)
1 Gallon U.S.=0·8327 gallons (imp.) (1·201)
1 Gallon U.S. weighs 8·327 lbs.
1 Gramme=0·03527 ounces (28·3)
1 Gramme per litre=0·1603 ounces per gallon (imp.) (6·236)
1 Gramme per square metre=0·0295 ounces per square yard (33·9)
1 Hectare=2·471 acres (0·4047)
1 Horse power=0·707 British Thermal Units per second (1·415)
1 Horse power=746 watts
1 Inch=25·4 millimetres (0·03937)
1 Joule=0·239 calories (4·18)
1 Kilogramme=2·204 lbs. (0·454)
1 Kilogramme per hectare=0·007968 cwt. per acre (125·5)
1 Kilogramme per hectare=0·0003984 tons per acre (2510)
1 Kilogramme per hectare=0·892 lbs. per acre (1·12)
1 Kilogramme per sq. cm.=14·22 lbs. per square inch (0·0703)
1 Kilojoule=0·947 British Thermal Units (1·055)
1 Kilojoule=0·278 watt hours (3600)
1 Kilometre=0·6213 miles (1·609)

Useful conversion factors

1 Kilometre per litre=2·825 miles per gallon (imp.) (0·354)
1 Kilometre per litre=2·352 miles per gallon U.S. (0·425)
1 Kilowatt=0·001344 horse power (746)
lb. see pound
1 Litre=0·0353 cubic feet (28·316)
1 Litre=0·220 gallons (imp.) (4·546)
1 Litre=0·2642 U.S. gallons (3·785)
1 Litre=1·76 pints (imp.) (0·568)
1 Litre=2·113 U.S. pints (0·473)
1 Metre=3·281 feet (0·3048)
1 Metre=39·37 inches (0·0254)
1 Metre=1·0936 yards (0·9144)
1 Millimetre=0·03937 inches (25·4)
1 Mile=1·609 kilometres (0·6213)
1 Mile per gallon (imp.)=0·354 kilometres per litre (2·825)
1 Mile per gallon U.S.=0·425 kilometres per litre (2·352)
1 Ounce=28·3 grammes (0·03527)
1 Ounce per gallon=6·236 grammes per litre (0·1604)
1 Ounce per square yard=33·9 grammes per sq. metre (0·0295)
1 Pica point=0·351 millimetres
1 Pint=0·568 litres (1·76)
1 Pound=0·454 kilogrammes (2·204)
1 lb. per acre=1·122 kilogrammes per hectare (0·8919)
1 lb. per square inch=0·0703 kilogrammes per sq. centimetre (14·22)
1 Square cm.=0·155 square inches (6·45)
1 Square foot=929 square centimetres (0·001076)
1 Square foot=0·0929 square metres (10·76)
1 Square inch=6·45 square centimetres (0·155)
1 Square metre=10·76 square feet (0·0929)
1 Square kilometre=0·3861 square miles (2·590)
1 Square metre=1·196 square yards (0·836)
1 Square mile=2·590 sq. kilometres (0·3861)
1 Square yard=0·836 square metres (1·196)
1 Ton=1·01605 (0·984)
1 Ton per acre=2510 kilogrammes per hectare (0·0003984)
1 Tonne=2204 lbs.
1 Tonne=0·984 tons (1·0165)
1 Watt hour=3·412 British Thermal Units (0·293)
1 Watt hour=3·60 kilojoules (278)
1 Yard=0·9144 metres (1·0936)

Conversion Table from £ s. d. to £p. using ½p

The shopper's table

1d	½p	4s 1d	20½p	8s 1d	40½p	12s 1d	60½p	16s 1d	80½p
2d	1p	4s 2d	21p	8s 2d	41p	12s 2d	61p	16s 2d	81p
3d	1p	4s 3d	21p	8s 3d	41p	12s 3d	61p	16s 3d	81p
4d	1½p	4s 4d	21½p	8s 4d	41½p	12s 4d	61½p	16s 4d	81½p
5d	2p	4s 5d	22p	8s 5d	42p	12s 5d	62p	16s 5d	82p
6d	**2½p**	**4s 6d**	**22½p**	**8s 6d**	**42½p**	**12s 6d**	**62½p**	**16s 6d**	**82½p**
7d	3p	4s 7d	23p	8s 7d	43p	12s 7d	63p	16s 7d	83p
8d	3½p	4s 8d	23½p	8s 8d	43½p	12s 8d	63½p	16s 8d	83½p
9d	4p	4s 9d	24p	8s 9d	44p	12s 9d	64p	16s 9d	84p
10d	4p	4s 10d	24p	8s 10d	44p	12s 10d	64p	16s 10d	84p
11d	4½p	4s 11d	24½p	8s 11d	44½p	12s 11d	64½p	16s 11d	84½p
1s	**5p**	**5s**	**25p**	**9s**	**45p**	**13s**	**65p**	**17s**	**85p**
1s 1d	5½	5s 1d	25½p	9s 1d	45½p	13s 1d	65½p	17s 1d	85½p
1s 2d	6p	5s 2d	26p	9s 2d	46p	13s 2d	66p	17s 2d	86p
1s 3d	6p	5s 3d	26p	9s 3d	46p	13s 3d	66p	17s 3d	86p
1s 4d	6½p	5s 4d	26½p	9s 4d	46½p	13s 4d	66½p	17s 4d	86½p
1s 5d	7p	5s 5d	27p	9s 5d	47p	13s 5d	67p	17s 5d	87p
1s 6d	**7½p**	**5s 6d**	**27½p**	**9s 6d**	**47½p**	**13s 6d**	**67½p**	**17s 6d**	**87½p**
1s 7d	8p	5s 7d	28p	9s 7d	48p	13s 7d	68p	17s 7d	88p
1s 8d	8½p	5s 8d	28½p	9s 8d	48½p	13s 8d	68½p	17s 8d	88½p
1s 9d	9p	5s 9d	29p	9s 9d	49p	13s 9d	69p	17s 9d	89P
1s 10d	9p	5s 10d	29p	9s 10d	49p	13s 10d	69p	17s 10d	89p
1s 11d	9½p	5s 11d	29½p	9s 11d	49½p	13s 11d	69½p	17s 11d	89½p
2s	**10p**	**6s**	**30p**	**10s**	**50p**	**14s**	**70p**	**18s**	**90p**
2s 1d	10½p	6s 1d	30½p	10s 1d	50½p	14s 1d	70½p	18s 1d	90½p
2s 2d	11p	6s 2d	31p	10s 2d	51p	14s 2d	71p	18s 2d	91p
2s 3d	11p	6s 3d	31p	10s 3d	51p	14s 3d	71p	18s 3d	91p
2s 4d	11½p	6s 4d	31½p	10s 4d	51½p	14s 4d	71½p	18s 4d	91½p
2s 5d	12p	6s 5d	32p	10s 5d	52p	14s 5d	72p	18s 5d	92p
2s 6d	**12½p**	**6s 6d**	**32½p**	**10s 6d**	**52½p**	**14s 6d**	**72½p**	**18s 6d**	**92½p**
2s 7d	13p	6s 7d	33p	10s 7d	53p	14s 7d	73p	18s 7d	93p
2s 8d	13½p	6s 8d	33½p	10s 8d	53½p	14s 8d	73½p	18s 8d	93½p
2s 9d	14p	6s 9d	34p	10s 9d	54p	14s 9d	74p	18s 9d	94p
2s 10d	14p	6s 10d	34p	10s 10d	54p	14s 10d	74p	18s 10d	94p
2s 11d	14½p	6s 11d	34½p	10s 11d	54½p	14s 11d	74½p	18s 11d	94½p
3s	**15p**	**7s**	**35p**	**11s**	**55p**	**15s**	**75p**	**19s**	**95p**
3s 1d	15½p	7s 1d	35½p	11s 1d	55½p	15s 1d	75½p	19s 1d	95½p
3s 2d	16p	7s 2d	36p	11s 2d	56p	15s 2d	76p	19s 2d	96p
3s 3d	16p	7s 3d	36p	11s 3d	56p	15s 3d	76p	19s 3d	96p
3s 4d	16½p	7s 4d	36½p	11s 4d	56½p	15s 4d	76½p	19s 4d	96½p
3s 5d	17p	7s 5d	37p	11s 5d	57p	15s 5d	77p	19s 5d	97p
3s 6d	**17½p**	**7s 6d**	**37½p**	**11s 6d**	**57½p**	**15s 6d**	**77½p**	**19s 6d**	**97½p**
3s 7d	18p	7s 7d	38p	11s 7d	58p	15s 7d	78p	19s 7d	98p
3s 8d	18½p	7s 8d	38½p	11s 8d	58½p	15s 8d	78½p	19s 8d	98½p
3s 9d	19p	7s 9d	39p	11s 9d	59p	15s 9d	79p	19s 9d	99p
3s 10d	19p	7s 10d	39p	11s 10d	59p	15s 10d	79p	19s 10d	99p
3s 11d	19½p	7s 11d	39½p	11s 11d	59½p	15s 11d	79½p	19s 11d	99½p
4s	**20p**	**8s**	**40p**	**12s**	**60p**	**16s**	**80p**	**20s**	**£1**

Conversion Table from £ s. d. to £p using whole New Pence only

The banker's table

1d	—	4s 1d	20p	8s 1d	40p	12s 1d	60p	16s 1d	80p
2d	1p	4s 2d	21p	8s 2d	41p	12s 2d	61p	16s 2d	81p
3d	1p	4s 3d	21p	8s 3d	41p	12s 3d	61p	16s 3d	81p
4d	2p	4s 4d	22p	8s 4d	42p	12s 4d	62p	16s 4d	82p
5d	2p	4s 5d	22p	8s 5d	42p	12s 5d	62p	16s 5d	82p
6d	3p	4s 6d	23p	8s 6d	43p	12s 6d	63p	16s 6d	83p
7d	3p	4s 7d	23p	8s 7d	43p	12s 7d	63p	16s 7d	83p
8d	3p	4s 8d	23p	8s 8d	43p	12s 8d	63p	16s 8d	83p
9d	4p	4s 9d	24p	8s 9d	44p	12s 9d	64p	16s 9d	84p
10d	4p	4s 10d	24p	8s 10d	44p	12s 10d	64p	16s 10d	84p
11d	5p	4s 11d	25p	8s 11d	45p	12s 11d	65p	16s 11d	85p
1s	**5p**	**5s**	**25p**	**9s**	**45p**	**13s**	**65p**	**17s**	**85p**
1s 1d	5p	5s 1d	25p	9s 1d	45p	13s 1d	65p	17s 1d	85p
1s 2d	6p	5s 2d	26p	9s 2d	46p	13s 2d	66p	17s 2d	86p
1s 3d	6p	5s 3d	26p	9s 3d	46p	13s 3d	66p	17s 3d	86p
1s 4d	7p	5s 4d	27p	9s 4d	47p	13s 4d	67p	17s 4d	87p
1s 5d	7p	5s 5d	27p	9s 5d	47p	13s 5d	67p	17s 5d	87p
1s 6d	7p	5s 6d	27p	9s 6d	47p	13s 6d	67p	17s 6d	87p
1s 7d	8p	5s 7d	28p	9s 7d	48p	13s 7d	68p	17s 7d	88p
1s 8d	8p	5s 8d	28p	9s 8d	48p	13s 8d	68p	17s 8d	88p
1s 9d	9p	5s 9d	29p	9s 9d	49p	13s 9d	69p	17s 9d	89p
1s 10d	9p	5s 10d	29p	9s 10d	49p	13s 10d	69p	17s 10d	89p
1s 11d	10p	5s 11d	30p	9s 11d	50p	13s 11d	70p	17s 11d	90p
2s	**10p**	**6s**	**30p**	**10s**	**50p**	**14s**	**70p**	**18s**	**90p**
2s 1d	10p	6s 1d	30p	10s 1d	50p	14s 1d	70p	18s 1d	90p
2s 2d	11p	6s 2d	31p	10s 2d	51p	14s 2d	71p	18s 2d	91p
2s 3d	11p	6s 3d	31p	10s 3d	51p	14s 3d	71p	18s 3d	91p
2s 4d	12p	6s 4d	32p	10s 4d	52p	14s 4d	72p	18s 4d	92p
2s 5d	12p	6s 5d	32p	10s 5d	52p	14s 5d	72p	18s 5d	92p
2s 6d	13p	6s 6d	33p	10s 6d	53p	14s 6d	73p	18s 6d	93p
2s 7d	13p	6s 7d	33p	10s 7d	53p	14s 7d	73p	18s 7d	93p
2s 8d	13p	6s 8d	33p	10s 8d	53p	14s 8d	73p	18s 8d	93p
2s 9d	14p	6s 9d	34p	10s 9d	54p	14s 9d	74p	18s 9d	94p
2s 10d	14p	6s 10d	34p	10s 10d	54p	14s 10d	74p	18s 10d	94p
2s 11d	15p	6s 11d	35p	10s 11d	55p	14s 11d	75p	18s 11d	95p
3s	**15p**	**7s**	**35p**	**11s**	**55p**	**15s**	**75p**	**19s**	**95p**
3s 1d	15p	7s 1d	35p	11s 1d	55p	15s 1d	75p	19s 1d	95p
3s 2d	16p	7s 2d	36p	11s 2d	56p	15s 2d	76p	10s 2d	96p
3s 3d	16p	7s 3d	36p	11s 3d	56p	15s 3d	76p	19s 3d	96p
3s 4d	17p	7s 4d	37p	11s 4d	57p	15s 4d	77p	19s 4d	97p
3s 5d	17p	7s 5d	37p	11s 5d	57p	15s 5d	77p	19s 5d	97p
3s 6d	17p	7s 6d	37p	11s 6d	57p	15s 6d	77p	19s 6d	97p
3s 7d	18p	7s 7d	38p	11s 7d	58p	15s 7d	78p	19s 7d	98p
3s 8d	18p	7s 8d	38p	11s 8d	58p	15s 8d	78p	19s 8d	98p
3s 9d	19p	7s 9d	39p	11s 9d	59p	15s 9d	79p	19s 9d	99p
3s 10d	19p	7s 10d	39p	11s 10d	59p	15s 10d	79p	19s 10d	99p
3s 11d	20p	7s 11d	40p	11s 11d	60p	15s 11d	80p	19s 11d	100p
4s	**20p**	**8s**	**40p**	**12s**	**60p**	**16s**	**80p**	**20s**	**£1**

40 extra pages are included for the following ratios

224(0)	lbs. per ton
239	sq. km.—sq. m.
2·47	acres—hectare
252	cals.—BThU.
254	mm.—inch
261	days per year (5 day week)
271	acre foot
282	km. per litre—miles per gallon
28·3	gr.—oz.
293	Kwh.—BThU.
·305	metre—ft.
313	days per year (6 day week)
3.14	π
3·22	g. ft.—sec^2.
3·28	ft.—metre
341	BThU—kwh.
·351	mm.—point pica
(·0)353	oz.—gr.
3·54	mpg.—km. per litre
365	days in 1 year
366	days in 1 leap year
39·4	ins.—metre
397	Kcal.—BThU.
·405	acres—hectare
·568	litres—pint
(·0)610	cu. inch—cc.
(0·)621	m.—km.
6·23	galls.—cu. ft.
62·5	oz./lb.—gm./kg.
703	lbs.—sq. in./Kgs.—sq. cm.
746	w/hp.
765	cu. m.—cu. yd.
836	sq. m.—sq. yd.
·914	m.—yd.
929	sq. m.—sq. ft.
981	g. cm./sec^2. (acceleration due to gravity)
·984	tons—tonne
41$\frac{1}{4}$	standard rate of income tax

Decimal parts of £1

	0d	1d	2d	3d	4d	5d	6d	7d	8d	9d	10d	11d	
0/–	—	·0042	·0083	·0125	·0167	·0208	·0250	·0292	·0333	·0375	·0417	·0458	0/–
1/–	·0500	·0542	·0583	·0625	·0667	·0708	·0750	·0792	·0833	·0875	·0917	·0958	1/–
2/–	·1000	·1042	·1083	·1125	·1167	·1208	·1250	·1292	·1333	·1375	·1417	·1458	2/–
3/–	·1500	·1542	·1583	·1625	·1667	·1708	·1750	·1792	·1833	·1875	·1917	·1958	3/–
4/–	·2000	·2042	·2083	·2125	·2167	·2208	·2250	·2292	·2333	·2375	·2417	·2458	4/–
5/–	·2500	·2542	·2583	·2625	·2667	·2708	·2750	·2792	·2833	·2875	·2917	·2958	5/–
6/–	·3000	·3042	·3083	·3125	·3167	·3208	·3250	·3292	·3333	·3375	·3417	·3458	6/–
7/–	·3500	·3542	·3583	·3625	·3667	·3708	·3750	·3792	·3833	·3875	·3917	·3958	7/–
8/–	·4000	·4042	·4083	·4125	·4167	·4208	·4250	·4292	·4333	·4375	·4417	·4458	8/–
9/–	·4500	·4542	·4583	·4625	·4667	·4708	·4750	·4792	·4833	·4875	·4917	·4958	9/–
10/–	·5000	·5042	·5033	·5125	·5167	·5208	·5250	·5292	·5333	·5375	·5417	·5458	10/–
11/–	·5500	·5542	·5583	·5625	·5667	·5708	·5750	·5792	·5833	·5875	·5917	·5958	11/–
12/–	·6000	·6042	·6083	·6125	·6167	·6208	·6250	·6292	·6333	·6375	·6417	·6458	12/–
13/–	·6500	·6542	·6583	·6625	·6667	·6708	·6750	·6792	·6833	·6875	·6917	·6958	13/–
14/–	·7000	·7042	·7083	·7125	·7167	·7208	·7250	·7292	·7333	·7375	·7417	·7458	14/–
15/–	·7500	·7542	·7583	·7625	·7667	·7708	·7750	·7792	·7833	·7875	·7917	·7958	15/–
16/–	·8000	·8042	·8083	·8125	·8167	·8208	·8250	·8292	·8333	·8375	·8417	·8458	16/–
17/–	·8500	·8542	·8583	·8625	·8667	·8708	·8750	·8792	·8833	·8875	·8917	·8958	17/–
18/–	·9000	·9042	·9083	·9125	·9167	·9208	·9250	·9292	·9333	·9375	·9417	·9458	18/–
19/–	·9500	·9542	·9583	·9625	·9667	·9708	·9750	·9792	·9833	·9875	·9917	·9958	19/–
	0d	1d	2d	3d	4d	5d	6d	7d	8d	9d	10d	11d	

(200.0=1£) **£0.00½**

1	0·00½	41	0·20½	81	0·40½	121	0·60½	161	0·80½
2	0·01	42	0·21	82	0·41	122	0·61	162	0·81
3	0·01½	43	0·21½	83	0·41½	123	0·61½	163	0·81½
4	0·02	44	0·22	84	0·42	124	0·62	164	0·82
5	0·02½	45	0·22½	85	0·42½	125	0·62½	165	0·82½
6	0·03	46	0·23	86	0·43	126	0·63	166	0·83
7	0·03½	47	0·23½	87	0·43½	127	0·63½	167	0·83½
8	0·04	48	0·24	88	0·44	128	0·64	168	0·84
9	0·04½	49	0·24½	89	0·44½	129	0·64½	169	0·84½
10	0·05	**50**	0·25	**90**	0·45	**130**	0·65	**170**	0·85
11	0·05½	51	0·25½	91	0·45½	131	0·65½	171	0·85½
12	0·06	52	0·26	92	0·46	132	0·66	172	0·86
13	0·06½	53	0·26½	93	0·46½	133	0·66½	173	0·86½
14	0·07	54	0·27	94	0·47	134	0·67	174	0·87
15	0·07½	55	0·27½	95	0·47½	135	0·67½	175	0·87½
16	0·08	56	0·28	96	0·48	136	0·68	176	0·88
17	0·08½	57	0·28½	97	0·48½	137	0·68½	177	0·88½
18	0·09	58	0·29	98	0·49	138	0·69	178	0·89
19	0·09½	59	0·29½	99	0·49½	139	0·69½	179	0·89½
20	0·10	**60**	0·30	**100**	0·50	**140**	0·70	**180**	0·90
21	0·10½	61	0·30½	101	0·50½	141	0·70½	181	0·90½
22	0·11	62	0·31	102	0·51	142	0·71	182	0·91
23	0·11½	63	0·31½	103	0·51½	143	0·71½	183	0·91½
24	0·12	64	0·32	104	0·52	144	0·72	184	0·92
25	0·12½	65	0·32½	105	0·52½	145	0·72½	185	0·92½
26	0·13	66	0·33	106	0·53	146	0·73	186	0·93
27	0·13½	67	0·33½	107	0·53½	147	0·73½	187	0·93½
28	0·14	68	0·34	108	0·54	148	0·74	188	0·94
29	0·14½	69	0·34½	109	0·54½	149	0·74½	189	0·94½
30	0·15	**70**	0·35	**110**	0·55	**150**	0·75	**190**	0·95
31	0·15½	71	0·35½	111	0·55½	151	0·75½	191	0·95½
32	0·16	72	0·36	112	0·56	152	0·76	192	0·96
33	0·16½	73	0·36½	113	0·56½	153	0·76½	193	0·96½
34	0·17	74	0·37	114	0·57	154	0·77	194	0·97
35	0·17½	75	0·37½	115	0·57½	155	0·77½	195	0·97½
36	0·18	76	0·38	116	0·58	156	0·78	196	0·98
37	0·18½	77	0·38½	117	0·58½	157	0·78½	197	0·98½
38	0·19	78	0·39	118	0·59	158	0·79	198	0·99
39	0·19½	79	0·39½	119	0·59½	159	0·79½	199	0·99½
40	0·20	**80**	0·40	**120**	0·60	**160**	0·80	**200**	1·00

$\frac{1}{16}=0{\cdot}00^0$	$\frac{1}{9}=0{\cdot}00^1$	$\frac{1}{8}=0{\cdot}00^1$	$\frac{1}{6}=0{\cdot}00^1$	$\frac{1}{5}=0{\cdot}00^1$	$\frac{1}{4}=0{\cdot}00^1$	$\frac{1}{3}=0{\cdot}00^2$
$\frac{3}{8}=0{\cdot}00^2$	$\frac{1}{2}=0{\cdot}00^3$	$\frac{5}{8}=0{\cdot}00^3$	$\frac{3}{4}=0{\cdot}00^4$	$\frac{7}{8}=0{\cdot}00^4$	$\frac{2}{3}=0{\cdot}00^3$	$\frac{5}{6}=0{\cdot}00^4$
$\frac{3}{16}=0{\cdot}00^1$	$\frac{5}{16}=0{\cdot}00^2$	$\frac{7}{16}=0{\cdot}00^2$	$\frac{9}{16}=0{\cdot}00^3$	$\frac{11}{16}=0{\cdot}00^3$	$\frac{13}{16}=0{\cdot}00^4$	$\frac{15}{16}=0{\cdot}00^5$

(100.0=1£) **£0.01**

1	0·01	41	0·41	81	0·81	121	1·21	161	1·61
2	0·02	42	0·42	82	0·82	122	1·22	162	1·62
3	0·03	43	0·43	83	0·83	123	1·23	163	1·63
4	0·04	44	0·44	84	0·84	124	1·24	164	1·64
5	0·05	45	0·45	85	0·85	125	1·25	165	1·65
6	0·06	46	0·46	86	0·86	126	1·26	166	1·66
7	0·07	47	0·47	87	0·87	127	1·27	167	1·67
8	0·08	48	0·48	88	0·88	128	1·28	168	1·68
9	0·09	49	0·49	89	0·89	129	1·29	169	1·69
10	0·10	**50**	0·50	**90**	0·90	**130**	1·30	**170**	1·70
11	0·11	51	0·51	91	0·91	131	1·31	171	1·71
12	0·12	52	0·52	92	0·92	132	1·32	172	1·72
13	0·13	53	0·53	93	0·93	133	1·33	173	1·73
14	0·14	54	0·54	94	0·94	134	1·34	174	1·74
15	0·15	55	0·55	95	0·95	135	1·35	175	1·75
16	0·16	56	0·56	96	0·96	136	1·36	176	1·76
17	0·17	57	0·57	97	0·97	137	1·37	177	1·77
18	0·18	58	0·58	98	0·98	138	1·38	178	1·78
19	0·19	59	0·59	99	0·99	139	1·39	179	1·79
20	0·20	**60**	0·60	**100**	1·00	**140**	1·40	**180**	1·80
21	0·21	61	0·61	101	1·01	141	1·41	181	1·81
22	0·22	62	0·62	102	1·02	142	1·42	182	1·82
23	0·23	63	0·63	103	1·03	143	1·43	183	1·83
24	0·24	64	0·64	104	1·04	144	1·44	184	1·84
25	0·25	65	0·65	105	1·05	145	1·45	185	1·85
26	0·26	66	0·66	106	1·06	146	1·46	186	1·86
27	0·27	67	0·67	107	1·07	147	1·47	187	1·87
28	0·28	68	0·68	108	1·08	148	1·48	188	1·88
29	0·29	69	0·69	109	1·09	149	1·49	189	1·89
30	0·30	**70**	0·70	**110**	1·10	**150**	1·50	**190**	1·90
31	0·31	71	0·71	111	1·11	151	1·51	191	1·91
32	0·32	72	0·72	112	1·12	152	1·52	192	1·92
33	0·33	73	0·73	113	1·13	153	1·53	193	1·93
34	0·34	74	0·74	114	1·14	154	1·54	194	1·94
35	0·35	75	0·75	115	1·15	155	1·55	195	1·95
36	0·36	76	0·76	116	1·16	156	1·56	196	1·96
37	0·37	77	0·77	117	1·17	157	1·57	197	1·97
38	0·38	78	0·78	118	1·18	158	1·58	198	1·98
39	0·39	79	0·79	119	1·19	159	1·59	199	1·99
40	0·40	**80**	0·80	**120**	1·20	**160**	1·60	**200**	2·00

$\frac{1}{16}=0{\cdot}00^{1}$ $\frac{1}{9}=0{\cdot}00^{1}$ $\frac{1}{8}=0{\cdot}00^{1}$ $\frac{1}{6}=0{\cdot}00^{2}$ $\frac{1}{5}=0{\cdot}00^{2}$ $\frac{1}{4}=0{\cdot}00^{3}$ $\frac{1}{3}=0{\cdot}00^{3}$

$\frac{3}{8}=0{\cdot}00^{4}$ $\frac{1}{2}=0{\cdot}00^{5}$ $\frac{5}{8}=0{\cdot}00^{6}$ $\frac{3}{4}=0{\cdot}00^{8}$ $\frac{7}{8}=0{\cdot}00^{9}$ $\frac{2}{3}=0{\cdot}00^{7}$ $\frac{5}{6}=0{\cdot}00^{8}$

$\frac{3}{16}=0{\cdot}00^{2}$ $\frac{5}{16}=0{\cdot}00^{3}$ $\frac{7}{16}=0{\cdot}00^{4}$ $\frac{9}{16}=0{\cdot}00^{6}$ $\frac{11}{16}=0{\cdot}00^{7}$ $\frac{13}{16}=0{\cdot}00^{8}$ $\frac{15}{16}=0{\cdot}00^{9}$

1	0·01½	41	0·61½	81	1·21½	121	1·81½	161	2·41½
2	0·03	42	0·63	82	1·23	122	1·83	162	2·43
3	0·04½	43	0·64½	83	1·24½	123	1·84½	163	2·44½
4	0·06	44	0·66	84	1·26	124	1·86	164	2·46
5	0·07½	45	0·67½	85	1·27½	125	1·87½	165	2·47½
6	0·09	46	0·69	86	1·29	126	1·89	166	2·49
7	0·10½	47	0·70½	87	1·30½	127	1·90½	167	2·50½
8	0·12	48	0·72	88	1·32	128	1·92	168	2·52
9	0·13½	49	0·73½	89	1·33½	129	1·93½	169	2·53½
10	0·15	**50**	0·75	**90**	1·35	**130**	1·95	**170**	2·55
11	0·16½	51	0·76½	91	1·36½	131	1·96½	171	2·56½
12	0·18	52	0·78	92	1·38	132	1·98	172	2·58
13	0·19½	53	0·79½	93	1·39½	133	1·99½	173	2·59½
14	0·21	54	0·81	94	1·41	134	2·01	174	2·61
15	0·22½	55	0·82½	95	1·42½	135	2·02½	175	2·62½
16	0·24	56	0·84	96	1·44	136	2·04	176	2·64
17	0·25½	57	0·85½	97	1·45½	137	2·05½	177	2·65½
18	0·27	58	0·87	98	1·47	138	2·07	178	2·67
19	0·28½	59	0·88½	99	1·48½	139	2·08½	179	2·68½
20	0·30	**60**	0·90	**100**	1·50	**140**	2·10	**180**	2·70
21	0·31½	61	0·91½	101	1·51½	141	2·11½	181	2·71½
22	0·33	62	0·93	102	1·53	142	2·13	182	2·73
23	0·34½	63	0·94½	103	1·54½	143	2·14½	183	2·74½
24	0·36	64	0·96	104	1·56	144	2·16	184	2·76
25	0·37½	65	0·97½	105	1·57½	145	2·17½	185	2·77½
26	0·39	66	0·99	106	1·59	146	2·19	186	2·79
27	0·40½	67	1·00½	107	1·60½	147	2·20½	187	2·80½
28	0·42	68	1·02	108	1·62	148	2·22	188	2·82
29	0·43½	69	1·03½	109	1·63½	149	2·23½	189	2·83½
30	0·45	**70**	1·05	**110**	1·65	**150**	2·25	**190**	2·85
31	0·46½	71	1·06½	111	1·66½	151	2·26½	191	2·86½
32	0·48	72	1·08	112	1·68	152	2·28	192	2·88
33	0·49½	73	1·09½	113	1·69½	153	2·29½	193	2·89½
34	0·51	74	1·11	114	1·71	154	2·31	194	2·91
35	0·52½	75	1·12½	115	1·72½	155	2·32½	195	2·92½
36	0·54	76	1·14	116	1·74	156	2·34	196	2·94
37	0·55½	77	1·15½	117	1·75½	157	2·35½	197	2·95½
38	0·57	78	1·17	118	1·77	158	2·37	198	2·97
39	0·58½	79	1·18½	119	1·78½	159	2·38½	199	2·98½
40	0·60	**80**	1·20	**120**	1·80	**160**	2·40	**200**	3·00

$\frac{1}{16}=0{\cdot}00^{1}$ $\frac{1}{9}=0{\cdot}00^{2}$ $\frac{1}{8}=0{\cdot}00^{2}$ $\frac{1}{6}=0{\cdot}00^{3}$ $\frac{1}{5}=0{\cdot}00^{3}$ $\frac{1}{4}=0{\cdot}00^{4}$ $\frac{1}{3}=0{\cdot}00^{5}$

$\frac{3}{8}=0{\cdot}00^{6}$ $\frac{1}{2}=0{\cdot}00^{8}$ $\frac{5}{8}=0{\cdot}00^{9}$ $\frac{3}{4}=0{\cdot}01^{1}$ $\frac{7}{8}=0{\cdot}01^{3}$ $\frac{2}{3}=0{\cdot}01^{0}$ $\frac{5}{6}=0{\cdot}01^{3}$

$\frac{3}{16}=0{\cdot}00^{3}$ $\frac{5}{16}=0{\cdot}00^{5}$ $\frac{7}{16}=0{\cdot}00^{7}$ $\frac{9}{16}=0{\cdot}00^{8}$ $\frac{11}{16}=0{\cdot}01^{0}$ $\frac{13}{16}=0{\cdot}01^{2}$ $\frac{15}{16}=0{\cdot}01^{4}$

(50.0=1£) **£0.02**

1	0·02	41	0·82	81	1·62	121	2·42	161	3·22
2	0·04	42	0·84	82	1·64	122	2·44	162	3·24
3	0·06	43	0·86	83	1·66	123	2·46	163	3·26
4	0·08	44	0·88	84	1·68	124	2·48	164	3·28
5	0·10	45	0·90	85	1·70	125	2·50	165	3·30
6	0·12	46	0·92	86	1·72	126	2·52	166	3·32
7	0·14	47	0·94	87	1·74	127	2·54	167	3·34
8	0·16	48	0·96	88	1·76	128	2·56	168	3·36
9	0·18	49	0·98	89	1·78	129	2·58	169	3·38
10	0·20	**50**	1·00	**90**	1·80	**130**	2·60	**170**	3·40
11	0·22	51	1·02	91	1·82	131	2·62	171	3·42
12	0·24	52	1·04	92	1·84	132	2·64	172	3·44
13	0·26	53	1·06	93	1·86	133	2·66	173	3·46
14	0·28	54	1·08	94	1·88	134	2·68	174	3·48
15	0·30	55	1·10	95	1·90	135	2·70	175	3·50
16	0·32	56	1·12	96	1·92	136	2·72	176	3·52
17	0·34	57	1·14	97	1·94	137	2·74	177	3·54
18	0·36	58	1·16	98	1·96	138	2·76	178	3·56
19	0·38	59	1·18	99	1·98	139	2·78	179	3·58
20	0·40	**60**	1·20	**100**	2·00	**140**	2·80	**180**	3·60
21	0·42	61	1·22	101	2·02	141	2·82	181	3·62
22	0·44	62	1·24	102	2·04	142	2·84	182	3·64
23	0·46	63	1·26	103	2·06	143	2·86	183	3·66
24	0·48	64	1·28	104	2·08	144	2·88	184	3·68
25	0·50	65	1·30	105	2·10	145	2·90	185	3·70
26	0·52	66	1·32	106	2·12	146	2·92	186	3·72
27	0·54	67	1·34	107	2·14	147	2·94	187	3·74
28	0·56	68	1·36	108	2·16	148	2·96	188	3·76
29	0·58	69	1·38	109	2·18	149	2·98	189	3·78
30	0·60	**70**	1·40	**110**	2·20	**150**	3·00	**190**	3·80
31	0·62	71	1·42	111	2·22	151	3·02	191	3·82
32	0·64	72	1·44	112	2·24	152	3·04	192	3·84
33	0·66	73	1·46	113	2·26	153	3·06	193	3·86
34	0·68	74	1·48	114	2·28	154	3·08	194	3·88
35	0·70	75	1·50	115	2·30	155	3·10	195	3·90
36	0·72	76	1·52	116	2·32	156	3·12	196	3·92
37	0·74	77	1·54	117	2·34	157	3·14	197	3·94
38	0·76	78	1·56	118	2·36	158	3·16	198	3·96
39	0·78	79	1·58	119	2·38	159	3·18	199	3·98
40	0·80	**80**	1·60	**120**	2·40	**160**	3·20	**200**	4·00

$\frac{1}{16}=0{\cdot}00^{1}$ $\frac{1}{9}=0{\cdot}00^{2}$ $\frac{1}{8}=0{\cdot}00^{3}$ $\frac{1}{6}=0{\cdot}00^{3}$ $\frac{1}{5}=0{\cdot}00^{4}$ $\frac{1}{4}=0{\cdot}00^{5}$ $\frac{1}{3}=0{\cdot}00^{7}$

$\frac{3}{8}=0{\cdot}00^{8}$ $\frac{1}{2}=0{\cdot}01^{0}$ $\frac{5}{8}=0{\cdot}01^{3}$ $\frac{3}{4}=0{\cdot}01^{5}$ $\frac{7}{8}=0{\cdot}01^{8}$ $\frac{2}{3}=0{\cdot}01^{3}$ $\frac{5}{6}=0{\cdot}01^{7}$

$\frac{3}{16}=0{\cdot}00^{4}$ $\frac{5}{16}=0{\cdot}00^{6}$ $\frac{7}{16}=0{\cdot}00^{9}$ $\frac{9}{16}=0{\cdot}01^{1}$ $\frac{11}{16}=0{\cdot}01^{4}$ $\frac{13}{16}=0{\cdot}01^{6}$ $\frac{15}{16}=0{\cdot}01^{9}$

1	0·02½	41	1·02½	81	2·02½	121	3·02½	161	4·02½
2	0·05	42	1·05	82	2·05	122	3·05	162	4·05
3	0·07½	43	1·07½	83	2·07½	123	3·07½	163	4·07½
4	0·10	44	1·10	84	2·10	124	3·10	164	4·10
5	0·12½	45	1·12½	85	2·12½	125	3·12½	165	4·12½
6	0·15	46	1·15	86	2·15	126	3·15	166	4·15
7	0·17½	47	1·17½	87	2·17½	127	3·17½	167	4·17½
8	0·20	48	1·20	88	2·20	128	3·20	168	4·20
9	0·22½	49	1·22½	89	2·22½	129	3·22½	169	4·22½
10	0·25	**50**	1·25	**90**	2·25	**130**	3·25	**170**	4·25
11	0·27½	51	1·27½	91	2·27½	131	3·27½	171	4·27½
12	0·30	52	1·30	92	2·30	132	3·30	172	4·30
13	0·32½	53	1·32½	93	2·32½	133	3·32½	173	4·32½
14	0·35	54	1·35	94	2·35	134	3·35	174	4·35
15	0·37½	55	1·37½	95	2·37½	135	3·37½	175	4·37½
16	0·40	56	1·40	96	2·40	136	3·40	176	4·40
17	0·42½	57	1·42½	97	2·42½	137	3·42½	177	4·42½
18	0·45	58	1·45	98	2·45	138	3·45	178	4·45
19	0·47½	59	1·47½	99	2·47½	139	3·47½	179	4·47½
20	0·50	**60**	1·50	**100**	2·50	**140**	3·50	**180**	4·50
21	0·52½	61	1·52½	101	2·52½	141	3·52½	181	4·52½
22	0·55	62	1·55	102	2·55	142	3·55	182	4·55
23	0·57½	63	1·57½	103	2·57½	143	3·57½	183	4·57½
24	0·60	64	1·60	104	2·60	144	3·60	184	4·60
25	0·62½	65	1·62½	105	2·62½	145	3·62½	185	4·62½
26	0·65	66	1·65	106	2·65	146	3·65	186	4·65
27	0·67½	67	1·67½	107	2·67½	147	3·67½	187	4·67½
28	0·70	68	1·70	108	2·70	148	3·70	188	4·70
29	0·72½	69	1·72½	109	2·72½	149	3·72½	189	4·72½
30	0·75	**70**	1·75	**110**	2·75	**150**	3·75	**190**	4·75
31	0·77½	71	1·77½	111	2·77½	151	3·77½	191	4·77½
32	0·80	72	1·80	112	2·80	152	3·80	192	4·80
33	0·82½	73	1·82½	113	2·82½	153	3·82½	193	4·82½
34	0·85	74	1·85	114	2·85	154	3·85	194	4·85
35	0·87½	75	1·87½	115	2·87½	155	3·87½	195	4·87½
36	0·90	76	1·90	116	2·90	156	3·90	196	4·90
37	0·92½	77	1·92½	117	2·92½	157	3·92½	197	4·92½
38	0·95	78	1·95	118	2·95	158	3·95	198	4·95
39	0·97½	79	1·97½	119	2·97½	159	3·97½	199	4·97½
40	1·00	**80**	2·00	**120**	3·00	**160**	4·00	**200**	5·00

$\frac{1}{16}$=0·00^2	$\frac{1}{9}$=0·00^3	$\frac{1}{8}$=0·00^3	$\frac{1}{6}$=0·00^4	$\frac{1}{5}$=0·00^5	$\frac{1}{4}$=0·00^6	$\frac{1}{3}$=0·00^8
$\frac{3}{8}$=0·00^9	$\frac{1}{2}$=0·01^3	$\frac{5}{8}$=0·01^6	$\frac{3}{4}$=0·01^9	$\frac{7}{8}$=0·02^2	$\frac{2}{3}$=0·01^7	$\frac{5}{6}$=0·02^1
$\frac{3}{16}$=0·00^5	$\frac{5}{16}$=0·00^8	$\frac{7}{16}$=0·01^1	$\frac{9}{16}$=0·01^4	$\frac{11}{16}$=0·01^7	$\frac{13}{16}$=0·02^0	$\frac{15}{16}$=0·02^3

(33.3=1£) £0.03

1	0·03	41	1·23	81	2·43	121	3·63	161	4·83
2	0·06	42	1·26	82	2·46	122	3·66	162	4·86
3	0·09	43	1·29	83	2·49	123	3·69	163	4·89
4	0·12	44	1·32	84	2·52	124	3·72	164	4·92
5	0·15	45	1·35	85	2·55	125	3·75	165	4·95
6	0·18	46	1·38	86	2·58	126	3·78	166	4·98
7	0·21	47	1·41	87	2·61	127	3·81	167	5·01
8	0·24	48	1·44	88	2·64	128	3·84	168	5·04
9	0·27	49	1·47	89	2·67	129	3·87	169	5·07
10	0·30	**50**	1·50	**90**	2·70	**130**	3·90	**170**	5·10
11	0·33	51	1·53	91	2·73	131	3·93	171	5·13
12	0·36	52	1·56	92	2·76	132	3·96	172	5·16
13	0·39	53	1·59	93	2·79	133	3·99	173	5·19
14	0·42	54	1·62	94	2·82	134	4·02	174	5·22
15	0·45	55	1·65	95	2·85	135	4·05	175	5·25
16	0·48	56	1·68	96	2·88	136	4·08	176	5·28
17	0·51	57	1·71	97	2·91	137	4·11	177	5·31
18	0·54	58	1·74	98	2·94	138	4·14	178	5·34
19	0·57	59	1·77	99	2·97	139	4·17	179	5·37
20	0·60	**60**	1·80	**100**	3·00	**140**	4·20	**180**	5·40
21	0·63	61	1·83	101	3·03	141	4·23	181	5·43
22	0·66	62	1·86	102	3·06	142	4·26	182	5·46
23	0·69	63	1·89	103	3·09	143	4·29	183	5·49
24	0·72	64	1·92	104	3·12	144	4·32	184	5·52
25	0·75	65	1·95	105	3·15	145	4·35	185	5·55
26	0·78	66	1·98	106	3·18	146	4·38	186	5·58
27	0·81	67	2·01	107	3·21	147	4·41	187	5·61
28	0·84	68	2·04	108	3·24	148	4·44	188	5·64
29	0·87	69	2·07	109	3·27	149	4·47	189	5·67
30	0·90	**70**	2·10	**110**	3·30	**150**	4·50	**190**	5·70
31	0·93	71	2·13	111	3·33	151	4·53	191	5·73
32	0·96	72	2·16	112	3·36	152	4·56	192	5·76
33	0·99	73	2·19	113	3·39	153	4·59	193	5·79
34	1·02	74	2·22	114	3·42	154	4·62	194	5·82
35	1·05	75	2·25	115	3·45	155	4·65	195	5·85
36	1·08	76	2·28	116	3·48	156	4·68	196	5·88
37	1·11	77	2·31	117	3·51	157	4·71	197	5·91
38	1·14	78	2·34	118	3·54	158	4·74	198	5·94
39	1·17	79	2·37	119	3·57	159	4·77	199	5·97
40	1·20	**80**	2·40	**120**	3·60	**160**	4·80	**200**	6·00

$\frac{1}{16}=0{\cdot}00^{2}$ $\frac{1}{9}=0{\cdot}00^{3}$ $\frac{1}{8}=0{\cdot}00^{4}$ $\frac{1}{6}=0{\cdot}00^{5}$ $\frac{1}{5}=0{\cdot}00^{6}$ $\frac{1}{4}=0{\cdot}00^{8}$ $\frac{1}{3}=0{\cdot}01^{0}$

$\frac{3}{8}=0{\cdot}01^{1}$ $\frac{1}{2}=0{\cdot}01^{5}$ $\frac{5}{8}=0{\cdot}01^{9}$ $\frac{3}{4}=0{\cdot}02^{3}$ $\frac{7}{8}=0{\cdot}02^{6}$ $\frac{2}{3}=0{\cdot}02^{0}$ $\frac{5}{6}=0{\cdot}02^{5}$

$\frac{3}{16}=0{\cdot}00^{6}$ $\frac{5}{16}=0{\cdot}00^{9}$ $\frac{7}{16}=0{\cdot}01^{3}$ $\frac{9}{16}=0{\cdot}01^{7}$ $\frac{11}{16}=0{\cdot}02^{1}$ $\frac{13}{16}=0{\cdot}02^{4}$ $\frac{15}{16}=0{\cdot}02^{8}$

1	0·03½	41	1·43½	81	2·83½	121	4·23½	161	5·63½
2	0·07	42	1·47	82	2·87	122	4·27	162	5·67
3	0·10½	43	1·50½	83	2·90½	123	4·30½	163	5·70½
4	0·14	44	1·54	84	2·94	124	4·34	164	5·74
5	0·17½	45	1·57½	85	2·97½	125	4·37½	165	5·77½
6	0·21	46	1·61	86	3·01	126	4·41	166	5·81
7	0·24½	47	1·64½	87	3·04½	127	4·44½	167	5·84½
8	0·28	48	1·68	88	3·08	128	4·48	168	5·88
9	0·31½	49	1·71½	89	3·11½	129	4·51½	169	5·91½
10	0·35	**50**	1·75	**90**	3·15	**130**	4·55	**170**	5·95
11	0·38½	51	1·78½	91	3·18½	131	4·58½	171	5·98½
12	0·42	52	1·82	92	3·22	132	4·62	172	6·02
13	0·45½	53	1·85½	93	3·25½	133	4·65½	173	6·05½
14	0·49	54	1·89	94	3·29	134	4·69	174	6·09
15	0·52½	55	1·92½	95	3·32½	135	4·72½	175	6·12½
16	0·56	56	1·96	96	3·36	136	4·76	176	6·16
17	0·59½	57	1·99½	97	3·39½	137	4·79½	177	6·19½
18	0·63	58	2·03	98	3·43	138	4·83	178	6·23
19	0·66½	59	2·06½	99	3·46½	139	4·86½	179	6·26½
20	0·70	**60**	2·10	**100**	3·50	**140**	4·90	**180**	6·30
21	0·73½	61	2·13½	101	3·53½	141	4·93½	181	6·33½
22	0·77	62	2·17	102	3·57	142	4·97	182	6·37
23	0·80½	63	2·20½	103	3·60½	143	5·00½	183	6·40½
24	0·84	64	2·24	104	3·64	144	5·04	184	6·44
25	0·87½	65	2·27½	105	3·67½	145	5·07½	185	6·47½
26	0·91	66	2·31	106	3·71	146	5·11	186	6·51
27	0·94½	67	2·34½	107	3·74½	147	5·14½	187	6·54½
28	0·98	68	2·38	108	3·78	148	5·18	188	6·58
29	1·01½	69	2·41½	109	3·81½	149	5·21½	189	6·61½
30	1·05	**70**	2·45	**110**	3·85	**150**	5·25	**190**	6·65
31	1·08½	71	2·48½	111	3·88½	151	5·28½	191	6·68½
32	1·12	72	2·52	112	3·92	152	5·32	192	6·72
33	1·15½	73	2·55½	113	3·95½	153	5·35½	193	6·75½
34	1·19	74	2·59	114	3·99	154	5·39	194	6·79
35	1·22½	75	2·62½	115	4·02½	155	5·42½	195	6·82½
36	1·26	76	2·66	116	4·06	156	5·46	196	6·86
37	1·29½	77	2·69½	117	4·09½	157	5·49½	197	6·89½
38	1·33	78	2·73	118	4·13	158	5·53	198	6·93
39	1·36½	79	2·76½	119	4·16½	159	5·56½	199	6·96½
40	1·40	**80**	2·80	**120**	4·20	**160**	5·60	**200**	7·00

$\frac{1}{16}$=0·00^2 $\frac{1}{9}$=0·00^4 $\frac{1}{8}$=0·00^4 $\frac{1}{6}$=0·00^6 $\frac{1}{5}$=0·00^7 $\frac{1}{4}$=0·00^9 $\frac{1}{3}$=0·01^2

$\frac{3}{8}$=0·01^3 $\frac{1}{2}$=0·01^8 $\frac{5}{8}$=0·02^2 $\frac{3}{4}$=0·02^6 $\frac{7}{8}$=0·03^1 $\frac{2}{3}$=0·02^3 $\frac{5}{6}$=0·02^9

$\frac{3}{16}$=0·00^7 $\frac{5}{16}$=0·01^1 $\frac{7}{16}$=0·01^5 $\frac{9}{16}$=0·02^0 $\frac{11}{16}$=0·02^4 $\frac{13}{16}$=0·02^8 $\frac{15}{16}$=0·03^3

(25.0=1£) **£0.04**

1	0·04	41	1·64	81	3·24	121	4·84	161	6·44
2	0·08	42	1·68	82	3·28	122	4·88	162	6·48
3	0·12	43	1·72	83	3·32	123	4·92	163	6·52
4	0·16	44	1·76	84	3·36	124	4·96	164	6·56
5	0·20	45	1·80	85	3·40	125	5·00	165	6·60
6	0·24	46	1·84	86	3·44	126	5·04	166	6·64
7	0·28	47	1·88	87	3·48	127	5·08	167	6·68
8	0·32	48	1·92	88	3·52	128	5·12	168	6·72
9	0·36	49	1·96	89	3·56	129	5·16	169	6·76
10	0·40	**50**	2·00	**90**	3·60	**130**	5·20	**170**	6·80
11	0·44	51	2·04	91	3·64	131	5·24	171	6·84
12	0·48	52	2·08	92	3·68	132	5·28	172	6·88
13	0·52	53	2·12	93	3·72	133	5·32	173	6·92
14	0·56	54	2·16	94	3·76	134	5·36	174	6·96
15	0·60	55	2·20	95	3·80	135	5·40	175	7·00
16	0·64	56	2·24	96	3·84	136	5·44	176	7·04
17	0·68	57	2·28	97	3·88	137	5·48	177	7·08
18	0·72	58	2·32	98	3·92	138	5·52	178	7·12
19	0·76	59	2·36	99	3·96	139	5·56	179	7·16
20	0·80	**60**	2·40	**100**	4·00	**140**	5·60	**180**	7·20
21	0·84	61	2·44	101	4·04	141	5·64	181	7·24
22	0·88	62	2·48	102	4·08	142	5·68	182	7·28
23	0·92	63	2·52	103	4·12	143	5·72	183	7·32
24	0·96	64	2·56	104	4·16	144	5·76	184	7·36
25	1·00	65	2·60	105	4·20	145	5·80	185	7·40
26	1·04	66	2·64	106	4·24	146	5·84	186	7·44
27	1·08	67	2·68	107	4·28	147	5·88	187	7·48
28	1·12	68	2·72	108	4·32	148	5·92	188	7·52
29	1·16	69	2·76	109	4·36	149	5·96	189	7·56
30	1·20	**70**	2·80	**110**	4·40	**150**	6·00	**190**	7·60
31	1·24	71	2·84	111	4·44	151	6·04	191	7·64
32	1·28	72	2·88	112	4·48	152	6·08	192	7·68
33	1·32	73	2·92	113	4·52	153	6·12	193	7·72
34	1·36	74	2·96	114	4·56	154	6·16	194	7·76
35	1·40	75	3·00	115	4·60	155	6·20	195	7·80
36	1·44	76	3·04	116	4·64	156	6·24	196	7·84
37	1·48	77	3·08	117	4·68	157	6·28	197	7·88
38	1·52	78	3·12	118	4·72	158	6·32	198	7·92
39	1·56	79	3·16	119	4·76	159	6·36	199	7·96
40	1·60	**80**	3·20	**120**	4·80	**160**	6·40	**200**	8·00

$\frac{1}{16}=0{\cdot}00^{3}$ $\frac{1}{9}=0{\cdot}00^{4}$ $\frac{1}{8}=0{\cdot}00^{5}$ $\frac{1}{6}=0{\cdot}00^{7}$ $\frac{1}{5}=0{\cdot}00^{8}$ $\frac{1}{4}=0{\cdot}01^{0}$ $\frac{1}{3}=0{\cdot}01^{3}$

$\frac{3}{8}=0{\cdot}01^{5}$ $\frac{1}{2}=0{\cdot}02^{0}$ $\frac{5}{8}=0{\cdot}02^{5}$ $\frac{3}{4}=0{\cdot}03^{0}$ $\frac{7}{8}=0{\cdot}03^{5}$ $\frac{2}{3}=0{\cdot}02^{7}$ $\frac{5}{6}=0{\cdot}03^{3}$

$\frac{3}{16}=0{\cdot}00^{8}$ $\frac{5}{16}=0{\cdot}01^{3}$ $\frac{7}{16}=0{\cdot}01^{8}$ $\frac{9}{16}=0{\cdot}02^{3}$ $\frac{11}{16}=0{\cdot}02^{8}$ $\frac{13}{16}=0{\cdot}03^{3}$ $\frac{15}{16}=0{\cdot}03^{8}$

1	0·04½	41	1·84½	81	3·64½	121	5·44½	161	7·24½
2	0·09	42	1·89	82	3·69	122	5·49	162	7·29
3	0·13½	43	1·93½	83	3·73½	123	5·53½	163	7·33½
4	0·18	44	1·98	84	3·78	124	5·58	164	7·38
5	0·22½	45	2·02½	85	3·82½	125	5·62½	165	7·42½
6	0·27	46	2·07	86	3·87	126	5·67	166	7·47
7	0·31½	47	2·11½	87	3·91½	127	5·71½	167	7·51½
8	0·36	48	2·16	88	3·96	128	5·76	168	7·56
9	0·40½	49	2·20½	89	4·00½	129	5·80½	169	7·60½
10	0·45	**50**	2·25	**90**	4·05	**130**	5·85	**170**	7·65
11	0·49½	51	2·29½	91	4·09½	131	5·89½	171	7·69½
12	0·54	52	2·34	92	4·14	132	5·94	172	7·74
13	0·58½	53	2·38½	93	4·18½	133	5·98½	173	7·78½
14	0·63	54	2·43	94	4·23	134	6·03	174	7·83
15	0·67½	55	2·47½	95	4·27½	135	6·07½	175	7·87½
16	0·72	56	2·52	96	4·32	136	6·12	176	7·92
17	0·76½	57	2·56½	97	4·36½	137	6·16½	177	7·96½
18	0·81	58	2·61	98	4·41	138	6·21	178	8·01
19	0·85½	59	2·65½	99	4·45½	139	6·25½	179	8·05½
20	0·90	**60**	2·70	**100**	4·50	**140**	6·30	**180**	8·10
21	0·94½	61	2·74½	101	4·54½	141	6·34½	181	8·14½
22	0·99	62	2·79	102	4·59	142	6·39	182	8·19
23	1·03½	63	2·83½	103	4·63½	143	6·43½	183	8·23½
24	1·08	64	2·88	104	4·68	144	6·48	184	8·28
25	1·12½	65	2·92½	105	4·72½	145	6·52½	185	8·32½
26	1·17	66	2·97	106	4·77	146	6·57	186	8·37
27	1·21½	67	3·01½	107	4·81½	147	6·61½	187	8·41½
28	1·26	68	3·06	108	4·86	148	6·66	188	8·46
29	1·30½	69	3·10½	109	4·90½	149	6·70½	189	8·50½
30	1·35	**70**	3·15	**110**	4·95	**150**	6·75	**190**	8·55
31	1·39½	71	3·19½	111	4·99½	151	6·79½	191	8·59½
32	1·44	72	3·24	112	5·04	152	6·84	192	8·64
33	1·48½	73	3·28½	113	5·08½	153	6·88½	193	8·68½
34	1·53	74	3·33	114	5·13	154	6·93	194	8·73
35	1·57½	75	3·37½	115	5·17½	155	6·97½	195	8·77½
36	1·62	76	3·42	116	5·22	156	7·02	196	8·82
37	1·66½	77	3·46½	117	5·26½	157	7·06½	197	8·86½
38	1·71	78	3·51	118	5·31	158	7·11	198	8·91
39	1·75½	79	3·55½	119	5·35½	159	7·15½	199	8·95½
40	1·80	**80**	3·60	**120**	5·40	**160**	7·20	**200**	9·00

$\frac{1}{16}=0{\cdot}00^3$	$\frac{1}{9}=0{\cdot}00^5$	$\frac{1}{8}=0{\cdot}00^6$	$\frac{1}{6}=0{\cdot}00^8$	$\frac{1}{5}=0{\cdot}00^9$	$\frac{1}{4}=0{\cdot}01^1$	$\frac{1}{3}=0{\cdot}01^5$
$\frac{3}{8}=0{\cdot}01^7$	$\frac{1}{2}=0{\cdot}02^3$	$\frac{5}{8}=0{\cdot}02^8$	$\frac{3}{4}=0{\cdot}03^4$	$\frac{7}{8}=0{\cdot}03^9$	$\frac{2}{3}=0{\cdot}03^0$	$\frac{5}{6}=0{\cdot}03^8$
$\frac{3}{16}=0{\cdot}00^8$	$\frac{5}{16}=0{\cdot}01^4$	$\frac{7}{16}=0{\cdot}02^0$	$\frac{9}{16}=0{\cdot}02^5$	$\frac{11}{16}=0{\cdot}03^1$	$\frac{13}{16}=0{\cdot}03^7$	$\frac{15}{16}=0{\cdot}04^2$

(20.0=1£) **£0.05**

1	0·05	41	2·05	81	4·05	121	6·05	161	8·05
2	0·10	42	2·10	82	4·10	122	6·10	162	8·10
3	0·15	43	2·15	83	4·15	123	6·15	163	8·15
4	0·20	44	2·20	84	4·20	124	6·20	164	8·20
5	0·25	45	2·25	85	4·25	125	6·25	165	8·25
6	0·30	46	2·30	86	4·30	126	6·30	166	8·30
7	0·35	47	2·35	87	4·35	127	6·35	167	8·35
8	0·40	48	2·40	88	4·40	128	6·40	168	8·40
9	0·45	49	2·45	89	4·45	129	6·45	169	8·45
10	0·50	**50**	2·50	**90**	4·50	**130**	6·50	**170**	8·50
11	0·55	51	2·55	91	4·55	131	6·55	171	8·55
12	0·60	52	2·60	92	4·60	132	6·60	172	8·60
13	0·65	53	2·65	93	4·65	133	6·65	173	8·65
14	0·70	54	2·70	94	4·70	134	6·70	174	8·70
15	0·75	55	2·75	95	4·75	135	6·75	175	8·75
16	0·80	56	2·80	96	4·80	136	6·80	176	8·80
17	0·85	57	2·85	97	4·85	137	6·85	177	8·85
18	0·90	58	2·90	98	4·90	138	6·90	178	8·90
19	0·95	59	2·95	99	4·95	139	6·95	179	8·95
20	1·00	**60**	3·00	**100**	5·00	**140**	7·00	**180**	9·00
21	1·05	61	3·05	101	5·05	141	7·05	181	9·05
22	1·10	62	3·10	102	5·10	142	7·10	182	9·10
23	1·15	63	3·15	103	5·15	143	7·15	183	9·15
24	1·20	64	3·20	104	5·20	144	7·20	184	9·20
25	1·25	65	3·25	105	5·25	145	7·25	185	9·25
26	1·30	66	3·30	106	5·30	146	7·30	186	9·30
27	1·35	67	3·35	107	5·35	147	7·35	187	9·35
28	1·40	68	3·40	108	5·40	148	7·40	188	9·40
29	1·45	69	3·45	109	5·45	149	7·45	189	9·45
30	1·50	**70**	3·50	**110**	5·50	**150**	7·50	**190**	9·50
31	1·55	71	3·55	111	5·55	151	7·55	191	9·55
32	1·60	72	3·60	112	5·60	152	7·60	192	9·60
33	1·65	73	3·65	113	5·65	153	7·65	193	9·65
34	1·70	74	3·70	114	5·70	154	7·70	194	9·70
35	1·75	75	3·75	115	5·75	155	7·75	195	9·75
36	1·80	76	3·80	116	5·80	156	7·80	196	9·80
37	1·85	77	3·85	117	5·85	157	7·85	197	9·85
38	1·90	78	3·90	118	5·90	158	7·90	198	9·90
39	1·95	79	3·95	119	5·95	159	7·95	199	9·95
40	2·00	**80**	4·00	**120**	6·00	**160**	8·00	**200**	10·00

$\frac{1}{16}=0{\cdot}00^{3}$	$\frac{1}{9}=0{\cdot}00^{6}$	$\frac{1}{8}=0{\cdot}00^{6}$	$\frac{1}{6}=0{\cdot}00^{8}$	$\frac{1}{5}=0{\cdot}01^{0}$	$\frac{1}{4}=0{\cdot}01^{3}$	$\frac{1}{3}=0{\cdot}01^{7}$
$\frac{3}{8}=0{\cdot}01^{9}$	$\frac{1}{2}=0{\cdot}02^{5}$	$\frac{5}{8}=0{\cdot}03^{1}$	$\frac{3}{4}=0{\cdot}03^{8}$	$\frac{7}{8}=0{\cdot}04^{4}$	$\frac{2}{3}=0{\cdot}03^{3}$	$\frac{5}{6}=0{\cdot}04^{2}$
$\frac{3}{16}=0{\cdot}00^{9}$	$\frac{5}{16}=0{\cdot}01^{6}$	$\frac{7}{16}=0{\cdot}02^{2}$	$\frac{9}{16}=0{\cdot}02^{8}$	$\frac{11}{16}=0{\cdot}03^{4}$	$\frac{13}{16}=0{\cdot}04^{1}$	$\frac{15}{16}=0{\cdot}04^{7}$

(18.2=1£) £0.05½

1	0·05½	41	2·25½	81	4·45½	121	6·65½	161	8·85½
2	0·11	42	2·31	82	4·51	122	6·71	162	8·91
3	0·16½	43	2·36½	83	4·56½	123	6·76½	163	8·96½
4	0·22	44	2·42	84	4·62	124	6·82	164	9·02
5	0·27½	45	2·47½	85	4·67½	125	6·87½	165	9·07½
6	0·33	46	2·53	86	4·73	126	6·93	166	9·13
7	0·38½	47	2·58½	87	4·78½	127	6·98½	167	9·18½
8	0·44	48	2·64	88	4·84	128	7·04	168	9·24
9	0·49½	49	2·69½	89	4·89½	129	7·09½	169	9·29½
10	0·55	**50**	2·75	**90**	4·95	**130**	7·15	**170**	9·35
11	0·60½	51	2·80½	91	5·00½	131	7·20½	171	9·40½
12	0·66	52	2·86	92	5·06	132	7·26	172	9·46
13	0·71½	53	2·91½	93	5·11½	133	7·31½	173	9·51½
14	0·77	54	2·97	94	5·17	134	7·37	174	9·57
15	0·82½	55	3·02½	95	5·22½	135	7·42½	175	9·62½
16	0·88	56	3·08	96	5·28	136	7·48	176	9·68
17	0·93½	57	3·13½	97	5·33½	137	7·53½	177	9·73½
18	0·99	58	3·19	98	5·39	138	7·59	178	9·79
19	1·04½	59	3·24½	99	5·44½	139	7·64½	179	9·84½
20	1·10	**60**	3·30	**100**	5·50	**140**	7·70	**180**	9·90
21	1·15½	61	3·35½	101	5·55½	141	7·75½	181	9·95½
22	1·21	62	3·41	102	5·61	142	7·81	182	10·01
23	1·26½	63	3·46½	103	5·66½	143	7·86½	183	10·06½
24	1·32	64	3·52	104	5·72	144	7·92	184	10·12
25	1·37½	65	3·57½	105	5·77½	145	7·97½	185	10·17½
26	1·43	66	3·63	106	5·83	146	8·03	186	10·23
27	1·48½	67	3·68½	107	5·88½	147	8·08½	187	10·28½
28	1·54	68	3·74	108	5·94	148	8·14	188	10·34
29	1·59½	69	3·79½	109	5·99½	149	8·19½	189	10·39½
30	1·65	**70**	3·85	**110**	6·05	**150**	8·25	**190**	10·45
31	1·70½	71	3·90½	111	6·10½	151	8·30½	191	10·50½
32	1·76	72	3·96	112	6·16	152	8·36	192	10·56
33	1·81½	73	4·01½	113	6·21½	153	8·41½	193	10·61½
34	1·87	74	4·07	114	6·27	154	8·47	194	10·67
35	1·92½	75	4·12½	115	6·32½	155	8·52½	195	10·72½
36	1·98	76	4·18	116	6·38	156	8·58	196	10·78
37	2·03½	77	4·23½	117	6·43½	157	8·63½	197	10·83½
38	2·09	78	4·29	118	6·49	158	8·69	198	10·89
39	2·14½	79	4·34½	119	6·54½	159	8·74½	199	10·94½
40	2·20	**80**	4·40	**120**	6·60	**160**	8·80	**200**	11·00

$\frac{1}{16}=0{\cdot}00^{3}$	$\frac{1}{9}=0{\cdot}00^{6}$	$\frac{1}{8}=0{\cdot}00^{7}$	$\frac{1}{6}=0{\cdot}00^{9}$	$\frac{1}{5}=0{\cdot}01^{1}$	$\frac{1}{4}=0{\cdot}01^{4}$	$\frac{1}{3}=0{\cdot}01^{8}$
$\frac{3}{8}=0{\cdot}02^{1}$	$\frac{1}{2}=0{\cdot}02^{8}$	$\frac{5}{8}=0{\cdot}03^{4}$	$\frac{3}{4}=0{\cdot}04^{1}$	$\frac{7}{8}=0{\cdot}04^{8}$	$\frac{2}{3}=0{\cdot}03^{7}$	$\frac{5}{6}=0{\cdot}04^{6}$
$\frac{3}{16}=0{\cdot}01^{0}$	$\frac{5}{16}=0{\cdot}01^{7}$	$\frac{7}{16}=0{\cdot}02^{4}$	$\frac{9}{16}=0{\cdot}03^{1}$	$\frac{11}{16}=0{\cdot}03^{8}$	$\frac{13}{16}=0{\cdot}04^{5}$	$\frac{15}{16}=0{\cdot}05^{2}$

1	0·06	41	2·46	81	4·86	121	7·26	161	9·66
2	0·12	42	2·52	82	4·92	122	7·32	162	9·72
3	0·18	43	2·58	83	4·98	123	7·38	163	9·78
4	0·24	44	2·64	84	5·04	124	7·44	164	9·84
5	0·30	45	2·70	85	5·10	125	7·50	165	9·90
6	0·36	46	2·76	86	5·16	126	7·56	166	9·96
7	0·42	47	2·82	87	5·22	127	7·62	167	10·02
8	0·48	48	2·88	88	5·28	128	7·68	168	10·08
9	0·54	49	2·94	89	5·34	129	7·74	169	10·14
10	0·60	**50**	3·00	**90**	5·40	**130**	7·80	**170**	10·20
11	0·66	51	3·06	91	5·46	131	7·86	171	10·26
12	0·72	52	3·12	92	5·52	132	7·92	172	10·32
13	0·78	53	3·18	93	5·58	133	7·98	173	10·38
14	0·84	54	3·24	94	5·64	134	8·04	174	10·44
15	0·90	55	3·30	95	5·70	135	8·10	175	10·50
16	0·96	56	3·36	96	5·76	136	8·16	176	10·56
17	1·02	57	3·42	97	5·82	137	8·22	177	10·62
18	1·08	58	3·48	98	5·88	138	8·28	178	10·68
19	1·14	59	3·54	99	5·94	139	8·34	179	10·74
20	1·20	**60**	3·60	**100**	6·00	**140**	8·40	**180**	10·80
21	1·26	61	3·66	101	6·06	141	8·46	181	10·86
22	1·32	62	3·72	102	6·12	142	8·52	182	10·92
23	1·38	63	3·78	103	6·18	143	8·58	183	10·98
24	1·44	64	3·84	104	6·24	144	8·64	184	11·04
25	1·50	65	3·90	105	6·30	145	8·70	185	11·10
26	1·56	66	3·96	106	6·36	146	8·76	186	11·16
27	1·62	67	4·02	107	6·42	147	8·82	187	11·22
28	1·68	68	4·08	108	6·48	148	8·88	188	11·28
29	1·74	69	4·14	109	6·54	149	8·94	189	11·34
30	1·80	**70**	4·20	**110**	6·60	**150**	9·00	**190**	11·40
31	1·86	71	4·26	111	6·66	151	9·06	191	11·46
32	1·92	72	4·32	112	6·72	152	9·12	192	11·52
33	1·98	73	4·38	113	6·78	153	9·18	193	11·58
34	2·04	74	4·44	114	6·84	154	9·24	194	11·64
35	2·10	75	4·50	115	6·90	155	9·30	195	11·70
36	2·16	76	4·56	116	6·96	156	9·36	196	11·76
37	2·22	77	4·62	117	7·02	157	9·42	197	11·82
38	2·28	78	4·68	118	7·08	158	9·48	198	11·88
39	2·34	79	4·74	119	7·14	159	9·54	199	11·94
40	2·40	**80**	4·80	**120**	7·20	**160**	9·60	**200**	12·00

$\frac{1}{16}=0{\cdot}00^{4}$ $\frac{1}{9}=0{\cdot}00^{7}$ $\frac{1}{8}=0{\cdot}00^{8}$ $\frac{1}{6}=0{\cdot}01^{0}$ $\frac{1}{5}=0{\cdot}01^{2}$ $\frac{1}{4}=0{\cdot}01^{5}$ $\frac{1}{3}=0{\cdot}02^{0}$

$\frac{3}{8}=0{\cdot}02^{3}$ $\frac{1}{2}=0{\cdot}03^{0}$ $\frac{5}{8}=0{\cdot}03^{8}$ $\frac{3}{4}=0{\cdot}04^{5}$ $\frac{7}{8}=0{\cdot}05^{3}$ $\frac{2}{3}=0{\cdot}04^{0}$ $\frac{5}{6}=0{\cdot}05^{0}$

$\frac{3}{16}=0{\cdot}01^{1}$ $\frac{5}{16}=0{\cdot}01^{9}$ $\frac{7}{16}=0{\cdot}02^{6}$ $\frac{9}{16}=0{\cdot}03^{4}$ $\frac{11}{16}=0{\cdot}04^{1}$ $\frac{13}{16}=0{\cdot}04^{9}$ $\frac{15}{16}=0{\cdot}05^{6}$

1	0·06½	41	2·66½	81	5·26½	121	7·86½	161	10·46½
2	0·13	42	2·73	82	5·33	122	7·93	162	10·53
3	0·19½	43	2·79½	83	5·39½	123	7·99½	163	10·59½
4	0·26	44	2·86	84	5·46	124	8·06	164	10·66
5	0·32½	45	2·92½	85	5·52½	125	8·12½	165	10·72½
6	0·39	46	2·99	86	5·59	126	8·19	166	10·79
7	0·45½	47	3·05½	87	5·65½	127	8·25½	167	10·85½
8	0·52	48	3·12	88	5·72	128	8·32	168	10·92
9	0·58½	49	3·18½	89	5·78½	129	8·38½	169	10·98½
10	0·65	**50**	3·25	**90**	5·85	**130**	8·45	**170**	11·05
11	0·71½	51	3·31½	91	5·91½	131	8·51½	171	11·11½
12	0·78	52	3·38	92	5·98	132	8·58	172	11·18
13	0·84½	53	3·44½	93	6·04½	133	8·64½	173	11·24½
14	0·91	54	3·51	94	6·11	134	8·71	174	11·31
15	0·97½	55	3·57½	95	6·17½	135	8·77½	175	11·37½
16	1·04	56	3·64	96	6·24	136	8·84	176	11·44
17	1·10½	57	3·70½	97	6·30½	137	8·90½	177	11·50½
18	1·17	58	3·77	98	6·37	138	8·97	178	11·57
19	1·23½	59	3·83½	99	6·43½	139	9·03½	179	11·63½
20	1·30	**60**	3·90	**100**	6·50	**140**	9·10	**180**	11·70
21	1·36½	61	3·96½	101	6·56½	141	9·16½	181	11·76½
22	1·43	62	4·03	102	6·63	142	9·23	182	11·83
23	1·49½	63	4·09½	103	6·69½	143	9·29½	183	11·89½
24	1·56	64	4·16	104	6·76	144	9·36	184	11·96
25	1·62½	65	4·22½	105	6·82½	145	9·42½	185	12·02½
26	1·69	66	4·29	106	6·89	146	9·49	186	12·09
27	1·75½	67	4·35½	107	6·95½	147	9·55½	187	12·15½
28	1·82	68	4·42	108	7·02	148	9·62	188	12·22
29	1·88½	69	4·48½	109	7·08½	149	9·68½	189	12·28½
30	1·95	**70**	4·55	**110**	7·15	**150**	9·75	**190**	12·35
31	2·01½	71	4·61½	111	7·21½	151	9·81½	191	12·41½
32	2·08	72	4·68	112	7·28	152	9·88	192	12·48
33	2·14½	73	4·74½	113	7·34½	153	9·94½	193	12·54½
34	2·21	74	4·81	114	7·41	154	10·01	194	12·61
35	2·27½	75	4·87½	115	7·47½	155	10·07½	195	12·67½
36	2·34	76	4·94	116	7·54	156	10·14	196	12·74
37	2·40½	77	5·00½	117	7·60½	157	10·20½	197	12·80½
38	2·47	78	5·07	118	7·67	158	10·27	198	12·87
39	2·53½	79	5·13½	119	7·73½	159	10·33½	199	12·93½
40	2·60	**80**	5·20	**120**	7·80	**160**	10·40	**200**	13·00

$\frac{1}{16}$=0·00^{4} $\frac{1}{9}$=0·00^{7} $\frac{1}{8}$=0·00^{8} $\frac{1}{6}$=0·01^{1} $\frac{1}{5}$=0·01^{3} $\frac{1}{4}$=0·01^{6} $\frac{1}{3}$=0·02^{2}

$\frac{3}{8}$=0·02^{4} $\frac{1}{2}$=0·03^{3} $\frac{5}{8}$=0·04^{1} $\frac{3}{4}$=0·04^{9} $\frac{7}{8}$=0·05^{7} $\frac{2}{3}$=0·04^{3} $\frac{5}{6}$=0·05^{4}

$\frac{3}{16}$=0·01^{2} $\frac{5}{16}$=0·02^{0} $\frac{7}{16}$=0·02^{8} $\frac{9}{16}$=0·03^{7} $\frac{11}{16}$=0·04^{5} $\frac{13}{16}$=0·05^{3} $\frac{15}{16}$=0·06^{1}

1	0·07	41	2·87	81	5·67	121	8·47	161	11·27
2	0·14	42	2·94	82	5·74	122	8·54	162	11·34
3	0·21	43	3·01	83	5·81	123	8·61	163	11·41
4	0·28	44	3·08	84	5·88	124	8·68	164	11·48
5	0·35	45	3·15	85	5·95	125	8·75	165	11·55
6	0·42	46	3·22	86	6·02	126	8·82	166	11·62
7	0·49	47	3·29	87	6·09	127	8·89	167	11·69
8	0·56	48	3·36	88	6·16	128	8·96	168	11·76
9	0·63	49	3·43	89	6·23	129	9·03	169	11·83
10	0·70	**50**	3·50	**90**	6·30	**130**	9·10	**170**	11·90
11	0·77	51	3·57	91	6·37	131	9·17	171	11·97
12	0·84	52	3·64	92	6·44	132	9·24	172	12·04
13	0·91	53	3·71	93	6·51	133	9·31	173	12·11
14	0·98	54	3·78	94	6·58	134	9·38	174	12·18
15	1·05	55	3·85	95	6·65	135	9·45	175	12·25
16	1·12	56	3·92	96	6·72	136	9·52	176	12·32
17	1·19	57	3·99	97	6·79	137	9·59	177	12·39
18	1·26	58	4·06	98	6·86	138	9·66	178	12·46
19	1·33	59	4·13	99	6·93	139	9·73	179	12·53
20	1·40	**60**	4·20	**100**	7·00	**140**	9·80	**180**	12·60
21	1·47	61	4·27	101	7·07	141	9·87	181	12·67
22	1·54	62	4·34	102	7·14	142	9·94	182	12·74
23	1·61	63	4·41	103	7·21	143	10·01	183	12·81
24	1·68	64	4·48	104	7·28	144	10·08	184	12·88
25	1·75	65	4·55	105	7·35	145	10·15	185	12·95
26	1·82	66	4·62	106	7·42	146	10·22	186	13·02
27	1·89	67	4·69	107	7·49	147	10·29	187	13·09
28	1·96	68	4·76	108	7·56	148	10·36	188	13·16
29	2·03	69	4·83	109	7·63	149	10·43	189	13·23
30	2·10	**70**	4·90	**110**	7·70	**150**	10·50	**190**	13·30
31	2·17	71	4·97	111	7·77	151	10·57	191	13·37
32	2·24	72	5·04	112	7·84	152	10·64	192	13·44
33	2·31	73	5·11	113	7·91	153	10·71	193	13·51
34	2·38	74	5·18	114	7·98	154	10·78	194	13·58
35	2·45	75	5·25	115	8·05	155	10·85	195	13·65
36	2·52	76	5·32	116	8·12	156	10·92	196	13·72
37	2·59	77	5·39	117	8·19	157	10·99	197	13·79
38	2·66	78	5·46	118	8·26	158	11·06	198	13·86
39	2·73	79	5·53	119	8·33	159	11·13	199	13·93
40	2·80	**80**	5·60	**120**	8·40	**160**	11·20	**200**	14·00

$\frac{1}{16}=0{\cdot}00^4$ $\frac{1}{9}=0{\cdot}00^8$ $\frac{1}{8}=0{\cdot}00^9$ $\frac{1}{6}=0{\cdot}01^2$ $\frac{1}{5}=0{\cdot}01^4$ $\frac{1}{4}=0{\cdot}01^8$ $\frac{1}{3}=0{\cdot}02^3$

$\frac{3}{8}=0{\cdot}02^6$ $\frac{1}{2}=0{\cdot}03^5$ $\frac{5}{8}=0{\cdot}04^4$ $\frac{3}{4}=0{\cdot}05^3$ $\frac{7}{8}=0{\cdot}06^1$ $\frac{2}{3}=0{\cdot}04^7$ $\frac{5}{6}=0{\cdot}05^8$

$\frac{3}{16}=0{\cdot}01^3$ $\frac{5}{16}=0{\cdot}02^2$ $\frac{7}{16}=0{\cdot}03^1$ $\frac{9}{16}=0{\cdot}03^9$ $\frac{11}{16}=0{\cdot}04^8$ $\frac{13}{16}=0{\cdot}05^7$ $\frac{15}{16}=0{\cdot}06^6$

(13.3=1£) £0.07½

1	0·07½	41	3·07½	81	6·07½	121	9·07½	161	12·07½
2	0·15	42	3·15	82	6·15	122	9·15	162	12·15
3	0·22½	43	3·22½	83	6·22½	123	9·22½	163	12·22½
4	0·30	44	3·30	84	6·30	124	9·30	164	12·30
5	0·37½	45	3·37½	85	6·37½	125	9·37½	165	12·37½
6	0·45	46	3·45	86	6·45	126	9·45	166	12·45
7	0·52½	47	3·52½	87	6·52½	127	9·52½	167	12·52½
8	0·60	48	3·60	88	6·60	128	9·60	168	12·60
9	0·67½	49	3·67½	89	6·67½	129	9·67½	169	12·67½
10	0·75	**50**	3·75	**90**	6·75	**130**	9·75	**170**	12·75
11	0·82½	51	3·82½	91	6·82½	131	9·82½	171	12·82½
12	0·90	52	3·90	92	6·90	132	9·90	172	12·90
13	0·97½	53	3·97½	93	6·97½	133	9·97½	173	12·97½
14	1·05	54	4·05	94	7·05	134	10·05	174	13·05
15	1·12½	55	4·12½	95	7·12½	135	10·12½	175	13·12½
16	1·20	56	4·20	96	7·20	136	10·20	176	13·20
17	1·27½	57	4·27½	97	7·27½	137	10·27½	177	13·27½
18	1·35	58	4·35	98	7·35	138	10·35	178	13·35
19	1·42½	59	4·42½	99	7·42½	139	10·42½	179	13·42½
20	1·50	**60**	4·50	**100**	7·50	**140**	10·50	**180**	13·50
21	1·57½	61	4·57½	101	7·57½	141	10·57½	181	13·57½
22	1·65	62	4·65	102	7·65	142	10·65	182	13·65
23	1·72½	63	4·72½	103	7·72½	143	10·72½	183	13·72½
24	1·80	64	4·80	104	7·80	144	10·80	184	13·80
25	1·87½	65	4·87½	105	7·87½	145	10·87½	185	13·87½
26	1·95	66	4·95	106	7·95	146	10·95	186	13·95
27	2·02½	67	5·02½	107	8·02½	147	11·02½	187	14·02½
28	2·10	68	5·10	108	8·10	148	11·10	188	14·10
29	2·17½	69	5·17½	109	8·17½	149	11·17½	189	14·17½
30	2·25	**70**	5·25	**110**	8·25	**150**	11·25	**190**	14·25
31	2·32½	71	5·32½	111	8·32½	151	11·32½	191	14·32½
32	2·40	72	5·40	112	8·40	152	11·40	192	14·40
33	2·47½	73	5·47½	113	8·47½	153	11·47½	193	14·47½
34	2·55	74	5·55	114	8·55	154	11·55	194	14·55
35	2·62½	75	5·62½	115	8·62½	155	11·62½	195	14·62½
36	2·70	76	5·70	116	8·70	156	11·70	196	14·70
37	2·77½	77	5·77½	117	8·77½	157	11·77½	197	14·77½
38	2·85	78	5·85	118	8·85	158	11·85	198	14·85
39	2·92½	79	5·92½	119	8·92½	159	11·92½	199	14·92½
40	3·00	**80**	6·00	**120**	9·00	**160**	12·00	**200**	15·00

$\frac{1}{16}=0{\cdot}00^5$ $\frac{1}{9}=0{\cdot}00^8$ $\frac{1}{8}=0{\cdot}00^9$ $\frac{1}{6}=0{\cdot}01^3$ $\frac{1}{5}=0{\cdot}01^5$ $\frac{1}{4}=0{\cdot}01^9$ $\frac{1}{3}=0{\cdot}02^5$

$\frac{3}{8}=0{\cdot}02^8$ $\frac{1}{2}=0{\cdot}03^8$ $\frac{5}{8}=0{\cdot}04^7$ $\frac{3}{4}=0{\cdot}05^6$ $\frac{7}{8}=0{\cdot}06^6$ $\frac{2}{3}=0{\cdot}05^0$ $\frac{5}{6}=0{\cdot}06^3$

$\frac{3}{16}=0{\cdot}01^4$ $\frac{5}{16}=0{\cdot}02^3$ $\frac{7}{16}=0{\cdot}03^3$ $\frac{9}{16}=0{\cdot}04^2$ $\frac{11}{16}=0{\cdot}05^2$ $\frac{13}{16}=0{\cdot}06^1$ $\frac{15}{16}=0{\cdot}07^0$

(12.5=1£) **£0.08**

1	0·08	41	3·28	81	6·48	121	9·68	161	12·88
2	0·16	42	3·36	82	6·56	122	9·76	162	12·96
3	0·24	43	3·44	83	6·64	123	9·84	163	13·04
4	0·32	44	3·52	84	6·72	124	9·92	164	13·12
5	0·40	45	3·60	85	6·80	125	10·00	165	13·20
6	0·48	46	3·68	86	6·88	126	10·08	166	13·28
7	0·56	47	3·76	87	6·96	127	10·16	167	13·36
8	0·64	48	3·84	88	7·04	128	10·24	168	13·44
9	0·72	49	3·92	89	7·12	129	10·32	169	13·52
10	0·80	**50**	4·00	**90**	7·20	**130**	10·40	**170**	13·60
11	0·88	51	4·08	91	7·28	131	10·48	171	13·68
12	0·96	52	4·16	92	7·36	132	10·56	172	13·76
13	1·04	53	4·24	93	7·44	133	10·64	173	13·84
14	1·12	54	4·32	94	7·52	134	10·72	174	13·92
15	1·20	55	4·40	95	7·60	135	10·80	175	14·00
16	1·28	56	4·48	96	7·68	136	10·88	176	14·08
17	1·36	57	4·56	97	7·76	137	10·96	177	14·16
18	1·44	58	4·64	98	7·84	138	11·04	178	14·24
19	1·52	59	4·72	99	7·92	139	11·12	179	14·32
20	1·60	**60**	4·80	**100**	8·00	**140**	11·20	**180**	14·40
21	1·68	61	4·88	101	8·08	141	11·28	181	14·48
22	1·76	62	4·96	102	8·16	142	11·36	182	14·56
23	1·84	63	5·04	103	8·24	143	11·44	183	14·64
24	1·92	64	5·12	104	8·32	144	11·52	184	14·72
25	2·00	65	5·20	105	8·40	145	11·60	185	14·80
26	2·08	66	5·28	106	8·48	146	11·68	186	14·88
27	2·16	67	5·36	107	8·56	147	11·76	187	14·96
28	2·24	68	5·44	108	8·64	148	11·84	188	15·04
29	2·32	69	5·52	109	8·72	149	11·92	189	15·12
30	2·40	**70**	5·60	**110**	8·80	**150**	12·00	**190**	15·20
31	2·48	71	5·68	111	8·88	151	12·08	191	15·28
32	2·56	72	5·76	112	8·96	152	12·16	192	15·36
33	2·64	73	5·84	113	9·04	153	12·24	193	15·44
34	2·72	74	5·92	114	9·12	154	12·32	194	15·52
35	2·80	75	6·00	115	9·20	155	12·40	195	15·60
36	2·88	76	6·08	116	9·28	156	12·48	196	15·68
37	2·96	77	6·16	117	9·36	157	12·56	197	15·76
38	3·04	78	6·24	118	9·44	158	12·64	198	15·84
39	3·12	79	6·32	119	9·52	159	12·72	199	15·92
40	3·20	**80**	6·40	**120**	9·60	**160**	12·80	**200**	16·00

$\frac{1}{16}=0{\cdot}00^{5}$ $\frac{1}{9}=0{\cdot}00^{9}$ $\frac{1}{8}=0{\cdot}01^{0}$ $\frac{1}{6}=0{\cdot}01^{3}$ $\frac{1}{5}=0{\cdot}01^{6}$ $\frac{1}{4}=0{\cdot}02^{0}$ $\frac{1}{3}=0{\cdot}02^{7}$

$\frac{3}{8}=0{\cdot}03^{0}$ $\frac{1}{2}=0{\cdot}04^{0}$ $\frac{5}{8}=0{\cdot}05^{0}$ $\frac{3}{4}=0{\cdot}06^{0}$ $\frac{7}{8}=0{\cdot}07^{0}$ $\frac{2}{3}=0{\cdot}05^{3}$ $\frac{5}{6}=0{\cdot}06^{7}$

$\frac{3}{16}=0{\cdot}01^{5}$ $\frac{5}{16}=0{\cdot}02^{5}$ $\frac{7}{16}=0{\cdot}03^{5}$ $\frac{9}{16}=0{\cdot}04^{5}$ $\frac{11}{16}=0{\cdot}05^{5}$ $\frac{13}{16}=0{\cdot}06^{5}$ $\frac{15}{16}=0{\cdot}07^{5}$

(11.8=1£) **£0.08½**

1	0·08½	41	3·48½	81	6·88½	121	10·28½	161	13·68½
2	0·17	42	3·57	82	6·97	122	10·37	162	13·77
3	0·25½	43	3·65½	83	7·05½	123	10·45½	163	13·85½
4	0·34	44	3·74	84	7·14	124	10·54	164	13·94
5	0·42½	45	3·82½	85	7·22½	125	10·62½	165	14·02½
6	0·51	46	3·91	86	7·31	126	10·71	166	14·11
7	0·59½	47	3·99½	87	7·39½	127	10·79½	167	14·19½
8	0·68	48	4·08	88	7·48	128	10·88	168	14·28
9	0·76½	49	4·16½	89	7·56½	129	10·96½	169	14·36½
10	0·85	**50**	4·25	**90**	7·65	**130**	11·05	**170**	14·45
11	0·93½	51	4·33½	91	7·73½	131	11·13½	171	14·53½
12	1·02	52	4·42	92	7·82	132	11·22	172	14·62
13	1·10½	53	4·50½	93	7·90½	133	11·30½	173	14·70½
14	1·19	54	4·59	94	7·99	134	11·39	174	14·79
15	1·27½	55	4·67½	95	8·07½	135	11·47½	175	14·87½
16	1·36	56	4·76	96	8·16	136	11·56	176	14·96
17	1·44½	57	4·84½	97	8·24½	137	11·64½	177	15·04½
18	1·53	58	4·93	98	8·33	138	11·73	178	15·13
19	1·61½	59	5·01½	99	8·41½	139	11·81½	179	15·21½
20	1·70	**60**	5·10	**100**	8·50	**140**	11·90	**180**	15·30
21	1·78½	61	5·18½	101	8·58½	141	11·98½	181	15·38½
22	1·87	62	5·27	102	8·67	142	12·07	182	15·47
23	1·95½	63	5·35½	103	8·75½	143	12·15½	183	15·55½
24	2·04	64	5·44	104	8·84	144	12·24	184	15·64
25	2·12½	65	5·52½	105	8·92½	145	12·32½	185	15·72½
26	2·21	66	5·61	106	9·01	146	12·41	186	15·81
27	2·29½	67	5·69½	107	9·09½	147	12·49½	187	15·89½
28	2·38	68	5·78	108	9·18	148	12·58	188	15·98
29	2·46½	69	5·86½	109	9·26½	149	12·66½	189	16·06½
30	2·55	**70**	5·95	**110**	9·35	**150**	12·75	**190**	16·15
31	2·63½	71	6·03½	111	9·43½	151	12·83½	191	16·23½
32	2·72	72	6·12	112	9·52	152	12·92	192	16·32
33	2·80½	73	6·20½	113	9·60½	153	13·00½	193	16·40½
34	2·89	74	6·29	114	9·69	154	13·09	194	16·49
35	2·97½	75	6·37½	115	9·77½	155	13·17½	195	16·57½
36	3·06	76	6·46	116	9·86	156	13·26	196	16·66
37	3·14½	77	6·54½	117	9·94½	157	13·34½	197	16·74½
38	3·23	78	6·63	118	10·03	158	13·43	198	16·83
39	3·31½	79	6·71½	119	10·11½	159	13·51½	199	16·91½
40	3·40	**80**	6·80	**120**	10·20	**160**	13·60	**200**	17·00

$\frac{1}{16}=0{\cdot}00^{5}$ $\frac{1}{9}=0{\cdot}00^{9}$ $\frac{1}{8}=0{\cdot}01^{1}$ $\frac{1}{6}=0{\cdot}01^{4}$ $\frac{1}{5}=0{\cdot}01^{7}$ $\frac{1}{4}=0{\cdot}02^{1}$ $\frac{1}{3}=0{\cdot}02^{8}$

$\frac{3}{8}=0{\cdot}03^{2}$ $\frac{1}{2}=0{\cdot}04^{3}$ $\frac{5}{8}=0{\cdot}05^{3}$ $\frac{3}{4}=0{\cdot}06^{4}$ $\frac{7}{8}=0{\cdot}07^{4}$ $\frac{2}{3}=0{\cdot}05^{7}$ $\frac{5}{6}=0{\cdot}07^{1}$

$\frac{3}{16}=0{\cdot}01^{6}$ $\frac{5}{16}=0{\cdot}02^{7}$ $\frac{7}{16}=0{\cdot}03^{7}$ $\frac{9}{16}=0{\cdot}04^{8}$ $\frac{11}{16}=0{\cdot}05^{8}$ $\frac{13}{16}=0{\cdot}06^{9}$ $\frac{15}{16}=0{\cdot}08^{0}$

(11.1=1£) **£0.09**

1	0·09	41	3·69	81	7·29	121	10·89	161	14·49
2	0·18	42	3·78	82	7·38	122	10·98	162	14·58
3	0·27	43	3·87	83	7·47	123	11·07	163	14·67
4	0·36	44	3·96	84	7·56	124	11·16	164	14·76
5	0·45	45	4·05	85	7·65	125	11·25	165	14·85
6	0·54	46	4·14	86	7·74	126	11·34	166	14·94
7	0·63	47	4·23	87	7·83	127	11·43	167	15·03
8	0·72	48	4·32	88	7·92	128	11·52	168	15·12
9	0·81	49	4·41	89	8·01	129	11·61	169	15·21
10	0·90	**50**	4·50	**90**	8·10	**130**	11·70	**170**	15·30
11	0·99	51	4·59	91	8·19	131	11·79	171	15·39
12	1·08	52	4·68	92	8·28	132	11·88	172	15·48
13	1·17	53	4·77	93	8·37	133	11·97	173	15·57
14	1·26	54	4·86	94	8·46	134	12·06	174	15·66
15	1·35	55	4·95	95	8·55	135	12·15	175	15·75
16	1·44	56	5·04	96	8·64	136	12·24	176	15·84
17	1·53	57	5·13	97	8·73	137	12·33	177	15·93
18	1·62	58	5·22	98	8·82	138	12·42	178	16·02
19	1·71	59	5·31	99	8·91	139	12·51	179	16·11
20	1·80	**60**	5·40	**100**	9·00	**140**	12·60	**180**	16·20
21	1·89	61	5·49	101	9·09	141	12·69	181	16·29
22	1·98	62	5·58	102	9·18	142	12·78	182	16·38
23	2·07	63	5·67	103	9·27	143	12·87	183	16·47
24	2·16	64	5·76	104	9·36	144	12·96	184	16·56
25	2·25	65	5·85	105	9·45	145	13·05	185	16·65
26	2·34	66	5·94	106	9·54	146	13·14	186	16·74
27	2·43	67	6·03	107	9·63	147	13·23	187	16·83
28	2·52	68	6·12	108	9·72	148	13·32	188	16·92
29	2·61	69	6·21	109	9·81	149	13·41	189	17·01
30	2·70	**70**	6·30	**110**	9·90	**150**	13·50	**190**	17·10
31	2·79	71	6·39	111	9·99	151	13·59	191	17·19
32	2·88	72	6·48	112	10·08	152	13·68	192	17·28
33	2·97	73	6·57	113	10·17	153	13·77	193	17·37
34	3·06	74	6·66	114	10·26	154	13·86	194	17·46
35	3·15	75	6·75	115	10·35	155	13·95	195	17·55
36	3·24	76	6·84	116	10·44	156	14·04	196	17·64
37	3·33	77	6·93	117	10·53	157	14·13	197	17·73
38	3·42	78	7·02	118	10·62	158	14·22	198	17·82
39	3·51	79	7·11	119	10·71	159	14·31	199	17·91
40	3·60	**80**	7·20	**120**	10·80	**160**	14·40	**200**	18·00

$\frac{1}{16}=0{\cdot}00^{6}$ $\frac{1}{9}=0{\cdot}01^{0}$ $\frac{1}{8}=0{\cdot}01^{1}$ $\frac{1}{6}=0{\cdot}01^{5}$ $\frac{1}{5}=0{\cdot}01^{8}$ $\frac{1}{4}=0{\cdot}02^{3}$ $\frac{1}{3}=0{\cdot}03^{0}$

$\frac{3}{8}=0{\cdot}03^{4}$ $\frac{1}{2}=0{\cdot}04^{5}$ $\frac{5}{8}=0{\cdot}05^{6}$ $\frac{3}{4}=0{\cdot}06^{8}$ $\frac{7}{8}=0{\cdot}07^{9}$ $\frac{2}{3}=0{\cdot}06^{0}$ $\frac{5}{6}=0{\cdot}07^{5}$

$\frac{3}{16}=0{\cdot}01^{7}$ $\frac{5}{16}=0{\cdot}02^{8}$ $\frac{7}{16}=0{\cdot}03^{9}$ $\frac{9}{16}=0{\cdot}05^{1}$ $\frac{11}{16}=0{\cdot}06^{2}$ $\frac{13}{16}=0{\cdot}07^{3}$ $\frac{15}{16}=0{\cdot}08^{4}$

(10.5=1£) **£0.09½**

1	0·09½	41	3·89½	81	7·69½	121	11·49½	161	15·29½
2	0·19	42	3·99	82	7·79	122	11·59	162	15·39
3	0·28½	43	4·08½	83	7·88½	123	11·68½	163	15·48½
4	0·38	44	4·18	84	7·98	124	11·78	164	15·58
5	0·47½	45	4·27½	85	8·07½	125	11·87½	165	15·67½
6	0·57	46	4·37	86	8·17	126	11·97	166	15·77
7	0·66½	47	4·46½	87	8·26½	127	12·06½	167	15·86½
8	0·76	48	4·56	88	8·36	128	12·16	168	15·96
9	0·85½	49	4·65½	89	8·45½	129	12·25½	169	16·05½
10	0·95	**50**	4·75	**90**	8·55	**130**	12·35	**170**	16·15
11	1·04½	51	4·84½	91	8·64½	131	12·44½	171	16·24½
12	1·14	52	4·94	92	8·74	132	12·54	172	16·34
13	1·23½	53	5·03½	93	8·83½	133	12·63½	173	16·43½
14	1·33	54	5·13	94	8·93	134	12·73	174	16·53
15	1·42½	55	5·22½	95	9·02½	135	12·82½	175	16·62½
16	1·52	56	5·32	96	9·12	136	12·92	176	16·72
17	1·61½	57	5·41½	97	9·21½	137	13·01½	177	16·81½
18	1·71	58	5·51	98	9·31	138	13·11	178	16·91
19	1·80½	59	5·60½	99	9·40½	139	13·20½	179	17·00½
20	1·90	**60**	5·70	**100**	9·50	**140**	13·30	**180**	17·10
21	1·99½	61	5·79½	101	9·59½	141	13·39½	181	17·19½
22	2·09	62	5·89	102	9·69	142	13·49	182	17·29
23	2·18½	63	5·98½	103	9·78½	143	13·58½	183	17·38½
24	2·28	64	6·08	104	9·88	144	13·68	184	17·48
25	2·37½	65	6·17½	105	9·97½	145	13·77½	185	17·57½
26	2·47	66	6·27	106	10·07	146	13·87	186	17·67
27	2·56½	67	6·36½	107	10·16½	147	13·96½	187	17·76½
28	2·66	68	6·46	108	10·26	148	14·06	188	17·86
29	2·75½	69	6·55½	109	10·35½	149	14·15½	189	17·95½
30	2·85	**70**	6·65	**110**	10·45	**150**	14·25	**190**	18·05
31	2·94½	71	6·74½	111	10·54½	151	14·34½	191	18·14½
32	3·04	72	6·84	112	10·64	152	14·44	192	18·24
33	3·13½	73	6·93½	113	10·73½	153	14·53½	193	18·33½
34	3·23	74	7·03	114	10·83	154	14·63	194	18·43
35	3·32½	75	7·12½	115	10·92½	155	14·72½	195	18·52½
36	3·42	76	7·22	116	11·02	156	14·82	196	18·62
37	3·51½	77	7·31½	117	11·11½	157	14·91½	197	18·71½
38	3·61	78	7·41	118	11·21	158	15·01	198	18·81
39	3·70½	79	7·50½	119	11·30½	159	15·10½	199	18·90½
40	3·80	**80**	7·60	**120**	11·40	**160**	15·20	**200**	19·00

$\frac{1}{16}=0{\cdot}00^6$ $\frac{1}{9}=0{\cdot}01^1$ $\frac{1}{8}=0{\cdot}01^2$ $\frac{1}{6}=0{\cdot}01^6$ $\frac{1}{5}=0{\cdot}01^9$ $\frac{1}{4}=0{\cdot}02^4$ $\frac{1}{3}=0{\cdot}03^2$

$\frac{3}{8}=0{\cdot}03^6$ $\frac{1}{2}=0{\cdot}04^8$ $\frac{5}{8}=0{\cdot}05^9$ $\frac{3}{4}=0{\cdot}07^1$ $\frac{7}{8}=0{\cdot}08^3$ $\frac{2}{3}=0{\cdot}06^3$ $\frac{5}{6}=0{\cdot}07^9$

$\frac{3}{16}=0{\cdot}01^8$ $\frac{5}{16}=0{\cdot}03^0$ $\frac{7}{16}=0{\cdot}04^2$ $\frac{9}{16}=0{\cdot}05^3$ $\frac{11}{16}=0{\cdot}06^5$ $\frac{13}{16}=0{\cdot}07^7$ $\frac{15}{16}=0{\cdot}08^9$

(10.0=1£) **£0.10**

1	0·10	41	4·10	81	8·10	121	12·10	161	16·10
2	0·20	42	4·20	82	8·20	122	12·20	162	16·20
3	0·30	43	4·30	83	8·30	123	12·30	163	16·30
4	0·40	44	4·40	84	8·40	124	12·40	164	16·40
5	0·50	45	4·50	85	8·50	125	12·50	165	16·50
6	0·60	46	4·60	86	8·60	126	12·60	166	16·60
7	0·70	47	4·70	87	8·70	127	12·70	167	16·70
8	0·80	48	4·80	88	8·80	128	12·80	168	16·80
9	0·90	49	4·90	89	8·90	129	12·90	169	16·90
10	1·00	**50**	5·00	**90**	9·00	**130**	13·00	**170**	17·00
11	1·10	51	5·10	91	9·10	131	13·10	171	17·10
12	1·20	52	5·20	92	9·20	132	13·20	172	17·20
13	1·30	53	5·30	93	9·30	133	13·30	173	17·30
14	1·40	54	5·40	94	9·40	134	13·40	174	17·40
15	1·50	55	5·50	95	9·50	135	13·50	175	17·50
16	1·60	56	5·60	96	9·60	136	13·60	176	17·60
17	1·70	57	5·70	97	9·70	137	13·70	177	17·70
18	1·80	58	5·80	98	9·80	138	13·80	178	17·80
19	1·90	59	5·90	99	9·90	139	13·90	179	17·90
20	2·00	**60**	6·00	**100**	10·00	**140**	14·00	**180**	18·00
21	2·10	61	6·10	101	10·10	141	14·10	181	18·10
22	2·20	62	6·20	102	10·20	142	14·20	182	18·20
23	2·30	63	6·30	103	10·30	143	14·30	183	18·30
24	2·40	64	6·40	104	10·40	144	14·40	184	18·40
25	2·50	65	6·50	105	10·50	145	14·50	185	18·50
26	2·60	66	6·60	106	10·60	146	14·60	186	18·60
27	2·70	67	6·70	107	10·70	147	14·70	187	18·70
28	2·80	68	6·80	108	10·80	148	14·80	188	18·80
29	2·90	69	6·90	109	10·90	149	14·90	189	18·90
30	3·00	**70**	7·00	**110**	11·00	**150**	15·00	**190**	19·00
31	3·10	71	7·10	111	11·10	151	15·10	191	19·10
32	3·20	72	7·20	112	11·20	152	15·20	192	19·20
33	3·30	73	7·30	113	11·30	153	15·30	193	19·30
34	3·40	74	7·40	114	11·40	154	15·40	194	19·40
35	3·50	75	7·50	115	11·50	155	15·50	195	19·50
36	3·60	76	7·60	116	11·60	156	15·60	196	19·60
37	3·70	77	7·70	117	11·70	157	15·70	197	19·70
38	3·80	78	7·80	118	11·80	158	15·80	198	19·80
39	3·90	79	7·90	119	11·90	159	15·90	199	19·90
40	4·00	**80**	8·00	**120**	12·00	**160**	16·00	**200**	20·00

$\frac{1}{16}=0{\cdot}00^{6}$ $\frac{1}{9}=0{\cdot}01^{1}$ $\frac{1}{8}=0{\cdot}01^{3}$ $\frac{1}{6}=0{\cdot}01^{7}$ $\frac{1}{5}=0{\cdot}02^{0}$ $\frac{1}{4}=0{\cdot}02^{5}$ $\frac{1}{3}=0{\cdot}03^{3}$

$\frac{3}{8}=0{\cdot}03^{8}$ $\frac{1}{2}=0{\cdot}05^{0}$ $\frac{5}{8}=0{\cdot}06^{3}$ $\frac{3}{4}=0{\cdot}07^{5}$ $\frac{7}{8}=0{\cdot}08^{8}$ $\frac{2}{3}=0{\cdot}06^{7}$ $\frac{5}{6}=0{\cdot}08^{3}$

$\frac{3}{16}=0{\cdot}01^{9}$ $\frac{5}{16}=0{\cdot}03^{1}$ $\frac{7}{16}=0{\cdot}04^{4}$ $\frac{9}{16}=0{\cdot}05^{6}$ $\frac{11}{16}=0{\cdot}06^{9}$ $\frac{13}{16}=0{\cdot}08^{1}$ $\frac{15}{16}=0{\cdot}09^{4}$

(9.52=1£) **£0.10½**

1	0·10½	41	4·30½	81	8·50½	121	12·70½	161	16·90½
2	0·21	42	4·41	82	8·61	122	12·81	162	17·01
3	0·31½	43	4·51½	83	8·71½	123	12·91½	163	17·11½
4	0·42	44	4·62	84	8·82	124	13·02	164	17·22
5	0·52½	45	4·72½	85	8·92½	125	13·12½	165	17·32½
6	0·63	46	4·83	86	9·03	126	13·23	166	17·43
7	0·73½	47	4·93½	87	9·13½	127	13·33½	167	17·53½
8	0·84	48	5·04	88	9·24	128	13·44	168	17·64
9	0·94½	49	5·14½	89	9·34½	129	13·54½	169	17·74½
10	1·05	**50**	5·25	**90**	9·45	**130**	13·65	**170**	17·85
11	1·15½	51	5·35½	91	9·55½	131	13·75½	171	17·95½
12	1·26	52	5·46	92	9·66	132	13·86	172	18·06
13	1·36½	53	5·56½	93	9·76½	133	13·96½	173	18·16½
14	1·47	54	5·67	94	9·87	134	14·07	174	18·27
15	1·57½	55	5·77½	95	9·97½	135	14·17½	175	18·37½
16	1·68	56	5·88	96	10·08	136	14·28	176	18·48
17	1·78½	57	5·98½	97	10·18½	137	14·38½	177	18·58½
18	1·89	58	6·09	98	10·29	138	14·49	178	18·69
19	1·99½	59	6·19½	99	10·39½	139	14·59½	179	18·79½
20	2·10	**60**	6·30	**100**	10·50	**140**	14·70	**180**	18·90
21	2·20½	61	6·40½	101	10·60½	141	14·80½	181	19·00½
22	2·31	62	6·51	102	10·71	142	14·91	182	19·11
23	2·41½	63	6·61½	103	10·81½	143	15·01½	183	19·21½
24	2·52	64	6·72	104	10·92	144	15·12	184	19·32
25	2·62½	65	6·82½	105	11·02½	145	15·22½	185	19·42½
26	2·73	66	6·93	106	11·13	146	15·33	186	19·53
27	2·83½	67	7·03½	107	11·23½	147	15·43½	187	19·63½
28	2·94	68	7·14	108	11·34	148	15·54	188	19·74
29	3·04½	69	7·24½	109	11·44½	149	15·64½	189	19·84½
30	3·15	**70**	7·35	**110**	11·55	**150**	15·75	**190**	19·95
31	3·25½	71	7·45½	111	11·65½	151	15·85½	191	20·05½
32	3·36	72	7·56	112	11·76	152	15·96	192	20·16
33	3·46½	73	7·66½	113	11·86½	153	16·06½	193	20·26½
34	3·57	74	7·77	114	11·97	154	16·17	194	20·37
35	3·67½	75	7·87½	115	12·07½	155	16·27½	195	20·47½
36	3·78	76	7·98	116	12·18	156	16·38	196	20·58
37	3·88½	77	8·08½	117	12·28½	157	16·48½	197	20·68½
38	3·99	78	8·19	118	12·39	158	16·59	198	20·79
39	4·09½	79	8·29½	119	12·49½	159	16·69½	199	20·89½
40	4·20	**80**	8·40	**120**	12·60	**160**	16·80	**200**	21·00

$\frac{1}{16}=0{\cdot}00^{7}$ $\frac{1}{9}=0{\cdot}01^{2}$ $\frac{1}{8}=0{\cdot}01^{3}$ $\frac{1}{6}=0{\cdot}01^{8}$ $\frac{1}{5}=0{\cdot}02^{1}$ $\frac{1}{4}=0{\cdot}02^{6}$ $\frac{1}{3}=0{\cdot}03^{5}$

$\frac{3}{8}=0{\cdot}03^{9}$ $\frac{1}{2}=0{\cdot}05^{3}$ $\frac{5}{8}=0{\cdot}06^{6}$ $\frac{3}{4}=0{\cdot}07^{9}$ $\frac{7}{8}=0{\cdot}09^{2}$ $\frac{2}{3}=0{\cdot}07^{0}$ $\frac{5}{6}=0{\cdot}08^{8}$

$\frac{3}{16}=0{\cdot}02^{0}$ $\frac{5}{16}=0{\cdot}03^{3}$ $\frac{7}{16}=0{\cdot}04^{6}$ $\frac{9}{16}=0{\cdot}05^{9}$ $\frac{11}{16}=0{\cdot}07^{2}$ $\frac{13}{16}=0{\cdot}08^{5}$ $\frac{15}{16}=0{\cdot}09^{8}$

(9.09=1£) **£0.11**

1	0·11	41	4·51	81	8·91	121	13·31	161	17·71
2	0·22	42	4·62	82	9·02	122	13·42	162	17·82
3	0·33	43	4·73	83	9·13	123	13·53	163	17·93
4	0·44	44	4·84	84	9·24	124	13·64	164	18·04
5	0·55	45	4·95	85	9·35	125	13·75	165	18·15
6	0·66	46	5·06	86	9·46	126	13·86	166	18·26
7	0·77	47	5·17	87	9·57	127	13·97	167	18·37
8	0·88	48	5·28	88	9·68	128	14·08	168	18·48
9	0·99	49	5·39	89	9·79	129	14·19	169	18·59
10	1·10	**50**	5·50	**90**	9·90	**130**	14·30	**170**	18·70
11	1·21	51	5·61	91	10·01	131	14·41	171	18·81
12	1·32	52	5·72	92	10·12	132	14·52	172	18·92
13	1·43	53	5·83	93	10·23	133	14·63	173	19·03
14	1·54	54	5·94	94	10·34	134	14·74	174	19·14
15	1·65	55	6·05	95	10·45	135	14·85	175	19·25
16	1·76	56	6·16	96	10·56	136	14·96	176	19·36
17	1·87	57	6·27	97	10·67	137	15·07	177	19·47
18	1·98	58	6·38	98	10·78	138	15·18	178	19·58
19	2·09	59	6·49	99	10·89	139	15·29	179	19·69
20	2·20	**60**	6·60	**100**	11·00	**140**	15·40	**180**	19·80
21	2·31	61	6·71	101	11·11	141	15·51	181	19·91
22	2·42	62	6·82	102	11·22	142	15·62	182	20·02
23	2·53	63	6·93	103	11·33	143	15·73	183	20·13
24	2·64	64	7·04	104	11·44	144	15·84	184	20·24
25	2·75	65	7·15	105	11·55	145	15·95	185	20·35
26	2·86	66	7·26	106	11·66	146	16·06	186	20·46
27	2·97	67	7·37	107	11·77	147	16·17	187	20·57
28	3·08	68	7·48	108	11·88	148	16·28	188	20·68
29	3·19	69	7·59	109	11·99	149	16·39	189	20·79
30	3·30	**70**	7·70	**110**	12·10	**150**	16·50	**190**	20·90
31	3·41	71	7·81	111	12·21	151	16·61	191	21·01
32	3·52	72	7·92	112	12·32	152	16·72	192	21·12
33	3·63	73	8·03	113	12·43	153	16·83	193	21·23
34	3·74	74	8·14	114	12·54	154	16·94	194	21·34
35	3·85	75	8·25	115	12·65	155	17·05	195	21·45
36	3·96	76	8·36	116	12·76	156	17·16	196	21·56
37	4·07	77	8·47	117	12·87	157	17·27	197	21·67
38	4·18	78	8·58	118	12·98	158	17·38	198	21·78
39	4·29	79	8·69	119	13·09	159	17·49	199	21·89
40	4·40	**80**	8·80	**120**	13·20	**160**	17·60	**200**	22·00

$\frac{1}{16}=0{\cdot}00^{7}$ $\frac{1}{9}=0{\cdot}01^{2}$ $\frac{1}{8}=0{\cdot}01^{4}$ $\frac{1}{6}=0{\cdot}01^{8}$ $\frac{1}{5}=0{\cdot}02^{2}$ $\frac{1}{4}=0{\cdot}02^{8}$ $\frac{1}{3}=0{\cdot}03^{7}$

$\frac{3}{8}=0{\cdot}04^{1}$ $\frac{1}{2}=0{\cdot}05^{5}$ $\frac{5}{8}=0{\cdot}06^{9}$ $\frac{3}{4}=0{\cdot}08^{3}$ $\frac{7}{8}=0{\cdot}09^{6}$ $\frac{2}{3}=0{\cdot}07^{3}$ $\frac{5}{6}=0{\cdot}09^{2}$

$\frac{3}{16}=0{\cdot}02^{1}$ $\frac{5}{16}=0{\cdot}03^{4}$ $\frac{7}{16}=0{\cdot}04^{8}$ $\frac{9}{16}=0{\cdot}06^{2}$ $\frac{11}{16}=0{\cdot}07^{6}$ $\frac{13}{16}=0{\cdot}08^{9}$ $\frac{15}{16}=0{\cdot}10^{3}$

1	0·11½	41	4·71½	81	9·31½	121	13·91½	161	18·51½
2	0·23	42	4·83	82	9·43	122	14·03	162	18·63
3	0·34½	43	4·94½	83	9·54½	123	14·14½	163	18·74½
4	0·46	44	5·06	84	9·66	124	14·26	164	18·86
5	0·57½	45	5·17½	85	9·77½	125	14·37½	165	18·97½
6	0·69	46	5·29	86	9·89	126	14·49	166	19·09
7	0·80½	47	5·40½	87	10·00½	127	14·60½	167	19·20½
8	0·92	48	5·52	88	10·12	128	14·72	168	19·32
9	1·03½	49	5·63½	89	10·23½	129	14·83½	169	19·43½
10	1·15	**50**	5·75	**90**	10·35	**130**	14·95	**170**	19·55
11	1·26½	51	5·86½	91	10·46½	131	15·06½	171	19·66½
12	1·38	52	5·98	92	10·58	132	15·18	172	19·78
13	1·49½	53	6·09½	93	10·69½	133	15·29½	173	19·89½
14	1·61	54	6·21	94	10·81	134	15·41	174	20·01
15	1·72½	55	6·32½	95	10·92½	135	15·52½	175	20·12½
16	1·84	56	6·44	96	11·04	136	15·64	176	20·24
17	1·95½	57	6·55½	97	11·15½	137	15·75½	177	20·35½
18	2·07	58	6·67	98	11·27	138	15·87	178	20·47
19	2·18½	59	6·78½	99	11·38½	139	15·98½	179	20·58½
20	2·30	**60**	6·90	**100**	11·50	**140**	16·10	**180**	20·70
21	2·41½	61	7·01½	101	11·61½	141	16·21½	181	20·81½
22	2·53	62	7·13	102	11·73	142	16·33	182	20·93
23	2·64½	63	7·24½	103	11·84½	143	16·44½	183	21·04½
24	2·76	64	7·36	104	11·96	144	16·56	184	21·16
25	2·87½	65	7·47½	105	12·07½	145	16·67½	185	21·27½
26	2·99	66	7·59	106	12·19	146	16·79	186	21·39
27	3·10½	67	7·70½	107	12·30½	147	16·90½	187	21·50½
28	3·22	68	7·82	108	12·42	148	17·02	188	21·62
29	3·33½	69	7·93½	109	12·53½	149	17·13½	189	21·73½
30	3·45	**70**	8·05	**110**	12·65	**150**	17·25	**190**	21·85
31	3·56½	71	8·16½	111	12·76½	151	17·36½	191	21·96½
32	3·68	72	8·28	112	12·88	152	17·48	192	22·08
33	3·79½	73	8·39½	113	12·99½	153	17·59½	193	22·19½
34	3·91	74	8·51	114	13·11	154	17·71	194	22·31
35	4·02½	75	8·62½	115	13·22½	155	17·82½	195	22·42½
36	4·14	76	8·74	116	13·34	156	17·94	196	22·54
37	4·25½	77	8·85½	117	13·45½	157	18·05½	197	22·65½
38	4·37	78	8·97	118	13·57	158	18·17	198	22·77
39	4·48½	79	9·08½	119	13·68½	159	18·28½	199	22·88½
40	4·60	**80**	9·20	**120**	13·80	**160**	18·40	**200**	23·00

$\frac{1}{16}=0{\cdot}00^{7}$ $\frac{1}{9}=0{\cdot}01^{3}$ $\frac{1}{8}=0{\cdot}01^{4}$ $\frac{1}{6}=0{\cdot}01^{9}$ $\frac{1}{5}=0{\cdot}02^{3}$ $\frac{1}{4}=0{\cdot}02^{9}$ $\frac{1}{3}=0{\cdot}03^{8}$

$\frac{3}{8}=0{\cdot}04^{3}$ $\frac{1}{2}=0{\cdot}05^{8}$ $\frac{5}{8}=0{\cdot}07^{2}$ $\frac{3}{4}=0{\cdot}08^{6}$ $\frac{7}{8}=0{\cdot}10^{1}$ $\frac{2}{3}=0{\cdot}07^{7}$ $\frac{5}{6}=0{\cdot}09^{6}$

$\frac{3}{16}=0{\cdot}02^{2}$ $\frac{5}{16}=0{\cdot}03^{6}$ $\frac{7}{16}=0{\cdot}05^{0}$ $\frac{9}{16}=0{\cdot}06^{5}$ $\frac{11}{16}=0{\cdot}07^{9}$ $\frac{13}{16}=0{\cdot}09^{3}$ $\frac{15}{16}=0{\cdot}10^{8}$

(8.33=1£) **£0.12**

1	0·12	41	4·92	81	9·72	121	14·52	161	19·32
2	0·24	42	5·04	82	9·84	122	14·64	162	19·44
3	0·36	43	5·16	83	9·96	123	14·76	163	19·56
4	0·48	44	5·28	84	10·08	124	14·88	164	19·68
5	0·60	45	5·40	85	10·20	125	15·00	165	19·80
6	0·72	46	5·52	86	10·32	126	15·12	166	19·92
7	0·84	47	5·64	87	10·44	127	15·24	167	20·04
8	0·96	48	5·76	88	10·56	128	15·36	168	20·16
9	1·08	49	5·88	89	10·68	129	15·48	169	20·28
10	1·20	**50**	6·00	**90**	10·80	**130**	15·60	**170**	20·40
11	1·32	51	6·12	91	10·92	131	15·72	171	20·52
12	1·44	52	6·24	92	11·04	132	15·84	172	20·64
13	1·56	53	6·36	93	11·16	133	15·96	173	20·76
14	1·68	54	6·48	94	11·28	134	16·08	174	20·88
15	1·80	55	6·60	95	11·40	135	16·20	175	21·00
16	1·92	56	6·72	96	11·52	136	16·32	176	21·12
17	2·04	57	6·84	97	11·64	137	16·44	177	21·24
18	2·16	58	6·96	98	11·76	138	16·56	178	21·36
19	2·28	59	7·08	99	11·88	139	16·68	179	21·48
20	2·40	**60**	7·20	**100**	12·00	**140**	16·80	**180**	21·60
21	2·52	61	7·32	101	12·12	141	16·92	181	21·72
22	2·64	62	7·44	102	12·24	142	17·04	182	21·84
23	2·76	63	7·56	103	12·36	143	17·16	183	21·96
24	2·88	64	7·68	104	12·48	144	17·28	184	22·08
25	3·00	65	7·80	105	12·60	145	17·40	185	22·20
26	3·12	66	7·92	106	12·72	146	17·52	186	22·32
27	3·24	67	8·04	107	12·84	147	17·64	187	22·44
28	3·36	68	8·16	108	12·96	148	17·76	188	22·56
29	3·48	69	8·28	109	13·08	149	17·88	189	22·68
30	3·60	**70**	8·40	**110**	13·20	**150**	18·00	**190**	22·80
31	3·72	71	8·52	111	13·32	151	18·12	191	22·92
32	3·84	72	8·64	112	13·44	152	18·24	192	23·04
33	3·96	73	8·76	113	13·56	153	18·36	193	23·16
34	4·08	74	8·88	114	13·68	154	18·48	194	23·28
35	4·20	75	9·00	115	13·80	155	18·60	195	23·40
36	4·32	76	9·12	116	13·92	156	18·72	196	23·52
37	4·44	77	9·24	117	14·04	157	18·84	197	23·64
38	4·56	78	9·36	118	14·16	158	18·96	198	23·76
39	4·68	79	9·48	119	14·28	159	19·08	199	23·88
40	4·80	**80**	9·60	**120**	14·40	**160**	19·20	**200**	24·00

$\frac{1}{16}=0{\cdot}00^{8}$	$\frac{1}{9}=0{\cdot}01^{3}$	$\frac{1}{8}=0{\cdot}01^{5}$	$\frac{1}{6}=0{\cdot}02^{0}$	$\frac{1}{5}=0{\cdot}02^{4}$	$\frac{1}{4}=0{\cdot}03^{0}$	$\frac{1}{3}=0{\cdot}04^{0}$
$\frac{3}{8}=0{\cdot}04^{5}$	$\frac{1}{2}=0{\cdot}06^{0}$	$\frac{5}{8}=0{\cdot}07^{5}$	$\frac{3}{4}=0{\cdot}09^{0}$	$\frac{7}{8}=0{\cdot}10^{5}$	$\frac{2}{3}=0{\cdot}08^{0}$	$\frac{5}{6}=0{\cdot}10^{0}$
$\frac{3}{16}=0{\cdot}02^{3}$	$\frac{5}{16}=0{\cdot}03^{8}$	$\frac{7}{16}=0{\cdot}05^{3}$	$\frac{9}{16}=0{\cdot}06^{8}$	$\frac{11}{16}=0{\cdot}08^{3}$	$\frac{13}{16}=0{\cdot}09^{8}$	$\frac{15}{16}=0{\cdot}11^{3}$

1	0·12½	41	5·12½	81	10·12½	121	15·12½	161	20·12½
2	0·25	42	5·25	82	10·25	122	15·25	162	20·25
3	0·37½	43	5·37½	83	10·37½	123	15·37½	163	20·37½
4	0·50	44	5·50	84	10·50	124	15·50	164	20·50
5	0·62½	45	5·62½	85	10·62½	125	15·62½	165	20·62½
6	0·75	46	5·75	86	10·75	126	15·75	166	20·75
7	0·87½	47	5·87½	87	10·87½	127	15·87½	167	20·87½
8	1·00	48	6·00	88	11·00	128	16·00	168	21·00
9	1·12½	49	6·12½	89	11·12½	129	16·12½	169	21·12½
10	1·25	**50**	6·25	**90**	11·25	**130**	16·25	**170**	21·25
11	1·37½	51	6·37½	91	11·37½	131	16·37½	171	21·37½
12	1·50	52	6·50	92	11·50	132	16·50	172	21·50
13	1·62½	53	6·62½	93	11·62½	133	16·62½	173	21·62½
14	1·75	54	6·75	94	11·75	134	16·75	174	21·75
15	1·87½	55	6·87½	95	11·87½	135	16·87½	175	21·87½
16	2·00	56	7·00	96	12·00	136	17·00	176	22·00
17	2·12½	57	7·12½	97	12·12½	137	17·12½	177	22·12½
18	2·25	58	7·25	98	12·25	138	17·25	178	22·25
19	2·37½	59	7·37½	99	12·37½	139	17·37½	179	22·37½
20	2·50	**60**	7·50	**100**	12·50	**140**	17·50	**180**	22·50
21	2·62½	61	7·62½	101	12·62½	141	17·62½	181	22·62½
22	2·75	62	7·75	102	12·75	142	17·75	182	22·75
23	2·87½	63	7·87½	103	12·87½	143	17·87½	183	22·87½
24	3·00	64	8·00	104	13·00	144	18·00	184	23·00
25	3·12½	65	8·12½	105	13·12½	145	18·12½	185	23·12½
26	3·25	66	8·25	106	13·25	146	18·25	186	23·25
27	3·37½	67	8·37½	107	13·37½	147	18·37½	187	23·37½
28	3·50	68	8·50	108	13·50	148	18·50	188	23·50
29	3·62½	69	8·62½	109	13·62½	149	18·62½	189	23·62½
30	3·75	**70**	8·75	**110**	13·75	**150**	18·75	**190**	23·75
31	3·87½	71	8·87½	111	13·87½	151	18·87½	191	23·87½
32	4·00	72	9·00	112	14·00	152	19·00	192	24·00
33	4·12½	73	9·12½	113	14·12½	153	19·12½	193	24·12½
34	4·25	74	9·25	114	14·25	154	19·25	194	24·25
35	4·37½	75	9·37½	115	14·37½	155	19·37½	195	24·37½
36	4·50	76	9·50	116	14·50	156	19·50	196	24·50
37	4·62½	77	9·62½	117	14·62½	157	19·62½	197	24·62½
38	4·75	78	9·75	118	14·75	158	19·75	198	24·75
39	4·87½	79	9·87½	119	14·87½	159	19·87½	199	24·87½
40	5·00	**80**	10·00	**120**	15·00	**160**	20·00	**200**	25·00

$\frac{1}{16}=0{\cdot}00^{8}$ $\frac{1}{9}=0{\cdot}01^{4}$ $\frac{1}{8}=0{\cdot}01^{6}$ $\frac{1}{6}=0{\cdot}02^{1}$ $\frac{1}{5}=0{\cdot}02^{5}$ $\frac{1}{4}=0{\cdot}03^{1}$ $\frac{1}{3}=0{\cdot}04^{2}$

$\frac{3}{8}=0{\cdot}04^{7}$ $\frac{1}{2}=0{\cdot}06^{3}$ $\frac{5}{8}=0{\cdot}07^{8}$ $\frac{3}{4}=0{\cdot}09^{4}$ $\frac{7}{8}=0{\cdot}10^{9}$ $\frac{2}{3}=0{\cdot}08^{3}$ $\frac{5}{6}=0{\cdot}10^{4}$

$\frac{3}{16}=0{\cdot}02^{3}$ $\frac{5}{16}=0{\cdot}03^{9}$ $\frac{7}{16}=0{\cdot}05^{5}$ $\frac{9}{16}=0{\cdot}07^{0}$ $\frac{11}{16}=0{\cdot}08^{6}$ $\frac{13}{16}=0{\cdot}10^{2}$ $\frac{15}{16}=0{\cdot}11^{7}$

(7.69=1£) **£0.13**

1	0·13	41	5·33	81	10·53	121	15·73	161	20·93
2	0·26	42	5·46	82	10·66	122	15·86	162	21·06
3	0·39	43	5·59	83	10·79	123	15·99	163	21·19
4	0·52	44	5·72	84	10·92	124	16·12	164	21·32
5	0·65	45	5·85	85	11·05	125	16·25	165	21·45
6	0·78	46	5·98	86	11·18	126	16·38	166	21·58
7	0·91	47	6·11	87	11·31	127	16·51	167	21·71
8	1·04	48	6·24	88	11·44	128	16·64	168	21·84
9	1·17	49	6·37	89	11·57	129	16·77	169	21·97
10	1·30	**50**	6·50	**90**	11·70	**130**	16·90	**170**	22·10
11	1·43	51	6·63	91	11·83	131	17·03	171	22·23
12	1·56	52	6·76	92	11·96	132	17·16	172	22·36
13	1·69	53	6·89	93	12·09	133	17·29	173	22·49
14	1·82	54	7·02	94	12·22	134	17·42	174	22·62
15	1·95	55	7·15	95	12·35	135	17·55	175	22·75
16	2·08	56	7·28	96	12·48	136	17·68	176	22·88
17	2·21	57	7·41	97	12·61	137	17·81	177	23·01
18	2·34	58	7·54	98	12·74	138	17·94	178	23·14
19	2·47	59	7·67	99	12·87	139	18·07	179	23·27
20	2·60	**60**	7·80	**100**	13·00	**140**	18·20	**180**	23·40
21	2·73	61	7·93	101	13·13	141	18·33	181	23·53
22	2·86	62	8·06	102	13·26	142	18·46	182	23·66
23	2·99	63	8·19	103	13·39	143	18·59	183	23·79
24	3·12	64	8·32	104	13·52	144	18·72	184	23·92
25	3·25	65	8·45	105	13·65	145	18·85	185	24·05
26	3·38	66	8·58	106	13·78	146	18·98	186	24·18
27	3·51	67	8·71	107	13·91	147	19·11	187	24·31
28	3·64	68	8·84	108	14·04	148	19·24	188	24·44
29	3·77	69	8·97	109	14·17	149	19·37	189	24·57
30	3·90	**70**	9·10	**110**	14·30	**150**	19·50	**190**	24·70
31	4·03	71	9·23	111	14·43	151	19·63	191	24·83
32	4·16	72	9·36	112	14·56	152	19·76	192	24·96
33	4·29	73	9·49	113	14·69	153	19·89	193	25·09
34	4·42	74	9·62	114	14·82	154	20·02	194	25·22
35	4·55	75	9·75	115	14·95	155	20·15	195	25·35
36	4·68	76	9·88	116	15·08	156	20·28	196	25·48
37	4·81	77	10·01	117	15·21	157	20·41	197	25·61
38	4·94	78	10·14	118	15·34	158	20·54	198	25·74
39	5·07	79	10·27	119	15·47	159	20·67	199	25·87
40	5·20	**80**	10·40	**120**	15·60	**160**	20·80	**200**	26·00

$\frac{1}{16}=0{\cdot}00^{8}$ $\frac{1}{9}=0{\cdot}01^{4}$ $\frac{1}{8}=0{\cdot}01^{6}$ $\frac{1}{6}=0{\cdot}02^{2}$ $\frac{1}{5}=0{\cdot}02^{6}$ $\frac{1}{4}=0{\cdot}03^{3}$ $\frac{1}{3}=0{\cdot}04^{3}$

$\frac{3}{8}=0{\cdot}04^{9}$ $\frac{1}{2}=0{\cdot}06^{5}$ $\frac{5}{8}=0{\cdot}08^{1}$ $\frac{3}{4}=0{\cdot}09^{8}$ $\frac{7}{8}=0{\cdot}11^{4}$ $\frac{2}{3}=0{\cdot}08^{7}$ $\frac{5}{6}=0{\cdot}10^{8}$

$\frac{3}{16}=0{\cdot}02^{4}$ $\frac{5}{16}=0{\cdot}04^{1}$ $\frac{7}{16}=0{\cdot}05^{7}$ $\frac{9}{16}=0{\cdot}07^{3}$ $\frac{11}{16}=0{\cdot}08^{9}$ $\frac{13}{16}=0{\cdot}10^{6}$ $\frac{15}{16}=0{\cdot}12^{2}$

(7.41=1£) **£0.13½**

1	0·13½	41	5·53½	81	10·93½	121	16·33½	161	21·73½
2	0·27	42	5·67	82	11·07	122	16·47	162	21·87
3	0·40½	43	5·80½	83	11·20½	123	16·60½	163	22·00½
4	0·54	44	5·94	84	11·34	124	16·74	164	22·14
5	0·67½	45	6·07½	85	11·47½	125	16·87½	165	22·27½
6	0·81	46	6·21	86	11·61	126	17·01	166	22·41
7	0·94½	47	6·34½	87	11·74½	127	17·14½	167	22·54½
8	1·08	48	6·48	88	11·88	128	17·28	168	22·68
9	1·21½	49	6·61½	89	12·01½	129	17·41½	169	22·81½
10	1·35	**50**	6·75	**90**	12·15	**130**	17·55	**170**	22·95
11	1·48½	51	6·88½	91	12·28½	131	17·68½	171	23·08½
12	1·62	52	7·02	92	12·42	132	17·82	172	23·22
13	1·75½	53	7·15½	93	12·55½	133	17·95½	173	23·35½
14	1·89	54	7·29	94	12·69	134	18·09	174	23·49
15	2·02½	55	7·42½	95	12·82½	135	18·22½	175	23·62½
16	2·16	56	7·56	96	12·96	136	18·36	176	23·76
17	2·29½	57	7·69½	97	13·09½	137	18·49½	177	23·89½
18	2·43	58	7·83	98	13·23	138	18·63	178	24·03
19	2·56½	59	7·96½	99	13·36½	139	18·76½	179	24·16½
20	2·70	**60**	8·10	**100**	13·50	**140**	18·90	**180**	24·30
21	2·83½	61	8·23½	101	13·63½	141	19·03½	181	24·43½
22	2·97	62	8·37	102	13·77	142	19·17	182	24·57
23	3·10½	63	8·50½	103	13·90½	143	19·30½	183	24·70½
24	3·24	64	8·64	104	14·04	144	19·44	184	24·84
25	3·37½	65	8·77½	105	14·17½	145	19·57½	185	24·97½
26	3·51	66	8·91	106	14·31	146	19·71	186	25·11
27	3·64½	67	9·04½	107	14·44½	147	19·84½	187	25·24½
28	3·78	68	9·18	108	14·58	148	19·98	188	25·38
29	3·91½	69	9·31½	109	14·71½	149	20·11½	189	25·51½
30	4·05	**70**	9·45	**110**	14·85	**150**	20·25	**190**	25·65
31	4·18½	71	9·58½	111	14·98½	151	20·38½	191	25·78½
32	4·32	72	9·72	112	15·12	152	20·52	192	25·92
33	4·45½	73	9·85½	113	15·25½	153	20·65½	193	26·05½
34	4·59	74	9·99	114	15·39	154	20·79	194	26·19
35	4·72½	75	10·12½	115	15·52½	155	20·92½	195	26·32½
36	4·86	76	10·26	116	15·66	156	21·06	196	26·46
37	4·99½	77	10·39½	117	15·79½	157	21·19½	197	26·59½
38	5·13	78	10·53	118	15·93	158	21·33	198	26·73
39	5·26½	79	10·66½	119	16·06½	159	21·46½	199	26·86½
40	5·40	**80**	10·80	**120**	16·20	**160**	21·60	**200**	27·00

$\frac{1}{16}=0{\cdot}00^{8}$	$\frac{1}{9}=0{\cdot}01^{5}$	$\frac{1}{8}=0{\cdot}01^{7}$	$\frac{1}{6}=0{\cdot}02^{3}$	$\frac{1}{5}=0{\cdot}02^{7}$	$\frac{1}{4}=0{\cdot}03^{4}$	$\frac{1}{3}=0{\cdot}04^{5}$
$\frac{3}{8}=0{\cdot}05^{1}$	$\frac{1}{2}=0{\cdot}06^{8}$	$\frac{5}{8}=0{\cdot}08^{4}$	$\frac{3}{4}=0{\cdot}10^{1}$	$\frac{7}{8}=0{\cdot}11^{8}$	$\frac{2}{3}=0{\cdot}09^{0}$	$\frac{5}{6}=0{\cdot}11^{3}$
$\frac{3}{16}=0{\cdot}02^{5}$	$\frac{5}{16}=0{\cdot}04^{2}$	$\frac{7}{16}=0{\cdot}05^{9}$	$\frac{9}{16}=0{\cdot}07^{6}$	$\frac{11}{16}=0{\cdot}09^{3}$	$\frac{13}{16}=0{\cdot}11^{0}$	$\frac{15}{16}=0{\cdot}12^{7}$

(7.14=1£) **£0.14**

1	0·14	41	5·74	81	11·34	121	16·94	161	22·54
2	0·28	42	5·88	82	11·48	122	17·08	162	22·68
3	0·42	43	6·02	83	11·62	123	17·22	163	22·82
4	0·56	44	6·16	84	11·76	124	17·36	164	22·96
5	0·70	45	6·30	85	11·90	125	17·50	165	23·10
6	0·84	46	6·44	86	12·04	126	17·64	166	23·24
7	0·98	47	6·58	87	12·18	127	17·78	167	23·38
8	1·12	48	6·72	88	12·32	128	17·92	168	23·52
9	1·26	49	6·86	89	12·46	129	18·06	169	23·66
10	1·40	**50**	7·00	**90**	12·60	**130**	18·20	**170**	23·80
11	1·54	51	7·14	91	12·74	131	18·34	171	23·94
12	1·68	52	7·28	92	12·88	132	18·48	172	24·08
13	1·82	53	7·42	93	13·02	133	18·62	173	24·22
14	1·96	54	7·56	94	13·16	134	18·76	174	24·36
15	2·10	55	7·70	95	13·30	135	18·90	175	24·50
16	2·24	56	7·84	96	13·44	136	19·04	176	24·64
17	2·38	57	7·98	97	13·58	137	19·18	177	24·78
18	2·52	58	8·12	98	13·72	138	19·32	178	24·92
19	2·66	59	8·26	99	13·86	139	19·46	179	25·06
20	2·80	**60**	8·40	**100**	14·00	**140**	19·60	**180**	25·20
21	2·94	61	8·54	101	14·14	141	19·74	181	25·34
22	3·08	62	8·68	102	14·28	142	19·88	182	25·48
23	3·22	63	8·82	103	14·42	143	20·02	183	25·62
24	3·36	64	8·96	104	14·56	144	20·16	184	25·76
25	3·50	65	9·10	105	14·70	145	20·30	185	25·90
26	3·64	66	9·24	106	14·84	146	20·44	186	26·04
27	3·78	67	9·38	107	14·98	147	20·58	187	26·18
28	3·92	68	9·52	108	15·12	148	20·72	188	26·32
29	4·06	69	9·66	109	15·26	149	20·86	189	26·46
30	4·20	**70**	9·80	**110**	15·40	**150**	21·00	**190**	26·60
31	4·34	71	9·94	111	15·54	151	21·14	191	26·74
32	4·48	72	10·08	112	15·68	152	21·28	192	26·88
33	4·62	73	10·22	113	15·82	153	21·42	193	27·02
34	4·76	74	10·36	114	15·96	154	21·56	194	27·16
35	4·90	75	10·50	115	16·10	155	21·70	195	27·30
36	5·04	76	10·64	116	16·24	156	21·84	196	27·44
37	5·18	77	10·78	117	16·38	157	21·98	197	27·58
38	5·32	78	10·92	118	16·52	158	22·12	198	27·72
39	5·46	79	11·06	119	16·66	159	22·26	199	27·86
40	5·60	**80**	11·20	**120**	16·80	**160**	22·40	**200**	28·00

$\frac{1}{16}=0{\cdot}00^{9}$	$\frac{1}{9}=0{\cdot}01^{6}$	$\frac{1}{8}=0{\cdot}01^{8}$	$\frac{1}{6}=0{\cdot}02^{3}$	$\frac{1}{5}=0{\cdot}02^{8}$	$\frac{1}{4}=0{\cdot}03^{5}$	$\frac{1}{3}=0{\cdot}04^{7}$
$\frac{3}{8}=0{\cdot}05^{3}$	$\frac{1}{2}=0{\cdot}07^{0}$	$\frac{5}{8}=0{\cdot}08^{8}$	$\frac{3}{4}=0{\cdot}10^{5}$	$\frac{7}{8}=0{\cdot}12^{3}$	$\frac{2}{3}=0{\cdot}09^{3}$	$\frac{5}{6}=0{\cdot}11^{7}$
$\frac{3}{16}=0{\cdot}02^{6}$	$\frac{5}{16}=0{\cdot}04^{4}$	$\frac{7}{16}=0{\cdot}06^{1}$	$\frac{9}{16}=0{\cdot}07^{9}$	$\frac{11}{16}=0{\cdot}09^{6}$	$\frac{13}{16}=0{\cdot}11^{4}$	$\frac{15}{16}=0{\cdot}13^{1}$

(6.90=1£) **£0.14½**

1	0·14½	41	5·94½	81	11·74½	121	17·54½	161	23·34½
2	0·29	42	6·09	82	11·89	122	17·69	162	23·49
3	0·43½	43	6·23½	83	12·03½	123	17·83½	163	23·63½
4	0·58	44	6·38	84	12·18	124	17·98	164	23·78
5	0·72½	45	6·52½	85	12·32½	125	18·12½	165	23·92½
6	0·87	46	6·67	86	12·47	126	18·27	166	24·07
7	1·01½	47	6·81½	87	12·61½	127	18·41½	167	24·21½
8	1·16	48	6·96	88	12·76	128	18·56	168	24·36
9	1·30½	49	7·10½	89	12·90½	129	18·70½	169	24·50½
10	1·45	**50**	7·25	**90**	13·05	**130**	18·85	**170**	24·65
11	1·59½	51	7·39½	91	13·19½	131	18·99½	171	24·79½
12	1·74	52	7·54	92	13·34	132	19·14	172	24·94
13	1·88½	53	7·68½	93	13·48½	133	19·28½	173	25·08½
14	2·03	54	7·83	94	13·63	134	19·43	174	25·23
15	2·17½	55	7·97½	95	13·77½	135	19·57½	175	25·37½
16	2·32	56	8·12	96	13·92	136	19·72	176	25·52
17	2·46½	57	8·26½	97	14·06½	137	19·86½	177	25·66½
18	2·61	58	8·41	98	14·21	138	20·01	178	25·81
19	2·75½	59	8·55½	99	14·35½	139	20·15½	179	25·95½
20	2·90	**60**	8·70	**100**	14·50	**140**	20·30	**180**	26·10
21	3·04½	61	8·84½	101	14·64½	141	20·44½	181	26·24½
22	3·19	62	8·99	102	14·79	142	20·59	182	26·39
23	3·33½	63	9·13½	103	14·93½	143	20·73½	183	26·53½
24	3·48	64	9·28	104	15·08	144	20·88	184	26·68
25	3·62½	65	9·42½	105	15·22½	145	21·02½	185	26·82½
26	3·77	66	9·57	106	15·37	146	21·17	186	26·97
27	3·91½	67	9·71½	107	15·51½	147	21·31½	187	27·11½
28	4·06	68	9·86	108	15·66	148	21·46	188	27·26
29	4·20½	69	10·00½	109	15·80½	149	21·60½	189	27·40½
30	4·35	**70**	10·15	**110**	15·95	**150**	21·75	**190**	27·55
31	4·49½	71	10·29½	111	16·09½	151	21·89½	191	27·69½
32	4·64	72	10·44	112	16·24	152	22·04	192	27·84
33	4·78½	73	10·58½	113	16·38½	153	22·18½	193	27·98½
34	4·93	74	10·73	114	16·53	154	22·33	194	28·13
35	5·07½	75	10·87½	115	16·67½	155	22·47½	195	28·27½
36	5·22	76	11·02	116	16·82	156	22·62	196	28·42
37	5·36½	77	11·16½	117	16·96½	157	22·76½	197	28·56½
38	5·51	78	11·31	118	17·11	158	22·91	198	28·71
39	5·65½	79	11·45½	119	17·25½	159	23·05½	199	28·85½
40	5·80	**80**	11·60	**120**	17·40	**160**	23·20	**200**	29·00

$\frac{1}{16}=0{\cdot}00^{9}$ $\frac{1}{9}=0{\cdot}01^{6}$ $\frac{1}{8}=0{\cdot}01^{8}$ $\frac{1}{6}=0{\cdot}02^{4}$ $\frac{1}{5}=0{\cdot}02^{9}$ $\frac{1}{4}=0{\cdot}03^{6}$ $\frac{1}{3}=0{\cdot}04^{8}$

$\frac{3}{8}=0{\cdot}05^{4}$ $\frac{1}{2}=0{\cdot}07^{3}$ $\frac{5}{8}=0{\cdot}09^{1}$ $\frac{3}{4}=0{\cdot}10^{9}$ $\frac{7}{8}=0{\cdot}12^{7}$ $\frac{2}{3}=0{\cdot}09^{7}$ $\frac{5}{6}=0{\cdot}12^{1}$

$\frac{3}{16}=0{\cdot}02^{7}$ $\frac{5}{16}=0{\cdot}04^{5}$ $\frac{7}{16}=0{\cdot}06^{3}$ $\frac{9}{16}=0{\cdot}08^{2}$ $\frac{11}{16}=0{\cdot}10^{0}$ $\frac{13}{16}=0{\cdot}11^{8}$ $\frac{15}{16}=0{\cdot}13^{6}$

(6.67=1£) **£0.15**

1	0·15	41	6·15	81	12·15	121	18·15	161	24·15
2	0·30	42	6·30	82	12·30	122	18·30	162	24·30
3	0·45	43	6·45	83	12·45	123	18·45	163	24·45
4	0·60	44	6·60	84	12·60	124	18·60	164	24·60
5	0·75	45	6·75	85	12·75	125	18·75	165	24·75
6	0·90	46	6·90	86	12·90	126	18·90	166	24·90
7	1·05	47	7·05	87	13·05	127	19·05	167	25·05
8	1·20	48	7·20	88	13·20	128	19·20	168	25·20
9	1·35	49	7·35	89	13·35	129	19·35	169	25·35
10	1·50	**50**	7·50	**90**	13·50	**130**	19·50	**170**	25·50
11	1·65	51	7·65	91	13·65	131	19·65	171	25·65
12	1·80	52	7·80	92	13·80	132	19·80	172	25·80
13	1·95	53	7·95	93	13·95	133	19·95	173	25·95
14	2·10	54	8·10	94	14·10	134	20·10	174	26·10
15	2·25	55	8·25	95	14·25	135	20·25	175	26·25
16	2·40	56	8·40	96	14·40	136	20·40	176	26·40
17	2·55	57	8·55	97	14·55	137	20·55	177	26·55
18	2·70	58	8·70	98	14·70	138	20·70	178	26·70
19	2·85	59	8·85	99	14·85	139	20·85	179	26·85
20	3·00	**60**	9·00	**100**	15·00	**140**	21·00	**180**	27·00
21	3·15	61	9·15	101	15·15	141	21·15	181	27·15
22	3·30	62	9·30	102	15·30	142	21·30	182	27·30
23	3·45	63	9·45	103	15·45	143	21·45	183	27·45
24	3·60	64	9·60	104	15·60	144	21·60	184	27·60
25	3·75	65	9·75	105	15·75	145	21·75	185	27·75
26	3·90	66	9·90	106	15·90	146	21·90	186	27·90
27	4·05	67	10·05	107	16·05	147	22·05	187	28·05
28	4·20	68	10·20	108	16·20	148	22·20	188	28·20
29	4·35	69	10·35	109	16·35	149	22·35	189	28·35
30	4·50	**70**	10·50	**110**	16·50	**150**	22·50	**190**	28·50
31	4·65	71	10·65	111	16·65	151	22·65	191	28·65
32	4·80	72	10·80	112	16·80	152	22·80	192	28·80
33	4·95	73	10·95	113	16·95	153	22·95	193	28·95
34	5·10	74	11·10	114	17·10	154	23·10	194	29·10
35	5·25	75	11·25	115	17·25	155	23·25	195	29·25
36	5·40	76	11·40	116	17·40	156	23·40	196	29·40
37	5·55	77	11·55	117	17·55	157	23·55	197	29·55
38	5·70	78	11·70	118	17·70	158	23·70	198	29·70
39	5·85	79	11·85	119	17·85	159	23·85	199	29·85
40	6·00	**80**	12·00	**120**	18·00	**160**	24·00	**200**	30·00

$\frac{1}{16}=0{\cdot}00^{9}$ $\frac{1}{9}=0{\cdot}01^{7}$ $\frac{1}{8}=0{\cdot}01^{9}$ $\frac{1}{6}=0{\cdot}02^{5}$ $\frac{1}{5}=0{\cdot}03^{0}$ $\frac{1}{4}=0{\cdot}03^{8}$ $\frac{1}{3}=0{\cdot}05^{0}$

$\frac{3}{8}=0{\cdot}05^{6}$ $\frac{1}{2}=0{\cdot}07^{5}$ $\frac{5}{8}=0{\cdot}09^{4}$ $\frac{3}{4}=0{\cdot}11^{3}$ $\frac{7}{8}=0{\cdot}13^{1}$ $\frac{2}{3}=0{\cdot}10^{0}$ $\frac{5}{6}=0{\cdot}12^{5}$

$\frac{3}{16}=0{\cdot}02^{8}$ $\frac{5}{16}=0{\cdot}04^{7}$ $\frac{7}{16}=0{\cdot}06^{6}$ $\frac{9}{16}=0{\cdot}08^{4}$ $\frac{11}{16}=0{\cdot}10^{3}$ $\frac{13}{16}=0{\cdot}12^{2}$ $\frac{15}{16}=0{\cdot}14^{1}$

(6.45=1£) **£0.15½**

1	0·15½	41	6·35½	81	12·55½	121	18·75½	161	24·95½
2	0·31	42	6·51	82	12·71	122	18·91	162	25·11
3	0·46½	43	6·66½	83	12·86½	123	19·06½	163	25·26½
4	0·62	44	6·82	84	13·02	124	19·22	164	25·42
5	0·77½	45	6·97½	85	13·17½	125	19·37½	165	25·57½
6	0·93	46	7·13	86	13·33	126	19·53	166	25·73
7	1·08½	47	7·28½	87	13·48½	127	19·68½	167	25·88½
8	1·24	48	7·44	88	13·64	128	19·84	168	26·04
9	1·39½	49	7·59½	89	13·79½	129	19·99½	169	26·19½
10	1·55	**50**	7·75	**90**	13·95	**130**	20·15	**170**	26·35
11	1·70½	51	7·90½	91	14·10½	131	20·30½	171	26·50½
12	1·86	52	8·06	92	14·26	132	20·46	172	26·66
13	2·01½	53	8·21½	93	14·41½	133	20·61½	173	26·81½
14	2·17	54	8·37	94	14·57	134	20·77	174	26·97
15	2·32½	55	8·52½	95	14·72½	135	20·92½	175	27·12½
16	2·48	56	8·68	96	14·88	136	21·08	176	27·28
17	2·63½	57	8·83½	97	15·03½	137	21·23½	177	27·43½
18	2·79	58	8·99	98	15·19	138	21·39	178	27·59
19	2·94½	59	9·14½	99	15·34½	139	21·54½	179	27·74½
20	3·10	**60**	9·30	**100**	15·50	**140**	21·70	**180**	27·90
21	3·25½	61	9·45½	101	15·65½	141	21·85½	181	28·05½
22	3·41	62	9·61	102	15·81	142	22·01	182	28·21
23	3·56½	63	9·76½	103	15·96½	143	22·16½	183	28·36½
24	3·72	64	9·92	104	16·12	144	22·32	184	28·52
25	3·87½	65	10·07½	105	16·27½	145	22·47½	185	28·67½
26	4·03	66	10·23	106	16·43	146	22·63	186	28·83
27	4·18½	67	10·38½	107	16·58½	147	22·78½	187	28·98½
28	4·34	68	10·54	108	16·74	148	22·94	188	29·14
29	4·49½	69	10·69½	109	16·89½	149	23·09½	189	29·29½
30	4·65	**70**	10·85	**110**	17·05	**150**	23·25	**190**	29·45
31	4·80½	71	11·00½	111	17·20½	151	23·40½	191	29·60½
32	4·96	72	11·16	112	17·36	152	23·56	192	29·76
33	5·11½	73	11·31½	113	17·51½	153	23·71½	193	29·91½
34	5·27	74	11·47	114	17·67	154	23·87	194	30·07
35	5·42½	75	11·62½	115	17·82½	155	24·02½	195	30·22½
36	5·58	76	11·78	116	17·98	156	24·18	196	30·38
37	5·73½	77	11·93½	117	18·13½	157	24·33½	197	30·53½
38	5·89	78	12·09	118	18·29	158	24·49	198	30·69
39	6·04½	79	12·24½	119	18·44½	159	24·64½	199	30·84½
40	6·20	**80**	12·40	**120**	18·60	**160**	24·80	**200**	31·00

$\frac{1}{16}=0{\cdot}01^{0}$ $\frac{1}{9}=0{\cdot}01^{7}$ $\frac{1}{8}=0{\cdot}01^{9}$ $\frac{1}{6}=0{\cdot}02^{6}$ $\frac{1}{5}=0{\cdot}03^{1}$ $\frac{1}{4}=0{\cdot}03^{9}$ $\frac{1}{3}=0{\cdot}05^{2}$

$\frac{3}{8}=0{\cdot}05^{8}$ $\frac{1}{2}=0{\cdot}07^{8}$ $\frac{5}{8}=0{\cdot}09^{7}$ $\frac{3}{4}=0{\cdot}11^{6}$ $\frac{7}{8}=0{\cdot}13^{6}$ $\frac{2}{3}=0{\cdot}10^{3}$ $\frac{5}{6}=0{\cdot}12^{9}$

$\frac{3}{16}=0{\cdot}02^{9}$ $\frac{5}{16}=0{\cdot}04^{8}$ $\frac{7}{16}=0{\cdot}06^{8}$ $\frac{9}{16}=0{\cdot}08^{7}$ $\frac{11}{16}=0{\cdot}10^{7}$ $\frac{13}{16}=0{\cdot}12^{6}$ $\frac{15}{16}=0{\cdot}14^{5}$

(6.25=1£) **£0.16**

1	0·16	41	6·56	81	12·96	121	19·36	161	25·76
2	0·32	42	6·72	82	13·12	122	19·52	162	25·92
3	0·48	43	6·88	83	13·28	123	19·68	163	26·08
4	0·64	44	7·04	84	13·44	124	19·84	164	26·24
5	0·80	45	7·20	85	13·60	125	20·00	165	26·40
6	0·96	46	7·36	86	13·76	126	20·16	166	26·56
7	1·12	47	7·52	87	13·92	127	20·32	167	26·72
8	1·28	48	7·68	88	14·08	128	20·48	168	26·88
9	1·44	49	7·84	89	14·24	129	20·64	169	27·04
10	1·60	**50**	8·00	**90**	14·40	**130**	20·80	**170**	27·20
11	1·76	51	8·16	91	14·56	131	20·96	171	27·36
12	1·92	52	8·32	92	14·72	132	21·12	172	27·52
13	2·08	53	8·48	93	14·88	133	21·28	173	27·68
14	2·24	54	8·64	94	15·04	134	21·44	174	27·84
15	2·40	55	8·80	95	15·20	135	21·60	175	28·00
16	2·56	56	8·96	96	15·36	136	21·76	176	28·16
17	2·72	57	9·12	97	15·52	137	21·92	177	28·32
18	2·88	58	9·28	98	15·68	138	22·08	178	28·48
19	3·04	59	9·44	99	15·84	139	22·24	179	28·64
20	3·20	**60**	9·60	**100**	16·00	**140**	22·40	**180**	28·80
21	3·36	61	9·76	101	16·16	141	22·56	181	28·96
22	3·52	62	9·92	102	16·32	142	22·72	182	29·12
23	3·68	63	10·08	103	16·48	143	22·88	183	29·28
24	3·84	64	10·24	104	16·64	144	23·04	184	29·44
25	4·00	65	10·40	105	16·80	145	23·20	185	29·60
26	4·16	66	10·56	106	16·96	146	23·36	186	29·76
27	4·32	67	10·72	107	17·12	147	23·52	187	29·92
28	4·48	68	10·88	108	17·28	148	23·68	188	30·08
29	4·64	69	11·04	109	17·44	149	23·84	189	30·24
30	4·80	**70**	11·20	**110**	17·60	**150**	24·00	**190**	30·40
31	4·96	71	11·36	111	17·76	151	24·16	191	30·56
32	5·12	72	11·52	112	17·92	152	24·32	192	30·72
33	5·28	73	11·68	113	18·08	153	24·48	193	30·88
34	5·44	74	11·84	114	18·24	154	24·64	194	31·04
35	5·60	75	12·00	115	18·40	155	24·80	195	31·20
36	5·76	76	12·16	116	18·56	156	24·96	196	31·36
37	5·92	77	12·32	117	18·72	157	25·12	197	31·52
38	6·08	78	12·48	118	18·88	158	25·28	198	31·68
39	6·24	79	12·64	119	19·04	159	25·44	199	31·84
40	6·40	**80**	12·80	**120**	19·20	**160**	25·60	**200**	32·00

$\frac{1}{16}$=0·01^{0} $\frac{1}{9}$=0·01^{8} $\frac{1}{8}$=0·02^{0} $\frac{1}{6}$=0·02^{7} $\frac{1}{5}$=0·03^{2} $\frac{1}{4}$=0·04^{0} $\frac{1}{3}$=0·05^{3}

$\frac{3}{8}$=0·06^{0} $\frac{1}{2}$=0·08^{0} $\frac{5}{8}$=0·10^{0} $\frac{3}{4}$=0·12^{0} $\frac{7}{8}$=0·14^{0} $\frac{2}{3}$=0·10^{7} $\frac{5}{6}$=0·13^{3}

$\frac{3}{16}$=0·03^{0} $\frac{5}{16}$=0·05^{0} $\frac{7}{16}$=0·07^{0} $\frac{9}{16}$=0·09^{0} $\frac{11}{16}$=0·11^{0} $\frac{13}{16}$=0·13^{0} $\frac{15}{16}$=0·15^{0}

1	0·16½	41	6·76½	81	13·36½	121	19·96½	161	26·56½
2	0·33	42	6·93	82	13·53	122	20·13	162	26·73
3	0·49½	43	7·09½	83	13·69½	123	20·29½	163	26·89½
4	0·66	44	7·26	84	13·86	124	20·46	164	27·06
5	0·82½	45	7·42½	85	14·02½	125	20·62½	165	27·22½
6	0·99	46	7·59	86	14·19	126	20·79	166	27·39
7	1·15½	47	7·75½	87	14·35½	127	20·95½	167	27·55½
8	1·32	48	7·92	88	14·52	128	21·12	168	27·72
9	1·48½	49	8·08½	89	14·68½	129	21·28½	169	27·88½
10	1·65	**50**	8·25	**90**	14·85	**130**	21·45	**170**	28·05
11	1·81½	51	8·41½	91	15·01½	131	21·61½	171	28·21½
12	1·98	52	8·58	92	15·18	132	21·78	172	28·38
13	2·14½	53	8·74½	93	15·34½	133	21·94½	173	28·54½
14	2·31	54	8·91	94	15·51	134	22·11	174	28·71
15	2·47½	55	9·07½	95	15·67½	135	22·27½	175	28·87½
16	2·64	56	9·24	96	15·84	136	22·44	176	29·04
17	2·80½	57	9·40½	97	16·00½	137	22·60½	177	29·20½
18	2·97	58	9·57	98	16·17	138	22·77	178	29·37
19	3·13½	59	9·73½	99	16·33½	139	22·93½	179	29·53½
20	3·30	**60**	9·90	**100**	16·50	**140**	23·10	**180**	29·70
21	3·46½	61	10·06½	101	16·66½	141	23·26½	181	29·86½
22	3·63	62	10·23	102	16·83	142	23·43	182	30·03
23	3·79½	63	10·39½	103	16·99½	143	23·59½	183	30·19½
24	3·96	64	10·56	104	17·16	144	23·76	184	30·36
25	4·12½	65	10·72½	105	17·32½	145	23·92½	185	30·52½
26	4·29	66	10·89	106	17·49	146	24·09	186	30·69
27	4·45½	67	11·05½	107	17·65½	147	24·25½	187	30·85½
28	4·62	68	11·22	108	17·82	148	24·42	188	31·02
29	4·78½	69	11·38½	109	17·98½	149	24·58½	189	31·18½
30	4·95	**70**	11·55	**110**	18·15	**150**	24·75	**190**	31·35
31	5·11½	71	11·71½	111	18·31½	151	24·91½	191	31·51½
32	5·28	72	11·88	112	18·48	152	25·08	192	31·68
33	5·44½	73	12·04½	113	18·64½	153	25·24½	193	31·84½
34	5·61	74	12·21	114	18·81	154	25·41	194	32·01
35	5·77½	75	12·37½	115	18·97½	155	25·57½	195	32·17½
36	5·94	76	12·54	116	19·14	156	25·74	196	32·34
37	6·10½	77	12·70½	117	19·30½	157	25·90½	197	32·50½
38	6·27	78	12·87	118	19·47	158	26·07	198	32·67
39	6·43½	79	13·03½	119	19·63½	159	26·23½	199	32·83½
40	6·60	**80**	13·20	**120**	19·80	**160**	26·40	**200**	33·00

$\frac{1}{16}=0{\cdot}01^{0}$	$\frac{1}{9}=0{\cdot}01^{8}$	$\frac{1}{8}=0{\cdot}02^{1}$	$\frac{1}{6}=0{\cdot}02^{8}$	$\frac{1}{5}=0{\cdot}03^{3}$	$\frac{1}{4}=0{\cdot}04^{1}$	$\frac{1}{3}=0{\cdot}05^{5}$
$\frac{3}{8}=0{\cdot}06^{2}$	$\frac{1}{2}=0{\cdot}08^{3}$	$\frac{5}{8}=0{\cdot}10^{3}$	$\frac{3}{4}=0{\cdot}12^{4}$	$\frac{7}{8}=0{\cdot}14^{4}$	$\frac{2}{3}=0{\cdot}11^{0}$	$\frac{5}{6}=0{\cdot}13^{8}$
$\frac{3}{16}=0{\cdot}03^{1}$	$\frac{5}{16}=0{\cdot}05^{2}$	$\frac{7}{16}=0{\cdot}07^{2}$	$\frac{9}{16}=0{\cdot}09^{3}$	$\frac{11}{16}=0{\cdot}11^{3}$	$\frac{13}{16}=0{\cdot}13^{4}$	$\frac{15}{16}=0{\cdot}15^{5}$

1	0·17	41	6·97	81	13·77	121	20·57	161	27·37
2	0·34	42	7·14	82	13·94	122	20·74	162	27·54
3	0·51	43	7·31	83	14·11	123	20·91	163	27·71
4	0·68	44	7·48	84	14·28	124	21·08	164	27·88
5	0·85	45	7·65	85	14·45	125	21·25	165	28·05
6	1·02	46	7·82	86	14·62	126	21·42	166	28·22
7	1·19	47	7·99	87	14·79	127	21·59	167	28·39
8	1·36	48	8·16	88	14·96	128	21·76	168	28·56
9	1·53	49	8·33	89	15·13	129	21·93	169	28·73
10	1·70	**50**	8·50	**90**	15·30	**130**	22·10	**170**	28·90
11	1·87	51	8·67	91	15·47	131	22·27	171	29·07
12	2·04	52	8·84	92	15·64	132	22·44	172	29·24
13	2·21	53	9·01	93	15·81	133	22·61	173	29·41
14	2·38	54	9·18	94	15·98	134	22·78	174	29·58
15	2·55	55	9·35	95	16·15	135	22·95	175	29·75
16	2·72	56	9·52	96	16·32	136	23·12	176	29·92
17	2·89	57	9·69	97	16·49	137	23·29	177	30·09
18	3·06	58	9·86	98	16·66	138	23·46	178	30·26
19	3·23	59	10·03	99	16·83	139	23·63	179	30·43
20	3·40	**60**	10·20	**100**	17·00	**140**	23·80	**180**	30·60
21	3·57	61	10·37	101	17·17	141	23·97	181	30·77
22	3·74	62	10·54	102	17·34	142	24·14	182	30·94
23	3·91	63	10·71	103	17·51	143	24·31	183	31·11
24	4·08	64	10·88	104	17·68	144	24·48	184	31·28
25	4·25	65	11·05	105	17·85	145	24·65	185	31·45
26	4·42	66	11·22	106	18·02	146	24·82	186	31·62
27	4·59	67	11·39	107	18·19	147	24·99	187	31·79
28	4·76	68	11·56	108	18·36	148	25·16	188	31·96
29	4·93	69	11·73	109	18·53	149	25·33	189	32·13
30	5·10	**70**	11·90	**110**	18·70	**150**	25·50	**190**	32·30
31	5·27	71	12·07	111	18·87	151	25·67	191	32·47
32	5·44	72	12·24	112	19·04	152	25·84	192	32·64
33	5·61	73	12·41	113	19·21	153	26·01	193	32·81
34	5·78	74	12·58	114	19·38	154	26·18	194	32·98
35	5·95	75	12·75	115	19·55	155	26·35	195	33·15
36	6·12	76	12·92	116	19·72	156	26·52	196	33·32
37	6·29	77	13·09	117	19·89	157	26·69	197	33·49
38	6·46	78	13·26	118	20·06	158	26·86	198	33·66
39	6·63	79	13·43	119	20·23	159	27·03	199	33·83
40	6·80	**80**	13·60	**120**	20·40	**160**	27·20	**200**	34·00

$\frac{1}{16}=0{\cdot}01^{1}$ $\frac{1}{9}=0{\cdot}01^{9}$ $\frac{1}{8}=0{\cdot}02^{1}$ $\frac{1}{6}=0{\cdot}02^{8}$ $\frac{1}{5}=0{\cdot}03^{4}$ $\frac{1}{4}=0{\cdot}04^{3}$ $\frac{1}{3}=0{\cdot}05^{7}$

$\frac{3}{8}=0{\cdot}06^{4}$ $\frac{1}{2}=0{\cdot}08^{5}$ $\frac{5}{8}=0{\cdot}10^{6}$ $\frac{3}{4}=0{\cdot}12^{8}$ $\frac{7}{8}=0{\cdot}14^{9}$ $\frac{2}{3}=0{\cdot}11^{3}$ $\frac{5}{6}=0{\cdot}14^{2}$

$\frac{3}{16}=0{\cdot}03^{2}$ $\frac{5}{16}=0{\cdot}05^{3}$ $\frac{7}{16}=0{\cdot}07^{4}$ $\frac{9}{16}=0{\cdot}09^{6}$ $\frac{11}{16}=0{\cdot}11^{7}$ $\frac{13}{16}=0{\cdot}13^{8}$ $\frac{15}{16}=0{\cdot}15^{9}$

(5.71=1£) **£0.17½**

1	0·17½	41	7·17½	81	14·17½	121	21·17½	161	28·17½
2	0·35	42	7·35	82	14·35	122	21·35	162	28·35
3	0·52½	43	7·52½	83	14·52½	123	21·52½	163	28·52½
4	0·70	44	7·70	84	14·70	124	21·70	164	28·70
5	0·87½	45	7·87½	85	14·87½	125	21·87½	165	28·87½
6	1·05	46	8·05	86	15·05	126	22·05	166	29·05
7	1·22½	47	8·22½	87	15·22½	127	22·22½	167	29·22½
8	1·40	48	8·40	88	15·40	128	22·40	168	29·40
9	1·57½	49	8·57½	89	15·57½	129	22·57½	169	29·57½
10	1·75	**50**	8·75	**90**	15·75	**130**	22·75	**170**	29·75
11	1·92½	51	8·92½	91	15·92½	131	22·92½	171	29·92½
12	2·10	52	9·10	92	16·10	132	23·10	172	30·10
13	2·27½	53	9·27½	93	16·27½	133	23·27½	173	30·27½
14	2·45	54	9·45	94	16·45	134	23·45	174	30·45
15	2·62½	55	9·62½	95	16·62½	135	23·62½	175	30·62½
16	2·80	56	9·80	96	16·80	136	23·80	176	30·80
17	2·97½	57	9·97½	97	16·97½	137	23·97½	177	30·97½
18	3·15	58	10·15	98	17·15	138	24·15	178	31·15
19	3·32½	59	10·32½	99	17·32½	139	24·32½	179	31·32½
20	3·50	**60**	10·50	**100**	17·50	**140**	24·50	**180**	31·50
21	3·67½	61	10·67½	101	17·67½	141	24·67½	181	31·67½
22	3·85	62	10·85	102	17·85	142	24·85	182	31·85
23	4·02½	63	11·02½	103	18·02½	143	25·02½	183	32·02½
24	4·20	64	11·20	104	18·20	144	25·20	184	32·20
25	4·37½	65	11·37½	105	18·37½	145	25·37½	185	32·37½
26	4·55	66	11·55	106	18·55	146	25·55	186	32·55
27	4·72½	67	11·72½	107	18·72½	147	25·72½	187	32·72½
28	4·90	68	11·90	108	18·90	148	25·90	188	32·90
29	5·07½	69	12·07½	109	19·07½	149	26·07½	189	33·07½
30	5·25	**70**	12·25	**110**	19·25	**150**	26·25	**190**	33·25
31	5·42½	71	12·42½	111	19·42½	151	26·42½	191	33·42½
32	5·60	72	12·60	112	19·60	152	26·60	192	33·60
33	5·77½	73	12·77½	113	19·77½	153	26·77½	193	33·77½
34	5·95	74	12·95	114	19·95	154	26·95	194	33·95
35	6·12½	75	13·12½	115	20·12½	155	27·12½	195	34·12½
36	6·30	76	13·30	116	20·30	156	27·30	196	34·30
37	6·47½	77	13·47½	117	20·47½	157	27·47½	197	34·47½
38	6·65	78	13·65	118	20·65	158	27·65	198	34·65
39	6·82½	79	13·82½	119	20·82½	159	27·82½	199	34·82½
40	7·00	**80**	14·00	**120**	21·00	**160**	28·00	**200**	35·00

$\frac{1}{16}$=0·01^{1} $\frac{1}{9}$=0·01^{9} $\frac{1}{8}$=0·02^{2} $\frac{1}{6}$=0·02^{9} $\frac{1}{5}$=0·03^{5} $\frac{1}{4}$=0·04^{4} $\frac{1}{3}$=0·05^{8}

$\frac{3}{8}$=0·06^{6} $\frac{1}{2}$=0·08^{8} $\frac{5}{8}$=0·10^{9} $\frac{3}{4}$=0·13^{1} $\frac{7}{8}$=0·15^{3} $\frac{2}{3}$=0·11^{7} $\frac{5}{6}$=0·14^{6}

$\frac{3}{16}$=0·03^{3} $\frac{5}{16}$=0·05^{5} $\frac{7}{16}$=0·07^{7} $\frac{9}{16}$=0·09^{8} $\frac{11}{16}$=0·12^{0} $\frac{13}{16}$=0·14^{2} $\frac{15}{16}$=0·16^{4}

(5.56=1£) **£0.18**

1	0·18	41	7·38	81	14·58	121	21·78	161	28·98
2	0·36	42	7·56	82	14·76	122	21·96	162	29·16
3	0·54	43	7·74	83	14·94	123	22·14	163	29·34
4	0·72	44	7·92	84	15·12	124	22·32	164	29·52
5	0·90	45	8·10	85	15·30	125	22·50	165	29·70
6	1·08	46	8·28	86	15·48	126	22·68	166	29·88
7	1·26	47	8·46	87	15·66	127	22·86	167	30·06
8	1·44	48	8·64	88	15·84	128	23·04	168	30·24
9	1·62	49	8·82	89	16·02	129	23·22	169	30·42
10	1·80	**50**	9·00	**90**	16·20	**130**	23·40	**170**	30·60
11	1·98	51	9·18	91	16·38	131	23·58	171	30·78
12	2·16	52	9·36	92	16·56	132	23·76	172	30·96
13	2·34	53	9·54	93	16·74	133	23·94	173	31·14
14	2·52	54	9·72	94	16·92	134	24·12	174	31·32
15	2·70	55	9·90	95	17·10	135	24·30	175	31·50
16	2·88	56	10·08	96	17·28	136	24·48	176	31·68
17	3·06	57	10·26	97	17·46	137	24·66	177	31·86
18	3·24	58	10·44	98	17·64	138	24·84	178	32·04
19	3·42	59	10·62	99	17·82	139	25·02	179	32·22
20	3·60	**60**	10·80	**100**	18·00	**140**	25·20	**180**	32·40
21	3·78	61	10·98	101	18·18	141	25·38	181	32·58
22	3·96	62	11·16	102	18·36	142	25·56	182	32·76
23	4·14	63	11·34	103	18·54	143	25·74	183	32·94
24	4·32	64	11·52	104	18·72	144	25·92	184	33·12
25	4·50	65	11·70	105	18·90	145	26·10	185	33·30
26	4·68	66	11·88	106	19·08	146	26·28	186	33·48
27	4·86	67	12·06	107	19·26	147	26·46	187	33·66
28	5·04	68	12·24	108	19·44	148	26·64	188	33·84
29	5·22	69	12·42	109	19·62	149	26·82	189	34·02
30	5·40	**70**	12·60	**110**	19·80	**150**	27·00	**190**	34·20
31	5·58	71	12·78	111	19·98	151	27·18	191	34·38
32	5·76	72	12·96	112	20·16	152	27·36	192	34·56
33	5·94	73	13·14	113	20·34	153	27·54	193	34·74
34	6·12	74	13·32	114	20·52	154	27·72	194	34·92
35	6·30	75	13·50	115	20·70	155	27·90	195	35·10
36	6·48	76	13·68	116	20·88	156	28·08	196	35·28
37	6·66	77	13·86	117	21·06	157	28·26	197	35·46
38	6·84	78	14·04	118	21·24	158	28·44	198	35·64
39	7·02	79	14·22	119	21·42	159	28·62	199	35·82
40	7·20	**80**	14·40	**120**	21·60	**160**	28·80	**200**	36·00

$\frac{1}{16}=0{\cdot}01^{1}$ $\frac{1}{9}=0{\cdot}02^{0}$ $\frac{1}{8}=0{\cdot}02^{3}$ $\frac{1}{6}=0{\cdot}03^{0}$ $\frac{1}{5}=0{\cdot}03^{6}$ $\frac{1}{4}=0{\cdot}04^{5}$ $\frac{1}{3}=0{\cdot}06^{0}$

$\frac{3}{8}=0{\cdot}06^{8}$ $\frac{1}{2}=0{\cdot}09^{0}$ $\frac{5}{8}=0{\cdot}11^{3}$ $\frac{3}{4}=0{\cdot}13^{5}$ $\frac{7}{8}=0{\cdot}15^{8}$ $\frac{2}{3}=0{\cdot}12^{0}$ $\frac{5}{6}=0{\cdot}15^{0}$

$\frac{3}{16}=0{\cdot}03^{4}$ $\frac{5}{16}=0{\cdot}05^{6}$ $\frac{7}{16}=0{\cdot}07^{9}$ $\frac{9}{16}=0{\cdot}10^{1}$ $\frac{11}{16}=0{\cdot}12^{4}$ $\frac{13}{16}=0{\cdot}14^{6}$ $\frac{15}{16}=0{\cdot}16^{9}$

(5.41=1£) **£0.18½**

1	0·18½	41	7·58½	81	14·98½	121	22·38½	161	29·78½
2	0·37	42	7·77	82	15·17	122	22·57	162	29·97
3	0·55½	43	7·95½	83	15·35½	123	22·75½	163	30·15½
4	0·74	44	8·14	84	15·54	124	22·94	164	30·34
5	0·92½	45	8·32½	85	15·72½	125	23·12½	165	30·52½
6	1·11	46	8·51	86	15·91	126	23·31	166	30·71
7	1·29½	47	8·69½	87	16·09½	127	23·49½	167	30·89½
8	1·48	48	8·88	88	16·28	128	23·68	168	31·08
9	1·66½	49	9·06½	89	16·46½	129	23·86½	169	31·26½
10	1·85	**50**	9·25	**90**	16·65	**130**	24·05	**170**	31·45
11	2·03½	51	9·43½	91	16·83½	131	24·23½	171	31·63½
12	2·22	52	9·62	92	17·02	132	24·42	172	31·82
13	2·40½	53	9·80½	93	17·20½	133	24·60½	173	32·00½
14	2·59	54	9·99	94	17·39	134	24·79	174	32·19
15	2·77½	55	10·17½	95	17·57½	135	24·97½	175	32·37½
16	2·96	56	10·36	96	17·76	136	25·16	176	32·56
17	3·14½	57	10·54½	97	17·94½	137	25·34½	177	32·74½
18	3·33	58	10·73	98	18·13	138	25·53	178	32·93
19	3·51½	59	10·91½	99	18·31½	139	25·71½	179	33·11½
20	3·70	**60**	11·10	**100**	18·50	**140**	25·90	**180**	33·30
21	3·88½	61	11·28½	101	18·68½	141	26·08½	181	33·48½
22	4·07	62	11·47	102	18·87	142	26·27	182	33·67
23	4·25½	63	11·65½	103	19·05½	143	26·45½	183	33·85½
24	4·44	64	11·84	104	19·24	144	26·64	184	34·04
25	4·62½	65	12·02½	105	19·42½	145	26·82½	185	34·22½
26	4·81	66	12·21	106	19·61	146	27·01	186	34·41
27	4·99½	67	12·39½	107	19·79½	147	27·19½	187	34·59½
28	5·18	68	12·58	108	19·98	148	27·38	188	34·78
29	5·36½	69	12·76½	109	20·16½	149	27·56½	189	34·96½
30	5·55	**70**	12·95	**110**	20·35	**150**	27·75	**190**	35·15
31	5·73½	71	13·13½	111	20·53½	151	27·93½	191	35·33½
32	5·92	72	13·32	112	20·72	152	28·12	192	35·52
33	6·10½	73	13·50½	113	20·90½	153	28·30½	193	35·70½
34	6·29	74	13·69	114	21·09	154	28·49	194	35·89
35	6·47½	75	13·87½	115	21·27½	155	28·67½	195	36·07½
36	6·66	76	14·06	116	21·46	156	28·86	196	36·26
37	6·84½	77	14·24½	117	21·64½	157	29·04½	197	36·44½
38	7·03	78	14·43	118	21·83	158	29·23	198	36·63
39	7·21½	79	14·61½	119	22·01½	159	29·41½	199	36·81½
40	7·40	**80**	14·80	**120**	22·20	**160**	29·60	**200**	37·00

$\frac{1}{16}$=0·01^{2}	$\frac{1}{9}$=0·02^{1}	$\frac{1}{8}$=0·02^{3}	$\frac{1}{6}$=0·03^{1}	$\frac{1}{5}$=0·03^{7}	$\frac{1}{4}$=0·04^{6}	$\frac{1}{3}$=0·06^{2}
$\frac{3}{8}$=0·06^{9}	$\frac{1}{2}$=0·09^{3}	$\frac{5}{8}$=0·11^{6}	$\frac{3}{4}$=0·13^{9}	$\frac{7}{8}$=0·16^{2}	$\frac{2}{3}$=0·12^{3}	$\frac{5}{6}$=0·15^{4}
$\frac{3}{16}$=0·03^{5}	$\frac{5}{16}$=0·05^{8}	$\frac{7}{16}$=0·08^{1}	$\frac{9}{16}$=0·10^{4}	$\frac{11}{16}$=0·12^{7}	$\frac{13}{16}$=0·15^{0}	$\frac{15}{16}$=0·17^{3}

(5.26=1£) **£0.19**

1	0·19	41	7·79	81	15·39	121	22·99	161	30·59
2	0·38	42	7·98	82	15·58	122	23·18	162	30·78
3	0·57	43	8·17	83	15·77	123	23·37	163	30·97
4	0·76	44	8·36	84	15·96	124	23·56	164	31·16
5	0·95	45	8·55	85	16·15	125	23·75	165	31·35
6	1·14	46	8·74	86	16·34	126	23·94	166	31·54
7	1·33	47	8·93	87	16·53	127	24·13	167	31·73
8	1·52	48	9·12	88	16·72	128	24·32	168	31·92
9	1·71	49	9·31	89	16·91	129	24·51	169	32·11
10	1·90	**50**	9·50	**90**	17·10	**130**	24·70	**170**	32·30
11	2·09	51	9·69	91	17·29	131	24·89	171	32·49
12	2·28	52	9·88	92	17·48	132	25·08	172	32·68
13	2·47	53	10·07	93	17·67	133	25·27	173	32·87
14	2·66	54	10·26	94	17·86	134	25·46	174	33·06
15	2·85	55	10·45	95	18·05	135	25·65	175	33·25
16	3·04	56	10·64	96	18·24	136	25·84	176	33·44
17	3·23	57	10·83	97	18·43	137	26·03	177	33·63
18	3·42	58	11·02	98	18·62	138	26·22	178	33·82
19	3·61	59	11·21	99	18·81	139	26·41	179	34·01
20	3·80	**60**	11·40	**100**	19·00	**140**	26·60	**180**	34·20
21	3·99	61	11·59	101	19·19	141	26·79	181	34·39
22	4·18	62	11·78	102	19·38	142	26·98	182	34·58
23	4·37	63	11·97	103	19·57	143	27·17	183	34·77
24	4·56	64	12·16	104	19·76	144	27·36	184	34·96
25	4·75	65	12·35	105	19·95	145	27·55	185	35·15
26	4·94	66	12·54	106	20·14	146	27·74	186	35·34
27	5·13	67	12·73	107	20·33	147	27·93	187	35·53
28	5·32	68	12·92	108	20·52	148	28·12	188	35·72
29	5·51	69	13·11	109	20·71	149	28·31	189	35·91
30	5·70	**70**	13·30	**110**	20·90	**150**	28·50	**190**	36·10
31	5·89	71	13·49	111	21·09	151	28·69	191	36·29
32	6·08	72	13·68	112	21·28	152	28·88	192	36·48
33	6·27	73	13·87	113	21·47	153	29·07	193	36·67
34	6·46	74	14·06	114	21·66	154	29·26	194	36·86
35	6·65	75	14·25	115	21·85	155	29·45	195	37·05
36	6·84	76	14·44	116	22·04	156	29·64	196	37·24
37	7·03	77	14·63	117	22·23	157	29·83	197	37·43
38	7·22	78	14·82	118	22·42	158	30·02	198	37·62
39	7·41	79	15·01	119	22·61	159	30·21	199	37·81
40	7·60	**80**	15·20	**120**	22·80	**160**	30·40	**200**	38·00

$\frac{1}{16}$=0·01^{2} $\frac{1}{9}$=0·02^{1} $\frac{1}{8}$=0·02^{4} $\frac{1}{6}$=0·03^{2} $\frac{1}{5}$=0·03^{8} $\frac{1}{4}$=0·04^{8} $\frac{1}{3}$=0·06^{3}

$\frac{3}{8}$=0·07^{1} $\frac{1}{2}$=0·09^{5} $\frac{5}{8}$=0·11^{9} $\frac{3}{4}$=0·14^{3} $\frac{7}{8}$=0·16^{6} $\frac{2}{3}$=0·12^{7} $\frac{5}{6}$=0·15^{8}

$\frac{3}{16}$=0·03^{6} $\frac{5}{16}$=0·05^{9} $\frac{7}{16}$=0·08^{3} $\frac{9}{16}$=0·10^{7} $\frac{11}{16}$=0·13^{1} $\frac{13}{16}$=0·15^{4} $\frac{15}{16}$=0·17^{8}

(5.13=1£) **£0.19½**

1	0·19½	41	7·99½	81	15·79½	121	23·59½	161	31·39½
2	0·39	42	8·19	82	15·99	122	23·79	162	31·59
3	0·58½	43	8·38½	83	16·18½	123	23·98½	163	31·78½
4	0·78	44	8·58	84	16·38	124	24·18	164	31·98
5	0·97½	45	8·77½	85	16·57½	125	24·37½	165	32·17½
6	1·17	46	8·97	86	16·77	126	24·57	166	32·37
7	1·36½	47	9·16½	87	16·96½	127	24·76½	167	32·56½
8	1·56	48	9·36	88	17·16	128	24·96	168	32·76
9	1·75½	49	9·55½	89	17·35½	129	25·15½	169	32·95½
10	1·95	**50**	9·75	**90**	17·55	**130**	25·35	**170**	33·15
11	2·14½	51	9·94½	91	17·74½	131	25·54½	171	33·34½
12	2·34	52	10·14	92	17·94	132	25·74	172	33·54
13	2·53½	53	10·33½	93	18·13½	133	25·93½	173	33·73½
14	2·73	54	10·53	94	18·33	134	26·13	174	33·93
15	2·92½	55	10·72½	95	18·52½	135	26·32½	175	34·12½
16	3·12	56	10·92	96	18·72	136	26·52	176	34·32
17	3·31½	57	11·11½	97	18·91½	137	26·71½	177	34·51½
18	3·51	58	11·31	98	19·11	138	26·91	178	34·71
19	3·70½	59	11·50½	99	19·30½	139	27·10½	179	34·90½
20	3·90	**60**	11·70	**100**	19·50	**140**	27·30	**180**	35·10
21	4·09½	61	11·89½	101	19·69½	141	27·49½	181	35·29½
22	4·29	62	12·09	102	19·89	142	27·69	182	35·49
23	4·48½	63	12·28½	103	20·08½	143	27·88½	183	35·68½
24	4·68	64	12·48	104	20·28	144	28·08	184	35·88
25	4·87½	65	12·67½	105	20·47½	145	28·27½	185	36·07½
26	5·07	66	12·87	106	20·67	146	28·47	186	36·27
27	5·26½	67	13·06½	107	20·86½	147	28·66½	187	36·46½
28	5·46	68	13·26	108	21·06	148	28·86	188	36·66
29	5·65½	69	13·45½	109	21·25½	149	29·05½	189	36·85½
30	5·85	**70**	13·65	**110**	21·45	**150**	29·25	**190**	37·05
31	6·04½	71	13·84½	111	21·64½	151	29·44½	191	37·24½
32	6·24	72	14·04	112	21·84	152	29·64	192	37·44
33	6·43½	73	14·23½	113	22·03½	153	29·83½	193	37·63½
34	6·63	74	14·43	114	22·23	154	30·03	194	37·83
35	6·82½	75	14·62½	115	22·42½	155	30·22½	195	38·02½
36	7·02	76	14·82	116	22·62	156	30·42	196	38·22
37	7·21½	77	15·01½	117	22·81½	157	30·61½	197	38·41½
38	7·41	78	15·21	118	23·01	158	30·81	198	38·61
39	7·60½	79	15·40½	119	23·20½	159	31·00½	199	38·80½
40	7·80	**80**	15·60	**120**	23·40	**160**	31·20	**200**	39·00

$\frac{1}{16}=0{\cdot}01^{2}$ $\frac{1}{9}=0{\cdot}02^{2}$ $\frac{1}{8}=0{\cdot}02^{4}$ $\frac{1}{6}=0{\cdot}03^{3}$ $\frac{1}{5}=0{\cdot}03^{9}$ $\frac{1}{4}=0{\cdot}04^{9}$ $\frac{1}{3}=0{\cdot}06^{5}$

$\frac{3}{8}=0{\cdot}07^{3}$ $\frac{1}{2}=0{\cdot}09^{8}$ $\frac{5}{8}=0{\cdot}12^{2}$ $\frac{3}{4}=0{\cdot}14^{6}$ $\frac{7}{8}=0{\cdot}17^{1}$ $\frac{2}{3}=0{\cdot}13^{0}$ $\frac{5}{6}=0{\cdot}16^{3}$

$\frac{3}{16}=0{\cdot}03^{7}$ $\frac{5}{16}=0{\cdot}06^{1}$ $\frac{7}{16}=0{\cdot}08^{5}$ $\frac{9}{16}=0{\cdot}11^{0}$ $\frac{11}{16}=0{\cdot}13^{4}$ $\frac{13}{16}=0{\cdot}15^{8}$ $\frac{15}{16}=0{\cdot}18^{3}$

(5.00=1£) **£0.20**

1	0·20	41	8·20	81	16·20	121	24·20	161	32·20
2	0·40	42	8·40	82	16·40	122	24·40	162	32·40
3	0·60	43	8·60	83	16·60	123	24·60	163	32·60
4	0·80	44	8·80	84	16·80	124	24·80	164	32·80
5	1·00	45	9·00	85	17·00	125	25·00	165	33·00
6	1·20	46	9·20	86	17·20	126	25·20	166	33·20
7	1·40	47	9·40	87	17·40	127	25·40	167	33·40
8	1·60	48	9·60	88	17·60	128	25·60	168	33·60
9	1·80	49	9·80	89	17·80	129	25·80	169	33·80
10	2·00	**50**	10·00	**90**	18·00	**130**	26·00	**170**	34·00
11	2·20	51	10·20	91	18·20	131	26·20	171	34·20
12	2·40	52	10·40	92	18·40	132	26·40	172	34·40
13	2·60	53	10·60	93	18·60	133	26·60	173	34·60
14	2·80	54	10·80	94	18·80	134	26·80	174	34·80
15	3·00	55	11·00	95	19·00	135	27·00	175	35·00
16	3·20	56	11·20	96	19·20	136	27·20	176	35·20
17	3·40	57	11·40	97	19·40	137	27·40	177	35·40
18	3·60	58	11·60	98	19·60	138	27·60	178	35·60
19	3·80	59	11·80	99	19·80	139	27·80	179	35·80
20	4·00	**60**	12·00	**100**	20·00	**140**	28·00	**180**	36·00
21	4·20	61	12·20	101	20·20	141	28·20	181	36·20
22	4·40	62	12·40	102	20·40	142	28·40	182	36·40
23	4·60	63	12·60	103	20·60	143	28·60	183	36·60
24	4·80	64	12·80	104	20·80	144	28·80	184	36·80
25	5·00	65	13·00	105	21·00	145	29·00	185	37·00
26	5·20	66	13·20	106	21·20	146	29·20	186	37·20
27	5·40	67	13·40	107	21·40	147	29·40	187	37·40
28	5·60	68	13·60	108	21·60	148	29·60	188	37·60
29	5·80	69	13·80	109	21·80	149	29·80	189	37·80
30	6·00	**70**	14·00	**110**	22·00	**150**	30·00	**190**	38·00
31	6·20	71	14·20	111	22·20	151	30·20	191	38·20
32	6·40	72	14·40	112	22·40	152	30·40	192	38·40
33	6·60	73	14·60	113	22·60	153	30·60	193	38·60
34	6·80	74	14·80	114	22·80	154	30·80	194	38·80
35	7·00	75	15·00	115	23·00	155	31·00	195	39·00
36	7·20	76	15·20	116	23·20	156	31·20	196	39·20
37	7·40	77	15·40	117	23·40	157	31·40	197	39·40
38	7·60	78	15·60	118	23·60	158	31·60	198	39·60
39	7·80	79	15·80	119	23·80	159	31·80	199	39·80
40	8·00	**80**	16·00	**120**	24·00	**160**	32·00	**200**	40·00

$\frac{1}{16}=0{\cdot}01^{3}$ $\frac{1}{9}=0{\cdot}02^{2}$ $\frac{1}{8}=0{\cdot}02^{5}$ $\frac{1}{6}=0{\cdot}03^{3}$ $\frac{1}{5}=0{\cdot}04^{0}$ $\frac{1}{4}=0{\cdot}05^{0}$ $\frac{1}{3}=0{\cdot}06^{7}$

$\frac{3}{8}=0{\cdot}07^{5}$ $\frac{1}{2}=0{\cdot}10^{0}$ $\frac{5}{8}=0{\cdot}12^{5}$ $\frac{3}{4}=0{\cdot}15^{0}$ $\frac{7}{8}=0{\cdot}17^{5}$ $\frac{2}{3}=0{\cdot}13^{3}$ $\frac{5}{6}=0{\cdot}16^{7}$

$\frac{3}{16}=0{\cdot}03^{8}$ $\frac{5}{16}=0{\cdot}06^{3}$ $\frac{7}{16}=0{\cdot}08^{8}$ $\frac{9}{16}=0{\cdot}11^{3}$ $\frac{11}{16}=0{\cdot}13^{8}$ $\frac{13}{16}=0{\cdot}16^{3}$ $\frac{15}{16}=0{\cdot}18^{8}$

(4.88=1£) **£0.20½**

1	0·20½	41	8·40½	81	16·60½	121	24·80½	161	33·00½
2	0·41	42	8·61	82	16·81	122	25·01	162	33·21
3	0·61½	43	8·81½	83	17·01½	123	25·21½	163	33·41½
4	0·82	44	9·02	84	17·22	124	25·42	164	33·62
5	1·02½	45	9·22½	85	17·42½	125	25·62½	165	33·82½
6	1·23	46	9·43	86	17·63	126	25·83	166	34·03
7	1·43½	47	9·63½	87	17·83½	127	26·03½	167	34·23½
8	1·64	48	9·84	88	18·04	128	26·24	168	34·44
9	1·84½	49	10·04½	89	18·24½	129	26·44½	169	34·64½
10	2·05	**50**	10·25	**90**	18·45	**130**	26·65	**170**	34·85
11	2·25½	51	10·45½	91	18·65½	131	26·85½	171	35·05½
12	2·46	52	10·66	92	18·86	132	27·06	172	35·26
13	2·66½	53	10·86½	93	19·06½	133	27·26½	173	35·46½
14	2·87	54	11·07	94	19·27	134	27·47	174	35·67
15	3·07½	55	11·27½	95	19·47½	135	27·67½	175	35·87½
16	3·28	56	11·48	96	19·68	136	27·88	176	36·08
17	3·48½	57	11·68½	97	19·88½	137	28·08½	177	36·28½
18	3·69	58	11·89	98	20·09	138	28·29	178	36·49
19	3·89½	59	12·09½	99	20·29½	139	28·49½	179	36·69½
20	4·10	**60**	12·30	**100**	20·50	**140**	28·70	**180**	36·90
21	4·30½	61	12·50½	101	20·70½	141	28·90½	181	37·10½
22	4·51	62	12·71	102	20·91	142	29·11	182	37·31
23	4·71½	63	12·91½	103	21·11½	143	29·31½	183	37·51½
24	4·92	64	13·12	104	21·32	144	29·52	184	37·72
25	5·12½	65	13·32½	105	21·52½	145	29·72½	185	37·92½
26	5·33	66	13·53	106	21·73	146	29·93	186	38·13
27	5·53½	67	13·73½	107	21·93½	147	30·13½	187	38·33½
28	5·74	68	13·94	108	22·14	148	30·34	188	38·54
29	5·94½	69	14·14½	109	22·34½	149	30·54½	189	38·74½
30	6·15	**70**	14·35	**110**	22·55	**150**	30·75	**190**	38·95
31	6·35½	71	14·55½	111	22·75½	151	30·95½	191	39·15½
32	6·56	72	14·76	112	22·96	152	31·16	192	39·36
33	6·76½	73	14·96½	113	23·16½	153	31·36½	193	39·56½
34	6·97	74	15·17	114	23·37	154	31·57	194	39·77
35	7·17½	75	15·37½	115	23·57½	155	31·77½	195	39·97½
36	7·38	76	15·58	116	23·78	156	31·98	196	40·18
37	7·58½	77	15·78½	117	23·98½	157	32·18½	197	40·38½
38	7·79	78	15·99	118	24·19	158	32·39	198	40·59
39	7·99½	79	16·19½	119	24·39½	159	32·59½	199	40·79½
40	8·20	**80**	16·40	**120**	24·60	**160**	32·80	**200**	41·00

$\frac{1}{16}=0{\cdot}01^{3}$ $\frac{1}{9}=0{\cdot}02^{3}$ $\frac{1}{8}=0{\cdot}02^{6}$ $\frac{1}{6}=0{\cdot}03^{4}$ $\frac{1}{5}=0{\cdot}04^{1}$ $\frac{1}{4}=0{\cdot}05^{1}$ $\frac{1}{3}=0{\cdot}06^{8}$

$\frac{3}{8}=0{\cdot}07^{7}$ $\frac{1}{2}=0{\cdot}10^{3}$ $\frac{5}{8}=0{\cdot}12^{8}$ $\frac{3}{4}=0{\cdot}15^{4}$ $\frac{7}{8}=0{\cdot}17^{9}$ $\frac{2}{3}=0{\cdot}13^{7}$ $\frac{5}{6}=0{\cdot}17^{1}$

$\frac{3}{16}=0{\cdot}03^{8}$ $\frac{5}{16}=0{\cdot}06^{4}$ $\frac{7}{16}=0{\cdot}09^{0}$ $\frac{9}{16}=0{\cdot}11^{5}$ $\frac{11}{16}=0{\cdot}14^{1}$ $\frac{13}{16}=0{\cdot}16^{7}$ $\frac{15}{16}=0{\cdot}19^{2}$

(4.76=1£) **£0.21**

1	0·21	41	8·61	81	17·01	121	25·41	161	33·81
2	0·42	42	8·82	82	17·22	122	25·62	162	34·02
3	0·63	43	9·03	83	17·43	123	25·83	163	34·23
4	0·84	44	9·24	84	17·64	124	26·04	164	34·44
5	1·05	45	9·45	85	17·85	125	26·25	165	34·65
6	1·26	46	9·66	86	18·06	126	26·46	166	34·86
7	1·47	47	9·87	87	18·27	127	26·67	167	35·07
8	1·68	48	10·08	88	18·48	128	26·88	168	35·28
9	1·89	49	10·29	89	18·69	129	27·09	169	35·49
10	2·10	**50**	10·50	**90**	18·90	**130**	27·30	**170**	35·70
11	2·31	51	10·71	91	19·11	131	27·51	171	35·91
12	2·52	52	10·92	92	19·32	132	27·72	172	36·12
13	2·73	53	11·13	93	19·53	133	27·93	173	36·33
14	2·94	54	11·34	94	19·74	134	28·14	174	36·54
15	3·15	55	11·55	95	19·95	135	28·35	175	36·75
16	3·36	56	11·76	96	20·16	136	28·56	176	36·96
17	3·57	57	11·97	97	20·37	137	28·77	177	37·17
18	3·78	58	12·18	98	20·58	138	28·98	178	37·38
19	3·99	59	12·39	99	20·79	139	29·19	179	37·59
20	4·20	**60**	12·60	**100**	21·00	**140**	29·40	**180**	37·80
21	4·41	61	12·81	101	21·21	141	29·61	181	38·01
22	4·62	62	13·02	102	21·42	142	29·82	182	38·22
23	4·83	63	13·23	103	21·63	143	30·03	183	38·43
24	5·04	64	13·44	104	21·84	144	30·24	184	38·64
25	5·25	65	13·65	105	22·05	145	30·45	185	38·85
26	5·46	66	13·86	106	22·26	146	30·66	186	39·06
27	5·67	67	14·07	107	22·47	147	30·87	187	39·27
28	5·88	68	14·28	108	22·68	148	31·08	188	39·48
29	6·09	69	14·49	109	22·89	149	31·29	189	39·69
30	6·30	**70**	14·70	**110**	23·10	**150**	31·50	**190**	39·90
31	6·51	71	14·91	111	23·31	151	31·71	191	40·11
32	6·72	72	15·12	112	23·52	152	31·92	192	40·32
33	6·93	73	15·33	113	23·73	153	32·13	193	40·53
34	7·14	74	15·54	114	23·94	154	32·34	194	40·74
35	7·35	75	15·75	115	24·15	155	32·55	195	40·95
36	7·56	76	15·96	116	24·36	156	32·76	196	41·16
37	7·77	77	16·17	117	24·57	157	32·97	197	41·37
38	7·98	78	16·38	118	24·78	158	33·18	198	41·58
39	8·19	79	16·59	119	24·99	159	33·39	199	41·79
40	8·40	**80**	16·80	**120**	25·20	**160**	33·60	**200**	42·00

$\frac{1}{16}=0{\cdot}01^3$ $\frac{1}{9}=0{\cdot}02^3$ $\frac{1}{8}=0{\cdot}02^6$ $\frac{1}{6}=0{\cdot}03^5$ $\frac{1}{5}=0{\cdot}04^2$ $\frac{1}{4}=0{\cdot}05^3$ $\frac{1}{3}=0{\cdot}07^0$

$\frac{3}{8}=0{\cdot}07^9$ $\frac{1}{2}=0{\cdot}10^5$ $\frac{5}{8}=0{\cdot}13^1$ $\frac{3}{4}=0{\cdot}15^8$ $\frac{7}{8}=0{\cdot}18^4$ $\frac{2}{3}=0{\cdot}14^0$ $\frac{5}{6}=0{\cdot}17^5$

$\frac{3}{16}=0{\cdot}03^9$ $\frac{5}{16}=0{\cdot}06^6$ $\frac{7}{16}=0{\cdot}09^2$ $\frac{9}{16}=0{\cdot}11^8$ $\frac{11}{16}=0{\cdot}14^4$ $\frac{13}{16}=0{\cdot}17^1$ $\frac{15}{16}=0{\cdot}19^7$

1 litre=2·113 US pints

1	0·21½	41	8·81½	81	17·41½	121	26·01½	161	34·61½
2	0·43	42	9·03	82	17·63	122	26·23	162	34·83
3	0·64½	43	9·24½	83	17·84½	123	26·44½	163	35·04½
4	0·86	44	9·46	84	18·06	124	26·66	164	35·26
5	1·07½	45	9·67½	85	18·27½	125	26·87½	165	35·47½
6	1·29	46	9·89	86	18·49	126	27·09	166	35·69
7	1·50½	47	10·10½	87	18·70½	127	27·30½	167	35·90½
8	1·72	48	10·32	88	18·92	128	27·52	168	36·12
9	1·93½	49	10·53½	89	19·13½	129	27·73½	169	36·33½
10	2·15	**50**	10·75	**90**	19·35	**130**	27·95	**170**	36·55
11	2·36½	51	10·96½	91	19·56½	131	28·16½	171	36·76½
12	2·58	52	11·18	92	19·78	132	28·38	172	36·98
13	2·79½	53	11·39½	93	19·99½	133	28·59½	173	37·19½
14	3·01	54	11·61	94	20·21	134	28·81	174	37·41
15	3·22½	55	11·82½	95	20·42½	135	29·02½	175	37·62½
16	3·44	56	12·04	96	20·64	136	29·24	176	37·84
17	3·65½	57	12·25½	97	20·85½	137	29·45½	177	38·05½
18	3·87	58	12·47	98	21·07	138	29·67	178	38·27
19	4·08½	59	12·68½	99	21·28½	139	29·88½	179	38·48½
20	4·30	**60**	12·90	**100**	21·50	**140**	30·10	**180**	38·70
21	4·51½	61	13·11½	101	21·71½	141	30·31½	181	38·91½
22	4·73	62	13·33	102	21·93	142	30·53	182	39·13
23	4·94½	63	13·54½	103	22·14½	143	30·74½	183	39·34½
24	5·16	64	13·76	104	22·36	144	30·96	184	39·56
25	5·37½	65	13·97½	105	22·57½	145	31·17½	185	39·77½
26	5·59	66	14·19	106	22·79	146	31·39	186	39·99
27	5·80½	67	14·40½	107	23·00½	147	31·60½	187	40·20½
28	6·02	68	14·62	108	23·22	148	31·82	188	40·42
29	6·23½	69	14·83½	109	23·43½	149	32·03½	189	40·63½
30	6·45	**70**	15·05	**110**	23·65	**150**	32·25	**190**	40·85
31	6·66½	71	15·26½	111	23·86½	151	32·46½	191	41·06½
32	6·88	72	15·48	112	24·08	152	32·68	192	41·28
33	7·09½	73	15·69½	113	24·29½	153	32·89½	193	41·49½
34	7·31	74	15·91	114	24·51	154	33·11	194	41·71
35	7·52½	75	16·12½	115	24·72½	155	33·32½	195	41·92½
36	7·74	76	16·34	116	24·94	156	33·54	196	42·14
37	7·95½	77	16·55½	117	25·15½	157	33·75½	197	42·35½
38	8·17	78	16·77	118	25·37	158	33·97	198	42·57
39	8·38½	79	16·98½	119	25·58½	159	34·18½	199	42·78½
40	8·60	**80**	17·20	**120**	25·80	**160**	34·40	**200**	43·00

$\frac{1}{16}=0{\cdot}01^{3}$ $\frac{1}{9}=0{\cdot}02^{4}$ $\frac{1}{8}=0{\cdot}02^{7}$ $\frac{1}{6}=0{\cdot}03^{6}$ $\frac{1}{5}=0{\cdot}04^{3}$ $\frac{1}{4}=0{\cdot}05^{4}$ $\frac{1}{3}=0{\cdot}07^{2}$

$\frac{3}{8}=0{\cdot}08^{1}$ $\frac{1}{2}=0{\cdot}10^{8}$ $\frac{5}{8}=0{\cdot}13^{4}$ $\frac{3}{4}=0{\cdot}16^{1}$ $\frac{7}{8}=0{\cdot}18^{8}$ $\frac{2}{3}=0{\cdot}14^{3}$ $\frac{5}{6}=0{\cdot}17^{9}$

$\frac{3}{16}=0{\cdot}04^{0}$ $\frac{5}{16}=0{\cdot}06^{7}$ $\frac{7}{16}=0{\cdot}09^{4}$ $\frac{9}{16}=0{\cdot}12^{1}$ $\frac{11}{16}=0{\cdot}14^{8}$ $\frac{13}{16}=0{\cdot}17^{5}$ $\frac{15}{16}=0{\cdot}20^{2}$

(4.55=1£) **£0.22**

1	0·22	41	9·02	81	17·82	121	26·62	161	35·42
2	0·44	42	9·24	82	18·04	122	26·84	162	35·64
3	0·66	43	9·46	83	18·26	123	27·06	163	35·86
4	0·88	44	9·68	84	18·48	124	27·28	164	36·08
5	1·10	45	9·90	85	18·70	125	27·50	165	36·30
6	1·32	46	10·12	86	18·92	126	27·72	166	36·52
7	1·54	47	10·34	87	19·14	127	27·94	167	36·74
8	1·76	48	10·56	88	19·36	128	28·16	168	36·96
9	1·98	49	10·78	89	19·58	129	28·38	169	37·18
10	2·20	**50**	11·00	**90**	19·80	**130**	28·60	**170**	37·40
11	2·42	51	11·22	91	20·02	131	28·82	171	37·62
12	2·64	52	11·44	92	20·24	132	29·04	172	37·84
13	2·86	53	11·66	93	20·46	133	29·26	173	38·06
14	3·08	54	11·88	94	20·68	134	29·48	174	38·28
15	3·30	55	12·10	95	20·90	135	29·70	175	38·50
16	3·52	56	12·32	96	21·12	136	29·92	176	38·72
17	3·74	57	12·54	97	21·34	137	30·14	177	38·94
18	3·96	58	12·76	98	21·56	138	30·36	178	39·16
19	4·18	59	12·98	99	21·78	139	30·58	179	39·38
20	4·40	**60**	13·20	**100**	22·00	**140**	30·80	**180**	39·60
21	4·62	61	13·42	101	22·22	141	31·02	181	39·82
22	4·84	62	13·64	102	22·44	142	31·24	182	40·04
23	5·06	63	13·86	103	22·66	143	31·46	183	40·26
24	5·28	64	14·08	104	22·88	144	31·68	184	40·48
25	5·50	65	14·30	105	23·10	145	31·90	185	40·70
26	5·72	66	14·52	106	23·32	146	32·12	186	40·92
27	5·94	67	14·74	107	23·54	147	32·34	187	41·14
28	6·16	68	14·96	108	23·76	148	32·56	188	41·36
29	6·38	69	15·18	109	23·98	149	32·78	189	41·58
30	6·60	**70**	15·40	**110**	24·20	**150**	33·00	**190**	41·80
31	6·82	71	15·62	111	24·42	151	33·22	191	42·02
32	7·04	72	15·84	112	24·64	152	33·44	192	42·24
33	7·26	73	16·06	113	24·86	153	33·66	193	42·46
34	7·48	74	16·28	114	25·08	154	33·88	194	42·68
35	7·70	75	16·50	115	25·30	155	34·10	195	42·90
36	7·92	76	16·72	116	25·52	156	34·32	196	43·12
37	8·14	77	16·94	117	25·74	157	34·54	197	43·34
38	8·36	78	17·16	118	25·96	158	34·76	198	43·56
39	8·58	79	17·38	119	26·18	159	34·98	199	43·78
40	8·80	**80**	17·60	**120**	26·40	**160**	35·20	**200**	44·00

$\frac{1}{16}=0{\cdot}01^{4}$ $\frac{1}{9}=0{\cdot}02^{4}$ $\frac{1}{8}=0{\cdot}02^{8}$ $\frac{1}{6}=0{\cdot}03^{7}$ $\frac{1}{5}=0{\cdot}04^{4}$ $\frac{1}{4}=0{\cdot}05^{5}$ $\frac{1}{3}=0{\cdot}07^{3}$

$\frac{3}{8}=0{\cdot}08^{3}$ $\frac{1}{2}=0{\cdot}11^{0}$ $\frac{5}{8}=0{\cdot}13^{8}$ $\frac{3}{4}=0{\cdot}16^{5}$ $\frac{7}{8}=0{\cdot}19^{3}$ $\frac{2}{3}=0{\cdot}14^{7}$ $\frac{5}{6}=0{\cdot}18^{3}$

$\frac{3}{16}=0{\cdot}04^{1}$ $\frac{5}{16}=0{\cdot}06^{9}$ $\frac{7}{16}=0{\cdot}09^{6}$ $\frac{9}{16}=0{\cdot}12^{4}$ $\frac{11}{16}=0{\cdot}15^{1}$ $\frac{13}{16}=0{\cdot}17^{9}$ $\frac{15}{16}=0{\cdot}20^{6}$

1 litre=0·220 gallons (imp.) 1 kilogramme=2·204 lbs 1 tonne=2204 lbs

1	0·22½	41	9·22½	81	18·22½	121	27·22½	161	36·22½
2	0·45	42	9·45	82	18·45	122	27·45	162	36·45
3	0·67½	43	9·67½	83	18·67½	123	27·67½	163	36·67½
4	0·90	44	9·90	84	18·90	124	27·90	164	36·90
5	1·12½	45	10·12½	85	19·12½	125	28·12½	165	37·12½
6	1·35	46	10·35	86	19·35	126	28·35	166	37·35
7	1·57½	47	10·57½	87	19·57½	127	28·57½	167	37·57½
8	1·80	48	10·80	88	19·80	128	28·80	168	37·80
9	2·02½	49	11·02½	89	20·02½	129	29·02½	169	38·02½
10	2·25	**50**	11·25	**90**	20·25	**130**	29·25	**170**	38·25
11	2·47½	51	11·47½	91	20·47½	131	29·47½	171	38·47½
12	2·70	52	11·70	92	20·70	132	29·70	172	38·70
13	2·92½	53	11·92½	93	20·92½	133	29·92½	173	38·92½
14	3·15	54	12·15	94	21·15	134	30·15	174	39·15
15	3·37½	55	12·37½	95	21·37½	135	30·37½	175	39·37½
16	3·60	56	12·60	96	21·60	136	30·60	176	39·60
17	3·82½	57	12·82½	97	21·82½	137	30·82½	177	39·82½
18	4·05	58	13·05	98	22·05	138	31·05	178	40·05
19	4·27½	59	13·27½	99	22·27½	139	31·27½	179	40·27½
20	4·50	**60**	13·50	**100**	22·50	**140**	31·50	**180**	40·50
21	4·72½	61	13·72½	101	22·72½	141	31·72½	181	40·72½
22	4·95	62	13·95	102	22·95	142	31·95	182	40·95
23	5·17½	63	14·17½	103	23·17½	143	32·17½	183	41·17½
24	5·40	64	14·40	104	23·40	144	32·40	184	41·40
25	5·62½	65	14·62½	105	23·62½	145	32·62½	185	41·62½
26	5·85	66	14·85	106	23·85	146	32·85	186	41·85
27	6·07½	67	15·07½	107	24·07½	147	33·07½	187	42·07½
28	6·30	68	15·30	108	24·30	148	33·30	188	42·30
29	6·52½	69	15·52½	109	24·52½	149	33·52½	189	42·52½
30	6·75	**70**	15·75	**110**	24·75	**150**	33·75	**190**	42·75
31	6·97½	71	15·97½	111	24·97½	151	33·97½	191	42·97½
32	7·20	72	16·20	112	25·20	152	34·20	192	43·20
33	7·42½	73	16·42½	113	25·42½	153	34·42½	193	43·42½
34	7·65	74	16·65	114	25·65	154	34·65	194	43·65
35	7·87½	75	16·87½	115	25·87½	155	34·87½	195	43·87½
36	8·10	76	17·10	116	26·10	156	35·10	196	44·10
37	8·32½	77	17·32½	117	26·32½	157	35·32½	197	44·32½
38	8·55	78	17·55	118	26·55	158	35·55	198	44·55
39	8·77½	79	17·77½	119	26·77½	159	35·77½	199	44·77½
40	9·00	**80**	18·00	**120**	27·00	**160**	36·00	**200**	45·00

$\frac{1}{16}=0{\cdot}01^{4}$ $\frac{1}{9}=0{\cdot}02^{5}$ $\frac{1}{8}=0{\cdot}02^{8}$ $\frac{1}{6}=0{\cdot}03^{8}$ $\frac{1}{5}=0{\cdot}04^{5}$ $\frac{1}{4}=0{\cdot}05^{6}$ $\frac{1}{3}=0{\cdot}07^{5}$

$\frac{3}{8}=0{\cdot}08^{4}$ $\frac{1}{2}=0{\cdot}11^{3}$ $\frac{5}{8}=0{\cdot}14^{1}$ $\frac{3}{4}=0{\cdot}16^{9}$ $\frac{7}{8}=0{\cdot}19^{7}$ $\frac{2}{3}=0{\cdot}15^{0}$ $\frac{5}{6}=0{\cdot}18^{8}$

$\frac{3}{16}=0{\cdot}04^{2}$ $\frac{5}{16}=0{\cdot}07^{0}$ $\frac{7}{16}=0{\cdot}09^{8}$ $\frac{9}{16}=0{\cdot}12^{7}$ $\frac{11}{16}=0{\cdot}15^{5}$ $\frac{13}{16}=0{\cdot}18^{3}$ $\frac{15}{16}=0{\cdot}21^{1}$

1	0·23	41	9·43	81	18·63	121	27·83	161	37·03
2	0·46	42	9·66	82	18·86	122	28·06	162	37·26
3	0·69	43	9·89	83	19·09	123	28·29	163	37·49
4	0·92	44	10·12	84	19·32	124	28·52	164	37·72
5	1·15	45	10·35	85	19·55	125	28·75	165	37·95
6	1·38	46	10·58	86	19·78	126	28·98	166	38·18
7	1·61	47	10·81	87	20·01	127	29·21	167	38·41
8	1·84	48	11·04	88	20·24	128	29·44	168	38·64
9	2·07	49	11·27	89	20·47	129	29·67	169	38·87
10	2·30	**50**	11·50	**90**	20·70	**130**	29·90	**170**	39·10
11	2·53	51	11·73	91	20·93	131	30·13	171	39·33
12	2·76	52	11·96	92	21·16	132	30·36	172	39·56
13	2·99	53	12·19	93	21·39	133	30·59	173	39·79
14	3·22	54	12·42	94	21·62	134	30·82	174	40·02
15	3·45	55	12·65	95	21·85	135	31·05	175	40·25
16	3·68	56	12·88	96	22·08	136	31·28	176	40·48
17	3·91	57	13·11	97	22·31	137	31·51	177	40·71
18	4·14	58	13·34	98	22·54	138	31·74	178	40·94
19	4·37	59	13·57	99	22·77	139	31·97	179	41·17
20	4·60	**60**	13·80	**100**	23·00	**140**	32·20	**180**	41·40
21	4·83	61	14·03	101	23·23	141	32·43	181	41·63
22	5·06	62	14·26	102	23·46	142	32·66	182	41·86
23	5·29	63	14·49	103	23·69	143	32·89	183	42·09
24	5·52	64	14·72	104	23·92	144	33·12	184	42·32
25	5·75	65	14·95	105	24·15	145	33·35	185	42·55
26	5·98	66	15·18	106	24·38	146	33·58	186	42·78
27	6·21	67	15·41	107	24·61	147	33·81	187	43·01
28	6·44	68	15·64	108	24·84	148	34·04	188	43·24
29	6·67	69	15·87	109	25·07	149	34·27	189	43·47
30	6·90	**70**	16·10	**110**	25·30	**150**	34·50	**190**	43·70
31	7·13	71	16·33	111	25·53	151	34·73	191	43·93
32	7·36	72	16·56	112	25·76	152	34·96	192	44·16
33	7·59	73	16·79	113	25·99	153	35·19	193	44·39
34	7·82	74	17·02	114	26·22	154	35·42	194	44·62
35	8·05	75	17·25	115	26·45	155	35·65	195	44·85
36	8·28	76	17·48	116	26·68	156	35·88	196	45·08
37	8·51	77	17·71	117	26·91	157	36·11	197	45·31
38	8·74	78	17·94	118	27·14	158	36·34	198	45·54
39	8·97	79	18·17	119	27·37	159	36·57	199	45·77
40	9·20	**80**	18·40	**120**	27·60	**160**	36·80	**200**	46·00

$\frac{1}{16}=0{\cdot}01^{4}$ $\frac{1}{9}=0{\cdot}02^{6}$ $\frac{1}{8}=0{\cdot}02^{9}$ $\frac{1}{6}=0{\cdot}03^{8}$ $\frac{1}{5}=0{\cdot}04^{6}$ $\frac{1}{4}=0{\cdot}05^{8}$ $\frac{1}{3}=0{\cdot}07^{7}$

$\frac{3}{8}=0{\cdot}08^{6}$ $\frac{1}{2}=0{\cdot}11^{5}$ $\frac{5}{8}=0{\cdot}14^{4}$ $\frac{3}{4}=0{\cdot}17^{3}$ $\frac{7}{8}=0{\cdot}20^{1}$ $\frac{2}{3}=0{\cdot}15^{3}$ $\frac{5}{6}=0{\cdot}19^{2}$

$\frac{3}{16}=0{\cdot}04^{3}$ $\frac{5}{16}=0{\cdot}07^{2}$ $\frac{7}{16}=0{\cdot}10^{1}$ $\frac{9}{16}=0{\cdot}12^{9}$ $\frac{11}{16}=0{\cdot}15^{8}$ $\frac{13}{16}=0{\cdot}18^{7}$ $\frac{15}{16}=0{\cdot}21^{6}$

(4.26=1£) **£0.23½**

1	0·23½	41	9·63½	81	19·03½	121	28·43½	161	37·83½
2	0·47	42	9·87	82	19·27	122	28·67	162	38·07
3	0·70½	43	10·10½	83	19·50½	123	28·90½	163	38·30½
4	0·94	44	10·34	84	19·74	124	29·14	164	38·54
5	1·17½	45	10·57½	85	19·97½	125	29·37½	165	38·77½
6	1·41	46	10·81	86	20·21	126	29·61	166	39·01
7	1·64½	47	11·04½	87	20·44½	127	29·84½	167	39·24½
8	1·88	48	11·28	88	20·68	128	30·08	168	39·48
9	2·11½	49	11·51½	89	20·91½	129	30·31½	169	39·71½
10	2·35	**50**	11·75	**90**	21·15	**130**	30·55	**170**	39·95
11	2·58½	51	11·98½	91	21·38½	131	30·78½	171	40·18½
12	2·82	52	12·22	92	21·62	132	31·02	172	40·42
13	3·05½	53	12·45½	93	21·85½	133	31·25½	173	40·65½
14	3·29	54	12·69	94	22·09	134	31·49	174	40·89
15	3·52½	55	12·92½	95	22·32½	135	31·72½	175	41·12½
16	3·76	56	13·16	96	22·56	136	31·96	176	41·36
17	3·99½	57	13·39½	97	22·79½	137	32·19½	177	41·59½
18	4·23	58	13·63	98	23·03	138	32·43	178	41·83
19	4·46½	59	13·86½	99	23·26½	139	32·66½	179	42·06½
20	4·70	**60**	14·10	**100**	23·50	**140**	32·90	**180**	42·30
21	4·93½	61	14·33½	101	23·73½	141	33·13½	181	42·53½
22	5·17	62	14·57	102	23·97	142	33·37	182	42·77
23	5·40½	63	14·80½	103	24·20½	143	33·60½	183	43·00½
24	5·64	64	15·04	104	24·44	144	33·84	184	43·24
25	5·87½	65	15·27½	105	24·67½	145	34·07½	185	43·47½
26	6·11	66	15·51	106	24·91	146	34·31	186	43·71
27	6·34½	67	15·74½	107	25·14½	147	34·54½	187	43·94½
28	6·58	68	15·98	108	25·38	148	34·78	188	44·18
29	6·81½	69	16·21½	109	25·61½	149	35·01½	189	44·41½
30	7·05	**70**	16·45	**110**	25·85	**150**	35·25	**190**	44·65
31	7·28½	71	16·68½	111	26·08½	151	35·48½	191	44·88½
32	7·52	72	16·92	112	26·32	152	35·72	192	45·12
33	7·75½	73	17·15½	113	26·55½	153	35·95½	193	45·35½
34	7·99	74	17·39	114	26·79	154	36·19	194	45·59
35	8·22½	75	17·62½	115	27·02½	155	36·42½	195	45·82½
36	8·46	76	17·86	116	27·26	156	36·66	196	46·06
37	8·69½	77	18·09½	117	27·49½	157	36·89½	197	46·29½
38	8·93	78	18·33	118	27·73	158	37·13	198	46·53
39	9·16½	79	18·56½	119	27·96½	159	37·36½	199	46·76½
40	9·40	**80**	18·80	**120**	28·20	**160**	37·60	**200**	47·00

$\frac{1}{16}=0{\cdot}01^{5}$ $\frac{1}{9}=0{\cdot}02^{6}$ $\frac{1}{8}=0{\cdot}02^{9}$ $\frac{1}{6}=0{\cdot}03^{9}$ $\frac{1}{5}=0{\cdot}04^{7}$ $\frac{1}{4}=0{\cdot}05^{9}$ $\frac{1}{3}=0{\cdot}07^{8}$

$\frac{3}{8}=0{\cdot}08^{8}$ $\frac{1}{2}=0{\cdot}11^{8}$ $\frac{5}{8}=0{\cdot}14^{7}$ $\frac{3}{4}=0{\cdot}17^{6}$ $\frac{7}{8}=0{\cdot}20^{6}$ $\frac{2}{3}=0{\cdot}15^{7}$ $\frac{5}{6}=0{\cdot}19^{6}$

$\frac{3}{16}=0{\cdot}04^{4}$ $\frac{5}{16}=0{\cdot}07^{3}$ $\frac{7}{16}=0{\cdot}10^{3}$ $\frac{9}{16}=0{\cdot}13^{2}$ $\frac{11}{16}=0{\cdot}16^{2}$ $\frac{13}{16}=0{\cdot}19^{1}$ $\frac{15}{16}=0{\cdot}22^{0}$

(4.17=1£) £0.24

1	0·24	41	9·84	81	19·44	121	29·04	161	38·64
2	0·48	42	10·08	82	19·68	122	29·28	162	38·88
3	0·72	43	10·32	83	19·92	123	29·52	163	39·12
4	0·96	44	10·56	84	20·16	124	29·76	164	39·36
5	1·20	45	10·80	85	20·40	125	30·00	165	39·60
6	1·44	46	11·04	86	20·64	126	30·24	166	39·84
7	1·68	47	11·28	87	20·88	127	30·48	167	40·08
8	1·92	48	11·52	88	21·12	128	30·72	168	40·32
9	2·16	49	11·76	89	21·36	129	30·96	169	40·56
10	2·40	**50**	12·00	**90**	21·60	**130**	31·20	**170**	40·80
11	2·64	51	12·24	91	21·84	131	31·44	171	41·04
12	2·88	52	12·48	92	22·08	132	31·68	172	41·28
13	3·12	53	12·72	93	22·32	133	31·92	173	41·52
14	3·36	54	12·96	94	22·56	134	32·16	174	41·76
15	3·60	55	13·20	95	22·80	135	32·40	175	42·00
16	3·84	56	13·44	96	23·04	136	32·64	176	42·24
17	4·08	57	13·68	97	23·28	137	32·88	177	42·48
18	4·32	58	13·92	98	23·52	138	33·12	178	42·72
19	4·56	59	14·16	99	23·76	139	33·36	179	42·96
20	4·80	**60**	14·40	**100**	24·00	**140**	33·60	**180**	43·20
21	5·04	61	14·64	101	24·24	141	33·84	181	43·44
22	5·28	62	14·88	102	24·48	142	34·08	182	43·68
23	5·52	63	15·12	103	24·72	143	34·32	183	43·92
24	5·76	64	15·36	104	24·96	144	34·56	184	44·16
25	6·00	65	15·60	105	25·20	145	34·80	185	44·40
26	6·24	66	15·84	106	25·44	146	35·04	186	44·64
27	6·48	67	16·08	107	25·68	147	35·28	187	44·88
28	6·72	68	16·32	108	25·92	148	35·52	188	45·12
29	6·96	69	16·56	109	26·16	149	35·76	189	45·36
30	7·20	**70**	16·80	**110**	26·40	**150**	36·00	**190**	45·60
31	7·44	71	17·04	111	26·64	151	36·24	191	45·84
32	7·68	72	17·28	112	26·88	152	36·48	192	46·08
33	7·92	73	17·52	113	27·12	153	36·72	193	46·32
34	8·16	74	17·76	114	27·36	154	36·96	194	46·56
35	8·40	75	18·00	115	27·60	155	37·20	195	46·80
36	8·64	76	18·24	116	27·84	156	37·44	196	47·04
37	8·88	77	18·48	117	28·08	157	37·68	197	47·28
38	9·12	78	18·72	118	28·32	158	37·92	198	47·52
39	9·36	79	18·96	119	28·56	159	38·16	199	47·76
40	9·60	**80**	19·20	**120**	28·80	**160**	38·40	**200**	48·00

$\frac{1}{16}=0{\cdot}01^5$ $\frac{1}{9}=0{\cdot}02^7$ $\frac{1}{8}=0{\cdot}03^0$ $\frac{1}{6}=0{\cdot}04^0$ $\frac{1}{5}=0{\cdot}04^8$ $\frac{1}{4}=0{\cdot}06^0$ $\frac{1}{3}=0{\cdot}08^0$

$\frac{3}{8}=0{\cdot}09^0$ $\frac{1}{2}=0{\cdot}12^0$ $\frac{5}{8}=0{\cdot}15^0$ $\frac{3}{4}=0{\cdot}18^0$ $\frac{7}{8}=0{\cdot}21^0$ $\frac{2}{3}=0{\cdot}16^0$ $\frac{5}{6}=0{\cdot}20^0$

$\frac{3}{16}=0{\cdot}04^5$ $\frac{5}{16}=0{\cdot}07^5$ $\frac{7}{16}=0{\cdot}10^5$ $\frac{9}{16}=0{\cdot}13^5$ $\frac{11}{16}=0{\cdot}16^5$ $\frac{13}{16}=0{\cdot}19^5$ $\frac{15}{16}=0{\cdot}22^5$

(4.08=1£) **£0.24½**

1	0·24½	41	10·04½	81	19·84½	121	29·64½	161	39·44½
2	0·49	42	10·29	82	20·09	122	29·89	162	39·69
3	0·73½	43	10·53½	83	20·33½	123	30·13½	163	39·93½
4	0·98	44	10·78	84	20·58	124	30·38	164	40·18
5	1·22½	45	11·02½	85	20·82½	125	30·62½	165	40·42½
6	1·47	46	11·27	86	21·07	126	30·87	166	40·67
7	1·71½	47	11·51½	87	21·31½	127	31·11½	167	40·91½
8	1·96	48	11·76	88	21·56	128	31·36	168	41·16
9	2·20½	49	12·00½	89	21·80½	129	31·60½	169	41·40½
10	2·45	**50**	12·25	**90**	22·05	**130**	31·85	**170**	41·65
11	2·69½	51	12·49½	91	22·29½	131	32·09½	171	41·89½
12	2·94	52	12·74	92	22·54	132	32·34	172	42·14
13	3·18½	53	12·98½	93	22·78½	133	32·58½	173	42·38½
14	3·43	54	13·23	94	23·03	134	32·83	174	42·63
15	3·67½	55	13·47½	95	23·27½	135	33·07½	175	42·87½
16	3·92	56	13·72	96	23·52	136	33·32	176	43·12
17	4·16½	57	13·96½	97	23·76½	137	33·56½	177	43·36½
18	4·41	58	14·21	98	24·01	138	33·81	178	43·61
19	4·65½	59	14·45½	99	24·25½	139	34·05½	179	43·85½
20	4·90	**60**	14·70	**100**	24·50	**140**	34·30	**180**	44·10
21	5·14½	61	14·94½	101	24·74½	141	34·54½	181	44·34½
22	5·39	62	15·19	102	24·99	142	34·79	182	44·59
23	5·63½	63	15·43½	103	25·23½	143	35·03½	183	44·83½
24	5·88	64	15·68	104	25·48	144	35·28	184	45·08
25	6·12½	65	15·92½	105	25·72½	145	35·52½	185	45·32½
26	6·37	66	16·17	106	25·97	146	35·77	186	45·57
27	6·61½	67	16·41½	107	26·21½	147	36·01½	187	45·81½
28	6·86	68	16·66	108	26·46	148	36·26	188	46·06
29	7·10½	69	16·90½	109	26·70½	149	36·50½	189	46·30½
30	7·35	**70**	17·15	**110**	26·95	**150**	36·75	**190**	46·55
31	7·59½	71	17·39½	111	27·19½	151	36·99½	191	46·79½
32	7·84	72	17·64	112	27·44	152	37·24	192	47·04
33	8·08½	73	17·88½	113	27·68½	153	37·48½	193	47·28½
34	8·33	74	18·13	114	27·93	154	37·73	194	47·53
35	8·57½	75	18·37½	115	28·17½	155	37·97½	195	47·77½
36	8·82	76	18·62	116	28·42	156	38·22	196	48·02
37	9·06½	77	18·86½	117	28·66½	157	38·46½	197	48·26½
38	9·31	78	19·11	118	28·91	158	38·71	198	48·51
39	9·55½	79	19·35½	119	29·15½	159	38·95½	199	48·75½
40	9·80	**80**	19·60	**120**	29·40	**160**	39·20	**200**	49·00

$\frac{1}{16}=0{\cdot}01^{5}$ $\frac{1}{9}=0{\cdot}02^{7}$ $\frac{1}{8}=0{\cdot}03^{1}$ $\frac{1}{6}=0{\cdot}04^{1}$ $\frac{1}{5}=0{\cdot}04^{9}$ $\frac{1}{4}=0{\cdot}06^{1}$ $\frac{1}{3}=0{\cdot}08^{2}$

$\frac{3}{8}=0{\cdot}09^{2}$ $\frac{1}{2}=0{\cdot}12^{3}$ $\frac{5}{8}=0{\cdot}15^{3}$ $\frac{3}{4}=0{\cdot}18^{4}$ $\frac{7}{8}=0{\cdot}21^{4}$ $\frac{2}{3}=0{\cdot}16^{3}$ $\frac{5}{6}=0{\cdot}20^{4}$

$\frac{3}{16}=0{\cdot}04^{6}$ $\frac{5}{16}=0{\cdot}07^{7}$ $\frac{7}{16}=0{\cdot}10^{7}$ $\frac{9}{16}=0{\cdot}13^{8}$ $\frac{11}{16}=0{\cdot}16^{8}$ $\frac{13}{16}=0{\cdot}19^{9}$ $\frac{15}{16}=0{\cdot}23^{0}$

(4.00=1£) **£0.25**

1	0·25	41	10·25	81	20·25	121	30·25	161	40·25
2	0·50	42	10·50	82	20·50	122	30·50	162	40·50
3	0·75	43	10·75	83	20·75	123	30·75	163	40·75
4	1·00	44	11·00	84	21·00	124	31·00	164	41·00
5	1·25	45	11·25	85	21·25	125	31·25	165	41·25
6	1·50	46	11·50	86	21·50	126	31·50	166	41·50
7	1·75	47	11·75	87	21·75	127	31·75	167	41·75
8	2·00	48	12·00	88	22·00	128	32·00	168	42·00
9	2·25	49	12·25	89	22·25	129	32·25	169	42·25
10	2·50	**50**	12·50	**90**	22·50	**130**	32·50	**170**	42·50
11	2·75	51	12·75	91	22·75	131	32·75	171	42·75
12	3·00	52	13·00	92	23·00	132	33·00	172	43·00
13	3·25	53	13·25	93	23·25	133	33·25	173	43·25
14	3·50	54	13·50	94	23·50	134	33·50	174	43·50
15	3·75	55	13·75	95	23·75	135	33·75	175	43·75
16	4·00	56	14·00	96	24·00	136	34·00	176	44·00
17	4·25	57	14·25	97	24·25	137	34·25	177	44·25
18	4·50	58	14·50	98	24·50	138	34·50	178	44·50
19	4·75	59	14·75	99	24·75	139	34·75	179	44·75
20	5·00	**60**	15·00	**100**	25·00	**140**	35·00	**180**	45·00
21	5·25	61	15·25	101	25·25	141	35·25	181	45·25
22	5·50	62	15·50	102	25·50	142	35·50	182	45·50
23	5·75	63	15·75	103	25·75	143	35·75	183	45·75
24	6·00	64	16·00	104	26·00	144	36·00	184	46·00
25	6·25	65	16·25	105	26·25	145	36·25	185	46·25
26	6·50	66	16·50	106	26·50	146	36·50	186	46·50
27	6·75	67	16·75	107	26·75	147	36·75	187	46·75
28	7·00	68	17·00	108	27·00	148	37·00	188	47·00
29	7·25	69	17·25	109	27·25	149	37·25	189	47·25
30	7·50	**70**	17·50	**110**	27·50	**150**	37·50	**190**	47·50
31	7·75	71	17·75	111	27·75	151	37·75	191	47·75
32	8·00	72	18·00	112	28·00	152	38·00	192	48·00
33	8·25	73	18·25	113	28·25	153	38·25	193	48·25
34	8·50	74	18·50	114	28·50	154	38·50	194	48·50
35	8·75	75	18·75	115	28·75	155	38·75	195	48·75
36	9·00	76	19·00	116	29·00	156	39·00	196	49·00
37	9·25	77	19·25	117	29·25	157	39·25	197	49·25
38	9·50	78	19·50	118	29·50	158	39·50	198	49·50
39	9·75	79	19·75	119	29·75	159	39·75	199	49·75
40	10·00	**80**	20·00	**120**	30·00	**160**	40·00	**200**	50·00

$\frac{1}{16}=0{\cdot}01^{6}$ $\frac{1}{9}=0{\cdot}02^{8}$ $\frac{1}{8}=0{\cdot}03^{1}$ $\frac{1}{6}=0{\cdot}04^{2}$ $\frac{1}{5}=0{\cdot}05^{0}$ $\frac{1}{4}=0{\cdot}06^{3}$ $\frac{1}{3}=0{\cdot}08^{3}$

$\frac{3}{8}=0{\cdot}09^{4}$ $\frac{1}{2}=0{\cdot}12^{5}$ $\frac{5}{8}=0{\cdot}15^{6}$ $\frac{3}{4}=0{\cdot}18^{8}$ $\frac{7}{8}=0{\cdot}21^{9}$ $\frac{2}{3}=0{\cdot}16^{7}$ $\frac{5}{6}=0{\cdot}20^{8}$

$\frac{3}{16}=0{\cdot}04^{7}$ $\frac{5}{16}=0{\cdot}07^{8}$ $\frac{7}{16}=0{\cdot}10^{9}$ $\frac{9}{16}=0{\cdot}14^{1}$ $\frac{11}{16}=0{\cdot}17^{2}$ $\frac{13}{16}=0{\cdot}20^{3}$ $\frac{15}{16}=0{\cdot}23^{4}$

1 ton per acre=2510 kilogrammes per hectare

(3.92=1£) **£0.25½**

1	0·25½	41	10·45½	81	20·65½	121	30·85½	161	41·05½
2	0·51	42	10·71	82	20·91	122	31·11	162	41·31
3	0·76½	43	10·96½	83	21·16½	123	31·36½	163	41·56½
4	1·02	44	11·22	84	21·42	124	31·62	164	41·82
5	1·27½	45	11·47½	85	21·67½	125	31·87½	165	42·07½
6	1·53	46	11·73	86	21·93	126	32·13	166	42·33
7	1·78½	47	11·98½	87	22·18½	127	32·38½	167	42·58½
8	2·04	48	12·24	88	22·44	128	32·64	168	42·84
9	2·29½	49	12·49½	89	22·69½	129	32·89½	169	43·09½
10	2·55	**50**	12·75	**90**	22·95	**130**	33·15	**170**	43·35
11	2·80½	51	13·00½	91	23·20½	131	33·40½	171	43·60½
12	3·06	52	13·26	92	23·46	132	33·66	172	43·86
13	3·31½	53	13·51½	93	23·71½	133	33·91½	173	44·11½
14	3·57	54	13·77	94	23·97	134	34·17	174	44·37
15	3·82½	55	14·02½	95	24·22½	135	34·42½	175	44·62½
16	4·08	56	14·28	96	24·48	136	34·68	176	44·88
17	4·33½	57	14·53½	97	24·73½	137	34·93½	177	45·13½
18	4·59	58	14·79	98	24·99	138	35·19	178	45·39
19	4·84½	59	15·04½	99	25·24½	139	35·44½	179	45·64½
20	5·10	**60**	15·30	**100**	25·50	**140**	35·70	**180**	45·90
21	5·35½	61	15·55½	101	25·75½	141	35·95½	181	46·15½
22	5·61	62	15·81	102	26·01	142	36·21	182	46·41
23	5·86½	63	16·06½	103	26·26½	143	36·46½	183	46·66½
24	6·12	64	16·32	104	26·52	144	36·72	184	46·92
25	6·37½	65	16·57½	105	26·77½	145	36·97½	185	47·17½
26	6·63	66	16·83	106	27·03	146	37·23	186	47·43
27	6·88½	67	17·08½	107	27·28½	147	37·48½	187	47·68½
28	7·14	68	17·34	108	27·54	148	37·74	188	47·94
29	7·39½	69	17·59½	109	27·79½	149	37·99½	189	48·19½
30	7·65	**70**	17·85	**110**	28·05	**150**	38·25	**190**	48·45
31	7·90½	71	18·10½	111	28·30½	151	38·50½	191	48·70½
32	8·16	72	18·36	112	28·56	152	38·76	192	48·96
33	8·41½	73	18·61½	113	28·81½	153	39·01½	193	49·21½
34	8·67	74	18·87	114	29·07	154	39·27	194	49·47
35	8·92½	75	19·12½	115	29·32½	155	39·52½	195	49·72½
36	9·18	76	19·38	116	29·58	156	39·78	196	49·98
37	9·43½	77	19·63½	117	29·83½	157	40·03½	197	50·23½
38	9·69	78	19·89	118	30·09	158	40·29	198	50·49
39	9·94½	79	20·14½	119	30·34½	159	40·54½	199	50·74½
40	10·20	**80**	20·40	**120**	30·60	**160**	40·80	**200**	51·00

$\frac{1}{16}=0{\cdot}01^{6}$ $\frac{1}{9}=0{\cdot}02^{8}$ $\frac{1}{8}=0{\cdot}03^{2}$ $\frac{1}{6}=0{\cdot}04^{3}$ $\frac{1}{5}=0{\cdot}05^{1}$ $\frac{1}{4}=0{\cdot}06^{4}$ $\frac{1}{3}=0{\cdot}08^{5}$

$\frac{3}{8}=0{\cdot}09^{6}$ $\frac{1}{2}=0{\cdot}12^{8}$ $\frac{5}{8}=0{\cdot}15^{9}$ $\frac{3}{4}=0{\cdot}19^{1}$ $\frac{7}{8}=0{\cdot}22^{3}$ $\frac{2}{3}=0{\cdot}17^{0}$ $\frac{5}{6}=0{\cdot}21^{3}$

$\frac{3}{16}=0{\cdot}04^{8}$ $\frac{5}{16}=0{\cdot}08^{0}$ $\frac{7}{16}=0{\cdot}11^{2}$ $\frac{9}{16}=0{\cdot}14^{3}$ $\frac{11}{16}=0{\cdot}17^{5}$ $\frac{13}{16}=0{\cdot}20^{7}$ $\frac{15}{16}=0{\cdot}23^{9}$

(3.85=1£) **£0.26**

1	0·26	41	10·66	81	21·06	121	31·46	161	41·86
2	0·52	42	10·92	82	21·32	122	31·72	162	42·12
3	0·78	43	11·18	83	21·58	123	31·98	163	42·38
4	1·04	44	11·44	84	21·84	124	32·24	164	42·64
5	1·30	45	11·70	85	22·10	125	32·50	165	42·90
6	1·56	46	11·96	86	22·36	126	32·76	166	43·16
7	1·82	47	12·22	87	22·62	127	33·02	167	43·42
8	2·08	48	12·48	88	22·88	128	33·28	168	43·68
9	2·34	49	12·74	89	23·14	129	33·54	169	43·94
10	2·60	**50**	13·00	**90**	23·40	**130**	33·80	**170**	44·20
11	2·86	51	13·26	91	23·66	131	34·06	171	44·46
12	3·12	52	13·52	92	23·92	132	34·32	172	44·72
13	3·38	53	13·78	93	24·18	133	34·58	173	44·98
14	3·64	54	14·04	94	24·44	134	34·84	174	45·24
15	3·90	55	14·30	95	24·70	135	35·10	175	45·50
16	4·16	56	14·56	96	24·96	136	35·36	176	45·76
17	4·42	57	14·82	97	25·22	137	35·62	177	46·02
18	4·68	58	15·08	98	25·48	138	35·88	178	46·28
19	4·94	59	15·34	99	25·74	139	36·14	179	46·54
20	5·20	**60**	15·60	**100**	26·00	**140**	36·40	**180**	46·80
21	5·46	61	15·86	101	26·26	141	36·66	181	47·06
22	5·72	62	16·12	102	26·52	142	36·92	182	47·32
23	5·98	63	16·38	103	26·78	143	37·18	183	47·58
24	6·24	64	16·64	104	27·04	144	37·44	184	47·84
25	6·50	65	16·90	105	27·30	145	37·70	185	48·10
26	6·76	66	17·16	106	27·56	146	37·96	186	48·36
27	7·02	67	17·42	107	27·82	147	38·22	187	48·62
28	7·28	68	17·68	108	28·08	148	38·48	188	48·88
29	7·54	69	17·94	109	28·34	149	38·74	189	49·14
30	7·80	**70**	18·20	**110**	28·60	**150**	39·00	**190**	49·40
31	8·06	71	18·46	111	28·86	151	39·26	191	49·66
32	8·32	72	18·72	112	29·12	152	39·52	192	49·92
33	8·58	73	18·98	113	29·38	153	39·78	193	50·18
34	8·84	74	19·24	114	29·64	154	40·04	194	50·44
35	9·10	75	19·50	115	29·90	155	40·30	195	50·70
36	9·36	76	19·76	116	30·16	156	40·56	196	50·96
37	9·62	77	20·02	117	30·42	157	40·82	197	51·22
38	9·88	78	20·28	118	30·68	158	41·08	198	51·48
39	10·14	79	20·54	119	30·94	159	41·34	199	51·74
40	10·40	**80**	20·80	**120**	31·20	**160**	41·60	**200**	52·00

$\frac{1}{16}=0{\cdot}01^{6}$ $\frac{1}{9}=0{\cdot}02^{9}$ $\frac{1}{8}=0{\cdot}03^{3}$ $\frac{1}{6}=0{\cdot}04^{3}$ $\frac{1}{5}=0{\cdot}05^{2}$ $\frac{1}{4}=0{\cdot}06^{5}$ $\frac{1}{3}=0{\cdot}08^{7}$

$\frac{3}{8}=0{\cdot}09^{8}$ $\frac{1}{2}=0{\cdot}13^{0}$ $\frac{5}{8}=0{\cdot}16^{3}$ $\frac{3}{4}=0{\cdot}19^{5}$ $\frac{7}{8}=0{\cdot}22^{8}$ $\frac{2}{3}=0{\cdot}17^{3}$ $\frac{5}{6}=0{\cdot}21^{7}$

$\frac{3}{16}=0{\cdot}04^{9}$ $\frac{5}{16}=0{\cdot}08^{1}$ $\frac{7}{16}=0{\cdot}11^{4}$ $\frac{9}{16}=0{\cdot}14^{6}$ $\frac{11}{16}=0{\cdot}17^{9}$ $\frac{13}{16}=0{\cdot}21^{1}$ $\frac{15}{16}=0{\cdot}24^{4}$

1 square mile=2·590 sq. kilometres

(3.77=1£) **£0.26½.**

1	0·26½	41	10·86½	81	21·46½	121	32·06½	161	42·66½
2	0·53	42	11·13	82	21·73	122	32·33	162	42·93
3	0·79½	43	11·39½	83	21·99½	123	32·59½	163	43·19½
4	1·06	44	11·66	84	22·26	124	32·86	164	43·46
5	1·32½	45	11·92½	85	22·52½	125	33·12½	165	43·72½
6	1·59	46	12·19	86	22·79	126	33·39	166	43·99
7	1·85½	47	12·45½	87	23·05½	127	33·65½	167	44·25½
8	2·12	48	12·72	88	23·32	128	33·92	168	44·52
9	2·38½	49	12·98½	89	23·58½	129	34·18½	169	44·78½
10	2·65	**50**	13·25	**90**	23·85	**130**	34·45	**170**	45·05
11	2·91½	51	13·51½	91	24·11½	131	34·71½	171	45·31½
12	3·18	52	13·78	92	24·38	132	34·98	172	45·58
13	3·44½	53	14·04½	93	24·64½	133	35·24½	173	45·84½
14	3·71	54	14·31	94	24·91	134	35·51	174	46·11
15	3·97½	55	14·57½	95	25·17½	135	35·77½	175	46·37½
16	4·24	56	14·84	96	25·44	136	36·04	176	46·64
17	4·50½	57	15·10½	97	25·70½	137	36·30½	177	46·90½
18	4·77	58	15·37	98	25·97	138	36·57	178	47·17
19	5·03½	59	15·63½	99	26·23½	139	36·83½	179	47·43½
20	5·30	**60**	15·90	**100**	26·50	**140**	37·10	**180**	47·70
21	5·56½	61	16·16½	101	26·76½	141	37·36½	181	47·96½
22	5·83	62	16·43	102	27·03	142	37·63	182	48·23
23	6·09½	63	16·69½	103	27·29½	143	37·89½	183	48·49½
24	6·36	64	16·96	104	27·56	144	38·16	184	48·76
25	6·62½	65	17·22½	105	27·82½	145	38·42½	185	49·02½
26	6·89	66	17·49	106	28·09	146	38·69	186	49·29
27	7·15½	67	17·75½	107	28·35½	147	38·95½	187	49·55½
28	7·42	68	18·02	108	28·62	148	39·22	188	49·82
29	7·68½	69	18·28½	109	28·88½	149	39·48½	189	50·08½
30	7·95	**70**	18·55	**110**	29·15	**150**	39·75	**190**	50·35
31	8·21½	71	18·81½	111	29·41½	151	40·01½	191	50·61½
32	8·48	72	19·08	112	29·68	152	40·28	192	50·88
33	8·74½	73	19·34½	113	29·94½	153	40·54½	193	51·14½
34	9·01	74	19·61	114	30·21	154	40·81	194	51·41
35	9·27½	75	19·87½	115	30·47½	155	41·07½	195	51·67½
36	9·54	76	20·14	116	30·74	156	41·34	196	51·94
37	9·80½	77	20·40½	117	31·00½	157	41·60½	197	52·20½
38	10·07	78	20·67	118	31·27	158	41·87	198	52·47
39	10·33½	79	20·93½	119	31·53½	159	42·13½	199	52·73½
40	10·60	**80**	21·20	**120**	31·80	**160**	42·40	**200**	53·00

$\frac{1}{16}=0{\cdot}01^{7}$ $\frac{1}{9}=0{\cdot}02^{9}$ $\frac{1}{8}=0{\cdot}03^{3}$ $\frac{1}{6}=0{\cdot}04^{4}$ $\frac{1}{5}=0{\cdot}05^{3}$ $\frac{1}{4}=0{\cdot}06^{6}$ $\frac{1}{3}=0{\cdot}08^{8}$

$\frac{3}{8}=0{\cdot}09^{9}$ $\frac{1}{2}=0{\cdot}13^{3}$ $\frac{5}{8}=0{\cdot}16^{6}$ $\frac{3}{4}=0{\cdot}19^{9}$ $\frac{7}{8}=0{\cdot}23^{2}$ $\frac{2}{3}=0{\cdot}17^{7}$ $\frac{5}{6}=0{\cdot}22^{1}$

$\frac{3}{16}=0{\cdot}05^{0}$ $\frac{5}{16}=0{\cdot}08^{3}$ $\frac{7}{16}=0{\cdot}11^{6}$ $\frac{9}{16}=0{\cdot}14^{9}$ $\frac{11}{16}=0{\cdot}18^{2}$ $\frac{13}{16}=0{\cdot}21^{5}$ $\frac{15}{16}=0{\cdot}24^{8}$

1 litre=0·2642 US gallons

1	0·27	41	11·07	81	21·87	121	32·67	161	43·47
2	0·54	42	11·34	82	22·14	122	32·94	162	43·74
3	0·81	43	11·61	83	22·41	123	33·21	163	44·01
4	1·08	44	11·88	84	22·68	124	33·48	164	44·28
5	1·35	45	12·15	85	22·95	125	33·75	165	44·55
6	1·62	46	12·42	86	23·22	126	34·02	166	44·82
7	1·89	47	12·69	87	23·49	127	34·29	167	45·09
8	2·16	48	12·96	88	23·76	128	34·56	168	45·36
9	2·43	49	13·23	89	24·03	129	34·83	169	45·63
10	2·70	**50**	13·50	**90**	24·30	**130**	35·10	**170**	45·90
11	2·97	51	13·77	91	24·57	131	35·37	171	46·17
12	3·24	52	14·04	92	24·84	132	35·64	172	46·44
13	3·51	53	14·31	93	25·11	133	35·91	173	46·71
14	3·78	54	14·58	94	25·38	134	36·18	174	46·98
15	4·05	55	14·85	95	25·65	135	36·45	175	47·25
16	4·32	56	15·12	96	25·92	136	36·72	176	47·52
17	4·59	57	15·39	97	26·19	137	36·99	177	47·79
18	4·86	58	15·66	98	26·46	138	37·26	178	48·06
19	5·13	59	15·93	99	26·73	139	37·53	179	48·33
20	5·40	**60**	16·20	**100**	27·00	**140**	37·80	**180**	48·60
21	5·67	61	16·47	101	27·27	141	38·07	181	48·87
22	5·94	62	16·74	102	27·54	142	38·34	182	49·14
23	6·21	63	17·01	103	27·81	143	38·61	183	49·41
24	6·48	64	17·28	104	28·08	144	38·88	184	49·68
25	6·75	65	17·55	105	28·35	145	39·15	185	49·95
26	7·02	66	17·82	106	28·62	146	39·42	186	50·22
27	7·29	67	18·09	107	28·89	147	39·69	187	50·49
28	7·56	68	18·36	108	29·16	148	39·96	188	50·76
29	7·83	69	18·63	109	29·43	149	40·23	189	51·03
30	8·10	**70**	18·90	**110**	29·70	**150**	40·50	**190**	51·30
31	8·37	71	19·17	111	29·97	151	40·77	191	51·57
32	8·64	72	19·44	112	30·24	152	41·04	192	51·84
33	8·91	73	19·71	113	30·51	153	41·31	193	52·11
34	9·18	74	19·98	114	30·78	154	41·58	194	52·38
35	9·45	75	20·25	115	31·05	155	41·85	195	52·65
36	9·72	76	20·52	116	31·32	156	42·12	196	52·92
37	9·99	77	20·79	117	31·59	157	42·39	197	53·19
38	10·26	78	21·06	118	31·86	158	42·66	198	53·46
39	10·53	79	21·33	119	32·13	159	42·93	199	53·73
40	10·80	**80**	21·60	**120**	32·40	**160**	43·20	**200**	54·00

$\frac{1}{16}=0{\cdot}01^{7}$	$\frac{1}{9}=0{\cdot}03^{0}$	$\frac{1}{8}=0{\cdot}03^{4}$	$\frac{1}{6}=0{\cdot}04^{5}$	$\frac{1}{5}=0{\cdot}05^{4}$	$\frac{1}{4}=0{\cdot}06^{8}$	$\frac{1}{3}=0{\cdot}09^{0}$
$\frac{3}{8}=0{\cdot}10^{1}$	$\frac{1}{2}=0{\cdot}13^{5}$	$\frac{5}{8}=0{\cdot}16^{9}$	$\frac{3}{4}=0{\cdot}20^{3}$	$\frac{7}{8}=0{\cdot}23^{6}$	$\frac{2}{3}=0{\cdot}18^{0}$	$\frac{5}{6}=0{\cdot}22^{5}$
$\frac{3}{16}=0{\cdot}05^{1}$	$\frac{5}{16}=0{\cdot}08^{4}$	$\frac{7}{16}=0{\cdot}11^{8}$	$\frac{9}{16}=0{\cdot}15^{2}$	$\frac{11}{16}=0{\cdot}18^{6}$	$\frac{13}{16}=0{\cdot}21^{9}$	$\frac{15}{16}=0{\cdot}25^{3}$

(3.64=1£) **£0.27½**

1	0·27½	41	11·27½	81	22·27½	121	33·27½	161	44·27½
2	0·55	42	11·55	82	22·55	122	33·55	162	44·55
3	0·82½	43	11·82½	83	22·82½	123	33·82½	163	44·82½
4	1·10	44	12·10	84	23·10	124	34·10	164	45·10
5	1·37½	45	12·37½	85	23·37½	125	34·37½	165	45·37½
6	1·65	46	12·65	86	23·65	126	34·65	166	45·65
7	1·92½	47	12·92½	87	23·92½	127	34·92½	167	45·92½
8	2·20	48	13·20	88	24·20	128	35·20	168	46·20
9	2·47½	49	13·47½	89	24·47½	129	35·47½	169	46·47½
10	2·75	**50**	13·75	**90**	24·75	**130**	35·75	**170**	46·75
11	3·02½	51	14·02½	91	25·02½	131	36·02½	171	47·02½
12	3·30	52	14·30	92	25·30	132	36·30	172	47·30
13	3·57½	53	14·57½	93	25·57½	133	36·57½	173	47·57½
14	3·85	54	14·85	94	25·85	134	36·85	174	47·85
15	4·12½	55	15·12½	95	26·12½	135	37·12½	175	48·12½
16	4·40	56	15·40	96	26·40	136	37·40	176	48·40
17	4·67½	57	15·67½	97	26·67½	137	37·67½	177	48·67½
18	4·95	58	15·95	98	26·95	138	37·95	178	48·95
19	5·22½	59	16·22½	99	27·22½	139	38·22½	179	49·22½
20	5·50	**60**	16·50	**100**	27·50	**140**	38·50	**180**	49·50
21	5·77½	61	16·77½	101	27·77½	141	38·77½	181	49·77½
22	6·05	62	17·05	102	28·05	142	39·05	182	50·05
23	6·32½	63	17·32½	103	28·32½	143	39·32½	183	50·32½
24	6·60	64	17·60	104	28·60	144	39·60	184	50·60
25	6·87½	65	17·87½	105	28·87½	145	39·87½	185	50·87½
26	7·15	66	18·15	106	29·15	146	40·15	186	51·15
27	7·42½	67	18·42½	107	29·42½	147	40·42½	187	51·42½
28	7·70	68	18·70	108	29·70	148	40·70	188	51·70
29	7·97½	69	18·97½	109	29·97½	149	40·97½	189	51·97½
30	8·25	**70**	19·25	**110**	30·25	**150**	41·25	**190**	52·25
31	8·52½	71	19·52½	111	30·52½	151	41·52½	191	52·52½
32	8·80	72	19·80	112	30·80	152	41·80	192	52·80
33	9·07½	73	20·07½	113	31·07½	153	42·07½	193	53·07½
34	9·35	74	20·35	114	31·35	154	42·35	194	53·35
35	9·62½	75	20·62½	115	31·62½	155	42·62½	195	53·62½
36	9·90	76	20·90	116	31·90	156	42·90	196	53·90
37	10·17½	77	21·17½	117	32·17½	157	43·17½	197	54·17½
38	10·45	78	21·45	118	32·45	158	43·45	198	54·45
39	10·72½	79	21·72½	119	32·72½	159	43·72½	199	54·72½
40	11·00	**80**	22·00	**120**	33·00	**160**	44·00	**200**	55·00

$\frac{1}{16}=0{\cdot}01^{7}$ $\frac{1}{9}=0{\cdot}03^{1}$ $\frac{1}{8}=0{\cdot}03^{4}$ $\frac{1}{6}=0{\cdot}04^{6}$ $\frac{1}{5}=0{\cdot}05^{5}$ $\frac{1}{4}=0{\cdot}06^{9}$ $\frac{1}{3}=0{\cdot}09^{2}$

$\frac{3}{8}=0{\cdot}10^{3}$ $\frac{1}{2}=0{\cdot}13^{8}$ $\frac{5}{8}=0{\cdot}17^{2}$ $\frac{3}{4}=0{\cdot}20^{6}$ $\frac{7}{8}=0{\cdot}24^{1}$ $\frac{2}{3}=0{\cdot}18^{3}$ $\frac{5}{6}=0{\cdot}22^{9}$

$\frac{3}{16}=0{\cdot}05^{2}$ $\frac{5}{16}=0{\cdot}08^{6}$ $\frac{7}{16}=0{\cdot}12^{0}$ $\frac{9}{16}=0{\cdot}15^{5}$ $\frac{11}{16}=0{\cdot}18^{9}$ $\frac{13}{16}=0{\cdot}22^{3}$ $\frac{15}{16}=0{\cdot}25^{8}$

(3.57=1£) **£0.28**

1	0·28	41	11·48	81	22·68	121	33·88	161	45·08
2	0·56	42	11·76	82	22·96	122	34·16	162	45·36
3	0·84	43	12·04	83	23·24	123	34·44	163	45·64
4	1·12	44	12·32	84	23·52	124	34·72	164	45·92
5	1·40	45	12·60	85	23·80	125	35·00	165	46·20
6	1·68	46	12·88	86	24·08	126	35·28	166	46·48
7	1·96	47	13·16	87	24·36	127	35·56	167	46·76
8	2·24	48	13·44	88	24·64	128	35·84	168	47·04
9	2·52	49	13·72	89	24·92	129	36·12	169	47·32
10	2·80	**50**	14·00	**90**	25·20	**130**	36·40	**170**	47·60
11	3·08	51	14·28	91	25·48	131	36·68	171	47·88
12	3·36	52	14·56	92	25·76	132	36·96	172	48·16
13	3·64	53	14·84	93	26·04	133	37·24	173	48·44
14	3·92	54	15·12	94	26·32	134	37·52	174	48·72
15	4·20	55	15·40	95	26·60	135	37·80	175	49·00
16	4·48	56	15·68	96	26·88	136	38·08	176	49·28
17	4·76	57	15·96	97	27·16	137	38·36	177	49·56
18	5·04	58	16·24	98	27·44	138	38·64	178	49·84
19	5·32	59	16·52	99	27·72	139	38·92	179	50·12
20	5·60	**60**	16·80	**100**	28·00	**140**	39·20	**180**	50·40
21	5·88	61	17·08	101	28·28	141	39·48	181	50·68
22	6·16	62	17·36	102	28·56	142	39·76	182	50·96
23	6·44	63	17·64	103	28·84	143	40·04	183	51·24
24	6·72	64	17·92	104	29·12	144	40·32	184	51·52
25	7·00	65	18·20	105	29·40	145	40·60	185	51·80
26	7·28	66	18·48	106	29·68	146	40·88	186	52·08
27	7·56	67	18·76	107	29·96	147	41·16	187	52·36
28	7·84	68	19·04	108	30·24	148	41·44	188	52·64
29	8·12	69	19·32	109	30·52	149	41·72	189	52·92
30	8·40	**70**	19·60	**110**	30·80	**150**	42·00	**190**	53·20
31	8·68	71	19·88	111	31·08	151	42·28	191	53·48
32	8·96	72	20·16	112	31·36	152	42·56	192	53·76
33	9·24	73	20·44	113	31·64	153	42·84	193	54·04
34	9·52	74	20·72	114	31·92	154	43·12	194	54·32
35	9·80	75	21·00	115	32·20	155	43·40	195	54·60
36	10·08	76	21·28	116	32·48	156	43·68	196	54·88
37	10·36	77	21·56	117	32·76	157	43·96	197	55·16
38	10·64	78	21·84	118	33·04	158	44·24	198	55·44
39	10·92	79	22·12	119	33·32	159	44·52	199	55·72
40	11·20	**80**	22·40	**120**	33·60	**160**	44·80	**200**	56·00

$\frac{1}{16}=0{\cdot}01^{8}$ $\frac{1}{9}=0{\cdot}03^{1}$ $\frac{1}{8}=0{\cdot}03^{5}$ $\frac{1}{6}=0{\cdot}04^{7}$ $\frac{1}{5}=0{\cdot}05^{6}$ $\frac{1}{4}=0{\cdot}07^{0}$ $\frac{1}{3}=0{\cdot}09^{3}$

$\frac{3}{8}=0{\cdot}10^{5}$ $\frac{1}{2}=0{\cdot}14^{0}$ $\frac{5}{8}=0{\cdot}17^{5}$ $\frac{3}{4}=0{\cdot}21^{0}$ $\frac{7}{8}=0{\cdot}24^{5}$ $\frac{2}{3}=0{\cdot}18^{7}$ $\frac{5}{6}=0{\cdot}23^{3}$

$\frac{3}{16}=0{\cdot}05^{3}$ $\frac{5}{16}=0{\cdot}08^{8}$ $\frac{7}{16}=0{\cdot}12^{3}$ $\frac{9}{16}=0{\cdot}15^{8}$ $\frac{11}{16}=0{\cdot}19^{3}$ $\frac{13}{16}=0{\cdot}22^{8}$ $\frac{15}{16}=0{\cdot}26^{3}$

1 kilojoule=0·278 Watt hours

(3.51=1£) **£0.28½**

1	0·28½	41	11·68½	81	23·08½	121	34·48½	161	45·88½
2	0·57	42	11·97	82	23·37	122	34·77	162	46·17
3	0·85½	43	12·25½	83	23·65½	123	35·05½	163	46·45½
4	1·14	44	12·54	84	23·94	124	35·34	164	46·74
5	1·42½	45	12·82½	85	24·22½	125	35·62½	165	47·02½
6	1·71	46	13·11	86	24·51	126	35·91	166	47·31
7	1·99½	47	13·39½	87	24·79½	127	36·19½	167	47·59½
8	2·28	48	13·68	88	25·08	128	36·48	168	47·88
9	2·56½	49	13·96½	89	25·36½	129	36·76½	169	48·16½
10	2·85	**50**	14·25	**90**	25·65	**130**	37·05	**170**	48·45
11	3·13½	51	14·53½	91	25·93½	131	37·33½	171	48·73½
12	3·42	52	14·82	92	26·22	132	37·62	172	49·02
13	3·70½	53	15·10½	93	26·50½	133	37·90½	173	49·30½
14	3·99	54	15·39	94	26·79	134	38·19	174	49·59
15	4·27½	55	15·67½	95	27·07½	135	38·47½	175	49·87½
16	4·56	56	15·96	96	27·36	136	38·76	176	50·16
17	4·84½	57	16·24½	97	27·64½	137	39·04½	177	50·44½
18	5·13	58	16·53	98	27·93	138	39·33	178	50·73
19	5·41½	59	16·81½	99	28·21½	139	39·61½	179	51·01½
20	5·70	**60**	17·10	**100**	28·50	**140**	39·90	**180**	51·30
21	5·98½	61	17·38½	101	28·78½	141	40·18½	181	51·58½
22	6·27	62	17·67	102	29·07	142	40·47	182	51·87
23	6·55½	63	17·95½	103	29·35½	143	40·75½	183	52·15½
24	6·84	64	18·24	104	29·64	144	41·04	184	52·44
25	7·12½	65	18·52½	105	29·92½	145	41·32½	185	52·72½
26	7·41	66	18·81	106	30·21	146	41·61	186	53·01
27	7·69½	67	19·09½	107	30·49½	147	41·89½	187	53·29½
28	7·98	68	19·38	108	30·78	148	42·18	188	53·58
29	8·26½	69	19·66½	109	31·06½	149	42·46½	189	53·86½
30	8·55	**70**	19·95	**110**	31·35	**150**	42·75	**190**	54·15
31	8·83½	71	20·23½	111	31·63½	151	43·03½	191	54·43½
32	9·12	72	20·52	112	31·92	152	43·32	192	54·72
33	9·40½	73	20·80½	113	32·20½	153	43·60½	193	55·00½
34	9·69	74	21·09	114	32·49	154	43·89	194	55·29
35	9·97½	75	21·37½	115	32·77½	155	44·17½	195	55·57½
36	10·26	76	21·66	116	33·06	156	44·46	196	55·86
37	10·54½	77	21·94½	117	33·34½	157	44·74½	197	56·14½
38	10·83	78	22·23	118	33·63	158	45·03	198	56·43
39	11·11½	79	22·51½	119	33·91½	159	45·31½	199	56·71½
40	11·40	**80**	22·80	**120**	34·20	**160**	45·60	**200**	57·00

$\frac{1}{16}=0{\cdot}01^{8}$ $\frac{1}{9}=0{\cdot}03^{2}$ $\frac{1}{8}=0{\cdot}03^{6}$ $\frac{1}{6}=0{\cdot}04^{8}$ $\frac{1}{5}=0{\cdot}05^{7}$ $\frac{1}{4}=0{\cdot}07^{1}$ $\frac{1}{3}=0{\cdot}09^{5}$

$\frac{3}{8}=0{\cdot}10^{7}$ $\frac{1}{2}=0{\cdot}14^{3}$ $\frac{5}{8}=0{\cdot}17^{8}$ $\frac{3}{4}=0{\cdot}21^{4}$ $\frac{7}{8}=0{\cdot}24^{9}$ $\frac{2}{3}=0{\cdot}19^{0}$ $\frac{5}{6}=0{\cdot}23^{8}$

$\frac{3}{16}=0{\cdot}05^{3}$ $\frac{5}{16}=0{\cdot}08^{9}$ $\frac{7}{16}=0{\cdot}12^{5}$ $\frac{9}{16}=0{\cdot}16^{0}$ $\frac{11}{16}=0{\cdot}19^{6}$ $\frac{13}{16}=0{\cdot}23^{2}$ $\frac{15}{16}=0{\cdot}26^{7}$

1 cubic foot=28·316 litres

(3.45=1£) **£0.29**

1	0·29	41	11·89	81	23·49	121	35·09	161	46·69
2	0·58	42	12·18	82	23·78	122	35·38	162	46·98
3	0·87	43	12·47	83	24·07	123	35·67	163	47·27
4	1·16	44	12·76	84	24·36	124	35·96	164	47·56
5	1·45	45	13·05	85	24·65	125	36·25	165	47·85
6	1·74	46	13·34	86	24·94	126	36·54	166	48·14
7	2·03	47	13·63	87	25·23	127	36·83	167	48·43
8	2·32	48	13·92	88	25·52	128	37·12	168	48·72
9	2·61	49	14·21	89	25·81	129	37·41	169	49·01
10	2·90	**50**	14·50	**90**	26·10	**130**	37·70	**170**	49·30
11	3·19	51	14·79	91	26·39	131	37·99	171	49·59
12	3·48	52	15·08	92	26·68	132	38·28	172	49·88
13	3·77	53	15·37	93	26·97	133	38·57	173	50·17
14	4·06	54	15·66	94	27·26	134	38·86	174	50·46
15	4·35	55	15·95	95	27·55	135	39·15	175	50·75
16	4·64	56	16·24	96	27·84	136	39·44	176	51·04
17	4·93	57	16·53	97	28·13	137	39·73	177	51·33
18	5·22	58	16·82	98	28·42	138	40·02	178	51·62
19	5·51	59	17·11	99	28·71	139	40·31	179	51·91
20	5·80	**60**	17·40	**100**	29·00	**140**	40·60	**180**	52·20
21	6·09	61	17·69	101	29·29	141	40·89	181	52·49
22	6·38	62	17·98	102	29·58	142	41·18	182	52·78
23	6·67	63	18·27	103	29·87	143	41·47	183	53·07
24	6·96	64	18·56	104	30·16	144	41·76	184	53·36
25	7·25	65	18·85	105	30·45	145	42·05	185	53·65
26	7·54	66	19·14	106	30·74	146	42·34	186	53·94
27	7·83	67	19·43	107	31·03	147	42·63	187	54·23
28	8·12	68	19·72	108	31·32	148	42·92	188	54·52
29	8·41	69	20·01	109	31·61	149	43·21	189	54·81
30	8·70	**70**	20·30	**110**	31·90	**150**	43·50	**190**	55·10
31	8·99	71	20·59	111	32·19	151	43·79	191	55·39
32	9·28	72	20·88	112	32·48	152	44·08	192	55·68
33	9·57	73	21·17	113	32·77	153	44·37	193	55·97
34	9·86	74	21·46	114	33·06	154	44·66	194	56·26
35	10·15	75	21·75	115	33·35	155	44·95	195	56·55
36	10·44	76	22·04	116	33·64	156	45·24	196	56·84
37	10·73	77	22·33	117	33·93	157	45·53	197	57·13
38	11·02	78	22·62	118	34·22	158	45·82	198	57·42
39	11·31	79	22·91	119	34·51	159	46·11	199	57·71
40	11·60	**80**	23·20	**120**	34·80	**160**	46·40	**200**	58·00

$\frac{1}{16}=0{\cdot}01^{8}$	$\frac{1}{9}=0{\cdot}03^{2}$	$\frac{1}{8}=0{\cdot}03^{6}$	$\frac{1}{6}=0{\cdot}04^{8}$	$\frac{1}{5}=0{\cdot}05^{8}$	$\frac{1}{4}=0{\cdot}07^{3}$	$\frac{1}{3}=0{\cdot}09^{7}$
$\frac{3}{8}=0{\cdot}10^{9}$	$\frac{1}{2}=0{\cdot}14^{5}$	$\frac{5}{8}=0{\cdot}18^{1}$	$\frac{3}{4}=0{\cdot}21^{8}$	$\frac{7}{8}=0{\cdot}25^{4}$	$\frac{2}{3}=0{\cdot}19^{3}$	$\frac{5}{6}=0{\cdot}24^{2}$
$\frac{3}{16}=0{\cdot}05^{4}$	$\frac{5}{16}=0{\cdot}09^{1}$	$\frac{7}{16}=0{\cdot}12^{7}$	$\frac{9}{16}=0{\cdot}16^{3}$	$\frac{11}{16}=0{\cdot}19^{9}$	$\frac{13}{16}=0{\cdot}23^{6}$	$\frac{15}{16}=0{\cdot}27^{2}$

(3.39=1£) **£0.29½**

1	0·29½	41	12·09½	81	23·89½	121	35·69½	161	47·49½
2	0·59	42	12·39	82	24·19	122	35·99	162	47·79
3	0·88½	43	12·68½	83	24·48½	123	36·28½	163	48·08½
4	1·18	44	12·98	84	24·78	124	36·58	164	48·38
5	1·47½	45	13·27½	85	25·07½	125	36·87½	165	48·67½
6	1·77	46	13·57	86	25·37	126	37·17	166	48·97
7	2·06½	47	13·86½	87	25·66½	127	37·46½	167	49·26½
8	2·36	48	14·16	88	25·96	128	37·76	168	49·56
9	2·65½	49	14·45½	89	26·25½	129	38·05½	169	49·85½
10	2·95	**50**	14·75	**90**	26·55	**130**	38·35	**170**	50·15
11	3·24½	51	15·04½	91	26·84½	131	38·64½	171	50·44½
12	3·54	52	15·34	92	27·14	132	38·94	172	50·74
13	3·83½	53	15·63½	93	27·43½	133	39·23½	173	51·03½
14	4·13	54	15·93	94	27·73	134	39·53	174	51·33
15	4·42½	55	16·22½	95	28·02½	135	39·82½	175	51·62½
16	4·72	56	16·52	96	28·32	136	40·12	176	51·92
17	5·01½	57	16·81½	97	28·61½	137	40·41½	177	52·21½
18	5·31	58	17·11	98	28·91	138	40·71	178	52·51
19	5·60½	59	17·40½	99	29·20½	139	41·00½	179	52·80½
20	5·90	**60**	17·70	**100**	29·50	**140**	41·30	**180**	53·10
21	6·19½	61	17·99½	101	29·79½	141	41·59½	181	53·39½
22	6·49	62	18·29	102	30·09	142	41·89	182	53·69
23	6·78½	63	18·58½	103	30·38½	143	42·18½	183	53·98½
24	7·08	64	18·88	104	30·68	144	42·48	184	54·28
25	7·37½	65	19·17½	105	30·97½	145	42·77½	185	54·57½
26	7·67	66	19·47	106	31·27	146	43·07	186	54·87
27	7·96½	67	19·76½	107	31·56½	147	43·36½	187	55·16½
28	8·26	68	20·06	108	31·86	148	43·66	188	55·46
29	8·55½	69	20·35½	109	32·15½	149	43·95½	189	55·75½
30	8·85	**70**	20·65	**110**	32·45	**150**	44·25	**190**	56·05
31	9·14½	71	20·94½	111	32·74½	151	44·54½	191	56·34½
32	9·44	72	21·24	112	33·04	152	44·84	192	56·64
33	9·73½	73	21·53½	113	33·33½	153	45·13½	193	56·93½
34	10·03	74	21·83	114	33·63	154	45·43	194	57·23
35	10·32½	75	22·12½	115	33·92½	155	45·72½	195	57·52½
36	10·62	76	22·42	116	34·22	156	46·02	196	57·82
37	10·91½	77	22·71½	117	34·51½	157	46·31½	197	58·11½
38	11·21	78	23·01	118	34·81	158	46·61	198	58·41
39	11·50½	79	23·30½	119	35·10½	159	46·90½	199	58·70½
40	11·80	**80**	23·60	**120**	35·40	**160**	47·20	**200**	59·00

$\frac{1}{16}=0{\cdot}01^{8}$ $\frac{1}{9}=0{\cdot}03^{3}$ $\frac{1}{8}=0{\cdot}03^{7}$ $\frac{1}{6}=0{\cdot}04^{9}$ $\frac{1}{5}=0{\cdot}05^{9}$ $\frac{1}{4}=0{\cdot}07^{4}$ $\frac{1}{3}=0{\cdot}09^{8}$

$\frac{3}{8}=0{\cdot}11^{1}$ $\frac{1}{2}=0{\cdot}14^{8}$ $\frac{5}{8}=0{\cdot}18^{4}$ $\frac{3}{4}=0{\cdot}22^{1}$ $\frac{7}{8}=0{\cdot}25^{8}$ $\frac{2}{3}=0{\cdot}19^{7}$ $\frac{5}{6}=0{\cdot}24^{6}$

$\frac{3}{16}=0{\cdot}05^{5}$ $\frac{5}{16}=0{\cdot}09^{2}$ $\frac{7}{16}=0{\cdot}12^{9}$ $\frac{9}{16}=0{\cdot}16^{6}$ $\frac{11}{16}=0{\cdot}20^{3}$ $\frac{13}{16}=0{\cdot}24^{0}$ $\frac{15}{16}=0{\cdot}27^{7}$

(3.33=1£) **£0.30**

1	0·30	41	12·30	81	24·30	121	36·30	161	48·30
2	0·60	42	12·60	82	24·60	122	36·60	162	48·60
3	0·90	43	12·90	83	24·90	123	36·90	163	48·90
4	1·20	44	13·20	84	25·20	124	37·20	164	49·20
5	1·50	45	13·50	85	25·50	125	37·50	165	49·50
6	1·80	46	13·80	86	25·80	126	37·80	166	49·80
7	2·10	47	14·10	87	26·10	127	38·10	167	50·10
8	2·40	48	14·40	88	26·40	128	38·40	168	50·40
9	2·70	49	14·70	89	26·70	129	38·70	169	50·70
10	3·00	**50**	15·00	**90**	27·00	**130**	39·00	**170**	51·00
11	3·30	51	15·30	91	27·30	131	39·30	171	51·30
12	3·60	52	15·60	92	27·60	132	39·60	172	51·60
13	3·90	53	15·90	93	27·90	133	39·90	173	51·90
14	4·20	54	16·20	94	28·20	134	40·20	174	52·20
15	4·50	55	16·50	95	28·50	135	40·50	175	52·50
16	4·80	56	16·80	96	28·80	136	40·80	176	52·80
17	5·10	57	17·10	97	29·10	137	41·10	177	53·10
18	5·40	58	17·40	98	29·40	138	41·40	178	53·40
19	5·70	59	17·70	99	29·70	139	41·70	179	53·70
20	6·00	**60**	18·00	**100**	30·00	**140**	42·00	**180**	54·00
21	6·30	61	18·30	101	30·30	141	42·30	181	54·30
22	6·60	62	18·60	102	30·60	142	42·60	182	54·60
23	6·90	63	18·90	103	30·90	143	42·90	183	54·90
24	7·20	64	19·20	104	31·20	144	43·20	184	55·20
25	7·50	65	19·50	105	31·50	145	43·50	185	55·50
26	7·80	66	19·80	106	31·80	146	43·80	186	55·80
27	8·10	67	20·10	107	32·10	147	44·10	187	56·10
28	8·40	68	20·40	108	32·40	148	44·40	188	56·40
29	8·70	69	20·70	109	32·70	149	44·70	189	56·70
30	9·00	**70**	21·00	**110**	33·00	**150**	45·00	**190**	57·00
31	9·30	71	21·30	111	33·30	151	45·30	191	57·30
32	9·60	72	21·60	112	33·60	152	45·60	192	57·60
33	9·90	73	21·90	113	33·90	153	45·90	193	57·90
34	10·20	74	22·20	114	34·20	154	46·20	194	58·20
35	10·50	75	22·50	115	34·50	155	46·50	195	58·50
36	10·80	76	22·80	116	34·80	156	46·80	196	58·80
37	11·10	77	23·10	117	35·10	157	47·10	197	59·10
38	11·40	78	23·40	118	35·40	158	47·40	198	59·40
39	11·70	79	23·70	119	35·70	159	47·70	199	59·70
40	12·00	**80**	24·00	**120**	36·00	**160**	48·00	**200**	60·00

$\frac{1}{16}=0{\cdot}01^{9}$ $\frac{1}{9}=0{\cdot}03^{3}$ $\frac{1}{8}=0{\cdot}03^{8}$ $\frac{1}{6}=0{\cdot}05^{0}$ $\frac{1}{5}=0{\cdot}06^{0}$ $\frac{1}{4}=0{\cdot}07^{5}$ $\frac{1}{3}=0{\cdot}10^{0}$

$\frac{3}{8}=0{\cdot}11^{3}$ $\frac{1}{2}=0{\cdot}15^{0}$ $\frac{5}{8}=0{\cdot}18^{8}$ $\frac{3}{4}=0{\cdot}22^{5}$ $\frac{7}{8}=0{\cdot}26^{3}$ $\frac{2}{3}=0{\cdot}20^{0}$ $\frac{5}{6}=0{\cdot}25^{0}$

$\frac{3}{16}=0{\cdot}05^{6}$ $\frac{5}{16}=0{\cdot}09^{4}$ $\frac{7}{16}=0{\cdot}13^{1}$ $\frac{9}{16}=0{\cdot}16^{9}$ $\frac{11}{16}=0{\cdot}20^{6}$ $\frac{13}{16}=0{\cdot}24^{4}$ $\frac{15}{16}=0{\cdot}28^{1}$

(3.28=1£) **£0.30½**

1	0·30½	41	12·50½	81	24·70½	121	36·90½	161	49·10½
2	0·61	42	12·81	82	25·01	122	37·21	162	49·41
3	0·91½	43	13·11½	83	25·31½	123	37·51½	163	49·71½
4	1·22	44	13·42	84	25·62	124	37·82	164	50·02
5	1·52½	45	13·72½	85	25·92½	125	38·12½	165	50·32½
6	1·83	46	14·03	86	26·23	126	38·43	166	50·63
7	2·13½	47	14·33½	87	26·53½	127	38·73½	167	50·93½
8	2·44	48	14·64	88	26·84	128	39·04	168	51·24
9	2·74½	49	14·94½	89	27·14½	129	39·34½	169	51·54½
10	3·05	**50**	15·25	**90**	27·45	**130**	39·65	**170**	51·85
11	3·35½	51	15·55½	91	27·75½	131	39·95½	171	52·15½
12	3·66	52	15·86	92	28·06	132	40·26	172	52·46
13	3·96½	53	16·16½	93	28·36½	133	40·56½	173	52·76½
14	4·27	54	16·47	94	28·67	134	40·87	174	53·07
15	4·57½	55	16·77½	95	28·97½	135	41·17½	175	53·37½
16	4·88	56	17·08	96	29·28	136	41·48	176	53·68
17	5·18½	57	17·38½	97	29·58½	137	41·78½	177	53·98½
18	5·49	58	17·69	98	29·89	138	42·09	178	54·29
19	5·79½	59	17·99½	99	30·19½	139	42·39½	179	54·59½
20	6·10	**60**	18·30	**100**	30·50	**140**	42·70	**180**	54·90
21	6·40½	61	18·60½	101	30·80½	141	43·00½	181	55·20½
22	6·71	62	18·91	102	31·11	142	43·31	182	55·51
23	7·01½	63	19·21½	103	31·41½	143	43·61½	183	55·81½
24	7·32	64	19·52	104	31·72	144	43·92	184	56·12
25	7·62½	65	19·82½	105	32·02½	145	44·22½	185	56·42½
26	7·93	66	20·13	106	32·33	146	44·53	186	56·73
27	8·23½	67	20·43½	107	32·63½	147	44·83½	187	57·03½
28	8·54	68	20·74	108	32·94	148	45·14	188	57·34
29	8·84½	69	21·04½	109	33·24½	149	45·44½	189	57·64½
30	9·15	**70**	21·35	**110**	33·55	**150**	45·75	**190**	57·95
31	9·45½	71	21·65½	111	33·85½	151	46·05½	191	58·25½
32	9·76	72	21·96	112	34·16	152	46·36	192	58·56
33	10·06½	73	22·26½	113	34·46½	153	46·66½	193	58·86½
34	10·37	74	22·57	114	34·77	154	46·97	194	59·17
35	10·67½	75	22·87½	115	35·07½	155	47·27½	195	59·47½
36	10·98	76	23·18	116	35·38	156	47·58	196	59·78
37	11·28½	77	23·48½	117	35·68½	157	47·88½	197	60·08½
38	11·59	78	23·79	118	35·99	158	48·19	198	60·39
39	11·89½	79	24·09½	119	36·29½	159	48·49½	199	60·69½
40	12·20	**80**	24·40	**120**	36·60	**160**	48·80	**200**	61·00

$\frac{1}{16}=0\cdot01^{9}$ $\frac{1}{9}=0\cdot03^{4}$ $\frac{1}{8}=0\cdot03^{8}$ $\frac{1}{6}=0\cdot05^{1}$ $\frac{1}{5}=0\cdot06^{1}$ $\frac{1}{4}=0\cdot07^{6}$ $\frac{1}{3}=0\cdot10^{2}$

$\frac{3}{8}=0\cdot11^{4}$ $\frac{1}{2}=0\cdot15^{3}$ $\frac{5}{8}=0\cdot19^{1}$ $\frac{3}{4}=0\cdot22^{9}$ $\frac{7}{8}=0\cdot26^{7}$ $\frac{2}{3}=0\cdot20^{3}$ $\frac{5}{6}=0\cdot25^{4}$

$\frac{3}{16}=0\cdot05^{7}$ $\frac{5}{16}=0\cdot09^{5}$ $\frac{7}{16}=0\cdot13^{3}$ $\frac{9}{16}=0\cdot17^{2}$ $\frac{11}{16}=0\cdot21^{0}$ $\frac{13}{16}=0\cdot24^{8}$ $\frac{15}{16}=0\cdot28^{6}$

1 foot=0·3048 metres

(3.23=1£) **£0.31**

1	0·31	41	12·71	81	25·11	121	37·51	161	49·91
2	0·62	42	13·02	82	25·42	122	37·82	162	50·22
3	0·93	43	13·33	83	25·73	123	38·13	163	50·53
4	1·24	44	13·64	84	26·04	124	38·44	164	50·84
5	1·55	45	13·95	85	26·35	125	38·75	165	51·15
6	1·86	46	14·26	86	26·66	126	39·06	166	51·46
7	2·17	47	14·57	87	26·97	127	39·37	167	51·77
8	2·48	48	14·88	88	27·28	128	39·68	168	52·08
9	2·79	49	15·19	89	27·59	129	39·99	169	52·39
10	3·10	**50**	15·50	**90**	27·90	**130**	40·30	**170**	52·70
11	3·41	51	15·81	91	28·21	131	40·61	171	53·01
12	3·72	52	16·12	92	28·52	132	40·92	172	53·32
13	4·03	53	16·43	93	28·83	133	41·23	173	53·63
14	4·34	54	16·74	94	29·14	134	41·54	174	53·94
15	4·65	55	17·05	95	29·45	135	41·85	175	54·25
16	4·96	56	17·36	96	29·76	136	42·16	176	54·56
17	5·27	57	17·67	97	30·07	137	42·47	177	54·87
18	5·58	58	17·98	98	30·38	138	42·78	178	55·18
19	5·89	59	18·29	99	30·69	139	43·09	179	55·49
20	6·20	**60**	18·60	**100**	31·00	**140**	43·40	**180**	55·80
21	6·51	61	18·91	101	31·31	141	43·71	181	56·11
22	6·82	62	19·22	102	31·62	142	44·02	182	56·42
23	7·13	63	19·53	103	31·93	143	44·33	183	56·73
24	7·44	64	19·84	104	32·24	144	44·64	184	57·04
25	7·75	65	20·15	105	32·55	145	44·95	185	57·35
26	8·06	66	20·46	106	32·86	146	45·26	186	57·66
27	8·37	67	20·77	107	33·17	147	45·57	187	57·97
28	8·68	68	21·08	108	33·48	148	45·88	188	58·28
29	8·99	69	21·39	109	33·79	149	46·19	189	58·59
30	9·30	**70**	21·70	**110**	34·10	**150**	46·50	**190**	58·90
31	9·61	71	22·01	111	34·41	151	46·81	191	59·21
32	9·92	72	22·32	112	34·72	152	47·12	192	59·52
33	10·23	73	22·63	113	35·03	153	47·43	193	59·83
34	10·54	74	22·94	114	35·34	154	47·74	194	60·14
35	10·85	75	23·25	115	35·65	155	48·05	195	60·45
36	11·16	76	23·56	116	35·96	156	48·36	196	60·76
37	11·47	77	23·87	117	36·27	157	48·67	197	61·07
38	11·78	78	24·18	118	36·58	158	48·98	198	61·38
39	12·09	79	24·49	119	36·89	159	49·29	199	61·69
40	12·40	**80**	24·80	**120**	37·20	**160**	49·60	**200**	62·00

$\frac{1}{16}=0{\cdot}01^{9}$ $\frac{1}{9}=0{\cdot}03^{4}$ $\frac{1}{8}=0{\cdot}03^{9}$ $\frac{1}{6}=0{\cdot}05^{2}$ $\frac{1}{5}=0{\cdot}06^{2}$ $\frac{1}{4}=0{\cdot}07^{8}$ $\frac{1}{3}=0{\cdot}10^{3}$

$\frac{3}{8}=0{\cdot}11^{6}$ $\frac{1}{2}=0{\cdot}15^{5}$ $\frac{5}{8}=0{\cdot}19^{4}$ $\frac{3}{4}=0{\cdot}23^{3}$ $\frac{7}{8}=0{\cdot}27^{1}$ $\frac{2}{3}=0{\cdot}20^{7}$ $\frac{5}{6}=0{\cdot}25^{8}$

$\frac{3}{16}=0{\cdot}05^{8}$ $\frac{5}{16}=0{\cdot}09^{7}$ $\frac{7}{16}=0{\cdot}13^{6}$ $\frac{9}{16}=0{\cdot}17^{4}$ $\frac{11}{16}=0{\cdot}21^{3}$ $\frac{13}{16}=0{\cdot}25^{2}$ $\frac{15}{16}=0{\cdot}29^{1}$

(3.17=1£) **£0.31½**

1	0·31½	41	12·91½	81	25·51½	121	38·11½	161	50·71½
2	0·63	42	13·23	82	25·83	122	38·43	162	51·03
3	0·94½	43	13·54½	83	26·14½	123	38·74½	163	51·34½
4	1·26	44	13·86	84	26·46	124	39·06	164	51·66
5	1·57½	45	14·17½	85	26·77½	125	39·37½	165	51·97½
6	1·89	46	14·49	86	27·09	126	39·69	166	52·29
7	2·20½	47	14·80½	87	27·40½	127	40·00½	167	52·60½
8	2·52	48	15·12	88	27·72	128	40·32	168	52·92
9	2·83½	49	15·43½	89	28·03½	129	40·63½	169	53·23½
10	3·15	**50**	15·75	**90**	28·35	**130**	40·95	**170**	53·55
11	3·46½	51	16·06½	91	28·66½	131	41·26½	171	53·86½
12	3·78	52	16·38	92	28·98	132	41·58	172	54·18
13	4·09½	53	16·69½	93	29·29½	133	41·89½	173	54·49½
14	4·41	54	17·01	94	29·61	134	42·21	174	54·81
15	4·72½	55	17·32½	95	29·92½	135	42·52½	175	55·12½
16	5·04	56	17·64	96	30·24	136	42·84	176	55·44
17	5·35½	57	17·95½	97	30·55½	137	43·15½	177	55·75½
18	5·67	58	18·27	98	30·87	138	43·47	178	56·07
19	5·98½	59	18·58½	99	31·18½	139	43·78½	179	56·38½
20	6·30	**60**	18·90	**100**	31·50	**140**	44·10	**180**	56·70
21	6·61½	61	19·21½	101	31·81½	141	44·41½	181	57·01½
22	6·93	62	19·53	102	32·13	142	44·73	182	57·33
23	7·24½	63	19·84½	103	32·44½	143	45·04½	183	57·64½
24	7·56	64	20·16	104	32·76	144	45·36	184	57·96
25	7·87½	65	20·47½	105	33·07½	145	45·67½	185	58·27½
26	8·19	66	20·79	106	33·39	146	45·99	186	58·59
27	8·50½	67	21·10½	107	33·70½	147	46·30½	187	58·90½
28	8·82	68	21·42	108	34·02	148	46·62	188	59·22
29	9·13½	69	21·73½	109	34·33½	149	46·93½	189	59·53½
30	9·45	**70**	22·05	**110**	34·65	**150**	47·25	**190**	59·85
31	9·76½	71	22·36½	111	34·96½	151	47·56½	191	60·16½
32	10·08	72	22·68	112	35·28	152	47·88	192	60·48
33	10·39½	73	22·99½	113	35·59½	153	48·19½	193	60·79½
34	10·71	74	23·31	114	35·91	154	48·51	194	61·11
35	11·02½	75	23·62½	115	36·22½	155	48·82½	195	61·42½
36	11·34	76	23·94	116	36·54	156	49·14	196	61·74
37	11·65½	77	24·25½	117	36·85½	157	49·45½	197	62·05½
38	11·97	78	24·57	118	37·17	158	49·77	198	62·37
39	12·28½	79	24·88½	119	37·48½	159	50·08½	199	62·68½
40	12·60	**80**	25·20	**120**	37·80	**160**	50·40	**200**	63·00

$\frac{1}{16}$=0·02^{0} $\frac{1}{9}$=0·03^{5} $\frac{1}{8}$=0·03^{9} $\frac{1}{6}$=0·05^{3} $\frac{1}{5}$=0·06^{3} $\frac{1}{4}$=0·07^{9} $\frac{1}{3}$=0·10^{5}

$\frac{3}{8}$=0·11^{8} $\frac{1}{2}$=0·15^{8} $\frac{5}{8}$=0·19^{7} $\frac{3}{4}$=0·23^{6} $\frac{7}{8}$=0·27^{6} $\frac{2}{3}$=0·21^{0} $\frac{5}{6}$=0·26^{3}

$\frac{3}{16}$=0·05^{9} $\frac{5}{16}$=0·09^{8} $\frac{7}{16}$=0·13^{8} $\frac{9}{16}$=0·17^{7} $\frac{11}{16}$=0·21^{7} $\frac{13}{16}$=0·25^{6} $\frac{15}{16}$=0·29^{5}

3·162 is the square root of ten

1	0·32	41	13·12	81	25·92	121	38·72	161	51·52
2	0·64	42	13·44	82	26·24	122	39·04	162	51·84
3	0·96	43	13·76	83	26·56	123	39·36	163	52·16
4	1·28	44	14·08	84	26·88	124	39·68	164	52·48
5	1·60	45	14·40	85	27·20	125	40·00	165	52·80
6	1·92	46	14·72	86	27·52	126	40·32	166	53·12
7	2·24	47	15·04	87	27·84	127	40·64	167	53·44
8	2·56	48	15·36	88	28·16	128	40·96	168	53·76
9	2·88	49	15·68	89	28·48	129	41·28	169	54·08
10	3·20	**50**	16·00	**90**	28·80	**130**	41·60	**170**	54·40
11	3·52	51	16·32	91	29·12	131	41·92	171	54·72
12	3·84	52	16·64	92	29·44	132	42·24	172	55·04
13	4·16	53	16·96	93	29·76	133	42·56	173	55·36
14	4·48	54	17·28	94	30·08	134	42·88	174	55·68
15	4·80	55	17·60	95	30·40	135	43·20	175	56·00
16	5·12	56	17·92	96	30·72	136	43·52	176	56·32
17	5·44	57	18·24	97	31·04	137	43·84	177	56·64
18	5·76	58	18·56	98	31·36	138	44·16	178	56·96
19	6·08	59	18·88	99	31·68	139	44·48	179	57·28
20	6·40	**60**	19·20	**100**	32·00	**140**	44·80	**180**	57·60
21	6·72	61	19·52	101	32·32	141	45·12	181	57·92
22	7·04	62	19·84	102	32·64	142	45·44	182	58·24
23	7·36	63	20·16	103	32·96	143	45·76	183	58·56
24	7·68	64	20·48	104	33·28	144	46·08	184	58·88
25	8·00	65	20·80	105	33·60	145	46·40	185	59·20
26	8·32	66	21·12	106	33·92	146	46·72	186	59·52
27	8·64	67	21·44	107	34·24	147	47·04	187	59·84
28	8·96	68	21·76	108	34·56	148	47·36	188	60·16
29	9·28	69	22·08	109	34·88	149	47·68	189	60·48
30	9·60	**70**	22·40	**110**	35·20	**150**	48·00	**190**	60·80
31	9·92	71	22·72	111	35·52	151	48·32	191	61·12
32	10·24	72	23·04	112	35·84	152	48·64	192	61·44
33	10·56	73	23·36	113	36·16	153	48·96	193	61·76
34	10·88	74	23·68	114	36·48	154	49·28	194	62·08
35	11·20	75	24·00	115	36·80	155	49·60	195	62·40
36	11·52	76	24·32	116	37·12	156	49·92	196	62·72
37	11·84	77	24·64	117	37·44	157	50·24	197	63·04
38	12·16	78	24·96	118	37·76	158	50·56	198	63·36
39	12·48	79	25·28	119	38·08	159	50·88	199	63·68
40	12·80	**80**	25·60	**120**	38·40	**160**	51·20	**200**	64·00

$\frac{1}{16}=0{\cdot}02^{0}$ $\frac{1}{9}=0{\cdot}03^{6}$ $\frac{1}{8}=0{\cdot}04^{0}$ $\frac{1}{6}=0{\cdot}05^{3}$ $\frac{1}{5}=0{\cdot}06^{4}$ $\frac{1}{4}=0{\cdot}08^{0}$ $\frac{1}{3}=0{\cdot}10^{7}$

$\frac{3}{8}=0{\cdot}12^{0}$ $\frac{1}{2}=0{\cdot}16^{0}$ $\frac{5}{8}=0{\cdot}20^{0}$ $\frac{3}{4}=0{\cdot}24^{0}$ $\frac{7}{8}=0{\cdot}28^{0}$ $\frac{2}{3}=0{\cdot}21^{3}$ $\frac{5}{6}=0{\cdot}26^{7}$

$\frac{3}{16}=0{\cdot}06^{0}$ $\frac{5}{16}=0{\cdot}10^{0}$ $\frac{7}{16}=0{\cdot}14^{0}$ $\frac{9}{16}=0{\cdot}18^{0}$ $\frac{11}{16}=0{\cdot}22^{0}$ $\frac{13}{16}=0{\cdot}26^{0}$ $\frac{15}{16}=0{\cdot}30^{0}$

(3.08=1£) **£0.32½**

1	0·32½	41	13·32½	81	26·32½	121	39·32½	161	52·32½
2	0·65	42	13·65	82	26·65	122	39·65	162	52·65
3	0·97½	43	13·97½	83	26·97½	123	39·97½	163	52·97½
4	1·30	44	14·30	84	27·30	124	40·30	164	53·30
5	1·62½	45	14·62½	85	27·62½	125	40·62½	165	53·62½
6	1·95	46	14·95	86	27·95	126	40·95	166	53·95
7	2·27½	47	15·27½	87	28·27½	127	41·27½	167	54·27½
8	2·60	48	15·60	88	28·60	128	41·60	168	54·60
9	2·92½	49	15·92½	89	28·92½	129	41·92½	169	54·92½
10	3·25	**50**	16·25	**90**	29·25	**130**	42·25	**170**	55·25
11	3·57½	51	16·57½	91	29·57½	131	42·57½	171	55·57½
12	3·90	52	16·90	92	29·90	132	42·90	172	55·90
13	4·22½	53	17·22½	93	30·22½	133	43·22½	173	56·22½
14	4·55	54	17·55	94	30·55	134	43·55	174	56·55
15	4·87½	55	17·87½	95	30·87½	135	43·87½	175	56·87½
16	5·20	56	18·20	96	31·20	136	44·20	176	57·20
17	5·52½	57	18·52½	97	31·52½	137	44·52½	177	57·52½
18	5·85	58	18·85	98	31·85	138	44·85	178	57·85
19	6·17½	59	19·17½	99	32·17½	139	45·17½	179	58·17½
20	6·50	**60**	19·50	**100**	32·50	**140**	45·50	**180**	58·50
21	6·82½	61	19·82½	101	32·82½	141	45·82½	181	58·82½
22	7·15	62	20·15	102	33·15	142	46·15	182	59·15
23	7·47½	63	20·47½	103	33·47½	143	46·47½	183	59·47½
24	7·80	64	20·80	104	33·80	144	46·80	184	59·80
25	8·12½	65	21·12½	105	34·12½	145	47·12½	185	60·12½
26	8·45	66	21·45	106	34·45	146	47·45	186	60·45
27	8·77½	67	21·77½	107	34·77½	147	47·77½	187	60·77½
28	9·10	68	22·10	108	35·10	148	48·10	188	61·10
29	9·42½	69	22·42½	109	35·42½	149	48·42½	189	61·42½
30	9·75	**70**	22·75	**110**	35·75	**150**	48·75	**190**	61·75
31	10·07½	71	23·07½	111	36·07½	151	49·07½	191	62·07½
32	10·40	72	23·40	112	36·40	152	49·40	192	62·40
33	10·72½	73	23·72½	113	36·72½	153	49·72½	193	62·72½
34	11·05	74	24·05	114	37·05	154	50·05	194	63·05
35	11·37½	75	24·37½	115	37·37½	155	50·37½	195	63·37½
36	11·70	76	24·70	116	37·70	156	50·70	196	63·70
37	12·02½	77	25·02½	117	38·02½	157	51·02½	197	64·02½
38	12·35	78	25·35	118	38·35	158	51·35	198	64·35
39	12·67½	79	25·67½	119	38·67½	159	51·67½	199	64·67½
40	13·00	**80**	26·00	**120**	39·00	**160**	52·00	**200**	65·00

$\frac{1}{16}=0{\cdot}02^{0}$	$\frac{1}{9}=0{\cdot}03^{6}$	$\frac{1}{8}=0{\cdot}04^{1}$	$\frac{1}{6}=0{\cdot}05^{4}$	$\frac{1}{5}=0{\cdot}06^{5}$	$\frac{1}{4}=0{\cdot}08^{1}$	$\frac{1}{3}=0{\cdot}10^{8}$
$\frac{3}{8}=0{\cdot}12^{2}$	$\frac{1}{2}=0{\cdot}16^{3}$	$\frac{5}{8}=0{\cdot}20^{3}$	$\frac{3}{4}=0{\cdot}24^{4}$	$\frac{7}{8}=0{\cdot}28^{4}$	$\frac{2}{3}=0{\cdot}21^{7}$	$\frac{5}{6}=0{\cdot}27^{1}$
$\frac{3}{16}=0{\cdot}06^{1}$	$\frac{5}{16}=0{\cdot}10^{2}$	$\frac{7}{16}=0{\cdot}14^{2}$	$\frac{9}{16}=0{\cdot}18^{3}$	$\frac{11}{16}=0{\cdot}22^{3}$	$\frac{13}{16}=0{\cdot}26^{4}$	$\frac{15}{16}=0{\cdot}30^{5}$

(3.03=1£) £0.33

1	0·33	41	13·53	81	26·73	121	39·93	161	53·13
2	0·66	42	13·86	82	27·06	122	40·26	162	53·46
3	0·99	43	14·19	83	27·39	123	40·59	163	53·79
4	1·32	44	14·52	84	27·72	124	40·92	164	54·12
5	1·65	45	14·85	85	28·05	125	41·25	165	54·45
6	1·98	46	15·18	86	28·38	126	41·58	166	54·78
7	2·31	47	15·51	87	28·71	127	41·91	167	55·11
8	2·64	48	15·84	88	29·04	128	42·24	168	55·44
9	2·97	49	16·17	89	29·37	129	42·57	169	55·77
10	3·30	**50**	16·50	**90**	29·70	**130**	42·90	**170**	56·10
11	3·63	51	16·83	91	30·03	131	43·23	171	56·43
12	3·96	52	17·16	92	30·36	132	43·56	172	56·76
13	4·29	53	17·49	93	30·69	133	43·89	173	57·09
14	4·62	54	17·82	94	31·02	134	44·22	174	57·42
15	4·95	55	18·15	95	31·35	135	44·55	175	57·75
16	5·28	56	18·48	96	31·68	136	44·88	176	58·08
17	5·61	57	18·81	97	32·01	137	45·21	177	58·41
18	5·94	58	19·14	98	32·34	138	45·54	178	58·74
19	6·27	59	19·47	99	32·67	139	45·87	179	59·07
20	6·60	**60**	19·80	**100**	33·00	**140**	46·20	**180**	59·40
21	6·93	61	20·13	101	33·33	141	46·53	181	59·73
22	7·26	62	20·46	102	33·66	142	46·86	182	60·06
23	7·59	63	20·79	103	33·99	143	47·19	183	60·39
24	7·92	64	21·12	104	34·32	144	47·52	184	60·72
25	8·25	65	21·45	105	34·65	145	47·85	185	61·05
26	8·58	66	21·78	106	34·98	146	48·18	186	61·38
27	8·91	67	22·11	107	35·31	147	48·51	187	61·71
28	9·24	68	22·44	108	35·64	148	48·84	188	62·04
29	9·57	69	22·77	109	35·97	149	49·17	189	62·37
30	9·90	**70**	23·10	**110**	36·30	**150**	49·50	**190**	62·70
31	10·23	71	23·43	111	36·63	151	49·83	191	63·03
32	10·56	72	23·76	112	36·96	152	50·16	192	63·36
33	10·89	73	24·09	113	37·29	153	50·49	193	63·69
34	11·22	74	24·42	114	37·62	154	50·82	194	64·02
35	11·55	75	24·75	115	37·95	155	51·15	195	64·35
36	11·88	76	25·08	116	38·28	156	51·48	196	64·68
37	12·21	77	25·41	117	38·61	157	51·81	197	65·01
38	12·54	78	25·74	118	38·94	158	52·14	198	65·34
39	12·87	79	26·07	119	39·27	159	52·47	199	65·67
40	13·20	**80**	26·40	**120**	39·60	**160**	52·80	**200**	66·00

$\frac{1}{16}=0{\cdot}02^1$ $\frac{1}{9}=0{\cdot}03^7$ $\frac{1}{8}=0{\cdot}04^1$ $\frac{1}{6}=0{\cdot}05^5$ $\frac{1}{5}=0{\cdot}06^6$ $\frac{1}{4}=0{\cdot}08^3$ $\frac{1}{3}=0{\cdot}11^0$

$\frac{3}{8}=0{\cdot}12^4$ $\frac{1}{2}=0{\cdot}16^5$ $\frac{5}{8}=0{\cdot}20^6$ $\frac{3}{4}=0{\cdot}24^8$ $\frac{7}{8}=0{\cdot}28^9$ $\frac{2}{3}=0{\cdot}22^0$ $\frac{5}{6}=0{\cdot}27^5$

$\frac{3}{16}=0{\cdot}06^2$ $\frac{5}{16}=0{\cdot}10^3$ $\frac{7}{16}=0{\cdot}14^4$ $\frac{9}{16}=0{\cdot}18^6$ $\frac{11}{16}=0{\cdot}22^7$ $\frac{13}{16}=0{\cdot}26^8$ $\frac{15}{16}=0{\cdot}30^9$

1 metre=3·28 feet

1	0·33½	41	13·73½	81	27·13½	121	40·53½	161	53·93½
2	0·67	42	14·07	82	27·47	122	40·87	162	54·27
3	1·00½	43	14·40½	83	27·80½	123	41·20½	163	54·60½
4	1·34	44	14·74	84	28·14	124	41·54	164	54·94
5	1·67½	45	15·07½	85	28·47½	125	41·87½	165	55·27½
6	2·01	46	15·41	86	28·81	126	42·21	166	55·61
7	2·34½	47	15·74½	87	29·14½	127	42·54½	167	55·94½
8	2·68	48	16·08	88	29·48	128	42·88	168	56·28
9	3·01½	49	16·41½	89	29·81½	129	43·21½	169	56·61½
10	3·35	**50**	16·75	**90**	30·15	**130**	43·55	**170**	56·95
11	3·68½	51	17·08½	91	30·48½	131	43·88½	171	57·28½
12	4·02	52	17·42	92	30·82	132	44·22	172	57·62
13	4·35½	53	17·75½	93	31·15½	133	44·55½	173	57·95½
14	4·69	54	18·09	94	31·49	134	44·89	174	58·29
15	5·02½	55	18·42½	95	31·82½	135	45·22½	175	58·62½
16	5·36	56	18·76	96	32·16	136	45·56	176	58·96
17	5·69½	57	19·09½	97	32·49½	137	45·89½	177	59·29½
18	6·03	58	19·43	98	32·83	138	46·23	178	59·63
19	6·36½	59	19·76½	99	33·16½	139	46·56½	179	59·96½
20	6·70	**60**	20·10	**100**	33·50	**140**	46·90	**180**	60·30
21	7·03½	61	20·43½	101	33·83½	141	47·23½	181	60·63½
22	7·37	62	20·77	102	34·17	142	47·57	182	60·97
23	7·70½	63	21·10½	103	34·50½	143	47·90½	183	61·30½
24	8·04	64	21·44	104	34·84	144	48·24	184	61·64
25	8·37½	65	21·77½	105	35·17½	145	48·57½	185	61·97½
26	8·71	66	22·11	106	35·51	146	48·91	186	62·31
27	9·04½	67	22·44½	107	35·84½	147	49·24½	187	62·64½
28	9·38	68	22·78	108	36·18	148	49·58	188	62·98
29	9·71½	69	23·11½	109	36·51½	149	49·91½	189	63·31½
30	10·05	**70**	23·45	**110**	36·85	**150**	50·25	**190**	63·65
31	10·38½	71	23·78½	111	37·18½	151	50·58½	191	63·98½
32	10·72	72	24·12	112	37·52	152	50·92	192	64·32
33	11·05½	73	24·45½	113	37·85½	153	51·25½	193	64·65½
34	11·39	74	24·79	114	38·19	154	51·59	194	64·99
35	11·72½	75	25·12½	115	38·52½	155	51·92½	195	65·32½
36	12·06	76	25·46	116	38·86	156	52·26	196	65·66
37	12·39½	77	25·79½	117	39·19½	157	52·59½	197	65·99½
38	12·73	78	26·13	118	39·53	158	52·93	198	66·33
39	13·06½	79	26·46½	119	39·86½	159	53·26½	199	66·66½
40	13·40	**80**	26·80	**120**	40·20	**160**	53·60	**200**	67·00

$\frac{1}{16}=0{\cdot}02^{1}$ $\frac{1}{9}=0{\cdot}03^{7}$ $\frac{1}{8}=0{\cdot}04^{2}$ $\frac{1}{6}=0{\cdot}05^{6}$ $\frac{1}{5}=0{\cdot}06^{7}$ $\frac{1}{4}=0{\cdot}08^{4}$ $\frac{1}{3}=0{\cdot}11^{2}$

$\frac{3}{8}=0{\cdot}12^{6}$ $\frac{1}{2}=0{\cdot}16^{8}$ $\frac{5}{8}=0{\cdot}20^{9}$ $\frac{3}{4}=0{\cdot}25^{1}$ $\frac{7}{8}=0{\cdot}29^{3}$ $\frac{2}{3}=0{\cdot}22^{3}$ $\frac{5}{6}=0{\cdot}27^{9}$

$\frac{3}{16}=0{\cdot}06^{3}$ $\frac{5}{16}=0{\cdot}10^{5}$ $\frac{7}{16}=0{\cdot}14^{7}$ $\frac{9}{16}=0{\cdot}18^{8}$ $\frac{11}{16}=0{\cdot}23^{0}$ $\frac{13}{16}=0{\cdot}27^{2}$ $\frac{15}{16}=0{\cdot}31^{4}$

(2.94=1£) **£0.34**

1	0·34	41	13·94	81	27·54	121	41·14	161	54·74
2	0·68	42	14·28	82	27·88	122	41·48	162	55·08
3	1·02	43	14·62	83	28·22	123	41·82	163	55·42
4	1·36	44	14·96	84	28·56	124	42·16	164	55·76
5	1·70	45	15·30	85	28·90	125	42·50	165	56·10
6	2·04	46	15·64	86	29·24	126	42·84	166	56·44
7	2·38	47	15·98	87	29·58	127	43·18	167	56·78
8	2·72	48	16·32	88	29·92	128	43·52	168	57·12
9	3·06	49	16·66	89	30·26	129	43·86	169	57·46
10	3·40	**50**	17·00	**90**	30·60	**130**	44·20	**170**	57·80
11	3·74	51	17·34	91	30·94	131	44·54	171	58·14
12	4·08	52	17·68	92	31·28	132	44·88	172	58·48
13	4·42	53	18·02	93	31·62	133	45·22	173	58·82
14	4·76	54	18·36	94	31·96	134	45·56	174	59·16
15	5·10	55	18·70	95	32·30	135	45·90	175	59·50
16	5·44	56	19·04	96	32·64	136	46·24	176	59·84
17	5·78	57	19·38	97	32·98	137	46·58	177	60·18
18	6·12	58	19·72	98	33·32	138	46·92	178	60·52
19	6·46	59	20·06	99	33·66	139	47·26	179	60·86
20	6·80	**60**	20·40	**100**	34·00	**140**	47·60	**180**	61·20
21	7·14	61	20·74	101	34·34	141	47·94	181	61·54
22	7·48	62	21·08	102	34·68	142	48·28	182	61·88
23	7·82	63	21·42	103	35·02	143	48·62	183	62·22
24	8·16	64	21·76	104	35·36	144	48·96	184	62·56
25	8·50	65	22·10	105	35·70	145	49·30	185	62·90
26	8·84	66	22·44	106	36·04	146	49·64	186	63·24
27	9·18	67	22·78	107	36·38	147	49·98	187	63·58
28	9·52	68	23·12	108	36·72	148	50·32	188	63·92
29	9·86	69	23·46	109	37·06	149	50·66	189	64·26
30	10·20	**70**	23·80	**110**	37·40	**150**	51·00	**190**	64·60
31	10·54	71	24·14	111	37·74	151	51·34	191	64·94
32	10·88	72	24·48	112	38·08	152	51·68	192	65·28
33	11·22	73	24·82	113	38·42	153	52·02	193	65·62
34	11·56	74	25·16	114	38·76	154	52·36	194	65·96
35	11·90	75	25·50	115	39·10	155	52·70	195	66·30
36	12·24	76	25·84	116	39·44	156	53·04	196	66·64
37	12·58	77	26·18	117	39·78	157	53·38	197	66·98
38	12·92	78	26·52	118	40·12	158	53·72	198	67·32
39	13·26	79	26·86	119	40·46	159	54·06	199	67·66
40	13·60	**80**	27·20	**120**	40·80	**160**	54·40	**200**	68·00

$\frac{1}{16}=0{\cdot}02^{1}$ $\frac{1}{9}=0{\cdot}03^{8}$ $\frac{1}{8}=0{\cdot}04^{3}$ $\frac{1}{6}=0{\cdot}05^{7}$ $\frac{1}{5}=0{\cdot}06^{8}$ $\frac{1}{4}=0{\cdot}08^{5}$ $\frac{1}{3}=0{\cdot}11^{3}$

$\frac{3}{8}=0{\cdot}12^{8}$ $\frac{1}{2}=0{\cdot}17^{0}$ $\frac{5}{8}=0{\cdot}21^{3}$ $\frac{3}{4}=0{\cdot}25^{5}$ $\frac{7}{8}=0{\cdot}29^{8}$ $\frac{2}{3}=0{\cdot}22^{7}$ $\frac{5}{6}=0{\cdot}28^{3}$

$\frac{3}{16}=0{\cdot}06^{4}$ $\frac{5}{16}=0{\cdot}10^{6}$ $\frac{7}{16}=0{\cdot}14^{9}$ $\frac{9}{16}=0{\cdot}19^{1}$ $\frac{11}{16}=0{\cdot}23^{4}$ $\frac{13}{16}=0{\cdot}27^{6}$ $\frac{15}{16}=0{\cdot}31^{9}$

1 ounce per square yard=33·9 grammes per sq. metre

(2.90=1£) **£0.34½**

1	0·34½	41	14·14½	81	27·94½	121	41·74½	161	55·54½
2	0·69	42	14·49	82	28·29	122	42·09	162	55·89
3	1·03½	43	14·83½	83	28·63½	123	42·43½	163	56·23½
4	1·38	44	15·18	84	28·98	124	42·78	164	56·58
5	1·72½	45	15·52½	85	29·32½	125	43·12½	165	56·92½
6	2·07	46	15·87	86	29·67	126	43·47	166	57·27
7	2·41½	47	16·21½	87	30·01½	127	43·81½	167	57·61½
8	2·76	48	16·56	88	30·36	128	44·16	168	57·96
9	3·10½	49	16·90½	89	30·70½	129	44·50½	169	58·30½
10	3·45	**50**	17·25	**90**	31·05	**130**	44·85	**170**	58·65
11	3·79½	51	17·59½	91	31·39½	131	45·19½	171	58·99½
12	4·14	52	17·94	92	31·74	132	45·54	172	59·34
13	4·48½	53	18·28½	93	32·08½	133	45·88½	173	59·68½
14	4·83	54	18·63	94	32·43	134	46·23	174	60·03
15	5·17½	55	18·97½	95	32·77½	135	46·57½	175	60·37½
16	5·52	56	19·32	96	33·12	136	46·92	176	60·72
17	5·86½	57	19·66½	97	33·46½	137	47·26½	177	61·06½
18	6·21	58	20·01	98	33·81	138	47·61	178	61·41
19	6·55½	59	20·35½	99	34·15½	139	47·95½	179	61·75½
20	6·90	**60**	20·70	**100**	34·50	**140**	48·30	**180**	62·10
21	7·24½	61	21·04½	101	34·84½	141	48·64½	181	62·44½
22	7·59	62	21·39	102	35·19	142	48·99	182	62·79
23	7·93½	63	21·73½	103	35·53½	143	49·33½	183	63·13½
24	8·28	64	22·08	104	35·88	144	49·68	184	63·48
25	8·62½	65	22·42½	105	36·22½	145	50·02½	185	63·82½
26	8·97	66	22·77	106	36·57	146	50·37	186	64·17
27	9·31½	67	23·11½	107	36·91½	147	50·71½	187	64·51½
28	9·66	68	23·46	108	37·26	148	51·06	188	64·86
29	10·00½	69	23·80½	109	37·60½	149	51·40½	189	65·20½
30	10·35	**70**	24·15	**110**	37·95	**150**	51·75	**190**	65·55
31	10·69½	71	24·49½	111	38·29½	151	52·09½	191	65·89½
32	11·04	72	24·84	112	38·64	152	52·44	192	66·24
33	11·38½	73	25·18½	113	38·98½	153	52·78½	193	66·58½
34	11·73	74	25·53	114	39·33	154	53·13	194	66·93
35	12·07½	75	25·87½	115	39·67½	155	53·47½	195	67·27½
36	12·42	76	26·22	116	40·02	156	53·82	196	67·62
37	12·76½	77	26·56½	117	40·36½	157	54·16½	197	67·96½
38	13·11	78	26·91	118	40·71	158	54·51	198	68·31
39	13·45½	79	27·25½	119	41·05½	159	54·85½	199	68·65½
40	13·80	**80**	27·60	**120**	41·40	**160**	55·20	**200**	69·00

$\frac{1}{16}=0{\cdot}02^{2}$ $\frac{1}{9}=0{\cdot}03^{8}$ $\frac{1}{8}=0{\cdot}04^{3}$ $\frac{1}{6}=0{\cdot}05^{8}$ $\frac{1}{5}=0{\cdot}06^{9}$ $\frac{1}{4}=0{\cdot}08^{6}$ $\frac{1}{3}=0{\cdot}11^{5}$

$\frac{3}{8}=0{\cdot}12^{9}$ $\frac{1}{2}=0{\cdot}17^{3}$ $\frac{5}{8}=0{\cdot}21^{6}$ $\frac{3}{4}=0{\cdot}25^{9}$ $\frac{7}{8}=0{\cdot}30^{2}$ $\frac{2}{3}=0{\cdot}23^{0}$ $\frac{5}{6}=0{\cdot}28^{8}$

$\frac{3}{16}=0{\cdot}06^{5}$ $\frac{5}{16}=0{\cdot}10^{8}$ $\frac{7}{16}=0{\cdot}15^{1}$ $\frac{9}{16}=0{\cdot}19^{4}$ $\frac{11}{16}=0{\cdot}23^{7}$ $\frac{13}{16}=0{\cdot}28^{0}$ $\frac{15}{16}=0{\cdot}32^{3}$

(2.86=1£) **£0.35**

1	0·35	41	14·35	81	28·35	121	42·35	161	56·35
2	0·70	42	14·70	82	28·70	122	42·70	162	56·70
3	1·05	43	15·05	83	29·05	123	43·05	163	57·05
4	1·40	44	15·40	84	29·40	124	43·40	164	57·40
5	1·75	45	15·75	85	29·75	125	43·75	165	57·75
6	2·10	46	16·10	86	30·10	126	44·10	166	58·10
7	2·45	47	16·45	87	30·45	127	44·45	167	58·45
8	2·80	48	16·80	88	30·80	128	44·80	168	58·80
9	3·15	49	17·15	89	31·15	129	45·15	169	59·15
10	3·50	**50**	17·50	**90**	31·50	**130**	45·50	**170**	59·50
11	3·85	51	17·85	91	31·85	131	45·85	171	59·85
12	4·20	52	18·20	92	32·20	132	46·20	172	60·20
13	4·55	53	18·55	93	32·55	133	46·55	173	60·55
14	4·90	54	18·90	94	32·90	134	46·90	174	60·90
15	5·25	55	19·25	95	33·25	135	47·25	175	61·25
16	5·60	56	19·60	96	33·60	136	47·60	176	61·60
17	5·95	57	19·95	97	33·95	137	47·95	177	61·95
18	6·30	58	20·30	98	34·30	138	48·30	178	62·30
19	6·65	59	20·65	99	34·65	139	48·65	179	62·65
20	7·00	**60**	21·00	**100**	35·00	**140**	49·00	**180**	63·00
21	7·35	61	21·35	101	35·35	141	49·35	181	63·35
22	7·70	62	21·70	102	35·70	142	49·70	182	63·70
23	8·05	63	22·05	103	36·05	143	50·05	183	64·05
24	8·40	64	22·40	104	36·40	144	50·40	184	64·40
25	8·75	65	22·75	105	36·75	145	50·75	185	64·75
26	9·10	66	23·10	106	37·10	146	51·10	186	65·10
27	9·45	67	23·45	107	37·45	147	51·45	187	65·45
28	9·80	68	23·80	108	37·80	148	51·80	188	65·80
29	10·15	69	24·15	109	38·15	149	52·15	189	66·15
30	10·50	**70**	24·50	**110**	38·50	**150**	52·50	**190**	66·50
31	10·85	71	24·85	111	38·85	151	52·85	191	66·85
32	11·20	72	25·20	112	39·20	152	53·20	192	67·20
33	11·55	73	25·55	113	39·55	153	53·55	193	67·55
34	11·90	74	25·90	114	39·90	154	53·90	194	67·90
35	12·25	75	26·25	115	40·25	155	54·25	195	68·25
36	12·60	76	26·60	116	40·60	156	54·60	196	68·60
37	12·95	77	26·95	117	40·95	157	54·95	197	68·95
38	13·30	78	27·30	118	41·30	158	55·30	198	69·30
39	13·65	79	27·65	119	41·65	159	55·65	199	69·65
40	14·00	**80**	28·00	**120**	42·00	**160**	56·00	**200**	70·00

$\frac{1}{16}$=0·02^{2} $\frac{1}{9}$=0·03^{9} $\frac{1}{8}$=0·04^{4} $\frac{1}{6}$=0·05^{8} $\frac{1}{5}$=0·07^{0} $\frac{1}{4}$=0·08^{8} $\frac{1}{3}$=0·11^{7}

$\frac{3}{8}$=0·13^{1} $\frac{1}{2}$=0·17^{5} $\frac{5}{8}$=0·21^{9} $\frac{3}{4}$=0·26^{3} $\frac{7}{8}$=0·30^{6} $\frac{2}{3}$=0·23^{3} $\frac{5}{6}$=0·29^{2}

$\frac{3}{16}$=0·06^{6} $\frac{5}{16}$=0·10^{9} $\frac{7}{16}$=0·15^{3} $\frac{9}{16}$=0·19^{7} $\frac{11}{16}$=0·24^{1} $\frac{13}{16}$=0·28^{4} $\frac{15}{16}$=0·32^{8}

(2.82=1£) **£0.35½**

1	0·35½	41	14·55½	81	28·75½	121	42·95½	161	57·15½
2	0·71	42	14·91	82	29·11	122	43·31	162	57·51
3	1·06½	43	15·26½	83	29·46½	123	43·66½	163	57·86½
4	1·42	44	15·62	84	29·82	124	44·02	164	58·22
5	1·77½	45	15·97½	85	30·17½	125	44·37½	165	58·57½
6	2·13	46	16·33	86	30·53	126	44·73	166	58·93
7	2·48½	47	16·68½	87	30·88½	127	45·08½	167	59·28½
8	2·84	48	17·04	88	31·24	128	45·44	168	59·64
9	3·19½	49	17·39½	89	31·59½	129	45·79½	169	59·99½
10	3·55	**50**	17·75	**90**	31·95	**130**	46·15	**170**	60·35
11	3·90½	51	18·10½	91	32·30½	131	46·50½	171	60·70½
12	4·26	52	18·46	92	32·66	132	46·86	172	61·06
13	4·61½	53	18·81½	93	33·01½	133	47·21½	173	61·41½
14	4·97	54	19·17	94	33·37	134	47·57	174	61·77
15	5·32½	55	19·52½	95	33·72½	135	47·92½	175	62·12½
16	5·68	56	19·88	96	34·08	136	48·28	176	62·48
17	6·03½	57	20·23½	97	34·43½	137	48·63½	177	62·83½
18	6·39	58	20·59	98	34·79	138	48·99	178	63·19
19	6·74½	59	20·94½	99	35·14½	139	49·34½	179	63·54½
20	7·10	**60**	21·30	**100**	35·50	**140**	49·70	**180**	63·90
21	7·45½	61	21·65½	101	35·85½	141	50·05½	181	64·25½
22	7·81	62	22·01	102	36·21	142	50·41	182	64·61
23	8·16½	63	22·36½	103	36·56½	143	50·76½	183	64·96½
24	8·52	64	22·72	104	36·92	144	51·12	184	65·32
25	8·87½	65	23·07½	105	37·27½	145	51·47½	185	65·67½
26	9·23	66	23·43	106	37·63	146	51·83	186	66·03
27	9·58½	67	23·78½	107	37·98½	147	52·18½	187	66·38½
28	9·94	68	24·14	108	38·34	148	52·54	188	66·74
29	10·29½	69	24·49½	109	38·69½	149	52·89½	189	67·09½
30	10·65	**70**	24·85	**110**	39·05	**150**	53·25	**190**	67·45
31	11·00½	71	25·20½	111	39·40½	151	53·60½	191	67·80½
32	11·36	72	25·56	112	39·76	152	53·96	192	68·16
33	11·71½	73	25·91½	113	40·11½	153	54·31½	193	68·51½
34	12·07	74	26·27	114	40·47	154	54·67	194	68·87
35	12·42½	75	26·62½	115	40·82½	155	55·02½	195	69·22½
36	12·78	76	26·98	116	41·18	156	55·38	196	69·58
37	13·13½	77	27·33½	117	41·53½	157	55·73½	197	69·93½
38	13·49	78	27·69	118	41·89	158	56·09	198	70·29
39	13·84½	79	28·04½	119	42·24½	159	56·44½	199	70·64½
40	14·20	**80**	28·40	**120**	42·60	**160**	56·80	**200**	71·00

$\frac{1}{16}$=0·02^{2} $\frac{1}{9}$=0·03^{9} $\frac{1}{8}$=0·04^{4} $\frac{1}{6}$=0·05^{9} $\frac{1}{5}$=0·07^{1} $\frac{1}{4}$=0·08^{9} $\frac{1}{3}$=0·11^{8}

$\frac{3}{8}$=0·13^{3} $\frac{1}{2}$=0·17^{8} $\frac{5}{8}$=0·22^{2} $\frac{3}{4}$=0·26^{6} $\frac{7}{8}$=0·31^{1} $\frac{2}{3}$=0·23^{7} $\frac{5}{6}$=0·29^{6}

$\frac{3}{16}$=0·06^{7} $\frac{5}{16}$=0·11^{1} $\frac{7}{16}$=0·15^{5} $\frac{9}{16}$=0·20^{0} $\frac{11}{16}$=0·24^{4} $\frac{13}{16}$=0·28^{8} $\frac{15}{16}$=0·33^{3}

1 litre=0·353 cubic feet

(2.78=1£) **£0.36**

1	0·36	41	14·76	81	29·16	121	43·56	161	57·96
2	0·72	42	15·12	82	29·52	122	43·92	162	58·32
3	1·08	43	15·48	83	29·88	123	44·28	163	58·68
4	1·44	44	15·84	84	30·24	124	44·64	164	59·04
5	1·80	45	16·20	85	30·60	125	45·00	165	59·40
6	2·16	46	16·56	86	30·96	126	45·36	166	59·76
7	2·52	47	16·92	87	31·32	127	45·72	167	60·12
8	2·88	48	17·28	88	31·68	128	46·08	168	60·48
9	3·24	49	17·64	89	32·04	129	46·44	169	60·84
10	3·60	**50**	18·00	**90**	32·40	**130**	46·80	**170**	61·20
11	3·96	51	18·36	91	32·76	131	47·16	171	61·56
12	4·32	52	18·72	92	33·12	132	47·52	172	61·92
13	4·68	53	19·08	93	33·48	133	47·88	173	62·28
14	5·04	54	19·44	94	33·84	134	48·24	174	62·64
15	5·40	55	19·80	95	34·20	135	48·60	175	63·00
16	5·76	56	20·16	96	34·56	136	48·96	176	63·36
17	6·12	57	20·52	97	34·92	137	49·32	177	63·72
18	6·48	58	20·88	98	35·28	138	49·68	178	64·08
19	6·84	59	21·24	99	35·64	139	50·04	179	64·44
20	7·20	**60**	21·60	**100**	36·00	**140**	50·40	**180**	64·80
21	7·56	61	21·96	101	36·36	141	50·76	181	65·16
22	7·92	62	22·32	102	36·72	142	51·12	182	65·52
23	8·28	63	22·68	103	37·08	143	51·48	183	65·88
24	8·64	64	23·04	104	37·44	144	51·84	184	66·24
25	9·00	65	23·40	105	37·80	145	52·20	185	66·60
26	9·36	66	23·76	106	38·16	146	52·56	186	66·96
27	9·72	67	24·12	107	38·52	147	52·92	187	67·32
28	10·08	68	24·48	108	38·88	148	53·28	188	67·68
29	10·44	69	24·84	109	39·24	149	53·64	189	68·04
30	10·80	**70**	25·20	**110**	39·60	**150**	54·00	**190**	68·40
31	11·16	71	25·56	111	39·96	151	54·36	191	68·76
32	11·52	72	25·92	112	40·32	152	54·72	192	69·12
33	11·88	73	26·28	113	40·68	153	55·08	193	69·48
34	12·24	74	26·64	114	41·04	154	55·44	194	69·84
35	12·60	75	27·00	115	41·40	155	55·80	195	70·20
36	12·96	76	27·36	116	41·76	156	56·16	196	70·56
37	13·32	77	27·72	117	42·12	157	56·52	197	70·92
38	13·68	78	28·08	118	42·48	158	56·88	198	71·28
39	14·04	79	28·44	119	42·84	159	57·24	199	71·64
40	14·40	**80**	28·80	**120**	43·20	**160**	57·60	**200**	72·00

$\frac{1}{16}=0{\cdot}02^{3}$ $\frac{1}{9}=0{\cdot}04^{0}$ $\frac{1}{8}=0{\cdot}04^{5}$ $\frac{1}{6}=0{\cdot}06^{0}$ $\frac{1}{5}=0{\cdot}07^{2}$ $\frac{1}{4}=0{\cdot}09^{0}$ $\frac{1}{3}=0{\cdot}12^{0}$

$\frac{3}{8}=0{\cdot}13^{5}$ $\frac{1}{2}=0{\cdot}18^{0}$ $\frac{5}{8}=0{\cdot}22^{5}$ $\frac{3}{4}=0{\cdot}27^{0}$ $\frac{7}{8}=0{\cdot}31^{5}$ $\frac{2}{3}=0{\cdot}24^{0}$ $\frac{5}{6}=0{\cdot}30^{0}$

$\frac{3}{16}=0{\cdot}06^{8}$ $\frac{5}{16}=0{\cdot}11^{3}$ $\frac{7}{16}=0{\cdot}15^{8}$ $\frac{9}{16}=0{\cdot}20^{3}$ $\frac{11}{16}=0{\cdot}24^{8}$ $\frac{13}{16}=0{\cdot}29^{3}$ $\frac{15}{16}=0{\cdot}33^{8}$

(2.74=1£) **£0.36½**

1	0·36½	41	14·96½	81	29·56½	121	44·16½	161	58·76½
2	0·73	42	15·33	82	29·93	122	44·53	162	59·13
3	1·09½	43	15·69½	83	30·29½	123	44·89½	163	59·49½
4	1·46	44	16·06	84	30·66	124	45·26	164	59·86
5	1·82½	45	16·42½	85	31·02½	125	45·62½	165	60·22½
6	2·19	46	16·79	86	31·39	126	45·99	166	60·59
7	2·55½	47	17·15½	87	31·75½	127	46·35½	167	60·95½
8	2·92	48	17·52	88	32·12	128	46·72	168	61·32
9	3·28½	49	17·88½	89	32·48½	129	47·08½	169	61·68½
10	3·65	**50**	18·25	**90**	32·85	**130**	47·45	**170**	62·05
11	4·01½	51	18·61½	91	33·21½	131	47·81½	171	62·41½
12	4·38	52	18·98	92	33·58	132	48·18	172	62·78
13	4·74½	53	19·34½	93	33·94½	133	48·54½	173	63·14½
14	5·11	54	19·71	94	34·31	134	48·91	174	63·51
15	5·47½	55	20·07½	95	34·67½	135	49·27½	175	63·87½
16	5·84	56	20·44	96	35·04	136	49·64	176	64·24
17	6·20½	57	20·80½	97	35·40½	137	50·00½	177	64·60½
18	6·57	58	21·17	98	35·77	138	50·37	178	64·97
19	6·93½	59	21·53½	99	36·13½	139	50·73½	179	65·33½
20	7·30	**60**	21·90	**100**	36·50	**140**	51·10	**180**	65·70
21	7·66½	61	22·26½	101	36·86½	141	51·46½	181	66·06½
22	8·03	62	22·63	102	37·23	142	51·83	182	66·43
23	8·39½	63	22·99½	103	37·59½	143	52·19½	183	66·79½
24	8·76	64	23·36	104	37·96	144	52·56	184	67·16
25	9·12½	65	23·72½	105	38·32½	145	52·92½	185	67·52½
26	9·49	66	24·09	106	38·69	146	53·29	186	67·89
27	9·85½	67	24·45½	107	39·05½	147	53·65½	187	68·25½
28	10·22	68	24·82	108	39·42	148	54·02	188	68·62
29	10·58½	69	25·18½	109	39·78½	149	54·38½	189	68·98½
30	10·95	**70**	25·55	**110**	40·15	**150**	54·75	**190**	69·35
31	11·31½	71	25·91½	111	40·51½	151	55·11½	191	69·71½
32	11·68	72	26·28	112	40·88	152	55·48	192	70·08
33	12·04½	73	26·64½	113	41·24½	153	55·84½	193	70·44½
34	12·41	74	27·01	114	41·61	154	56·21	194	70·81
35	12·77½	75	27·37½	115	41·97½	155	56·57½	195	71·17½
36	13·14	76	27·74	116	42·34	156	56·94	196	71·54
37	13·50½	77	28·10½	117	42·70½	157	57·30½	197	71·90½
38	13·87	78	28·47	118	43·07	158	57·67	198	72·27
39	14·23½	79	28·83½	119	43·43½	159	58·03½	199	72·63½
40	14·60	**80**	29·20	**120**	43·80	**160**	58·40	**200**	73·00

$\frac{1}{16}=0{\cdot}02^{3}$ $\frac{1}{9}=0{\cdot}04^{1}$ $\frac{1}{8}=0{\cdot}04^{6}$ $\frac{1}{6}=0{\cdot}06^{1}$ $\frac{1}{5}=0{\cdot}07^{3}$ $\frac{1}{4}=0{\cdot}09^{1}$ $\frac{1}{3}=0{\cdot}12^{2}$

$\frac{3}{8}=0{\cdot}13^{7}$ $\frac{1}{2}=0{\cdot}18^{3}$ $\frac{5}{8}=0{\cdot}22^{8}$ $\frac{3}{4}=0{\cdot}27^{4}$ $\frac{7}{8}=0{\cdot}31^{9}$ $\frac{2}{3}=0{\cdot}24^{3}$ $\frac{5}{6}=0{\cdot}30^{4}$

$\frac{3}{16}=0{\cdot}06^{8}$ $\frac{5}{16}=0{\cdot}11^{4}$ $\frac{7}{16}=0{\cdot}16^{0}$ $\frac{9}{16}=0{\cdot}20^{5}$ $\frac{11}{16}=0{\cdot}25^{1}$ $\frac{13}{16}=0{\cdot}29^{7}$ $\frac{15}{16}=0{\cdot}34^{2}$

1	0·37	41	15·17	81	29·97	121	44·77	161	59·57
2	0·74	42	15·54	82	30·34	122	45·14	162	59·94
3	1·11	43	15·91	83	30·71	123	45·51	163	60·31
4	1·48	44	16·28	84	31·08	124	45·88	164	60·68
5	1·85	45	16·65	85	31·45	125	46·25	165	61·05
6	2·22	46	17·02	86	31·82	126	46·62	166	61·42
7	2·59	47	17·39	87	32·19	127	46·99	167	61·79
8	2·96	48	17·76	88	32·56	128	47·36	168	62·16
9	3·33	49	18·13	89	32·93	129	47·73	169	62·53
10	3·70	**50**	18·50	**90**	33·30	**130**	48·10	**170**	62·90
11	4·07	51	18·87	91	33·67	131	48·47	171	63·27
12	4·44	52	19·24	92	34·04	132	48·84	172	63·64
13	4·81	53	19·61	93	34·41	133	49·21	173	64·01
14	5·18	54	19·98	94	34·78	134	49·58	174	64·38
15	5·55	55	20·35	95	35·15	135	49·95	175	64·75
16	5·92	56	20·72	96	35·52	136	50·32	176	65·12
17	6·29	57	21·09	97	35·89	137	50·69	177	65·49
18	6·66	58	21·46	98	36·26	138	51·06	178	65·86
19	7·03	59	21·83	99	36·63	139	51·43	179	66·23
20	7·40	**60**	22·20	**100**	37·00	**140**	51·80	**180**	66·60
21	7·77	61	22·57	101	37·37	141	52·17	181	66·97
22	8·14	62	22·94	102	37·74	142	52·54	182	67·34
23	8·51	63	23·31	103	38·11	143	52·91	183	67·71
24	8·88	64	23·68	104	38·48	144	53·28	184	68·08
25	9·25	65	24·05	105	38·85	145	53·65	185	68·45
26	9·62	66	24·42	106	39·22	146	54·02	186	68·82
27	9·99	67	24·79	107	39·59	147	54·39	187	69·19
28	10·36	68	25·16	108	39·96	148	54·76	188	69·56
29	10·73	69	25·53	109	40·33	149	55·13	189	69·93
30	11·10	**70**	25·90	**110**	40·70	**150**	55·50	**190**	70·30
31	11·47	71	26·27	111	41·07	151	55·87	191	70·67
32	11·84	72	26·64	112	41·44	152	56·24	192	71·04
33	12·21	73	27·01	113	41·81	153	56·61	193	71·41
34	12·58	74	27·38	114	42·18	154	56·98	194	71·78
35	12·95	75	27·75	115	42·55	155	57·35	195	72·15
36	13·32	76	28·12	116	42·92	156	57·72	196	72·52
37	13·69	77	28·49	117	43·29	157	58·09	197	72·89
38	14·06	78	28·86	118	43·66	158	58·46	198	73·26
39	14·43	79	29·23	119	44·03	159	58·83	199	73·63
40	14·80	**80**	29·60	**120**	44·40	**160**	59·20	**200**	74·00

$\frac{1}{16}$=0·02^{3} $\frac{1}{9}$=0·04^{1} $\frac{1}{8}$=0·04^{6} $\frac{1}{6}$=0·06^{2} $\frac{1}{5}$=0·07^{4} $\frac{1}{4}$=0·09^{3} $\frac{1}{3}$=0·12^{3}

$\frac{3}{8}$=0·13^{9} $\frac{1}{2}$=0·18^{5} $\frac{5}{8}$=0·23^{1} $\frac{3}{4}$=0·27^{8} $\frac{7}{8}$=0·32^{4} $\frac{2}{3}$=0·24^{7} $\frac{5}{6}$=0·30^{8}

$\frac{3}{16}$=0·06^{9} $\frac{5}{16}$=0·11^{6} $\frac{7}{16}$=0·16^{2} $\frac{9}{16}$=0·20^{8} $\frac{11}{16}$=0·25^{4} $\frac{13}{16}$=0·30^{1} $\frac{15}{16}$=0·34^{7}

(2.67=1£) **£0.37½**

1	0·37½	41	15·37½	81	30·37½	121	45·37½	161	60·37½
2	0·75	42	15·75	82	30·75	122	45·75	162	60·75
3	1·12½	43	16·12½	83	31·12½	123	46·12½	163	61·12½
4	1·50	44	16·50	84	31·50	124	46·50	164	61·50
5	1·87½	45	16·87½	85	31·87½	125	46·87½	165	61·87½
6	2·25	46	17·25	86	32·25	126	47·25	166	62·25
7	2·62½	47	17·62½	87	32·62½	127	47·62½	167	62·62½
8	3·00	48	18·00	88	33·00	128	48·00	168	63·00
9	3·37½	49	18·37½	89	33·37½	129	48·37½	169	63·37½
10	3·75	**50**	18·75	**90**	33·75	**130**	48·75	**170**	63·75
11	4·12½	51	19·12½	91	34·12½	131	49·12½	171	64·12½
12	4·50	52	19·50	92	34·50	132	49·50	172	64·50
13	4·87½	53	19·87½	93	34·87½	133	49·87½	173	64·87½
14	5·25	54	20·25	94	35·25	134	50·25	174	65·25
15	5·62½	55	20·62½	95	35·62½	135	50·62½	175	65·62½
16	6·00	56	21·00	96	36·00	136	51·00	176	66·00
17	6·37½	57	21·37½	97	36·37½	137	51·37½	177	66·37½
18	6·75	58	21·75	98	36·75	138	51·75	178	66·75
19	7·12½	59	22·12½	99	37·12½	139	52·12½	179	67·12½
20	7·50	**60**	22·50	**100**	37·50	**140**	52·50	**180**	67·50
21	7·87½	61	22·87½	101	37·87½	141	52·87½	181	67·87½
22	8·25	62	23·25	102	38·25	142	53·25	182	68·25
23	8·62½	63	23·62½	103	38·62½	143	53·62½	183	68·62½
24	9·00	64	24·00	104	39·00	144	54·00	184	69·00
25	9·37½	65	24·37½	105	39·37½	145	54·37½	185	69·37½
26	9·75	66	24·75	106	39·75	146	54·75	186	69·75
27	10·12½	67	25·12½	107	40·12½	147	55·12½	187	70·12½
28	10·50	68	25·50	108	40·50	148	55·50	188	70·50
29	10·87½	69	25·87½	109	40·87½	149	55·87½	189	70·87½
30	11·25	**70**	26·25	**110**	41·25	**150**	56·25	**190**	71·25
31	11·62½	71	26·62½	111	41·62½	151	56·62½	191	71·62½
32	12·00	72	27·00	112	42·00	152	57·00	192	72·00
33	12·37½	73	27·37½	113	42·37½	153	57·37½	193	72·37½
34	12·75	74	27·75	114	42·75	154	57·75	194	72·75
35	13·12½	75	28·12½	115	43·12½	155	58·12½	195	73·12½
36	13·50	76	28·50	116	43·50	156	58·50	196	73·50
37	13·87½	77	28·87½	117	43·87½	157	58·87½	197	73·87½
38	14·25	78	29·25	118	44·25	158	59·25	198	74·25
39	14·62½	79	29·62½	119	44·62½	159	59·62½	199	74·62½
40	15·00	**80**	30·00	**120**	45·00	**160**	60·00	**200**	75·00

$\frac{1}{16}=0{\cdot}02^{3}$ $\frac{1}{9}=0{\cdot}04^{2}$ $\frac{1}{8}=0{\cdot}04^{7}$ $\frac{1}{6}=0{\cdot}06^{3}$ $\frac{1}{5}=0{\cdot}07^{5}$ $\frac{1}{4}=0{\cdot}09^{4}$ $\frac{1}{3}=0{\cdot}12^{5}$

$\frac{3}{8}=0{\cdot}14^{1}$ $\frac{1}{2}=0{\cdot}18^{8}$ $\frac{5}{8}=0{\cdot}23^{4}$ $\frac{3}{4}=0{\cdot}28^{1}$ $\frac{7}{8}=0{\cdot}32^{8}$ $\frac{2}{3}=0{\cdot}25^{0}$ $\frac{5}{6}=0{\cdot}31^{3}$

$\frac{3}{16}=0{\cdot}07^{0}$ $\frac{5}{16}=0{\cdot}11^{7}$ $\frac{7}{16}=0{\cdot}16^{4}$ $\frac{9}{16}=0{\cdot}21^{1}$ $\frac{11}{16}=0{\cdot}25^{8}$ $\frac{13}{16}=0{\cdot}30^{5}$ $\frac{15}{16}=0{\cdot}35^{2}$

(2.63=1£) £0.38

1	0·38	41	15·58	81	30·78	121	45·98	161	61·18
2	0·76	42	15·96	82	31·16	122	46·36	162	61·56
3	1·14	43	16·34	83	31·54	123	46·74	163	61·94
4	1·52	44	16·72	84	31·92	124	47·12	164	62·32
5	1·90	45	17·10	85	32·30	125	47·50	165	62·70
6	2·28	46	17·48	86	32·68	126	47·88	166	63·08
7	2·66	47	17·86	87	33·06	127	48·26	167	63·46
8	3·04	48	18·24	88	33·44	128	48·64	168	63·84
9	3·42	49	18·62	89	33·82	129	49·02	169	64·22
10	3·80	**50**	19·00	**90**	34·20	**130**	49·40	**170**	64·60
11	4·18	51	19·38	91	34·58	131	49·78	171	64·98
12	4·56	52	19·76	92	34·96	132	50·16	172	65·36
13	4·94	53	20·14	93	35·34	133	50·54	173	65·74
14	5·32	54	20·52	94	35·72	134	50·92	174	66·12
15	5·70	55	20·90	95	36·10	135	51·30	175	66·50
16	6·08	56	21·28	96	36·48	136	51·68	176	66·88
17	6·46	57	21·66	97	36·86	137	52·06	177	67·26
18	6·84	58	22·04	98	37·24	138	52·44	178	67·64
19	7·22	59	22·42	99	37·62	139	52·82	179	68·02
20	7·60	**60**	22·80	**100**	38·00	**140**	53·20	**180**	68·40
21	7·98	61	23·18	101	38·38	141	53·58	181	68·78
22	8·36	62	23·56	102	38·76	142	53·96	182	69·16
23	8·74	63	23·94	103	39·14	143	54·34	183	69·54
24	9·12	64	24·32	104	39·52	144	54·72	184	69·92
25	9·50	65	24·70	105	39·90	145	55·10	185	70·30
26	9·88	66	25·08	106	40·28	146	55·48	186	70·68
27	10·26	67	25·46	107	40·66	147	55·86	187	71·06
28	10·64	68	25·84	108	41·04	148	56·24	188	71·44
29	11·02	69	26·22	109	41·42	149	56·62	189	71·82
30	11·40	**70**	26·60	**110**	41·80	**150**	57·00	**190**	72·20
31	11·78	71	26·98	111	42·18	151	57·38	191	72·58
32	12·16	72	27·36	112	42·56	152	57·76	192	72·96
33	12·54	73	27·74	113	42·94	153	58·14	193	73·34
34	12·92	74	28·12	114	43·32	154	58·52	194	73·72
35	13·30	75	28·50	115	43·70	155	58·90	195	74·10
36	13·68	76	28·88	116	44·08	156	59·28	196	74·48
37	14·06	77	29·26	117	44·46	157	59·66	197	74·86
38	14·44	78	29·64	118	44·84	158	60·04	198	75·24
39	14·82	79	30·02	119	45·22	159	60·42	199	75·62
40	15·20	**80**	30·40	**120**	45·60	**160**	60·80	**200**	76·00

$\frac{1}{16}=0{\cdot}02^{4}$ $\frac{1}{9}=0{\cdot}04^{2}$ $\frac{1}{8}=0{\cdot}04^{8}$ $\frac{1}{6}=0{\cdot}06^{3}$ $\frac{1}{5}=0{\cdot}07^{6}$ $\frac{1}{4}=0{\cdot}09^{5}$ $\frac{1}{3}=0{\cdot}12^{7}$

$\frac{3}{8}=0{\cdot}14^{3}$ $\frac{1}{2}=0{\cdot}19^{0}$ $\frac{5}{8}=0{\cdot}23^{8}$ $\frac{3}{4}=0{\cdot}28^{5}$ $\frac{7}{8}=0{\cdot}33^{3}$ $\frac{2}{3}=0{\cdot}25^{3}$ $\frac{5}{6}=0{\cdot}31^{7}$

$\frac{3}{16}=0{\cdot}07^{1}$ $\frac{5}{16}=0{\cdot}11^{9}$ $\frac{7}{16}=0{\cdot}16^{6}$ $\frac{9}{16}=0{\cdot}21^{4}$ $\frac{11}{16}=0{\cdot}26^{1}$ $\frac{13}{16}=0{\cdot}30^{9}$ $\frac{15}{16}=0{\cdot}35^{6}$

1 US gallon=3·785 litres

(2.60=1£) **£0.38½**

1	0·38½	41	15·78½	81	31·18½	121	46·58½	161	61·98½
2	0·77	42	16·17	82	31·57	122	46·97	162	62·37
3	1·15½	43	16·55½	83	31·95½	123	47·35½	163	62·75½
4	1·54	44	16·94	84	32·34	124	47·74	164	63·14
5	1·92½	45	17·32½	85	32·72½	125	48·12½	165	63·52½
6	2·31	46	17·71	86	33·11	126	48·51	166	63·91
7	2·69½	47	18·09½	87	33·49½	127	48·89½	167	64·29½
8	3·08	48	18·48	88	33·88	128	49·28	168	64·68
9	3·46½	49	18·86½	89	34·26½	129	49·66½	169	65·06½
10	3·85	**50**	19·25	**90**	34·65	**130**	50·05	**170**	65·45
11	4·23½	51	19·63½	91	35·03½	131	50·43½	171	65·83½
12	4·62	52	20·02	92	35·42	132	50·82	172	66·22
13	5·00½	53	20·40½	93	35·80½	133	51·20½	173	66·60½
14	5·39	54	20·79	94	36·19	134	51·59	174	66·99
15	5·77½	55	21·17½	95	36·57½	135	51·97½	175	67·37½
16	6·16	56	21·56	96	36·96	136	52·36	176	67·76
17	6·54½	57	21·94½	97	37·34½	137	52·74½	177	68·14½
18	6·93	58	22·33	98	37·73	138	53·13	178	68·53
19	7·31½	59	22·71½	99	38·11½	139	53·51½	179	68·91½
20	7·70	**60**	23·10	**100**	38·50	**140**	53·90	**180**	69·30
21	8·08½	61	23·48½	101	38·88½	141	54·28½	181	69·68½
22	8·47	62	23·87	102	39·27	142	54·67	182	70·07
23	8·85½	63	24·25½	103	39·65½	143	55·05½	183	70·45½
24	9·24	64	24·64	104	40·04	144	55·44	184	70·84
25	9·62½	65	25·02½	105	40·42½	145	55·82½	185	71·22½
26	10·01	66	25·41	106	40·81	146	56·21	186	71·61
27	10·39½	67	25·79½	107	41·19½	147	56·59½	187	71·99½
28	10·78	68	26·18	108	41·58	148	56·98	188	72·38
29	11·16½	69	26·56½	109	41·96½	149	57·36½	189	72·76½
30	11·55	**70**	26·95	**110**	42·35	**150**	57·75	**190**	73·15
31	11·93½	71	27·33½	111	42·73½	151	58·13½	191	73·53½
32	12·32	72	27·72	112	43·12	152	58·52	192	73·92
33	12·70½	73	28·10½	113	43·50½	153	58·90½	193	74·30½
34	13·09	74	28·49	114	43·89	154	59·29	194	74·69
35	13·47½	75	28·87½	115	44·27½	155	59·67½	195	75·07½
36	13·86	76	29·26	116	44·66	156	60·06	196	75·46
37	14·24½	77	29·64½	117	45·04½	157	60·44½	197	75·84½
38	14·63	78	30·03	118	45·43	158	60·83	198	76·23
39	15·01½	79	30·41½	119	45·81½	159	61·21½	199	76·61½
40	15·40	**80**	30·80	**120**	46·20	**160**	61·60	**200**	77·00

$\frac{1}{16}$=0·02^{4} $\frac{1}{9}$=0·04^{3} $\frac{1}{8}$=0·04^{8} $\frac{1}{6}$=0·06^{4} $\frac{1}{5}$=0·07^{7} $\frac{1}{4}$=0·09^{6} $\frac{1}{3}$=0·12^{8}

$\frac{3}{8}$=0·14^{4} $\frac{1}{2}$=0·19^{3} $\frac{5}{8}$=0·24^{1} $\frac{3}{4}$=0·28^{9} $\frac{7}{8}$=0·33^{7} $\frac{2}{3}$=0·25^{7} $\frac{5}{6}$=0·32^{1}

$\frac{3}{16}$=0·07^{2} $\frac{5}{16}$=0·12^{0} $\frac{7}{16}$=0·16^{8} $\frac{9}{16}$=0·21^{7} $\frac{11}{16}$=0·26^{5} $\frac{13}{16}$=0·31^{3} $\frac{15}{16}$=0·36^{1}

1	0·39	41	15·99	81	31·59	121	47·19	161	62·79
2	0·78	42	16·38	82	31·98	122	47·58	162	63·18
3	1·17	43	16·77	83	32·37	123	47·97	163	63·57
4	1·56	44	17·16	84	32·76	124	48·36	164	63·96
5	1·95	45	17·55	85	33·15	125	48·75	165	64·35
6	2·34	46	17·94	86	33·54	126	49·14	166	64·74
7	2·73	47	18·33	87	33·93	127	49·53	167	65·13
8	3·12	48	18·72	88	34·32	128	49·92	168	65·52
9	3·51	49	19·11	89	34·71	129	50·31	169	65·91
10	3·90	**50**	19·50	**90**	35·10	**130**	50·70	**170**	66·30
11	4·29	51	19·89	91	35·49	131	51·09	171	66·69
12	4·68	52	20·28	92	35·88	132	51·48	172	67·08
13	5·07	53	20·67	93	36·27	133	51·87	173	67·47
14	5·46	54	21·06	94	36·66	134	52·26	174	67·86
15	5·85	55	21·45	95	37·05	135	52·65	175	68·25
16	6·24	56	21·84	96	37·44	136	53·04	176	68·64
17	6·63	57	22·23	97	37·83	137	53·43	177	69·03
18	7·02	58	22·62	98	38·22	138	53·82	178	69·42
19	7·41	59	23·01	99	38·61	139	54·21	179	69·81
20	7·80	**60**	23·40	**100**	39·00	**140**	54·60	**180**	70·20
21	8·19	61	23·79	101	39·39	141	54·99	181	70·59
22	8·58	62	24·18	102	39·78	142	55·38	182	70·98
23	8·97	63	24·57	103	40·17	143	55·77	183	71·37
24	9·36	64	24·96	104	40·56	144	56·16	184	71·76
25	9·75	65	25·35	105	40·95	145	56·55	185	72·15
26	10·14	66	25·74	106	41·34	146	56·94	186	72·54
27	10·53	67	26·13	107	41·73	147	57·33	187	72·93
28	10·92	68	26·52	108	42·12	148	57·72	188	73·32
29	11·31	69	26·91	109	42·51	149	58·11	189	73·71
30	11·70	**70**	27·30	**110**	42·90	**150**	58·50	**190**	74·10
31	12·09	71	27·69	111	43·29	151	58·89	191	74·49
32	12·48	72	28·08	112	43·68	152	59·28	192	74·88
33	12·87	73	28·47	113	44·07	153	59·67	193	75·27
34	13·26	74	28·86	114	44·46	154	60·06	194	75·66
35	13·65	75	29·25	115	44·85	155	60·45	195	76·05
36	14·04	76	29·64	116	45·24	156	60·84	196	76·44
37	14·43	77	30·03	117	45·63	157	61·23	197	76·83
38	14·82	78	30·42	118	46·02	158	61·62	198	77·22
39	15·21	79	30·81	119	46·41	159	62·01	199	77·61
40	15·60	**80**	31·20	**120**	46·80	**160**	62·40	**200**	78·00

$\frac{1}{16}=0{\cdot}02^{4}$ $\frac{1}{9}=0{\cdot}04^{3}$ $\frac{1}{8}=0{\cdot}04^{9}$ $\frac{1}{6}=0{\cdot}06^{5}$ $\frac{1}{5}=0{\cdot}07^{8}$ $\frac{1}{4}=0{\cdot}09^{8}$ $\frac{1}{3}=0{\cdot}13^{0}$

$\frac{3}{8}=0{\cdot}14^{6}$ $\frac{1}{2}=0{\cdot}19^{5}$ $\frac{5}{8}=0{\cdot}24^{4}$ $\frac{3}{4}=0{\cdot}29^{3}$ $\frac{7}{8}=0{\cdot}34^{1}$ $\frac{2}{3}=0{\cdot}26^{0}$ $\frac{5}{6}=0{\cdot}32^{5}$

$\frac{3}{16}=0{\cdot}07^{3}$ $\frac{5}{16}=0{\cdot}12^{2}$ $\frac{7}{16}=0{\cdot}17^{1}$ $\frac{9}{16}=0{\cdot}21^{9}$ $\frac{11}{16}=0{\cdot}26^{8}$ $\frac{13}{16}=0{\cdot}31^{7}$ $\frac{15}{16}=0{\cdot}36^{6}$

(2.53=1£) **£0.39½**

1	0·39½	41	16·19½	81	31·99½	121	47·79½	161	63·59½
2	0·79	42	16·59	82	32·39	122	48·19	162	63·99
3	1·18½	43	16·98½	83	32·78½	123	48·58½	163	64·38½
4	1·58	44	17·38	84	33·18	124	48·98	164	64·78
5	1·97½	45	17·77½	85	33·57½	125	49·37½	165	65·17½
6	2·37	46	18·17	86	33·97	126	49·77	166	65·57
7	2·76½	47	18·56½	87	34·36½	127	50·16½	167	65·96½
8	3·16	48	18·96	88	34·76	128	50·56	168	66·36
9	3·55½	49	19·35½	89	35·15½	129	50·95½	169	66·75½
10	3·95	**50**	19·75	**90**	35·55	**130**	51·35	**170**	67·15
11	4·34½	51	20·14½	91	35·94½	131	51·74½	171	67·54½
12	4·74	52	20·54	92	36·34	132	52·14	172	67·94
13	5·13½	53	20·93½	93	36·73½	133	52·53½	173	68·33½
14	5·53	54	21·33	94	37·13	134	52·93	174	68·73
15	5·92½	55	21·72½	95	37·52½	135	53·32½	175	69·12½
16	6·32	56	22·12	96	37·92	136	53·72	176	69·52
17	6·71½	57	22·51½	97	38·31½	137	54·11½	177	69·91½
18	7·11	58	22·91	98	38·71	138	54·51	178	70·31
19	7·50½	59	23·30½	99	39·10½	139	54·90½	179	70·70½
20	7·90	**60**	23·70	**100**	39·50	**140**	55·30	**180**	71·10
21	8·29½	61	24·09½	101	39·89½	141	55·69½	181	71·49½
22	8·69	62	24·49	102	40·29	142	56·09	182	71·89
23	9·08½	63	24·88½	103	40·68½	143	56·48½	183	72·28½
24	9·48	64	25·28	104	41·08	144	56·88	184	72·68
25	9·87½	65	25·67½	105	41·47½	145	57·27½	185	73·07½
26	10·27	66	26·07	106	41·87	146	57·67	186	73·47
27	10·66½	67	26·46½	107	42·26½	147	58·06½	187	73·86½
28	11·06	68	26·86	108	42·66	148	58·46	188	74·26
29	11·45½	69	27·25½	109	43·05½	149	58·85½	189	74·65½
30	11·85	**70**	27·65	**110**	43·45	**150**	59·25	**190**	75·05
31	12·24½	71	28·04½	111	43·84½	151	59·64½	191	75·44½
32	12·64	72	28·44	112	44·24	152	60·04	192	75·84
33	13·03½	73	28·83½	113	44·63½	153	60·43½	193	76·23½
34	13·43	74	29·23	114	45·03	154	60·83	194	76·63
35	13·82½	75	29·62½	115	45·42½	155	61·22½	195	77·02½
36	14·22	76	30·02	116	45·82	156	61·62	196	77·42
37	14·61½	77	30·41½	117	46·21½	157	62·01½	197	77·81½
38	15·01	78	30·81	118	46·61	158	62·41	198	78·21
39	15·40½	79	31·20½	119	47·00½	159	62·80½	199	78·60½
40	15·80	**80**	31·60	**120**	47·40	**160**	63·20	**200**	79·00

$\frac{1}{16}=0{\cdot}02^{5}$ $\frac{1}{9}=0{\cdot}04^{4}$ $\frac{1}{8}=0{\cdot}04^{9}$ $\frac{1}{6}=0{\cdot}06^{6}$ $\frac{1}{5}=0{\cdot}07^{9}$ $\frac{1}{4}=0{\cdot}09^{9}$ $\frac{1}{3}=0{\cdot}13^{2}$

$\frac{3}{8}=0{\cdot}14^{8}$ $\frac{1}{2}=0{\cdot}19^{8}$ $\frac{5}{8}=0{\cdot}24^{7}$ $\frac{3}{4}=0{\cdot}29^{6}$ $\frac{7}{8}=0{\cdot}34^{6}$ $\frac{2}{3}=0{\cdot}26^{3}$ $\frac{5}{6}=0{\cdot}32^{9}$

$\frac{3}{16}=0{\cdot}07^{4}$ $\frac{5}{16}=0{\cdot}12^{3}$ $\frac{7}{16}=0{\cdot}17^{3}$ $\frac{9}{16}=0{\cdot}22^{2}$ $\frac{11}{16}=0{\cdot}27^{2}$ $\frac{13}{16}=0{\cdot}32^{1}$ $\frac{15}{16}=0{\cdot}37^{0}$

(2.50=1£) **£0.40**

1	0·40	41	16·40	81	32·40	121	48·40	161	64·40
2	0·80	42	16·80	82	32·80	122	48·80	162	64·80
3	1·20	43	17·20	83	33·20	123	49·20	163	65·20
4	1·60	44	17·60	84	33·60	124	49·60	164	65·60
5	2·00	45	18·00	85	34·00	125	50·00	165	66·00
6	2·40	46	18·40	86	34·40	126	50·40	166	66·40
7	2·80	47	18·80	87	34·80	127	50·80	167	66·80
8	3·20	48	19·20	88	35·20	128	51·20	168	67·20
9	3·60	49	19·60	89	35·60	129	51·60	169	67·60
10	4·00	**50**	20·00	**90**	36·00	**130**	52·00	**170**	68·00
11	4·40	51	20·40	91	36·40	131	52·40	171	68·40
12	4·80	52	20·80	92	36·80	132	52·80	172	68·80
13	5·20	53	21·20	93	37·20	133	53·20	173	69·20
14	5·60	54	21·60	94	37·60	134	53·60	174	69·60
15	6·00	55	22·00	95	38·00	135	54·00	175	70·00
16	6·40	56	22·40	96	38·40	136	54·40	176	70·40
17	6·80	57	22·80	97	38·80	137	54·80	177	70·80
18	7·20	58	23·20	98	39·20	138	55·20	178	71·20
19	7·60	59	23·60	99	39·60	139	55·60	179	71·60
20	8·00	**60**	24·00	**100**	40·00	**140**	56·00	**180**	72·00
21	8·40	61	24·40	101	40·40	141	56·40	181	72·40
22	8·80	62	24·80	102	40·80	142	56·80	182	72·80
23	9·20	63	25·20	103	41·20	143	57·20	183	73·20
24	9·60	64	25·60	104	41·60	144	57·60	184	73·60
25	10·00	65	26·00	105	42·00	145	58·00	185	74·00
26	10·40	66	26·40	106	42·40	146	58·40	186	74·40
27	10·80	67	26·80	107	42·80	147	58·80	187	74·80
28	11·20	68	27·20	108	43·20	148	59·20	188	75·20
29	11·60	69	27·60	109	43·60	149	59·60	189	75·60
30	12·00	**70**	28·00	**110**	44·00	**150**	60·00	**190**	76·00
31	12·40	71	28·40	111	44·40	151	60·40	191	76·40
32	12·80	72	28·80	112	44·80	152	60·80	192	76·80
33	13·20	73	29·20	113	45·20	153	61·20	193	77·20
34	13·60	74	29·60	114	45·60	154	61·60	194	77·60
35	14·00	75	30·00	115	46·00	155	62·00	195	78·00
36	14·40	76	30·40	116	46·40	156	62·40	196	78·40
37	14·80	77	30·80	117	46·80	157	62·80	197	78·80
38	15·20	78	31·20	118	47·20	158	63·20	198	79·20
39	15·60	79	31·60	119	47·60	159	63·60	199	79·60
40	16·00	**80**	32·00	**120**	48·00	**160**	64·00	**200**	80·00

$\frac{1}{16}=0{\cdot}02^5$ $\frac{1}{9}=0{\cdot}04^4$ $\frac{1}{8}=0{\cdot}05^0$ $\frac{1}{6}=0{\cdot}06^7$ $\frac{1}{5}=0{\cdot}08^0$ $\frac{1}{4}=0{\cdot}10^0$ $\frac{1}{3}=0{\cdot}13^3$

$\frac{3}{8}=0{\cdot}15^0$ $\frac{1}{2}=0{\cdot}20^0$ $\frac{5}{8}=0{\cdot}25^0$ $\frac{3}{4}=0{\cdot}30^0$ $\frac{7}{8}=0{\cdot}35^0$ $\frac{2}{3}=0{\cdot}26^7$ $\frac{5}{6}=0{\cdot}33^3$

$\frac{3}{16}=0{\cdot}07^5$ $\frac{5}{16}=0{\cdot}12^5$ $\frac{7}{16}=0{\cdot}17^5$ $\frac{9}{16}=0{\cdot}22^5$ $\frac{11}{16}=0{\cdot}27^5$ $\frac{13}{16}=0{\cdot}32^5$ $\frac{15}{16}=0{\cdot}37^5$

(2.47=1£) **£0.40½**

1	0·40½	41	16·60½	81	32·80½	121	49·00½	161	65·20½
2	0·81	42	17·01	82	33·21	122	49·41	162	65·61
3	1·21½	43	17·41½	83	33·61½	123	49·81½	163	66·01½
4	1·62	44	17·82	84	34·02	124	50·22	164	66·42
5	2·02½	45	18·22½	85	34·42½	125	50·62½	165	66·82½
6	2·43	46	18·63	86	34·83	126	51·03	166	67·23
7	2·83½	47	19·03½	87	35·23½	127	51·43½	167	67·63½
8	3·24	48	19·44	88	35·64	128	51·84	168	68·04
9	3·64½	49	19·84½	89	36·04½	129	52·24½	169	68·44½
10	4·05	**50**	20·25	**90**	36·45	**130**	52·65	**170**	68·85
11	4·45½	51	20·65½	91	36·85½	131	53·05½	171	69·25½
12	4·86	52	21·06	92	37·26	132	53·46	172	69·66
13	5·26½	53	21·46½	93	37·66½	133	53·86½	173	70·06½
14	5·67	54	21·87	94	38·07	134	54·27	174	70·47
15	6·07½	55	22·27½	95	38·47½	135	54·67½	175	70·87½
16	6·48	56	22·68	96	38·88	136	55·08	176	71·28
17	6·88½	57	23·08½	97	39·28½	137	55·48½	177	71·68½
18	7·29	58	23·49	98	39·69	138	55·89	178	72·09
19	7·69½	59	23·89½	99	40·09½	139	56·29½	179	72·49½
20	8·10	**60**	24·30	**100**	40·50	**140**	56·70	**180**	72·90
21	8·50½	61	24·70½	101	40·90½	141	57·10½	181	73·30½
22	8·91	62	25·11	102	41·31	142	57·51	182	73·71
23	9·31½	63	25·51½	103	41·71½	143	57·91½	183	74·11½
24	9·72	64	25·92	104	42·12	144	58·32	184	74·52
25	10·12½	65	26·32½	105	42·52½	145	58·72½	185	74·92½
26	10·53	66	26·73	106	42·93	146	59·13	186	75·33
27	10·93½	67	27·13½	107	43·33½	147	59·53½	187	75·73½
28	11·34	68	27·54	108	43·74	148	59·94	188	76·14
29	11·74½	69	27·94½	109	44·14½	149	60·34½	189	76·54½
30	12·15	**70**	28·35	**110**	44·55	**150**	60·75	**190**	76·95
31	12·55½	71	28·75½	111	44·95½	151	61·15½	191	77·35½
32	12·96	72	29·16	112	45·36	152	61·56	192	77·76
33	13·36½	73	29·56½	113	45·76½	153	61·96½	193	78·16½
34	13·77	74	29·97	114	46·17	154	62·37	194	78·57
35	14·17½	75	30·37½	115	46·57½	155	62·77½	195	78·97½
36	14·58	76	30·78	116	46·98	156	63·18	196	79·38
37	14·98½	77	31·18½	117	47·38½	157	63·58½	197	79·78½
38	15·39	78	31·59	118	47·79	158	63·99	198	80·19
39	15·79½	79	31·99½	119	48·19½	159	64·39½	199	80·59½
40	16·20	**80**	32·40	**120**	48·60	**160**	64·80	**200**	81·00

$\frac{1}{16}=0{\cdot}02^{5}$ $\frac{1}{9}=0{\cdot}04^{5}$ $\frac{1}{8}=0{\cdot}05^{1}$ $\frac{1}{6}=0{\cdot}06^{8}$ $\frac{1}{5}=0{\cdot}08^{1}$ $\frac{1}{4}=0{\cdot}10^{1}$ $\frac{1}{3}=0{\cdot}13^{5}$

$\frac{3}{8}=0{\cdot}15^{2}$ $\frac{1}{2}=0{\cdot}20^{3}$ $\frac{5}{8}=0{\cdot}25^{3}$ $\frac{3}{4}=0{\cdot}30^{4}$ $\frac{7}{8}=0{\cdot}35^{4}$ $\frac{2}{3}=0{\cdot}27^{0}$ $\frac{5}{6}=0{\cdot}33^{8}$

$\frac{3}{16}=0{\cdot}07^{6}$ $\frac{5}{16}=0{\cdot}12^{7}$ $\frac{7}{16}=0{\cdot}17^{7}$ $\frac{9}{16}=0{\cdot}22^{8}$ $\frac{11}{16}=0{\cdot}27^{8}$ $\frac{13}{16}=0{\cdot}32^{9}$ $\frac{15}{16}=0{\cdot}38^{0}$

1	0·41	41	16·81	81	33·21	121	49·61	161	66·01
2	0·82	42	17·22	82	33·62·	122	50·02	162	66·42
3	1·23	43	17·63	83	34·03	123	50·43	163	66·83
4	1·64	44	18·04	84	34·44	124	50·84	164	67·24
5	2·05	45	18·45	85	34·85	125	51·25	165	67·65
6	2·46	46	18·86	86	35·26	126	51·66	166	68·06
7	2·87	47	19·27	87	35·67	127	52·07	167	68·47
8	3·28	48	19·68	88	36·08	128	52·48	168	68·88
9	3·69	49	20·09	89	36·49	129	52·89	169	69·29
10	4·10	**50**	20·50	**90**	36·90	**130**	53·30	**170**	69·70
11	4·51	51	20·91	91	37·31	131	53·71	171	70·11
12	4·92	52	21·32	92	37·72	132	54·12	172	70·52
13	5·33	53	21·73	93	38·13	133	54·53	173	70·93
14	5·74	54	22·14	94	38·54	134	54·94	174	71·34
15	6·15	55	22·55	95	38·95	135	55·35	175	71·75
16	6·56	56	22·96	96	39·36	136	55·76	176	72·16
17	6·97	57	23·37	97	39·77	137	56·17	177	72·57
18	7·38	58	23·78	98	40·18	138	56·58	178	72·98
19	7·79	59	24·19	99	40·59	139	56·99	179	73·39
20	8·20	**60**	24·60	**100**	41·00	**140**	57·40	**180**	73·80
21	8·61	61	25·01	101	41·41	141	57·81	181	74·21
22	9·02	62	25·42	102	41·82	142	58·22	182	74·62
23	9·43	63	25·83	103	42·23	143	58·63	183	75·03
24	9·84	64	26·24	104	42·64	144	59·04	184	75·44
25	10·25	65	26·65	105	43·05	145	59·45	185	75·85
26	10·66	66	27·06	106	43·46	146	59·86	186	76·26
27	11·07	67	27·47	107	43·87	147	60·27	187	76·67
28	11·48	68	27·88	108	44·28	148	60·68	188	77·08
29	11·89	69	28·29	109	44·69	149	61·09	189	77·49
30	12·30	**70**	28·70	**110**	45·10	**150**	61·50	**190**	77·90
31	12·71	71	29·11	111	45·51	151	61·91	191	78·31
32	13·12	72	29·52	112	45·92	152	62·32	192	78·72
33	13·53	73	29·93	113	46·33	153	62·73	193	79·13
34	13·94	74	30·34	114	46·74	154	63·14	194	79·54
35	14·35	75	30·75	115	47·15	155	63·55	195	79·95
36	14·76	76	31·16	116	47·56	156	63·96	196	80·36
37	15·17	77	31·57	117	47·97	157	64·37	197	80·77
38	15·58	78	31·98	118	48·38	158	64·78	198	81·18
39	15·99	79	32·39	119	48·79	159	65·19	199	81·59
40	16·40	**80**	32·80	**120**	49·20	**160**	65·60	**200**	82·00

$\frac{1}{16}=0{\cdot}02^{6}$ $\frac{1}{9}=0{\cdot}04^{6}$ $\frac{1}{8}=0{\cdot}05^{1}$ $\frac{1}{6}=0{\cdot}06^{8}$ $\frac{1}{5}=0{\cdot}08^{2}$ $\frac{1}{4}=0{\cdot}10^{3}$ $\frac{1}{3}=0{\cdot}13^{7}$

$\frac{3}{8}=0{\cdot}15^{4}$ $\frac{1}{2}=0{\cdot}20^{5}$ $\frac{5}{8}=0{\cdot}25^{6}$ $\frac{3}{4}=0{\cdot}30^{8}$ $\frac{7}{8}=0{\cdot}35^{9}$ $\frac{2}{3}=0{\cdot}27^{3}$ $\frac{5}{6}=0{\cdot}34^{2}$

$\frac{3}{16}=0{\cdot}07^{7}$ $\frac{5}{16}=0{\cdot}12^{8}$ $\frac{7}{16}=0{\cdot}17^{9}$ $\frac{9}{16}=0{\cdot}23^{1}$ $\frac{11}{16}=0{\cdot}28^{2}$ $\frac{13}{16}=0{\cdot}33^{3}$ $\frac{15}{16}=0{\cdot}38^{4}$

1	0·41½	41	17·01½	81	33·61½	121	50·21½	161	66·81½
2	0·83	42	17·43	82	34·03	122	50·63	162	67·23
3	1·24½	43	17·84½	83	34·44½	123	51·04½	163	67·64½
4	1·66	44	18·26	84	34·86	124	51·46	164	68·06
5	2·07½	45	18·67½	85	35·27½	125	51·87½	165	68·47½
6	2·49	46	19·09	86	35·69	126	52·29	166	68·89
7	2·90½	47	19·50½	87	36·10½	127	52·70½	167	69·30½
8	3·32	48	19·92	88	36·52	128	53·12	168	69·72
9	3·73½	49	20·33½	89	36·93½	129	53·53½	169	70·13½
10	4·15	**50**	20·75	**90**	37·35	**130**	53·95	**170**	70·55
11	4·56½	51	21·16½	91	37·76½	131	54·36½	171	70·96½
12	4·98	52	21·58	92	38·18	132	54·78	172	71·38
13	5·39½	53	21·99½	93	38·59½	133	55·19½	173	71·79½
14	5·81	54	22·41	94	39·01	134	55·61	174	72·21
15	6·22½	55	22·82½	95	39·42½	135	56·02½	175	72·62½
16	6·64	56	23·24	96	39·84	136	56·44	176	73·04
17	7·05½	57	23·65½	97	40·25½	137	56·85½	177	73·45½
18	7·47	58	24·07	98	40·67	138	57·27	178	73·87
19	7·88½	59	24·48½	99	41·08½	139	57·68½	179	74·28½
20	8·30	**60**	24·90	**100**	41·50	**140**	58·10	**180**	74·70
21	8·71½	61	25·31½	101	41·91½	141	58·51½	181	75·11½
22	9·13	62	25·73	102	42·33	142	58·93	182	75·53
23	9·54½	63	26·14½	103	42·74½	143	59·34½	183	75·94½
24	9·96	64	26·56	104	43·16	144	59·76	184	76·36
25	10·37½	65	26·97½	105	43·57½	145	60·17½	185	76·77½
26	10·79	66	27·39	106	43·99	146	60·59	186	77·19
27	11·20½	67	27·80½	107	44·40½	147	61·00½	187	77·60½
28	11·62	68	28·22	108	44·82	148	61·42	188	78·02
29	12·03½	69	28·63½	109	45·23½	149	61·83½	189	78·43½
30	12·45	**70**	29·05	**110**	45·65	**150**	62·25	**190**	78·85
31	12·86½	71	29·46½	111	46·06½	151	62·66½	191	79·26½
32	13·28	72	29·88	112	46·48	152	63·08	192	79·68
33	13·69½	73	30·29½	113	46·89½	153	63·49½	193	80·09½
34	14·11	74	30·71	114	47·31	154	63·91	194	80·51
35	14·52½	75	31·12½	115	47·72½	155	64·32½	195	80·92½
36	14·94	76	31·54	116	48·14	156	64·74	196	81·34
37	15·35½	77	31·95½	117	48·55½	157	65·15½	197	81·75½
38	15·77	78	32·37	118	48·97	158	65·57	198	82·17
39	16·18½	79	32·78½	119	49·38½	159	65·98½	199	82·58½
40	16·60	**80**	33·20	**120**	49·80	**160**	66·40	**200**	83·00

$\frac{1}{16}=0{\cdot}02^{6}$	$\frac{1}{9}=0{\cdot}04^{6}$	$\frac{1}{8}=0{\cdot}05^{2}$	$\frac{1}{6}=0{\cdot}06^{9}$	$\frac{1}{5}=0{\cdot}08^{3}$	$\frac{1}{4}=0{\cdot}10^{4}$	$\frac{1}{3}=0{\cdot}13^{8}$
$\frac{3}{8}=0{\cdot}15^{6}$	$\frac{1}{2}=0{\cdot}20^{8}$	$\frac{5}{8}=0{\cdot}25^{9}$	$\frac{3}{4}=0{\cdot}31^{1}$	$\frac{7}{8}=0{\cdot}36^{3}$	$\frac{2}{3}=0{\cdot}27^{7}$	$\frac{5}{6}=0{\cdot}34^{6}$
$\frac{3}{16}=0{\cdot}07^{8}$	$\frac{5}{16}=0{\cdot}13^{0}$	$\frac{7}{16}=0{\cdot}18^{2}$	$\frac{9}{16}=0{\cdot}23^{3}$	$\frac{11}{16}=0{\cdot}28^{5}$	$\frac{13}{16}=0{\cdot}33^{7}$	$\frac{15}{16}=0{\cdot}38^{9}$

(2.38=1£) **£0.42**

1	0·42	41	17·22	81	34·02	121	50·82	161	67·62
2	0·84	42	17·64	82	34·44	122	51·24	162	68·04
3	1·26	43	18·06	83	34·86	123	51·66	163	68·46
4	1·68	44	18·48	84	35·28	124	52·08	164	68·88
5	2·10	45	18·90	85	35·70	125	52·50	165	69·30
6	2·52	46	19·32	86	36·12	126	52·92	166	69·72
7	2·94	47	19·74	87	36·54	127	53·34	167	70·14
8	3·36	48	20·16	88	36·96	128	53·76	168	70·56
9	3·78	49	20·58	89	37·38	129	54·18	169	70·98
10	4·20	**50**	21·00	**90**	37·80	**130**	54·60	**170**	71·40
11	4·62	51	21·42	91	38·22	131	55·02	171	71·82
12	5·04	52	21·84	92	38·64	132	55·44	172	72·24
13	5·46	53	22·26	93	39·06	133	55·86	173	72·66
14	5·88	54	22·68	94	39·48	134	56·28	174	73·08
15	6·30	55	23·10	95	39·90	135	56·70	175	73·50
16	6·72	56	23·52	96	40·32	136	57·12	176	73·92
17	7·14	57	23·94	97	40·74	137	57·54	177	74·34
18	7·56	58	24·36	98	41·16	138	57·96	178	74·76
19	7·98	59	24·78	99	41·58	139	58·38	179	75·18
20	8·40	**60**	25·20	**100**	42·00	**140**	58·80	**180**	75·60
21	8·82	61	25·62	101	42·42	141	59·22	181	76·02
22	9·24	62	26·04	102	42·84	142	59·64	182	76·44
23	9·66	63	26·46	103	43·26	143	60·06	183	76·86
24	10·08	64	26·88	104	43·68	144	60·48	184	77·28
25	10·50	65	27·30	105	44·10	145	60·90	185	77·70
26	10·92	66	27·72	106	44·52	146	61·32	186	78·12
27	11·34	67	28·14	107	44·94	147	61·74	187	78·54
28	11·76	68	28·56	108	45·36	148	62·16	188	78·96
29	12·18	69	28·98	109	45·78	149	62·58	189	79·38
30	12·60	**70**	29·40	**110**	46·20	**150**	63·00	**190**	79·80
31	13·02	71	29·82	111	46·62	151	63·42	191	80·22
32	13·44	72	30·24	112	47·04	152	63·84	192	80·64
33	13·86	73	30·66	113	47·46	153	64·26	193	81·06
34	14·28	74	31·08	114	47·88	154	64·68	194	81·48
35	14·70	75	31·50	115	48·30	155	65·10	195	81·90
36	15·12	76	31·92	116	48·72	156	65·52	196	82·32
37	15·54	77	32·34	117	49·14	157	65·94	197	82·74
38	15·96	78	32·76	118	49·56	158	66·36	198	83·16
39	16·38	79	33·18	119	49·98	159	66·78	199	83·58
40	16·80	**80**	33·60	**120**	50·40	**160**	67·20	**200**	84·00

$\frac{1}{16}=0{\cdot}02^{6}$ $\frac{1}{9}=0{\cdot}04^{7}$ $\frac{1}{8}=0{\cdot}05^{3}$ $\frac{1}{6}=0{\cdot}07^{0}$ $\frac{1}{5}=0{\cdot}08^{4}$ $\frac{1}{4}=0{\cdot}10^{5}$ $\frac{1}{3}=0{\cdot}14^{0}$

$\frac{3}{8}=0{\cdot}15^{8}$ $\frac{1}{2}=0{\cdot}21^{0}$ $\frac{5}{8}=0{\cdot}26^{3}$ $\frac{3}{4}=0{\cdot}31^{5}$ $\frac{7}{8}=0{\cdot}36^{8}$ $\frac{2}{3}=0{\cdot}28^{0}$ $\frac{5}{6}=0{\cdot}35^{0}$

$\frac{3}{16}=0{\cdot}07^{9}$ $\frac{5}{16}=0{\cdot}13^{1}$ $\frac{7}{16}=0{\cdot}18^{4}$ $\frac{9}{16}=0{\cdot}23^{6}$ $\frac{11}{16}=0{\cdot}28^{9}$ $\frac{13}{16}=0{\cdot}34^{1}$ $\frac{15}{16}=0{\cdot}39^{4}$

1 calorie=4·186 joules

1	0·42½	41	17·42½	81	34·42½	121	51·42½	161	68·42½
2	0·85	42	17·85	82	34·85	122	51·85	162	68·85
3	1·27½	43	18·27½	83	35·27½	123	52·27½	163	69·27½
4	1·70	44	18·70	84	35·70	124	52·70	164	69·70
5	2·12½	45	19·12½	85	36·12½	125	53·12½	165	70·12½
6	2·55	46	19·55	86	36·55	126	53·55	166	70·55
7	2·97½	47	19·97½	87	36·97½	127	53·97½	167	70·97½
8	3·40	48	20·40	88	37·40	128	54·40	168	71·40
9	3·82½	49	20·82½	89	37·82½	129	54·82½	169	71·82½
10	4·25	**50**	21·25	**90**	38·25	**130**	55·25	**170**	72·25
11	4·67½	51	21·67½	91	38·67½	131	55·67½	171	72·67½
12	5·10	52	22·10	92	39·10	132	56·10	172	73·10
13	5·52½	53	22·52½	93	39·52½	133	56·52½	173	73·52½
14	5·95	54	22·95	94	39·95	134	56·95	174	73·95
15	6·37½	55	23·37½	95	40·37½	135	57·37½	175	74·37½
16	6·80	56	23·80	96	40·80	136	57·80	176	74·80
17	7·22½	57	24·22½	97	41·22½	137	58·22½	177	75·22½
18	7·65	58	24·65	98	41·65	138	58·65	178	75·65
19	8·07½	59	25·07½	99	42·07½	139	59·07½	179	76·07½
20	8·50	**60**	25·50	**100**	42·50	**140**	59·50	**180**	76·50
21	8·92½	61	25·92½	101	42·92½	141	59·92½	181	76·92½
22	9·35	62	26·35	102	43·35	142	60·35	182	77·35
23	9·77½	63	26·77½	103	43·77½	143	60·77½	183	77·77½
24	10·20	64	27·20	104	44·20	144	61·20	184	78·20
25	10·62½	65	27·62½	105	44·62½	145	61·62½	185	78·62½
26	11·05	66	28·05	106	45·05	146	62·05	186	79·05
27	11·47½	67	28·47½	107	45·47½	147	62·47½	187	79·47½
28	11·90	68	28·90	108	45·90	148	62·90	188	79·90
29	12·32½	69	29·32½	109	46·32½	149	63·32½	189	80·32½
30	12·75	**70**	29·75	**110**	46·75	**150**	63·75	**190**	80·75
31	13·17½	71	30·17½	111	47·17½	151	64·17½	191	81·17½
32	13·60	72	30·60	112	47·60	152	64·60	192	81·60
33	14·02½	73	31·02½	113	48·02½	153	65·02½	193	82·02½
34	14·45	74	31·45	114	48·45	154	65·45	194	82·45
35	14·87½	75	31·87½	115	48·87½	155	65·87½	195	82·87½
36	15·30	76	32·30	116	49·30	156	66·30	196	83·30
37	15·72½	77	32·72½	117	49·72½	157	66·72½	197	83·72½
38	16·15	78	33·15	118	50·15	158	67·15	198	84·15
39	16·57½	79	33·57½	119	50·57½	159	67·57½	199	84·57½
40	17·00	**80**	34·00	**120**	51·00	**160**	68·00	**200**	85·00

$\frac{1}{16}$=0·02⁷	$\frac{1}{9}$=0·04⁷	$\frac{1}{8}$=0·05³	$\frac{1}{6}$=0·07¹	$\frac{1}{5}$=0·08⁵	$\frac{1}{4}$=0·10⁶	$\frac{1}{3}$=0·14²
$\frac{3}{8}$=0·15⁹	$\frac{1}{2}$=0·21³	$\frac{5}{8}$=0·26⁶	$\frac{3}{4}$=0·31⁹	$\frac{7}{8}$=0·37²	$\frac{2}{3}$=0·28³	$\frac{5}{6}$=0·35⁴
$\frac{3}{16}$=0·08⁰	$\frac{5}{16}$=0·13³	$\frac{7}{16}$=0·18⁶	$\frac{9}{16}$=0·23⁹	$\frac{11}{16}$=0·29²	$\frac{13}{16}$=0·34⁵	$\frac{15}{16}$=0·39⁸

1	0·43	41	17·63	81	34·83	121	52·03	161	69·23
2	0·86	42	18·06	82	35·26	122	52·46	162	69·66
3	1·29	43	18·49	83	35·69	123	52·89	163	70·09
4	1·72	44	18·92	84	36·12	124	53·32	164	70·52
5	2·15	45	19·35	85	36·55	125	53·75	165	70·95
6	2·58	46	19·78	86	36·98	126	54·18	166	71·38
7	3·01	47	20·21	87	37·41	127	54·61	167	71·81
8	3·44	48	20·64	88	37·84	128	55·04	168	72·24
9	3·87	49	21·07	89	38·27	129	55·47	169	72·67
10	4·30	**50**	21·50	**90**	38·70	**130**	55·90	**170**	73·10
11	4·73	51	21·93	91	39·13	131	56·33	171	73·53
12	5·16	52	22·36	92	39·56	132	56·76	172	73·96
13	5·59	53	22·79	93	39·99	133	57·19	173	74·39
14	6·02	54	23·22	94	40·42	134	57·62	174	74·82
15	6·45	55	23·65	95	40·85	135	58·05	175	75·25
16	6·88	56	24·08	96	41·28	136	58·48	176	75·68
17	7·31	57	24·51	97	41·71	137	58·91	177	76·11
18	7·74	58	24·94	98	42·14	138	59·34	178	76·54
19	8·17	59	25·37	99	42·57	139	59·77	179	76·97
20	8·60	**60**	25·80	**100**	43·00	**140**	60·20	**180**	77·40
21	9·03	61	26·23	101	43·43	141	60·63	181	77·83
22	9·46	62	26·66	102	43·86	142	61·06	182	78·26
23	9·89	63	27·09	103	44·29	143	61·49	183	78·69
24	10·32	64	27·52	104	44·72	144	61·92	184	79·12
25	10·75	65	27·95	105	45·15	145	62·35	185	79·55
26	11·18	66	28·38	106	45·58	146	62·78	186	79·98
27	11·61	67	28·81	107	46·01	147	63·21	187	80·41
28	12·04	68	29·24	108	46·44	148	63·64	188	80·84
29	12·47	69	29·67	109	46·87	149	64·07	189	81·27
30	12·90	**70**	30·10	**110**	47·30	**150**	64·50	**190**	81·70
31	13·33	71	30·53	111	47·73	151	64·93	191	82·13
32	13·76	72	30·96	112	48·16	152	65·36	192	82·56
33	14·19	73	31·39	113	48·59	153	65·79	193	82·99
34	14·62	74	31·82	114	49·02	154	66·22	194	83·42
35	15·05	75	32·25	115	49·45	155	66·65	195	83·85
36	15·48	76	32·68	116	49·88	156	67·08	196	84·28
37	15·91	77	33·11	117	50·31	157	67·51	197	84·71
38	16·34	78	33·54	118	50·74	158	67·94	198	85·14
39	16·77	79	33·97	119	51·17	159	68·37	199	85·57
40	17·20	**80**	34·40	**120**	51·60	**160**	68·80	**200**	86·00

$\frac{1}{16}=0{\cdot}02^{7}$ $\frac{1}{9}=0{\cdot}04^{8}$ $\frac{1}{8}=0{\cdot}05^{4}$ $\frac{1}{6}=0{\cdot}07^{2}$ $\frac{1}{5}=0{\cdot}08^{6}$ $\frac{1}{4}=0{\cdot}10^{8}$ $\frac{1}{3}=0{\cdot}14^{3}$

$\frac{3}{8}=0{\cdot}16^{1}$ $\frac{1}{2}=0{\cdot}21^{5}$ $\frac{5}{8}=0{\cdot}26^{9}$ $\frac{3}{4}=0{\cdot}32^{3}$ $\frac{7}{8}=0{\cdot}37^{6}$ $\frac{2}{3}=0{\cdot}28^{7}$ $\frac{5}{6}=0{\cdot}35^{8}$

$\frac{3}{16}=0{\cdot}08^{1}$ $\frac{5}{16}=0{\cdot}13^{4}$ $\frac{7}{16}=0{\cdot}18^{8}$ $\frac{9}{16}=0{\cdot}24^{2}$ $\frac{11}{16}=0{\cdot}29^{6}$ $\frac{13}{16}=0{\cdot}34^{9}$ $\frac{15}{16}=0{\cdot}40^{3}$

(2.30=1£) **£0.43½**

1	0·43½	41	17·83½	81	35·23½	121	52·63½	161	70·03½
2	0·87	42	18·27	82	35·67	122	53·07	162	70·47
3	1·30½	43	18·70½	83	36·10½	123	53·50½	163	70·90½
4	1·74	44	19·14	84	36·54	124	53·94	164	71·34
5	2·17½	45	19·57½	85	36·97½	125	54·37½	165	71·77½
6	2·61	46	20·01	86	37·41	126	54·81	166	72·21
7	3·04½	47	20·44½	87	37·84½	127	55·24½	167	72·64½
8	3·48	48	20·88	88	38·28	128	55·68	168	73·08
9	3·91½	49	21·31½	89	38·71½	129	56·11½	169	73·51½
10	4·35	**50**	21·75	**90**	39·15	**130**	56·55	**170**	73·95
11	4·78½	51	22·18½	91	39·58½	131	56·98½	171	74·38½
12	5·22	52	22·62	92	40·02	132	57·42	172	74·82
13	5·65½	53	23·05½	93	40·45½	133	57·85½	173	75·25½
14	6·09	54	23·49	94	40·89	134	58·29	174	75·69
15	6·52½	55	23·92½	95	41·32½	135	58·72½	175	76·12½
16	6·96	56	24·36	96	41·76	136	59·16	176	76·56
17	7·39½	57	24·79½	97	42·19½	137	59·59½	177	76·99½
18	7·83	58	25·23	98	42·63	138	60·03	178	77·43
19	8·26½	59	25·66½	99	43·06½	139	60·46½	179	77·86½
20	8·70	**60**	26·10	**100**	43·50	**140**	60·90	**180**	78·30
21	9·13½	61	26·53½	101	43·93½	141	61·33½	181	78·73½
22	9·57	62	26·97	102	44·37	142	61·77	182	79·17
23	10·00½	63	27·40½	103	44·80½	143	62·20½	183	79·60½
24	10·44	64	27·84	104	45·24	144	62·64	184	80·04
25	10·87½	65	28·27½	105	45·67½	145	63·07½	185	80·47½
26	11·31	66	28·71	106	46·11	146	63·51	186	80·91
27	11·74½	67	29·14½	107	46·54½	147	63·94½	187	81·34½
28	12·18	68	29·58	108	46·98	148	64·38	188	81·78
29	12·61½	69	30·01½	109	47·41½	149	64·81½	189	82·21½
30	13·05	**70**	30·45	**110**	47·85	**150**	65·25	**190**	82·65
31	13·48½	71	30·88½	111	48·28½	151	65·68½	191	83·08½
32	13·92	72	31·32	112	48·72	152	66·12	192	83·52
33	14·35½	73	31·75½	113	49·15½	153	66·55½	193	83·95½
34	14·79	74	32·19	114	49·59	154	66·99	194	84·39
35	15·22½	75	32·62½	115	50·02½	155	67·42½	195	84·82½
36	15·66	76	33·06	116	50·46	156	67·86	196	85·26
37	16·09½	77	33·49½	117	50·89½	157	68·29½	197	85·69½
38	16·53	78	33·93	118	51·33	158	68·73	198	86·13
39	16·96½	79	34·36½	119	51·76½	159	69·16½	199	86·56½
40	17·40	**80**	34·80	**120**	52·20	**160**	69·60	**200**	87·00

$\frac{1}{16}=0{\cdot}02^{7}$ $\frac{1}{9}=0{\cdot}04^{8}$ $\frac{1}{8}=0{\cdot}05^{4}$ $\frac{1}{6}=0{\cdot}07^{3}$ $\frac{1}{5}=0{\cdot}08^{7}$ $\frac{1}{4}=0{\cdot}10^{9}$ $\frac{1}{3}=0{\cdot}14^{5}$

$\frac{3}{8}=0{\cdot}16^{3}$ $\frac{1}{2}=0{\cdot}21^{8}$ $\frac{5}{8}=0{\cdot}27^{2}$ $\frac{3}{4}=0{\cdot}32^{6}$ $\frac{7}{8}=0{\cdot}38^{1}$ $\frac{2}{3}=0{\cdot}29^{0}$ $\frac{5}{6}=0{\cdot}36^{3}$

$\frac{3}{16}=0{\cdot}08^{2}$ $\frac{5}{16}=0{\cdot}13^{6}$ $\frac{7}{16}=0{\cdot}19^{0}$ $\frac{9}{16}=0{\cdot}24^{5}$ $\frac{11}{16}=0{\cdot}29^{9}$ $\frac{13}{16}=0{\cdot}35^{3}$ $\frac{15}{16}=0{\cdot}40^{8}$

1	0·44	41	18·04	81	35·64	121	53·24	161	70·84
2	0·88	42	18·48	82	36·08	122	53·68	162	71·28
3	1·32	43	18·92	83	36·52	123	54·12	163	71·72
4	1·76	44	19·36	84	36·96	124	54·56	164	72·16
5	2·20	45	19·80	85	37·40	125	55·00	165	72·60
6	2·64	46	20·24	86	37·84	126	55·44	166	73·04
7	3·08	47	20·68	87	38·28	127	55·88	167	73·48
8	3·52	48	21·12	88	38·72	128	56·32	168	73·92
9	3·96	49	21·56	89	39·16	129	56·76	169	74·36
10	4·40	**50**	22·00	**90**	39·60	**130**	57·20	**170**	74·80
11	4·84	51	22·44	91	40·04	131	57·64	171	75·24
12	5·28	52	22·88	92	40·48	132	58·08	172	75·68
13	5·72	53	23·32	93	40·92	133	58·52	173	76·12
14	6·16	54	23·76	94	41·36	134	58·96	174	76·56
15	6·60	55	24·20	95	41·80	135	59·40	175	77·00
16	7·04	56	24·64	96	42·24	136	59·84	176	77·44
17	7·48	57	25·08	97	42·68	137	60·28	177	77·88
18	7·92	58	25·52	98	43·12	138	60·72	178	78·32
19	8·36	59	25·96	99	43·56	139	61·16	179	78·76
20	8·80	**60**	26·40	**100**	44·00	**140**	61·60	**180**	79·20
21	9·24	61	26·84	101	44·44	141	62·04	181	79·64
22	9·68	62	27·28	102	44·88	142	62·48	182	80·08
23	10·12	63	27·72	103	45·32	143	62·92	183	80·52
24	10·56	64	28·16	104	45·76	144	63·36	184	80·96
25	11·00	65	28·60	105	46·20	145	63·80	185	81·40
26	11·44	66	29·04	106	46·64	146	64·24	186	81·84
27	11·88	67	29·48	107	47·08	147	64·68	187	82·28
28	12·32	68	29·92	108	47·52	148	65·12	188	82·72
29	12·76	69	30·36	109	47·96	149	65·56	189	83·16
30	13·20	**70**	30·80	**110**	48·40	**150**	66·00	**190**	83·60
31	13·64	71	31·24	111	48·84	151	66·44	191	84·04
32	14·08	72	31·68	112	49·28	152	66·88	192	84·48
33	14·52	73	32·12	113	49·72	153	67·32	193	84·92
34	14·96	74	32·56	114	50·16	154	67·76	194	85·36
35	15·40	75	33·00	115	50·60	155	68·20	195	85·80
36	15·84	76	33·44	116	51·04	156	68·64	196	86·24
37	16·28	77	33·88	117	51·48	157	69·08	197	86·68
38	16·72	78	34·32	118	51·92	158	69·52	198	87·12
39	17·16	79	34·76	119	52·36	159	69·96	199	87·56
40	17·60	**80**	35·20	**120**	52·80	**160**	70·40	**200**	88·00

$\frac{1}{16}=0{\cdot}02^8$ $\frac{1}{9}=0{\cdot}04^9$ $\frac{1}{8}=0{\cdot}05^5$ $\frac{1}{6}=0{\cdot}07^3$ $\frac{1}{5}=0{\cdot}08^8$ $\frac{1}{4}=0{\cdot}11^0$ $\frac{1}{3}=0{\cdot}14^7$

$\frac{3}{8}=0{\cdot}16^5$ $\frac{1}{2}=0{\cdot}22^0$ $\frac{5}{8}=0{\cdot}27^5$ $\frac{3}{4}=0{\cdot}33^0$ $\frac{7}{8}=0{\cdot}38^5$ $\frac{2}{3}=0{\cdot}29^3$ $\frac{5}{6}=0{\cdot}36^7$

$\frac{3}{16}=0{\cdot}08^3$ $\frac{5}{16}=0{\cdot}13^8$ $\frac{7}{16}=0{\cdot}19^3$ $\frac{9}{16}=0{\cdot}24^8$ $\frac{11}{16}=0{\cdot}30^3$ $\frac{13}{16}=0{\cdot}35^8$ $\frac{15}{16}=0{\cdot}41^3$

1	0·44½	41	18·24½	81	36·04½	121	53·84½	161	71·64½
2	0·89	42	18·69	82	36·49	122	54·29	162	72·09
3	1·33½	43	19·13½	83	36·93½	123	54·73½	163	72·53½
4	1·78	44	19·58	84	37·38	124	55·18	164	72·98
5	2·22½	45	20·02½	85	37·82½	125	55·62½	165	73·42½
6	2·67	46	20·47	86	38·27	126	56·07	166	73·87
7	3·11½	47	20·91½	87	38·71½	127	56·51½	167	74·31½
8	3·56	48	21·36	88	39·16	128	56·96	168	74·76
9	4·00½	49	21·80½	89	39·60½	129	57·40½	169	75·20½
10	4·45	**50**	22·25	**90**	40·05	**130**	57·85	**170**	75·65
11	4·89½	51	22·69½	91	40·49½	131	58·29½	171	76·09½
12	5·34	52	23·14	92	40·94	132	58·74	172	76·54
13	5·78½	53	23·58½	93	41·38½	133	59·18½	173	76·98½
14	6·23	54	24·03	94	41·83	134	59·63	174	77·43
15	6·67½	55	24·47½	95	42·27½	135	60·07½	175	77·87½
16	7·12	56	24·92	96	42·72	136	60·52	176	78·32
17	7·56½	57	25·36½	97	43·16½	137	60·96½	177	78·76½
18	8·01	58	25·81	98	43·61	138	61·41	178	79·21
19	8·45½	59	26·25½	99	44·05½	139	61·85½	179	79·65½
20	8·90	**60**	26·70	**100**	44·50	**140**	62·30	**180**	80·10
21	9·34½	61	27·14½	101	44·94½	141	62·74½	181	80·54½
22	9·79	62	27·59	102	45·39	142	63·19	182	80·99
23	10·23½	63	28·03½	103	45·83½	143	63·63½	183	81·43½
24	10·68	64	28·48	104	46·28	144	64·08	184	81·88
25	11·12½	65	28·92½	105	46·72½	145	64·52½	185	82·32½
26	11·57	66	29·37	106	47·17	146	64·97	186	82·77
27	12·01½	67	29·81½	107	47·61½	147	65·41½	187	83·21½
28	12·46	68	30·26	108	48·06	148	65·86	188	83·66
29	12·90½	69	30·70½	109	48·50½	149	66·30½	189	84·10½
30	13·35	**70**	31·15	**110**	48·95	**150**	66·75	**190**	84·55
31	13·79½	71	31·59½	111	49·39½	151	67·19½	191	84·99½
32	14·24	72	32·04	112	49·84	152	67·64	192	85·44
33	14·68½	73	32·48½	113	50·28½	153	68·08½	193	85·88½
34	15·13	74	32·93	114	50·73	154	68·53	194	86·33
35	15·57½	75	33·37½	115	51·17½	155	68·97½	195	86·77½
36	16·02	76	33·82	116	51·62	156	69·42	196	87·22
37	16·46½	77	34·26½	117	52·06½	157	69·86½	197	87·66½
38	16·91	78	34·71	118	52·51	158	70·31	198	88·11
39	17·35½	79	35·15½	119	52·95½	159	70·75½	199	88·55½
40	17·80	**80**	35·60	**120**	53·40	**160**	71·20	**200**	89·00

$\frac{1}{16}=0{\cdot}02^8$	$\frac{1}{9}=0{\cdot}04^9$	$\frac{1}{8}=0{\cdot}05^6$	$\frac{1}{6}=0{\cdot}07^4$	$\frac{1}{5}=0{\cdot}08^9$	$\frac{1}{4}=0{\cdot}11^1$	$\frac{1}{3}=0{\cdot}14^8$
$\frac{3}{8}=0{\cdot}16^7$	$\frac{1}{2}=0{\cdot}22^3$	$\frac{5}{8}=0{\cdot}27^8$	$\frac{3}{4}=0{\cdot}33^4$	$\frac{7}{8}=0{\cdot}38^9$	$\frac{2}{3}=0{\cdot}29^7$	$\frac{5}{6}=0{\cdot}37^1$
$\frac{3}{16}=0{\cdot}08^3$	$\frac{5}{16}=0{\cdot}13^9$	$\frac{7}{16}=0{\cdot}19^5$	$\frac{9}{16}=0{\cdot}25^0$	$\frac{11}{16}=0{\cdot}30^6$	$\frac{13}{16}=0{\cdot}36^2$	$\frac{15}{16}=0{\cdot}41^7$

(2.22=1£) **£0.45**

1	0·45	41	18·45	81	36·45	121	54·45	161	72·45
2	0·90	42	18·90	82	36·90	122	54·90	162	72·90
3	1·35	43	19·35	83	37·35	123	55·35	163	73·35
4	1·80	44	19·80	84	37·80	124	55·80	164	73·80
5	2·25	45	20·25	85	38·25	125	56·25	165	74·25
6	2·70	46	20·70	86	38·70	126	56·70	166	74·70
7	3·15	47	21·15	87	39·15	127	57·15	167	75·15
8	3·60	48	21·60	88	39·60	128	57·60	168	75·60
9	4·05	49	22·05	89	40·05	129	58·05	169	76·05
10	4·50	**50**	22·50	**90**	40·50	**130**	58·50	**170**	76·50
11	4·95	51	22·95	91	40·95	131	58·95	171	76·95
12	5·40	52	23·40	92	41·40	132	59·40	172	77·40
13	5·85	53	23·85	93	41·85	133	59·85	173	77·85
14	6·30	54	24·30	94	42·30	134	60·30	174	78·30
15	6·75	55	24·75	95	42·75	135	60·75	175	78·75
16	7·20	56	25·20	96	43·20	136	61·20	176	79·20
17	7·65	57	25·65	97	43·65	137	61·65	177	79·65
18	8·10	58	26·10	98	44·10	138	62·10	178	80·10
19	8·55	59	26·55	99	44·55	139	62·55	179	80·55
20	9·00	**60**	27·00	**100**	45·00	**140**	63·00	**180**	81·00
21	9·45	61	27·45	101	45·45	141	63·45	181	81·45
22	9·90	62	27·90	102	45·90	142	63·90	182	81·90
23	10·35	63	28·35	103	46·35	143	64·35	183	82·35
24	10·80	64	28·80	104	46·80	144	64·80	184	82·80
25	11·25	65	29·25	105	47·25	145	65·25	185	83·25
26	11·70	66	29·70	106	47·70	146	65·70	186	83·70
27	12·15	67	30·15	107	48·15	147	66·15	187	84·15
28	12·60	68	30·60	108	48·60	148	66·60	188	84·60
29	13·05	69	31·05	109	49·05	149	67·05	189	85·05
30	13·50	**70**	31·50	**110**	49·50	**150**	67·50	**190**	85·50
31	13·95	71	31·95	111	49·95	151	67·95	191	85·95
32	14·40	72	32·40	112	50·40	152	68·40	192	86·40
33	14·85	73	32·85	113	50·85	153	68·85	193	86·85
34	15·30	74	33·30	114	51·30	154	69·30	194	87·30
35	15·75	75	33·75	115	51·75	155	69·75	195	87·75
36	16·20	76	34·20	116	52·20	156	70·20	196	88·20
37	16·65	77	34·65	117	52·65	157	70·65	197	88·65
38	17·10	78	35·10	118	53·10	158	71·10	198	89·10
39	17·55	79	35·55	119	53·55	159	71·55	199	89·55
40	18·00	**80**	36·00	**120**	54·00	**160**	72·00	**200**	90·00

$\frac{1}{16}=0{\cdot}02^{8}$ $\frac{1}{9}=0{\cdot}05^{0}$ $\frac{1}{8}=0{\cdot}05^{6}$ $\frac{1}{6}=0{\cdot}07^{5}$ $\frac{1}{5}=0{\cdot}09^{0}$ $\frac{1}{4}=0{\cdot}11^{3}$ $\frac{1}{3}=0{\cdot}15^{0}$

$\frac{3}{8}=0{\cdot}16^{9}$ $\frac{1}{2}=0{\cdot}22^{5}$ $\frac{5}{8}=0{\cdot}28^{1}$ $\frac{3}{4}=0{\cdot}33^{8}$ $\frac{7}{8}=0{\cdot}39^{4}$ $\frac{2}{3}=0{\cdot}30^{0}$ $\frac{5}{6}=0{\cdot}37^{5}$

$\frac{3}{16}=0{\cdot}08^{4}$ $\frac{5}{16}=0{\cdot}14^{1}$ $\frac{7}{16}=0{\cdot}19^{7}$ $\frac{9}{16}=0{\cdot}25^{3}$ $\frac{11}{16}=0{\cdot}30^{9}$ $\frac{13}{16}=0{\cdot}36^{6}$ $\frac{15}{16}=0{\cdot}42^{2}$

(2.20=1£) **£0.45½**

1	0·45½	41	18·65½	81	36·85½	121	55·05½	161	73·25½
2	0·91	42	19·11	82	37·31	122	55·51	162	73·71
3	1·36½	43	19·56½	83	37·76½	123	55·96½	163	74·16½
4	1·82	44	20·02	84	38·22	124	56·42	164	74·62
5	2·27½	45	20·47½	85	38·67½	125	56·87½	165	75·07½
6	2·73	46	20·93	86	39·13	126	57·33	166	75·53
7	3·18½	47	21·38½	87	39·58½	127	57·78½	167	75·98½
8	3·64	48	21·84	88	40·04	128	58·24	168	76·44
9	4·09½	49	22·29½	89	40·49½	129	58·69½	169	76·89½
10	4·55	**50**	22·75	**90**	40·95	**130**	59·15	**170**	77·35
11	5·00½	51	23·20½	91	41·40½	131	59·60½	171	77·80½
12	5·46	52	23·66	92	41·86	132	60·06	172	78·26
13	5·91½	53	24·11½	93	42·31½	133	60·51½	173	78·71½
14	6·37	54	24·57	94	42·77	134	60·97	174	79·17
15	6·82½	55	25·02½	95	43·22½	135	61·42½	175	79·62½
16	7·28	56	25·48	96	43·68	136	61·88	176	80·08
17	7·73½	57	25·93½	97	44·13½	137	62·33½	177	80·53½
18	8·19	58	26·39	98	44·59	138	62·79	178	80·99
19	8·64½	59	26·84½	99	45·04½	139	63·24½	179	81·44½
20	9·10	**60**	27·30	**100**	45·50	**140**	63·70	**180**	81·90
21	9·55½	61	27·75½	101	45·95½	141	64·15½	181	82·35½
22	10·01	62	28·21	102	46·41	142	64·61	182	82·81
23	10·46½	63	28·66½	103	46·86½	143	65·06½	183	83·26½
24	10·92	64	29·12	104	47·32	144	65·52	184	83·72
25	11·37½	65	29·57½	105	47·77½	145	65·97½	185	84·17½
26	11·83	66	30·03	106	48·23	146	66·43	186	84·63
27	12·28½	67	30·48½	107	48·68½	147	66·88½	187	85·08½
28	12·74	68	30·94	108	49·14	148	67·34	188	85·54
29	13·19½	69	31·39½	109	49·59½	149	67·79½	189	85·99½
30	13·65	**70**	31·85	**110**	50·05	**150**	68·25	**190**	86·45
31	14·10½	71	32·30½	111	50·50½	151	68·70½	191	86·90½
32	14·56	72	32·76	112	50·96	152	69·16	192	87·36
33	15·01½	73	33·21½	113	51·41½	153	69·61½	193	87·81½
34	15·47	74	33·67	114	51·87	154	70·07	194	88·27
35	15·92½	75	34·12½	115	52·32½	155	70·52½	195	88·72½
36	16·38	76	34·58	116	52·78	156	70·98	196	89·18
37	16·83½	77	35·03½	117	53·23½	157	71·43½	197	89·63½
38	17·29	78	35·49	118	53·69	158	71·89	198	90·09
39	17·74½	79	35·94½	119	54·14½	159	72·34½	199	90·54½
40	18·20	**80**	36·40	**120**	54·60	**160**	72·80	**200**	91·00

$\frac{1}{16}=0{\cdot}02^{8}$	$\frac{1}{9}=0{\cdot}05^{1}$	$\frac{1}{8}=0{\cdot}05^{7}$	$\frac{1}{6}=0{\cdot}07^{6}$	$\frac{1}{5}=0{\cdot}09^{1}$	$\frac{1}{4}=0{\cdot}11^{4}$	$\frac{1}{3}=0{\cdot}15^{2}$
$\frac{3}{8}=0{\cdot}17^{1}$	$\frac{1}{2}=0{\cdot}22^{8}$	$\frac{5}{8}=0{\cdot}28^{4}$	$\frac{3}{4}=0{\cdot}34^{1}$	$\frac{7}{8}=0{\cdot}39^{8}$	$\frac{2}{3}=0{\cdot}30^{3}$	$\frac{5}{6}=0{\cdot}37^{9}$
$\frac{3}{16}=0{\cdot}08^{5}$	$\frac{5}{16}=0{\cdot}14^{2}$	$\frac{7}{16}=0{\cdot}19^{9}$	$\frac{9}{16}=0{\cdot}25^{6}$	$\frac{11}{16}=0{\cdot}31^{3}$	$\frac{13}{16}=0{\cdot}37^{0}$	$\frac{15}{16}=0{\cdot}42^{7}$

1 imperial gallon=4·546 litres 1 lb=454 grammes

(2.17=1£) **£0.46**

1	0·46	41	18·86	81	37·26	121	55·66	161	74·06
2	0·92	42	19·32	82	37·72	122	56·12	162	74·52
3	1·38	43	19·78	83	38·18	123	56·58	163	74·98
4	1·84	44	20·24	84	38·64	124	57·04	164	75·44
5	2·30	45	20·70	85	39·10	125	57·50	165	75·90
6	2·76	46	21·16	86	39·56	126	57·96	166	76·36
7	3·22	47	21·62	87	40·02	127	58·42	167	76·82
8	3·68	48	22·08	88	40·48	128	58·88	168	77·28
9	4·14	49	22·54	89	40·94	129	59·34	169	77·74
10	4·60	**50**	23·00	**90**	41·40	**130**	59·80	**170**	78·20
11	5·06	51	23·46	91	41·86	131	60·26	171	78·66
12	5·52	52	23·92	92	42·32	132	60·72	172	79·12
13	5·98	53	24·38	93	42·78	133	61·18	173	79·58
14	6·44	54	24·84	94	43·24	134	61·64	174	80·04
15	6·90	55	25·30	95	43·70	135	62·10	175	80·50
16	7·36	56	25·76	96	44·16	136	62·56	176	80·96
17	7·82	57	26·22	97	44·62	137	63·02	177	81·42
18	8·28	58	26·68	98	45·08	138	63·48	178	81·88
19	8·74	59	27·14	99	45·54	139	63·94	179	82·34
20	9·20	**60**	27·60	**100**	46·00	**140**	64·40	**180**	82·80
21	9·66	61	28·06	101	46·46	141	64·86	181	83·26
22	10·12	62	28·52	102	46·92	142	65·32	182	83·72
23	10·58	63	28·98	103	47·38	143	65·78	183	84·18
24	11·04	64	29·44	104	47·84	144	66·24	184	84·64
25	11·50	65	29·90	105	48·30	145	66·70	185	85·10
26	11·96	66	30·36	106	48·76	146	67·16	186	85·56
27	12·42	67	30·82	107	49·22	147	67·62	187	86·02
28	12·88	68	31·28	108	49·68	148	68·08	188	86·48
29	13·34	69	31·74	109	50·14	149	68·54	189	86·94
30	13·80	**70**	32·20	**110**	50·60	**150**	69·00	**190**	87·40
31	14·26	71	32·66	111	51·06	151	69·46	191	87·86
32	14·72	72	33·12	112	51·52	152	69·92	192	88·32
33	15·18	73	33·58	113	51·98	153	70·38	193	88·78
34	15·64	74	34·04	114	52·44	154	70·84	194	89·24
35	16·10	75	34·50	115	52·90	155	71·30	195	89·70
36	16·56	76	34·96	116	53·36	156	71·76	196	90·16
37	17·02	77	35·42	117	53·82	157	72·22	197	90·62
38	17·48	78	35·88	118	54·28	158	72·68	198	91·08
39	17·94	79	36·34	119	54·74	159	73·14	199	91·54
40	18·40	**80**	36·80	**120**	55·20	**160**	73·60	**200**	92·00

$\frac{1}{16}=0{\cdot}02^{9}$ $\frac{1}{9}=0{\cdot}05^{1}$ $\frac{1}{8}=0{\cdot}05^{8}$ $\frac{1}{6}=0{\cdot}07^{7}$ $\frac{1}{5}=0{\cdot}09^{2}$ $\frac{1}{4}=0{\cdot}11^{5}$ $\frac{1}{3}=0{\cdot}15^{3}$

$\frac{3}{8}=0{\cdot}17^{3}$ $\frac{1}{2}=0{\cdot}23^{0}$ $\frac{5}{8}=0{\cdot}28^{8}$ $\frac{3}{4}=0{\cdot}34^{5}$ $\frac{7}{8}=0{\cdot}40^{3}$ $\frac{2}{3}=0{\cdot}30^{7}$ $\frac{5}{6}=0{\cdot}38^{3}$

$\frac{3}{16}=0{\cdot}08^{6}$ $\frac{5}{16}=0{\cdot}14^{4}$ $\frac{7}{16}=0{\cdot}20^{1}$ $\frac{9}{16}=0{\cdot}25^{9}$ $\frac{11}{16}=0{\cdot}31^{6}$ $\frac{13}{16}=0{\cdot}37^{4}$ $\frac{15}{16}=0{\cdot}43^{1}$

(2.15=1£) **£0.46½**

1	0·46½	41	19·06½	81	37·66½	121	56·26½	161	74·86½
2	0·93	42	19·53	82	38·13	122	56·73	162	75·33
3	1·39½	43	19·99½	83	38·59½	123	57·19½	163	75·79½
4	1·86	44	20·46	84	39·06	124	57·66	164	76·26
5	2·32½	45	20·92½	85	39·52½	125	58·12½	165	76·72½
6	2·79	46	21·39	86	39·99	126	58·59	166	77·19
7	3·25½	47	21·85½	87	40·45½	127	59·05½	167	77·65½
8	3·72	48	22·32	88	40·92	128	59·52	168	78·12
9	4·18½	49	22·78½	89	41·38½	129	59·98½	169	78·58½
10	4·65	**50**	23·25	**90**	41·85	**130**	60·45	**170**	79·05
11	5·11½	51	23·71½	91	42·31½	131	60·91½	171	79·51½
12	5·58	52	24·18	92	42·78	132	61·38	172	79·98
13	6·04½	53	24·64½	93	43·24½	133	61·84½	173	80·44½
14	6·51	54	25·11	94	43·71	134	62·31	174	80·91
15	6·97½	55	25·57½	95	44·17½	135	62·77½	175	81·37½
16	7·44	56	26·04	96	44·64	136	63·24	176	81·84
17	7·90½	57	26·50½	97	45·10½	137	63·70½	177	82·30½
18	8·37	58	26·97	98	45·57	138	64·17	178	82·77
19	8·83½	59	27·43½	99	46·03½	139	64·63½	179	83·23½
20	9·30	**60**	27·90	**100**	46·50	**140**	65·10	**180**	83·70
21	9·76½	61	28·36½	101	46·96½	141	65·56½	181	84·16½
22	10·23	62	28·83	102	47·43	142	66·03	182	84·63
23	10·69½	63	29·29½	103	47·89½	143	66·49½	183	85·09½
24	11·16	64	29·76	104	48·36	144	66·96	184	85·56
25	11·62½	65	30·22½	105	48·82½	145	67·42½	185	86·02½
26	12·09	66	30·69	106	49·29	146	67·89	186	86·49
27	12·55½	67	31·15½	107	49·75½	147	68·35½	187	86·95½
28	13·02	68	31·62	108	50·22	148	68·82	188	87·42
29	13·48½	69	32·08½	109	50·68½	149	69·28½	189	87·88½
30	13·95	**70**	32·55	**110**	51·15	**150**	69·75	**190**	88·35
31	14·41½	71	33·01½	111	51·61½	151	70·21½	191	88·81½
32	14·88	72	33·48	112	52·08	152	70·68	192	89·28
33	15·34½	73	33·94½	113	52·54½	153	71·14½	193	89·74½
34	15·81	74	34·41	114	53·01	154	71·61	194	90·21
35	16·27½	75	34·87½	115	53·47½	155	72·07½	195	90·67½
36	16·74	76	35·34	116	53·94	156	72·54	196	91·14
37	17·20½	77	35·80½	117	54·40½	157	73·00½	197	91·60½
38	17·67	78	36·27	118	54·87	158	73·47	198	92·07
39	18·13½	79	36·73½	119	55·33½	159	73·93½	199	92·53½
40	18·60	**80**	37·20	**120**	55·80	**160**	74·40	**200**	93·00

$\frac{1}{16}=0{\cdot}02^{9}$	$\frac{1}{9}=0{\cdot}05^{2}$	$\frac{1}{8}=0{\cdot}05^{8}$	$\frac{1}{6}=0{\cdot}07^{8}$	$\frac{1}{5}=0{\cdot}09^{3}$	$\frac{1}{4}=0{\cdot}11^{6}$	$\frac{1}{3}=0{\cdot}15^{5}$
$\frac{3}{8}=0{\cdot}17^{4}$	$\frac{1}{2}=0{\cdot}23^{3}$	$\frac{5}{8}=0{\cdot}29^{1}$	$\frac{3}{4}=0{\cdot}34^{9}$	$\frac{7}{8}=0{\cdot}40^{7}$	$\frac{2}{3}=0{\cdot}31^{0}$	$\frac{5}{6}=0{\cdot}38^{8}$
$\frac{3}{16}=0{\cdot}08^{7}$	$\frac{5}{16}=0{\cdot}14^{5}$	$\frac{7}{16}=0{\cdot}20^{3}$	$\frac{9}{16}=0{\cdot}26^{2}$	$\frac{11}{16}=0{\cdot}32^{0}$	$\frac{13}{16}=0{\cdot}37^{8}$	$\frac{15}{16}=0{\cdot}43^{6}$

1	0·47	41	19·27	81	38·07	121	56·87	161	75·67
2	0·94	42	19·74	82	38·54	122	57·34	162	76·14
3	1·41	43	20·21	83	39·01	123	57·81	163	76·61
4	1·88	44	20·68	84	39·48	124	58·28	164	77·08
5	2·35	45	21·15	85	39·95	125	58·75	165	77·55
6	2·82	46	21·62	86	40·42	126	59·22	166	78·02
7	3·29	47	22·09	87	40·89	127	59·69	167	78·49
8	3·76	48	22·56	88	41·36	128	60·16	168	78·96
9	4·23	49	23·03	89	41·83	129	60·63	169	79·43
10	4·70	**50**	23·50	**90**	42·30	**130**	61·10	**170**	79·90
11	5·17	51	23·97	91	42·77	131	61·57	171	80·37
12	5·64	52	24·44	92	43·24	132	62·04	172	80·84
13	6·11	53	24·91	93	43·71	133	62·51	173	81·31
14	6·58	54	25·38	94	44·18	134	62·98	174	81·78
15	7·05	55	25·85	95	44·65	135	63·45	175	82·25
16	7·52	56	26·32	96	45·12	136	63·92	176	82·72
17	7·99	57	26·79	97	45·59	137	64·39	177	83·19
18	8·46	58	27·26	98	46·06	138	64·86	178	83·66
19	8·93	59	27·73	99	46·53	139	65·33	179	84·13
20	9·40	**60**	28·20	**100**	47·00	**140**	65·80	**180**	84·60
21	9·87	61	28·67	101	47·47	141	66·27	181	85·07
22	10·34	62	29·14	102	47·94	142	66·74	182	85·54
23	10·81	63	29·61	103	48·41	143	67·21	183	86·01
24	11·28	64	30·08	104	48·88	144	67·68	184	86·48
25	11·75	65	30·55	105	49·35	145	68·15	185	86·95
26	12·22	66	31·02	106	49·82	146	68·62	186	87·42
27	12·69	67	31·49	107	50·29	147	69·09	187	87·89
28	13·16	68	31·96	108	50·76	148	69·56	188	88·36
29	13·63	69	32·43	109	51·23	149	70·03	189	88·83
30	14·10	**70**	32·90	**110**	51·70	**150**	70·50	**190**	89·30
31	14·57	71	33·37	111	52·17	151	70·97	191	89·77
32	15·04	72	33·84	112	52·64	152	71·44	192	90·24
33	15·51	73	34·31	113	53·11	153	71·91	193	90·71
34	15·98	74	34·78	114	53·58	154	72·38	194	91·18
35	16·45	75	35·25	115	54·05	155	72·85	195	91·65
36	16·92	76	35·72	116	54·52	156	73·32	196	92·12
37	17·39	77	36·19	117	54·99	157	73·79	197	92·59
38	17·86	78	36·66	118	55·46	158	74·26	198	93·06
39	18·33	79	37·13	119	55·93	159	74·73	199	93·53
40	18·80	**80**	37·60	**120**	56·40	**160**	75·20	**200**	94·00

$\frac{1}{16}=0{\cdot}02^9$ $\frac{1}{9}=0{\cdot}05^2$ $\frac{1}{8}=0{\cdot}05^9$ $\frac{1}{6}=0{\cdot}07^8$ $\frac{1}{5}=0{\cdot}09^4$ $\frac{1}{4}=0{\cdot}11^8$ $\frac{1}{3}=0{\cdot}15^7$

$\frac{3}{8}=0{\cdot}17^6$ $\frac{1}{2}=0{\cdot}23^5$ $\frac{5}{8}=0{\cdot}29^4$ $\frac{3}{4}=0{\cdot}35^3$ $\frac{7}{8}=0{\cdot}41^1$ $\frac{2}{3}=0{\cdot}31^3$ $\frac{5}{6}=0{\cdot}39^2$

$\frac{3}{16}=0{\cdot}08^8$ $\frac{5}{16}=0{\cdot}14^7$ $\frac{7}{16}=0{\cdot}20^6$ $\frac{9}{16}=0{\cdot}26^4$ $\frac{11}{16}=0{\cdot}32^3$ $\frac{13}{16}=0{\cdot}38^2$ $\frac{15}{16}=0{\cdot}44^1$

(2.11=1£) **£0.47½**

1	0·47½	41	19·47½	81	38·47½	121	57·47½	161	76·47½
2	0·95	42	19·95	82	38·95	122	57·95	162	76·95
3	1·42½	43	20·42½	83	39·42½	123	58·42½	163	77·42½
4	1·90	44	20·90	84	39·90	124	58·90	164	77·90
5	2·37½	45	21·37½	85	40·37½	125	59·37½	165	78·37½
6	2·85	46	21·85	86	40·85	126	59·85	166	78·85
7	3·32½	47	22·32½	87	41·32½	127	60·32½	167	79·32½
8	3·80	48	22·80	88	41·80	128	60·80	168	79·80
9	4·27½	49	23·27½	89	42·27½	129	61·27½	169	80·27½
10	4·75	**50**	23·75	**90**	42·75	**130**	61·75	**170**	80·75
11	5·22½	51	24·22½	91	43·22½	131	62·22½	171	81·22½
12	5·70	52	24·70	92	43·70	132	62·70	172	81·70
13	6·17½	53	25·17½	93	44·17½	133	63·17½	173	82·17½
14	6·65	54	25·65	94	44·65	134	63·65	174	82·65
15	7·12½	55	26·12½	95	45·12½	135	64·12½	175	83·12½
16	7·60	56	26·60	96	45·60	136	64·60	176	83·60
17	8·07½	57	27·07½	97	46·07½	137	65·07½	177	84·07½
18	8·55	58	27·55	98	46·55	138	65·55	178	84·55
19	9·02½	59	28·02½	99	47·02½	139	66·02½	179	85·02½
20	9·50	**60**	28·50	**100**	47·50	**140**	66·50	**180**	85·50
21	9·97½	61	28·97½	101	47·97½	141	66·97½	181	85·97½
22	10·45	62	29·45	102	48·45	142	67·45	182	86·45
23	10·92½	63	29·92½	103	48·92½	143	67·92½	183	86·92½
24	11·40	64	30·40	104	49·40	144	68·40	184	87·40
25	11·87½	65	30·87½	105	49·87½	145	68·87½	185	87·87½
26	12·35	66	31·35	106	50·35	146	69·35	186	88·35
27	12·82½	67	31·82½	107	50·82½	147	69·82½	187	88·82½
28	13·30	68	32·30	108	51·30	148	70·30	188	89·30
29	13·77½	69	32·77½	109	51·77½	149	70·77½	189	89·77½
30	14·25	**70**	33·25	**110**	52·25	**150**	71·25	**190**	90·25
31	14·72½	71	33·72½	111	52·72½	151	71·72½	191	90·72½
32	15·20	72	34·20	112	53·20	152	72·20	192	91·20
33	15·67½	73	34·67½	113	53·67½	153	72·67½	193	91·67½
34	16·15	74	35·15	114	54·15	154	73·15	194	92·15
35	16·62½	75	35·62½	115	54·62½	155	73·62½	195	92·62½
36	17·10	76	36·10	116	55·10	156	74·10	196	93·10
37	17·57½	77	36·57½	117	55·57½	157	74·57½	197	93·57½
38	18·05	78	37·05	118	56·05	158	75·05	198	94·05
39	18·52½	79	37·52½	119	56·52½	159	75·52½	199	94·52½
40	19·00	**80**	38·00	**120**	57·00	**160**	76·00	**200**	95·00

$\frac{1}{16}=0{\cdot}03^{0}$ $\frac{1}{9}=0{\cdot}05^{3}$ $\frac{1}{8}=0{\cdot}05^{9}$ $\frac{1}{6}=0{\cdot}07^{9}$ $\frac{1}{5}=0{\cdot}09^{5}$ $\frac{1}{4}=0{\cdot}11^{9}$ $\frac{1}{3}=0{\cdot}15^{8}$

$\frac{3}{8}=0{\cdot}17^{8}$ $\frac{1}{2}=0{\cdot}23^{8}$ $\frac{5}{8}=0{\cdot}29^{7}$ $\frac{3}{4}=0{\cdot}35^{6}$ $\frac{7}{8}=0{\cdot}41^{6}$ $\frac{2}{3}=0{\cdot}31^{7}$ $\frac{5}{6}=0{\cdot}39^{6}$

$\frac{3}{16}=0{\cdot}08^{9}$ $\frac{5}{16}=0{\cdot}14^{8}$ $\frac{7}{16}=0{\cdot}20^{8}$ $\frac{9}{16}=0{\cdot}26^{7}$ $\frac{11}{16}=0{\cdot}32^{7}$ $\frac{13}{16}=0{\cdot}38^{6}$ $\frac{15}{16}=0{\cdot}44^{5}$

1	0·48	41	19·68	81	38·88	121	58·08	161	77·28
2	0·96	42	20·16	82	39·36	122	58·56	162	77·76
3	1·44	43	20·64	83	39·84	123	59·04	163	78·24
4	1·92	44	21·12	84	40·32	124	59·52	164	78·72
5	2·40	45	21·60	85	40·80	125	60·00	165	79·20
6	2·88	46	22·08	86	41·28	126	60·48	166	79·68
7	3·36	47	22·56	87	41·76	127	60·96	167	80·16
8	3·84	48	23·04	88	42·24	128	61·44	168	80·64
9	4·32	49	23·52	89	42·72	129	61·92	169	81·12
10	4·80	**50**	24·00	**90**	43·20	**130**	62·40	**170**	81·60
11	5·28	51	24·48	91	43·68	131	62·88	171	82·08
12	5·76	52	24·96	92	44·16	132	63·36	172	82·56
13	6·24	53	25·44	93	44·64	133	63·84	173	83·04
14	6·72	54	25·92	94	45·12	134	64·32	174	83·52
15	7·20	55	26·40	95	45·60	135	64·80	175	84·00
16	7·68	56	26·88	96	46·08	136	65·28	176	84·48
17	8·16	57	27·36	97	46·56	137	65·76	177	84·96
18	8·64	58	27·84	98	47·04	138	66·24	178	85·44
19	9·12	59	28·32	99	47·52	139	66·72	179	85·92
20	9·60	**60**	28·80	**100**	48·00	**140**	67·20	**180**	86·40
21	10·08	61	29·28	101	48·48	141	67·68	181	86·88
22	10·56	62	29·76	102	48·96	142	68·16	182	87·36
23	11·04	63	30·24	103	49·44	143	68·64	183	87·84
24	11·52	64	30·72	104	49·92	144	69·12	184	88·32
25	12·00	65	31·20	105	50·40	145	69·60	185	88·80
26	12·48	66	31·68	106	50·88	146	70·08	186	89·28
27	12·96	67	32·16	107	51·36	147	70·56	187	89·76
28	13·44	68	32·64	108	51·84	148	71·04	188	90·24
29	13·92	69	33·12	109	52·32	149	71·52	189	90·72
30	14·40	**70**	33·60	**110**	52·80	**150**	72·00	**190**	91·20
31	14·88	71	34·08	111	53·28	151	72·48	191	91·68
32	15·36	72	34·56	112	53·76	152	72·96	192	92·16
33	15·84	73	35·04	113	54·24	153	73·44	193	92·64
34	16·32	74	35·52	114	54·72	154	73·92	194	93·12
35	16·80	75	36·00	115	55·20	155	74·40	195	93·60
36	17·28	76	36·48	116	55·68	156	74·88	196	94·08
37	17·76	77	36·96	117	56·16	157	75·36	197	94·56
38	18·24	78	37·44	118	56·64	158	75·84	198	95·04
39	18·72	79	37·92	119	57·12	159	76·32	199	95·52
40	19·20	**80**	38·40	**120**	57·60	**160**	76·80	**200**	96·00

$\frac{1}{16}=0{\cdot}03^{0}$ $\frac{1}{9}=0{\cdot}05^{3}$ $\frac{1}{8}=0{\cdot}06^{0}$ $\frac{1}{6}=0{\cdot}08^{0}$ $\frac{1}{5}=0{\cdot}09^{6}$ $\frac{1}{4}=0{\cdot}12^{0}$ $\frac{1}{3}=0{\cdot}16^{0}$

$\frac{3}{8}=0{\cdot}18^{0}$ $\frac{1}{2}=0{\cdot}24^{0}$ $\frac{5}{8}=0{\cdot}30^{0}$ $\frac{3}{4}=0{\cdot}36^{0}$ $\frac{7}{8}=0{\cdot}42^{0}$ $\frac{2}{3}=0{\cdot}32^{0}$ $\frac{5}{6}=0{\cdot}40^{0}$

$\frac{3}{16}=0{\cdot}09^{0}$ $\frac{5}{16}=0{\cdot}15^{0}$ $\frac{7}{16}=0{\cdot}21^{0}$ $\frac{9}{16}=0{\cdot}27^{0}$ $\frac{11}{16}=0{\cdot}33^{0}$ $\frac{13}{16}=0{\cdot}39^{0}$ $\frac{15}{16}=0{\cdot}45^{0}$

(2.06=1£) **£0.48½**

1	0·48½	41	19·88½	81	39·28½	121	58·68½	161	78·08½
2	0·97	42	20·37	82	39·77	122	59·17	162	78·57
3	1·45½	43	20·85½	83	40·25½	123	59·65½	163	79·05½
4	1·94	44	21·34	84	40·74	124	60·14	164	79·54
5	2·42½	45	21·82½	85	41·22½	125	60·62½	165	80·02½
6	2·91	46	22·31	86	41·71	126	61·11	166	80·51
7	3·39½	47	22·79½	87	42·19½	127	61·59½	167	80·99½
8	3·88	48	23·28	88	42·68	128	62·08	168	81·48
9	4·36½	49	23·76½	89	43·16½	129	62·56½	169	81·96½
10	4·85	**50**	24·25	**90**	43·65	**130**	63·05	**170**	82·45
11	5·33½	51	24·73½	91	44·13½	131	63·53½	171	82·93½
12	5·82	52	25·22	92	44·62	132	64·02	172	83·42
13	6·30½	53	25·70½	93	45·10½	133	64·50½	173	83·90½
14	6·79	54	26·19	94	45·59	134	64·99	174	84·39
15	7·27½	55	26·67½	95	46·07½	135	65·47½	175	84·87½
16	7·76	56	27·16	96	46·56	136	65·96	176	85·36
17	8·24½	57	27·64½	97	47·04½	137	66·44½	177	85·84½
18	8·73	58	28·13	98	47·53	138	66·93	178	86·33
19	9·21½	59	28·61½	99	48·01½	139	67·41½	179	86·81½
20	9·70	**60**	29·10	**100**	48·50	**140**	67·90	**180**	87·30
21	10·18½	61	29·58½	101	48·98½	141	68·38½	181	87·78½
22	10·67	62	30·07	102	49·47	142	68·87	182	88·27
23	11·15½	63	30·55½	103	49·95½	143	69·35½	183	88·75½
24	11·64	64	31·04	104	50·44	144	69·84	184	89·24
25	12·12½	65	31·52½	105	50·92½	145	70·32½	185	89·72½
26	12·61	66	32·01	106	51·41	146	70·81	186	90·21
27	13·09½	67	32·49½	107	51·89½	147	71·29½	187	90·69½
28	13·58	68	32·98	108	52·38	148	71·78	188	91·18
29	14·06½	69	33·46½	109	52·86½	149	72·26½	189	91·66½
30	14·55	**70**	33·95	**110**	53·35	**150**	72·75	**190**	92·15
31	15·03½	71	34·43½	111	53·83½	151	73·23½	191	92·63½
32	15·52	72	34·92	112	54·32	152	73·72	192	93·12
33	16·00½	73	35·40½	113	54·80½	153	74·20½	193	93·60½
34	16·49	74	35·89	114	55·29	154	74·69	194	94·09
35	16·97½	75	36·37½	115	55·77½	155	75·17½	195	94·57½
36	17·46	76	36·86	116	56·26	156	75·66	196	95·06
37	17·94½	77	37·34½	117	56·74½	157	76·14½	197	95·54½
38	18·43	78	37·83	118	57·23	158	76·63	198	96·03
39	18·91½	79	38·31½	119	57·71½	159	77·11½	199	96·51½
40	19·40	**80**	38·80	**120**	58·20	**160**	77·60	**200**	97·00

$\frac{1}{16}=0{\cdot}03^{0}$ $\frac{1}{9}=0{\cdot}05^{4}$ $\frac{1}{8}=0{\cdot}06^{1}$ $\frac{1}{6}=0{\cdot}08^{1}$ $\frac{1}{5}=0{\cdot}09^{7}$ $\frac{1}{4}=0{\cdot}12^{1}$ $\frac{1}{3}=0{\cdot}16^{2}$

$\frac{3}{8}=0{\cdot}18^{2}$ $\frac{1}{2}=0{\cdot}24^{3}$ $\frac{5}{8}=0{\cdot}30^{3}$ $\frac{3}{4}=0{\cdot}36^{4}$ $\frac{7}{8}=0{\cdot}42^{4}$ $\frac{2}{3}=0{\cdot}32^{3}$ $\frac{5}{6}=0{\cdot}40^{4}$

$\frac{3}{16}=0{\cdot}09^{1}$ $\frac{5}{16}=0{\cdot}15^{2}$ $\frac{7}{16}=0{\cdot}21^{2}$ $\frac{9}{16}=0{\cdot}27^{3}$ $\frac{11}{16}=0{\cdot}33^{3}$ $\frac{13}{16}=0{\cdot}39^{4}$ $\frac{15}{16}=0{\cdot}45^{5}$

(2.04=1£) **£0.49**

1	0·49	41	20·09	81	39·69	121	59·29	161	78·89
2	0·98	42	20·58	82	40·18	122	59·78	162	79·38
3	1·47	43	21·07	83	40·67	123	60·27	163	79·87
4	1·96	44	21·56	84	41·16	124	60·76	164	80·36
5	2·45	45	22·05	85	41·65	125	61·25	165	80·85
6	2·94	46	22·54	86	42·14	126	61·74	166	81·34
7	3·43	47	23·03	87	42·63	127	62·23	167	81·83
8	3·92	48	23·52	88	43·12	128	62·72	168	82·32
9	4·41	49	24·01	89	43·61	129	63·21	169	82·81
10	4·90	**50**	24·50	**90**	44·10	**130**	63·70	**170**	83·30
11	5·39	51	24·99	91	44·59	131	64·19	171	83·79
12	5·88	52	25·48	92	45·08	132	64·68	172	84·28
13	6·37	53	25·97	93	45·57	133	65·17	173	84·77
14	6·86	54	26·46	94	46·06	134	65·66	174	85·26
15	7·35	55	26·95	95	46·55	135	66·15	175	85·75
16	7·84	56	27·44	96	47·04	136	66·64	176	86·24
17	8·33	57	27·93	97	47·53	137	67·13	177	86·73
18	8·82	58	28·42	98	48·02	138	67·62	178	87·22
19	9·31	59	28·91	99	48·51	139	68·11	179	87·71
20	9·80	**60**	29·40	**100**	49·00	**140**	68·60	**180**	88·20
21	10·29	61	29·89	101	49·49	141	69·09	181	88·69
22	10·78	62	30·38	102	49·98	142	69·58	182	89·18
23	11·27	63	30·87	103	50·47	143	70·07	183	89·67
24	11·76	64	31·36	104	50·96	144	70·56	184	90·16
25	12·25	65	31·85	105	51·45	145	71·05	185	90·65
26	12·74	66	32·34	106	51·94	146	71·54	186	91·14
27	13·23	67	32·83	107	52·43	147	72·03	187	91·63
28	13·72	68	33·32	108	52·92	148	72·52	188	92·12
29	14·21	69	33·81	109	53·41	149	73·01	189	92·61
30	14·70	**70**	34·30	**110**	53·90	**150**	73·50	**190**	93·10
31	15·19	71	34·79	111	54·39	151	73·99	191	93·59
32	15·68	72	35·28	112	54·88	152	74·48	192	94·08
33	16·17	73	35·77	113	55·37	153	74·97	193	94·57
34	16·66	74	36·26	114	55·86	154	75·46	194	95·06
35	17·15	75	36·75	115	56·35	155	75·95	195	95·55
36	17·64	76	37·24	116	56·84	156	76·44	196	96·04
37	18·13	77	37·73	117	57·33	157	76·93	197	96·53
38	18·62	78	38·22	118	57·82	158	77·42	198	97·02
39	19·11	79	38·71	119	58·31	159	77·91	199	97·51
40	19·60	**80**	39·20	**120**	58·80	**160**	78·40	**200**	98·00

$\frac{1}{16}$=0·03^{1} $\frac{1}{9}$=0·05^{4} $\frac{1}{8}$=0·06^{1} $\frac{1}{6}$=0·08^{2} $\frac{1}{5}$=0·09^{8} $\frac{1}{4}$=0·12^{3} $\frac{1}{3}$=0·16^{3}

$\frac{3}{8}$=0·18^{4} $\frac{1}{2}$=0·24^{5} $\frac{5}{8}$=0·30^{6} $\frac{3}{4}$=0·36^{8} $\frac{7}{8}$=0·42^{9} $\frac{2}{3}$=0·32^{7} $\frac{5}{6}$=0·40^{8}

$\frac{3}{16}$=0·09^{2} $\frac{5}{16}$=0·15^{3} $\frac{7}{16}$=0·21^{4} $\frac{9}{16}$=0·27^{6} $\frac{11}{16}$=0·33^{7} $\frac{13}{16}$=0·39^{8} $\frac{15}{16}$=0·45^{9}

(2.02=1£) **£0.49½**

1	0·49½	41	20·29½	81	40·09½	121	59·89½	161	79·69½
2	0·99	42	20·79	82	40·59	122	60·39	162	80·19
3	1·48½	43	21·28½	83	41·08½	123	60·88½	163	80·68½
4	1·98	44	21·78	84	41·58	124	61·38	164	81·18
5	2·47½	45	22·27½	85	42·07½	125	61·87½	165	81·67½
6	2·97	46	22·77	86	42·57	126	62·37	166	82·17
7	3·46½	47	23·26½	87	43·06½	127	62·86½	167	82·66½
8	3·96	48	23·76	88	43·56	128	63·36	168	83·16
9	4·45½	49	24·25½	89	44·05½	129	63·85½	169	83·65½
10	4·95	**50**	24·75	**90**	44·55	**130**	64·35	**170**	84·15
11	5·44½	51	25·24½	91	45·04½	131	64·84½	171	84·64½
12	5·94	52	25·74	92	45·54	132	65·34	172	85·14
13	6·43½	53	26·23½	93	46·03½	133	65·83½	173	85·63½
14	6·93	54	26·73	94	46·53	134	66·33	174	86·13
15	7·42½	55	27·22½	95	47·02½	135	66·82½	175	86·62½
16	7·92	56	27·72	96	47·52	136	67·32	176	87·12
17	8·41½	57	28·21½	97	48·01½	137	67·81½	177	87·61½
18	8·91	58	28·71	98	48·51	138	68·31	178	88·11
19	9·40½	59	29·20½	99	49·00½	139	68·80½	179	88·60½
20	9·90	**60**	29·70	**100**	49·50	**140**	69·30	**180**	89·10
21	10·39½	61	30·19½	101	49·99½	141	69·79½	181	89·59½
22	10·89	62	30·69	102	50·49	142	70·29	182	90·09
23	11·38½	63	31·18½	103	50·98½	143	70·78½	183	90·58½
24	11·88	64	31·68	104	51·48	144	71·28	184	91·08
25	12·37½	65	32·17½	105	51·97½	145	71·77½	185	91·57½
26	12·87	66	32·67	106	52·47	146	72·27	186	92·07
27	13·36½	67	33·16½	107	52·96½	147	72·76½	187	92·56½
28	13·86	68	33·66	108	53·46	148	73·26	188	93·06
29	14·35½	69	34·15½	109	53·95½	149	73·75½	189	93·55½
30	14·85	**70**	34·65	**110**	54·45	**150**	74·25	**190**	94·05
31	15·34½	71	35·14½	111	54·94½	151	74·74½	191	94·54½
32	15·84	72	35·64	112	55·44	152	75·24	192	95·04
33	16·33½	73	36·13½	113	55·93½	153	75·73½	193	95·53½
34	16·83	74	36·63	114	56·43	154	76·23	194	96·03
35	17·32½	75	37·12½	115	56·92½	155	76·72½	195	96·52½
36	17·82	76	37·62	116	57·42	156	77·22	196	97·02
37	18·31½	77	38·11½	117	57·91½	157	77·71½	197	97·51½
38	18·81	78	38·61	118	58·41	158	78·21	198	98·01
39	19·30½	79	39·10½	119	58·90½	159	78·70½	199	98·50½
40	19·80	**80**	39·60	**120**	59·40	**160**	79·20	**200**	99·00

$\frac{1}{16}$=0·03^{1}	$\frac{1}{9}$=0·05^{5}	$\frac{1}{8}$=0·06^{2}	$\frac{1}{6}$=0·08^{3}	$\frac{1}{5}$=0·09^{9}	$\frac{1}{4}$=0·12^{4}	$\frac{1}{3}$=0·16^{5}
$\frac{3}{8}$=0·18^{6}	$\frac{1}{2}$=0·24^{8}	$\frac{5}{8}$=0·30^{9}	$\frac{3}{4}$=0·37^{1}	$\frac{7}{8}$=0·43^{3}	$\frac{2}{3}$=0·33^{0}	$\frac{5}{6}$=0·41^{3}
$\frac{3}{16}$=0·09^{3}	$\frac{5}{16}$=0·15^{5}	$\frac{7}{16}$=0·21^{7}	$\frac{9}{16}$=0·27^{8}	$\frac{11}{16}$=0·34^{0}	$\frac{13}{16}$=0·40^{2}	$\frac{15}{16}$=0·46^{4}

(2.00=1£) **£0.50**

1	0·50	41	20·50	81	40·50	121	60·50	161	80·50
2	1·00	42	21·00	82	41·00	122	61·00	162	81·00
3	1·50	43	21·50	83	41·50	123	61·50	163	81·50
4	2·00	44	22·00	84	42·00	124	62·00	164	82·00
5	2·50	45	22·50	85	42·50	125	62·50	165	82·50
6	3·00	46	23·00	86	43·00	126	63·00	166	83·00
7	3·50	47	23·50	87	43·50	127	63·50	167	83·50
8	4·00	48	24·00	88	44·00	128	64·00	168	84·00
9	4·50	49	24·50	89	44·50	129	64·50	169	84·50
10	5·00	**50**	25·00	**90**	45·00	**130**	65·00	**170**	85·00
11	5·50	51	25·50	91	45·50	131	65·50	171	85·50
12	6·00	52	26·00	92	46·00	132	66·00	172	86·00
13	6·50	53	26·50	93	46·50	133	66·50	173	86·50
14	7·00	54	27·00	94	47·00	134	67·00	174	87·00
15	7·50	55	27·50	95	47·50	135	67·50	175	87·50
16	8·00	56	28·00	96	48·00	136	68·00	176	88·00
17	8·50	57	28·50	97	48·50	137	68·50	177	88·50
18	9·00	58	29·00	98	49·00	138	69·00	178	89·00
19	9·50	59	29·50	99	49·50	139	69·50	179	89·50
20	10·00	**60**	30·00	**100**	50·00	**140**	70·00	**180**	90·00
21	10·50	61	30·50	101	50·50	141	70·50	181	90·50
22	11·00	62	31·00	102	51·00	142	71·00	182	91·00
23	11·50	63	31·50	103	51·50	143	71·50	183	91·50
24	12·00	64	32·00	104	52·00	144	72·00	184	92·00
25	12·50	65	32·50	105	52·50	145	72·50	185	92·50
26	13·00	66	33·00	106	53·00	146	73·00	186	93·00
27	13·50	67	33·50	107	53·50	147	73·50	187	93·50
28	14·00	68	34·00	108	54·00	148	74·00	188	94·00
29	14·50	69	34·50	109	54·50	149	74·50	189	94·50
30	15·00	**70**	35·00	**110**	55·00	**150**	75·00	**190**	95·00
31	15·50	71	35·50	111	55·50	151	75·50	191	95·50
32	16·00	72	36·00	112	56·00	152	76·00	192	96·00
33	16·50	73	36·50	113	56·50	153	76·50	193	96·50
34	17·00	74	37·00	114	57·00	154	77·00	194	97·00
35	17·50	75	37·50	115	57·50	155	77·50	195	97·50
36	18·00	76	38·00	116	58·00	156	78·00	196	98·00
37	18·50	77	38·50	117	58·50	157	78·50	197	98·50
38	19·00	78	39·00	118	59·00	158	79·00	198	99·00
39	19·50	79	39·50	119	59·50	159	79·50	199	99·50
40	20·00	**80**	40·00	**120**	60·00	**160**	80·00	**200**	100·00

$\frac{1}{16}=0{\cdot}03^{1}$ $\frac{1}{9}=0{\cdot}05^{6}$ $\frac{1}{8}=0{\cdot}06^{3}$ $\frac{1}{6}=0{\cdot}08^{3}$ $\frac{1}{5}=0{\cdot}10^{0}$ $\frac{1}{4}=0{\cdot}12^{5}$ $\frac{1}{3}=0{\cdot}16^{7}$

$\frac{3}{8}=0{\cdot}18^{8}$ $\frac{1}{2}=0{\cdot}25^{0}$ $\frac{5}{8}=0{\cdot}31^{3}$ $\frac{3}{4}=0{\cdot}37^{5}$ $\frac{7}{8}=0{\cdot}43^{8}$ $\frac{2}{3}=0{\cdot}33^{3}$ $\frac{5}{6}=0{\cdot}41^{7}$

$\frac{3}{16}=0{\cdot}09^{4}$ $\frac{5}{16}=0{\cdot}15^{6}$ $\frac{7}{16}=0{\cdot}21^{9}$ $\frac{9}{16}=0{\cdot}28^{1}$ $\frac{11}{16}=0{\cdot}34^{4}$ $\frac{13}{16}=0{\cdot}40^{6}$ $\frac{15}{16}=0{\cdot}46^{9}$

(1.98=1£) **£0.50½**

1	0·50½	41	20·70½	81	40·90½	121	61·10½	161	81·30½
2	1·01	42	21·21	82	41·41	122	61·61	162	81·81
3	1·51½	43	21·71½	83	41·91½	123	62·11½	163	82·31½
4	2·02	44	22·22	84	42·42	124	62·62	164	82·82
5	2·52½	45	22·72½	85	42·92½	125	63·12½	165	83·32½
6	3·03	46	23·23	86	43·43	126	63·63	166	83·83
7	3·53½	47	23·73½	87	43·93½	127	64·13½	167	84·33½
8	4·04	48	24·24	88	44·44	128	64·64	168	84·84
9	4·54½	49	24·74½	89	44·94½	129	65·14½	169	85·34½
10	5·05	**50**	25·25	**90**	45·45	**130**	65·65	**170**	85·85
11	5·55½	51	25·75½	91	45·95½	131	66·15½	171	86·35½
12	6·06	52	26·26	92	46·46	132	66·66	172	86·86
13	6·56½	53	26·76½	93	46·96½	133	67·16½	173	87·36½
14	7·07	54	27·27	94	47·47	134	67·67	174	87·87
15	7·57½	55	27·77½	95	47·97½	135	68·17½	175	88·37½
16	8·08	56	28·28	96	48·48	136	68·68	176	88·88
17	8·58½	57	28·78½	97	48·98½	137	69·18½	177	89·38½
18	9·09	58	29·29	98	49·49	138	69·69	178	89·89
19	9·59½	59	29·79½	99	49·99½	139	70·19½	179	90·39½
20	10·10	**60**	30·30	**100**	50·50	**140**	70·70	**180**	90·90
21	10·60½	61	30·80½	101	51·00½	141	71·20½	181	91·40½
22	11·11	62	31·31	102	51·51	142	71·71	182	91·91
23	11·61½	63	31·81½	103	52·01½	143	72·21½	183	92·41½
24	12·12	64	32·32	104	52·52	144	72·72	184	92·92
25	12·62½	65	32·82½	105	53·02½	145	73·22½	185	93·42½
26	13·13	66	33·33	106	53·53	146	73·73	186	93·93
27	13·63½	67	33·83½	107	54·03½	147	74·23½	187	94·43½
28	14·14	68	34·34	108	54·54	148	74·74	188	94·94
29	14·64½	69	34·84½	109	55·04½	149	75·24½	189	95·44½
30	15·15	**70**	35·35	**110**	55·55	**150**	75·75	**190**	95·95
31	15·65½	71	35·85½	111	56·05½	151	76·25½	191	96·45½
32	16·16	72	36·36	112	56·56	152	76·76	192	96·96
33	16·66½	73	36·86½	113	57·06½	153	77·26½	193	97·46½
34	17·17	74	37·37	114	57·57	154	77·77	194	97·97
35	17·67½	75	37·87½	115	58·07½	155	78·27½	195	98·47½
36	18·18	76	38·38	116	58·58	156	78·78	196	98·98
37	18·68½	77	38·88½	117	59·08½	157	79·28½	197	99·48½
38	19·19	78	39·39	118	59·59	158	79·79	198	99·99
39	19·69½	79	39·89½	119	60·09½	159	80·29½	199	100·49½
40	20·20	**80**	40·40	**120**	60·60	**160**	80·80	**200**	101·00

$\frac{1}{16}=0{\cdot}03^2$	$\frac{1}{9}=0{\cdot}05^6$	$\frac{1}{8}=0{\cdot}06^3$	$\frac{1}{6}=0{\cdot}08^4$	$\frac{1}{5}=0{\cdot}10^1$	$\frac{1}{4}=0{\cdot}12^6$	$\frac{1}{3}=0{\cdot}16^8$
$\frac{3}{8}=0{\cdot}18^9$	$\frac{1}{2}=0{\cdot}25^3$	$\frac{5}{8}=0{\cdot}31^6$	$\frac{3}{4}=0{\cdot}37^9$	$\frac{7}{8}=0{\cdot}44^2$	$\frac{2}{3}=0{\cdot}33^7$	$\frac{5}{6}=0{\cdot}42^1$
$\frac{3}{16}=0{\cdot}09^5$	$\frac{5}{16}=0{\cdot}15^8$	$\frac{7}{16}=0{\cdot}22^1$	$\frac{9}{16}=0{\cdot}28^4$	$\frac{11}{16}=0{\cdot}34^7$	$\frac{13}{16}=0{\cdot}41^0$	$\frac{15}{16}=0{\cdot}47^3$

(1.96=1£) **£0.51**

1	0·51	41	20·91	81	41·31	121	61·71	161	82·11
2	1·02	42	21·42	82	41·82	122	62·22	162	82·62
3	1·53	43	21·93	83	42·33	123	62·73	163	83·13
4	2·04	44	22·44	84	42·84	124	63·24	164	83·64
5	2·55	45	22·95	85	43·35	125	63·75	165	84·15
6	3·06	46	23·46	86	43·86	126	64·26	166	84·66
7	3·57	47	23·97	87	44·37	127	64·77	167	85·17
8	4·08	48	24·48	88	44·88	128	65·28	168	85·68
9	4·59	49	24·99	89	45·39	129	65·79	169	86·19
10	5·10	**50**	25·50	**90**	45·90	**130**	66·30	**170**	86·70
11	5·61	51	26·01	91	46·41	131	66·81	171	87·21
12	6·12	52	26·52	92	46·92	132	67·32	172	87·72
13	6·63	53	27·03	93	47·43	133	67·83	173	88·23
14	7·14	54	27·54	94	47·94	134	68·34	174	88·74
15	7·65	55	28·05	95	48·45	135	68·85	175	89·25
16	8·16	56	28·56	96	48·96	136	69·36	176	89·76
17	8·67	57	29·07	97	49·47	137	69·87	177	90·27
18	9·18	58	29·58	98	49·98	138	70·38	178	90·78
19	9·69	59	30·09	99	50·49	139	70·89	179	91·29
20	10·20	**60**	30·60	**100**	51·00	**140**	71·40	**180**	91·80
21	10·71	61	31·11	101	51·51	141	71·91	181	92·31
22	11·22	62	31·62	102	52·02	142	72·42	182	92·82
23	11·73	63	32·13	103	52·53	143	72·93	183	93·33
24	12·24	64	32·64	104	53·04	144	73·44	184	93·84
25	12·75	65	33·15	105	53·55	145	73·95	185	94·35
26	13·26	66	33·66	106	54·06	146	74·46	186	94·86
27	13·77	67	34·17	107	54·57	147	74·97	187	95·37
28	14·28	68	34·68	108	55·08	148	75·48	188	95·88
29	14·79	69	35·19	109	55·59	149	75·99	189	96·39
30	15·30	**70**	35·70	**110**	56·10	**150**	76·50	**190**	96·90
31	15·81	71	36·21	111	56·61	151	77·01	191	97·41
32	16·32	72	36·72	112	57·12	152	77·52	192	97·92
33	16·83	73	37·23	113	57·63	153	78·03	193	98·43
34	17·34	74	37·74	114	58·14	154	78·54	194	98·94
35	17·85	75	38·25	115	58·65	155	79·05	195	99·45
36	18·36	76	38·76	116	59·16	156	79·56	196	99·96
37	18·87	77	39·27	117	59·67	157	80·07	197	100·47
38	19·38	78	39·78	118	60·18	158	80·58	198	100·98
39	19·89	79	40·29	119	60·69	159	81·09	199	101·49
40	20·40	**80**	40·80	**120**	61·20	**160**	81·60	**200**	102·00

$\frac{1}{16}$=0·03^{2} $\frac{1}{9}$=0·05^{7} $\frac{1}{8}$=0·06^{4} $\frac{1}{6}$=0·08^{5} $\frac{1}{5}$=0·10^{2} $\frac{1}{4}$=0·12^{8} $\frac{1}{3}$=0·17^{0}

$\frac{3}{8}$=0·19^{1} $\frac{1}{2}$=0·25^{5} $\frac{5}{8}$=0·31^{9} $\frac{3}{4}$=0·38^{3} $\frac{7}{8}$=0·44^{6} $\frac{2}{3}$=0·34^{0} $\frac{5}{6}$=0·42^{5}

$\frac{3}{16}$=0·09^{6} $\frac{5}{16}$=0·15^{9} $\frac{7}{16}$=0·22^{3} $\frac{9}{16}$=0·28^{7} $\frac{11}{16}$=0·35^{1} $\frac{13}{16}$=0·41^{4} $\frac{15}{16}$=0·47^{8}

(1.94=1£) £0.51½

1	0·51½	41	21·11½	81	41·71½	121	62·31½	161	82·91½
2	1·03	42	21·63	82	42·23	122	62·83	162	83·43
3	1·54½	43	22·14½	83	42·74½	123	63·34½	163	83·94½
4	2·06	44	22·66	84	43·26	124	63·86	164	84·46
5	2·57½	45	23·17½	85	43·77½	125	64·37½	165	84·97½
6	3·09	46	23·69	86	44·29	126	64·89	166	85·49
7	3·60½	47	24·20½	87	44·80½	127	65·40½	167	86·00½
8	4·12	48	24·72	88	45·32	128	65·92	168	86·52
9	4·63½	49	25·23½	89	45·83½	129	66·43½	169	87·03½
10	5·15	**50**	25·75	**90**	46·35	**130**	66·95	**170**	87·55
11	5·66½	51	26·26½	91	46·86½	131	67·46½	171	88·06½
12	6·18	52	26·78	92	47·38	132	67·98	172	88·58
13	6·69½	53	27·29½	93	47·89½	133	68·49½	173	89·09½
14	7·21	54	27·81	94	48·41	134	69·01	174	89·61
15	7·72½	55	28·32½	95	48·92½	135	69·52½	175	90·12½
16	8·24	56	28·84	96	49·44	136	70·04	176	90·64
17	8·75½	57	29·35½	97	49·95½	137	70·55½	177	91·15½
18	9·27	58	29·87	98	50·47	138	71·07	178	91·67
19	9·78½	59	30·38½	99	50·98½	139	71·58½	179	92·18½
20	10·30	**60**	30·90	**100**	51·50	**140**	72·10	**180**	92·70
21	10·81½	61	31·41½	101	52·01½	141	72·61½	181	93·21½
22	11·33	62	31·93	102	52·53	142	73·13	182	93·73
23	11·84½	63	32·44½	103	53·04½	143	73·64½	183	94·24½
24	12·36	64	32·96	104	53·56	144	74·16	184	94·76
25	12·87½	65	33·47½	105	54·07½	145	74·67½	185	95·27½
26	13·39	66	33·99	106	54·59	146	75·19	186	95·79
27	13·90½	67	34·50½	107	55·10½	147	75·70½	187	96·30½
28	14·42	68	35·02	108	55·62	148	76·22	188	96·82
29	14·93½	69	35·53½	109	56·13½	149	76·73½	189	97·33½
30	15·45	**70**	36·05	**110**	56·65	**150**	77·25	**190**	97·85
31	15·96½	71	36·56½	111	57·16½	151	77·76½	191	98·36½
32	16·48	72	37·08	112	57·68	152	78·28	192	98·88
33	16·99½	73	37·59½	113	58·19½	153	78·79½	193	99·39½
34	17·51	74	38·11	114	58·71	154	79·31	194	99·91
35	18·02½	75	38·62½	115	59·22½	155	79·82½	195	100·42½
36	18·54	76	39·14	116	59·74	156	80·34	196	100·94
37	19·05½	77	39·65½	117	60·25½	157	80·85½	197	101·45½
38	19·57	78	40·17	118	60·77	158	81·37	198	101·97
39	20·08½	79	40·68½	119	61·28½	159	81·88½	199	102·48½
40	20·60	**80**	41·20	**120**	61·80	**160**	82·40	**200**	103·00

$\frac{1}{16}$=0·03²	$\frac{1}{9}$=0·05⁷	$\frac{1}{8}$=0·06⁴	$\frac{1}{6}$=0·08⁶	$\frac{1}{5}$=0·10³	$\frac{1}{4}$=0·12⁹	$\frac{1}{3}$=0·17²
$\frac{3}{8}$=0·19³	$\frac{1}{2}$=0·25⁸	$\frac{5}{8}$=0·32²	$\frac{3}{4}$=0·38⁶	$\frac{7}{8}$=0·45¹	$\frac{2}{3}$=0·34³	$\frac{5}{6}$=0·42⁹
$\frac{3}{16}$=0·09⁷	$\frac{5}{16}$=0·16¹	$\frac{7}{16}$=0·22⁵	$\frac{9}{16}$=0·29⁰	$\frac{11}{16}$=0·35⁴	$\frac{13}{16}$=0·41⁸	$\frac{15}{16}$=0·48³

1	0·52	41	21·32	81	42·12	121	62·92	161	83·72
2	1·04	42	21·84	82	42·64	122	63·44	162	84·24
3	1·56	43	22·36	83	43·16	123	63·96	163	84·76
4	2·08	44	22·88	84	43·68	124	64·48	164	85·28
5	2·60	45	23·40	85	44·20	125	65·00	165	85·80
6	3·12	46	23·92	86	44·72	126	65·52	166	86·32
7	3·64	47	24·44	87	45·24	127	66·04	167	86·84
8	4·16	48	24·96	88	45·76	128	66·56	168	87·36
9	4·68	49	25·48	89	46·28	129	67·08	169	87·88
10	5·20	**50**	26·00	**90**	46·80	**130**	67·60	**170**	88·40
11	5·72	51	26·52	91	47·32	131	68·12	171	88·92
12	6·24	52	27·04	92	47·84	132	68·64	172	89·44
13	6·76	53	27·56	93	48·36	133	69·16	173	89·96
14	7·28	54	28·08	94	48·88	134	69·68	174	90·48
15	7·80	55	28·60	95	49·40	135	70·20	175	91·00
16	8·32	56	29·12	96	49·92	136	70·72	176	91·52
17	8·84	57	29·64	97	50·44	137	71·24	177	92·04
18	9·36	58	30·16	98	50·96	138	71·76	178	92·56
19	9·88	59	30·68	99	51·48	139	72·28	179	93·08
20	10·40	**60**	31·20	**100**	52·00	**140**	72·80	**180**	93·60
21	10·92	61	31·72	101	52·52	141	73·32	181	94·12
22	11·44	62	32·24	102	53·04	142	73·84	182	94·64
23	11·96	63	32·76	103	53·56	143	74·36	183	95·16
24	12·48	64	33·28	104	54·08	144	74·88	184	95·68
25	13·00	65	33·80	105	54·60	145	75·40	185	96·20
26	13·52	66	34·32	106	55·12	146	75·92	186	96·72
27	14·04	67	34·84	107	55·64	147	76·44	187	97·24
28	14·56	68	35·36	108	56·16	148	76·96	188	97·76
29	15·08	69	35·88	109	56·68	149	77·48	189	98·28
30	15·60	**70**	36·40	**110**	57·20	**150**	78·00	**190**	98·80
31	16·12	71	36·92	111	57·72	151	78·52	191	99·32
32	16·64	72	37·44	112	58·24	152	79·04	192	99·84
33	17·16	73	37·96	113	58·76	153	79·56	193	100·36
34	17·68	74	38·48	114	59·28	154	80·08	194	100·88
35	18·20	75	39·00	115	59·80	155	80·60	195	101·40
36	18·72	76	39·52	116	60·32	156	81·12	196	101·92
37	19·24	77	40·04	117	60·84	157	81·64	197	102·44
38	19·76	78	40·56	118	61·36	158	82·16	198	102·96
39	20·28	79	41·08	119	61·88	159	82·68	199	103·48
40	20·80	**80**	41·60	**120**	62·40	**160**	83·20	**200**	104·00

$\frac{1}{16}=0{\cdot}03^{3}$ $\frac{1}{9}=0{\cdot}05^{8}$ $\frac{1}{8}=0{\cdot}06^{5}$ $\frac{1}{6}=0{\cdot}08^{7}$ $\frac{1}{5}=0{\cdot}10^{4}$ $\frac{1}{4}=0{\cdot}13^{0}$ $\frac{1}{3}=0{\cdot}17^{3}$

$\frac{3}{8}=0{\cdot}19^{5}$ $\frac{1}{2}=0{\cdot}26^{0}$ $\frac{5}{8}=0{\cdot}32^{5}$ $\frac{3}{4}=0{\cdot}39^{0}$ $\frac{7}{8}=0{\cdot}45^{5}$ $\frac{2}{3}=0{\cdot}34^{7}$ $\frac{5}{6}=0{\cdot}43^{3}$

$\frac{3}{16}=0{\cdot}09^{8}$ $\frac{5}{16}=0{\cdot}16^{3}$ $\frac{7}{16}=0{\cdot}22^{8}$ $\frac{9}{16}=0{\cdot}29^{3}$ $\frac{11}{16}=0{\cdot}35^{8}$ $\frac{13}{16}=0{\cdot}42^{3}$ $\frac{15}{16}=0{\cdot}48^{8}$

(1.90=1£) **£0.52½**

1	0·52½	41	21·52½	81	42·52½	121	63·52½	161	84·52½
2	1·05	42	22·05	82	43·05	122	64·05	162	85·05
3	1·57½	43	22·57½	83	43·57½	123	64·57½	163	85·57½
4	2·10	44	23·10	84	44·10	124	65·10	164	86·10
5	2·62½	45	23·62½	85	44·62½	125	65·62½	165	86·62½
6	3·15	46	24·15	86	45·15	126	66·15	166	87·15
7	3·67½	47	24·67½	87	45·67½	127	66·67½	167	87·67½
8	4·20	48	25·20	88	46·20	128	67·20	168	88·20
9	4·72½	49	25·72½	89	46·72½	129	67·72½	169	88·72½
10	5·25	**50**	26·25	**90**	47·25	**130**	68·25	**170**	89·25
11	5·77½	51	26·77½	91	47·77½	131	68·77½	171	89·77½
12	6·30	52	27·30	92	48·30	132	69·30	172	90·30
13	6·82½	53	27·82½	93	48·82½	133	69·82½	173	90·82½
14	7·35	54	28·35	94	49·35	134	70·35	174	91·35
15	7·87½	55	28·87½	95	49·87½	135	70·87½	175	91·87½
16	8·40	56	29·40	96	50·40	136	71·40	176	92·40
17	8·92½	57	29·92½	97	50·92½	137	71·92½	177	92·92½
18	9·45	58	30·45	98	51·45	138	72·45	178	93·45
19	9·97½	59	30·97½	99	51·97½	139	72·97½	179	93·97½
20	10·50	**60**	31·50	**100**	52·50	**140**	73·50	**180**	94·50
21	11·02½	61	32·02½	101	53·02½	141	74·02½	181	95·02½
22	11·55	62	32·55	102	53·55	142	74·55	182	95·55
23	12·07½	63	33·07½	103	54·07½	143	75·07½	183	96·07½
24	12·60	64	33·60	104	54·60	144	75·60	184	96·60
25	13·12½	65	34·12½	105	55·12½	145	76·12½	185	97·12½
26	13·65	66	34·65	106	55·65	146	76·65	186	97·65
27	14·17½	67	35·17½	107	56·17½	147	77·17½	187	98·17½
28	14·70	68	35·70	108	56·70	148	77·70	188	98·70
29	15·22½	69	36·22½	109	57·22½	149	78·22½	189	99·22½
30	15·75	**70**	36·75	**110**	57·75	**150**	78·75	**190**	99·75
31	16·27½	71	37·27½	111	58·27½	151	79·27½	191	100·27½
32	16·80	72	37·80	112	58·80	152	79·80	192	100·80
33	17·32½	73	38·32½	113	59·32½	153	80·32½	193	101·32½
34	17·85	74	38·85	114	59·85	154	80·85	194	101·85
35	18·37½	75	39·37½	115	60·37½	155	81·37½	195	102·37½
36	18·90	76	39·90	116	60·90	156	81·90	196	102·90
37	19·42½	77	40·42½	117	61·42½	157	82·42½	197	103·42½
38	19·95	78	40·95	118	61·95	158	82·95	198	103·95
39	20·47½	79	41·47½	119	62·47½	159	83·47½	199	104·47½
40	21·00	**80**	42·00	**120**	63·00	**160**	84·00	**200**	105·00

$\frac{1}{16}=0{\cdot}03^{3}$ $\frac{1}{9}=0{\cdot}05^{8}$ $\frac{1}{8}=0{\cdot}06^{6}$ $\frac{1}{6}=0{\cdot}08^{8}$ $\frac{1}{5}=0{\cdot}10^{5}$ $\frac{1}{4}=0{\cdot}13^{1}$ $\frac{1}{3}=0{\cdot}17^{5}$

$\frac{3}{8}=0{\cdot}19^{7}$ $\frac{1}{2}=0{\cdot}26^{3}$ $\frac{5}{8}=0{\cdot}32^{8}$ $\frac{3}{4}=0{\cdot}39^{4}$ $\frac{7}{8}=0{\cdot}45^{9}$ $\frac{2}{3}=0{\cdot}35^{0}$ $\frac{5}{6}=0{\cdot}43^{8}$

$\frac{3}{16}=0{\cdot}09^{8}$ $\frac{5}{16}=0{\cdot}16^{4}$ $\frac{7}{16}=0{\cdot}23^{0}$ $\frac{9}{16}=0{\cdot}29^{5}$ $\frac{11}{16}=0{\cdot}36^{1}$ $\frac{13}{16}=0{\cdot}42^{7}$ $\frac{15}{16}=0{\cdot}49^{2}$

(1.89=1£) **£0.53**

1	0·53	41	21·73	81	42·93	121	64·13	161	85·33
2	1·06	42	22·26	82	43·46	122	64·66	162	85·86
3	1·59	43	22·79	83	43·99	123	65·19	163	86·39
4	2·12	44	23·32	84	44·52	124	65·72	164	86·92
5	2·65	45	23·85	85	45·05	125	66·25	165	87·45
6	3·18	46	24·38	86	45·58	126	66·78	166	87·98
7	3·71	47	24·91	87	46·11	127	67·31	167	88·51
8	4·24	48	25·44	88	46·64	128	67·84	168	89·04
9	4·77	49	25·97	89	47·17	129	68·37	169	89·57
10	5·30	**50**	26·50	**90**	47·70	**130**	68·90	**170**	90·10
11	5·83	51	27·03	91	48·23	131	69·43	171	90·63
12	6·36	52	27·56	92	48·76	132	69·96	172	91·16
13	6·89	53	28·09	93	49·29	133	70·49	173	91·69
14	7·42	54	28·62	94	49·82	134	71·02	174	92·22
15	7·95	55	29·15	95	50·35	135	71·55	175	92·75
16	8·48	56	29·68	96	50·88	136	72·08	176	93·28
17	9·01	57	30·21	97	51·41	137	72·61	177	93·81
18	9·54	58	30·74	98	51·94	138	73·14	178	94·34
19	10·07	59	31·27	99	52·47	139	73·67	179	94·87
20	10·60	**60**	31·80	**100**	53·00	**140**	74·20	**180**	95·40
21	11·13	61	32·33	101	53·53	141	74·73	181	95·93
22	11·66	62	32·86	102	54·06	142	75·26	182	96·46
23	12·19	63	33·39	103	54·59	143	75·79	183	96·99
24	12·72	64	33·92	104	55·12	144	76·32	184	97·52
25	13·25	65	34·45	105	55·65	145	76·85	185	98·05
26	13·78	66	34·98	106	56·18	146	77·38	186	98·58
27	14·31	67	35·51	107	56·71	147	77·91	187	99·11
28	14·84	68	36·04	108	57·24	148	78·44	188	99·64
29	15·37	69	36·57	109	57·77	149	78·97	189	100·17
30	15·90	**70**	37·10	**110**	58·30	**150**	79·50	**190**	100·70
31	16·43	71	37·63	111	58·83	151	80·03	191	101·23
32	16·96	72	38·16	112	59·36	152	80·56	192	101·76
33	17·49	73	38·69	113	59·89	153	81·09	193	102·29
34	18·02	74	39·22	114	60·42	154	81·62	194	102·82
35	18·55	75	39·75	115	60·95	155	82·15	195	103·35
36	19·08	76	40·28	116	61·48	156	82·68	196	103·88
37	19·61	77	40·81	117	62·01	157	83·21	197	104·41
38	20·14	78	41·34	118	62·54	158	83·74	198	104·94
39	20·67	79	41·87	119	63·07	159	84·27	199	105·47
40	21·20	**80**	42·40	**120**	63·60	**160**	84·80	**200**	106·00

$\frac{1}{16}=0{\cdot}03^{3}$ $\frac{1}{9}=0{\cdot}05^{9}$ $\frac{1}{8}=0{\cdot}06^{6}$ $\frac{1}{6}=0{\cdot}08^{8}$ $\frac{1}{5}=0{\cdot}10^{6}$ $\frac{1}{4}=0{\cdot}13^{3}$ $\frac{1}{3}=0{\cdot}17^{7}$

$\frac{3}{8}=0{\cdot}19^{9}$ $\frac{1}{2}=0{\cdot}26^{5}$ $\frac{5}{8}=0{\cdot}33^{1}$ $\frac{3}{4}=0{\cdot}39^{8}$ $\frac{7}{8}=0{\cdot}46^{4}$ $\frac{2}{3}=0{\cdot}35^{3}$ $\frac{5}{6}=0{\cdot}44^{2}$

$\frac{3}{16}=0{\cdot}09^{9}$ $\frac{5}{16}=0{\cdot}16^{6}$ $\frac{7}{16}=0{\cdot}23^{2}$ $\frac{9}{16}=0{\cdot}29^{8}$ $\frac{11}{16}=0{\cdot}36^{4}$ $\frac{13}{16}=0{\cdot}43^{1}$ $\frac{15}{16}=0{\cdot}49^{7}$

(1.87=1£) **£0.53½**

1	0·53½	41	21·93½	81	43·33½	121	64·73½	161	86·13½
2	1·07	42	22·47	82	43·87	122	65·27	162	86·67
3	1·60½	43	23·00½	83	44·40½	123	65·80½	163	87·20½
4	2·14	44	23·54	84	44·94	124	66·34	164	87·74
5	2·67½	45	24·07½	85	45·47½	125	66·87½	165	88·27½
6	3·21	46	24·61	86	46·01	126	67·41	166	88·81
7	3·74½	47	25·14½	87	46·54½	127	67·94½	167	89·34½
8	4·28	48	25·68	88	47·08	128	68·48	168	89·88
9	4·81½	49	26·21½	89	47·61½	129	69·01½	169	90·41½
10	5·35	**50**	26·75	**90**	48·15	**130**	69·55	**170**	90·95
11	5·88½	51	27·28½	91	48·68½	131	70·08½	171	91·48½
12	6·42	52	27·82	92	49·22	132	70·62	172	92·02
13	6·95½	53	28·35½	93	49·75½	133	71·15½	173	92·55½
14	7·49	54	28·89	94	50·29	134	71·69	174	93·09
15	8·02½	55	29·42½	95	50·82½	135	72·22½	175	93·62½
16	8·56	56	29·96	96	51·36	136	72·76	176	94·16
17	9·09½	57	30·49½	97	51·89½	137	73·29½	177	94·69½
18	9·63	58	31·03	98	52·43	138	73·83	178	95·23
19	10·16½	59	31·56½	99	52·96½	139	74·36½	179	95·76½
20	10·70	**60**	32·10	**100**	53·50	**140**	74·90	**180**	96·30
21	11·23½	61	32·63½	101	54·03½	141	75·43½	181	96·83½
22	11·77	62	33·17	102	54·57	142	75·97	182	97·37
23	12·30½	63	33·70½	103	55·10½	143	76·50½	183	97·90½
24	12·84	64	34·24	104	55·64	144	77·04	184	98·44
25	13·37½	65	34·77½	105	56·17½	145	77·57½	185	98·97½
26	13·91	66	35·31	106	56·71	146	78·11	186	99·51
27	14·44½	67	35·84½	107	57·24½	147	78·64½	187	100·04½
28	14·98	68	36·38	108	57·78	148	79·18	188	100·58
29	15·51½	69	36·91½	109	58·31½	149	79·71½	189	101·11½
30	16·05	**70**	37·45	**110**	58·85	**150**	80·25	**190**	101·65
31	16·58½	71	37·98½	111	59·38½	151	80·78½	191	102·18½
32	17·12	72	38·52	112	59·92	152	81·32	192	102·72
33	17·65½	73	39·05½	113	60·45½	153	81·85½	193	103·25½
34	18·19	74	39·59	114	60·99	154	82·39	194	103·79
35	18·72½	75	40·12½	115	61·52½	155	82·92½	195	104·32½
36	19·26	76	40·66	116	62·06	156	83·46	196	104·86
37	19·79½	77	41·19½	117	62·59½	157	83·99½	197	105·39½
38	20·33	78	41·73	118	63·13	158	84·53	198	105·93
39	20·86½	79	42·26½	119	63·66½	159	85·06½	199	106·46½
40	21·40	**80**	42·80	**120**	64·20	**160**	85·60	**200**	107·00

$\frac{1}{16}=0{\cdot}03^{3}$ $\frac{1}{9}=0{\cdot}05^{9}$ $\frac{1}{8}=0{\cdot}06^{7}$ $\frac{1}{6}=0{\cdot}08^{9}$ $\frac{1}{5}=0{\cdot}10^{7}$ $\frac{1}{4}=0{\cdot}13^{4}$ $\frac{1}{3}=0{\cdot}17^{8}$

$\frac{3}{8}=0{\cdot}20^{1}$ $\frac{1}{2}=0{\cdot}26^{8}$ $\frac{5}{8}=0{\cdot}33^{4}$ $\frac{3}{4}=0{\cdot}40^{1}$ $\frac{7}{8}=0{\cdot}46^{8}$ $\frac{2}{3}=0{\cdot}35^{7}$ $\frac{5}{6}=0{\cdot}44^{6}$

$\frac{3}{16}=0{\cdot}10^{0}$ $\frac{5}{16}=0{\cdot}16^{7}$ $\frac{7}{16}=0{\cdot}23^{4}$ $\frac{9}{16}=0{\cdot}30^{1}$ $\frac{11}{16}=0{\cdot}36^{8}$ $\frac{13}{16}=0{\cdot}43^{5}$ $\frac{15}{16}=0{\cdot}50^{2}$

(1.85=1£) **£0.54**

1	0·54	41	22·14	81	43·74	121	65·34	161	86·94
2	1·08	42	22·68	82	44·28	122	65·88	162	87·48
3	1·62	43	23·22	83	44·82	123	66·42	163	88·02
4	2·16	44	23·76	84	45·36	124	66·96	164	88·56
5	2·70	45	24·30	85	45·90	125	67·50	165	89·10
6	3·24	46	24·84	86	46·44	126	68·04	166	89·64
7	3·78	47	25·38	87	46·98	127	68·58	167	90·18
8	4·32	48	25·92	88	47·52	128	69·12	168	90·72
9	4·86	49	26·46	89	48·06	129	69·66	169	91·26
10	5·40	**50**	27·00	**90**	48·60	**130**	70·20	**170**	91·80
11	5·94	51	27·54	91	49·14	131	70·74	171	92·34
12	6·48	52	28·08	92	49·68	132	71·28	172	92·88
13	7·02	53	28·62	93	50·22	133	71·82	173	93·42
14	7·56	54	29·16	94	50·76	134	72·36	174	93·96
15	8·10	55	29·70	95	51·30	135	72·90	175	94·50
16	8·64	56	30·24	96	51·84	136	73·44	176	95·04
17	9·18	57	30·78	97	52·38	137	73·98	177	95·58
18	9·72	58	31·32	98	52·92	138	74·52	178	96·12
19	10·26	59	31·86	99	53·46	139	75·06	179	96·66
20	10·80	**60**	32·40	**100**	54·00	**140**	75·60	**180**	97·20
21	11·34	61	32·94	101	54·54	141	76·14	181	97·74
22	11·88	62	33·48	102	55·08	142	76·68	182	98·28
23	12·42	63	34·02	103	55·62	143	77·22	183	98·82
24	12·96	64	34·56	104	56·16	144	77·76	184	99·36
25	13·50	65	35·10	105	56·70	145	78·30	185	99·90
26	14·04	66	35·64	106	57·24	146	78·84	186	100·44
27	14·58	67	36·18	107	57·78	147	79·38	187	100·98
28	15·12	68	36·72	108	58·32	148	79·92	188	101·52
29	15·66	69	37·26	109	58·86	149	80·46	189	102·06
30	16·20	**70**	37·80	**110**	59·40	**150**	81·00	**190**	102·60
31	16·74	71	38·34	111	59·94	151	81·54	191	103·14
32	17·28	72	38·88	112	60·48	152	82·08	192	103·68
33	17·82	73	39·42	113	61·02	153	82·62	193	104·22
34	18·36	74	39·96	114	61·56	154	83·16	194	104·76
35	18·90	75	40·50	115	62·10	155	83·70	195	105·30
36	19·44	76	41·04	116	62·64	156	84·24	196	105·84
37	19·98	77	41·58	117	63·18	157	84·78	197	106·38
38	20·52	78	42·12	118	63·72	158	85·32	198	106·92
39	21·06	79	42·66	119	64·26	159	85·86	199	107·46
40	21·60	**80**	43·20	**120**	64·80	**160**	86·40	**200**	108·00

$\frac{1}{16}=0{\cdot}03^{4}$ $\frac{1}{9}=0{\cdot}06^{0}$ $\frac{1}{8}=0{\cdot}06^{8}$ $\frac{1}{6}=0{\cdot}09^{0}$ $\frac{1}{5}=0{\cdot}10^{8}$ $\frac{1}{4}=0{\cdot}13^{5}$ $\frac{1}{3}=0{\cdot}18^{0}$

$\frac{3}{8}=0{\cdot}20^{3}$ $\frac{1}{2}=0{\cdot}27^{0}$ $\frac{5}{8}=0{\cdot}33^{8}$ $\frac{3}{4}=0{\cdot}40^{5}$ $\frac{7}{8}=0{\cdot}47^{3}$ $\frac{2}{3}=0{\cdot}36^{0}$ $\frac{5}{6}=0{\cdot}45^{0}$

$\frac{3}{16}=0{\cdot}10^{1}$ $\frac{5}{16}=0{\cdot}16^{9}$ $\frac{7}{16}=0{\cdot}23^{6}$ $\frac{9}{16}=0{\cdot}30^{4}$ $\frac{11}{16}=0{\cdot}37^{1}$ $\frac{13}{16}=0{\cdot}43^{9}$ $\frac{15}{16}=0{\cdot}50^{6}$

(1.83=1£) **£0.54½**

1	0·54½	41	22·34½	81	44·14½	121	65·94½	161	87·74½
2	1·09	42	22·89	82	44·69	122	66·49	162	88·29
3	1·63½	43	23·43½	83	45·23½	123	67·03½	163	88·83½
4	2·18	44	23·98	84	45·78	124	67·58	164	89·38
5	2·72½	45	24·52½	85	46·32½	125	68·12½	165	89·92½
6	3·27	46	25·07	86	46·87	126	68·67	166	90·47
7	3·81½	47	25·61½	87	47·41½	127	69·21½	167	91·01½
8	4·36	48	26·16	88	47·96	128	69·76	168	91·56
9	4·90½	49	26·70½	89	48·50½	129	70·30½	169	92·10½
10	5·45	**50**	27·25	**90**	49·05	**130**	70·85	**170**	92·65
11	5·99½	51	27·79½	91	49·59½	131	71·39½	171	93·19½
12	6·54	52	28·34	92	50·14	132	71·94	172	93·74
13	7·08½	53	28·88½	93	50·68½	133	72·48½	173	94·28½
14	7·63	54	29·43	94	51·23	134	73·03	174	94·83
15	8·17½	55	29·97½	95	51·77½	135	73·57½	175	95·37½
16	8·72	56	30·52	96	52·32	136	74·12	176	95·92
17	9·26½	57	31·06½	97	52·86½	137	74·66½	177	96·46½
18	9·81	58	31·61	98	53·41	138	75·21	178	97·01
19	10·35½	59	32·15½	99	53·95½	139	75·75½	179	97·55½
20	10·90	**60**	32·70	**100**	54·50	**140**	76·30	**180**	98·10
21	11·44½	61	33·24½	101	55·04½	141	76·84½	181	98·64½
22	11·99	62	33·79	102	55·59	142	77·39	182	99·19
23	12·53½	63	34·33½	103	56·13½	143	77·93½	183	99·73½
24	13·08	64	34·88	104	56·68	144	78·48	184	100·28
25	13·62½	65	35·42½	105	57·22½	145	79·02½	185	100·82½
26	14·17	66	35·97	106	57·77	146	79·57	186	101·37
27	14·71½	67	36·51½	107	58·31½	147	80·11½	187	101·91½
28	15·26	68	37·06	108	58·86	148	80·66	188	102·46
29	15·80½	69	37·60½	109	59·40½	149	81·20½	189	103·00½
30	16·35	**70**	38·15	**110**	59·95	**150**	81·75	**190**	103·55
31	16·89½	71	38·69½	111	60·49½	151	82·29½	191	104·09½
32	17·44	72	39·24	112	61·04	152	82·84	192	104·64
33	17·98½	73	39·78½	113	61·58½	153	83·38½	193	105·18½
34	18·53	74	40·33	114	62·13	154	83·93	194	105·73
35	19·07½	75	40·87½	115	62·67½	155	84·47½	195	106·27½
36	19·62	76	41·42	116	63·22	156	85·02	196	106·82
37	20·16½	77	41·96½	117	63·76½	157	85·56½	197	107·36½
38	20·71	78	42·51	118	64·31	158	86·11	198	107·91
39	21·25½	79	43·05½	119	64·85½	159	86·65½	199	108·45½
40	21·80	**80**	43·60	**120**	65·40	**160**	87·20	**200**	109·00

$\frac{1}{16}=0{\cdot}03^{4}$ $\frac{1}{9}=0{\cdot}06^{1}$ $\frac{1}{8}=0{\cdot}06^{8}$ $\frac{1}{6}=0{\cdot}09^{1}$ $\frac{1}{5}=0{\cdot}10^{9}$ $\frac{1}{4}=0{\cdot}13^{6}$ $\frac{1}{3}=0{\cdot}18^{2}$

$\frac{3}{8}=0{\cdot}20^{4}$ $\frac{1}{2}=0{\cdot}27^{3}$ $\frac{5}{8}=0{\cdot}34^{1}$ $\frac{3}{4}=0{\cdot}40^{9}$ $\frac{7}{8}=0{\cdot}47^{7}$ $\frac{2}{3}=0{\cdot}36^{3}$ $\frac{5}{6}=0{\cdot}45^{4}$

$\frac{3}{16}=0{\cdot}10^{2}$ $\frac{5}{16}=0{\cdot}17^{0}$ $\frac{7}{16}=0{\cdot}23^{8}$ $\frac{9}{16}=0{\cdot}30^{7}$ $\frac{11}{16}=0{\cdot}37^{5}$ $\frac{13}{16}=0{\cdot}44^{3}$ $\frac{15}{16}=0{\cdot}51^{1}$

(1.82=1£) **£0.55**

1	0·55	41	22·55	81	44·55	121	66·55	161	88·55
2	1·10	42	23·10	82	45·10	122	67·10	162	89·10
3	1·65	43	23·65	83	45·65	123	67·65	163	89·65
4	2·20	44	24·20	84	46·20	124	68·20	164	90·20
5	2·75	45	24·75	85	46·75	125	68·75	165	90·75
6	3·30	46	25·30	86	47·30	126	69·30	166	91·30
7	3·85	47	25·85	87	47·85	127	69·85	167	91·85
8	4·40	48	26·40	88	48·40	128	70·40	168	92·40
9	4·95	49	26·95	89	48·95	129	70·95	169	92·95
10	5·50	**50**	27·50	**90**	49·50	**130**	71·50	**170**	93·50
11	6·05	51	28·05	91	50·05	131	72·05	171	94·05
12	6·60	52	28·60	92	50·60	132	72·60	172	94·60
13	7·15	53	29·15	93	51·15	133	73·15	173	95·15
14	7·70	54	29·70	94	51·70	134	73·70	174	95·70
15	8·25	55	30·25	95	52·25	135	74·25	175	96·25
16	8·80	56	30·80	96	52·80	136	74·80	176	96·80
17	9·35	57	31·35	97	53·35	137	75·35	177	97·35
18	9·90	58	31·90	98	53·90	138	75·90	178	97·90
19	10·45	59	32·45	99	54·45	139	76·45	179	98·45
20	11·00	**60**	33·00	**100**	55·00	**140**	77·00	**180**	99·00
21	11·55	61	33·55	101	55·55	141	77·55	181	99·55
22	12·10	62	34·10	102	56·10	142	78·10	182	100·10
23	12·65	63	34·65	103	56·65	143	78·65	183	100·65
24	13·20	64	35·20	104	57·20	144	79·20	184	101·20
25	13·75	65	35·75	105	57·75	145	79·75	185	101·75
26	14·30	66	36·30	106	58·30	146	80·30	186	102·30
27	14·85	67	36·85	107	58·85	147	80·85	187	102·85
28	15·40	68	37·40	108	59·40	148	81·40	188	103·40
29	15·95	69	37·95	109	59·95	149	81·95	189	103·95
30	16·50	**70**	38·50	**110**	60·50	**150**	82·50	**190**	104·50
31	17·05	71	39·05	111	61·05	151	83·05	191	105·05
32	17·60	72	39·60	112	61·60	152	83·60	192	105·60
33	18·15	73	40·15	113	62·15	153	84·15	193	106·15
34	18·70	74	40·70	114	62·70	154	84·70	194	106·70
35	19·25	75	41·25	115	63·25	155	85·25	195	107·25
36	19·80	76	41·80	116	63·80	156	85·80	196	107·80
37	20·35	77	42·35	117	64·35	157	86·35	197	108·35
38	20·90	78	42·90	118	64·90	158	86·90	198	108·90
39	21·45	79	43·45	119	65·45	159	87·45	199	109·45
40	22·00	**80**	44·00	**120**	66·00	**160**	88·00	**200**	110·00

$\frac{1}{16}=0{\cdot}03^4$ $\frac{1}{9}=0{\cdot}06^1$ $\frac{1}{8}=0{\cdot}06^9$ $\frac{1}{6}=0{\cdot}09^2$ $\frac{1}{5}=0{\cdot}11^0$ $\frac{1}{4}=0{\cdot}13^8$ $\frac{1}{3}=0{\cdot}18^3$

$\frac{3}{8}=0{\cdot}20^6$ $\frac{1}{2}=0{\cdot}27^5$ $\frac{5}{8}=0{\cdot}34^4$ $\frac{3}{4}=0{\cdot}41^3$ $\frac{7}{8}=0{\cdot}48^1$ $\frac{2}{3}=0{\cdot}36^7$ $\frac{5}{6}=0{\cdot}45^8$

$\frac{3}{16}=0{\cdot}10^3$ $\frac{5}{16}=0{\cdot}17^2$ $\frac{7}{16}=0{\cdot}24^1$ $\frac{9}{16}=0{\cdot}30^9$ $\frac{11}{16}=0{\cdot}37^8$ $\frac{13}{16}=0{\cdot}44^7$ $\frac{15}{16}=0{\cdot}51^6$

(1.80=1£) **£0.55½**

1	0·55½	41	22·75½	81	44·95½	121	67·15½	161	89·35½
2	1·11	42	23·31	82	45·51	122	67·71	162	89·91
3	1·66½	43	23·86½	83	46·06½	123	68·26½	163	90·46½
4	2·22	44	24·42	84	46·62	124	68·82	164	91·02
5	2·77½	45	24·97½	85	47·17½	125	69·37½	165	91·57½
6	3·33	46	25·53	86	47·73	126	69·93	166	92·13
7	3·88½	47	26·08½	87	48·28½	127	70·48½	167	92·68½
8	4·44	48	26·64	88	48·84	128	71·04	168	93·24
9	4·99½	49	27·19½	89	49·39½	129	71·59½	169	93·79½
10	5·55	**50**	27·75	**90**	49·95	**130**	72·15	**170**	94·35
11	6·10½	51	28·30½	91	50·50½	131	72·70½	171	94·90½
12	6·66	52	28·86	92	51·06	132	73·26	172	95·46
13	7·21½	53	29·41½	93	51·61½	133	73·81½	173	96·01½
14	7·77	54	29·97	94	52·17	134	74·37	174	96·57
15	8·32½	55	30·52½	95	52·72½	135	74·92½	175	97·12½
16	8·88	56	31·08	96	53·28	136	75·48	176	97·68
17	9·43½	57	31·63½	97	53·83½	137	76·03½	177	98·23½
18	9·99	58	32·19	98	54·39	138	76·59	178	98·79
19	10·54½	59	32·74½	99	54·94½	139	77·14½	179	99·34½
20	11·10	**60**	33·30	**100**	55·50	**140**	77·70	**180**	99·90
21	11·65½	61	33·85½	101	56·05½	141	78·25½	181	100·45½
22	12·21	62	34·41	102	56·61	142	78·81	182	101·01
23	12·76½	63	34·96½	103	57·16½	143	79·36½	183	101·56½
24	13·32	64	35·52	104	57·72	144	79·92	184	102·12
25	13·87½	65	36·07½	105	58·27½	145	80·47½	185	102·67½
26	14·43	66	36·63	106	58·83	146	81·03	186	103·23
27	14·98½	67	37·18½	107	59·38½	147	81·58½	187	103·78½
28	15·54	68	37·74	108	59·94	148	82·14	188	104·34
29	16·09½	69	38·29½	109	60·49½	149	82·69½	189	104·89½
30	16·65	**70**	38·85	**110**	61·05	**150**	83·25	**190**	105·45
31	17·20½	71	39·40½	111	61·60½	151	83·80½	191	106·00½
32	17·76	72	39·96	112	62·16	152	84·36	192	106·56
33	18·31½	73	40·51½	113	62·71½	153	84·91½	193	107·11½
34	18·87	74	41·07	114	63·27	154	85·47	194	107·67
35	19·42½	75	41·62½	115	63·82½	155	86·02½	195	108·22½
36	19·98	76	42·18	116	64·38	156	86·58	196	108·78
37	20·53½	77	42·73½	117	64·93½	157	87·13½	197	109·33½
38	21·09	78	43·29	118	65·49	158	87·69	198	109·89
39	21·64½	79	43·84½	119	66·04½	159	88·24½	199	110·44½
40	22·20	**80**	44·40	**120**	66·60	**160**	88·80	**200**	111·00

$\frac{1}{16}=0{\cdot}03^{5}$ $\frac{1}{9}=0{\cdot}06^{2}$ $\frac{1}{8}=0{\cdot}06^{9}$ $\frac{1}{6}=0{\cdot}09^{3}$ $\frac{1}{5}=0{\cdot}11^{1}$ $\frac{1}{4}=0{\cdot}13^{9}$ $\frac{1}{3}=0{\cdot}18^{5}$

$\frac{3}{8}=0{\cdot}20^{8}$ $\frac{1}{2}=0{\cdot}27^{8}$ $\frac{5}{8}=0{\cdot}34^{7}$ $\frac{3}{4}=0{\cdot}41^{6}$ $\frac{7}{8}=0{\cdot}48^{6}$ $\frac{2}{3}=0{\cdot}37^{0}$ $\frac{5}{6}=0{\cdot}46^{3}$

$\frac{3}{16}=0{\cdot}10^{4}$ $\frac{5}{16}=0{\cdot}17^{3}$ $\frac{7}{16}=0{\cdot}24^{3}$ $\frac{9}{16}=0{\cdot}31^{2}$ $\frac{11}{16}=0{\cdot}38^{2}$ $\frac{13}{16}=0{\cdot}45^{1}$ $\frac{15}{16}=0{\cdot}52^{0}$

1	0·56	41	22·96	81	45·36	121	67·76	161	90·16
2	1·12	42	23·52	82	45·92	122	68·32	162	90·72
3	1·68	43	24·08	83	46·48	123	68·88	163	91·28
4	2·24	44	24·64	84	47·04	124	69·44	164	91·84
5	2·80	45	25·20	85	47·60	125	70·00	165	92·40
6	3·36	46	25·76	86	48·16	126	70·56	166	92·96
7	3·92	47	26·32	87	48·72	127	71·12	167	93·52
8	4·48	48	26·88	88	49·28	128	71·68	168	94·08
9	5·04	49	27·44	89	49·84	129	72·24	169	94·64
10	5·60	**50**	28·00	**90**	50·40	**130**	72·80	**170**	95·20
11	6·16	51	28·56	91	50·96	131	73·36	171	95·76
12	6·72	52	29·12	92	51·52	132	73·92	172	96·32
13	7·28	53	29·68	93	52·08	133	74·48	173	96·88
14	7·84	54	30·24	94	52·64	134	75·04	174	97·44
15	8·40	55	30·80	95	53·20	135	75·60	175	98·00
16	8·96	56	31·36	96	53·76	136	76·16	176	98·56
17	9·52	57	31·92	97	54·32	137	76·72	177	99·12
18	10·08	58	32·48	98	54·88	138	77·28	178	99·68
19	10·64	59	33·04	99	55·44	139	77·84	179	100·24
20	11·20	**60**	33·60	**100**	56·00	**140**	78·40	**180**	100·80
21	11·76	61	34·16	101	56·56	141	78·96	181	101·36
22	12·32	62	34·72	102	57·12	142	79·52	182	101·92
23	12·88	63	35·28	103	57·68	143	80·08	183	102·48
24	13·44	64	35·84	104	58·24	144	80·64	184	103·04
25	14·00	65	36·40	105	58·80	145	81·20	185	103·60
26	14·56	66	36·96	106	59·36	146	81·76	186	104·16
27	15·12	67	37·52	107	59·92	147	82·32	187	104·72
28	15·68	68	38·08	108	60·48	148	82·88	188	105·28
29	16·24	69	38·64	109	61·04	149	83·44	189	105·84
30	16·80	**70**	39·20	**110**	61·60	**150**	84·00	**190**	106·40
31	17·36	71	39·76	111	62·16	151	84·56	191	106·96
32	17·92	72	40·32	112	62·72	152	85·12	192	107·52
33	18·48	73	40·88	113	63·28	153	85·68	193	108·08
34	19·04	74	41·44	114	63·84	154	86·24	194	108·64
35	19·60	75	42·00	115	64·40	155	86·80	195	109·20
36	20·16	76	42·56	116	64·96	156	87·36	196	109·76
37	20·72	77	43·12	117	65·52	157	87·92	197	110·32
38	21·28	78	43·68	118	66·08	158	88·48	198	110·88
39	21·84	79	44·24	119	66·64	159	89·04	199	111·44
40	22·40	**80**	44·80	**120**	67·20	**160**	89·60	**200**	112·00

$\frac{1}{16}=0{\cdot}03^5$ $\frac{1}{9}=0{\cdot}06^2$ $\frac{1}{8}=0{\cdot}07^0$ $\frac{1}{6}=0{\cdot}09^3$ $\frac{1}{5}=0{\cdot}11^2$ $\frac{1}{4}=0{\cdot}14^0$ $\frac{1}{3}=0{\cdot}18^7$

$\frac{3}{8}=0{\cdot}21^0$ $\frac{1}{2}=0{\cdot}28^0$ $\frac{5}{8}=0{\cdot}35^0$ $\frac{3}{4}=0{\cdot}42^0$ $\frac{7}{8}=0{\cdot}49^0$ $\frac{2}{3}=0{\cdot}37^3$ $\frac{5}{6}=0{\cdot}46^7$

$\frac{3}{16}=0{\cdot}10^5$ $\frac{5}{16}=0{\cdot}17^5$ $\frac{7}{16}=0{\cdot}24^5$ $\frac{9}{16}=0{\cdot}31^5$ $\frac{11}{16}=0{\cdot}38^5$ $\frac{13}{16}=0{\cdot}45^5$ $\frac{15}{16}=0{\cdot}52^5$

(1.77=1£) **£0.56½**

1	0·56½	41	23·16½	81	45·76½	121	68·36½	161	90·96½
2	1·13	42	23·73	82	46·33	122	68·93	162	91·53
3	1·69½	43	24·29½	83	46·89½	123	69·49½	163	92·09½
4	2·26	44	24·86	84	47·46	124	70·06	164	92·66
5	2·82½	45	25·42½	85	48·02½	125	70·62½	165	93·22½
6	3·39	46	25·99	86	48·59	126	71·19	166	93·79
7	3·95½	47	26·55½	87	49·15½	127	71·75½	167	94·35½
8	4·52	48	27·12	88	49·72	128	72·32	168	94·92
9	5·08½	49	27·68½	89	50·28½	129	72·88½	169	95·48½
10	5·65	**50**	28·25	**90**	50·85	**130**	73·45	**170**	96·05
11	6·21½	51	28·81½	91	51·41½	131	74·01½	171	96·61½
12	6·78	52	29·38	92	51·98	132	74·58	172	97·18
13	7·34½	53	29·94½	93	52·54½	133	75·14½	173	97·74½
14	7·91	54	30·51	94	53·11	134	75·71	174	98·31
15	8·47½	55	31·07½	95	53·67½	135	76·27½	175	98·87½
16	9·04	56	31·64	96	54·24	136	76·84	176	99·44
17	9·60½	57	32·20½	97	54·80½	137	77·40½	177	100·00½
18	10·17	58	32·77	98	55·37	138	77·97	178	100·57
19	10·73½	59	33·33½	99	55·93½	139	78·53½	179	101·13½
20	11·30	**60**	33·90	**100**	56·50	**140**	79·10	**180**	101·70
21	11·86½	61	34·46½	101	57·06½	141	79·66½	181	102·26½
22	12·43	62	35·03	102	57·63	142	80·23	182	102·83
23	12·99½	63	35·59½	103	58·19½	143	80·79½	183	103·39½
24	13·56	64	36·16	104	58·76	144	81·36	184	103·96
25	14·12½	65	36·72½	105	59·32½	145	81·92½	185	104·52½
26	14·69	66	37·29	106	59·89	146	82·49	186	105·09
27	15·25½	67	37·85½	107	60·45½	147	83·05½	187	105·65½
28	15·82	68	38·42	108	61·02	148	83·62	188	106·22
29	16·38½	69	38·98½	109	61·58½	149	84·18½	189	106·78½
30	16·95	**70**	39·55	**110**	62·15	**150**	84·75	**190**	107·35
31	17·51½	71	40·11½	111	62·71½	151	85·31½	191	107·91½
32	18·08	72	40·68	112	63·28	152	85·88	192	108·48
33	18·64½	73	41·24½	113	63·84½	153	86·44½	193	109·04½
34	19·21	74	41·81	114	64·41	154	87·01	194	109·61
35	19·77½	75	42·37½	115	64·97½	155	87·57½	195	110·17½
36	20·34	76	42·94	116	65·54	156	88·14	196	110·74
37	20·90½	77	43·50½	117	66·10½	157	88·70½	197	111·30½
38	21·47	78	44·07	118	66·67	158	89·27	198	111·87
39	22·03½	79	44·63½	119	67·23½	159	89·83½	199	112·43½
40	22·60	**80**	45·20	**120**	67·80	**160**	90·40	**200**	113·00

$\frac{1}{16}=0{\cdot}03^{5}$ $\frac{1}{9}=0{\cdot}06^{3}$ $\frac{1}{8}=0{\cdot}07^{1}$ $\frac{1}{6}=0{\cdot}09^{4}$ $\frac{1}{5}=0{\cdot}11^{3}$ $\frac{1}{4}=0{\cdot}14^{1}$ $\frac{1}{3}=0{\cdot}18^{8}$

$\frac{3}{8}=0{\cdot}21^{2}$ $\frac{1}{2}=0{\cdot}28^{3}$ $\frac{5}{8}=0{\cdot}35^{3}$ $\frac{3}{4}=0{\cdot}42^{4}$ $\frac{7}{8}=0{\cdot}49^{4}$ $\frac{2}{3}=0{\cdot}37^{7}$ $\frac{5}{6}=0{\cdot}47^{1}$

$\frac{3}{16}=0{\cdot}10^{6}$ $\frac{5}{16}=0{\cdot}17^{7}$ $\frac{7}{16}=0{\cdot}24^{7}$ $\frac{9}{16}=0{\cdot}31^{8}$ $\frac{11}{16}=0{\cdot}38^{8}$ $\frac{13}{16}=0{\cdot}45^{9}$ $\frac{15}{16}=0{\cdot}53^{0}$

(1.75=1£) **£0.57**

1	0·57	41	23·37	81	46·17	121	68·97	161	91·77
2	1·14	42	23·94	82	46·74	122	69·54	162	92·34
3	1·71	43	24·51	83	47·31	123	70·11	163	92·91
4	2·28	44	25·08	84	47·88	124	70·68	164	93·48
5	2·85	45	25·65	85	48·45	125	71·25	165	94·05
6	3·42	46	26·22	86	49·02	126	71·82	166	94·62
7	3·99	47	26·79	87	49·59	127	72·39	167	95·19
8	4·56	48	27·36	88	50·16	128	72·96	168	95·76
9	5·13	49	27·93	89	50·73	129	73·53	169	96·33
10	5·70	**50**	28·50	**90**	51·30	**130**	74·10	**170**	96·90
11	6·27	51	29·07	91	51·87	131	74·67	171	97·47
12	6·84	52	29·64	92	52·44	132	75·24	172	98·04
13	7·41	53	30·21	93	53·01	133	75·81	173	98·61
14	7·98	54	30·78	94	53·58	134	76·38	174	99·18
15	8·55	55	31·35	95	54·15	135	76·95	175	99·75
16	9·12	56	31·92	96	54·72	136	77·52	176	100·32
17	9·69	57	32·49	97	55·29	137	78·09	177	100·89
18	10·26	58	33·06	98	55·86	138	78·66	178	101·46
19	10·83	59	33·63	99	56·43	139	79·23	179	102·03
20	11·40	**60**	34·20	**100**	57·00	**140**	79·80	**180**	102·60
21	11·97	61	34·77	101	57·57	141	80·37	181	103·17
22	12·54	62	35·34	102	58·14	142	80·94	182	103·74
23	13·11	63	35·91	103	58·71	143	81·51	183	104·31
24	13·68	64	36·48	104	59·28	144	82·08	184	104·88
25	14·25	65	37·05	105	59·85	145	82·65	185	105·45
26	14·82	66	37·62	106	60·42	146	83·22	186	106·02
27	15·39	67	38·19	107	60·99	147	83·79	187	106·59
28	15·96	68	38·76	108	61·56	148	84·36	188	107·16
29	16·53	69	39·33	109	62·13	149	84·93	189	107·73
30	17·10	**70**	39·90	**110**	62·70	**150**	85·50	**190**	108·30
31	17·67	71	40·47	111	63·27	151	86·07	191	108·87
32	18·24	72	41·04	112	63·84	152	86·64	192	109·44
33	18·81	73	41·61	113	64·41	153	87·21	193	110·01
34	19·38	74	42·18	114	64·98	154	87·78	194	110·58
35	19·95	75	42·75	115	65·55	155	88·35	195	111·15
36	20·52	76	43·32	116	66·12	156	88·92	196	111·72
37	21·09	77	43·89	117	66·69	157	89·49	197	112·29
38	21·66	78	44·46	118	67·26	158	90·06	198	112·86
39	22·23	79	45·03	119	67·83	159	90·63	199	113·43
40	22·80	**80**	45·60	**120**	68·40	**160**	91·20	**200**	114·00

$\frac{1}{16}=0{\cdot}03^{6}$ $\frac{1}{9}=0{\cdot}06^{3}$ $\frac{1}{8}=0{\cdot}07^{1}$ $\frac{1}{6}=0{\cdot}09^{5}$ $\frac{1}{5}=0{\cdot}11^{4}$ $\frac{1}{4}=0{\cdot}14^{3}$ $\frac{1}{3}=0{\cdot}19^{0}$

$\frac{3}{8}=0{\cdot}21^{4}$ $\frac{1}{2}=0{\cdot}28^{5}$ $\frac{5}{8}=0{\cdot}35^{6}$ $\frac{3}{4}=0{\cdot}42^{8}$ $\frac{7}{8}=0{\cdot}49^{9}$ $\frac{2}{3}=0{\cdot}38^{0}$ $\frac{5}{6}=0{\cdot}47^{5}$

$\frac{3}{16}=0{\cdot}10^{7}$ $\frac{5}{16}=0{\cdot}17^{8}$ $\frac{7}{16}=0{\cdot}24^{9}$ $\frac{9}{16}=0{\cdot}32^{1}$ $\frac{11}{16}=0{\cdot}39^{2}$ $\frac{13}{16}=0{\cdot}46^{3}$ $\frac{15}{16}=0{\cdot}53^{4}$

1 pint=0·568 litres

(1.74=1£) **£0.57½**

1	0·57½	41	23·57½	81	46·57½	121	69·57½	161	92·57½
2	1·15	42	24·15	82	47·15	122	70·15	162	93·15
3	1·72½	43	24·72½	83	47·72½	123	70·72½	163	93·72½
4	2·30	44	25·30	84	48·30	124	71·30	164	94·30
5	2·87½	45	25·87½	85	48·87½	125	71·87½	165	94·87½
6	3·45	46	26·45	86	49·45	126	72·45	166	95·45
7	4·02½	47	27·02½	87	50·02½	127	73·02½	167	96·02½
8	4·60	48	27·60	88	50·60	128	73·60	168	96·60
9	5·17½	49	28·17½	89	51·17½	129	74·17½	169	97·17½
10	5·75	**50**	28·75	**90**	51·75	**130**	74·75	**170**	97·75
11	6·32½	51	29·32½	91	52·32½	131	75·32½	171	98·32½
12	6·90	52	29·90	92	52·90	132	75·90	172	98·90
13	7·47½	53	30·47½	93	53·47½	133	76·47½	173	99·47½
14	8·05	54	31·05	94	54·05	134	77·05	174	100·05
15	8·62½	55	31·62½	95	54·62½	135	77·62½	175	100·62½
16	9·20	56	32·20	96	55·20	136	78·20	176	101·20
17	9·77½	57	32·77½	97	55·77½	137	78·77½	177	101·77½
18	10·35	58	33·35	98	56·35	138	79·35	178	102·35
19	10·92½	59	33·92½	99	56·92½	139	79·92½	179	102·92½
20	11·50	**60**	34·50	**100**	57·50	**140**	80·50	**180**	103·50
21	12·07½	61	35·07½	101	58·07½	141	81·07½	181	104·07½
22	12·65	62	35·65	102	58·65	142	81·65	182	104·65
23	13·22½	63	36·22½	103	59·22½	143	82·22½	183	105·22½
24	13·80	64	36·80	104	59·80	144	82·80	184	105·80
25	14·37½	65	37·37½	105	60·37½	145	83·37½	185	106·37½
26	14·95	66	37·95	106	60·95	146	83·95	186	106·95
27	15·52½	67	38·52½	107	61·52½	147	84·52½	187	107·52½
28	16·10	68	39·10	108	62·10	148	85·10	188	108·10
29	16·67½	69	39·67½	109	62·67½	149	85·67½	189	108·67½
30	17·25	**70**	40·25	**110**	63·25	**150**	86·25	**190**	109·25
31	17·82½	71	40·82½	111	63·82½	151	86·82½	191	109·82½
32	18·40	72	41·40	112	64·40	152	87·40	192	110·40
33	18·97½	73	41·97½	113	64·97½	153	87·97½	193	110·97½
34	19·55	74	42·55	114	65·55	154	88·55	194	111·55
35	20·12½	75	43·12½	115	66·12½	155	89·12½	195	112·12½
36	20·70	76	43·70	116	66·70	156	89·70	196	112·70
37	21·27½	77	44·27½	117	67·27½	157	90·27½	197	113·27½
38	21·85	78	44·85	118	67·85	158	90·85	198	113·85
39	22·42½	79	45·42½	119	68·42½	159	91·42½	199	114·42½
40	23·00	**80**	46·00	**120**	69·00	**160**	92·00	**200**	115·00

$\frac{1}{16}$=0·03^{6}	$\frac{1}{9}$=0·06^{4}	$\frac{1}{8}$=0·07^{2}	$\frac{1}{6}$=0·09^{6}	$\frac{1}{5}$=0·11^{5}	$\frac{1}{4}$=0·14^{4}	$\frac{1}{3}$=0·19^{2}
$\frac{3}{8}$=0·21^{6}	$\frac{1}{2}$=0·28^{8}	$\frac{5}{8}$=0·35^{9}	$\frac{3}{4}$=0·43^{1}	$\frac{7}{8}$=0·50^{3}	$\frac{2}{3}$=0·38^{3}	$\frac{5}{6}$=0·47^{9}
$\frac{3}{16}$=0·10^{8}	$\frac{5}{16}$=0·18^{0}	$\frac{7}{16}$=0·25^{2}	$\frac{9}{16}$=0·32^{3}	$\frac{11}{16}$=0·39^{5}	$\frac{13}{16}$=0·46^{7}	$\frac{15}{16}$=0·53^{9}

1	0·58	41	23·78	81	46·98	121	70·18	161	93·38
2	1·16	42	24·36	82	47·56	122	70·76	162	93·96
3	1·74	43	24·94	83	48·14	123	71·34	163	94·54
4	2·32	44	25·52	84	48·72	124	71·92	164	95·12
5	2·90	45	26·10	85	49·30	125	72·50	165	95·70
6	3·48	46	26·68	86	49·88	126	73·08	166	96·28
7	4·06	47	27·26	87	50·46	127	73·66	167	96·86
8	4·64	48	27·84	88	51·04	128	74·24	168	97·44
9	5·22	49	28·42	89	51·62	129	74·82	169	98·02
10	5·80	**50**	29·00	**90**	52·20	**130**	75·40	**170**	98·60
11	6·38	51	29·58	91	52·78	131	75·98	171	99·18
12	6·96	52	30·16	92	53·36	132	76·56	172	99·76
13	7·54	53	30·74	93	53·94	133	77·14	173	100·34
14	8·12	54	31·32	94	54·52	134	77·72	174	100·92
15	8·70	55	31·90	95	55·10	135	78·30	175	101·50
16	9·28	56	32·48	96	55·68	136	78·88	176	102·08
17	9·86	57	33·06	97	56·26	137	79·46	177	102·66
18	10·44	58	33·64	98	56·84	138	80·04	178	103·24
19	11·02	59	34·22	99	57·42	139	80·62	179	103·82
20	11·60	**60**	34·80	**100**	58·00	**140**	81·20	**180**	104·40
21	12·18	61	35·38	101	58·58	141	81·78	181	104·98
22	12·76	62	35·96	102	59·16	142	82·36	182	105·56
23	13·34	63	36·54	103	59·74	143	82·94	183	106·14
24	13·92	64	37·12	104	60·32	144	83·52	184	106·72
25	14·50	65	37·70	105	60·90	145	84·10	185	107·30
26	15·08	66	38·28	106	61·48	146	84·68	186	107·88
27	15·66	67	38·86	107	62·06	147	85·26	187	108·46
28	16·24	68	39·44	108	62·64	148	85·84	188	109·04
29	16·82	69	40·02	109	63·22	149	86·42	189	109·62
30	17·40	**70**	40·60	**110**	63·80	**150**	87·00	**190**	110·20
31	17·98	71	41·18	111	64·38	151	87·58	191	110·78
32	18·56	72	41·76	112	64·96	152	88·16	192	111·36
33	19·14	73	42·34	113	65·54	153	88·74	193	111·94
34	19·72	74	42·92	114	66·12	154	89·32	194	112·52
35	20·30	75	43·50	115	66·70	155	89·90	195	113·10
36	20·88	76	44·08	116	67·28	156	90·48	196	113·68
37	21·46	77	44·66	117	67·86	157	91·06	197	114·26
38	22·04	78	45·24	118	68·44	158	91·64	198	114·84
39	22·62	79	45·82	119	69·02	159	92·22	199	115·42
40	23·20	**80**	46·40	**120**	69·60	**160**	92·80	**200**	116·00

$\frac{1}{16}=0{\cdot}03^{6}$ $\frac{1}{9}=0{\cdot}06^{4}$ $\frac{1}{8}=0{\cdot}07^{3}$ $\frac{1}{6}=0{\cdot}09^{7}$ $\frac{1}{5}=0{\cdot}11^{6}$ $\frac{1}{4}=0{\cdot}14^{5}$ $\frac{1}{3}=0{\cdot}19^{3}$

$\frac{3}{8}=0{\cdot}21^{8}$ $\frac{1}{2}=0{\cdot}29^{0}$ $\frac{5}{8}=0{\cdot}36^{3}$ $\frac{3}{4}=0{\cdot}43^{5}$ $\frac{7}{8}=0{\cdot}50^{8}$ $\frac{2}{3}=0{\cdot}38^{7}$ $\frac{5}{6}=0{\cdot}48^{3}$

$\frac{3}{16}=0{\cdot}10^{9}$ $\frac{5}{16}=0{\cdot}18^{1}$ $\frac{7}{16}=0{\cdot}25^{4}$ $\frac{9}{16}=0{\cdot}32^{6}$ $\frac{11}{16}=0{\cdot}39^{9}$ $\frac{13}{16}=0{\cdot}47^{1}$ $\frac{15}{16}=0{\cdot}54^{4}$

(1.71=1£) **£0.58½**

1	0·58½	41	23·98½	81	47·38½	121	70·78½	161	94·18½
2	1·17	42	24·57	82	47·97	122	71·37	162	94·77
3	1·75½	43	25·15½	83	48·55½	123	71·95½	163	95·35½
4	2·34	44	25·74	84	49·14	124	72·54	164	95·94
5	2·92½	45	26·32½	85	49·72½	125	73·12½	165	96·52½
6	3·51	46	26·91	86	50·31	126	73·71	166	97·11
7	4·09½	47	27·49½	87	50·89½	127	74·29½	167	97·69½
8	4·68	48	28·08	88	51·48	128	74·88	168	98·28
9	5·26½	49	28·66½	89	52·06½	129	75·46½	169	98·86½
10	5·85	**50**	29·25	**90**	52·65	**130**	76·05	**170**	99·45
11	6·43½	51	29·83½	91	53·23½	131	76·63½	171	100·03½
12	7·02	52	30·42	92	53·82	132	77·22	172	100·62
13	7·60½	53	31·00½	93	54·40½	133	77·80½	173	101·20½
14	8·19	54	31·59	94	54·99	134	78·39	174	101·79
15	8·77½	55	32·17½	95	55·57½	135	78·97½	175	102·37½
16	9·36	56	32·76	96	56·16	136	79·56	176	102·96
17	9·94½	57	33·34½	97	56·74½	137	80·14½	177	103·54½
18	10·53	58	33·93	98	57·33	138	80·73	178	104·13
19	11·11½	59	34·51½	99	57·91½	139	81·31½	179	104·71½
20	11·70	**60**	35·10	**100**	58·50	**140**	81·90	**180**	105·30
21	12·28½	61	35·68½	101	59·08½	141	82·48½	181	105·88½
22	12·87	62	36·27	102	59·67	142	83·07	182	106·47
23	13·45½	63	36·85½	103	60·25½	143	83·65½	183	107·05½
24	14·04	64	37·44	104	60·84	144	84·24	184	107·64
25	14·62½	65	38·02½	105	61·42½	145	84·82½	185	108·22½
26	15·21	66	38·61	106	62·01	146	85·41	186	108·81
27	15·79½	67	39·19½	107	62·59½	147	85·99½	187	109·39½
28	16·38	68	39·78	108	63·18	148	86·58	188	109·98
29	16·96½	69	40·36½	109	63·76½	149	87·16½	189	110·56½
30	17·55	**70**	40·95	**110**	64·35	**150**	87·75	**190**	111·15
31	18·13½	71	41·53½	111	64·93½	151	88·33½	191	111·73½
32	18·72	72	42·12	112	65·52	152	88·92	192	112·32
33	19·30½	73	42·70½	113	66·10½	153	89·50½	193	112·90½
34	19·89	74	43·29	114	66·69	154	90·09	194	113·49
35	20·47½	75	43·87½	115	67·27½	155	90·67½	195	114·07½
36	21·06	76	44·46	116	67·86	156	91·26	196	114·66
37	21·64½	77	45·04½	117	68·44½	157	91·84½	197	115·24½
38	22·23	78	45·63	118	69·03	158	92·43	198	115·83
39	22·81½	79	46·21½	119	69·61½	159	93·01½	199	116·41½
40	23·40	**80**	46·80	**120**	70·20	**160**	93·60	**200**	117·00

$\frac{1}{16}=0{\cdot}03^{7}$	$\frac{1}{9}=0{\cdot}06^{5}$	$\frac{1}{8}=0{\cdot}07^{3}$	$\frac{1}{6}=0{\cdot}09^{8}$	$\frac{1}{5}=0{\cdot}11^{7}$	$\frac{1}{4}=0{\cdot}14^{6}$	$\frac{1}{3}=0{\cdot}19^{5}$
$\frac{3}{8}=0{\cdot}21^{9}$	$\frac{1}{2}=0{\cdot}29^{3}$	$\frac{5}{8}=0{\cdot}36^{6}$	$\frac{3}{4}=0{\cdot}43^{9}$	$\frac{7}{8}=0{\cdot}51^{2}$	$\frac{2}{3}=0{\cdot}39^{0}$	$\frac{5}{6}=0{\cdot}48^{8}$
$\frac{3}{16}=0{\cdot}11^{0}$	$\frac{5}{16}=0{\cdot}18^{3}$	$\frac{7}{16}=0{\cdot}25^{6}$	$\frac{9}{16}=0{\cdot}32^{9}$	$\frac{11}{16}=0{\cdot}40^{2}$	$\frac{13}{16}=0{\cdot}47^{5}$	$\frac{15}{16}=0{\cdot}54^{8}$

(1.69=1£) £0.59

1	0·59	41	24·19	81	47·79	121	71·39	161	94·99
2	1·18	42	24·78	82	48·38	122	71·98	162	95·58
3	1·77	43	25·37	83	48·97	123	72·57	163	96·17
4	2·36	44	25·96	84	49·56	124	73·16	164	96·76
5	2·95	45	26·55	85	50·15	125	73·75	165	97·35
6	3·54	46	27·14	86	50·74	126	74·34	166	97·94
7	4·13	47	27·73	87	51·33	127	74·93	167	98·53
8	4·72	48	28·32	88	51·92	128	75·52	168	99·12
9	5·31	49	28·91	89	52·51	129	76·11	169	99·71
10	5·90	**50**	29·50	**90**	53·10	**130**	76·70	**170**	100·30
11	6·49	51	30·09	91	53·69	131	77·29	171	100·89
12	7·08	52	30·68	92	54·28	132	77·88	172	101·48
13	7·67	53	31·27	93	54·87	133	78·47	173	102·07
14	8·26	54	31·86	94	55·46	134	79·06	174	102·66
15	8·85	55	32·45	95	56·05	135	79·65	175	103·25
16	9·44	56	33·04	96	56·64	136	80·24	176	103·84
17	10·03	57	33·63	97	57·23	137	80·83	177	104·43
18	10·62	58	34·22	98	57·82	138	81·42	178	105·02
19	11·21	59	34·81	99	58·41	139	82·01	179	105·61
20	11·80	**60**	35·40	**100**	59·00	**140**	82·60	**180**	106·20
21	12·39	61	35·99	101	59·59	141	83·19	181	106·79
22	12·98	62	36·58	102	60·18	142	83·78	182	107·38
23	13·57	63	37·17	103	60·77	143	84·37	183	107·97
24	14·16	64	37·76	104	61·36	144	84·96	184	108·56
25	14·75	65	38·35	105	61·95	145	85·55	185	109·15
26	15·34	66	38·94	106	62·54	146	86·14	186	109·74
27	15·93	67	39·53	107	63·13	147	86·73	187	110·33
28	16·52	68	40·12	108	63·72	148	87·32	188	110·92
29	17·11	69	40·71	109	64·31	149	87·91	189	111·51
30	17·70	**70**	41·30	**110**	64·90	**150**	88·50	**190**	112·10
31	18·29	71	41·89	111	65·49	151	89·09	191	112·69
32	18·88	72	42·48	112	66·08	152	89·68	192	113·28
33	19·47	73	43·07	113	66·67	153	90·27	193	113·87
34	20·06	74	43·66	114	67·26	154	90·86	194	114·46
35	20·65	75	44·25	115	67·85	155	91·45	195	115·05
36	21·24	76	44·84	116	68·44	156	92·04	196	115·64
37	21·83	77	45·43	117	69·03	157	92·63	197	116·23
38	22·42	78	46·02	118	69·62	158	93·22	198	116·82
39	23·01	79	46·61	119	70·21	159	93·81	199	117·41
40	23·60	**80**	47·20	**120**	70·80	**160**	94·40	**200**	118·00

$\frac{1}{16}=0{\cdot}03^{7}$ $\frac{1}{9}=0{\cdot}06^{6}$ $\frac{1}{8}=0{\cdot}07^{4}$ $\frac{1}{6}=0{\cdot}09^{8}$ $\frac{1}{5}=0{\cdot}11^{8}$ $\frac{1}{4}=0{\cdot}14^{8}$ $\frac{1}{3}=0{\cdot}19^{7}$

$\frac{3}{8}=0{\cdot}22^{1}$ $\frac{1}{2}=0{\cdot}29^{5}$ $\frac{5}{8}=0{\cdot}36^{9}$ $\frac{3}{4}=0{\cdot}44^{3}$ $\frac{7}{8}=0{\cdot}51^{6}$ $\frac{2}{3}=0{\cdot}39^{3}$ $\frac{5}{6}=0{\cdot}49^{2}$

$\frac{3}{16}=0{\cdot}11^{1}$ $\frac{5}{16}=0{\cdot}18^{4}$ $\frac{7}{16}=0{\cdot}25^{8}$ $\frac{9}{16}=0{\cdot}33^{2}$ $\frac{11}{16}=0{\cdot}40^{6}$ $\frac{13}{16}=0{\cdot}47^{9}$ $\frac{15}{16}=0{\cdot}55^{3}$

(1.68=1£) **£0.59½**

1	0·59½	41	24·39½	81	48·19½	121	71·99½	161	95·79½
2	1·19	42	24·99	82	48·79	122	72·59	162	96·39
3	1·78½	43	25·58½	83	49·38½	123	73·18½	163	96·98½
4	2·38	44	26·18	84	49·98	124	73·78	164	97·58
5	2·97½	45	26·77½	85	50·57½	125	74·37½	165	98·17½
6	3·57	46	27·37	86	51·17	126	74·97	166	98·77
7	4·16½	47	27·96½	87	51·76½	127	75·56½	167	99·36½
8	4·76	48	28·56	88	52·36	128	76·16	168	99·96
9	5·35½	49	29·15½	89	52·95½	129	76·75½	169	100·55½
10	5·95	**50**	29·75	**90**	53·55	**130**	77·35	**170**	101·15
11	6·54½	51	30·34½	91	54·14½	131	77·94½	171	101·74½
12	7·14	52	30·94	92	54·74	132	78·54	172	102·34
13	7·73½	53	31·53½	93	55·33½	133	79·13½	173	102·93½
14	8·33	54	32·13	94	55·93	134	79·73	174	103·53
15	8·92½	55	32·72½	95	56·52½	135	80·32½	175	104·12½
16	9·52	56	33·32	96	57·12	136	80·92	176	104·72
17	10·11½	57	33·91½	97	57·71½	137	81·51½	177	105·31½
18	10·71	58	34·51	98	58·31	138	82·11	178	105·91
19	11·30½	59	35·10½	99	58·90½	139	82·70½	179	106·50½
20	11·90	**60**	35·70	**100**	59·50	**140**	83·30	**180**	107·10
21	12·49½	61	36·29½	101	60·09½	141	83·89½	181	107·69½
22	13·09	62	36·89	102	60·69	142	84·49	182	108·29
23	13·68½	63	37·48½	103	61·28½	143	85·08½	183	108·88½
24	14·28	64	38·08	104	61·88	144	85·68	184	109·48
25	14·87½	65	38·67½	105	62·47½	145	86·27½	185	110·07½
26	15·47	66	39·27	106	63·07	146	86·87	186	110·67
27	16·06½	67	39·86½	107	63·66½	147	87·46½	187	111·26½
28	16·66	68	40·46	108	64·26	148	88·06	188	111·86
29	17·25½	69	41·05½	109	64·85½	149	88·65½	189	112·45½
30	17·85	**70**	41·65	**110**	65·45	**150**	89·25	**190**	113·05
31	18·44½	71	42·24½	111	66·04½	151	89·84½	191	113·64½
32	19·04	72	42·84	112	66·64	152	90·44	192	114·24
33	19·63½	73	43·43½	113	67·23½	153	91·03½	193	114·83½
34	20·23	74	44·03	114	67·83	154	91·63	194	115·43
35	20·82½	75	44·62½	115	68·42½	155	92·22½	195	116·02½
36	21·42	76	45·22	116	69·02	156	92·82	196	116·62
37	22·01½	77	45·81½	117	69·61½	157	93·41½	197	117·21½
38	22·61	78	46·41	118	70·21	158	94·01	198	117·81
39	23·20½	79	47·00½	119	70·80½	159	94·60½	199	118·40½
40	23·80	**80**	47·60	**120**	71·40	**160**	95·20	**200**	119·00

$\frac{1}{16}=0{\cdot}03^{7}$ $\frac{1}{9}=0{\cdot}06^{6}$ $\frac{1}{8}=0{\cdot}07^{4}$ $\frac{1}{6}=0{\cdot}09^{9}$ $\frac{1}{5}=0{\cdot}11^{9}$ $\frac{1}{4}=0{\cdot}14^{9}$ $\frac{1}{3}=0{\cdot}19^{8}$

$\frac{3}{8}=0{\cdot}22^{3}$ $\frac{1}{2}=0{\cdot}29^{8}$ $\frac{5}{8}=0{\cdot}37^{2}$ $\frac{3}{4}=0{\cdot}44^{6}$ $\frac{7}{8}=0{\cdot}52^{1}$ $\frac{2}{3}=0{\cdot}39^{7}$ $\frac{5}{6}=0{\cdot}49^{6}$

$\frac{3}{16}=0{\cdot}11^{2}$ $\frac{5}{16}=0{\cdot}18^{6}$ $\frac{7}{16}=0{\cdot}26^{0}$ $\frac{9}{16}=0{\cdot}33^{5}$ $\frac{11}{16}=0{\cdot}40^{9}$ $\frac{13}{16}=0{\cdot}48^{3}$ $\frac{15}{16}=0{\cdot}55^{8}$

(1.67=1£) **£0.60**

1	0·60	41	24·60	81	48·60	121	72·60	161	96·60
2	1·20	42	25·20	82	49·20	122	73·20	162	97·20
3	1·80	43	25·80	83	49·80	123	73·80	163	97·80
4	2·40	44	26·40	84	50·40	124	74·40	164	98·40
5	3·00	45	27·00	85	51·00	125	75·00	165	99·00
6	3·60	46	27·60	86	51·60	126	75·60	166	99·60
7	4·20	47	28·20	87	52·20	127	76·20	167	100·20
8	4·80	48	28·80	88	52·80	128	76·80	168	100·80
9	5·40	49	29·40	89	53·40	129	77·40	169	101·40
10	6·00	**50**	30·00	**90**	54·00	**130**	78·00	**170**	102·00
11	6·60	51	30·60	91	54·60	131	78·60	171	102·60
12	7·20	52	31·20	92	55·20	132	79·20	172	103·20
13	7·80	53	31·80	93	55·80	133	79·80	173	103·80
14	8·40	54	32·40	94	56·40	134	80·40	174	104·40
15	9·00	55	33·00	95	57·00	135	81·00	175	105·00
16	9·60	56	33·60	96	57·60	136	81·60	176	105·60
17	10·20	57	34·20	97	58·20	137	82·20	177	106·20
18	10·80	58	34·80	98	58·80	138	82·80	178	106·80
19	11·40	59	35·40	99	59·40	139	83·40	179	107·40
20	12·00	**60**	36·00	**100**	60·00	**140**	84·00	**180**	108·00
21	12·60	61	36·60	101	60·60	141	84·60	181	108·60
22	13·20	62	37·20	102	61·20	142	85·20	182	109·20
23	13·80	63	37·80	103	61·80	143	85·80	183	109·80
24	14·40	64	38·40	104	62·40	144	86·40	184	110·40
25	15·00	65	39·00	105	63·00	145	87·00	185	111·00
26	15·60	66	39·60	106	63·60	146	87·60	186	111·60
27	16·20	67	40·20	107	64·20	147	88·20	187	112·20
28	16·80	68	40·80	108	64·80	148	88·80	188	112·80
29	17·40	69	41·40	109	65·40	149	89·40	189	113·40
30	18·00	**70**	42·00	**110**	66·00	**150**	90·00	**190**	114·00
31	18·60	71	42·60	111	66·60	151	90·60	191	114·60
32	19·20	72	43·20	112	67·20	152	91·20	192	115·20
33	19·80	73	43·80	113	67·80	153	91·80	193	115·80
34	20·40	74	44·40	114	68·40	154	92·40	194	116·40
35	21·00	75	45·00	115	69·00	155	93·00	195	117·00
36	21·60	76	45·60	116	69·60	156	93·60	196	117·60
37	22·20	77	46·20	117	70·20	157	94·20	197	118·20
38	22·80	78	46·80	118	70·80	158	94·80	198	118·80
39	23·40	79	47·40	119	71·40	159	95·40	199	119·40
40	24·00	**80**	48·00	**120**	72·00	**160**	96·00	**200**	120·00

$\frac{1}{16}=0{\cdot}03^8$ $\frac{1}{9}=0{\cdot}06^7$ $\frac{1}{8}=0{\cdot}07^5$ $\frac{1}{6}=0{\cdot}10^0$ $\frac{1}{5}=0{\cdot}12^0$ $\frac{1}{4}=0{\cdot}15^0$ $\frac{1}{3}=0{\cdot}20^0$

$\frac{3}{8}=0{\cdot}22^5$ $\frac{1}{2}=0{\cdot}30^0$ $\frac{5}{8}=0{\cdot}37^5$ $\frac{3}{4}=0{\cdot}45^0$ $\frac{7}{8}=0{\cdot}52^5$ $\frac{2}{3}=0{\cdot}40^0$ $\frac{5}{6}=0{\cdot}50^0$

$\frac{3}{16}=0{\cdot}11^3$ $\frac{5}{16}=0{\cdot}18^8$ $\frac{7}{16}=0{\cdot}26^3$ $\frac{9}{16}=0{\cdot}33^8$ $\frac{11}{16}=0{\cdot}41^3$ $\frac{13}{16}=0{\cdot}48^8$ $\frac{15}{16}=0{\cdot}56^3$

1	0·60½	41	24·80½	81	49·00½	121	73·20½	161	97·40½
2	1·21	42	25·41	82	49·61	122	73·81	162	98·01
3	1·81½	43	26·01½	83	50·21½	123	74·41½	163	98·61½
4	2·42	44	26·62	84	50·82	124	75·02	164	99·22
5	3·02½	45	27·22½	85	51·42½	125	75·62½	165	99·82½
6	3·63	46	27·83	86	52·03	126	76·23	166	100·43
7	4·23½	47	28·43½	87	52·63½	127	76·83½	167	101·03½
8	4·84	48	29·04	88	53·24	128	77·44	168	101·64
9	5·44½	49	29·64½	89	53·84½	129	78·04½	169	102·24½
10	6·05	**50**	30·25	**90**	54·45	**130**	78·65	**170**	102·85
11	6·65½	51	30·85½	91	55·05½	131	79·25½	171	103·45½
12	7·26	52	31·46	92	55·66	132	79·86	172	104·06
13	7·86½	53	32·06½	93	56·26½	133	80·46½	173	104·66½
14	8·47	54	32·67	94	56·87	134	81·07	174	105·27
15	9·07½	55	33·27½	95	57·47½	135	81·67½	175	105·87½
16	9·68	56	33·88	96	58·08	136	82·28	176	106·48
17	10·28½	57	34·48½	97	58·68½	137	82·88½	177	107·08½
18	10·89	58	35·09	98	59·29	138	83·49	178	107·69
19	11·49½	59	35·69½	99	59·89½	139	84·09½	179	108·29½
20	12·10	**60**	36·30	**100**	60·50	**140**	84·70	**180**	108·90
21	12·70½	61	36·90½	101	61·10½	141	85·30½	181	109·50½
22	13·31	62	37·51	102	61·71	142	85·91	182	110·11
23	13·91½	63	38·11½	103	62·31½	143	86·51½	183	110·71½
24	14·52	64	38·72	104	62·92	144	87·12	184	111·32
25	15·12½	65	39·32½	105	63·52½	145	87·72½	185	111·92½
26	15·73	66	39·93	106	64·13	146	88·33	186	112·53
27	16·33½	67	40·53½	107	64·73½	147	88·93½	187	113·13½
28	16·94	68	41·14	108	65·34	148	89·54	188	113·74
29	17·54½	69	41·74½	109	65·94½	149	90·14½	189	114·34½
30	18·15	**70**	42·35	**110**	66·55	**150**	90·75	**190**	114·95
31	18·75½	71	42·95½	111	67·15½	151	91·35½	191	115·55½
32	19·36	72	43·56	112	67·76	152	91·96	192	116·16
33	19·96½	73	44·16½	113	68·36½	153	92·56½	193	116·76½
34	20·57	74	44·77	114	68·97	154	93·17	194	117·37
35	21·17½	75	45·37½	115	69·57½	155	93·77½	195	117·97½
36	21·78	76	45·98	116	70·18	156	94·38	196	118·58
37	22·38½	77	46·58½	117	70·78½	157	94·98½	197	119·18½
38	22·99	78	47·19	118	71·39	158	95·59	198	119·79
39	23·59½	79	47·79½	119	71·99½	159	96·19½	199	120·39½
40	24·20	**80**	48·40	**120**	72·60	**160**	96·80	**200**	121·00

$\frac{1}{16}$=0·03^{8} $\frac{1}{9}$=0·06^{7} $\frac{1}{8}$=0·07^{6} $\frac{1}{6}$=0·10^{1} $\frac{1}{5}$=0·12^{1} $\frac{1}{4}$=0·15^{1} $\frac{1}{3}$=0·20^{2}

$\frac{3}{8}$=0·22^{7} $\frac{1}{2}$=0·30^{3} $\frac{5}{8}$=0·37^{8} $\frac{3}{4}$=0·45^{4} $\frac{7}{8}$=0·52^{9} $\frac{2}{3}$=0·40^{3} $\frac{5}{6}$=0·50^{4}

$\frac{3}{16}$=0·11^{3} $\frac{5}{16}$=0·18^{9} $\frac{7}{16}$=0·26^{5} $\frac{9}{16}$=0·34^{0} $\frac{11}{16}$=0·41^{6} $\frac{13}{16}$=0·49^{2} $\frac{15}{16}$=0·56^{7}

(1.64=1£) **£0.61**

1	0·61	41	25·01	81	49·41	121	73·81	161	98·21
2	1·22	42	25·62	82	50·02	122	74·42	162	98·82
3	1·83	43	26·23	83	50·63	123	75·03	163	99·43
4	2·44	44	26·84	84	51·24	124	75·64	164	100·04
5	3·05	45	27·45	85	51·85	125	76·25	165	100·65
6	3·66	46	28·06	86	52·46	126	76·86	166	101·26
7	4·27	47	28·67	87	53·07	127	77·47	167	101·87
8	4·88	48	29·28	88	53·68	128	78·08	168	102·48
9	5·49	49	29·89	89	54·29	129	78·69	169	103·09
10	6·10	**50**	30·50	**90**	54·90	**130**	79·30	**170**	103·70
11	6·71	51	31·11	91	55·51	131	79·91	171	104·31
12	7·32	52	31·72	92	56·12	132	80·52	172	104·92
13	7·93	53	32·33	93	56·73	133	81·13	173	105·53
14	8·54	54	32·94	94	57·34	134	81·74	174	106·14
15	9·15	55	33·55	95	57·95	135	82·35	175	106·75
16	9·76	56	34·16	96	58·56	136	82·96	176	107·36
17	10·37	57	34·77	97	59·17	137	83·57	177	107·97
18	10·98	58	35·38	98	59·78	138	84·18	178	108·58
19	11·59	59	35·99	99	60·39	139	84·79	179	109·19
20	12·20	**60**	36·60	**100**	61·00	**140**	85·40	**180**	109·80
21	12·81	61	37·21	101	61·61	141	86·01	181	110·41
22	13·42	62	37·82	102	62·22	142	86·62	182	111·02
23	14·03	63	38·43	103	62·83	143	87·23	183	111·63
24	14·64	64	39·04	104	63·44	144	87·84	184	112·24
25	15·25	65	39·65	105	64·05	145	88·45	185	112·85
26	15·86	66	40·26	106	64·66	146	89·06	186	113·46
27	16·47	67	40·87	107	65·27	147	89·67	187	114·07
28	17·08	68	41·48	108	65·88	148	90·28	188	114·68
29	17·69	69	42·09	109	66·49	149	90·89	189	115·29
30	18·30	**70**	42·70	**110**	67·10	**150**	91·50	**190**	115·90
31	18·91	71	43·31	111	67·71	151	92·11	191	116·51
32	19·52	72	43·92	112	68·32	152	92·72	192	117·12
33	20·13	73	44·53	113	68·93	153	93·33	193	117·73
34	20·74	74	45·14	114	69·54	154	93·94	194	118·34
35	21·35	75	45·75	115	70·15	155	94·55	195	118·95
36	21·96	76	46·36	116	70·76	156	95·16	196	119·56
37	22·57	77	46·97	117	71·37	157	95·77	197	120·17
38	23·18	78	47·58	118	71·98	158	96·38	198	120·78
39	23·79	79	48·19	119	72·59	159	96·99	199	121·39
40	24·40	**80**	48·80	**120**	73·20	**160**	97·60	**200**	122·00

$\frac{1}{16}=0{\cdot}03^{8}$ $\frac{1}{9}=0{\cdot}06^{8}$ $\frac{1}{8}=0{\cdot}07^{6}$ $\frac{1}{6}=0{\cdot}10^{2}$ $\frac{1}{5}=0{\cdot}12^{2}$ $\frac{1}{4}=0{\cdot}15^{3}$ $\frac{1}{3}=0{\cdot}20^{3}$

$\frac{3}{8}=0{\cdot}22^{9}$ $\frac{1}{2}=0{\cdot}30^{5}$ $\frac{5}{8}=0{\cdot}38^{1}$ $\frac{3}{4}=0{\cdot}45^{8}$ $\frac{7}{8}=0{\cdot}53^{4}$ $\frac{2}{3}=0{\cdot}40^{7}$ $\frac{5}{6}=0{\cdot}50^{8}$

$\frac{3}{16}=0{\cdot}11^{4}$ $\frac{5}{16}=0{\cdot}19^{1}$ $\frac{7}{16}=0{\cdot}26^{7}$ $\frac{9}{16}=0{\cdot}34^{3}$ $\frac{11}{16}=0{\cdot}41^{9}$ $\frac{13}{16}=0{\cdot}49^{6}$ $\frac{15}{16}=0{\cdot}57^{2}$

(1.63=1£) **£0.61½**

1	0·61½	41	25·21½	81	49·81½	121	74·41½	161	99·01½
2	1·23	42	25·83	82	50·43	122	75·03	162	99·63
3	1·84½	43	26·44½	83	51·04½	123	75·64½	163	100·24½
4	2·46	44	27·06	84	51·66	124	76·26	164	100·86
5	3·07½	45	27·67½	85	52·27½	125	76·87½	165	101·47½
6	3·69	46	28·29	86	52·89	126	77·49	166	102·09
7	4·30½	47	28·90½	87	53·50½	127	78·10½	167	102·70½
8	4·92	48	29·52	88	54·12	128	78·72	168	103·32
9	5·53½	49	30·13½	89	54·73½	129	79·33½	169	103·93½
10	6·15	**50**	30·75	**90**	55·35	**130**	79·95	**170**	104·55
11	6·76½	51	31·36½	91	55·96½	131	80·56½	171	105·16½
12	7·38	52	31·98	92	56·58	132	81·18	172	105·78
13	7·99½	53	32·59½	93	57·19½	133	81·79½	173	106·39½
14	8·61	54	33·21	94	57·81	134	82·41	174	107·01
15	9·22½	55	33·82½	95	58·42½	135	83·02½	175	107·62½
16	9·84	56	34·44	96	59·04	136	83·64	176	108·24
17	10·45½	57	35·05½	97	59·65½	137	84·25½	177	108·85½
18	11·07	58	35·67	98	60·27	138	84·87	178	109·47
19	11·68½	59	36·28½	99	60·88½	139	85·48½	179	110·08½
20	12·30	**60**	36·90	**100**	61·50	**140**	86·10	**180**	110·70
21	12·91½	61	37·51½	101	62·11½	141	86·71½	181	111·31½
22	13·53	62	38·13	102	62·73	142	87·33	182	111·93
23	14·14½	63	38·74½	103	63·34½	143	87·94½	183	112·54½
24	14·76	64	39·36	104	63·96	144	88·56	184	113·16
25	15·37½	65	39·97½	105	64·57½	145	89·17½	185	113·77½
26	15·99	66	40·59	106	65·19	146	89·79	186	114·39
27	16·60½	67	41·20½	107	65·80½	147	90·40½	187	115·00½
28	17·22	68	41·82	108	66·42	148	91·02	188	115·62
29	17·83½	69	42·43½	109	67·03½	149	91·63½	189	116·23½
30	18·45	**70**	43·05	**110**	67·65	**150**	92·25	**190**	116·85
31	19·06½	71	43·66½	111	68·26½	151	92·86½	191	117·46½
32	19·68	72	44·28	112	68·88	152	93·48	192	118·08
33	20·29½	73	44·89½	113	69·49½	153	94·09½	193	118·69½
34	20·91	74	45·51	114	70·11	154	94·71	194	119·31
35	21·52½	75	46·12½	115	70·72½	155	95·32½	195	119·92½
36	22·14	76	46·74	116	71·34	156	95·94	196	120·54
37	22·75½	77	47·35½	117	71·95½	157	96·55½	197	121·15½
38	23·37	78	47·97	118	72·57	158	97·17	198	121·77
39	23·98½	79	48·58½	119	73·18½	159	97·78½	199	122·38½
40	24·60	**80**	49·20	**120**	73·80	**160**	98·40	**200**	123·00

$\frac{1}{16}$=0·03^{8} $\frac{1}{9}$=0·06^{8} $\frac{1}{8}$=0·07^{7} $\frac{1}{6}$=0·10^{3} $\frac{1}{5}$=0·12^{3} $\frac{1}{4}$=0·15^{4} $\frac{1}{3}$=0·20^{5}

$\frac{3}{8}$=0·23^{1} $\frac{1}{2}$=0·30^{8} $\frac{5}{8}$=0·38^{4} $\frac{3}{4}$=0·46^{1} $\frac{7}{8}$=0·53^{8} $\frac{2}{3}$=0·41^{0} $\frac{5}{6}$=0·51^{3}

$\frac{3}{16}$=0·11^{5} $\frac{5}{16}$=0·19^{2} $\frac{7}{16}$=0·26^{9} $\frac{9}{16}$=0·34^{6} $\frac{11}{16}$=0·42^{3} $\frac{13}{16}$=0·50^{0} $\frac{15}{16}$=0·57^{7}

(1.61=1£) **£0.62**

1	0·62	41	25·42	81	50·22	121	75·02	161	99·82
2	1·24	42	26·04	82	50·84	122	75·64	162	100·44
3	1·86	43	26·66	83	51·46	123	76·26	163	101·06
4	2·48	44	27·28	84	52·08	124	76·88	164	101·68
5	3·10	45	27·90	85	52·70	125	77·50	165	102·30
6	3·72	46	28·52	86	53·32	126	78·12	166	102·92
7	4·34	47	29·14	87	53·94	127	78·74	167	103·54
8	4·96	48	29·76	88	54·56	128	79·36	168	104·16
9	5·58	49	30·38	89	55·18	129	79·98	169	104·78
10	6·20	**50**	31·00	**90**	55·80	**130**	80·60	**170**	105·40
11	6·82	51	31·62	91	56·42	131	81·22	171	106·02
12	7·44	52	32·24	92	57·04	132	81·84	172	106·64
13	8·06	53	32·86	93	57·66	133	82·46	173	107·26
14	8·68	54	33·48	94	58·28	134	83·08	174	107·88
15	9·30	55	34·10	95	58·90	135	83·70	175	108·50
16	9·92	56	34·72	96	59·52	136	84·32	176	109·12
17	10·54	57	35·34	97	60·14	137	84·94	177	109·74
18	11·16	58	35·96	98	60·76	138	85·56	178	110·36
19	11·78	59	36·58	99	61·38	139	86·18	179	110·98
20	12·40	**60**	37·20	**100**	62·00	**140**	86·80	**180**	111·60
21	13·02	61	37·82	101	62·62	141	87·42	181	112·22
22	13·64	62	38·44	102	63·24	142	88·04	182	112·84
23	14·26	63	39·06	103	63·86	143	88·66	183	113·46
24	14·88	64	39·68	104	64·48	144	89·28	184	114·08
25	15·50	65	40·30	105	65·10	145	89·90	185	114·70
26	16·12	66	40·92	106	65·72	146	90·52	186	115·32
27	16·74	67	41·54	107	66·34	147	91·14	187	115·94
28	17·36	68	42·16	108	66·96	148	91·76	188	116·56
29	17·98	69	42·78	109	67·58	149	92·38	189	117·18
30	18·60	**70**	43·40	**110**	68·20	**150**	93·00	**190**	117·80
31	19·22	71	44·02	111	68·82	151	93·62	191	118·42
32	19·84	72	44·64	112	69·44	152	94·24	192	119·04
33	20·46	73	45·26	113	70·06	153	94·86	193	119·66
34	21·08	74	45·88	114	70·68	154	95·48	194	120·28
35	21·70	75	46·50	115	71·30	155	96·10	195	120·90
36	22·32	76	47·12	116	71·92	156	96·72	196	121·52
37	22·94	77	47·74	117	72·54	157	97·34	197	122·14
38	23·56	78	48·36	118	73·16	158	97·96	198	122·76
39	24·18	79	48·98	119	73·78	159	98·58	199	123·38
40	24·80	**80**	49·60	**120**	74·40	**160**	99·20	**200**	124·00

$\frac{1}{16}=0{\cdot}03^{9}$ $\frac{1}{9}=0{\cdot}06^{9}$ $\frac{1}{8}=0{\cdot}07^{8}$ $\frac{1}{6}=0{\cdot}10^{3}$ $\frac{1}{5}=0{\cdot}12^{4}$ $\frac{1}{4}=0{\cdot}15^{5}$ $\frac{1}{3}=0{\cdot}20^{7}$

$\frac{3}{8}=0{\cdot}23^{3}$ $\frac{1}{2}=0{\cdot}31^{0}$ $\frac{5}{8}=0{\cdot}38^{8}$ $\frac{3}{4}=0{\cdot}46^{5}$ $\frac{7}{8}=0{\cdot}54^{3}$ $\frac{2}{3}=0{\cdot}41^{3}$ $\frac{5}{6}=0{\cdot}51^{7}$

$\frac{3}{16}=0{\cdot}11^{6}$ $\frac{5}{16}=0{\cdot}19^{4}$ $\frac{7}{16}=0{\cdot}27^{1}$ $\frac{9}{16}=0{\cdot}34^{9}$ $\frac{11}{16}=0{\cdot}42^{6}$ $\frac{13}{16}=0{\cdot}50^{4}$ $\frac{15}{16}=0{\cdot}58^{1}$

(1.60=1£) **£0.62½**

1	0·62½	41	25·62½	81	50·62½	121	75·62½	161	100·62½
2	1·25	42	26·25	82	51·25	122	76·25	162	101·25
3	1·87½	43	26·87½	83	51·87½	123	76·87½	163	101·87½
4	2·50	44	27·50	84	52·50	124	77·50	164	102·50
5	3·12½	45	28·12½	85	53·12½	125	78·12½	165	103·12½
6	3·75	46	28·75	86	53·75	126	78·75	166	103·75
7	4·37½	47	29·37½	87	54·37½	127	79·37½	167	104·37½
8	5·00	48	30·00	88	55·00	128	80·00	168	105·00
9	5·62½	49	30·62½	89	55·62½	129	80·62½	169	105·62½
10	6·25	**50**	31·25	**90**	56·25	**130**	81·25	**170**	106·25
11	6·87½	51	31·87½	91	56·87½	131	81·87½	171	106·87½
12	7·50	52	32·50	92	57·50	132	82·50	172	107·50
13	8·12½	53	33·12½	93	58·12½	133	83·12½	173	108·12½
14	8·75	54	33·75	94	58·75	134	83·75	174	108·75
15	9·37½	55	34·37½	95	59·37½	135	84·37½	175	109·37½
16	10·00	56	35·00	96	60·00	136	85·00	176	110·00
17	10·62½	57	35·62½	97	60·62½	137	85·62½	177	110·62½
18	11·25	58	36·25	98	61·25	138	86·25	178	111·25
19	11·87½	59	36·87½	99	61·87½	139	86·87½	179	111·87½
20	12·50	**60**	37·50	**100**	62·50	**140**	87·50	**180**	112·50
21	13·12½	61	38·12½	101	63·12½	141	88·12½	181	113·12½
22	13·75	62	38·75	102	63·75	142	88·75	182	113·75
23	14·37½	63	39·37½	103	64·37½	143	89·37½	183	114·37½
24	15·00	64	40·00	104	65·00	144	90·00	184	115·00
25	15·62½	65	40·62½	105	65·62½	145	90·62½	185	115·62½
26	16·25	66	41·25	106	66·25	146	91·25	186	116·25
27	16·87½	67	41·87½	107	66·87½	147	91·87½	187	116·87½
28	17·50	68	42·50	108	67·50	148	92·50	188	117·50
29	18·12½	69	43·12½	109	68·12½	149	93·12½	189	118·12½
30	18·75	**70**	43·75	**110**	68·75	**150**	93·75	**190**	118·75
31	19·37½	71	44·37½	111	69·37½	151	94·37½	191	119·37½
32	20·00	72	45·00	112	70·00	152	95·00	192	120·00
33	20·62½	73	45·62½	113	70·62½	153	95·62½	193	120·62½
34	21·25	74	46·25	114	71·25	154	96·25	194	121·25
35	21·87½	75	46·87½	115	71·87½	155	96·87½	195	121·87½
36	22·50	76	47·50	116	72·50	156	97·50	196	122·50
37	23·12½	77	48·12½	117	73·12½	157	98·12½	197	123·12½
38	23·75	78	48·75	118	73·75	158	98·75	198	123·75
39	24·37½	79	49·37½	119	74·37½	159	99·37½	199	124·37½
40	25·00	**80**	50·00	**120**	75·00	**160**	100·00	**200**	125·00

$\frac{1}{16}=0{\cdot}03^{9}$ $\frac{1}{9}=0{\cdot}06^{9}$ $\frac{1}{8}=0{\cdot}07^{8}$ $\frac{1}{6}=0{\cdot}10^{4}$ $\frac{1}{5}=0{\cdot}12^{5}$ $\frac{1}{4}=0{\cdot}15^{6}$ $\frac{1}{3}=0{\cdot}20^{8}$

$\frac{3}{8}=0{\cdot}23^{4}$ $\frac{1}{2}=0{\cdot}31^{3}$ $\frac{5}{8}=0{\cdot}39^{1}$ $\frac{3}{4}=0{\cdot}46^{9}$ $\frac{7}{8}=0{\cdot}54^{7}$ $\frac{2}{3}=0{\cdot}41^{7}$ $\frac{5}{6}=0{\cdot}52^{1}$

$\frac{3}{16}=0{\cdot}11^{7}$ $\frac{5}{16}=0{\cdot}19^{5}$ $\frac{7}{16}=0{\cdot}27^{3}$ $\frac{9}{16}=0{\cdot}35^{2}$ $\frac{11}{16}=0{\cdot}43^{0}$ $\frac{13}{16}=0{\cdot}50^{8}$ $\frac{15}{16}=0{\cdot}58^{6}$

1 cubic foot contains 6·23 imperial gallons 1 ounce per gallon=6·236 grammes per litre

(1.59=1£) **£0.63**

1	0·63	41	25·83	81	51·03	121	76·23	161	101·43
2	1·26	42	26·46	82	51·66	122	76·86	162	102·06
3	1·89	43	27·09	83	52·29	123	77·49	163	102·69
4	2·52	44	27·72	84	52·92	124	78·12	164	103·32
5	3·15	45	28·35	85	53·55	125	78·75	165	103·95
6	3·78	46	28·98	86	54·18	126	79·38	166	104·58
7	4·41	47	29·61	87	54·81	127	80·01	167	105·21
8	5·04	48	30·24	88	55·44	128	80·64	168	105·84
9	5·67	49	30·87	89	56·07	129	81·27	169	106·47
10	6·30	**50**	31·50	**90**	56·70	**130**	81·90	**170**	107·10
11	6·93	51	32·13	91	57·33	131	82·53	171	107·73
12	7·56	52	32·76	92	57·96	132	83·16	172	108·36
13	8·19	53	33·39	93	58·59	133	83·79	173	108·99
14	8·82	54	34·02	94	59·22	134	84·42	174	109·62
15	9·45	55	34·65	95	59·85	135	85·05	175	110·25
16	10·08	56	35·28	96	60·48	136	85·68	176	110·88
17	10·71	57	35·91	97	61·11	137	86·31	177	111·51
18	11·34	58	36·54	98	61·74	138	86·94	178	112·14
19	11·97	59	37·17	99	62·37	139	87·57	179	112·77
20	12·60	**60**	37·80	**100**	63·00	**140**	88·20	**180**	113·40
21	13·23	61	38·43	101	63·63	141	88·83	181	114·03
22	13·86	62	39·06	102	64·26	142	89·46	182	114·66
23	14·49	63	39·69	103	64·89	143	90·09	183	115·29
24	15·12	64	40·32	104	65·52	144	90·72	184	115·92
25	15·75	65	40·95	105	66·15	145	91·35	185	116·55
26	16·38	66	41·58	106	66·78	146	91·98	186	117·18
27	17·01	67	42·21	107	67·41	147	92·61	187	117·81
28	17·64	68	42·84	108	68·04	148	93·24	188	118·44
29	18·27	69	43·47	109	68·67	149	93·87	189	119·07
30	18·90	**70**	44·10	**110**	69·30	**150**	94·50	**190**	119·70
31	19·53	71	44·73	111	69·93	151	95·13	191	120·33
32	20·16	72	45·36	112	70·56	152	95·76	192	120·96
33	20·79	73	45·99	113	71·19	153	96·39	193	121·59
34	21·42	74	46·62	114	71·82	154	97·02	194	122·22
35	22·05	75	47·25	115	72·45	155	97·65	195	122·85
36	22·68	76	47·88	116	73·08	156	98·28	196	123·48
37	23·31	77	48·51	117	73·71	157	98·91	197	124·11
38	23·94	78	49·14	118	74·34	158	99·54	198	124·74
39	24·57	79	49·77	119	74·97	159	100·17	199	125·37
40	25·20	**80**	50·40	**120**	75·60	**160**	100·80	**200**	126·00

$\frac{1}{16}=0{\cdot}03^{9}$ $\frac{1}{9}=0{\cdot}07^{0}$ $\frac{1}{8}=0{\cdot}07^{9}$ $\frac{1}{6}=0{\cdot}10^{5}$ $\frac{1}{5}=0{\cdot}12^{6}$ $\frac{1}{4}=0{\cdot}15^{8}$ $\frac{1}{3}=0{\cdot}21^{0}$

$\frac{3}{8}=0{\cdot}23^{6}$ $\frac{1}{2}=0{\cdot}31^{5}$ $\frac{5}{8}=0{\cdot}39^{4}$ $\frac{3}{4}=0{\cdot}47^{3}$ $\frac{7}{8}=0{\cdot}55^{1}$ $\frac{2}{3}=0{\cdot}42^{0}$ $\frac{5}{6}=0{\cdot}52^{5}$

$\frac{3}{16}=0{\cdot}11^{8}$ $\frac{5}{16}=0{\cdot}19^{7}$ $\frac{7}{16}=0{\cdot}27^{6}$ $\frac{9}{16}=0{\cdot}35^{4}$ $\frac{11}{16}=0{\cdot}43^{3}$ $\frac{13}{16}=0{\cdot}51^{2}$ $\frac{15}{16}=0{\cdot}59^{1}$

1	0·63½	41	26·03½	81	51·43½	121	76·83½	161	102·23½
2	1·27	42	26·67	82	52·07	122	77·47	162	102·87
3	1·90½	43	27·30½	83	52·70½	123	78·10½	163	103·50½
4	2·54	44	27·94	84	53·34	124	78·74	164	104·14
5	3·17½	45	28·57½	85	53·97½	125	79·37½	165	104·77½
6	3·81	46	29·21	86	54·61	126	80·01	166	105·41
7	4·44½	47	29·84½	87	55·24½	127	80·64½	167	106·04½
8	5·08	48	30·48	88	55·88	128	81·28	168	106·68
9	5·71½	49	31·11½	89	56·51½	129	81·91½	169	107·31½
10	6·35	**50**	31·75	**90**	57·15	**130**	82·55	**170**	107·95
11	6·98½	51	32·38½	91	57·78½	131	83·18½	171	108·58½
12	7·62	52	33·02	92	58·42	132	83·82	172	109·22
13	8·25½	53	33·65½	93	59·05½	133	84·45½	173	109·85½
14	8·89	54	34·29	94	59·69	134	85·09	174	110·49
15	9·52½	55	34·92½	95	60·32½	135	85·72½	175	111·12½
16	10·16	56	35·56	96	60·96	136	86·36	176	111·76
17	10·79½	57	36·19½	97	61·59½	137	86·99½	177	112·39½
18	11·43	58	36·83	98	62·23	138	87·63	178	113·03
19	12·06½	59	37·46½	99	62·86½	139	88·26½	179	113·66½
20	12·70	**60**	38·10	**100**	63·50	**140**	88·90	**180**	114·30
21	13·33½	61	38·73½	101	64·13½	141	89·53½	181	114·93½
22	13·97	62	39·37	102	64·77	142	90·17	182	115·57
23	14·60½	63	40·00½	103	65·40½	143	90·80½	183	116·20½
24	15·24	64	40·64	104	66·04	144	91·44	184	116·84
25	15·87½	65	41·27½	105	66·67½	145	92·07½	185	117·47½
26	16·51	66	41·91	106	67·31	146	92·71	186	118·11
27	17·14½	67	42·54½	107	67·94½	147	93·34½	187	118·74½
28	17·78	68	43·18	108	68·58	148	93·98	188	119·38
29	18·41½	69	43·81½	109	69·21½	149	94·61½	189	120·01½
30	19·05	**70**	44·45	**110**	69·85	**150**	95·25	**190**	120·65
31	19·68½	71	45·08½	111	70·48½	151	95·88½	191	121·28½
32	20·32	72	45·72	112	71·12	152	96·52	192	121·92
33	20·95½	73	46·35½	113	71·75½	153	97·15½	193	122·55½
34	21·59	74	46·99	114	72·39	154	97·79	194	123·19
35	22·22½	75	47·62½	115	73·02½	155	98·42½	195	123·82½
36	22·86	76	48·26	116	73·66	156	99·06	196	124·46
37	23·49½	77	48·89½	117	74·29½	157	99·69½	197	125·09½
38	24·13	78	49·53	118	74·93	158	100·33	198	125·73
39	24·76½	79	50·16½	119	75·56½	159	100·96½	199	126·36½
40	25·40	**80**	50·80	**120**	76·20	**160**	101·60	**200**	127·00

$\frac{1}{16}=0{\cdot}04^{0}$ $\frac{1}{9}=0{\cdot}07^{1}$ $\frac{1}{8}=0{\cdot}07^{9}$ $\frac{1}{6}=0{\cdot}10^{6}$ $\frac{1}{5}=0{\cdot}12^{7}$ $\frac{1}{4}=0{\cdot}15^{9}$ $\frac{1}{3}=0{\cdot}21^{2}$

$\frac{3}{8}=0{\cdot}23^{8}$ $\frac{1}{2}=0{\cdot}31^{8}$ $\frac{5}{8}=0{\cdot}39^{7}$ $\frac{3}{4}=0{\cdot}47^{6}$ $\frac{7}{8}=0{\cdot}55^{6}$ $\frac{2}{3}=0{\cdot}42^{3}$ $\frac{5}{6}=0{\cdot}52^{9}$

$\frac{3}{16}=0{\cdot}11^{9}$ $\frac{5}{16}=0{\cdot}19^{8}$ $\frac{7}{16}=0{\cdot}27^{8}$ $\frac{9}{16}=0{\cdot}35^{7}$ $\frac{11}{16}=0{\cdot}43^{7}$ $\frac{13}{16}=0{\cdot}51^{6}$ $\frac{15}{16}=0{\cdot}59^{5}$

(1.56=1£) **£0.64**

1	0·64	41	26·24	81	51·84	121	77·44	161	103·04
2	1·28	42	26·88	82	52·48	122	78·08	162	103·68
3	1·92	43	27·52	83	53·12	123	78·72	163	104·32
4	2·56	44	28·16	84	53·76	124	79·36	164	104·96
5	3·20	45	28·80	85	54·40	125	80·00	165	105·60
6	3·84	46	29·44	86	55·04	126	80·64	166	106·24
7	4·48	47	30·08	87	55·68	127	81·28	167	106·88
8	5·12	48	30·72	88	56·32	128	81·92	168	107·52
9	5·76	49	31·36	89	56·96	129	82·56	169	108·16
10	6·40	**50**	32·00	**90**	57·60	**130**	83·20	**170**	108·80
11	7·04	51	32·64	91	58·24	131	83·84	171	109·44
12	7·68	52	33·28	92	58·88	132	84·48	172	110·08
13	8·32	53	33·92	93	59·52	133	85·12	173	110·72
14	8·96	54	34·56	94	60·16	134	85·76	174	111·36
15	9·60	55	35·20	95	60·80	135	86·40	175	112·00
16	10·24	56	35·84	96	61·44	136	87·04	176	112·64
17	10·88	57	36·48	97	62·08	137	87·68	177	113·28
18	11·52	58	37·12	98	62·72	138	88·32	178	113·92
19	12·16	59	37·76	99	63·36	139	88·96	179	114·56
20	12·80	**60**	38·40	**100**	64·00	**140**	89·60	**180**	115·20
21	13·44	61	39·04	101	64·64	141	90·24	181	115·84
22	14·08	62	39·68	102	65·28	142	90·88	182	116·48
23	14·72	63	40·32	103	65·92	143	91·52	183	117·12
24	15·36	64	40·96	104	66·56	144	92·16	184	117·76
25	16·00	65	41·60	105	67·20	145	92·80	185	118·40
26	16·64	66	42·24	106	67·84	146	93·44	186	119·04
27	17·28	67	42·88	107	68·48	147	94·08	187	119·68
28	17·92	68	43·52	108	69·12	148	94·72	188	120·32
29	18·56	69	44·16	109	69·76	149	95·36	189	120·96
30	19·20	**70**	44·80	**110**	70·40	**150**	96·00	**190**	121·60
31	19·84	71	45·44	111	71·04	151	96·64	191	122·24
32	20·48	72	46·08	112	71·68	152	97·28	192	122·88
33	21·12	73	46·72	113	72·32	153	97·92	193	123·52
34	21·76	74	47·36	114	72·96	154	98·56	194	124·16
35	22·40	75	48·00	115	73·60	155	99·20	195	124·80
36	23·04	76	48·64	116	74·24	156	99·84	196	125·44
37	23·68	77	49·28	117	74·88	157	100·48	197	126·08
38	24·32	78	49·92	118	75·52	158	101·12	198	126·72
39	24·96	79	50·56	119	76·16	159	101·76	199	127·36
40	25·60	**80**	51·20	**120**	76·80	**160**	102·40	**200**	128·00

$\frac{1}{16}=0{\cdot}04^{0}$ $\frac{1}{9}=0{\cdot}07^{1}$ $\frac{1}{8}=0{\cdot}08^{0}$ $\frac{1}{6}=0{\cdot}10^{7}$ $\frac{1}{5}=0{\cdot}12^{8}$ $\frac{1}{4}=0{\cdot}16^{0}$ $\frac{1}{3}=0{\cdot}21^{3}$

$\frac{3}{8}=0{\cdot}24^{0}$ $\frac{1}{2}=0{\cdot}32^{0}$ $\frac{5}{8}=0{\cdot}40^{0}$ $\frac{3}{4}=0{\cdot}48^{0}$ $\frac{7}{8}=0{\cdot}56^{0}$ $\frac{2}{3}=0{\cdot}42^{7}$ $\frac{5}{6}=0{\cdot}53^{3}$

$\frac{3}{16}=0{\cdot}12^{0}$ $\frac{5}{16}=0{\cdot}20^{0}$ $\frac{7}{16}=0{\cdot}28^{0}$ $\frac{9}{16}=0{\cdot}36^{0}$ $\frac{11}{16}=0{\cdot}44^{0}$ $\frac{13}{16}=0{\cdot}52^{0}$ $\frac{15}{16}=0{\cdot}60^{0}$

(1.55=1£) **£0.64½**

1	0·64½	41	26·44½	81	52·24½	121	78·04½	161	103·84½
2	1·29	42	27·09	82	52·89	122	78·69	162	104·49
3	1·93½	43	27·73½	83	53·53½	123	79·33½	163	105·13½
4	2·58	44	28·38	84	54·18	124	79·98	164	105·78
5	3·22½	45	29·02½	85	54·82½	125	80·62½	165	106·42½
6	3·87	46	29·67	86	55·47	126	81·27	166	107·07
7	4·51½	47	30·31½	87	56·11½	127	81·91½	167	107·71½
8	5·16	48	30·96	88	56·76	128	82·56	168	108·36
9	5·80½	49	31·60½	89	57·40½	129	83·20½	169	109·00½
10	6·45	**50**	32·25	**90**	58·05	**130**	83·85	**170**	109·65
11	7·09½	51	32·89½	91	58·69½	131	84·49½	171	110·29½
12	7·74	52	33·54	92	59·34	132	85·14	172	110·94
13	8·38½	53	34·18½	93	59·98½	133	85·78½	173	111·58½
14	9·03	54	34·83	94	60·63	134	86·43	174	112·23
15	9·67½	55	35·47½	95	61·27½	135	87·07½	175	112·87½
16	10·32	56	36·12	96	61·92	136	87·72	176	113·52
17	10·96½	57	36·76½	97	62·56½	137	88·36½	177	114·16½
18	11·61	58	37·41	98	63·21	138	89·01	178	114·81
19	12·25½	59	38·05½	99	63·85½	139	89·65½	179	115·45½
20	12·90	**60**	38·70	**100**	64·50	**140**	90·30	**180**	116·10
21	13·54½	61	39·34½	101	65·14½	141	90·94½	181	116·74½
22	14·19	62	39·99	102	65·79	142	91·59	182	117·39
23	14·83½	63	40·63½	103	66·43½	143	92·23½	183	118·03½
24	15·48	64	41·28	104	67·08	144	92·88	184	118·68
25	16·12½	65	41·92½	105	67·72½	145	93·52½	185	119·32½
26	16·77	66	42·57	106	68·37	146	94·17	186	119·97
27	17·41½	67	43·21½	107	69·01½	147	94·81½	187	120·61½
28	18·06	68	43·86	108	69·66	148	95·46	188	121·26
29	18·70½	69	44·50½	109	70·30½	149	96·10½	189	121·90½
30	19·35	**70**	45·15	**110**	70·95	**150**	96·75	**190**	122·55
31	19·99½	71	45·79½	111	71·59½	151	97·39½	191	123·19½
32	20·64	72	46·44	112	72·24	152	98·04	192	123·84
33	21·28½	73	47·08½	113	72·88½	153	98·68½	193	124·48½
34	21·93	74	47·73	114	73·53	154	99·33	194	125·13
35	22·57½	75	48·37½	115	74·17½	155	99·97½	195	125·77½
36	23·22	76	49·02	116	74·82	156	100·62	196	126·42
37	23·86½	77	49·66½	117	75·46½	157	101·26½	197	127·06½
38	24·51	78	50·31	118	76·11	158	101·91	198	127·71
39	25·15½	79	50·95½	119	76·75½	159	102·55½	199	128·35½
40	25·80	**80**	51·60	**120**	77·40	**160**	103·20	**200**	129·00

$\frac{1}{16}=0{\cdot}04^{0}$	$\frac{1}{9}=0{\cdot}07^{2}$	$\frac{1}{8}=0{\cdot}08^{1}$	$\frac{1}{6}=0{\cdot}10^{8}$	$\frac{1}{5}=0{\cdot}12^{9}$	$\frac{1}{4}=0{\cdot}16^{1}$	$\frac{1}{3}=0{\cdot}21^{5}$
$\frac{3}{8}=0{\cdot}24^{2}$	$\frac{1}{2}=0{\cdot}32^{3}$	$\frac{5}{8}=0{\cdot}40^{3}$	$\frac{3}{4}=0{\cdot}48^{4}$	$\frac{7}{8}=0{\cdot}56^{4}$	$\frac{2}{3}=0{\cdot}43^{0}$	$\frac{5}{6}=0{\cdot}53^{8}$
$\frac{3}{16}=0{\cdot}12^{1}$	$\frac{5}{16}=0{\cdot}20^{2}$	$\frac{7}{16}=0{\cdot}28^{2}$	$\frac{9}{16}=0{\cdot}36^{3}$	$\frac{11}{16}=0{\cdot}44^{3}$	$\frac{13}{16}=0{\cdot}52^{4}$	$\frac{15}{16}=0{\cdot}60^{5}$

(1.54=1£) **£0.65**

1	0·65	41	26·65	81	52·65	121	78·65	161	104·65
2	1·30	42	27·30	82	53·30	122	79·30	162	105·30
3	1·95	43	27·95	83	53·95	123	79·95	163	105·95
4	2·60	44	28·60	84	54·60	124	80·60	164	106·60
5	3·25	45	29·25	85	55·25	125	81·25	165	107·25
6	3·90	46	29·90	86	55·90	126	81·90	166	107·90
7	4·55	47	30·55	87	56·55	127	82·55	167	108·55
8	5·20	48	31·20	88	57·20	128	83·20	168	109·20
9	5·85	49	31·85	89	57·85	129	83·85	169	109·85
10	6·50	**50**	32·50	**90**	58·50	**130**	84·50	**170**	110·50
11	7·15	51	33·15	91	59·15	131	85·15	171	111·15
12	7·80	52	33·80	92	59·80	132	85·80	172	111·80
13	8·45	53	34·45	93	60·45	133	86·45	173	112·45
14	9·10	54	35·10	94	61·10	134	87·10	174	113·10
15	9·75	55	35·75	95	61·75	135	87·75	175	113·75
16	10·40	56	36·40	96	62·40	136	88·40	176	114·40
17	11·05	57	37·05	97	63·05	137	89·05	177	115·05
18	11·70	58	37·70	98	63·70	138	89·70	178	115·70
19	12·35	59	38·35	99	64·35	139	90·35	179	116·35
20	13·00	**60**	39·00	**100**	65·00	**140**	91·00	**180**	117·00
21	13·65	61	39·65	101	65·65	141	91·65	181	117·65
22	14·30	62	40·30	102	66·30	142	92·30	182	118·30
23	14·95	63	40·95	103	66·95	143	92·95	183	118·95
24	15·60	64	41·60	104	67·60	144	93·60	184	119·60
25	16·25	65	42·25	105	68·25	145	94·25	185	120·25
26	16·90	66	42·90	106	68·90	146	94·90	186	120·90
27	17·55	67	43·55	107	69·55	147	95·55	187	121·55
28	18·20	68	44·20	108	70·20	148	96·20	188	122·20
29	18·85	69	44·85	109	70·85	149	96·85	189	122·85
30	19·50	**70**	45·50	**110**	71·50	**150**	97·50	**190**	123·50
31	20·15	71	46·15	111	72·15	151	98·15	191	124·15
32	20·80	72	46·80	112	72·80	152	98·80	192	124·80
33	21·45	73	47·45	113	73·45	153	99·45	193	125·45
34	22·10	74	48·10	114	74·10	154	100·10	194	126·10
35	22·75	75	48·75	115	74·75	155	100·75	195	126·75
36	23·40	76	49·40	116	75·40	156	101·40	196	127·40
37	24·05	77	50·05	117	76·05	157	102·05	197	128·05
38	24·70	78	50·70	118	76·70	158	102·70	198	128·70
39	25·35	79	51·35	119	77·35	159	103·35	199	129·35
40	26·00	**80**	52·00	**120**	78·00	**160**	104·00	**200**	130·00

$\frac{1}{16}$=0·04^{1} $\frac{1}{9}$=0·07^{2} $\frac{1}{8}$=0·08^{1} $\frac{1}{6}$=0·10^{8} $\frac{1}{5}$=0·13^{0} $\frac{1}{4}$=0·16^{3} $\frac{1}{3}$=0·21^{7}

$\frac{3}{8}$=0·24^{4} $\frac{1}{2}$=0·32^{5} $\frac{5}{8}$=0·40^{6} $\frac{3}{4}$=0·48^{8} $\frac{7}{8}$=0·56^{9} $\frac{2}{3}$=0·43^{3} $\frac{5}{6}$=0·54^{2}

$\frac{3}{16}$=0·12^{2} $\frac{5}{16}$=0·20^{3} $\frac{7}{16}$=0·28^{4} $\frac{9}{16}$=0·36^{6} $\frac{11}{16}$=0·44^{7} $\frac{13}{16}$=0·52^{8} $\frac{15}{16}$=0·60^{9}

(1.53=1£) **£0.65½**

1	0·65½	41	26·85½	81	53·05½	121	79·25½	161	105·45½
2	1·31	42	27·51	82	53·71	122	79·91	162	106·11
3	1·96½	43	28·16½	83	54·36½	123	80·56½	163	106·76½
4	2·62	44	28·82	84	55·02	124	81·22	164	107·42
5	3·27½	45	29·47½	85	55·67½	125	81·87½	165	108·07½
6	3·93	46	30·13	86	56·33	126	82·53	166	108·73
7	4·58½	47	30·78½	87	56·98½	127	83·18½	167	109·38½
8	5·24	48	31·44	88	57·64	128	83·84	168	110·04
9	5·89½	49	32·09½	89	58·29½	129	84·49½	169	110·69½
10	6·55	**50**	32·75	**90**	58·95	**130**	85·15	**170**	111·35
11	7·20½	51	33·40½	91	59·60½	131	85·80½	171	112·00½
12	7·86	52	34·06	92	60·26	132	86·46	172	112·66
13	8·51½	53	34·71½	93	60·91½	133	87·11½	173	113·31½
14	9·17	54	35·37	94	61·57	134	87·77	174	113·97
15	9·82½	55	36·02½	95	62·22½	135	88·42½	175	114·62½
16	10·48	56	36·68	96	62·88	136	89·08	176	115·28
17	11·13½	57	37·33½	97	63·53½	137	89·73½	177	115·93½
18	11·79	58	37·99	98	64·19	138	90·39	178	116·59
19	12·44½	59	38·64½	99	64·84½	139	91·04½	179	117·24½
20	13·10	**60**	39·30	**100**	65·50	**140**	91·70	**180**	117·90
21	13·75½	61	39·95½	101	66·15½	141	92·35½	181	118·55½
22	14·41	62	40·61	102	66·81	142	93·01	182	119·21
23	15·06½	63	41·26½	103	67·46½	143	93·66½	183	119·86½
24	15·72	64	41·92	104	68·12	144	94·32	184	120·52
25	16·37½	65	42·57½	105	68·77½	145	94·97½	185	121·17½
26	17·03	66	43·23	106	69·43	146	95·63	186	121·83
27	17·68½	67	43·88½	107	70·08½	147	96·28½	187	122·48½
28	18·34	68	44·54	108	70·74	148	96·94	188	123·14
29	18·99½	69	45·19½	109	71·39½	149	97·59½	189	123·79½
30	19·65	**70**	45·85	**110**	72·05	**150**	98·25	**190**	124·45
31	20·30½	71	46·50½	111	72·70½	151	98·90½	191	125·10½
32	20·96	72	47·16	112	73·36	152	99·56	192	125·76
33	21·61½	73	47·81½	113	74·01½	153	100·21½	193	126·41½
34	22·27	74	48·47	114	74·67	154	100·87	194	127·07
35	22·92½	75	49·12½	115	75·32½	155	101·52½	195	127·72½
36	23·58	76	49·78	116	75·98	156	102·18	196	128·38
37	24·23½	77	50·43½	117	76·63½	157	102·83½	197	129·03½
38	24·89	78	51·09	118	77·29	158	103·49	198	129·69
39	25·54½	79	51·74½	119	77·94½	159	104·14½	199	130·34½
40	26·20	**80**	52·40	**120**	78·60	**160**	104·80	**200**	131·00

$\frac{1}{16}=0{\cdot}04^{1}$ $\frac{1}{9}=0{\cdot}07^{3}$ $\frac{1}{8}=0{\cdot}08^{2}$ $\frac{1}{6}=0{\cdot}10^{9}$ $\frac{1}{5}=0{\cdot}13^{1}$ $\frac{1}{4}=0{\cdot}16^{4}$ $\frac{1}{3}=0{\cdot}21^{8}$

$\frac{3}{8}=0{\cdot}24^{6}$ $\frac{1}{2}=0{\cdot}32^{8}$ $\frac{5}{8}=0{\cdot}40^{9}$ $\frac{3}{4}=0{\cdot}49^{1}$ $\frac{7}{8}=0{\cdot}57^{3}$ $\frac{2}{3}=0{\cdot}43^{7}$ $\frac{5}{6}=0{\cdot}54^{6}$

$\frac{3}{16}=0{\cdot}12^{3}$ $\frac{5}{16}=0{\cdot}20^{5}$ $\frac{7}{16}=0{\cdot}28^{7}$ $\frac{9}{16}=0{\cdot}36^{8}$ $\frac{11}{16}=0{\cdot}45^{0}$ $\frac{13}{16}=0{\cdot}53^{2}$ $\frac{15}{16}=0{\cdot}61^{4}$

(1.52=1£) **£0.66**

1	0·66	41	27·06	81	53·46	121	79·86	161	106·26
2	1·32	42	27·72	82	54·12	122	80·52	162	106·92
3	1·98	43	28·38	83	54·78	123	81·18	163	107·58
4	2·64	44	29·04	84	55·44	124	81·84	164	108·24
5	3·30	45	29·70	85	56·10	125	82·50	165	108·90
6	3·96	46	30·36	86	56·76	126	83·16	166	109·56
7	4·62	47	31·02	87	57·42	127	83·82	167	110·22
8	5·28	48	31·68	88	58·08	128	84·48	168	110·88
9	5·94	49	32·34	89	58·74	129	85·14	169	111·54
10	6·60	**50**	33·00	**90**	59·40	**130**	85·80	**170**	112·20
11	7·26	51	33·66	91	60·06	131	86·46	171	112·86
12	7·92	52	34·32	92	60·72	132	87·12	172	113·52
13	8·58	53	34·98	93	61·38	133	87·78	173	114·18
14	9·24	54	35·64	94	62·04	134	88·44	174	114·84
15	9·90	55	36·30	95	62·70	135	89·10	175	115·50
16	10·56	56	36·96	96	63·36	136	89·76	176	116·16
17	11·22	57	37·62	97	64·02	137	90·42	177	116·82
18	11·88	58	38·28	98	64·68	138	91·08	178	117·48
19	12·54	59	38·94	99	65·34	139	91·74	179	118·14
20	13·20	**60**	39·60	**100**	66·00	**140**	92·40	**180**	118·80
21	13·86	61	40·26	101	66·66	141	93·06	181	119·46
22	14·52	62	40·92	102	67·32	142	93·72	182	120·12
23	15·18	63	41·58	103	67·98	143	94·38	183	120·78
24	15·84	64	42·24	104	68·64	144	95·04	184	121·44
25	16·50	65	42·90	105	69·30	145	95·70	185	122·10
26	17·16	66	43·56	106	69·96	146	96·36	186	122·76
27	17·82	67	44·22	107	70·62	147	97·02	187	123·42
28	18·48	68	44·88	108	71·28	148	97·68	188	124·08
29	19·14	69	45·54	109	71·94	149	98·34	189	124·74
30	19·80	**70**	46·20	**110**	72·60	**150**	99·00	**190**	125·40
31	20·46	71	46·86	111	73·26	151	99·66	191	126·06
32	21·12	72	47·52	112	73·92	152	100·32	192	126·72
33	21·78	73	48·18	113	74·58	153	100·98	193	127·38
34	22·44	74	48·84	114	75·24	154	101·64	194	128·04
35	23·10	75	49·50	115	75·90	155	102·30	195	128·70
36	23·76	76	50·16	116	76·56	156	102·96	196	129·36
37	24·42	77	50·82	117	77·22	157	103·62	197	130·02
38	25·08	78	51·48	118	77·88	158	104·28	198	130·68
39	25·74	79	52·14	119	78·54	159	104·94	199	131·34
40	26·40	**80**	52·80	**120**	79·20	**160**	105·60	**200**	132·00

$\frac{1}{16}=0{\cdot}04^{1}$ $\frac{1}{9}=0{\cdot}07^{3}$ $\frac{1}{8}=0{\cdot}08^{3}$ $\frac{1}{6}=0{\cdot}11^{0}$ $\frac{1}{5}=0{\cdot}13^{2}$ $\frac{1}{4}=0{\cdot}16^{5}$ $\frac{1}{3}=0{\cdot}22^{0}$

$\frac{3}{8}=0{\cdot}24^{8}$ $\frac{1}{2}=0{\cdot}33^{0}$ $\frac{5}{8}=0{\cdot}41^{3}$ $\frac{3}{4}=0{\cdot}49^{5}$ $\frac{7}{8}=0{\cdot}57^{8}$ $\frac{2}{3}=0{\cdot}44^{0}$ $\frac{5}{6}=0{\cdot}55^{0}$

$\frac{3}{16}=0{\cdot}12^{4}$ $\frac{5}{16}=0{\cdot}20^{6}$ $\frac{7}{16}=0{\cdot}28^{9}$ $\frac{9}{16}=0{\cdot}37^{1}$ $\frac{11}{16}=0{\cdot}45^{4}$ $\frac{13}{16}=0{\cdot}53^{6}$ $\frac{15}{16}=0{\cdot}61^{9}$

(1.50=1£) **£0.66½**

1	0·66½	41	27·26½	81	53·86½	121	80·46½	161	107·06½
2	1·33	42	27·93	82	54·53	122	81·13	162	107·73
3	1·99½	43	28·59½	83	55·19½	123	81·79½	163	108·39½
4	2·66	44	29·26	84	55·86	124	82·46	164	109·06
5	3·32½	45	29·92½	85	56·52½	125	83·12½	165	109·72½
6	3·99	46	30·59	86	57·19	126	83·79	166	110·39
7	4·65½	47	31·25½	87	57·85½	127	84·45½	167	111·05½
8	5·32	48	31·92	88	58·52	128	85·12	168	111·72
9	5·98½	49	32·58½	89	59·18½	129	85·78½	169	112·38½
10	6·65	**50**	33·25	**90**	59·85	**130**	86·45	**170**	113·05
11	7·31½	51	33·91½	91	60·51½	131	87·11½	171	113·71½
12	7·98	52	34·58	92	61·18	132	87·78	172	114·38
13	8·64½	53	35·24½	93	61·84½	133	88·44½	173	115·04½
14	9·31	54	35·91	94	62·51	134	89·11	174	115·71
15	9·97½	55	36·57½	95	63·17½	135	89·77½	175	116·37½
16	10·64	56	37·24	96	63·84	136	90·44	176	117·04
17	11·30½	57	37·90½	97	64·50½	137	91·10½	177	117·70½
18	11·97	58	38·57	98	65·17	138	91·77	178	118·37
19	12·63½	59	39·23½	99	65·83½	139	92·43½	179	119·03½
20	13·30	**60**	39·90	**100**	66·50	**140**	93·10	**180**	119·70
21	13·96½	61	40·56½	101	67·16½	141	93·76½	181	120·36½
22	14·63	62	41·23	102	67·83	142	94·43	182	121·03
23	15·29½	63	41·89½	103	68·49½	143	95·09½	183	121·69½
24	15·96	64	42·56	104	69·16	144	95·76	184	122·36
25	16·62½	65	43·22½	105	69·82½	145	96·42½	185	123·02½
26	17·29	66	43·89	106	70·49	146	97·09	186	123·69
27	17·95½	67	44·55½	107	71·15½	147	97·75½	187	124·35½
28	18·62	68	45·22	108	71·82	148	98·42	188	125·02
29	19·28½	69	45·88½	109	72·48½	149	99·08½	189	125·68½
30	19·95	**70**	46·55	**110**	73·15	**150**	99·75	**190**	126·35
31	20·61½	71	47·21½	111	73·81½	151	100·41½	191	127·01½
32	21·28	72	47·88	112	74·48	152	101·08	192	127·68
33	21·94½	73	48·54½	113	75·14½	153	101·74½	193	128·34½
34	22·61	74	49·21	114	75·81	154	102·41	194	129·01
35	23·27½	75	49·87½	115	76·47½	155	103·07½	195	129·67½
36	23·94	76	50·54	116	77·14	156	103·74	196	130·34
37	24·60½	77	51·20½	117	77·80½	157	104·40½	197	131·00½
38	25·27	78	51·87	118	78·47	158	105·07	198	131·67
39	25·93½	79	52·53½	119	79·13½	159	105·73½	199	132·33½
40	26·60	**80**	53·20	**120**	79·80	**160**	106·40	**200**	133·00

$\frac{1}{16}=0{\cdot}04^{2}$ $\frac{1}{9}=0{\cdot}07^{4}$ $\frac{1}{8}=0{\cdot}08^{3}$ $\frac{1}{6}=0{\cdot}11^{1}$ $\frac{1}{5}=0{\cdot}13^{3}$ $\frac{1}{4}=0{\cdot}16^{6}$ $\frac{1}{3}=0{\cdot}22^{2}$

$\frac{3}{8}=0{\cdot}24^{9}$ $\frac{1}{2}=0{\cdot}33^{3}$ $\frac{5}{8}=0{\cdot}41^{6}$ $\frac{3}{4}=0{\cdot}49^{9}$ $\frac{7}{8}=0{\cdot}58^{2}$ $\frac{2}{3}=0{\cdot}44^{3}$ $\frac{5}{6}=0{\cdot}55^{4}$

$\frac{3}{16}=0{\cdot}12^{5}$ $\frac{5}{16}=0{\cdot}20^{8}$ $\frac{7}{16}=0{\cdot}29^{1}$ $\frac{9}{16}=0{\cdot}37^{4}$ $\frac{11}{16}=0{\cdot}45^{7}$ $\frac{13}{16}=0{\cdot}54^{0}$ $\frac{15}{16}=0{\cdot}62^{3}$

(1.49=1£) **£0.67**

1	0·67	41	27·47	81	54·27	121	81·07	161	107·87
2	1·34	42	28·14	82	54·94	122	81·74	162	108·54
3	2·01	43	28·81	83	55·61	123	82·41	163	109·21
4	2·68	44	29·48	84	56·28	124	83·08	164	109·88
5	3·35	45	30·15	85	56·95	125	83·75	165	110·55
6	4·02	46	30·82	86	57·62	126	84·42	166	111·22
7	4·69	47	31·49	87	58·29	127	85·09	167	111·89
8	5·36	48	32·16	88	58·96	128	85·76	168	112·56
9	6·03	49	32·83	89	59·63	129	86·43	169	113·23
10	6·70	**50**	33·50	**90**	60·30	**130**	87·10	**170**	113·90
11	7·37	51	34·17	91	60·97	131	87·77	171	114·57
12	8·04	52	34·84	92	61·64	132	88·44	172	115·24
13	8·71	53	35·51	93	62·31	133	89·11	173	115·91
14	9·38	54	36·18	94	62·98	134	89·78	174	116·58
15	10·05	55	36·85	95	63·65	135	90·45	175	117·25
16	10·72	56	37·52	96	64·32	136	91·12	176	117·92
17	11·39	57	38·19	97	64·99	137	91·79	177	118·59
18	12·06	58	38·86	98	65·66	138	92·46	178	119·26
19	12·73	59	39·53	99	66·33	139	93·13	179	119·93
20	13·40	**60**	40·20	**100**	67·00	**140**	93·80	**180**	120·60
21	14·07	61	40·87	101	67·67	141	94·47	181	121·27
22	14·74	62	41·54	102	68·34	142	95·14	182	121·94
23	15·41	63	42·21	103	69·01	143	95·81	183	122·61
24	16·08	64	42·88	104	69·68	144	96·48	184	123·28
25	16·75	65	43·55	105	70·35	145	97·15	185	123·95
26	17·42	66	44·22	106	71·02	146	97·82	186	124·62
27	18·09	67	44·89	107	71·69	147	98·49	187	125·29
28	18·76	68	45·56	108	72·36	148	99·16	188	125·96
29	19·43	69	46·23	109	73·03	149	99·83	189	126·63
30	20·10	**70**	46·90	**110**	73·70	**150**	100·50	**190**	127·30
31	20·77	71	47·57	111	74·37	151	101·17	191	127·97
32	21·44	72	48·24	112	75·04	152	101·84	192	128·64
33	22·11	73	48·91	113	75·71	153	102·51	193	129·31
34	22·78	74	49·58	114	76·38	154	103·18	194	129·98
35	23·45	75	50·25	115	77·05	155	103·85	195	130·65
36	24·12	76	50·92	116	77·72	156	104·52	196	131·32
37	24·79	77	51·59	117	78·39	157	105·19	197	131·99
38	25·46	78	52·26	118	79·06	158	105·86	198	132·66
39	26·13	79	52·93	119	79·73	159	106·53	199	133·33
40	26·80	**80**	53·60	**120**	80·40	**160**	107·20	**200**	134·00

$\frac{1}{16}=0{\cdot}04^2$ $\frac{1}{9}=0{\cdot}07^4$ $\frac{1}{8}=0{\cdot}08^4$ $\frac{1}{6}=0{\cdot}11^2$ $\frac{1}{5}=0{\cdot}13^4$ $\frac{1}{4}=0{\cdot}16^8$ $\frac{1}{3}=0{\cdot}22^3$

$\frac{3}{8}=0{\cdot}25^1$ $\frac{1}{2}=0{\cdot}33^5$ $\frac{5}{8}=0{\cdot}41^9$ $\frac{3}{4}=0{\cdot}50^3$ $\frac{7}{8}=0{\cdot}58^6$ $\frac{2}{3}=0{\cdot}44^7$ $\frac{5}{6}=0{\cdot}55^8$

$\frac{3}{16}=0{\cdot}12^6$ $\frac{5}{16}=0{\cdot}20^9$ $\frac{7}{16}=0{\cdot}29^3$ $\frac{9}{16}=0{\cdot}37^7$ $\frac{11}{16}=0{\cdot}46^1$ $\frac{13}{16}=0{\cdot}54^4$ $\frac{15}{16}=0{\cdot}62^8$

(1.48=1£) **£0.67½**

1	0·67½	41	27·67½	81	54·67½	121	81·67½	161	108·67½
2	1·35	42	28·35	82	55·35	122	82·35	162	109·35
3	2·02½	43	29·02½	83	56·02½	123	83·02½	163	110·02½
4	2·70	44	29·70	84	56·70	124	83·70	164	110·70
5	3·37½	45	30·37½	85	57·37½	125	84·37½	165	111·37½
6	4·05	46	31·05	86	58·05	126	85·05	166	112·05
7	4·72½	47	31·72½	87	58·72½	127	85·72½	167	112·72½
8	5·40	48	32·40	88	59·40	128	86·40	168	113·40
9	6·07½	49	33·07½	89	60·07½	129	87·07½	169	114·07½
10	6·75	**50**	33·75	**90**	60·75	**130**	87·75	**170**	114·75
11	7·42½	51	34·42½	91	61·42½	131	88·42½	171	115·42½
12	8·10	52	35·10	92	62·10	132	89·10	172	116·10
13	8·77½	53	35·77½	93	62·77½	133	89·77½	173	116·77½
14	9·45	54	36·45	94	63·45	134	90·45	174	117·45
15	10·12½	55	37·12½	95	64·12½	135	91·12½	175	118·12½
16	10·80	56	37·80	96	64·80	136	91·80	176	118·80
17	11·47½	57	38·47½	97	65·47½	137	92·47½	177	119·47½
18	12·15	58	39·15	98	66·15	138	93·15	178	120·15
19	12·82½	59	39·82½	99	66·82½	139	93·82½	179	120·82½
20	13·50	**60**	40·50	**100**	67·50	**140**	94·50	**180**	121·50
21	14·17½	61	41·17½	101	68·17½	141	95·17½	181	122·17½
22	14·85	62	41·85	102	68·85	142	95·85	182	122·85
23	15·52½	63	42·52½	103	69·52½	143	96·52½	183	123·52½
24	16·20	64	43·20	104	70·20	144	97·20	184	124·20
25	16·87½	65	43·87½	105	70·87½	145	97·87½	185	124·87½
26	17·55	66	44·55	106	71·55	146	98·55	186	125·55
27	18·22½	67	45·22½	107	72·22½	147	99·22½	187	126·22½
28	18·90	68	45·90	108	72·90	148	99·90	188	126·90
29	19·57½	69	46·57½	109	73·57½	149	100·57½	189	127·57½
30	20·25	**70**	47·25	**110**	74·25	**150**	101·25	**190**	128·25
31	20·92½	71	47·92½	111	74·92½	151	101·92½	191	128·92½
32	21·60	72	48·60	112	75·60	152	102·60	192	129·60
33	22·27½	73	49·27½	113	76·27½	153	103·27½	193	130·27½
34	22·95	74	49·95	114	76·95	154	103·95	194	130·95
35	23·62½	75	50·62½	115	77·62½	155	104·62½	195	131·62½
36	24·30	76	51·30	116	78·30	156	105·30	196	132·30
37	24·97½	77	51·97½	117	78·97½	157	105·97½	197	132·97½
38	25·65	78	52·65	118	79·65	158	106·65	198	133·65
39	26·32½	79	53·32½	119	80·32½	159	107·32½	199	134·32½
40	27·00	**80**	54·00	**120**	81·00	**160**	108·00	**200**	135·00

$\frac{1}{16}=0{\cdot}04^{2}$ $\frac{1}{9}=0{\cdot}07^{5}$ $\frac{1}{8}=0{\cdot}08^{4}$ $\frac{1}{6}=0{\cdot}11^{3}$ $\frac{1}{5}=0{\cdot}13^{5}$ $\frac{1}{4}=0{\cdot}16^{9}$ $\frac{1}{3}=0{\cdot}22^{5}$

$\frac{3}{8}=0{\cdot}25^{3}$ $\frac{1}{2}=0{\cdot}33^{8}$ $\frac{5}{8}=0{\cdot}42^{2}$ $\frac{3}{4}=0{\cdot}50^{6}$ $\frac{7}{8}=0{\cdot}59^{1}$ $\frac{2}{3}=0{\cdot}45^{0}$ $\frac{5}{6}=0{\cdot}56^{3}$

$\frac{3}{16}=0{\cdot}12^{7}$ $\frac{5}{16}=0{\cdot}21^{1}$ $\frac{7}{16}=0{\cdot}29^{5}$ $\frac{9}{16}=0{\cdot}38^{0}$ $\frac{11}{16}=0{\cdot}46^{4}$ $\frac{13}{16}=0{\cdot}54^{8}$ $\frac{15}{16}=0{\cdot}63^{3}$

(1.47=1£) **£0.68**

1	0·68	41	27·88	81	55·08	121	82·28	161	109·48
2	1·36	42	28·56	82	55·76	122	82·96	162	110·16
3	2·04	43	29·24	83	56·44	123	83·64	163	110·84
4	2·72	44	29·92	84	57·12	124	84·32	164	111·52
5	3·40	45	30·60	85	57·80	125	85·00	165	112·20
6	4·08	46	31·28	86	58·48	126	85·68	166	112·88
7	4·76	47	31·96	87	59·16	127	86·36	167	113·56
8	5·44	48	32·64	88	59·84	128	87·04	168	114·24
9	6·12	49	33·32	89	60·52	129	87·72	169	114·92
10	6·80	**50**	34·00	**90**	61·20	**130**	88·40	**170**	115·60
11	7·48	51	34·68	91	61·88	131	89·08	171	116·28
12	8·16	52	35·36	92	62·56	132	89·76	172	116·96
13	8·84	53	36·04	93	63·24	133	90·44	173	117·64
14	9·52	54	36·72	94	63·92	134	91·12	174	118·32
15	10·20	55	37·40	95	64·60	135	91·80	175	119·00
16	10·88	56	38·08	96	65·28	136	92·48	176	119·68
17	11·56	57	38·76	97	65·96	137	93·16	177	120·36
18	12·24	58	39·44	98	66·64	138	93·84	178	121·04
19	12·92	59	40·12	99	67·32	139	94·52	179	121·72
20	13·60	**60**	40·80	**100**	68·00	**140**	95·20	**180**	122·40
21	14·28	61	41·48	101	68·68	141	95·88	181	123·08
22	14·96	62	42·16	102	69·36	142	96·56	182	123·76
23	15·64	63	42·84	103	70·04	143	97·24	183	124·44
24	16·32	64	43·52	104	70·72	144	97·92	184	125·12
25	17·00	65	44·20	105	71·40	145	98·60	185	125·80
26	17·68	66	44·88	106	72·08	146	99·28	186	126·48
27	18·36	67	45·56	107	72·76	147	99·96	187	127·16
28	19·04	68	46·24	108	73·44	148	100·64	188	127·84
29	19·72	69	46·92	109	74·12	149	101·32	189	128·52
30	20·40	**70**	47·60	**110**	74·80	**150**	102·00	**190**	129·20
31	21·08	71	48·28	111	75·48	151	102·68	191	129·88
32	21·76	72	48·96	112	76·16	152	103·36	192	130·56
33	22·44	73	49·64	113	76·84	153	104·04	193	131·24
34	23·12	74	50·32	114	77·52	154	104·72	194	131·92
35	23·80	75	51·00	115	78·20	155	105·40	195	132·60
36	24·48	76	51·68	116	78·88	156	106·08	196	133·28
37	25·16	77	52·36	117	79·56	157	106·76	197	133·96
38	25·84	78	53·04	118	80·24	158	107·44	198	134·64
39	26·52	79	53·72	119	80·92	159	108·12	199	135·32
40	27·20	**80**	54·40	**120**	81·60	**160**	108·80	**200**	136·00

$\frac{1}{16}=0{\cdot}04^{3}$ $\frac{1}{9}=0{\cdot}07^{6}$ $\frac{1}{8}=0{\cdot}08^{5}$ $\frac{1}{6}=0{\cdot}11^{3}$ $\frac{1}{5}=0{\cdot}13^{6}$ $\frac{1}{4}=0{\cdot}17^{0}$ $\frac{1}{3}=0{\cdot}22^{7}$

$\frac{3}{8}=0{\cdot}25^{5}$ $\frac{1}{2}=0{\cdot}34^{0}$ $\frac{5}{8}=0{\cdot}42^{5}$ $\frac{3}{4}=0{\cdot}51^{0}$ $\frac{7}{8}=0{\cdot}59^{5}$ $\frac{2}{3}=0{\cdot}45^{3}$ $\frac{5}{6}=0{\cdot}56^{7}$

$\frac{3}{16}=0{\cdot}12^{8}$ $\frac{5}{16}=0{\cdot}21^{3}$ $\frac{7}{16}=0{\cdot}29^{8}$ $\frac{9}{16}=0{\cdot}38^{3}$ $\frac{11}{16}=0{\cdot}46^{8}$ $\frac{13}{16}=0{\cdot}55^{3}$ $\frac{15}{16}=0{\cdot}63^{8}$

(1.46=1£) £0.68½

1	0·68½	41	28·08½	81	55·48½	121	82·88½	161	110·28½
2	1·37	42	28·77	82	56·17	122	83·57	162	110·97
3	2·05½	43	29·45½	83	56·85½	123	84·25½	163	111·65½
4	2·74	44	30·14	84	57·54	124	84·94	164	112·34
5	3·42½	45	30·82½	85	58·22½	125	85·62½	165	113·02½
6	4·11	46	31·51	86	58·91	126	86·31	166	113·71
7	4·79½	47	32·19½	87	59·59½	127	86·99½	167	114·39½
8	5·48	48	32·88	88	60·28	128	87·68	168	115·08
9	6·16½	49	33·56½	89	60·96½	129	88·36½	169	115·76½
10	6·85	**50**	34·25	**90**	61·65	**130**	89·05	**170**	116·45
11	7·53½	51	34·93½	91	62·33½	131	89·73½	171	117·13½
12	8·22	52	35·62	92	63·02	132	90·42	172	117·82
13	8·90½	53	36·30½	93	63·70½	133	91·10½	173	118·50½
14	9·59	54	36·99	94	64·39	134	91·79	174	119·19
15	10·27½	55	37·67½	95	65·07½	135	92·47½	175	119·87½
16	10·96	56	38·36	96	65·76	136	93·16	176	120·56
17	11·64½	57	39·04½	97	66·44½	137	93·84½	177	121·24½
18	12·33	58	39·73	98	67·13	138	94·53	178	121·93
19	13·01½	59	40·41½	99	67·81½	139	95·21½	179	122·61½
20	13·70	**60**	41·10	**100**	68·50	**140**	95·90	**180**	123·30
21	14·38½	61	41·78½	101	69·18½	141	96·58½	181	123·98½
22	15·07	62	42·47	102	69·87	142	97·27	182	124·67
23	15·75½	63	43·15½	103	70·55½	143	97·95½	183	125·35½
24	16·44	64	43·84	104	71·24	144	98·64	184	126·04
25	17·12½	65	44·52½	105	71·92½	145	99·32½	185	126·72½
26	17·81	66	45·21	106	72·61	146	100·01	186	127·41
27	18·49½	67	45·89½	107	73·29½	147	100·69½	187	128·09½
28	19·18	68	46·58	108	73·98	148	101·38	188	128·78
29	19·86½	69	47·26½	109	74·66½	149	102·06½	189	129·46½
30	20·55	**70**	47·95	**110**	75·35	**150**	102·75	**190**	130·15
31	21·23½	71	48·63½	111	76·03½	151	103·43½	191	130·83½
32	21·92	72	49·32	112	76·72	152	104·12	192	131·52
33	22·60½	73	50·00½	113	77·40½	153	104·80½	193	132·20½
34	23·29	74	50·69	114	78·09	154	105·49	194	132·89
35	23·97½	75	51·37½	115	78·77½	155	106·17½	195	133·57½
36	24·66	76	52·06	116	79·46	156	106·86	196	134·26
37	25·34½	77	52·74½	117	80·14½	157	107·54½	197	134·94½
38	26·03	78	53·43	118	80·83	158	108·23	198	135·63
39	26·71½	79	54·11½	119	81·51½	159	108·91½	199	136·31½
40	27·40	**80**	54·80	**120**	82·20	**160**	109·60	**200**	137·00

$\frac{1}{16}=0{\cdot}04^{3}$ $\frac{1}{9}=0{\cdot}07^{6}$ $\frac{1}{8}=0{\cdot}08^{6}$ $\frac{1}{6}=0{\cdot}11^{4}$ $\frac{1}{5}=0{\cdot}13^{7}$ $\frac{1}{4}=0{\cdot}17^{1}$ $\frac{1}{3}=0{\cdot}22^{8}$

$\frac{3}{8}=0{\cdot}25^{7}$ $\frac{1}{2}=0{\cdot}34^{3}$ $\frac{5}{8}=0{\cdot}42^{8}$ $\frac{3}{4}=0{\cdot}51^{4}$ $\frac{7}{8}=0{\cdot}59^{9}$ $\frac{2}{3}=0{\cdot}45^{7}$ $\frac{5}{6}=0{\cdot}57^{1}$

$\frac{3}{16}=0{\cdot}12^{8}$ $\frac{5}{16}=0{\cdot}21^{4}$ $\frac{7}{16}=0{\cdot}30^{0}$ $\frac{9}{16}=0{\cdot}38^{5}$ $\frac{11}{16}=0{\cdot}47^{1}$ $\frac{13}{16}=0{\cdot}55^{7}$ $\frac{15}{16}=0{\cdot}64^{2}$

(1.45=1£) **£0.69**

1	0·69	41	28·29	81	55·89	121	83·49	161	111·09
2	1·38	42	28·98	82	56·58	122	84·18	162	111·78
3	2·07	43	29·67	83	57·27	123	84·87	163	112·47
4	2·76	44	30·36	84	57·96	124	85·56	164	113·16
5	3·45	45	31·05	85	58·65	125	86·25	165	113·85
6	4·14	46	31·74	86	59·34	126	86·94	166	114·54
7	4·83	47	32·43	87	60·03	127	87·63	167	115·23
8	5·52	48	33·12	88	60·72	128	88·32	168	115·92
9	6·21	49	33·81	89	61·41	129	89·01	169	116·61
10	6·90	**50**	34·50	**90**	62·10	**130**	89·70	**170**	117·30
11	7·59	51	35·19	91	62·79	131	90·39	171	117·99
12	8·28	52	35·88	92	63·48	132	91·08	172	118·68
13	8·97	53	36·57	93	64·17	133	91·77	173	119·37
14	9·66	54	37·26	94	64·86	134	92·46	174	120·06
15	10·35	55	37·95	95	65·55	135	93·15	175	120·75
16	11·04	56	38·64	96	66·24	136	93·84	176	121·44
17	11·73	57	39·33	97	66·93	137	94·53	177	122·13
18	12·42	58	40·02	98	67·62	138	95·22	178	122·82
19	13·11	59	40·71	99	68·31	139	95·91	179	123·51
20	13·80	**60**	41·40	**100**	69·00	**140**	96·60	**180**	124·20
21	14·49	61	42·09	101	69·69	141	97·29	181	124·89
22	15·18	62	42·78	102	70·38	142	97·98	182	125·58
23	15·87	63	43·47	103	71·07	143	98·67	183	126·27
24	16·56	64	44·16	104	71·76	144	99·36	184	126·96
25	17·25	65	44·85	105	72·45	145	100·05	185	127·65
26	17·94	66	45·54	106	73·14	146	100·74	186	128·34
27	18·63	67	46·23	107	73·83	147	101·43	187	129·03
28	19·32	68	46·92	108	74·52	148	102·12	188	129·72
29	20·01	69	47·61	109	75·21	149	102·81	189	130·41
30	20·70	**70**	48·30	**110**	75·90	**150**	103·50	**190**	131·10
31	21·39	71	48·99	111	76·59	151	104·19	191	131·79
32	22·08	72	49·68	112	77·28	152	104·88	192	132·48
33	22·77	73	50·37	113	77·97	153	105·57	193	133·17
34	23·46	74	51·06	114	78·66	154	106·26	194	133·86
35	24·15	75	51·75	115	79·35	155	106·95	195	134·55
36	24·84	76	52·44	116	80·04	156	107·64	196	135·24
37	25·53	77	53·13	117	80·73	157	108·33	197	135·93
38	26·22	78	53·82	118	81·42	158	109·02	198	136·62
39	26·91	79	54·51	119	82·11	159	109·71	199	137·31
40	27·60	**80**	55·20	**120**	82·80	**160**	110·40	**200**	138·00

$\frac{1}{16}=0{\cdot}04^{3}$ $\frac{1}{9}=0{\cdot}07^{7}$ $\frac{1}{8}=0{\cdot}08^{6}$ $\frac{1}{6}=0{\cdot}11^{5}$ $\frac{1}{5}=0{\cdot}13^{8}$ $\frac{1}{4}=0{\cdot}17^{3}$ $\frac{1}{3}=0{\cdot}23^{0}$

$\frac{3}{8}=0{\cdot}25^{9}$ $\frac{1}{2}=0{\cdot}34^{5}$ $\frac{5}{8}=0{\cdot}43^{1}$ $\frac{3}{4}=0{\cdot}51^{8}$ $\frac{7}{8}=0{\cdot}60^{4}$ $\frac{2}{3}=0{\cdot}46^{0}$ $\frac{5}{6}=0{\cdot}57^{5}$

$\frac{3}{16}=0{\cdot}12^{9}$ $\frac{5}{16}=0{\cdot}21^{6}$ $\frac{7}{16}=0{\cdot}30^{2}$ $\frac{9}{16}=0{\cdot}38^{8}$ $\frac{11}{16}=0{\cdot}47^{4}$ $\frac{13}{16}=0{\cdot}56^{1}$ $\frac{15}{16}=0{\cdot}64^{7}$

1	0·69½	41	28·49½	81	56·29½	121	84·09½	161	111·89½
2	1·39	42	29·19	82	56·99	122	84·79	162	112·59
3	2·08½	43	29·88½	83	57·68½	123	85·48½	163	113·28½
4	2·78	44	30·58	84	58·38	124	86·18	164	113·98
5	3·47½	45	31·27½	85	59·07½	125	86·87½	165	114·67½
6	4·17	46	31·97	86	59·77	126	87·57	166	115·37
7	4·86½	47	32·66½	87	60·46½	127	88·26½	167	116·06½
8	5·56	48	33·36	88	61·16	128	88·96	168	116·76
9	6·25½	49	34·05½	89	61·85½	129	89·65½	169	117·45½
10	6·95	**50**	34·75	**90**	62·55	**130**	90·35	**170**	118·15
11	7·64½	51	35·44½	91	63·24½	131	91·04½	171	118·84½
12	8·34	52	36·14	92	63·94	132	91·74	172	119·54
13	9·03½	53	36·83½	93	64·63½	133	92·43½	173	120·23½
14	9·73	54	37·53	94	65·33	134	93·13	174	120·93
15	10·42½	55	38·22½	95	66·02½	135	93·82½	175	121·62½
16	11·12	56	38·92	96	66·72	136	94·52	176	122·32
17	11·81½	57	39·61½	97	67·41½	137	95·21½	177	123·01½
18	12·51	58	40·31	98	68·11	138	95·91	178	123·71
19	13·20½	59	41·00½	99	68·80½	139	96·60½	179	124·40½
20	13·90	**60**	41·70	**100**	69·50	**140**	97·30	**180**	125·10
21	14·59½	61	42·39½	101	70·19½	141	97·99½	181	125·79½
22	15·29	62	43·09	102	70·89	142	98·69	182	126·49
23	15·98½	63	43·78½	103	71·58½	143	99·38½	183	127·18½
24	16·68	64	44·48	104	72·28	144	100·08	184	127·88
25	17·37½	65	45·17½	105	72·97½	145	100·77½	185	128·57½
26	18·07	66	45·87	106	73·67	146	101·47	186	129·27
27	18·76½	67	46·56½	107	74·36½	147	102·16½	187	129·96½
28	19·46	68	47·26	108	75·06	148	102·86	188	130·66
29	20·15½	69	47·95½	109	75·75½	149	103·55½	189	131·35½
30	20·85	**70**	48·65	**110**	76·45	**150**	104·25	**190**	132·05
31	21·54½	71	49·34½	111	77·14½	151	104·94½	191	132·74½
32	22·24	72	50·04	112	77·84	152	105·64	192	133·44
33	22·93½	73	50·73½	113	78·53½	153	106·33½	193	134·13½
34	23·63	74	51·43	114	79·23	154	107·03	194	134·83
35	24·32½	75	52·12½	115	79·92½	155	107·72½	195	135·52½
36	25·02	76	52·82	116	80·62	156	108·42	196	136·22
37	25·71½	77	53·51½	117	81·31½	157	109·11½	197	136·91½
38	26·41	78	54·21	118	82·01	158	109·81	198	137·61
39	27·10½	79	54·90½	119	82·70½	159	110·50½	199	138·30½
40	27·80	**80**	55·60	**120**	83·40	**160**	111·20	**200**	139·00

$\frac{1}{16}$=0·04^{3} $\frac{1}{9}$=0·07^{7} $\frac{1}{8}$=0·08^{7} $\frac{1}{6}$=0·11^{6} $\frac{1}{5}$=0·13^{9} $\frac{1}{4}$=0·17^{4} $\frac{1}{3}$=0·23^{2}

$\frac{3}{8}$=0·26^{1} $\frac{1}{2}$=0·34^{8} $\frac{5}{8}$=0·43^{4} $\frac{3}{4}$=0·52^{1} $\frac{7}{8}$=0·60^{8} $\frac{2}{3}$=0·46^{3} $\frac{5}{6}$=0·57^{9}

$\frac{3}{16}$=0·13^{0} $\frac{5}{16}$=0·21^{7} $\frac{7}{16}$=0·30^{4} $\frac{9}{16}$=0·39^{1} $\frac{11}{16}$=0·47^{8} $\frac{13}{16}$=0·56^{5} $\frac{15}{16}$=0·65^{2}

(1.43=1£) **£0.70**

1	0·70	41	28·70	81	56·70	121	84·70	161	112·70
2	1·40	42	29·40	82	57·40	122	85·40	162	113·40
3	2·10	43	30·10	83	58·10	123	86·10	163	114·10
4	2·80	44	30·80	84	58·80	124	86·80	164	114·80
5	3·50	45	31·50	85	59·50	125	87·50	165	115·50
6	4·20	46	32·20	86	60·20	126	88·20	166	116·20
7	4·90	47	32·90	87	60·90	127	88·90	167	116·90
8	5·60	48	33·60	88	61·60	128	89·60	168	117·60
9	6·30	49	34·30	89	62·30	129	90·30	169	118·30
10	7·00	**50**	35·00	**90**	63·00	**130**	91·00	**170**	119·00
11	7·70	51	35·70	91	63·70	131	91·70	171	119·70
12	8·40	52	36·40	92	64·40	132	92·40	172	120·40
13	9·10	53	37·10	93	65·10	133	93·10	173	121·10
14	9·80	54	37·80	94	65·80	134	93·80	174	121·80
15	10·50	55	38·50	95	66·50	135	94·50	175	122·50
16	11·20	56	39·20	96	67·20	136	95·20	176	123·20
17	11·90	57	39·90	97	67·90	137	95·90	177	123·90
18	12·60	58	40·60	98	68·60	138	96·60	178	124·60
19	13·30	59	41·30	99	69·30	139	97·30	179	125·30
20	14·00	**60**	42·00	**100**	70·00	**140**	98·00	**180**	126·00
21	14·70	61	42·70	101	70·70	141	98·70	181	126·70
22	15·40	62	43·40	102	71·40	142	99·40	182	127·40
23	16·10	63	44·10	103	72·10	143	100·10	183	128·10
24	16·80	64	44·80	104	72·80	144	100·80	184	128·80
25	17·50	65	45·50	105	73·50	145	101·50	185	129·50
26	18·20	66	46·20	106	74·20	146	102·20	186	130·20
27	18·90	67	46·90	107	74·90	147	102·90	187	130·90
28	19·60	68	47·60	108	75·60	148	103·60	188	131·60
29	20·30	69	48·30	109	76·30	149	104·30	189	132·30
30	21·00	**70**	49·00	**110**	77·00	**150**	105·00	**190**	133·00
31	21·70	71	49·70	111	77·70	151	105·70	191	133·70
32	22·40	72	50·40	112	78·40	152	106·40	192	134·40
33	23·10	73	51·10	113	79·10	153	107·10	193	135·10
34	23·80	74	51·80	114	79·80	154	107·80	194	135·80
35	24·50	75	52·50	115	80·50	155	108·50	195	136·50
36	25·20	76	53·20	116	81·20	156	109·20	196	137·20
37	25·90	77	53·90	117	81·90	157	109·90	197	137·90
38	26·60	78	54·60	118	82·60	158	110·60	198	138·60
39	27·30	79	55·30	119	83·30	159	111·30	199	139·30
40	28·00	**80**	56·00	**120**	84·00	**160**	112·00	**200**	140·00

$\frac{1}{16}=0{\cdot}04^{4}$ $\frac{1}{9}=0{\cdot}07^{8}$ $\frac{1}{8}=0{\cdot}08^{8}$ $\frac{1}{6}=0{\cdot}11^{7}$ $\frac{1}{5}=0{\cdot}14^{0}$ $\frac{1}{4}=0{\cdot}17^{5}$ $\frac{1}{3}=0{\cdot}23^{3}$

$\frac{3}{8}=0{\cdot}26^{3}$ $\frac{1}{2}=0{\cdot}35^{0}$ $\frac{5}{8}=0{\cdot}43^{8}$ $\frac{3}{4}=0{\cdot}52^{5}$ $\frac{7}{8}=0{\cdot}61^{3}$ $\frac{2}{3}=0{\cdot}46^{7}$ $\frac{5}{6}=0{\cdot}58^{3}$

$\frac{3}{16}=0{\cdot}13^{1}$ $\frac{5}{16}=0{\cdot}21^{9}$ $\frac{7}{16}=0{\cdot}30^{6}$ $\frac{9}{16}=0{\cdot}39^{4}$ $\frac{11}{16}=0{\cdot}48^{1}$ $\frac{13}{16}=0{\cdot}56^{9}$ $\frac{15}{16}=0{\cdot}65^{6}$

(1.42=1£) £0.70½

1	0·70½	41	28·90½	81	57·10½	121	85·30½	161	113·50½
2	1·41	42	29·61	82	57·81	122	86·01	162	114·21
3	2·11½	43	30·31½	83	58·51½	123	86·71½	163	114·91½
4	2·82	44	31·02	84	59·22	124	87·42	164	115·62
5	3·52½	45	31·72½	85	59·92½	125	88·12½	165	116·32½
6	4·23	46	32·43	86	60·63	126	88·83	166	117·03
7	4·93½	47	33·13½	87	61·33½	127	89·53½	167	117·73½
8	5·64	48	33·84	88	62·04	128	90·24	168	118·44
9	6·34½	49	34·54½	89	62·74½	129	90·94½	169	119·14½
10	7·05	**50**	35·25	**90**	63·45	**130**	91·65	**170**	119·85
11	7·75½	51	35·95½	91	64·15½	131	92·35½	171	120·55½
12	8·46	52	36·66	92	64·86	132	93·06	172	121·26
13	9·16½	53	37·36½	93	65·56½	133	93·76½	173	121·96½
14	9·87	54	38·07	94	66·27	134	94·47	174	122·67
15	10·57½	55	38·77½	95	66·97½	135	95·17½	175	123·37½
16	11·28	56	39·48	96	67·68	136	95·88	176	124·08
17	11·98½	57	40·18½	97	68·38½	137	96·58½	177	124·78½
18	12·69	58	40·89	98	69·09	138	97·29	178	125·49
19	13·39½	59	41·59½	99	69·79½	139	97·99½	179	126·19½
20	14·10	**60**	42·30	**100**	70·50	**140**	98·70	**180**	126·90
21	14·80½	61	43·00½	101	71·20½	141	99·40½	181	127·60½
22	15·51	62	43·71	102	71·91	142	100·11	182	128·31
23	16·21½	63	44·41½	103	72·61½	143	100·81½	183	129·01½
24	16·92	64	45·12	104	73·32	144	101·52	184	129·72
25	17·62½	65	45·82½	105	74·02½	145	102·22½	185	130·42½
26	18·33	66	46·53	106	74·73	146	102·93	186	131·13
27	19·03½	67	47·23½	107	75·43½	147	103·63½	187	131·83½
28	19·74	68	47·94	108	76·14	148	104·34	188	132·54
29	20·44½	69	48·64½	109	76·84½	149	105·04½	189	133·24½
30	21·15	**70**	49·35	**110**	77·55	**150**	105·75	**190**	133·95
31	21·85½	71	50·05½	111	78·25½	151	106·45½	191	134·65½
32	22·56	72	50·76	112	78·96	152	107·16	192	135·36
33	23·26½	73	51·46½	113	79·66½	153	107·86½	193	136·06½
34	23·97	74	52·17	114	80·37	154	108·57	194	136·77
35	24·67½	75	52·87½	115	81·07½	155	109·27½	195	137·47½
36	25·38	76	53·58	116	81·78	156	109·98	196	138·18
37	26·08½	77	54·28½	117	82·48½	157	110·68½	197	138·88½
38	26·79	78	54·99	118	83·19	158	111·39	198	139·59
39	27·49½	79	55·69½	119	83·89½	159	112·09½	199	140·29½
40	28·20	**80**	56·40	**120**	84·60	**160**	112·80	**200**	141·00

$\frac{1}{16}=0{\cdot}04^{4}$ $\frac{1}{9}=0{\cdot}07^{8}$ $\frac{1}{8}=0{\cdot}08^{8}$ $\frac{1}{6}=0{\cdot}11^{8}$ $\frac{1}{5}=0{\cdot}14^{1}$ $\frac{1}{4}=0{\cdot}17^{6}$ $\frac{1}{3}=0{\cdot}23^{5}$

$\frac{3}{8}=0{\cdot}26^{4}$ $\frac{1}{2}=0{\cdot}35^{3}$ $\frac{5}{8}=0{\cdot}44^{1}$ $\frac{3}{4}=0{\cdot}52^{9}$ $\frac{7}{8}=0{\cdot}61^{7}$ $\frac{2}{3}=0{\cdot}47^{0}$ $\frac{5}{6}=0{\cdot}58^{8}$

$\frac{3}{16}=0{\cdot}13^{2}$ $\frac{5}{16}=0{\cdot}22^{0}$ $\frac{7}{16}=0{\cdot}30^{8}$ $\frac{9}{16}=0{\cdot}39^{7}$ $\frac{11}{16}=0{\cdot}48^{5}$ $\frac{13}{16}=0{\cdot}57^{3}$ $\frac{15}{16}=0{\cdot}66^{1}$

·7071 is the square root of ½ 1 lb per square inch=0·703 kilogrammes per square centimetre

(1.41=1£) £0.71

1	0·71	41	29·11	81	57·51	121	85·91	161	114·31
2	1·42	42	29·82	82	58·22	122	86·62	162	115·02
3	2·13	43	30·53	83	58·93	123	87·33	163	115·73
4	2·84	44	31·24	84	59·64	124	88·04	164	116·44
5	3·55	45	31·95	85	60·35	125	88·75	165	117·15
6	4·26	46	32·66	86	61·06	126	89·46	166	117·86
7	4·97	47	33·37	87	61·77	127	90·17	167	118·57
8	5·68	48	34·08	88	62·48	128	90·88	168	119·28
9	6·39	49	34·79	89	63·19	129	91·59	169	119·99
10	7·10	**50**	35·50	**90**	63·90	**130**	92·30	**170**	120·70
11	7·81	51	36·21	91	64·61	131	93·01	171	121·41
12	8·52	52	36·92	92	65·32	132	93·72	172	122·12
13	9·23	53	37·63	93	66·03	133	94·43	173	122·83
14	9·94	54	38·34	94	66·74	134	95·14	174	123·54
15	10·65	55	39·05	95	67·45	135	95·85	175	124·25
16	11·36	56	39·76	96	68·16	136	96·56	176	124·96
17	12·07	57	40·47	97	68·87	137	97·27	177	125·67
18	12·78	58	41·18	98	69·58	138	97·98	178	126·38
19	13·49	59	41·89	99	70·29	139	98·69	179	127·09
20	14·20	**60**	42·60	**100**	71·00	**140**	99·40	**180**	127·80
21	14·91	61	43·31	101	71·71	141	100·11	181	128·51
22	15·62	62	44·02	102	72·42	142	100·82	182	129·22
23	16·33	63	44·73	103	73·13	143	101·53	183	129·93
24	17·04	64	45·44	104	73·84	144	102·24	184	130·64
25	17·75	65	46·15	105	74·55	145	102·95	185	131·35
26	18·46	66	46·86	106	75·26	146	103·66	186	132·06
27	19·17	67	47·57	107	75·97	147	104·37	187	132·77
28	19·88	68	48·28	108	76·68	148	105·08	188	133·48
29	20·59	69	48·99	109	77·39	149	105·79	189	134·19
30	21·30	**70**	49·70	**110**	78·10	**150**	106·50	**190**	134·90
31	22·01	71	50·41	111	78·81	151	107·21	191	135·61
32	22·72	72	51·12	112	79·52	152	107·92	192	136·32
33	23·43	73	51·83	113	80·23	153	108·63	193	137·03
34	24·14	74	52·54	114	80·94	154	109·34	194	137·74
35	24·85	75	53·25	115	81·65	155	110·05	195	138·45
36	25·56	76	53·96	116	82·36	156	110·76	196	139·16
37	26·27	77	54·67	117	83·07	157	111·47	197	139·87
38	26·98	78	55·38	118	83·78	158	112·18	198	140·58
39	27·69	79	56·09	119	84·49	159	112·89	199	141·29
40	28·40	**80**	56·80	**120**	85·20	**160**	113·60	**200**	142·00

$\frac{1}{16}=0{\cdot}04^{4}$ $\frac{1}{9}=0{\cdot}07^{9}$ $\frac{1}{8}=0{\cdot}08^{9}$ $\frac{1}{6}=0{\cdot}11^{8}$ $\frac{1}{5}=0{\cdot}14^{2}$ $\frac{1}{4}=0{\cdot}17^{8}$ $\frac{1}{3}=0{\cdot}23^{7}$

$\frac{3}{8}=0{\cdot}26^{6}$ $\frac{1}{2}=0{\cdot}35^{5}$ $\frac{5}{8}=0{\cdot}44^{4}$ $\frac{3}{4}=0{\cdot}53^{3}$ $\frac{7}{8}=0{\cdot}62^{1}$ $\frac{2}{3}=0{\cdot}47^{3}$ $\frac{5}{6}=0{\cdot}59^{2}$

$\frac{3}{16}=0{\cdot}13^{3}$ $\frac{5}{16}=0{\cdot}22^{2}$ $\frac{7}{16}=0{\cdot}31^{1}$ $\frac{9}{16}=0{\cdot}39^{9}$ $\frac{11}{16}=0{\cdot}48^{8}$ $\frac{13}{16}=0{\cdot}57^{7}$ $\frac{15}{16}=0{\cdot}66^{6}$

(1.40=1£) **£0.71½**

1	0·71½	41	29·31½	81	57·91½	121	86·51½	161	115·11½
2	1·43	42	30·03	82	58·63	122	87·23	162	115·83
3	2·14½	43	30·74½	83	59·34½	123	87·94½	163	116·54½
4	2·86	44	31·46	84	60·06	124	88·66	164	117·26
5	3·57½	45	32·17½	85	60·77½	125	89·37½	165	117·97½
6	4·29	46	32·89	86	61·49	126	90·09	166	118·69
7	5·00½	47	33·60½	87	62·20½	127	90·80½	167	119·40½
8	5·72	48	34·32	88	62·92	128	91·52	168	120·12
9	6·43½	49	35·03½	89	63·63½	129	92·23½	169	120·83½
10	7·15	**50**	35·75	**90**	64·35	**130**	92·95	**170**	121·55
11	7·86½	51	36·46½	91	65·06½	131	93·66½	171	122·26½
12	8·58	52	37·18	92	65·78	132	94·38	172	122·98
13	9·29½	53	37·89½	93	66·49½	133	95·09½	173	123·69½
14	10·01	54	38·61	94	67·21	134	95·81	174	124·41
15	10·72½	55	39·32½	95	67·92½	135	96·52½	175	125·12½
16	11·44	56	40·04	96	68·64	136	97·24	176	125·84
17	12·15½	57	40·75½	97	69·35½	137	97·95½	177	126·55½
18	12·87	58	41·47	98	70·07	138	98·67	178	127·27
19	13·58½	59	42·18½	99	70·78½	139	99·38½	179	127·98½
20	14·30	**60**	42·90	**100**	71·50	**140**	100·10	**180**	128·70
21	15·01½	61	43·61½	101	72·21½	141	100·81½	181	129·41½
22	15·73	62	44·33	102	72·93	142	101·53	182	130·13
23	16·44½	63	45·04½	103	73·64½	143	102·24½	183	130·84½
24	17·16	64	45·76	104	74·36	144	102·96	184	131·56
25	17·87½	65	46·47½	105	75·07½	145	103·67½	185	132·27½
26	18·59	66	47·19	106	75·79	146	104·39	186	132·99
27	19·30½	67	47·90½	107	76·50½	147	105·10½	187	133·70½
28	20·02	68	48·62	108	77·22	148	105·82	188	134·42
29	20·73½	69	49·33½	109	77·93½	149	106·53½	189	135·13½
30	21·45	**70**	50·05	**110**	78·65	**150**	107·25	**190**	135·85
31	22·16½	71	50·76½	111	79·36½	151	107·96½	191	136·56½
32	22·88	72	51·48	112	80·08	152	108·68	192	137·28
33	23·59½	73	52·19½	113	80·79½	153	109·39½	193	137·99½
34	24·31	74	52·91	114	81·51	154	110·11	194	138·71
35	25·02½	75	53·62½	115	82·22½	155	110·82½	195	139·42½
36	25·74	76	54·34	116	82·94	156	111·54	196	140·14
37	26·45½	77	55·05½	117	83·65½	157	112·25½	197	140·85½
38	27·17	78	55·77	118	84·37	158	112·97	198	141·57
39	27·88½	79	56·48½	119	85·08½	159	113·68½	199	142·28½
40	28·60	**80**	57·20	**120**	85·80	**160**	114·40	**200**	143·00

$\frac{1}{16}=0{\cdot}04^{5}$	$\frac{1}{9}=0{\cdot}07^{9}$	$\frac{1}{8}=0{\cdot}08^{9}$	$\frac{1}{6}=0{\cdot}11^{9}$	$\frac{1}{5}=0{\cdot}14^{3}$	$\frac{1}{4}=0{\cdot}17^{9}$	$\frac{1}{3}=0{\cdot}23^{8}$
$\frac{3}{8}=0{\cdot}26^{8}$	$\frac{1}{2}=0{\cdot}35^{8}$	$\frac{5}{8}=0{\cdot}44^{7}$	$\frac{3}{4}=0{\cdot}53^{6}$	$\frac{7}{8}=0{\cdot}62^{6}$	$\frac{2}{3}=0{\cdot}47^{7}$	$\frac{5}{6}=0{\cdot}59^{6}$
$\frac{3}{16}=0{\cdot}13^{4}$	$\frac{5}{16}=0{\cdot}22^{3}$	$\frac{7}{16}=0{\cdot}31^{3}$	$\frac{9}{16}=0{\cdot}40^{2}$	$\frac{11}{16}=0{\cdot}49^{2}$	$\frac{13}{16}=0{\cdot}58^{1}$	$\frac{15}{16}=0{\cdot}67^{0}$

(1.39=1£) **£0.72**

1	0·72	41	29·52	81	58·32	121	87·12	161	115·92
2	1·44	42	30·24	82	59·04	122	87·84	162	116·64
3	2·16	43	30·96	83	59·76	123	88·56	163	117·36
4	2·88	44	31·68	84	60·48	124	89·28	164	118·08
5	3·60	45	32·40	85	61·20	125	90·00	165	118·80
6	4·32	46	33·12	86	61·92	126	90·72	166	119·52
7	5·04	47	33·84	87	62·64	127	91·44	167	120·24
8	5·76	48	34·56	88	63·36	128	92·16	168	120·96
9	6·48	49	35·28	89	64·08	129	92·88	169	121·68
10	7·20	**50**	36·00	**90**	64·80	**130**	93·60	**170**	122·40
11	7·92	51	36·72	91	65·52	131	94·32	171	123·12
12	8·64	52	37·44	92	66·24	132	95·04	172	123·84
13	9·36	53	38·16	93	66·96	133	95·76	173	124·56
14	10·08	54	38·88	94	67·68	134	96·48	174	125·28
15	10·80	55	39·60	95	68·40	135	97·20	175	126·00
16	11·52	56	40·32	96	69·12	136	97·92	176	126·72
17	12·24	57	41·04	97	69·84	137	98·64	177	127·44
18	12·96	58	41·76	98	70·56	138	99·36	178	128·16
19	13·68	59	42·48	99	71·28	139	100·08	179	128·88
20	14·40	**60**	43·20	**100**	72·00	**140**	100·80	**180**	129·60
21	15·12	61	43·92	101	72·72	141	101·52	181	130·32
22	15·84	62	44·64	102	73·44	142	102·24	182	131·04
23	16·56	63	45·36	103	74·16	143	102·96	183	131·76
24	17·28	64	46·08	104	74·88	144	103·68	184	132·48
25	18·00	65	46·80	105	75·60	145	104·40	185	133·20
26	18·72	66	47·52	106	76·32	146	105·12	186	133·92
27	19·44	67	48·24	107	77·04	147	105·84	187	134·64
28	20·16	68	48·96	108	77·76	148	106·56	188	135·36
29	20·88	69	49·68	109	78·48	149	107·28	189	136·08
30	21·60	**70**	50·40	**110**	79·20	**150**	108·00	**190**	136·80
31	22·32	71	51·12	111	79·92	151	108·72	191	137·52
32	23·04	72	51·84	112	80·64	152	109·44	192	138·24
33	23·76	73	52·56	113	81·36	153	110·16	193	138·96
34	24·48	74	53·28	114	82·08	154	110·88	194	139·68
35	25·20	75	54·00	115	82·80	155	111·60	195	140·40
36	25·92	76	54·72	116	83·52	156	112·32	196	141·12
37	26·64	77	55·44	117	84·24	157	113·04	197	141·84
38	27·36	78	56·16	118	84·96	158	113·76	198	142·56
39	28·08	79	56·88	119	85·68	159	114·48	199	143·28
40	28·80	**80**	57·60	**120**	86·40	**160**	115·20	**200**	144·00

$\frac{1}{16}=0{\cdot}04^5$	$\frac{1}{9}=0{\cdot}08^0$	$\frac{1}{8}=0{\cdot}09^0$	$\frac{1}{6}=0{\cdot}12^0$	$\frac{1}{5}=0{\cdot}14^4$	$\frac{1}{4}=0{\cdot}18^0$	$\frac{1}{3}=0{\cdot}24^0$
$\frac{3}{8}=0{\cdot}27^0$	$\frac{1}{2}=0{\cdot}36^0$	$\frac{5}{8}=0{\cdot}45^0$	$\frac{3}{4}=0{\cdot}54^0$	$\frac{7}{8}=0{\cdot}63^0$	$\frac{2}{3}=0{\cdot}48^0$	$\frac{5}{6}=0{\cdot}60^0$
$\frac{3}{16}=0{\cdot}13^5$	$\frac{5}{16}=0{\cdot}22^5$	$\frac{7}{16}=0{\cdot}31^5$	$\frac{9}{16}=0{\cdot}40^5$	$\frac{11}{16}=0{\cdot}49^5$	$\frac{13}{16}=0{\cdot}58^5$	$\frac{15}{16}=0{\cdot}67^5$

(1.38=1£) **£0.72½**

1	0·72½	41	29·72½	81	58·72½	121	87·72½	161	116·72½
2	1·45	42	30·45	82	59·45	122	88·45	162	117·45
3	2·17½	43	31·17½	83	60·17½	123	89·17½	163	118·17½
4	2·90	44	31·90	84	60·90	124	89·90	164	118·90
5	3·62½	45	32·62½	85	61·62½	125	90·62½	165	119·62½
6	4·35	46	33·35	86	62·35	126	91·35	166	120·35
7	5·07½	47	34·07½	87	63·07½	127	92·07½	167	121·07½
8	5·80	48	34·80	88	63·80	128	92·80	168	121·80
9	6·52½	49	35·52½	89	64·52½	129	93·52½	169	122·52½
10	7·25	**50**	36·25	**90**	65·25	**130**	94·25	**170**	123·25
11	7·97½	51	36·97½	91	65·97½	131	94·97½	171	123·97½
12	8·70	52	37·70	92	66·70	132	95·70	172	124·70
13	9·42½	53	38·42½	93	67·42½	133	96·42½	173	125·42½
14	10·15	54	39·15	94	68·15	134	97·15	174	126·15
15	10·87½	55	39·87½	95	68·87½	135	97·87½	175	126·87½
16	11·60	56	40·60	96	69·60	136	98·60	176	127·60
17	12·32½	57	41·32½	97	70·32½	137	99·32½	177	128·32½
18	13·05	58	42·05	98	71·05	138	100·05	178	129·05
19	13·77½	59	42·77½	99	71·77½	139	100·77½	179	129·77½
20	14·50	**60**	43·50	**100**	72·50	**140**	101·50	**180**	130·50
21	15·22½	61	44·22½	101	73·22½	141	102·22½	181	131·22½
22	15·95	62	44·95	102	73·95	142	102·95	182	131·95
23	16·67½	63	45·67½	103	74·67½	143	103·67½	183	132·67½
24	17·40	64	46·40	104	75·40	144	104·40	184	133·40
25	18·12½	65	47·12½	105	76·12½	145	105·12½	185	134·12½
26	18·85	66	47·85	106	76·85	146	105·85	186	134·85
27	19·57½	67	48·57½	107	77·57½	147	106·57½	187	135·57½
28	20·30	68	49·30	108	78·30	148	107·30	188	136·30
29	21·02½	69	50·02½	109	79·02½	149	108·02½	189	137·02½
30	21·75	**70**	50·75	**110**	79·75	**150**	108·75	**190**	137·75
31	22·47½	71	51·47½	111	80·47½	151	109·47½	191	138·47½
32	23·20	72	52·20	112	81·20	152	110·20	192	139·20
33	23·92½	73	52·92½	113	81·92½	153	110·92½	193	139·92½
34	24·65	74	53·65	114	82·65	154	111·65	194	140·65
35	25·37½	75	54·37½	115	83·37½	155	112·37½	195	141·37½
36	26·10	76	55·10	116	84·10	156	113·10	196	142·10
37	26·82½	77	55·82½	117	84·82½	157	113·82½	197	142·82½
38	27·55	78	56·55	118	85·55	158	114·55	198	143·55
39	28·27½	79	57·27½	119	86·27½	159	115·27½	199	144·27½
40	29·00	**80**	58·00	**120**	87·00	**160**	116·00	**200**	145·00

$\frac{1}{16}=0{\cdot}04^5$ $\frac{1}{9}=0{\cdot}08^1$ $\frac{1}{8}=0{\cdot}09^1$ $\frac{1}{6}=0{\cdot}12^1$ $\frac{1}{5}=0{\cdot}14^5$ $\frac{1}{4}=0{\cdot}18^1$ $\frac{1}{3}=0{\cdot}24^2$

$\frac{3}{8}=0{\cdot}27^2$ $\frac{1}{2}=0{\cdot}36^3$ $\frac{5}{8}=0{\cdot}45^3$ $\frac{3}{4}=0{\cdot}54^4$ $\frac{7}{8}=0{\cdot}63^4$ $\frac{2}{3}=0{\cdot}48^3$ $\frac{5}{6}=0{\cdot}60^4$

$\frac{3}{16}=0{\cdot}13^6$ $\frac{5}{16}=0{\cdot}22^7$ $\frac{7}{16}=0{\cdot}31^7$ $\frac{9}{16}=0{\cdot}40^8$ $\frac{11}{16}=0{\cdot}49^8$ $\frac{13}{16}=0{\cdot}58^9$ $\frac{15}{16}=0{\cdot}68^0$

(1.37=1£) **£0.73**

1	0·73	41	29·93	81	59·13	121	88·33	161	117·53
2	1·46	42	30·66	82	59·86	122	89·06	162	118·26
3	2·19	43	31·39	83	60·59	123	89·79	163	118·99
4	2·92	44	32·12	84	61·32	124	90·52	164	119·72
5	3·65	45	32·85	85	62·05	125	91·25	165	120·45
6	4·38	46	33·58	86	62·78	126	91·98	166	121·18
7	5·11	47	34·31	87	63·51	127	92·71	167	121·91
8	5·84	48	35·04	88	64·24	128	93·44	168	122·64
9	6·57	49	35·77	89	64·97	129	94·17	169	123·37
10	7·30	**50**	36·50	**90**	65·70	**130**	94·90	**170**	124·10
11	8·03	51	37·23	91	66·43	131	95·63	171	124·83
12	8·76	52	37·96	92	67·16	132	96·36	172	125·56
13	9·49	53	38·69	93	67·89	133	97·09	173	126·29
14	10·22	54	39·42	94	68·62	134	97·82	174	127·02
15	10·95	55	40·15	95	69·35	135	98·55	175	127·75
16	11·68	56	40·88	96	70·08	136	99·28	176	128·48
17	12·41	57	41·61	97	70·81	137	100·01	177	129·21
18	13·14	58	42·34	98	71·54	138	100·74	178	129·94
19	13·87	59	43·07	99	72·27	139	101·47	179	130·67
20	14·60	**60**	43·80	**100**	73·00	**140**	102·20	**180**	131·40
21	15·33	61	44·53	101	73·73	141	102·93	181	132·13
22	16·06	62	45·26	102	74·46	142	103·66	182	132·86
23	16·79	63	45·99	103	75·19	143	104·39	183	133·59
24	17·52	64	46·72	104	75·92	144	105·12	184	134·32
25	18·25	65	47·45	105	76·65	145	105·85	185	135·05
26	18·98	66	48·18	106	77·38	146	106·58	186	135·78
27	19·71	67	48·91	107	78·11	147	107·31	187	136·51
28	20·44	68	49·64	108	78·84	148	108·04	188	137·24
29	21·17	69	50·37	109	79·57	149	108·77	189	137·97
30	21·90	**70**	51·10	**110**	80·30	**150**	109·50	**190**	138·70
31	22·63	71	51·83	111	81·03	151	110·23	191	139·43
32	23·36	72	52·56	112	81·76	152	110·96	192	140·16
33	24·09	73	53·29	113	82·49	153	111·69	193	140·89
34	24·82	74	54·02	114	83·22	154	112·42	194	141·62
35	25·55	75	54·75	115	83·95	155	113·15	195	142·35
36	26·28	76	55·48	116	84·68	156	113·88	196	143·08
37	27·01	77	56·21	117	85·41	157	114·61	197	143·81
38	27·74	78	56·94	118	86·14	158	115·34	198	144·54
39	28·47	79	57·67	119	86·87	159	116·07	199	145·27
40	29·20	**80**	58·40	**120**	87·60	**160**	116·80	**200**	146·00

$\frac{1}{16}=0{\cdot}04^{6}$ $\frac{1}{9}=0{\cdot}08^{1}$ $\frac{1}{8}=0{\cdot}09^{1}$ $\frac{1}{6}=0{\cdot}12^{2}$ $\frac{1}{5}=0{\cdot}14^{6}$ $\frac{1}{4}=0{\cdot}18^{3}$ $\frac{1}{3}=0{\cdot}24^{3}$

$\frac{3}{8}=0{\cdot}27^{4}$ $\frac{1}{2}=0{\cdot}36^{5}$ $\frac{5}{8}=0{\cdot}45^{6}$ $\frac{3}{4}=0{\cdot}54^{8}$ $\frac{7}{8}=0{\cdot}63^{9}$ $\frac{2}{3}=0{\cdot}48^{7}$ $\frac{5}{6}=0{\cdot}60^{8}$

$\frac{3}{16}=0{\cdot}13^{7}$ $\frac{5}{16}=0{\cdot}22^{8}$ $\frac{7}{16}=0{\cdot}31^{9}$ $\frac{9}{16}=0{\cdot}41^{1}$ $\frac{11}{16}=0{\cdot}50^{2}$ $\frac{13}{16}=0{\cdot}59^{3}$ $\frac{15}{16}=0{\cdot}68^{4}$

(1.36=1£) **£0.73½**

1	0·73½	41	30·13½	81	59·53½	121	88·93½	161	118·33½
2	1·47	42	30·87	82	60·27	122	89·67	162	119·07
3	2·20½	43	31·60½	83	61·00½	123	90·40½	163	119·80½
4	2·94	44	32·34	84	61·74	124	91·14	164	120·54
5	3·67½	45	33·07½	85	62·47½	125	91·87½	165	121·27½
6	4·41	46	33·81	86	63·21	126	92·61	166	122·01
7	5·14½	47	34·54½	87	63·94½	127	93·34½	167	122·74½
8	5·88	48	35·28	88	64·68	128	94·08	168	123·48
9	6·61½	49	36·01½	89	65·41½	129	94·81½	169	124·21½
10	7·35	**50**	36·75	**90**	66·15	**130**	95·55	**170**	124·95
11	8·08½	51	37·48½	91	66·88½	131	96·28½	171	125·68½
12	8·82	52	38·22	92	67·62	132	97·02	172	126·42
13	9·55½	53	38·95½	93	68·35½	133	97·75½	173	127·15½
14	10·29	54	39·69	94	69·09	134	98·49	174	127·89
15	11·02½	55	40·42½	95	69·82½	135	99·22½	175	128·62½
16	11·76	56	41·16	96	70·56	136	99·96	176	129·36
17	12·49½	57	41·89½	97	71·29½	137	100·69½	177	130·09½
18	13·23	58	42·63	98	72·03	138	101·43	178	130·83
19	13·96½	59	43·36½	99	72·76½	139	102·16½	179	131·56½
20	14·70	**60**	44·10	**100**	73·50	**140**	102·90	**180**	132·30
21	15·43½	61	44·83½	101	74·23½	141	103·63½	181	133·03½
22	16·17	62	45·57	102	74·97	142	104·37	182	133·77
23	16·90½	63	46·30½	103	75·70½	143	105·10½	183	134·50½
24	17·64	64	47·04	104	76·44	144	105·84	184	135·24
25	18·37½	65	47·77½	105	77·17½	145	106·57½	185	135·97½
26	19·11	66	48·51	106	77·91	146	107·31	186	136·71
27	19·84½	67	49·24½	107	78·64½	147	108·04½	187	137·44½
28	20·58	68	49·98	108	79·38	148	108·78	188	138·18
29	21·31½	69	50·71½	109	80·11½	149	109·51½	189	138·91½
30	22·05	**70**	51·45	**110**	80·85	**150**	110·25	**190**	139·65
31	22·78½	71	52·18½	111	81·58½	151	110·98½	191	140·38½
32	23·52	72	52·92	112	82·32	152	111·72	192	141·12
33	24·25½	73	53·65½	113	83·05½	153	112·45½	193	141·85½
34	24·99	74	54·39	114	83·79	154	113·19	194	142·59
35	25·72½	75	55·12½	115	84·52½	155	113·92½	195	143·32½
36	26·46	76	55·86	116	85·26	156	114·66	196	144·06
37	27·19½	77	56·59½	117	85·99½	157	115·39½	197	144·79½
38	27·93	78	57·33	118	86·73	158	116·13	198	145·53
39	28·66½	79	58·06½	119	87·46½	159	116·86½	199	146·26½
40	29·40	**80**	58·80	**120**	88·20	**160**	117·60	**200**	147·00

$\frac{1}{16}=0{\cdot}04^{6}$ $\frac{1}{9}=0{\cdot}08^{2}$ $\frac{1}{8}=0{\cdot}09^{2}$ $\frac{1}{6}=0{\cdot}12^{3}$ $\frac{1}{5}=0{\cdot}14^{7}$ $\frac{1}{4}=0{\cdot}18^{4}$ $\frac{1}{3}=0{\cdot}24^{5}$

$\frac{3}{8}=0{\cdot}27^{6}$ $\frac{1}{2}=0{\cdot}36^{8}$ $\frac{5}{8}=0{\cdot}45^{9}$ $\frac{3}{4}=0{\cdot}55^{1}$ $\frac{7}{8}=0{\cdot}64^{3}$ $\frac{2}{3}=0{\cdot}49^{0}$ $\frac{5}{6}=0{\cdot}61^{3}$

$\frac{3}{16}=0{\cdot}13^{8}$ $\frac{5}{16}=0{\cdot}23^{0}$ $\frac{7}{16}=0{\cdot}32^{2}$ $\frac{9}{16}=0{\cdot}41^{3}$ $\frac{11}{16}=0{\cdot}50^{5}$ $\frac{13}{16}=0{\cdot}59^{7}$ $\frac{15}{16}=0{\cdot}68^{9}$

(1.35=1£) **£0.74**

1	0·74	41	30·34	81	59·94	121	89·54	161	119·14
2	1·48	42	31·08	82	60·68	122	90·28	162	119·88
3	2·22	43	31·82	83	61·42	123	91·02	163	120·62
4	2·96	44	32·56	84	62·16	124	91·76	164	121·36
5	3·70	45	33·30	85	62·90	125	92·50	165	122·10
6	4·44	46	34·04	86	63·64	126	93·24	166	122·84
7	5·18	47	34·78	87	64·38	127	93·98	167	123·58
8	5·92	48	35·52	88	65·12	128	94·72	168	124·32
9	6·66	49	36·26	89	65·86	129	95·46	169	125·06
10	7·40	**50**	37·00	**90**	66·60	**130**	96·20	**170**	125·80
11	8·14	51	37·74	91	67·34	131	96·94	171	126·54
12	8·88	52	38·48	92	68·08	132	97·68	172	127·28
13	9·62	53	39·22	93	68·82	133	98·42	173	128·02
14	10·36	54	39·96	94	69·56	134	99·16	174	128·76
15	11·10	55	40·70	95	70·30	135	99·90	175	129·50
16	11·84	56	41·44	96	71·04	136	100·64	176	130·24
17	12·58	57	42·18	97	71·78	137	101·38	177	130·98
18	13·32	58	42·92	98	72·52	138	102·12	178	131·72
19	14·06	59	43·66	99	73·26	139	102·86	179	132·46
20	14·80	**60**	44·40	**100**	74·00	**140**	103·60	**180**	133·20
21	15·54	61	45·14	101	74·74	141	104·34	181	133·94
22	16·28	62	45·88	102	75·48	142	105·08	182	134·68
23	17·02	63	46·62	103	76·22	143	105·82	183	135·42
24	17·76	64	47·36	104	76·96	144	106·56	184	136·16
25	18·50	65	48·10	105	77·70	145	107·30	185	136·90
26	19·24	66	48·84	106	78·44	146	108·04	186	137·64
27	19·98	67	49·58	107	79·18	147	108·78	187	138·38
28	20·72	68	50·32	108	79·92	148	109·52	188	139·12
29	21·46	69	51·06	109	80·66	149	110·26	189	139·86
30	22·20	**70**	51·80	**110**	81·40	**150**	111·00	**190**	140·60
31	22·94	71	52·54	111	82·14	151	111·74	191	141·34
32	23·68	72	53·28	112	82·88	152	112·48	192	142·08
33	24·42	73	54·02	113	83·62	153	113·22	193	142·82
34	25·16	74	54·76	114	84·36	154	113·96	194	143·56
35	25·90	75	55·50	115	85·10	155	114·70	195	144·30
36	26·64	76	56·24	116	85·84	156	115·44	196	145·04
37	27·38	77	56·98	117	86·58	157	116·18	197	145·78
38	28·12	78	57·72	118	87·32	158	116·92	198	146·52
39	28·86	79	58·46	119	88·06	159	117·66	199	147·26
40	29·60	**80**	59·20	**120**	88·80	**160**	118·40	**200**	148·00

$\frac{1}{16}=0{\cdot}04^{6}$ $\frac{1}{9}=0{\cdot}08^{2}$ $\frac{1}{8}=0{\cdot}09^{3}$ $\frac{1}{6}=0{\cdot}12^{3}$ $\frac{1}{5}=0{\cdot}14^{8}$ $\frac{1}{4}=0{\cdot}18^{5}$ $\frac{1}{3}=0{\cdot}24^{7}$

$\frac{3}{8}=0{\cdot}27^{8}$ $\frac{1}{2}=0{\cdot}37^{0}$ $\frac{5}{8}=0{\cdot}46^{3}$ $\frac{3}{4}=0{\cdot}55^{5}$ $\frac{7}{8}=0{\cdot}64^{8}$ $\frac{2}{3}=0{\cdot}49^{3}$ $\frac{5}{6}=0{\cdot}61^{7}$

$\frac{3}{16}=0{\cdot}13^{9}$ $\frac{5}{16}=0{\cdot}23^{1}$ $\frac{7}{16}=0{\cdot}32^{4}$ $\frac{9}{16}=0{\cdot}41^{6}$ $\frac{11}{16}=0{\cdot}50^{9}$ $\frac{13}{16}=0{\cdot}60^{1}$ $\frac{15}{16}=0{\cdot}69^{4}$

(1.34=1£) **£0.74½**

1	0·74½	41	30·54½	81	60·34½	121	90·14½	161	119·94½
2	1·49	42	31·29	82	61·09	122	90·89	162	120·69
3	2·23½	43	32·03½	83	61·83½	123	91·63½	163	121·43½
4	2·98	44	32·78	84	62·58	124	92·38	164	122·18
5	3·72½	45	33·52½	85	63·32½	125	93·12½	165	122·92½
6	4·47	46	34·27	86	64·07	126	93·87	166	123·67
7	5·21½	47	35·01½	87	64·81½	127	94·61½	167	124·41½
8	5·96	48	35·76	88	65·56	128	95·36	168	125·16
9	6·70½	49	36·50½	89	66·30½	129	96·10½	169	125·90½
10	7·45	**50**	37·25	**90**	67·05	**130**	96·85	**170**	126·65
11	8·19½	51	37·99½	91	67·79½	131	97·59½	171	127·39½
12	8·94	52	38·74	92	68·54	132	98·34	172	128·14
13	9·68½	53	39·48½	93	69·28½	133	99·08½	173	128·88½
14	10·43	54	40·23	94	70·03	134	99·83	174	129·63
15	11·17½	55	40·97½	95	70·77½	135	100·57½	175	130·37½
16	11·92	56	41·72	96	71·52	136	101·32	176	131·12
17	12·66½	57	42·46½	97	72·26½	137	102·06½	177	131·86½
18	13·41	58	43·21	98	73·01	138	102·81	178	132·61
19	14·15½	59	43·95½	99	73·75½	139	103·55½	179	133·35½
20	14·90	**60**	44·70	**100**	74·50	**140**	104·30	**180**	134·10
21	15·64½	61	45·44½	101	75·24½	141	105·04½	181	134·84½
22	16·39	62	46·19	102	75·99	142	105·79	182	135·59
23	17·13½	63	46·93½	103	76·73½	143	106·53½	183	136·33½
24	17·88	64	47·68	104	77·48	144	107·28	184	137·08
25	18·62½	65	48·42½	105	78·22½	145	108·02½	185	137·82½
26	19·37	66	49·17	106	78·97	146	108·77	186	138·57
27	20·11½	67	49·91½	107	79·71½	147	109·51½	187	139·31½
28	20·86	68	50·66	108	80·46	148	110·26	188	140·06
29	21·60½	69	51·40½	109	81·20½	149	111·00½	189	140·80½
30	22·35	**70**	52·15	**110**	81·95	**150**	111·75	**190**	141·55
31	23·09½	71	52·89½	111	82·69½	151	112·49½	191	142·29½
32	23·84	72	53·64	112	83·44	152	113·24	192	143·04
33	24·58½	73	54·38½	113	84·18½	153	113·98½	193	143·78½
34	25·33	74	55·13	114	84·93	154	114·73	194	144·53
35	26·07½	75	55·87½	115	85·67½	155	115·47½	195	145·27½
36	26·82	76	56·62	116	86·42	156	116·22	196	146·02
37	27·56½	77	57·36½	117	87·16½	157	116·96½	197	146·76½
38	28·31	78	58·11	118	87·91	158	117·71	198	147·51
39	29·05½	79	58·85½	119	88·65½	159	118·45½	199	148·25½
40	29·80	**80**	59·60	**120**	89·40	**160**	119·20	**200**	149·00

$\frac{1}{16}=0{\cdot}04^{7}$ $\frac{1}{9}=0{\cdot}08^{3}$ $\frac{1}{8}=0{\cdot}09^{3}$ $\frac{1}{6}=0{\cdot}12^{4}$ $\frac{1}{5}=0{\cdot}14^{9}$ $\frac{1}{4}=0{\cdot}18^{6}$ $\frac{1}{3}=0{\cdot}24^{8}$

$\frac{3}{8}=0{\cdot}27^{9}$ $\frac{1}{2}=0{\cdot}37^{3}$ $\frac{5}{8}=0{\cdot}46^{6}$ $\frac{3}{4}=0{\cdot}55^{9}$ $\frac{7}{8}=0{\cdot}65^{2}$ $\frac{2}{3}=0{\cdot}49^{7}$ $\frac{5}{6}=0{\cdot}62^{1}$

$\frac{3}{16}=0{\cdot}14^{0}$ $\frac{5}{16}=0{\cdot}23^{3}$ $\frac{7}{16}=0{\cdot}32^{6}$ $\frac{9}{16}=0{\cdot}41^{9}$ $\frac{11}{16}=0{\cdot}51^{2}$ $\frac{13}{16}=0{\cdot}60^{5}$ $\frac{15}{16}=0{\cdot}69^{8}$

(1.333=1£) **£0.75**

1	0·75	41	30·75	81	60·75	121	90·75	161	120·75
2	1·50	42	31·50	82	61·50	122	91·50	162	121·50
3	2·25	43	32·25	83	62·25	123	92·25	163	122·25
4	3·00	44	33·00	84	63·00	124	93·00	164	123·00
5	3·75	45	33·75	85	63·75	125	93·75	165	123·75
6	4·50	46	34·50	86	64·50	126	94·50	166	124·50
7	5·25	47	35·25	87	65·25	127	95·25	167	125·25
8	6·00	48	36·00	88	66·00	128	96·00	168	126·00
9	6·75	49	36·75	89	66·75	129	96·75	169	126·75
10	7·50	**50**	37·50	**90**	67·50	**130**	97·50	**170**	127·50
11	8·25	51	38·25	91	68·25	131	98·25	171	128·25
12	9·00	52	39·00	92	69·00	132	99·00	172	129·00
13	9·75	53	39·75	93	69·75	133	99·75	173	129·75
14	10·50	54	40·50	94	70·50	134	100·50	174	130·50
15	11·25	55	41·25	95	71·25	135	101·25	175	131·25
16	12·00	56	42·00	96	72·00	136	102·00	176	132·00
17	12·75	57	42·75	97	72·75	137	102·75	177	132·75
18	13·50	58	43·50	98	73·50	138	103·50	178	133·50
19	14·25	59	44·25	99	74·25	139	104·25	179	134·25
20	15·00	**60**	45·00	**100**	75·00	**140**	105·00	**180**	135·00
21	15·75	61	45·75	101	75·75	141	105·75	181	135·75
22	16·50	62	46·50	102	76·50	142	106·50	182	136·50
23	17·25	63	47·25	103	77·25	143	107·25	183	137·25
24	18·00	64	48·00	104	78·00	144	108·00	184	138·00
25	18·75	65	48·75	105	78·75	145	108·75	185	138·75
26	19·50	66	49·50	106	79·50	146	109·50	186	139·50
27	20·25	67	50·25	107	80·25	147	110·25	187	140·25
28	21·00	68	51·00	108	81·00	148	111·00	188	141·00
29	21·75	69	51·75	109	81·75	149	111·75	189	141·75
30	22·50	**70**	52·50	**110**	82·50	**150**	112·50	**190**	142·50
31	23·25	71	53·25	111	83·25	151	113·25	191	143·25
32	24·00	72	54·00	112	84·00	152	114·00	192	144·00
33	24·75	73	54·75	113	84·75	153	114·75	193	144·75
34	25·50	74	55·50	114	85·50	154	115·50	194	145·50
35	26·25	75	56·25	115	86·25	155	116·25	195	146·25
36	27·00	76	57·00	116	87·00	156	117·00	196	147·00
37	27·75	77	57·75	117	87·75	157	117·75	197	147·75
38	28·50	78	58·50	118	88·50	158	118·50	198	148·50
39	29·25	79	59·25	119	89·25	159	119·25	199	149·25
40	30·00	**80**	60·00	**120**	90·00	**160**	120·00	**200**	150·00

$\frac{1}{16}=0{\cdot}04^{7}$ $\frac{1}{9}=0{\cdot}08^{3}$ $\frac{1}{8}=0{\cdot}09^{4}$ $\frac{1}{6}=0{\cdot}12^{5}$ $\frac{1}{5}=0{\cdot}15^{0}$ $\frac{1}{4}=0{\cdot}18^{8}$ $\frac{1}{3}=0{\cdot}25^{0}$

$\frac{3}{8}=0{\cdot}28^{1}$ $\frac{1}{2}=0{\cdot}37^{5}$ $\frac{5}{8}=0{\cdot}46^{9}$ $\frac{3}{4}=0{\cdot}56^{3}$ $\frac{7}{8}=0{\cdot}65^{6}$ $\frac{2}{3}=0{\cdot}50^{0}$ $\frac{5}{6}=0{\cdot}62^{5}$

$\frac{3}{16}=0{\cdot}14^{1}$ $\frac{5}{16}=0{\cdot}23^{4}$ $\frac{7}{16}=0{\cdot}32^{8}$ $\frac{9}{16}=0{\cdot}42^{2}$ $\frac{11}{16}=0{\cdot}51^{6}$ $\frac{13}{16}=0{\cdot}60^{9}$ $\frac{15}{16}=0{\cdot}70^{3}$

1	0·75½	41	30·95½	81	61·15½	121	91·35½	161	121·55½
2	1·51	42	31·71	82	61·91	122	92·11	162	122·31
3	2·26½	43	32·46½	83	62·66½	123	92·86½	163	123·06½
4	3·02	44	33·22	84	63·42	124	93·62	164	123·82
5	3·77½	45	33·97½	85	64·17½	125	94·37½	165	124·57½
6	4·53	46	34·73	86	64·93	126	95·13	166	125·33
7	5·28½	47	35·48½	87	65·68½	127	95·88½	167	126·08½
8	6·04	48	36·24	88	66·44	128	96·64	168	126·84
9	6·79½	49	36·99½	89	67·19½	129	97·39½	169	127·59½
10	7·55	**50**	37·75	**90**	67·95	**130**	98·15	**170**	128·35
11	8·30½	51	38·50½	91	68·70½	131	98·90½	171	129·10½
12	9·06	52	39·26	92	69·46	132	99·66	172	129·86
13	9·81½	53	40·01½	93	70·21½	133	100·41½	173	130·61½
14	10·57	54	40·77	94	70·97	134	101·17	174	131·37
15	11·32½	55	41·52½	95	71·72½	135	101·92½	175	132·12½
16	12·08	56	42·28	96	72·48	136	102·68	176	132·88
17	12·83½	57	43·03½	97	73·23½	137	103·43½	177	133·63½
18	13·59	58	43·79	98	73·99	138	104·19	178	134·39
19	14·34½	59	44·54½	99	74·74½	139	104·94½	179	135·14½
20	15·10	**60**	45·30	**100**	75·50	**140**	105·70	**180**	135·90
21	15·85½	61	46·05½	101	76·25½	141	106·45½	181	136·65½
22t	16·61	62	46·81	102	77·01	142	107·21	182	137·41
23	17·36½	63	47·56½	103	77·76½	143	107·96½	183	138·16½
24	18·12	64	48·32	104	78·52	144	108·72	184	138·92
25	18·87½	65	49·07½	105	79·27½	145	109·47½	185	139·67½
26	19·63	66	49·83	106	80·03	146	110·23	186	140·43
27	20·38½	67	50·58½	107	80·78½	147	110·98½	187	141·18½
28	21·14	68	51·34	108	81·54	148	111·74	188	141·94
29	21·89½	69	52·09½	109	82·29½	149	112·49½	189	142·69½
30	22·65	**70**	52·85	**110**	83·05	**150**	113·25	**190**	143·45
31	23·40½	71	53·60½	111	83·80½	151	114·00½	191	144·20½
32	24·16	72	54·36	112	84·56	152	114·76	192	144·96
33	24·91½	73	55·11½	113	85·31½	153	115·51½	193	145·71½
34	25·67	74	55·87	114	86·07	154	116·27	194	146·47
35	26·42½	75	56·62½	115	86·82½	155	117·02½	195	147·22½
36	27·18	76	57·38	116	87·58	156	117·78	196	147·98
37	27·93½	77	58·13½	117	88·33½	157	118·53½	197	148·73½
38	28·69	78	58·89	118	89·09	158	119·29	198	149·49
39	29·44½	79	59·64½	119	89·84½	159	120·04½	199	150·24½
40	30·20	**80**	60·40	**120**	90·60	**160**	120·80	**200**	151·00

$\frac{1}{16}=0{\cdot}04^{7}$ $\frac{1}{9}=0{\cdot}08^{4}$ $\frac{1}{8}=0{\cdot}09^{4}$ $\frac{1}{6}=0{\cdot}12^{6}$ $\frac{1}{5}=0{\cdot}15^{1}$ $\frac{1}{4}=0{\cdot}18^{9}$ $\frac{1}{3}=0{\cdot}25^{2}$

$\frac{3}{8}=0{\cdot}28^{3}$ $\frac{1}{2}=0{\cdot}37^{8}$ $\frac{5}{8}=0{\cdot}47^{2}$ $\frac{3}{4}=0{\cdot}56^{6}$ $\frac{7}{8}=0{\cdot}66^{1}$ $\frac{2}{3}=0{\cdot}50^{3}$ $\frac{5}{6}=0{\cdot}62^{9}$

$\frac{3}{16}=0{\cdot}14^{2}$ $\frac{5}{16}=0{\cdot}23^{6}$ $\frac{7}{16}=0{\cdot}33^{0}$ $\frac{9}{16}=0{\cdot}42^{5}$ $\frac{11}{16}=0{\cdot}51^{9}$ $\frac{13}{16}=0{\cdot}61^{3}$ $\frac{15}{16}=0{\cdot}70^{8}$

(1.316=1£) **£0.76**

1	0·76	41	31·16	81	61·56	121	91·96	161	122·36
2	1·52	42	31·92	82	62·32	122	92·72	162	123·12
3	2·28	43	32·68	83	63·08	123	93·48	163	123·88
4	3·04	44	33·44	84	63·84	124	94·24	164	124·64
5	3·80	45	34·20	85	64·60	125	95·00	165	125·40
6	4·56	46	34·96	86	65·36	126	95·76	166	126·16
7	5·32	47	35·72	87	66·12	127	96·52	167	126·92
8	6·08	48	36·48	88	66·88	128	97·28	168	127·68
9	6·84	49	37·24	89	67·64	129	98·04	169	128·44
10	7·60	**50**	38·00	**90**	68·40	**130**	98·80	**170**	129·20
11	8·36	51	38·76	91	69·16	131	99·56	171	129·96
12	9·12	52	39·52	92	69·92	132	100·32	172	130·72
13	9·88	53	40·28	93	70·68	133	101·08	173	131·48
14	10·64	54	41·04	94	71·44	134	101·84	174	132·24
15	11·40	55	41·80	95	72·20	135	102·60	175	133·00
16	12·16	56	42·56	96	72·96	136	103·36	176	133·76
17	12·92	57	43·32	97	73·72	137	104·12	177	134·52
18	13·68	58	44·08	98	74·48	138	104·88	178	135·28
19	14·44	59	44·84	99	75·24	139	105·64	179	136·04
20	15·20	**60**	45·60	**100**	76·00	**140**	106·40	**180**	136·80
21	15·96	61	46·36	101	76·76	141	107·16	181	137·56
22	16·72	62	47·12	102	77·52	142	107·92	182	138·32
23	17·48	63	47·88	103	78·28	143	108·68	183	139·08
24	18·24	64	48·64	104	79·04	144	109·44	184	139·84
25	19·00	65	49·40	105	79·80	145	110·20	185	140·60
26	19·76	66	50·16	106	80·56	146	110·96	186	141·36
27	20·52	67	50·92	107	81·32	147	111·72	187	142·12
28	21·28	68	51·68	108	82·08	148	112·48	188	142·88
29	22·04	69	52·44	109	82·84	149	113·24	189	143·64
30	22·80	**70**	53·20	**110**	83·60	**150**	114·00	**190**	144·40
31	23·56	71	53·96	111	84·36	151	114·76	191	145·16
32	24·32	72	54·72	112	85·12	152	115·52	192	145·92
33	25·08	73	55·48	113	85·88	153	116·28	193	146·68
34	25·84	74	56·24	114	86·64	154	117·04	194	147·44
35	26·60	75	57·00	115	87·40	155	117·80	195	148·20
36	27·36	76	57·76	116	88·16	156	118·56	196	148·96
37	28·12	77	58·52	117	88·92	157	119·32	197	149·72
38	28·88	78	59·28	118	89·68	158	120·08	198	150·48
39	29·64	79	60·04	119	90·44	159	120·84	199	151·24
40	30·40	**80**	60·80	**120**	91·20	**160**	121·60	**200**	152·00

$\frac{1}{16}=0{\cdot}04^{8}$ $\frac{1}{9}=0{\cdot}08^{4}$ $\frac{1}{8}=0{\cdot}09^{5}$ $\frac{1}{6}=0{\cdot}12^{7}$ $\frac{1}{5}=0{\cdot}15^{2}$ $\frac{1}{4}=0{\cdot}19^{0}$ $\frac{1}{3}=0{\cdot}25^{3}$

$\frac{3}{8}=0{\cdot}28^{5}$ $\frac{1}{2}=0{\cdot}38^{0}$ $\frac{5}{8}=0{\cdot}47^{5}$ $\frac{3}{4}=0{\cdot}57^{0}$ $\frac{7}{8}=0{\cdot}66^{5}$ $\frac{2}{3}=0{\cdot}50^{7}$ $\frac{5}{6}=0{\cdot}63^{3}$

$\frac{3}{16}=0{\cdot}14^{3}$ $\frac{5}{16}=0{\cdot}23^{8}$ $\frac{7}{16}=0{\cdot}33^{3}$ $\frac{9}{16}=0{\cdot}42^{8}$ $\frac{11}{16}=0{\cdot}52^{3}$ $\frac{13}{16}=0{\cdot}61^{8}$ $\frac{15}{16}=0{\cdot}71^{3}$

(1.308=1£) **£0.76½**

1	0·76½	41	31·36½	81	61·96j]	121	92·56½	161	123·16½
2	1·53	42	32·13	82	62·73	122	93·33	162	123·93
3	2·29½	43	32·89½	83	63·49½	123	94·09½	163	124·69½
4	3·06	44	33·66	84	64·26	124	94·86	164	125·46
5	3·82½	45	34·42½	85	65·02½	125	95·62½	165	126·22½
6	4·59	46	35·19	86	65·79	126	96·39	166	126·99
7	5·35½	47	35·95½	87	66·55½	127	97·15½	167	127·75½
8	6·12	48	36·72	88	67·32	128	97·92	168	128·52
9	6·88½	49	37·48½	89	68·08½	129	98·68½	169	129·28½
10	7·65	**50**	38·25	**90**	68·85	**130**	99·45	**170**	130·05
11	8·41½	51	39·01½	91	69·61½	131	100·21½	171	130·81½
12	9·18	52	39·78	92	70·38	132	100·98	172	131·58
13	9·94½	53	40·54½	93	71·14½	133	101·74½	173	132·34½
14	10·71	54	41·31	94	71·91	134	102·51	174	133·11
15	11·47½	55	42·07½	95	72·67½	135	103·27½	175	133·87½
16	12·24	56	42·84	96	73·44	136	104·04	176	134·64
17	13·00½	57	43·60½	97	74·20½	137	104·80½	177	135·40½
18	13·77	58	44·37	98	74·97	138	105·57	178	136·17
19	14·53½	59	45·13½	99	75·73½	139	106·33½	179	136·93½
20	15·30	**60**	45·90	**100**	76·50	**140**	107·10	**180**	137·70
21	16·06½	61	46·66½	101	77·26½	141	107·86½	181	138·46½
22	16·83	62	47·43	102	78·03	142	108·63	182	139·23
23	17·59½	63	48·19½	103	78·79½	143	109·39½	183	139·99½
24	18·36	64	48·96	104	79·56	144	110·16	184	140·76
25	19·12½	65	49·72½	105	80·32½	145	110·92½	185	141·52½
26	19·89	66	50·49	106	81·09	146	111·69	186	142·29
27	20·65½	67	51·25½	107	81·85½	147	112·45½	187	143·05½
28	21·42	68	52·02	108	82·62	148	113·22	188	143·82
29	22·18½	69	52·78½	109	83·38½	149	113·98½	189	144·58½
30	22·95	**70**	53·55	**110**	84·15	**150**	114·75	**190**	145·35
31	23·71½	71	54·31½	111	84·91½	151	115·51½	191	146·11½
32	24·48	72	55·08	112	85·68	152	116·28	192	146·88
33	25·24½	73	55·84½	113	86·44½	153	117·04½	193	147·64½
34	26·01	74	56·61	114	87·21	154	117·81	194	148·41
35	26·77½	75	57·37½	115	87·97½	155	118·57½	195	149·17½
36	27·54	76	58·14	116	88·74	156	119·34	196	149·94
37	28·30½	77	58·90½	117	89·50½	157	120·10½	197	150·70½
38	29·07	78	59·67	118	90·27	158	120·87	198	151·47
39	29·83½	79	60·43½	119	91·03½	159	121·63½	199	152·23½
40	30·60	**80**	61·20	**120**	91·80	**160**	122·40	**200**	153·00

$\frac{1}{16}=0{\cdot}04^{8}$ $\frac{1}{9}=0{\cdot}08^{5}$ $\frac{1}{8}=0{\cdot}09^{6}$ $\frac{1}{6}=0{\cdot}12^{8}$ $\frac{1}{5}=0{\cdot}15^{3}$ $\frac{1}{4}=0{\cdot}19^{1}$ $\frac{1}{3}=0{\cdot}25^{5}$

$\frac{3}{8}=0{\cdot}28^{7}$ $\frac{1}{2}=0{\cdot}38^{3}$ $\frac{5}{8}=0{\cdot}47^{8}$ $\frac{3}{4}=0{\cdot}57^{4}$ $\frac{7}{8}=0{\cdot}66^{9}$ $\frac{2}{3}=0{\cdot}51^{0}$ $\frac{5}{6}=0{\cdot}63^{8}$

$\frac{3}{16}=0{\cdot}14^{3}$ $\frac{5}{16}=0{\cdot}23^{9}$ $\frac{7}{16}=0{\cdot}33^{5}$ $\frac{9}{16}=0{\cdot}43^{0}$ $\frac{11}{16}=0{\cdot}52^{6}$ $\frac{13}{16}=0{\cdot}62^{2}$ $\frac{15}{16}=0{\cdot}71^{7}$

1 cubic yard=0·7645 cubic metres

1	0·77	41	31·57	81	62·37	121	93·17	161	123·97
2	1·54	42	32·34	82	63·14	122	93·94	162	124·74
3	2·31	43	33·11	83	63·91	123	94·71	163	125·51
4	3·08	44	33·88	84	64·68	124	95·48	164	126·28
5	3·85	45	34·65	85	65·45	125	96·25	165	127·05
6	4·62	46	35·42	86	66·22	126	97·02	166	127·82
7	5·39	47	36·19	87	66·99	127	97·79	167	128·59
8	6·16	48	36·96	88	67·76	128	98·56	168	129·36
9	6·93	49	37·73	89	68·53	129	99·33	169	130·13
10	7·70	**50**	38·50	**90**	69·30	**130**	100·10	**170**	130·90
11	8·47	51	39·27	91	70·07	131	100·87	171	131·67
12	9·24	52	40·04	92	70·84	132	101·64	172	132·44
13	10·01	53	40·81	93	71·61	133	102·41	173	133·21
14	10·78	54	41·58	94	72·38	134	103·18	174	133·98
15	11·55	55	42·35	95	73·15	135	103·95	175	134·75
16	12·32	56	43·12	96	73·92	136	104·72	176	135·52
17	13·09	57	43·89	97	74·69	137	105·49	177	136·29
18	13·86	58	44·66	98	75·46	138	106·26	178	137·06
19	14·63	59	45·43	99	76·23	139	107·03	179	137·83
20	15·40	**60**	46·20	**100**	77·00	**140**	107·80	**180**	138·60
21	16·17	61	46·97	101	77·77	141	108·57	181	139·37
22	16·94	62	47·74	102	78·54	142	109·34	182	140·14
23	17·71	63	48·51	103	79·31	143	110·11	183	140·91
24	18·48	64	49·28	104	80·08	144	110·88	184	141·68
25	19·25	65	50·05	105	80·85	145	111·65	185	142·45
26	20·02	66	50·82	106	81·62	146	112·42	186	143·22
27	20·79	67	51·59	107	82·39	147	113·19	187	143·99
28	21·56	68	52·36	108	83·16	148	113·96	188	144·76
29	22·33	69	53·13	109	83·93	149	114·73	189	145·53
30	23·10	**70**	53·90	**110**	84·70	**150**	115·50	**190**	146·30
31	23·87	71	54·67	111	85·47	151	116·27	191	147·07
32	24·64	72	55·44	112	86·24	152	117·04	192	147·84
33	25·41	73	56·21	113	87·01	153	117·81	193	148·61
34	26·18	74	56·98	114	87·78	154	118·58	194	149·38
35	26·95	75	57·75	115	88·55	155	119·35	195	150·15
36	27·72	76	58·52	116	89·32	156	120·12	196	150·92
37	28·49	77	59·29	117	90·09	157	120·89	197	151·69
38	29·26	78	60·06	118	90·86	158	121·66	198	152·46
39	30·03	79	60·83	119	91·63	159	122·43	199	153·23
40	30·80	**80**	61·60	**120**	92·40	**160**	123·20	**200**	154·00

$\frac{1}{16}=0{\cdot}04^8$ $\frac{1}{9}=0{\cdot}08^6$ $\frac{1}{8}=0{\cdot}09^6$ $\frac{1}{6}=0{\cdot}12^8$ $\frac{1}{5}=0{\cdot}15^4$ $\frac{1}{4}=0{\cdot}19^3$ $\frac{1}{3}=0{\cdot}25^7$

$\frac{3}{8}=0{\cdot}28^9$ $\frac{1}{2}=0{\cdot}38^5$ $\frac{5}{8}=0{\cdot}48^1$ $\frac{3}{4}=0{\cdot}57^8$ $\frac{7}{8}=0{\cdot}67^4$ $\frac{2}{3}=0{\cdot}51^3$ $\frac{5}{6}=0{\cdot}64^2$

$\frac{3}{16}=0{\cdot}14^4$ $\frac{5}{16}=0{\cdot}24^1$ $\frac{7}{16}=0{\cdot}33^7$ $\frac{9}{16}=0{\cdot}43^3$ $\frac{11}{16}=0{\cdot}52^9$ $\frac{13}{16}=0{\cdot}62^6$ $\frac{15}{16}=0{\cdot}72^2$

(1.290=1£) **£0.77½**

1	0·77½	41	31·77½	81	62·77½	121	93·77½	161	124·77½
2	1·55	42	32·55	82	63·55	122	94·55	162	125·55
3	2·32½	43	33·32½	83	64·32½	123	95·32½	163	126·32½
4	3·10	44	34·10	84	65·10	124	96·10	164	127·10
5	3·87½	45	34·87½	85	65·87½	125	96·87½	165	127·87½
6	4·65	46	35·65	86	66·65	126	97·65	166	128·65
7	5·42½	47	36·42½	87	67·42½	127	98·42½	167	129·42½
8	6·20	48	37·20	88	68·20	128	99·20	168	130·20
9	6·97½	49	37·97½	89	68·97½	129	99·97½	169	130·97½
10	7·75	**50**	38·75	**90**	69·75	**130**	100·75	**170**	131·75
11	8·52½	51	39·52½	91	70·52½	131	101·52½	171	132·52½
12	9·30	52	40·30	92	71·30	132	102·30	172	133·30
13	10·07½	53	41·07½	93	72·07½	133	103·07½	173	134·07½
14	10·85	54	41·85	94	72·85	134	103·85	174	134·85
15	11·62½	55	42·62½	95	73·62½	135	104·62½	175	135·62½
16	12·40	56	43·40	96	74·40	136	105·40	176	136·40
17	13·17½	57	44·17½	97	75·17½	137	106·17½	177	137·17½
18	13·95	58	44·95	98	75·95	138	106·95	178	137·95
19	14·72½	59	45·72½	99	76·72½	139	107·72½	179	138·72½
20	15·50	**60**	46·50	**100**	77·50	**140**	108·50	**180**	139·50
21	16·27½	61	47·27½	101	78·27½	141	109·27½	181	140·27½
22	17·05	62	48·05	102	79·05	142	110·05	182	141·05
23	17·82½	63	48·82½	103	79·82½	143	110·82½	183	141·82½
24	18·60	64	49·60	104	80·60	144	111·60	184	142·60
25	19·37½	65	50·37½	105	81·37½	145	112·37½	185	143·37½
26	20·15	66	51·15	106	82·15	146	113·15	186	144·15
27	20·92½	67	51·92½	107	82·92½	147	113·92½	187	144·92½
28	21·70	68	52·70	108	83·70	148	114·70	188	145·70
29	22·47½	69	53·47½	109	84·47½	149	115·47½	189	146·47½
30	23·25	**70**	54·25	**110**	85·25	**150**	116·25	**190**	147·25
31	24·02½	71	55·02½	111	86·02½	151	117·02½	191	148·02½
32	24·80	72	55·80	112	86·80	152	117·80	192	148·80
33	25·57½	73	56·57½	113	87·57½	153	118·57½	193	149·57½
34	26·35	74	57·35	114	88·35	154	119·35	194	150·35
35	27·12½	75	58·12½	115	89·12½	155	120·12½	195	151·12½
36	27·90	76	58·90	116	89·90	156	120·90	196	151·90
37	28·67½	77	59·67½	117	90·67½	157	121·67½	197	152·67½
38	29·45	78	60·45	118	91·45	158	122·45	198	153·45
39	30·22½	79	61·22½	119	92·22½	159	123·22½	199	154·22½
40	31·00	**80**	62·00	**120**	93·00	**160**	124·00	**200**	155·00

$\frac{1}{16}$=0·04^{8} $\frac{1}{9}$=0·08^{6} $\frac{1}{8}$=0·09^{7} $\frac{1}{6}$=0·12^{9} $\frac{1}{5}$=0·15^{5} $\frac{1}{4}$=0·19^{4} $\frac{1}{3}$=0·25^{8}

$\frac{3}{8}$=0·29^{1} $\frac{1}{2}$=0·38^{8} $\frac{5}{8}$=0·48^{4} $\frac{3}{4}$=0·58^{1} $\frac{7}{8}$=0·67^{8} $\frac{2}{3}$=0·51^{7} $\frac{5}{6}$=0·64^{6}

$\frac{3}{16}$=0·14^{5} $\frac{5}{16}$=0·24^{2} $\frac{7}{16}$=0·33^{9} $\frac{9}{16}$=0·43^{6} $\frac{11}{16}$=0·53^{3} $\frac{13}{16}$=0·63^{0} $\frac{15}{16}$=0·72^{7}

1	0·78	41	31·98	81	63·18	121	94·38	161	125·58
2	1·56	42	32·76	82	63·96	122	95·16	162	126·36
3	2·34	43	33·54	83	64·74	123	95·94	163	127·14
4	3·12	44	34·32	84	65·52	124	96·72	164	127·92
5	3·90	45	35·10	85	66·30	125	97·50	165	128·70
6	4·68	46	35·88	86	67·08	126	98·28	166	129·48
7	5·46	47	36·66	87	67·86	127	99·06	167	130·26
8	6·24	48	37·44	88	68·64	128	99·84	168	131·04
9	7·02	49	38·22	89	69·42	129	100·62	169	131·82
10	7·80	**50**	39·00	**90**	70·20	**130**	101·40	**170**	132·60
11	8·58	51	39·78	91	70·98	131	102·18	171	133·38
12	9·36	52	40·56	92	71·76	132	102·96	172	134·16
13	10·14	53	41·34	93	72·54	133	103·74	173	134·94
14	10·92	54	42·12	94	73·32	134	104·52	174	135·72
15	11·70	55	42·90	95	74·10	135	105·30	175	136·50
16	12·48	56	43·68	96	74·88	136	106·08	176	137·28
17	13·26	57	44·46	97	75·66	137	106·86	177	138·06
18	14·04	58	45·24	98	76·44	138	107·64	178	138·84
19	14·82	59	46·02	99	77·22	139	108·42	179	139·62
20	15·60	**60**	46·80	**100**	78·00	**140**	109·20	**180**	140·40
21	16·38	61	47·58	101	78·78	141	109·98	181	141·18
22	17·16	62	48·36	102	79·56	142	110·76	182	141·96
23	17·94	63	49·14	103	80·34	143	111·54	183	142·74
24	18·72	64	49·92	104	81·12	144	112·32	184	143·52
25	19·50	65	50·70	105	81·90	145	113·10	185	144·30
26	20·28	66	51·48	106	82·68	146	113·88	186	145·08
27	21·06	67	52·26	107	83·46	147	114·66	187	145·86
28	21·84	68	53·04	108	84·24	148	115·44	188	146·64
29	22·62	69	53·82	109	85·02	149	116·22	189	147·42
30	23·40	**70**	54·60	**110**	85·80	**150**	117·00	**190**	148·20
31	24·18	71	55·38	111	86·58	151	117·78	191	148·98
32	24·96	72	56·16	112	87·36	152	118·56	192	149·76
33	25·74	73	56·94	113	88·14	153	119·34	193	150·54
34	26·52	74	57·72	114	88·92	154	120·12	194	151·32
35	27·30	75	58·50	115	89·70	155	120·90	195	152·10
36	28·08	76	59·28	116	90·48	156	121·68	196	152·88
37	28·86	77	60·06	117	91·26	157	122·46	197	153·66
38	29·64	78	60·84	118	92·04	158	123·24	198	154·44
39	30·42	79	61·62	119	92·82	159	124·02	199	155·22
40	31·20	**80**	62·40	**120**	93·60	**160**	124·80	**200**	156·00

=$0{\cdot}04^{9}$ $\frac{1}{9}=0{\cdot}08^{7}$ $\frac{1}{8}=0{\cdot}09^{8}$ $\frac{1}{6}=0{\cdot}13^{0}$ $\frac{1}{5}=0{\cdot}15^{6}$ $\frac{1}{4}=0{\cdot}19^{5}$ $\frac{1}{3}=0{\cdot}26^{0}$

=$0{\cdot}29^{3}$ $\frac{1}{2}=0{\cdot}39^{0}$ $\frac{5}{8}=0{\cdot}48^{8}$ $\frac{3}{4}=0{\cdot}58^{5}$ $\frac{7}{8}=0{\cdot}68^{3}$ $\frac{2}{3}=0{\cdot}52^{0}$ $\frac{5}{6}=0{\cdot}65^{0}$

=$0{\cdot}14^{6}$ $\frac{5}{16}=0{\cdot}24^{4}$ $\frac{7}{16}=0{\cdot}34^{1}$ $\frac{9}{16}=0{\cdot}43^{9}$ $\frac{11}{16}=0{\cdot}53^{6}$ $\frac{13}{16}=0{\cdot}63^{4}$ $\frac{15}{16}=0{\cdot}73^{1}$

(1.274=1£) **£0.78½**

1	0·78½	41	32·18½	81	63·58½	121	94·98½	161	126·38½
2	1·57	42	32·97	82	64·37	122	95·77	162	127·17
3	2·35½	43	33·75½	83	65·15½	123	96·55½	163	127·95½
4	3·14	44	34·54	84	65·94	124	97·34	164	128·74
5	3·92½	45	35·32½	85	66·72½	125	98·12½	165	129·52½
6	4·71	46	36·11	86	67·51	126	98·91	166	130·31
7	5·49½	47	36·89½	87	68·29½	127	99·69½	167	131·09½
8	6·28	48	37·68	88	69·08	128	100·48	168	131·88
9	7·06½	49	38·46½	89	69·86½	129	101·26½	169	132·66½
10	7·85	**50**	39·25	**90**	70·65	**130**	102·05	**170**	133·45
11	8·63½	51	40·03½	91	71·43½	131	102·83½	171	134·23½
12	9·42	52	40·82	92	72·22	132	103·62	172	135·02
13	10·20½	53	41·60½	93	73·00½	133	104·40½	173	135·80½
14	10·99	54	42·39	94	73·79	134	105·19	174	136·59
15	11·77½	55	43·17½	95	74·57½	135	105·97½	175	137·37½
16	12·56	56	43·96	96	75·36	136	106·76	176	138·16
17	13·34½	57	44·74½	97	76·14½	137	107·54½	177	138·94½
18	14·13	58	45·53	98	76·93	138	108·33	178	139·73
19	14·91½	59	46·31½	99	77·71½	139	109·11½	179	140·51½
20	15·70	**60**	47·10	**100**	78·50	**140**	109·90	**180**	141·30
21	16·48½	61	47·88½	101	79·28½	141	110·68½	181	142·08½
22	17·27	62	48·67	102	80·07	142	111·47	182	142·87
23	18·05½	63	49·45½	103	80·85½	143	112·25½	183	143·65½
24	18·84	64	50·24	104	81·64	144	113·04	184	144·44
25	19·62½	65	51·02½	105	82·42½	145	113·82½	185	145·22½
26	20·41	66	51·81	106	83·21	146	114·61	186	146·01
27	21·19½	67	52·59½	107	83·99½	147	115·39½	187	146·79½
28	21·98	68	53·38	108	84·78	148	116·18	188	147·58
29	22·76½	69	54·16½	109	85·56½	149	116·96½	189	148·36½
30	23·55	**70**	54·95	**110**	86·35	**150**	117·75	**190**	149·15
31	24·33½	71	55·73½	111	87·13½	151	118·53½	191	149·93½
32	25·12	72	56·52	112	87·92	152	119·32	192	150·72
33	25·90½	73	57·30½	113	88·70½	153	120·10½	193	151·50½
34	26·69	74	58·09	114	89·49	154	120·89	194	152·29
35	27·47½	75	58·87½	115	90·27½	155	121·67½	195	153·07½
36	28·26	76	59·66	116	91·06	156	122·46	196	153·86
37	29·04½	77	60·44½	117	91·84½	157	123·24½	197	154·64½
38	29·83	78	61·23	118	92·63	158	124·03	198	155·43
39	30·61½	79	62·01½	119	93·41½	159	124·81½	199	156·21½
40	31·40	**80**	62·80	**120**	94·20	**160**	125·60	**200**	157·00

$\frac{1}{16}=0{\cdot}04^{9}$ $\frac{1}{9}=0{\cdot}08^{7}$ $\frac{1}{8}=0{\cdot}09^{8}$ $\frac{1}{6}=0{\cdot}13^{1}$ $\frac{1}{5}=0{\cdot}15^{7}$ $\frac{1}{4}=0{\cdot}19^{6}$ $\frac{1}{3}=0{\cdot}26^{2}$

$\frac{3}{8}=0{\cdot}29^{4}$ $\frac{1}{2}=0{\cdot}39^{3}$ $\frac{5}{8}=0{\cdot}49^{1}$ $\frac{3}{4}=0{\cdot}58^{9}$ $\frac{7}{8}=0{\cdot}68^{7}$ $\frac{2}{3}=0{\cdot}52^{3}$ $\frac{5}{6}=0{\cdot}65$

$\frac{3}{16}=0{\cdot}14^{7}$ $\frac{5}{16}=0{\cdot}24^{5}$ $\frac{7}{16}=0{\cdot}34^{3}$ $\frac{9}{16}=0{\cdot}44^{2}$ $\frac{11}{16}=0{\cdot}54^{0}$ $\frac{13}{16}=0{\cdot}63^{8}$ $\frac{15}{16}=0{\cdot}73$

1.266=1£) £0.79

1	0·79	41	32·39	81	63·99	121	95·59	161	127·19
2	1·58	42	33·18	82	64·78	122	96·38	162	127·98
3	2·37	43	33·97	83	65·57	123	97·17	163	128·77
4	3·16	44	34·76	84	66·36	124	97·96	164	129·56
5	3·95	45	35·55	85	67·15	125	98·75	165	130·35
6	4·74	46	36·34	86	67·94	126	99·54	166	131·14
7	5·53	47	37·13	87	68·73	127	100·33	167	131·93
8	6·32	48	37·92	88	69·52	128	101·12	168	132·72
9	7·11	49	38·71	89	70·31	129	101·91	169	133·51
0	7·90	**50**	39·50	**90**	71·10	**130**	102·70	**170**	134·30
1	8·69	51	40·29	91	71·89	131	103·49	171	135·09
2	9·48	52	41·08	92	72·68	132	104·28	172	135·88
3	10·27	53	41·87	93	73·47	133	105·07	173	136·67
4	11·06	54	42·66	94	74·26	134	105·86	174	137·46
5	11·85	55	43·45	95	75·05	135	106·65	175	138·25
6	12·64	56	44·24	96	75·84	136	107·44	176	139·04
7	13·43	57	45·03	97	76·63	137	108·23	177	139·83
8	14·22	58	45·82	98	77·42	138	109·02	178	140·62
9	15·01	59	46·61	99	78·21	139	109·81	179	141·41
0	15·80	**60**	47·40	**100**	79·00	**140**	110·60	**180**	142·20
1	16·59	61	48·19	101	79·79	141	111·39	181	142·99
2	17·38	62	48·98	102	80·58	142	112·18	182	143·78
3	18·17	63	49·77	103	81·37	143	112·97	183	144·57
4	18·96	64	50·56	104	82·16	144	113·76	184	145·36
5	19·75	65	51·35	105	82·95	145	114·55	185	146·15
6	20·54	66	52·14	106	83·74	146	115·34	186	146·94
7	21·33	67	52·93	107	84·53	147	116·13	187	147·73
8	22·12	68	53·72	108	85·32	148	116·92	188	148·52
9	22·91	69	54·51	109	86·11	149	117·71	189	149·31
0	23·70	**70**	55·30	**110**	86·90	**150**	118·50	**190**	150·10
1	24·49	71	56·09	111	87·69	151	119·29	191	150·89
2	25·28	72	56·88	112	88·48	152	120·08	192	151·68
3	26·07	73	57·67	113	89·27	153	120·87	193	152·47
4	26·86	74	58·46	114	90·06	154	121·66	194	153·26
5	27·65	75	59·25	115	90·85	155	122·45	195	154·05
6	28·44	76	60·04	116	91·64	156	123·24	196	154·84
7	29·23	77	60·83	117	92·43	157	124·03	197	155·63
8	30·02	78	61·62	118	93·22	158	124·82	198	156·42
9	30·81	79	62·41	119	94·01	159	125·61	199	157·21
0	31·60	**80**	63·20	**120**	94·80	**160**	126·40	**200**	158·00

=0·04^{9} $\frac{1}{9}$=0·08^{8} $\frac{1}{8}$=0·09^{9} $\frac{1}{6}$=0·13^{2} $\frac{1}{5}$=0·15^{8} $\frac{1}{4}$=0·19^{8} $\frac{1}{3}$=0·26^{3}

=0·29^{6} $\frac{1}{2}$=0·39^{5} $\frac{5}{8}$=0·49^{4} $\frac{3}{4}$=0·59^{3} $\frac{7}{8}$=0·69^{1} $\frac{2}{3}$=0·52^{7} $\frac{5}{6}$=0·65^{8}

=0·14^{8} $\frac{5}{16}$=0·24^{7} $\frac{7}{16}$=0·34^{6} $\frac{9}{16}$=0·44^{4} $\frac{11}{16}$=0·54^{3} $\frac{13}{16}$=0·64^{2} $\frac{15}{16}$=0·74^{1}

(1.258=1£) **£0.79½**

1	0·79½	41	32·59½	81	64·39½	121	96·19½	161	127·99½
2	1·59	42	33·39	82	65·19	122	96·99	162	128·79
3	2·38½	43	34·18½	83	65·98½	123	97·78½	163	129·58½
4	3·18	44	34·98	84	66·78	124	98·58	164	130·38
5	3·97½	45	35·77½	85	67·57½	125	99·37½	165	131·17½
6	4·77	46	36·57	86	68·37	126	100·17	166	131·97
7	5·56½	47	37·36½	87	69·16½	127	100·96½	167	132·76½
8	6·36	48	38·16	88	69·96	128	101·76	168	133·56
9	7·15½	49	38·95½	89	70·75½	129	102·55½	169	134·35½
10	7·95	**50**	39·75	**90**	71·55	**130**	103·35	**170**	135·15
11	8·74½	51	40·54½	91	72·34½	131	104·14½	171	135·94½
12	9·54	52	41·34	92	73·14	132	104·94	172	136·74
13	10·33½	53	42·13½	93	73·93½	133	105·73½	173	137·53½
14	11·13	54	42·93	94	74·73	134	106·53	174	138·33
15	11·92½	55	43·72½	95	75·52½	135	107·32½	175	139·12½
16	12·72	56	44·52	96	76·32	136	108·12	176	139·92
17	13·51½	57	45·31½	97	77·11½	137	108·91½	177	140·71½
18	14·31	58	46·11	98	77·91	138	109·71	178	141·51
19	15·10½	59	46·90½	99	78·70½	139	110·50½	179	142·30½
20	15·90	**60**	47·70	**100**	79·50	**140**	111·30	**180**	143·10
21	16·69½	61	48·49½	101	80·29½	141	112·09½	181	143·89½
22	17·49	62	49·29	102	81·09	142	112·89	182	144·69
23	18·28½	63	50·08½	103	81·88½	143	113·68½	183	145·48½
24	19·08	64	50·88	104	82·68	144	114·48	184	146·28
25	19·87½	65	51·67½	105	83·47½	145	115·27½	185	147·07½
26	20·67	66	52·47	106	84·27	146	116·07	186	147·87
27	21·46½	67	53·26½	107	85·06½	147	116·86½	187	148·66½
28	22·26	68	54·06	108	85·86	148	117·66	188	149·46
29	23·05½	69	54·85½	109	86·65½	149	118·45½	189	150·25½
30	23·85	**70**	55·65	**110**	87·45	**150**	119·25	**190**	151·05
31	24·64½	71	56·44½	111	88·24½	151	120·04½	191	151·84½
32	25·44	72	57·24	112	89·04	152	120·84	192	152·64
33	26·23½	73	58·03½	113	89·83½	153	121·63½	193	153·43½
34	27·03	74	58·83	114	90·63	154	122·43	194	154·23
35	27·82½	75	59·62½	115	91·42½	155	123·22½	195	155·02½
36	28·62	76	60·42	116	92·22	156	124·02	196	155·82
37	29·41½	77	61·21½	117	93·01½	157	124·81½	197	156·61½
38	30·21	78	62·01	118	93·81	158	125·61	198	157·41
39	31·00½	79	62·80½	119	94·60½	159	126·40½	199	158·20½
40	31·80	**80**	63·60	**120**	95·40	**160**	127·20	**200**	159·00

$\frac{1}{16}=0{\cdot}05^{0}$ $\frac{1}{9}=0{\cdot}08^{8}$ $\frac{1}{8}=0{\cdot}09^{9}$ $\frac{1}{6}=0{\cdot}13^{3}$ $\frac{1}{5}=0{\cdot}15^{9}$ $\frac{1}{4}=0{\cdot}19^{9}$ $\frac{1}{3}=0{\cdot}26^{5}$

$\frac{3}{8}=0{\cdot}29^{8}$ $\frac{1}{2}=0{\cdot}39^{8}$ $\frac{5}{8}=0{\cdot}49^{7}$ $\frac{3}{4}=0{\cdot}59^{6}$ $\frac{7}{8}=0{\cdot}69^{6}$ $\frac{2}{3}=0{\cdot}53^{0}$ $\frac{5}{6}=0{\cdot}66^{3}$

$\frac{3}{16}=0{\cdot}14^{9}$ $\frac{5}{16}=0{\cdot}24^{8}$ $\frac{7}{16}=0{\cdot}34^{8}$ $\frac{9}{16}=0{\cdot}44^{7}$ $\frac{11}{16}=0{\cdot}54^{7}$ $\frac{13}{16}=0{\cdot}64^{6}$ $\frac{15}{16}=0{\cdot}74^{5}$

(1.250=1£) **£0.80**

1	0·80	41	32·80	81	64·80	121	96·80	161	128·80
2	1·60	42	33·60	82	65·60	122	97·60	162	129·60
3	2·40	43	34·40	83	66·40	123	98·40	163	130·40
4	3·20	44	35·20	84	67·20	124	99·20	164	131·20
5	4·00	45	36·00	85	68·00	125	100·00	165	132·00
6	4·80	46	36·80	86	68·80	126	100·80	166	132·80
7	5·60	47	37·60	87	69·60	127	101·60	167	133·60
8	6·40	48	38·40	88	70·40	128	102·40	168	134·40
9	7·20	49	39·20	89	71·20	129	103·20	169	135·20
10	8·00	**50**	40·00	**90**	72·00	**130**	104·00	**170**	136·00
11	8·80	51	40·80	91	72·80	131	104·80	171	136·80
12	9·60	52	41·60	92	73·60	132	105·60	172	137·60
13	10·40	53	42·40	93	74·40	133	106·40	173	138·40
14	11·20	54	43·20	94	75·20	134	107·20	174	139·20
15	12·00	55	44·00	95	76·00	135	108·00	175	140·00
16	12·80	56	44·80	96	76·80	136	108·80	176	140·80
17	13·60	57	45·60	97	77·60	137	109·60	177	141·60
18	14·40	58	46·40	98	78·40	138	110·40	178	142·40
19	15·20	59	47·20	99	79·20	139	111·20	179	143·20
20	16·00	**60**	48·00	**100**	80·00	**140**	112·00	**180**	144·00
21	16·80	61	48·80	101	80·80	141	112·80	181	144·80
22	17·60	62	49·60	102	81·60	142	113·60	182	145·60
23	18·40	63	50·40	103	82·40	143	114·40	183	146·40
24	19·20	64	51·20	104	83·20	144	115·20	184	147·20
25	20·00	65	52·00	105	84·00	145	116·00	185	148·00
26	20·80	66	52·80	106	84·80	146	116·80	186	148·80
27	21·60	67	53·60	107	85·60	147	117·60	187	149·60
28	22·40	68	54·40	108	86·40	148	118·40	188	150·40
29	23·20	69	55·20	109	87·20	149	119·20	189	151·20
30	24·00	**70**	56·00	**110**	88·00	**150**	120·00	**190**	152·00
31	24·80	71	56·80	111	88·80	151	120·80	191	152·80
32	25·60	72	57·60	112	89·60	152	121·60	192	153·60
33	26·40	73	58·40	113	90·40	153	122·40	193	154·40
34	27·20	74	59·20	114	91·20	154	123·20	194	155·20
35	28·00	75	60·00	115	92·00	155	124·00	195	156·00
36	28·80	76	60·80	116	92·80	156	124·80	196	156·80
37	29·60	77	61·60	117	93·60	157	125·60	197	157·60
38	30·40	78	62·40	118	94·40	158	126·40	198	158·40
39	31·20	79	63·20	119	95·20	159	127·20	199	159·20
40	32·00	**80**	64·00	**120**	96·00	**160**	128·00	**200**	160·00

$\frac{1}{16}$=0·05^{0}	$\frac{1}{9}$=0·08^{9}	$\frac{1}{8}$=0·10^{0}	$\frac{1}{6}$=0·13^{3}	$\frac{1}{5}$=0·16^{0}	$\frac{1}{4}$=0·20^{0}	$\frac{1}{3}$=0·26^{7}
$\frac{3}{8}$=0·30^{0}	$\frac{1}{2}$=0·40^{0}	$\frac{5}{8}$=0·50^{0}	$\frac{3}{4}$=0·60^{0}	$\frac{7}{8}$=0·70^{0}	$\frac{2}{3}$=0·53^{3}	$\frac{5}{6}$=0·66^{7}
$\frac{3}{16}$=0·15^{0}	$\frac{5}{16}$=0·25^{0}	$\frac{7}{16}$=0·35^{0}	$\frac{9}{16}$=0·45^{0}	$\frac{11}{16}$=0·55^{0}	$\frac{13}{16}$=0·65^{0}	$\frac{15}{16}$=0·75^{0}

1	0·80½	41	33·00½	81	65·20½	121	97·40½	161	129·60½
2	1·61	42	33·81	82	66·01	122	98·21	162	130·41
3	2·41½	43	34·61½	83	66·81½	123	99·01½	163	131·21½
4	3·22	44	35·42	84	67·62	124	99·82	164	132·02
5	4·02½	45	36·22½	85	68·42½	125	100·62½	165	132·82½
6	4·83	46	37·03	86	69·23	126	101·43	166	133·63
7	5·63½	47	37·83½	87	70·03½	127	102·23½	167	134·43½
8	6·44	48	38·64	88	70·84	128	103·04	168	135·24
9	7·24½	49	39·44½	89	71·64½	129	103·84½	169	136·04½
10	8·05	**50**	40·25	**90**	72·45	**130**	104·65	**170**	136·85
11	8·85½	51	41·05½	91	73·25½	131	105·45½	171	137·65½
12	9·66	52	41·86	92	74·06	132	106·26	172	138·46
13	10·46½	53	42·66½	93	74·86½	133	107·06½	173	139·26½
14	11·27	54	43·47	94	75·67	134	107·87	174	140·07
15	12·07½	55	44·27½	95	76·47½	135	108·67½	175	140·87½
16	12·88	56	45·08	96	77·28	136	109·48	176	141·68
17	13·68½	57	45·88½	97	78·08½	137	110·28½	177	142·48½
18	14·49	58	46·69	98	78·89	138	111·09	178	143·29
19	15·29½	59	47·49½	99	79·69½	139	111·89½	179	144·09½
20	16·10	**60**	48·30	**100**	80·50	**140**	112·70	**180**	144·90
21	16·90½	61	49·10½	101	81·30½	141	113·50½	181	145·70½
22	17·71	62	49·91	102	82·11	142	114·31	182	146·51
23	18·51½	63	50·71½	103	82·91½	143	115·11½	183	147·31½
24	19·32	64	51·52	104	83·72	144	115·92	184	148·12
25	20·12½	65	52·32½	105	84·52½	145	116·72½	185	148·92½
26	20·93	66	53·13	106	85·33	146	117·53	186	149·73
27	21·73½	67	53·93½	107	86·13½	147	118·33½	187	150·53½
28	22·54	68	54·74	108	86·94	148	119·14	188	151·34
29	23·34½	69	55·54½	109	87·74½	149	119·94½	189	152·14½
30	24·15	**70**	56·35	**110**	88·55	**150**	120·75	**190**	152·95
31	24·95½	71	57·15½	111	89·35½	151	121·55½	191	153·75½
32	25·76	72	57·96	112	90·16	152	122·36	192	154·56
33	26·56½	73	58·76½	113	90·96½	153	123·16½	193	155·36½
34	27·37	74	59·57	114	91·77	154	123·97	194	156·17
35	28·17½	75	60·37½	115	92·57½	155	124·77½	195	156·97½
36	28·98	76	61·18	116	93·38	156	125·58	196	157·78
37	29·78½	77	61·98½	117	94·18½	157	126·38½	197	158·58½
38	30·59	78	62·79	118	94·99	158	127·19	198	159·39
39	31·39½	79	63·59½	119	95·79½	159	127·99½	199	160·19½
40	32·20	**80**	64·40	**120**	96·60	**160**	128·80	**200**	161·00

$\frac{1}{16}=0{\cdot}05^{0}$ $\frac{1}{9}=0{\cdot}08^{9}$ $\frac{1}{8}=0{\cdot}10^{1}$ $\frac{1}{6}=0{\cdot}13^{4}$ $\frac{1}{5}=0{\cdot}16^{1}$ $\frac{1}{4}=0{\cdot}20^{1}$ $\frac{1}{3}=0{\cdot}26^{8}$

$\frac{3}{8}=0{\cdot}30^{2}$ $\frac{1}{2}=0{\cdot}40^{3}$ $\frac{5}{8}=0{\cdot}50^{3}$ $\frac{3}{4}=0{\cdot}60^{4}$ $\frac{7}{8}=0{\cdot}70^{4}$ $\frac{2}{3}=0{\cdot}53^{7}$ $\frac{5}{6}=0{\cdot}67^{1}$

$\frac{3}{16}=0{\cdot}15^{1}$ $\frac{5}{16}=0{\cdot}25^{2}$ $\frac{7}{16}=0{\cdot}35^{2}$ $\frac{9}{16}=0{\cdot}45^{3}$ $\frac{11}{16}=0{\cdot}55^{3}$ $\frac{13}{16}=0{\cdot}65^{4}$ $\frac{15}{16}=0{\cdot}75^{5}$

(1.235=1£) **£0.81**

1	0·81	41	33·21	81	65·61	121	98·01	161	130·41
2	1·62	42	34·02	82	66·42	122	98·82	162	131·22
3	2·43	43	34·83	83	67·23	123	99·63	163	132·03
4	3·24	44	35·64	84	68·04	124	100·44	164	132·84
5	4·05	45	36·45	85	68·85	125	101·25	165	133·65
6	4·86	46	37·26	86	69·66	126	102·06	166	134·46
7	5·67	47	38·07	87	70·47	127	102·87	167	135·27
8	6·48	48	38·88	88	71·28	128	103·68	168	136·08
9	7·29	49	39·69	89	72·09	129	104·49	169	136·89
10	8·10	**50**	40·50	**90**	72·90	**130**	105·30	**170**	137·70
11	8·91	51	41·31	91	73·71	131	106·11	171	138·51
12	9·72	52	42·12	92	74·52	132	106·92	172	139·32
13	10·53	53	42·93	93	75·33	133	107·73	173	140·13
14	11·34	54	43·74	94	76·14	134	108·54	174	140·94
15	12·15	55	44·55	95	76·95	135	109·35	175	141·75
16	12·96	56	45·36	96	77·76	136	110·16	176	142·56
17	13·77	57	46·17	97	78·57	137	110·97	177	143·37
18	14·58	58	46·98	98	79·38	138	111·78	178	144·18
19	15·39	59	47·79	99	80·19	139	112·59	179	144·99
20	16·20	**60**	48·60	**100**	81·00	**140**	113·40	**180**	145·80
21	17·01	61	49·41	101	81·81	141	114·21	181	146·61
22	17·82	62	50·22	102	82·62	142	115·02	182	147·42
23	18·63	63	51·03	103	83·43	143	115·83	183	148·23
24	19·44	64	51·84	104	84·24	144	116·64	184	149·04
25	20·25	65	52·65	105	85·05	145	117·45	185	149·85
26	21·06	66	53·46	106	85·86	146	118·26	186	150·66
27	21·87	67	54·27	107	86·67	147	119·07	187	151·47
28	22·68	68	55·08	108	87·48	148	119·88	188	152·28
29	23·49	69	55·89	109	88·29	149	120·69	189	153·09
30	24·30	**70**	56·70	**110**	89·10	**150**	121·50	**190**	153·90
31	25·11	71	57·51	111	89·91	151	122·31	191	154·71
32	25·92	72	58·32	112	90·72	152	123·12	192	155·52
33	26·73	73	59·13	113	91·53	153	123·93	193	156·33
34	27·54	74	59·94	114	92·34	154	124·74	194	157·14
35	28·35	75	60·75	115	93·15	155	125·55	195	157·95
36	29·16	76	61·56	116	93·96	156	126·36	196	158·76
37	29·97	77	62·37	117	94·77	157	127·17	197	159·57
38	30·78	78	63·18	118	95·58	158	127·98	198	160·38
39	31·59	79	63·99	119	96·39	159	128·79	199	161·19
40	32·40	**80**	64·80	**120**	97·20	**160**	129·60	**200**	162·00

$\frac{1}{16}=0{\cdot}05^{1}$ $\frac{1}{9}=0{\cdot}09^{0}$ $\frac{1}{8}=0{\cdot}10^{1}$ $\frac{1}{6}=0{\cdot}13^{5}$ $\frac{1}{5}=0{\cdot}16^{2}$ $\frac{1}{4}=0{\cdot}20^{3}$ $\frac{1}{3}=0{\cdot}27^{0}$

$\frac{3}{8}=0{\cdot}30^{4}$ $\frac{1}{2}=0{\cdot}40^{5}$ $\frac{5}{8}=0{\cdot}50^{6}$ $\frac{3}{4}=0{\cdot}60^{8}$ $\frac{7}{8}=0{\cdot}70^{9}$ $\frac{2}{3}=0{\cdot}54^{0}$ $\frac{5}{6}=0{\cdot}67^{5}$

$\frac{3}{16}=0{\cdot}15^{2}$ $\frac{5}{16}=0{\cdot}25^{3}$ $\frac{7}{16}=0{\cdot}35^{4}$ $\frac{9}{16}=0{\cdot}45^{6}$ $\frac{11}{16}=0{\cdot}55^{7}$ $\frac{13}{16}=0{\cdot}65^{8}$ $\frac{15}{16}=0{\cdot}75^{9}$

(1.227=1£) **£0.81½**

1	0·81½	41	33·41½	81	66·01½	121	98·61½	161	131·21½
2	1·63	42	34·23	82	66·83	122	99·43	162	132·03
3	2·44½	43	35·04½	83	67·64½	123	100·24½	163	132·84½
4	3·26	44	35·86	84	68·46	124	101·06	164	133·66
5	4·07½	45	36·67½	85	69·27½	125	101·87½	165	134·47½
6	4·89	46	37·49	86	70·09	126	102·69	166	135·29
7	5·70½	47	38·30½	87	70·90½	127	103·50½	167	136·10½
8	6·52	48	39·12	88	71·72	128	104·32	168	136·92
9	7·33½	49	39·93½	89	72·53½	129	105·13½	169	137·73½
10	8·15	**50**	40·75	**90**	73·35	**130**	105·95	**170**	138·55
11	8·96½	51	41·56½	91	74·16½	131	106·76½	171	139·36½
12	9·78	52	42·38	92	74·98	132	107·58	172	140·18
13	10·59½	53	43·19½	93	75·79½	133	108·39½	173	140·99½
14	11·41	54	44·01	94	76·61	134	109·21	174	141·81
15	12·22½	55	44·82½	95	77·42½	135	110·02½	175	142·62½
16	13·04	56	45·64	96	78·24	136	110·84	176	143·44
17	13·85½	57	46·45½	97	79·05½	137	111·65½	177	144·25½
18	14·67	58	47·27	98	79·87	138	112·47	178	145·07
19	15·48½	59	48·08½	99	80·68½	139	113·28½	179	145·88½
20	16·30	**60**	48·90	**100**	81·50	**140**	114·10	**180**	146·70
21	17·11½	61	49·71½	101	82·31½	141	114·91½	181	147·51½
22	17·93	62	50·53	102	83·13	142	115·73	182	148·33
23	18·74½	63	51·34½	103	83·94½	143	116·54½	183	149·14½
24	19·56	64	52·16	104	84·76	144	117·36	184	149·96
25	20·37½	65	52·97½	105	85·57½	145	118·17½	185	150·77½
26	21·19	66	53·79	106	86·39	146	118·99	186	151·59
27	22·00½	67	54·60½	107	87·20½	147	119·80½	187	152·40½
28	22·82	68	55·42	108	88·02	148	120·62	188	153·22
29	23·63½	69	56·23½	109	88·83½	149	121·43½	189	154·03½
30	24·45	**70**	57·05	**110**	89·65	**150**	122·25	**190**	154·85
31	25·26½	71	57·86½	111	90·46½	151	123·06½	191	155·66½
32	26·08	72	58·68	112	91·28	152	123·88	192	156·48
33	26·89½	73	59·49½	113	92·09½	153	124·69½	193	157·29½
34	27·71	74	60·31	114	92·91	154	125·51	194	158·11
35	28·52½	75	61·12½	115	93·72½	155	126·32½	195	158·92½
36	29·34	76	61·94	116	94·54	156	127·14	196	159·74
37	30·15½	77	62·75½	117	95·35½	157	127·95½	197	160·55½
38	30·97	78	63·57	118	96·17	158	128·77	198	161·37
39	31·78½	79	64·38½	119	96·98½	159	129·58½	199	162·18½
40	32·60	**80**	65·20	**120**	97·80	**160**	130·40	**200**	163·00

$\frac{1}{16}=0{\cdot}05^{1}$ $\frac{1}{9}=0{\cdot}09^{1}$ $\frac{1}{8}=0{\cdot}10^{2}$ $\frac{1}{6}=0{\cdot}13^{6}$ $\frac{1}{5}=0{\cdot}16^{3}$ $\frac{1}{4}=0{\cdot}20^{4}$ $\frac{1}{3}=0{\cdot}27^{2}$

$\frac{3}{8}=0{\cdot}30^{6}$ $\frac{1}{2}=0{\cdot}40^{8}$ $\frac{5}{8}=0{\cdot}50^{9}$ $\frac{3}{4}=0{\cdot}61^{1}$ $\frac{7}{8}=0{\cdot}71^{3}$ $\frac{2}{3}=0{\cdot}54^{3}$ $\frac{5}{6}=0{\cdot}67^{9}$

$\frac{3}{16}=0{\cdot}15^{3}$ $\frac{5}{16}=0{\cdot}25^{5}$ $\frac{7}{16}=0{\cdot}35^{7}$ $\frac{9}{16}=0{\cdot}45^{8}$ $\frac{11}{16}=0{\cdot}56^{0}$ $\frac{13}{16}=0{\cdot}66^{2}$ $\frac{15}{16}=0{\cdot}76^{4}$

(1.220=1£) **£0.82**

1	0·82	41	33·62	81	66·42	121	99·22	161	132·02
2	1·64	42	34·44	82	67·24	122	100·04	162	132·84
3	2·46	43	35·26	83	68·06	123	100·86	163	133·66
4	3·28	44	36·08	84	68·88	124	101·68	164	134·48
5	4·10	45	36·90	85	69·70	125	102·50	165	135·30
6	4·92	46	37·72	86	70·52	126	103·32	166	136·12
7	5·74	47	38·54	87	71·34	127	104·14	167	136·94
8	6·56	48	39·36	88	72·16	128	104·96	168	137·76
9	7·38	49	40·18	89	72·98	129	105·78	169	138·58
10	8·20	**50**	41·00	**90**	73·80	**130**	106·60	**170**	139·40
11	9·02	51	41·82	91	74·62	131	107·42	171	140·22
12	9·84	52	42·64	92	75·44	132	108·24	172	141·04
13	10·66	53	43·46	93	76·26	133	109·06	173	141·86
14	11·48	54	44·28	94	77·08	134	109·88	174	142·68
15	12·30	55	45·10	95	77·90	135	110·70	175	143·50
16	13·12	56	45·92	96	78·72	136	111·52	176	144·32
17	13·94	57	46·74	97	79·54	137	112·34	177	145·14
18	14·76	58	47·56	98	80·36	138	113·16	178	145·96
19	15·58	59	48·38	99	81·18	139	113·98	179	146·78
20	16·40	**60**	49·20	**100**	82·00	**140**	114·80	**180**	147·60
21	17·22	61	50·02	101	82·82	141	115·62	181	148·42
22	18·04	62	50·84	102	83·64	142	116·44	182	149·24
23	18·86	63	51·66	103	84·46	143	117·26	183	150·06
24	19·68	64	52·48	104	85·28	144	118·08	184	150·88
25	20·50	65	53·30	105	86·10	145	118·90	185	151·70
26	21·32	66	54·12	106	86·92	146	119·72	186	152·52
27	22·14	67	54·94	107	87·74	147	120·54	187	153·34
28	22·96	68	55·76	108	88·56	148	121·36	188	154·16
29	23·78	69	56·58	109	89·38	149	122·18	189	154·98
30	24·60	**70**	57·40	**110**	90·20	**150**	123·00	**190**	155·80
31	25·42	71	58·22	111	91·02	151	123·82	191	156·62
32	26·24	72	59·04	112	91·84	152	124·64	192	157·44
33	27·06	73	59·86	113	92·66	153	125·46	193	158·26
34	27·88	74	60·68	114	93·48	154	126·28	194	159·08
35	28·70	75	61·50	115	94·30	155	127·10	195	159·90
36	29·52	76	62·32	116	95·12	156	127·92	196	160·72
37	30·34	77	63·14	117	95·94	157	128·74	197	161·54
38	31·16	78	63·96	118	96·76	158	129·56	198	162·36
39	31·98	79	64·78	119	97·58	159	130·38	199	163·18
40	32·80	**80**	65·60	**120**	98·40	**160**	131·20	**200**	164·00

$\frac{1}{16}=0{\cdot}05^{1}$ $\frac{1}{9}=0{\cdot}09^{1}$ $\frac{1}{8}=0{\cdot}10^{3}$ $\frac{1}{6}=0{\cdot}13^{7}$ $\frac{1}{5}=0{\cdot}16^{4}$ $\frac{1}{4}=0{\cdot}20^{5}$ $\frac{1}{3}=0{\cdot}27^{3}$

$\frac{3}{8}=0{\cdot}30^{8}$ $\frac{1}{2}=0{\cdot}41^{0}$ $\frac{5}{8}=0{\cdot}51^{3}$ $\frac{3}{4}=0{\cdot}61^{5}$ $\frac{7}{8}=0{\cdot}71^{8}$ $\frac{2}{3}=0{\cdot}54^{7}$ $\frac{5}{6}=0{\cdot}68^{3}$

$\frac{3}{16}=0{\cdot}15^{4}$ $\frac{5}{16}=0{\cdot}25^{6}$ $\frac{7}{16}=0{\cdot}35^{9}$ $\frac{9}{16}=0{\cdot}46^{1}$ $\frac{11}{16}=0{\cdot}56^{4}$ $\frac{13}{16}=0{\cdot}66^{6}$ $\frac{15}{16}=0{\cdot}76^{9}$

(1.212=1£) **£0.82½**

1	0·82½	41	33·82½	81	66·82½	121	99·82½	161	132·82½
2	1·65	42	34·65	82	67·65	122	100·65	162	133·65
3	2·47½	43	35·47½	83	68·47½	123	101·47½	163	134·47½
4	3·30	44	36·30	84	69·30	124	102·30	164	135·30
5	4·12½	45	37·12½	85	70·12½	125	103·12½	165	136·12½
6	4·95	46	37·95	86	70·95	126	103·95	166	136·95
7	5·77½	47	38·77½	87	71·77½	127	104·77½	167	137·77½
8	6·60	48	39·60	88	72·60	128	105·60	168	138·60
9	7·42½	49	40·42½	89	73·42½	129	106·42½	169	139·42½
10	8·25	**50**	41·25	**90**	74·25	**130**	107·25	**170**	140·25
11	9·07½	51	42·07½	91	75·07½	131	108·07½	171	141·07½
12	9·90	52	42·90	92	75·90	132	108·90	172	141·90
13	10·72½	53	43·72½	93	76·72½	133	109·72½	173	142·72½
14	11·55	54	44·55	94	77·55	134	110·55	174	143·55
15	12·37½	55	45·37½	95	78·37½	135	111·37½	175	144·37½
16	13·20	56	46·20	96	79·20	136	112·20	176	145·20
17	14·02½	57	47·02½	97	80·02½	137	113·02½	177	146·02½
18	14·85	58	47·85	98	80·85	138	113·85	178	146·85
19	15·67½	59	48·67½	99	81·67½	139	114·67½	179	147·67½
20	16·50	**60**	49·50	**100**	82·50	**140**	115·50	**180**	148·50
21	17·32½	61	50·32½	101	83·32½	141	116·32½	181	149·32½
22	18·15	62	51·15	102	84·15	142	117·15	182	150·15
23	18·97½	63	51·97½	103	84·97½	143	117·97½	183	150·97½
24	19·80	64	52·80	104	85·80	144	118·80	184	151·80
25	20·62½	65	53·62½	105	86·62½	145	119·62½	185	152·62½
26	21·45	66	54·45	106	87·45	146	120·45	186	153·45
27	22·27½	67	55·27½	107	88·27½	147	121·27½	187	154·27½
28	23·10	68	56·10	108	89·10	148	122·10	188	155·10
29	23·92½	69	56·92½	109	89·92½	149	122·92½	189	155·92½
30	24·75	**70**	57·75	**110**	90·75	**150**	123·75	**190**	156·75
31	25·57½	71	58·57½	111	91·57½	151	124·57½	191	157·57½
32	26·40	72	59·40	112	92·40	152	125·40	192	158·40
33	27·22½	73	60·22½	113	93·22½	153	126·22½	193	159·22½
34	28·05	74	61·05	114	94·05	154	127·05	194	160·05
35	28·87½	75	61·87½	115	94·87½	155	127·87½	195	160·87½
36	29·70	76	62·70	116	95·70	156	128·70	196	161·70
37	30·52½	77	63·52½	117	96·52½	157	129·52½	197	162·52½
38	31·35	78	64·35	118	97·35	158	130·35	198	163·35
39	32·17½	79	65·17½	119	98·17½	159	131·17½	199	164·17½
40	33·00	**80**	66·00	**120**	99·00	**160**	132·00	**200**	165·00

$\frac{1}{16}=0{\cdot}05^{2}$ $\frac{1}{9}=0{\cdot}09^{2}$ $\frac{1}{8}=0{\cdot}10^{3}$ $\frac{1}{6}=0{\cdot}13^{8}$ $\frac{1}{5}=0{\cdot}16^{5}$ $\frac{1}{4}=0{\cdot}20^{6}$ $\frac{1}{3}=0{\cdot}27^{5}$

$\frac{3}{8}=0{\cdot}30^{9}$ $\frac{1}{2}=0{\cdot}41^{3}$ $\frac{5}{8}=0{\cdot}51^{6}$ $\frac{3}{4}=0{\cdot}61^{9}$ $\frac{7}{8}=0{\cdot}72^{2}$ $\frac{2}{3}=0{\cdot}55^{0}$ $\frac{5}{6}=0{\cdot}68^{8}$

$\frac{3}{16}=0{\cdot}15^{5}$ $\frac{5}{16}=0{\cdot}25^{8}$ $\frac{7}{16}=0{\cdot}36^{1}$ $\frac{9}{16}=0{\cdot}46^{4}$ $\frac{11}{16}=0{\cdot}56^{7}$ $\frac{13}{16}=0{\cdot}67^{0}$ $\frac{15}{16}=0{\cdot}77^{3}$

(1.205=1£) **£0.83**

1	0·83	41	34·03	81	67·23	121	100·43	161	133·63
2	1·66	42	34·86	82	68·06	122	101·26	162	134·46
3	2·49	43	35·69	83	68·89	123	102·09	163	135·29
4	3·32	44	36·52	84	69·72	124	102·92	164	136·12
5	4·15	45	37·35	85	70·55	125	103·75	165	136·95
6	4·98	46	38·18	86	71·38	126	104·58	166	137·78
7	5·81	47	39·01	87	72·21	127	105·41	167	138·61
8	6·64	48	39·84	88	73·04	128	106·24	168	139·44
9	7·47	49	40·67	89	73·87	129	107·07	169	140·27
10	8·30	**50**	41·50	**90**	74·70	**130**	107·90	**170**	141·10
11	9·13	51	42·33	91	75·53	131	108·73	171	141·93
12	9·96	52	43·16	92	76·36	132	109·56	172	142·76
13	10·79	53	43·99	93	77·19	133	110·39	173	143·59
14	11·62	54	44·82	94	78·02	134	111·22	174	144·42
15	12·45	55	45·65	95	78·85	135	112·05	175	145·25
16	13·28	56	46·48	96	79·68	136	112·88	176	146·08
17	14·11	57	47·31	97	80·51	137	113·71	177	146·91
18	14·94	58	48·14	98	81·34	138	114·54	178	147·74
19	15·77	59	48·97	99	82·17	139	115·37	179	148·57
20	16·60	**60**	49·80	**100**	83·00	**140**	116·20	**180**	149·40
21	17·43	61	50·63	101	83·83	141	117·03	181	150·23
22	18·26	62	51·46	102	84·66	142	117·86	182	151·06
23	19·09	63	52·29	103	85·49	143	118·69	183	151·89
24	19·92	64	53·12	104	86·32	144	119·52	184	152·72
25	20·75	65	53·95	105	87·15	145	120·35	185	153·55
26	21·58	66	54·78	106	87·98	146	121·18	186	154·38
27	22·41	67	55·61	107	88·81	147	122·01	187	155·21
28	23·24	68	56·44	108	89·64	148	122·84	188	156·04
29	24·07	69	57·27	109	90·47	149	123·67	189	156·87
30	24·90	**70**	58·10	**110**	91·30	**150**	124·50	**190**	157·70
31	25·73	71	58·93	111	92·13	151	125·33	191	158·53
32	26·56	72	59·76	112	92·96	152	126·16	192	159·36
33	27·39	73	60·59	113	93·79	153	126·99	193	160·19
34	28·22	74	61·42	114	94·62	154	127·82	194	161·02
35	29·05	75	62·25	115	95·45	155	128·65	195	161·85
36	29·88	76	63·08	116	96·28	156	129·48	196	162·68
37	30·71	77	63·91	117	97·11	157	130·31	197	163·51
38	31·54	78	64·74	118	97·94	158	131·14	198	164·34
39	32·37	79	65·57	119	98·77	159	131·97	199	165·17
40	33·20	**80**	66·40	**120**	99·60	**160**	132·80	**200**	166·00

$\frac{1}{16}=0{\cdot}05^{2}$ $\frac{1}{9}=0{\cdot}09^{2}$ $\frac{1}{8}=0{\cdot}10^{4}$ $\frac{1}{6}=0{\cdot}13^{8}$ $\frac{1}{5}=0{\cdot}16^{6}$ $\frac{1}{4}=0{\cdot}20^{8}$ $\frac{1}{3}=0{\cdot}27^{7}$

$\frac{3}{8}=0{\cdot}31^{1}$ $\frac{1}{2}=0{\cdot}41^{5}$ $\frac{5}{8}=0{\cdot}51^{9}$ $\frac{3}{4}=0{\cdot}62^{3}$ $\frac{7}{8}=0{\cdot}72^{6}$ $\frac{2}{3}=0{\cdot}55^{3}$ $\frac{5}{6}=0{\cdot}69^{2}$

$\frac{3}{16}=0{\cdot}15^{6}$ $\frac{5}{16}=0{\cdot}25^{9}$ $\frac{7}{16}=0{\cdot}36^{3}$ $\frac{9}{16}=0{\cdot}46^{7}$ $\frac{11}{16}=0{\cdot}57^{1}$ $\frac{13}{16}=0{\cdot}67^{4}$ $\frac{15}{16}=0{\cdot}77^{8}$

1	0·83½	41	34·23½	81	67·63½	121	101·03½	161	134·43½
2	1·67	42	35·07	82	68·47	122	101·87	162	135·27
3	2·50½	43	35·90½	83	69·30½	123	102·70½	163	136·10½
4	3·34	44	36·74	84	70·14	124	103·54	164	136·94
5	4·17½	45	37·57½	85	70·97½	125	104·37½	165	137·77½
6	5·01	46	38·41	86	71·81	126	105·21	166	138·61
7	5·84½	47	39·24½	87	72·64½	127	106·04½	167	139·44½
8	6·68	48	40·08	88	73·48	128	106·88	168	140·28
9	7·51½	49	40·91½	89	74·31½	129	107·71½	169	141·11½
10	8·35	**50**	41·75	**90**	75·15	**130**	108·55	**170**	141·95
11	9·18½	51	42·58½	91	75·98½	131	109·38½	171	142·78½
12	10·02	52	43·42	92	76·82	132	110·22	172	143·62
13	10·85½	53	44·25½	93	77·65½	133	111·05½	173	144·45½
14	11·69	54	45·09	94	78·49	134	111·89	174	145·29
15	12·52½	55	45·92½	95	79·32½	135	112·72½	175	146·12½
16	13·36	56	46·76	96	80·16	136	113·56	176	146·96
17	14·19½	57	47·59½	97	80·99½	137	114·39½	177	147·79½
18	15·03	58	48·43	98	81·83	138	115·23	178	148·63
19	15·86½	59	49·26½	99	82·66½	139	116·06½	179	149·46½
20	16·70	**60**	50·10	**100**	83·50	**140**	116·90	**180**	150·30
21	17·53½	61	50·93½	101	84·33½	141	117·73½	181	151·13½
22	18·37	62	51·77	102	85·17	142	118·57	182	151·97
23	19·20½	63	52·60½	103	86·00½	143	119·40½	183	152·80½
24	20·04	64	53·44	104	86·84	144	120·24	184	153·64
25	20·87½	65	54·27½	105	87·67½	145	121·07½	185	154·47½
26	21·71	66	55·11	106	88·51	146	121·91	186	155·31
27	22·54½	67	55·94½	107	89·34½	147	122·74½	187	156·14½
28	23·38	68	56·78	108	90·18	148	123·58	188	156·98
29	24·21½	69	57·61½	109	91·01½	149	124·41½	189	157·81½
30	25·05	**70**	58·45	**110**	91·85	**150**	125·25	**190**	158·65
31	25·88½	71	59·28½	111	92·68½	151	126·08½	191	159·48½
32	26·72	72	60·12	112	93·52	152	126·92	192	160·32
33	27·55½	73	60·95½	113	94·35½	153	127·75½	193	161·15½
34	28·39	74	61·79	114	95·19	154	128·59	194	161·99
35	29·22½	75	62·62½	115	96·02½	155	129·42½	195	162·82½
36	30·06	76	63·46	116	96·86	156	130·26	196	163·66
37	30·89½	77	64·29½	117	97·69½	157	131·09½	197	164·49½
38	31·73	78	65·13	118	98·53	158	131·93	198	165·33
39	32·56½	79	65·96½	119	99·36½	159	132·76½	199	166·16½
40	33·40	**80**	66·80	**120**	100·20	**160**	133·60	**200**	167·00

$\frac{1}{16}=0\cdot05^{2}$ $\frac{1}{9}=0\cdot09^{3}$ $\frac{1}{8}=0\cdot10^{4}$ $\frac{1}{6}=0\cdot13^{9}$ $\frac{1}{5}=0\cdot16^{7}$ $\frac{1}{4}=0\cdot20^{9}$ $\frac{1}{3}=0\cdot27^{8}$

$\frac{3}{8}=0\cdot31^{3}$ $\frac{1}{2}=0\cdot41^{8}$ $\frac{5}{8}=0\cdot52^{2}$ $\frac{3}{4}=0\cdot62^{6}$ $\frac{7}{8}=0\cdot73^{1}$ $\frac{2}{3}=0\cdot55^{7}$ $\frac{5}{6}=0\cdot69^{6}$

$\frac{3}{16}=0\cdot15^{7}$ $\frac{5}{16}=0\cdot26^{1}$ $\frac{7}{16}=0\cdot36^{5}$ $\frac{9}{16}=0\cdot47^{0}$ $\frac{11}{16}=0\cdot57^{4}$ $\frac{13}{16}=0\cdot67^{8}$ $\frac{15}{16}=0\cdot78^{3}$

1 US gallon weighs 8·327 lbs

(1.190=1£) **£0.84**

1	0·84	41	34·44	81	68·04	121	101·64	161	135·24
2	1·68	42	35·28	82	68·88	122	102·48	162	136·08
3	2·52	43	36·12	83	69·72	123	103·32	163	136·92
4	3·36	44	36·96	84	70·56	124	104·16	164	137·76
5	4·20	45	37·80	85	71·40	125	105·00	165	138·60
6	5·04	46	38·64	86	72·24	126	105·84	166	139·44
7	5·88	47	39·48	87	73·08	127	106·68	167	140·28
8	6·72	48	40·32	88	73·92	128	107·52	168	141·12
9	7·56	49	41·16	89	74·76	129	108·36	169	141·96
10	8·40	**50**	42·00	**90**	75·60	**130**	109·20	**170**	142·80
11	9·24	51	42·84	91	76·44	131	110·04	171	143·64
12	10·08	52	43·68	92	77·28	132	110·88	172	144·48
13	10·92	53	44·52	93	78·12	133	111·72	173	145·32
14	11·76	54	45·36	94	78·96	134	112·56	174	146·16
15	12·60	55	46·20	95	79·80	135	113·40	175	147·00
16	13·44	56	47·04	96	80·64	136	114·24	176	147·84
17	14·28	57	47·88	97	81·48	137	115·08	177	148·68
18	15·12	58	48·72	98	82·32	138	115·92	178	149·52
19	15·96	59	49·56	99	83·16	139	116·76	179	150·36
20	16·80	**60**	50·40	**100**	84·00	**140**	117·60	**180**	151·20
21	17·64	61	51·24	101	84·84	141	118·44	181	152·04
22	18·48	62	52·08	102	85·68	142	119·28	182	152·88
23	19·32	63	52·92	103	86·52	143	120·12	183	153·72
24	20·16	64	53·76	104	87·36	144	120·96	184	154·56
25	21·00	65	54·60	105	88·20	145	121·80	185	155·40
26	21·84	66	55·44	106	89·04	146	122·64	186	156·24
27	22·68	67	56·28	107	89·88	147	123·48	187	157·08
28	23·52	68	57·12	108	90·72	148	124·32	188	157·92
29	24·36	69	57·96	109	91·56	149	125·16	189	158·76
30	25·20	**70**	58·80	**110**	92·40	**150**	126·00	**190**	159·60
31	26·04	71	59·64	111	93·24	151	126·84	191	160·44
32	26·88	72	60·48	112	94·08	152	127·68	192	161·28
33	27·72	73	61·32	113	94·92	153	128·52	193	162·12
34	28·56	74	62·16	114	95·76	154	129·36	194	162·96
35	29·40	75	63·00	115	96·60	155	130·20	195	163·80
36	30·24	76	63·84	116	97·44	156	131·04	196	164·64
37	31·08	77	64·68	117	98·28	157	131·88	197	165·48
38	31·92	78	65·52	118	99·12	158	132·72	198	166·32
39	32·76	79	66·36	119	99·96	159	133·56	199	167·16
40	33·60	**80**	67·20	**120**	100·80	**160**	134·40	**200**	168·00

$\frac{1}{16}$=0·05^{3} $\frac{1}{9}$=0·09^{3} $\frac{1}{8}$=0·10^{5} $\frac{1}{6}$=0·14^{0} $\frac{1}{5}$=0·16^{8} $\frac{1}{4}$=0·21^{0} $\frac{1}{3}$=0·28^{0}

$\frac{3}{8}$=0·31^{5} $\frac{1}{2}$=0·42^{0} $\frac{5}{8}$=0·52^{5} $\frac{3}{4}$=0·63^{0} $\frac{7}{8}$=0·73^{5} $\frac{2}{3}$=0·56^{0} $\frac{5}{6}$=0·70^{0}

$\frac{3}{16}$=0·15^{8} $\frac{5}{16}$=0·26^{3} $\frac{7}{16}$=0·36^{8} $\frac{9}{16}$=0·47^{3} $\frac{11}{16}$=0·57^{8} $\frac{13}{16}$=0·68^{3} $\frac{15}{16}$=0·78^{8}

(1.183=1£) **£0.84½**

1	0·84½	41	34·64½	81	68·44½	121	102·24½	161	136·04½
2	1·69	42	35·49	82	69·29	122	103·09	162	136·89
3	2·53½	43	36·33½	83	70·13½	123	103·93½	163	137·73½
4	3·38	44	37·18	84	70·98	124	104·78	164	138·58
5	4·22½	45	38·02½	85	71·82½	125	105·62½	165	139·42½
6	5·07	46	38·87	86	72·67	126	106·47	166	140·27
7	5·91½	47	39·71½	87	73·51½	127	107·31½	167	141·11½
8	6·76	48	40·56	88	74·36	128	108·16	168	141·96
9	7·60½	49	41·40½	89	75·20½	129	109·00½	169	142·80½
10	8·45	**50**	42·25	**90**	76·05	**130**	109·85	**170**	143·65
11	9·29½	51	43·09½	91	76·89½	131	110·69½	171	144·49½
12	10·14	52	43·94	92	77·74	132	111·54	172	145·34
13	10·98½	53	44·78½	93	78·58½	133	112·38½	173	146·18½
14	11·83	54	45·63	94	79·43	134	113·23	174	147·03
15	12·67½	55	46·47½	95	80·27½	135	114·07½	175	147·87½
16	13·52	56	47·32	96	81·12	136	114·92	176	148·72
17	14·36½	57	48·16½	97	81·96½	137	115·76½	177	149·56½
18	15·21	58	49·01	98	82·81	138	116·61	178	150·41
19	16·05½	59	49·85½	99	83·65½	139	117·45½	179	151·25½
20	16·90	**60**	50·70	**100**	84·50	**140**	118·30	**180**	152·10
21	17·74½	61	51·54½	101	85·34½	141	119·14½	181	152·94½
22	18·59	62	52·39	102	86·19	142	119·99	182	153·79
23	19·43½	63	53·23½	103	87·03½	143	120·83½	183	154·63½
24	20·28	64	54·08	104	87·88	144	121·68	184	155·48
25	21·12½	65	54·92½	105	88·72½	145	122·52½	185	156·32½
26	21·97	66	55·77	106	89·57	146	123·37	186	157·17
27	22·81½	67	56·61½	107	90·41½	147	124·21½	187	158·01½
28	23·66	68	57·46	108	91·26	148	125·06	188	158·86
29	24·50½	69	58·30½	109	92·10½	149	125·90½	189	159·70½
30	25·35	**70**	59·15	**110**	92·95	**150**	126·75	**190**	160·55
31	26·19½	71	59·99½	111	93·79½	151	127·59½	191	161·39½
32	27·04	72	60·84	112	94·64	152	128·44	192	162·24
33	27·88½	73	61·68½	113	95·48½	153	129·28½	193	163·08½
34	28·73	74	62·53	114	96·33	154	130·13	194	163·93
35	29·57½	75	63·37½	115	97·17½	155	130·97½	195	164·77½
36	30·42	76	64·22	116	98·02	156	131·82	196	165·62
37	31·26½	77	65·06½	117	98·86½	157	132·66½	197	166·46½
38	32·11	78	65·91	118	99·71	158	133·51	198	167·31
39	32·95½	79	66·75½	119	100·55½	159	134·35½	199	168·15½
40	33·80	**80**	67·60	**120**	101·40	**160**	135·20	**200**	169·00

$\frac{1}{16}$=0·05^{3} $\frac{1}{9}$=0·09^{4} $\frac{1}{8}$=0·10^{6} $\frac{1}{6}$=0·14^{1} $\frac{1}{5}$=0·16^{9} $\frac{1}{4}$=0·21^{1} $\frac{1}{3}$=0·28^{2}

$\frac{3}{8}$=0·31^{7} $\frac{1}{2}$=0·42^{3} $\frac{5}{8}$=0·52^{8} $\frac{3}{4}$=0·63^{4} $\frac{7}{8}$=0·73^{9} $\frac{2}{3}$=0·56^{3} $\frac{5}{6}$=0·70^{4}

$\frac{3}{16}$=0·15^{8} $\frac{5}{16}$=0·26^{4} $\frac{7}{16}$=0·37^{0} $\frac{9}{16}$=0·47^{5} $\frac{11}{16}$=0·58^{1} $\frac{13}{16}$=0·68^{7} $\frac{15}{16}$=0·79^{2}

(1.175=1£) **£0.85**

1	0·85	41	34·85	81	68·85	121	102·85	161	136·85
2	1·70	42	35·70	82	69·70	122	103·70	162	137·70
3	2·55	43	36·55	83	70·55	123	104·55	163	138·55
4	3·40	44	37·40	84	71·40	124	105·40	164	139·40
5	4·25	45	38·25	85	72·25	125	106·25	165	140·25
6	5·10	46	39·10	86	73·10	126	107·10	166	141·10
7	5·95	47	39·95	87	73·95	127	107·95	167	141·95
8	6·80	48	40·80	88	74·80	128	108·80	168	142·80
9	7·65	49	41·65	89	75·65	129	109·65	169	143·65
10	8·50	**50**	42·50	**90**	76·50	**130**	110·50	**170**	144·50
11	9·35	51	43·35	91	77·35	131	111·35	171	145·35
12	10·20	52	44·20	92	78·20	132	112·20	172	146·20
13	11·05	53	45·05	93	79·05	133	113·05	173	147·05
14	11·90	54	45·90	94	79·90	134	113·90	174	147·90
15	12·75	55	46·75	95	80·75	135	114·75	175	148·75
16	13·60	56	47·60	96	81·60	136	115·60	176	149·60
17	14·45	57	48·45	97	82·45	137	116·45	177	150·45
18	15·30	58	49·30	98	83·30	138	117·30	178	151·30
19	16·15	59	50·15	99	84·15	139	118·15	179	152·15
20	17·00	**60**	51·00	**100**	85·00	**140**	119·00	**180**	153·00
21	17·85	61	51·85	101	85·85	141	119·85	181	153·85
22	18·70	62	52·70	102	86·70	142	120·70	182	154·70
23	19·55	63	53·55	103	87·55	143	121·55	183	155·55
24	20·40	64	54·40	104	88·40	144	122·40	184	156·40
25	21·25	65	55·25	105	89·25	145	123·25	185	157·25
26	22·10	66	56·10	106	90·10	146	124·10	186	158·10
27	22·95	67	56·95	107	90·95	147	124·95	187	158·95
28	23·80	68	57·80	108	91·80	148	125·80	188	159·80
29	24·65	69	58·65	109	92·65	149	126·65	189	160·65
30	25·50	**70**	59·50	**110**	93·50	**150**	127·50	**190**	161·50
31	26·35	71	60·35	111	94·35	151	128·35	191	162·35
32	27·20	72	61·20	112	95·20	152	129·20	192	163·20
33	28·05	73	62·05	113	96·05	153	130·05	193	164·05
34	28·90	74	62·90	114	96·90	154	130·90	194	164·90
35	29·75	75	63·75	115	97·75	155	131·75	195	165·75
36	30·60	76	64·60	116	98·60	156	132·60	196	166·60
37	31·45	77	65·45	117	99·45	157	133·45	197	167·45
38	32·30	78	66·30	118	100·30	158	134·30	198	168·30
39	33·15	79	67·15	119	101·15	159	135·15	199	169·15
40	34·00	**80**	68·00	**120**	102·00	**160**	136·00	**200**	170·00

$\frac{1}{16}=0{\cdot}05^{3}$ $\frac{1}{9}=0{\cdot}09^{4}$ $\frac{1}{8}=0{\cdot}10^{6}$ $\frac{1}{6}=0{\cdot}14^{2}$ $\frac{1}{5}=0{\cdot}17^{0}$ $\frac{1}{4}=0{\cdot}21^{3}$ $\frac{1}{3}=0{\cdot}28^{3}$

$\frac{3}{8}=0{\cdot}31^{9}$ $\frac{1}{2}=0{\cdot}42^{5}$ $\frac{5}{8}=0{\cdot}53^{1}$ $\frac{3}{4}=0{\cdot}63^{8}$ $\frac{7}{8}=0{\cdot}74^{4}$ $\frac{2}{3}=0{\cdot}56^{7}$ $\frac{5}{6}=0{\cdot}70^{8}$

$\frac{3}{16}=0{\cdot}15^{9}$ $\frac{5}{16}=0{\cdot}26^{6}$ $\frac{7}{16}=0{\cdot}37^{2}$ $\frac{9}{16}=0{\cdot}47^{8}$ $\frac{11}{16}=0{\cdot}58^{4}$ $\frac{13}{16}=0{\cdot}69^{1}$ $\frac{15}{16}=0{\cdot}79^{7}$

(1.170=1£) **£0.85½**

1	0·85½	41	35·05½	81	69·25½	121	103·45½	161	137·65½
2	1·71	42	35·91	82	70·11	122	104·31	162	138·51
3	2·56½	43	36·76½	83	70·96½	123	105·16½	163	139·36½
4	3·42	44	37·62	84	71·82	124	106·02	164	140·22
5	4·27½	45	38·47½	85	72·67½	125	106·87½	165	141·07½
6	5·13	46	39·33	86	73·53	126	107·73	166	141·93
7	5·98½	47	40·18½	87	74·38½	127	108·58½	167	142·78½
8	6·84	48	41·04	88	75·24	128	109·44	168	143·64
9	7·69½	49	41·89½	89	76·09½	129	110·29½	169	144·49½
10	8·55	**50**	42·75	**90**	76·95	**130**	111·15	**170**	145·35
11	9·40½	51	43·60½	91	77·80½	131	112·00½	171	146·20½
12	10·26	52	44·46	92	78·66	132	112·86	172	147·06
13	11·11½	53	45·31½	93	79·51½	133	113·71½	173	147·91½
14	11·97	54	46·17	94	80·37	134	114·57	174	148·77
15	12·82½	55	47·02½	95	81·22½	135	115·42½	175	149·62½
16	13·68	56	47·88	96	82·08	136	116·28	176	150·48
17	14·53½	57	48·73½	97	82·93½	137	117·13½	177	151·33½
18	15·39	58	49·59	98	83·79	138	117·99	178	152·19
19	16·24½	59	50·44½	99	84·64½	139	118·84½	179	153·04½
20	17·10	**60**	51·30	**100**	85·50	**140**	119·70	**180**	153·90
21	17·95½	61	52·15½	101	86·35½	141	120·55½	181	154·75½
22	18·81	62	53·01	102	87·21	142	121·41	182	155·61
23	19·66½	63	53·86½	103	88·06½	143	122·26½	183	156·46½
24	20·52	64	54·72	104	88·92	144	123·12	184	157·32
25	21·37½	65	55·57½	105	89·77½	145	123·97½	185	158·17½
26	22·23	66	56·43	106	90·63	146	124·83	186	159·03
27	23·08½	67	57·28½	107	91·48½	147	125·68½	187	159·88½
28	23·94	68	58·14	108	92·34	148	126·54	188	160·74
29	24·79½	69	58·99½	109	93·19½	149	127·39½	189	161·59½
30	25·65	**70**	59·85	**110**	94·05	**150**	128·25	**190**	162·45
31	26·50½	71	60·70½	111	94·90½	151	129·10½	191	163·30½
32	27·36	72	61·56	112	95·76	152	129·96	192	164·16
33	28·21½	73	62·41½	113	96·61½	153	130·81½	193	165·01½
34	29·07	74	63·27	114	97·47	154	131·67	194	165·87
35	29·92½	75	64·12½	115	98·32½	155	132·52½	195	166·72½
36	30·78	76	64·98	116	99·18	156	133·38	196	167·58
37	31·63½	77	65·83½	117	100·03½	157	134·23½	197	168·43½
38	32·49	78	66·69	118	100·89	158	135·09	198	169·29
39	33·34½	79	67·54½	119	101·74½	159	135·94½	199	170·14½
40	34·20	**80**	68·40	**120**	102·60	**160**	136·80	**200**	171·00

$\frac{1}{16}=0{\cdot}05^{3}$	$\frac{1}{9}=0{\cdot}09^{5}$	$\frac{1}{8}=0{\cdot}10^{7}$	$\frac{1}{6}=0{\cdot}14^{3}$	$\frac{1}{5}=0{\cdot}17^{1}$	$\frac{1}{4}=0{\cdot}21^{4}$	$\frac{1}{3}=0{\cdot}28^{5}$
$\frac{3}{8}=0{\cdot}32^{1}$	$\frac{1}{2}=0{\cdot}42^{8}$	$\frac{5}{8}=0{\cdot}53^{4}$	$\frac{3}{4}=0{\cdot}64^{1}$	$\frac{7}{8}=0{\cdot}74^{8}$	$\frac{2}{3}=0{\cdot}57^{0}$	$\frac{5}{6}=0{\cdot}71^{3}$
$\frac{3}{16}=0{\cdot}16^{0}$	$\frac{5}{16}=0{\cdot}26^{7}$	$\frac{7}{16}=0{\cdot}37^{4}$	$\frac{9}{16}=0{\cdot}48^{1}$	$\frac{11}{16}=0{\cdot}58^{8}$	$\frac{13}{16}=0{\cdot}69^{5}$	$\frac{15}{16}=0{\cdot}80^{2}$

1	0·86	41	35·26	81	69·66	121	104·06	161	138·46
2	1·72	42	36·12	82	70·52	122	104·92	162	139·32
3	2·58	43	36·98	83	71·38	123	105·78	163	140·18
4	3·44	44	37·84	84	72·24	124	106·64	164	141·04
5	4·30	45	38·70	85	73·10	125	107·50	165	141·90
6	5·16	46	39·56	86	73·96	126	108·36	166	142·76
7	6·02	47	40·42	87	74·82	127	109·22	167	143·62
8	6·88	48	41·28	88	75·68	128	110·08	168	144·48
9	7·74	49	42·14	89	76·54	129	110·94	169	145·34
10	8·60	**50**	43·00	**90**	77·40	**130**	111·80	**170**	146·20
11	9·46	51	43·86	91	78·26	131	112·66	171	147·06
12	10·32	52	44·72	92	79·12	132	113·52	172	147·92
13	11·18	53	45·58	93	79·98	133	114·38	173	148·78
14	12·04	54	46·44	94	80·84	134	115·24	174	149·64
15	12·90	55	47·30	95	81·70	135	116·10	175	150·50
16	13·76	56	48·16	96	82·56	136	116·96	176	151·36
17	14·62	57	49·02	97	83·42	137	117·82	177	152·22
18	15·48	58	49·88	98	84·28	138	118·68	178	153·08
19	16·34	59	50·74	99	85·14	139	119·54	179	153·94
20	17·20	**60**	51·60	**100**	86·00	**140**	120·40	**180**	154·80
21	18·06	61	52·46	101	86·86	141	121·26	181	155·66
22	18·92	62	53·32	102	87·72	142	122·12	182	156·52
23	19·78	63	54·18	103	88·58	143	122·98	183	157·38
24	20·64	64	55·04	104	89·44	144	123·84	184	158·24
25	21·50	65	55·90	105	90·30	145	124·70	185	159·10
26	22·36	66	56·76	106	91·16	146	125·56	186	159·96
27	23·22	67	57·62	107	92·02	147	126·42	187	160·82
28	24·08	68	58·48	108	92·88	148	127·28	188	161·68
29	24·94	69	59·34	109	93·74	149	128·14	189	162·54
30	25·80	**70**	60·20	**110**	94·60	**150**	129·00	**190**	163·40
31	26·66	71	61·06	111	95·46	151	129·86	191	164·26
32	27·52	72	61·92	112	96·32	152	130·72	192	165·12
33	28·38	73	62·78	113	97·18	153	131·58	193	165·98
34	29·24	74	63·64	114	98·04	154	132·44	194	166·84
35	30·10	75	64·50	115	98·90	155	133·30	195	167·70
36	30·96	76	65·36	116	99·76	156	134·16	196	168·56
37	31·82	77	66·22	117	100·62	157	135·02	197	169·42
38	32·68	78	67·08	118	101·48	158	135·88	198	170·28
39	33·54	79	67·94	119	102·34	159	136·74	199	171·14
40	34·40	**80**	68·80	**120**	103·20	**160**	137·60	**200**	172·00

$\frac{1}{16}=0{\cdot}05^{4}$ $\frac{1}{9}=0{\cdot}09^{6}$ $\frac{1}{8}=0{\cdot}10^{8}$ $\frac{1}{6}=0{\cdot}14^{3}$ $\frac{1}{5}=0{\cdot}17^{2}$ $\frac{1}{4}=0{\cdot}21^{5}$ $\frac{1}{3}=0{\cdot}28^{7}$

$\frac{3}{8}=0{\cdot}32^{3}$ $\frac{1}{2}=0{\cdot}43^{0}$ $\frac{5}{8}=0{\cdot}53^{8}$ $\frac{3}{4}=0{\cdot}64^{5}$ $\frac{7}{8}=0{\cdot}75^{3}$ $\frac{2}{3}=0{\cdot}57^{3}$ $\frac{5}{6}=0{\cdot}71^{7}$

$\frac{3}{16}=0{\cdot}16^{1}$ $\frac{5}{16}=0{\cdot}26^{9}$ $\frac{7}{16}=0{\cdot}37^{6}$ $\frac{9}{16}=0{\cdot}48^{4}$ $\frac{11}{16}=0{\cdot}59^{1}$ $\frac{13}{16}=0{\cdot}69^{9}$ $\frac{15}{16}=0{\cdot}80^{6}$

(1.156=1£) **£0.86½**

1	0·86½	41	35·46½	81	70·06½	121	104·66½	161	139·26½
2	1·73	42	36·33	82	70·93	122	105·53	162	140·13
3	2·59½	43	37·19½	83	71·79½	123	106·39½	163	140·99½
4	3·46	44	38·06	84	72·66	124	107·26	164	141·86
5	4·32½	45	38·92½	85	73·52½	125	108·12½	165	142·72½
6	5·19	46	39·79	86	74·39	126	108·99	166	143·59
7	6·05½	47	40·65½	87	75·25½	127	109·85½	167	144·45½
8	6·92	48	41·52	88	76·12	128	110·72	168	145·32
9	7·78½	49	42·38½	89	76·98½	129	111·58½	169	146·18½
10	8·65	**50**	43·25	**90**	77·85	**130**	112·45	**170**	147·05
11	9·51½	51	44·11½	91	78·71½	131	113·31½	171	147·91½
12	10·38	52	44·98	92	79·58	132	114·18	172	148·78
13	11·24½	53	45·84½	93	80·44½	133	115·04½	173	149·64½
14	12·11	54	46·71	94	81·31	134	115·91	174	150·51
15	12·97½	55	47·57½	95	82·17½	135	116·77½	175	151·37½
16	13·84	56	48·44	96	83·04	136	117·64	176	152·24
17	14·70½	57	49·30½	97	83·90½	137	118·50½	177	153·10½
18	15·57	58	50·17	98	84·77	138	119·37	178	153·97
19	16·43½	59	51·03½	99	85·63½	139	120·23½	179	154·83½
20	17·30	**60**	51·90	**100**	86·50	**140**	121·10	**180**	155·70
21	18·16½	61	52·76½	101	87·36½	141	121·96½	181	156·56½
22	19·03	62	53·63	102	88·23	142	122·83	182	157·43
23	19·89½	63	54·49½	103	89·09½	143	123·69½	183	158·29½
24	20·76	64	55·36	104	89·96	144	124·56	184	159·16
25	21·62½	65	56·22½	105	90·82½	145	125·42½	185	160·02½
26	22·49	66	57·09	106	91·69	146	126·29	186	160·89
27	23·35½	67	57·95½	107	92·55½	147	127·15½	187	161·75½
28	24·22	68	58·82	108	93·42	148	128·02	188	162·62
29	25·08½	69	59·68½	109	94·28½	149	128·88½	189	163·48½
30	25·95	**70**	60·55	**110**	95·15	**150**	129·75	**190**	164·35
31	26·81½	71	61·41½	111	96·01½	151	130·61½	191	165·21½
32	27·68	72	62·28	112	96·88	152	131·48	192	166·08
33	28·54½	73	63·14½	113	97·74½	153	132·34½	193	166·94½
34	29·41	74	64·01	114	98·61	154	133·21	194	167·81
35	30·27½	75	64·87½	115	99·47½	155	134·07½	195	168·67½
36	31·14	76	65·74	116	100·34	156	134·94	196	169·54
37	32·00½	77	66·60½	117	101·20½	157	135·80½	197	170·40½
38	32·87	78	67·47	118	102·07	158	136·67	198	171·27
39	33·73½	79	68·33½	119	102·93½	159	137·53½	199	172·13½
40	34·60	**80**	69·20	**120**	103·80	**160**	138·40	**200**	173·00

$\frac{1}{16}$=0·05^{4} $\frac{1}{9}$=0·09^{6} $\frac{1}{8}$=0·10^{8} $\frac{1}{6}$=0·14^{4} $\frac{1}{5}$=0·17^{3} $\frac{1}{4}$=0·21^{6} $\frac{1}{3}$=0·28^{8}

$\frac{3}{8}$=0·32^{4} $\frac{1}{2}$=0·43^{3} $\frac{5}{8}$=0·54^{1} $\frac{3}{4}$=0·64^{9} $\frac{7}{8}$=0·75^{7} $\frac{2}{3}$=0·57^{7} $\frac{5}{6}$=0·72^{1}

$\frac{3}{16}$=0·16^{2} $\frac{5}{16}$=0·27^{0} $\frac{7}{16}$=0·37^{8} $\frac{9}{16}$=0·48^{7} $\frac{11}{16}$=0·59^{5} $\frac{13}{16}$=0·70^{3} $\frac{15}{16}$=0·81^{1}

·8660 is the sine of 60°

(1.149=1£) **£0.87**

1	0·87	41	35·67	81	70·47	121	105·27	161	140·07
2	1·74	42	36·54	82	71·34	122	106·14	162	140·94
3	2·61	43	37·41	83	72·21	123	107·01	163	141·81
4	3·48	44	38·28	84	73·08	124	107·88	164	142·68
5	4·35	45	39·15	85	73·95	125	108·75	165	143·55
6	5·22	46	40·02	86	74·82	126	109·62	166	144·42
7	6·09	47	40·89	87	75·69	127	110·49	167	145·29
8	6·96	48	41·76	88	76·56	128	111·36	168	146·16
9	7·83	49	42·63	89	77·43	129	112·23	169	147·03
10	8·70	**50**	43·50	**90**	78·30	**130**	113·10	**170**	147·90
11	9·57	51	44·37	91	79·17	131	113·97	171	148·77
12	10·44	52	45·24	92	80·04	132	114·84	172	149·64
13	11·31	53	46·11	93	80·91	133	115·71	173	150·51
14	12·18	54	46·98	94	81·78	134	116·58	174	151·38
15	13·05	55	47·85	95	82·65	135	117·45	175	152·25
16	13·92	56	48·72	96	83·52	136	118·32	176	153·12
17	14·79	57	49·59	97	84·39	137	119·19	177	153·99
18	15·66	58	50·46	98	85·26	138	120·06	178	154·86
19	16·53	59	51·33	99	86·13	139	120·93	179	155·73
20	17·40	**60**	52·20	**100**	87·00	**140**	121·80	**180**	156·60
21	18·27	61	53·07	101	87·87	141	122·67	181	157·47
22	19·14	62	53·94	102	88·74	142	123·54	182	158·34
23	20·01	63	54·81	103	89·61	143	124·41	183	159·21
24	20·88	64	55·68	104	90·48	144	125·28	184	160·08
25	21·75	65	56·55	105	91·35	145	126·15	185	160·95
26	22·62	66	57·42	106	92·22	146	127·02	186	161·82
27	23·49	67	58·29	107	93·09	147	127·89	187	162·69
28	24·36	68	59·16	108	93·96	148	128·76	188	163·56
29	25·23	69	60·03	109	94·83	149	129·63	189	164·43
30	26·10	**70**	60·90	**110**	95·70	**150**	130·50	**190**	165·30
31	26·97	71	61·77	111	96·57	151	131·37	191	166·17
32	27·84	72	62·64	112	97·44	152	132·24	192	167·04
33	28·71	73	63·51	113	98·31	153	133·11	193	167·91
34	29·58	74	64·38	114	99·18	154	133·98	194	168·78
35	30·45	75	65·25	115	100·05	155	134·85	195	169·65
36	31·32	76	66·12	116	100·92	156	135·72	196	170·52
37	32·19	77	66·99	117	101·79	157	136·59	197	171·39
38	33·06	78	67·86	118	102·66	158	137·46	198	172·26
39	33·93	79	68·73	119	103·53	159	138·33	199	173·13
40	34·80	**80**	69·60	**120**	104·40	**160**	139·20	**200**	174·00

$\frac{1}{16}=0{\cdot}05^4$ $\frac{1}{9}=0{\cdot}09^7$ $\frac{1}{8}=0{\cdot}10^9$ $\frac{1}{6}=0{\cdot}14^5$ $\frac{1}{5}=0{\cdot}17^4$ $\frac{1}{4}=0{\cdot}21^8$ $\frac{1}{3}=0{\cdot}29^0$

$\frac{3}{8}=0{\cdot}32^6$ $\frac{1}{2}=0{\cdot}43^5$ $\frac{5}{8}=0{\cdot}54^4$ $\frac{3}{4}=0{\cdot}65^3$ $\frac{7}{8}=0{\cdot}76^1$ $\frac{2}{3}=0{\cdot}58^0$ $\frac{5}{6}=0{\cdot}72^5$

$\frac{3}{16}=0{\cdot}16^3$ $\frac{5}{16}=0{\cdot}27^2$ $\frac{7}{16}=0{\cdot}38^1$ $\frac{9}{16}=0{\cdot}48^9$ $\frac{11}{16}=0{\cdot}59^8$ $\frac{13}{16}=0{\cdot}70^7$ $\frac{15}{16}=0{\cdot}81^6$

(1.143=1£) **£0.87½**

1	0·87½	41	35·87½	81	70·87½	121	105·87½	161	140·87½
2	1·75	42	36·75	82	71·75	122	106·75	162	141·75
3	2·62½	43	37·62½	83	72·62½	123	107·62½	163	142·62½
4	3·50	44	38·50	84	73·50	124	108·50	164	143·50
5	4·37½	45	39·37½	85	74·37½	125	109·37½	165	144·37½
6	5·25	46	40·25	86	75·25	126	110·25	166	145·25
7	6·12½	47	41·12½	87	76·12½	127	111·12½	167	146·12½
8	7·00	48	42·00	88	77·00	128	112·00	168	147·00
9	7·87½	49	42·87½	89	77·87½	129	112·87½	169	147·87½
10	8·75	**50**	43·75	**90**	78·75	**130**	113·75	**170**	148·75
11	9·62½	51	44·62½	91	79·62½	131	114·62½	171	149·62½
12	10·50	52	45·50	92	80·50	132	115·50	172	150·50
13	11·37½	53	46·37½	93	81·37½	133	116·37½	173	151·37½
14	12·25	54	47·25	94	82·25	134	117·25	174	152·25
15	13·12½	55	48·12½	95	83·12½	135	118·12½	175	153·12½
16	14·00	56	49·00	96	84·00	136	119·00	176	154·00
17	14·87½	57	49·87½	97	84·87½	137	119·87½	177	154·87½
18	15·75	58	50·75	98	85·75	138	120·75	178	155·75
19	16·62½	59	51·62½	99	86·62½	139	121·62½	179	156·62½
20	17·50	**60**	52·50	**100**	87·50	**140**	122·50	**180**	157·50
21	18·37½	61	53·37½	101	88·37½	141	123·37½	181	158·37½
22	19·25	62	54·25	102	89·25	142	124·25	182	159·25
23	20·12½	63	55·12½	103	90·12½	143	125·12½	183	160·12½
24	21·00	64	56·00	104	91·00	144	126·00	184	161·00
25	21·87½	65	56·87½	105	91·87½	145	126·87½	185	161·87½
26	22·75	66	57·75	106	92·75	146	127·75	186	162·75
27	23·62½	67	58·62½	107	93·62½	147	128·62½	187	163·62½
28	24·50	68	59·50	108	94·50	148	129·50	188	164·50
29	25·37½	69	60·37½	109	95·37½	149	130·37½	189	165·37½
30	26·25	**70**	61·25	**110**	96·25	**150**	131·25	**190**	166·25
31	27·12½	71	62·12½	111	97·12½	151	132·12½	191	167·12½
32	28·00	72	63·00	112	98·00	152	133·00	192	168·00
33	28·87½	73	63·87½	113	98·87½	153	133·87½	193	168·87½
34	29·75	74	64·75	114	99·75	154	134·75	194	169·75
35	30·62½	75	65·62½	115	100·62½	155	135·62½	195	170·62½
36	31·50	76	66·50	116	101·50	156	136·50	196	171·50
37	32·37½	77	67·37½	117	102·37½	157	137·37½	197	172·37½
38	33·25	78	68·25	118	103·25	158	138·25	198	173·25
39	34·12½	79	69·12½	119	104·12½	159	139·12½	199	174·12½
40	35·00	**80**	70·00	**120**	105·00	**160**	140·00	**200**	175·00

$\frac{1}{16}=0{\cdot}05^5$	$\frac{1}{9}=0{\cdot}09^7$	$\frac{1}{8}=0{\cdot}10^9$	$\frac{1}{6}=0{\cdot}14^6$	$\frac{1}{5}=0{\cdot}17^5$	$\frac{1}{4}=0{\cdot}21^9$	$\frac{1}{3}=0{\cdot}29^2$
$\frac{3}{8}=0{\cdot}32^8$	$\frac{1}{2}=0{\cdot}43^8$	$\frac{5}{8}=0{\cdot}54^7$	$\frac{3}{4}=0{\cdot}65^6$	$\frac{7}{8}=0{\cdot}76^6$	$\frac{2}{3}=0{\cdot}58^3$	$\frac{5}{6}=0{\cdot}72^9$
$\frac{3}{16}=0{\cdot}16^4$	$\frac{5}{16}=0{\cdot}27^3$	$\frac{7}{16}=0{\cdot}38^3$	$\frac{9}{16}=0{\cdot}49^2$	$\frac{11}{16}=0{\cdot}60^2$	$\frac{13}{16}=0{\cdot}71^1$	$\frac{15}{16}=0{\cdot}82^0$

1	0·88	41	36·08	81	71·28	121	106·48	161	141·68
2	1·76	42	36·96	82	72·16	122	107·36	162	142·56
3	2·64	43	37·84	83	73·04	123	108·24	163	143·44
4	3·52	44	38·72	84	73·92	124	109·12	164	144·32
5	4·40	45	39·60	85	74·80	125	110·00	165	145·20
6	5·28	46	40·48	86	75·68	126	110·88	166	146·08
7	6·16	47	41·36	87	76·56	127	111·76	167	146·96
8	7·04	48	42·24	88	77·44	128	112·64	168	147·84
9	7·92	49	43·12	89	78·32	129	113·52	169	148·72
10	8·80	**50**	44·00	**90**	79·20	**130**	114·40	**170**	149·60
11	9·68	51	44·88	91	80·08	131	115·28	171	150·48
12	10·56	52	45·76	92	80·96	132	116·16	172	151·36
13	11·44	53	46·64	93	81·84	133	117·04	173	152·24
14	12·32	54	47·52	94	82·72	134	117·92	174	153·12
15	13·20	55	48·40	95	83·60	135	118·80	175	154·00
16	14·08	56	49·28	96	84·48	136	119·68	176	154·88
17	14·96	57	50·16	97	85·36	137	120·56	177	155·76
18	15·84	58	51·04	98	86·24	138	121·44	178	156·64
19	16·72	59	51·92	99	87·12	139	122·32	179	157·52
20	17·60	**60**	52·80	**100**	88·00	**140**	123·20	**180**	158·40
21	18·48	61	53·68	101	88·88	141	124·08	181	159·28
22	19·36	62	54·56	102	89·76	142	124·96	182	160·16
23	20·24	63	55·44	103	90·64	143	125·84	183	161·04
24	21·12	64	56·32	104	91·52	144	126·72	184	161·92
25	22·00	65	57·20	105	92·40	145	127·60	185	162·80
26	22·88	66	58·08	106	93·28	146	128·48	186	163·68
27	23·76	67	58·96	107	94·16	147	129·36	187	164·56
28	24·64	68	59·84	108	95·04	148	130·24	188	165·44
29	25·52	69	60·72	109	95·92	149	131·12	189	166·32
30	26·40	**70**	61·60	**110**	96·80	**150**	132·00	**190**	167·20
31	27·28	71	62·48	111	97·68	151	132·88	191	168·08
32	28·16	72	63·36	112	98·56	152	133·76	192	168·96
33	29·04	73	64·24	113	99·44	153	134·64	193	169·84
34	29·92	74	65·12	114	100·32	154	135·52	194	170·72
35	30·80	75	66·00	115	101·20	155	136·40	195	171·60
36	31·68	76	66·88	116	102·08	156	137·28	196	172·48
37	32·56	77	67·76	117	102·96	157	138·16	197	173·36
38	33·44	78	68·64	118	103·84	158	139·04	198	174·24
39	34·32	79	69·52	119	104·72	159	139·92	199	175·12
40	35·20	**80**	70·40	**120**	105·60	**160**	140·80	**200**	176·00

$\frac{1}{16}$=0·05^{5} $\frac{1}{9}$=0·09^{8} $\frac{1}{8}$=0·11^{0} $\frac{1}{6}$=0·14^{7} $\frac{1}{5}$=0·17^{6} $\frac{1}{4}$=0·22^{0} $\frac{1}{3}$=0·29^{3}

$\frac{3}{8}$=0·33^{0} $\frac{1}{2}$=0·44^{0} $\frac{5}{8}$=0·55^{0} $\frac{3}{4}$=0·66^{0} $\frac{7}{8}$=0·77^{0} $\frac{2}{3}$=0·58^{7} $\frac{5}{6}$=0·73^{3}

$\frac{3}{16}$=0·16^{5} $\frac{5}{16}$=0·27^{5} $\frac{7}{16}$=0·38^{5} $\frac{9}{16}$=0·49^{5} $\frac{11}{16}$=0·60^{5} $\frac{13}{16}$=0·71^{5} $\frac{15}{16}$=0·82^{5}

(1.130=1£) £0.88½

1	0·88½	41	36·28½	81	71·68½	121	107·08½	161	142·48½
2	1·77	42	37·17	82	72·57	122	107·97	162	143·37
3	2·65½	43	38·05½	83	73·45½	123	108·85½	163	144·25½
4	3·54	44	38·94	84	74·34	124	109·74	164	145·14
5	4·42½	45	39·82½	85	75·22½	125	110·62½	165	146·02½
6	5·31	46	40·71	86	76·11	126	111·51	166	146·91
7	6·19½	47	41·59½	87	76·99½	127	112·39½	167	147·79½
8	7·08	48	42·48	88	77·88	128	113·28	168	148·68
9	7·96½	49	43·36½	89	78·76½	129	114·16½	169	149·56½
10	8·85	**50**	44·25	**90**	79·65	**130**	115·05	**170**	150·45
11	9·73½	51	45·13½	91	80·53½	131	115·93½	171	151·33½
12	10·62	52	46·02	92	81·42	132	116·82	172	152·22
13	11·50½	53	46·90½	93	82·30½	133	117·70½	173	153·10½
14	12·39	54	47·79	94	83·19	134	118·59	174	153·99
15	13·27½	55	48·67½	95	84·07½	135	119·47½	175	154·87½
16	14·16	56	49·56	96	84·96	136	120·36	176	155·76
17	15·04½	57	50·44½	97	85·84½	137	121·24½	177	156·64½
18	15·93	58	51·33	98	86·73	138	122·13	178	157·53
19	16·81½	59	52·21½	99	87·61½	139	123·01½	179	158·41½
20	17·70	**60**	53·10	**100**	88·50	**140**	123·90	**180**	159·30
21	18·58½	61	53·98½	101	89·38½	141	124·78½	181	160·18½
22	19·47	62	54·87	102	90·27	142	125·67	182	161·07
23	20·35½	63	55·75½	103	91·15½	143	126·55½	183	161·95½
24	21·24	64	56·64	104	92·04	144	127·44	184	162·84
25	22·12½	65	57·52½	105	92·92½	145	128·32½	185	163·72½
26	23·01	66	58·41	106	93·81	146	129·21	186	164·61
27	23·89½	67	59·29½	107	94·69½	147	130·09½	187	165·49½
28	24·78	68	60·18	108	95·58	148	130·98	188	166·38
29	25·66½	69	61·06½	109	96·46½	149	131·86½	189	167·26½
30	26·55	**70**	61·95	**110**	97·35	**150**	132·75	**190**	168·15
31	27·43½	71	62·83½	111	98·23½	151	133·63½	191	169·03½
32	28·32	72	63·72	112	99·12	152	134·52	192	169·92
33	29·20½	73	64·60½	113	100·00½	153	135·40½	193	170·80½
34	30·09	74	65·49	114	100·89	154	136·29	194	171·69
35	30·97½	75	66·37½	115	101·77½	155	137·17½	195	172·57½
36	31·86	76	67·26	116	102·66	156	138·06	196	173·46
37	32·74½	77	68·14½	117	103·54½	157	138·94½	197	174·34½
38	33·63	78	69·03	118	104·43	158	139·83	198	175·23
39	34·51½	79	69·91½	119	105·31½	159	140·71½	199	176·11½
40	35·40	**80**	70·80	**120**	106·20	**160**	141·60	**200**	177·00

$\frac{1}{16}=0{\cdot}05^{5}$ $\frac{1}{9}=0{\cdot}09^{8}$ $\frac{1}{8}=0{\cdot}11^{1}$ $\frac{1}{6}=0{\cdot}14^{8}$ $\frac{1}{5}=0{\cdot}17^{7}$ $\frac{1}{4}=0{\cdot}22^{1}$ $\frac{1}{3}=0{\cdot}29^{5}$

$\frac{3}{8}=0{\cdot}33^{2}$ $\frac{1}{2}=0{\cdot}44^{3}$ $\frac{5}{8}=0{\cdot}55^{3}$ $\frac{3}{4}=0{\cdot}66^{4}$ $\frac{7}{8}=0{\cdot}77^{4}$ $\frac{2}{3}=0{\cdot}59^{0}$ $\frac{5}{6}=0{\cdot}73^{8}$

$\frac{3}{16}=0{\cdot}16^{6}$ $\frac{5}{16}=0{\cdot}27^{7}$ $\frac{7}{16}=0{\cdot}38^{7}$ $\frac{9}{16}=0{\cdot}49^{8}$ $\frac{11}{16}=0{\cdot}60^{8}$ $\frac{13}{16}=0{\cdot}71^{9}$ $\frac{15}{16}=0{\cdot}83^{0}$

(1.124=1£) **£0.89**

1	0·89	41	36·49	81	72·09	121	107·69	161	143·29
2	1·78	42	37·38	82	72·98	122	108·58	162	144·18
3	2·67	43	38·27	83	73·87	123	109·47	163	145·07
4	3·56	44	39·16	84	74·76	124	110·36	164	145·96
5	4·45	45	40·05	85	75·65	125	111·25	165	146·85
6	5·34	46	40·94	86	76·54	126	112·14	166	147·74
7	6·23	47	41·83	87	77·43	127	113·03	167	148·63
8	7·12	48	42·72	88	78·32	128	113·92	168	149·52
9	8·01	49	43·61	89	79·21	129	114·81	169	150·41
10	8·90	**50**	44·50	**90**	80·10	**130**	115·70	**170**	151·30
11	9·79	51	45·39	91	80·99	131	116·59	171	152·19
12	10·68	52	46·28	92	81·88	132	117·48	172	153·08
13	11·57	53	47·17	93	82·77	133	118·37	173	153·97
14	12·46	54	48·06	94	83·66	134	119·26	174	154·86
15	13·35	55	48·95	95	84·55	135	120·15	175	155·75
16	14·24	56	49·84	96	85·44	136	121·04	176	156·64
17	15·13	57	50·73	97	86·33	137	121·93	177	157·53
18	16·02	58	51·62	98	87·22	138	122·82	178	158·42
19	16·91	59	52·51	99	88·11	139	123·71	179	159·31
20	17·80	**60**	53·40	**100**	89·00	**140**	124·60	**180**	160·20
21	18·69	61	54·29	101	89·89	141	125·49	181	161·09
22	19·58	62	55·18	102	90·78	142	126·38	182	161·98
23	20·47	63	56·07	103	91·67	143	127·27	183	162·87
24	21·36	64	56·96	104	92·56	144	128·16	184	163·76
25	22·25	65	57·85	105	93·45	145	129·05	185	164·65
26	23·14	66	58·74	106	94·34	146	129·94	186	165·54
27	24·03	67	59·63	107	95·23	147	130·83	187	166·43
28	24·92	68	60·52	108	96·12	148	131·72	188	167·32
29	25·81	69	61·41	109	97·01	149	132·61	189	168·21
30	26·70	**70**	62·30	**110**	97·90	**150**	133·50	**190**	169·10
31	27·59	71	63·19	111	98·79	151	134·39	191	169·99
32	28·48	72	64·08	112	99·68	152	135·28	192	170·88
33	29·37	73	64·97	113	100·57	153	136·17	193	171·77
34	30·26	74	65·86	114	101·46	154	137·06	194	172·66
35	31·15	75	66·75	115	102·35	155	137·95	195	173·55
36	32·04	76	67·64	116	103·24	156	138·84	196	174·44
37	32·93	77	68·53	117	104·13	157	139·73	197	175·33
38	33·82	78	69·42	118	105·02	158	140·62	198	176·22
39	34·71	79	70·31	119	105·91	159	141·51	199	177·11
40	35·60	**80**	71·20	**120**	106·80	**160**	142·40	**200**	178·00

$\frac{1}{16}=0{\cdot}05^{6}$ $\frac{1}{9}=0{\cdot}09^{9}$ $\frac{1}{8}=0{\cdot}11^{1}$ $\frac{1}{6}=0{\cdot}14^{8}$ $\frac{1}{5}=0{\cdot}17^{8}$ $\frac{1}{4}=0{\cdot}22^{3}$ $\frac{1}{3}=0{\cdot}29^{7}$

$\frac{3}{8}=0{\cdot}33^{4}$ $\frac{1}{2}=0{\cdot}44^{5}$ $\frac{5}{8}=0{\cdot}55^{6}$ $\frac{3}{4}=0{\cdot}66^{8}$ $\frac{7}{8}=0{\cdot}77^{9}$ $\frac{2}{3}=0{\cdot}59^{3}$ $\frac{5}{6}=0{\cdot}74^{2}$

$\frac{3}{16}=0{\cdot}16^{7}$ $\frac{5}{16}=0{\cdot}27^{8}$ $\frac{7}{16}=0{\cdot}38^{9}$ $\frac{9}{16}=0{\cdot}50^{1}$ $\frac{11}{16}=0{\cdot}61^{2}$ $\frac{13}{16}=0{\cdot}72^{3}$ $\frac{15}{16}=0{\cdot}83^{4}$

(1.117=1£) **£0.89½**

1	0·89½	41	36·69½	81	72·49½	121	108·29½	161	144·09½
2	1·79	42	37·59	82	73·39	122	109·19	162	144·99
3	2·68½	43	38·48½	83	74·28½	123	110·08½	163	145·88½
4	3·58	44	39·38	84	75·18	124	110·98	164	146·78
5	4·47½	45	40·27½	85	76·07½	125	111·87½	165	147·67½
6	5·37	46	41·17	86	76·97	126	112·77	166	148·57
7	6·26½	47	42·06½	87	77·86½	127	113·66½	167	149·46½
8	7·16	48	42·96	88	78·76	128	114·56	168	150·36
9	8·05½	49	43·85½	89	79·65½	129	115·45½	169	151·25½
10	8·95	**50**	44·75	**90**	80·55	**130**	116·35	**170**	152·15
11	9·84½	51	45·64½	91	81·44½	131	117·24½	171	153·04½
12	10·74	52	46·54	92	82·34	132	118·14	172	153·94
13	11·63½	53	47·43½	93	83·23½	133	119·03½	173	154·83½
14	12·53	54	48·33	94	84·13	134	119·93	174	155·73
15	13·42½	55	49·22½	95	85·02½	135	120·82½	175	156·62½
16	14·32	56	50·12	96	85·92	136	121·72	176	157·52
17	15·21½	57	51·01½	97	86·81½	137	122·61½	177	158·41½
18	16·11	58	51·91	98	87·71	138	123·51	178	159·31
19	17·00½	59	52·80½	99	88·60½	139	124·40½	179	160·20½
20	17·90	**60**	53·70	**100**	89·50	**140**	125·30	**180**	161·10
21	18·79½	61	54·59½	101	90·39½	141	126·19½	181	161·99½
22	19·69	62	55·49	102	91·29	142	127·09	182	162·89
23	20·58½	63	56·38½	103	92·18½	143	127·98½	183	163·78½
24	21·48	64	57·28	104	93·08	144	128·88	184	164·68
25	22·37½	65	58·17½	105	93·97½	145	129·77½	185	165·57½
26	23·27	66	59·07	106	94·87	146	130·67	186	166·47
27	24·16½	67	59·96½	107	95·76½	147	131·56½	187	167·36½
28	25·06	68	60·86	108	96·66	148	132·46	188	168·26
29	25·95½	69	61·75½	109	97·55½	149	133·35½	189	169·15½
30	26·85	**70**	62·65	**110**	98·45	**150**	134·25	**190**	170·05
31	27·74½	71	63·54½	111	99·34½	151	135·14½	191	170·94½
32	28·64	72	64·44	112	100·24	152	136·04	192	171·84
33	29·53½	73	65·33½	113	101·13½	153	136·93½	193	172·73½
34	30·43	74	66·23	114	102·03	154	137·83	194	173·63
35	31·32½	75	67·12½	115	102·92½	155	138·72½	195	174·52½
36	32·22	76	68·02	116	103·82	156	139·62	196	175·42
37	33·11½	77	68·91½	117	104·71½	157	140·51½	197	176·31½
38	34·01	78	69·81	118	105·61	158	141·41	198	177·21
39	34·90½	79	70·70½	119	106·50½	159	142·30½	199	178·10½
40	35·80	**80**	71·60	**120**	107·40	**160**	143·20	**200**	179·00

$\frac{1}{16}=0\cdot05^{6}$ $\frac{1}{9}=0\cdot09^{9}$ $\frac{1}{8}=0\cdot11^{2}$ $\frac{1}{6}=0\cdot14^{9}$ $\frac{1}{5}=0\cdot17^{9}$ $\frac{1}{4}=0\cdot22^{4}$ $\frac{1}{3}=0\cdot29^{8}$

$\frac{3}{8}=0\cdot33^{6}$ $\frac{1}{2}=0\cdot44^{8}$ $\frac{5}{8}=0\cdot55^{9}$ $\frac{3}{4}=0\cdot67^{1}$ $\frac{7}{8}=0\cdot78^{3}$ $\frac{2}{3}=0\cdot59^{7}$ $\frac{5}{6}=0\cdot74^{6}$

$\frac{3}{16}=0\cdot16^{8}$ $\frac{5}{16}=0\cdot28^{0}$ $\frac{7}{16}=0\cdot39^{2}$ $\frac{9}{16}=0\cdot50^{3}$ $\frac{11}{16}=0\cdot61^{5}$ $\frac{13}{16}=0\cdot72^{7}$ $\frac{15}{16}=0\cdot83^{9}$

(1.111=1£) **£0.90**

1	0·90	41	36·90	81	72·90	121	108·90	161	144·90
2	1·80	42	37·80	82	73·80	122	109·80	162	145·80
3	2·70	43	38·70	83	74·70	123	110·70	163	146·70
4	3·60	44	39·60	84	75·60	124	111·60	164	147·60
5	4·50	45	40·50	85	76·50	125	112·50	165	148·50
6	5·40	46	41·40	86	77·40	126	113·40	166	149·40
7	6·30	47	42·30	87	78·30	127	114·30	167	150·30
8	7·20	48	43·20	88	79·20	128	115·20	168	151·20
9	8·10	49	44·10	89	80·10	129	116·10	169	152·10
10	9·00	**50**	45·00	**90**	81·00	**130**	117·00	**170**	153·00
11	9·90	51	45·90	91	81·90	131	117·90	171	153·90
12	10·80	52	46·80	92	82·80	132	118·80	172	154·80
13	11·70	53	47·70	93	83·70	133	119·70	173	155·70
14	12·60	54	48·60	94	84·60	134	120·60	174	156·60
15	13·50	55	49·50	95	85·50	135	121·50	175	157·50
16	14·40	56	50·40	96	86·40	136	122·40	176	158·40
17	15·30	57	51·30	97	87·30	137	123·30	177	159·30
18	16·20	58	52·20	98	88·20	138	124·20	178	160·20
19	17·10	59	53·10	99	89·10	139	125·10	179	161·10
20	18·00	**60**	54·00	**100**	90·00	**140**	126·00	**180**	162·00
21	18·90	61	54·90	101	90·90	141	126·90	181	162·90
22	19·80	62	55·80	102	91·80	142	127·80	182	163·80
23	20·70	63	56·70	103	92·70	143	128·70	183	164·70
24	21·60	64	57·60	104	93·60	144	129·60	184	165·60
25	22·50	65	58·50	105	94·50	145	130·50	185	166·50
26	23·40	66	59·40	106	95·40	146	131·40	186	167·40
27	24·30	67	60·30	107	96·30	147	132·30	187	168·30
28	25·20	68	61·20	108	97·20	148	133·20	188	169·20
29	26·10	69	62·10	109	98·10	149	134·10	189	170·10
30	27·00	**70**	63·00	**110**	99·00	**150**	135·00	**190**	171·00
31	27·90	71	63·90	111	99·90	151	135·90	191	171·90
32	28·80	72	64·80	112	100·80	152	136·80	192	172·80
33	29·70	73	65·70	113	101·70	153	137·70	193	173·70
34	30·60	74	66·60	114	102·60	154	138·60	194	174·60
35	31·50	75	67·50	115	103·50	155	139·50	195	175·50
36	32·40	76	68·40	116	104·40	156	140·40	196	176·40
37	33·30	77	69·30	117	105·30	157	141·30	197	177·30
38	34·20	78	70·20	118	106·20	158	142·20	198	178·20
39	35·10	79	71·10	119	107·10	159	143·10	199	179·10
40	36·00	**80**	72·00	**120**	108·00	**160**	144·00	**200**	180·00

$\frac{1}{16}=0{\cdot}05^{6}$ $\frac{1}{9}=0{\cdot}10^{0}$ $\frac{1}{8}=0{\cdot}11^{3}$ $\frac{1}{6}=0{\cdot}15^{0}$ $\frac{1}{5}=0{\cdot}18^{0}$ $\frac{1}{4}=0{\cdot}22^{5}$ $\frac{1}{3}=0{\cdot}30^{0}$

$\frac{3}{8}=0{\cdot}33^{8}$ $\frac{1}{2}=0{\cdot}45^{0}$ $\frac{5}{8}=0{\cdot}56^{3}$ $\frac{3}{4}=0{\cdot}67^{5}$ $\frac{7}{8}=0{\cdot}78^{8}$ $\frac{2}{3}=0{\cdot}60^{0}$ $\frac{5}{6}=0{\cdot}75^{0}$

$\frac{3}{16}=0{\cdot}16^{9}$ $\frac{5}{16}=0{\cdot}28^{1}$ $\frac{7}{16}=0{\cdot}39^{4}$ $\frac{9}{16}=0{\cdot}50^{6}$ $\frac{11}{16}=0{\cdot}61^{9}$ $\frac{13}{16}=0{\cdot}73^{1}$ $\frac{15}{16}=0{\cdot}84^{4}$

1	0·90½	41	37·10½	81	73·30½	121	109·50½	161	145·70½
2	1·81	42	38·01	82	74·21	122	110·41	162	146·61
3	2·71½	43	38·91½	83	75·11½	123	111·31½	163	147·51½
4	3·62	44	39·82	84	76·02	124	112·22	164	148·42
5	4·52½	45	40·72½	85	76·92½	125	113·12½	165	149·32½
6	5·43	46	41·63	86	77·83	126	114·03	166	150·23
7	6·33½	47	42·53½	87	78·73½	127	114·93½	167	151·13½
8	7·24	48	43·44	88	79·64	128	115·84	168	152·04
9	8·14½	49	44·34½	89	80·54½	129	116·74½	169	152·94½
10	9·05	**50**	45·25	**90**	81·45	**130**	117·65	**170**	153·85
11	9·95½	51	46·15½	91	82·35½	131	118·55½	171	154·75½
12	10·86	52	47·06	92	83·26	132	119·46	172	155·66
13	11·76½	53	47·96½	93	84·16½	133	120·36½	173	156·56½
14	12·67	54	48·87	94	85·07	134	121·27	174	157·47
15	13·57½	55	49·77½	95	85·97½	135	122·17½	175	158·37½
16	14·48	56	50·68	96	86·88	136	123·08	176	159·28
17	15·38½	57	51·58½	97	87·78½	137	123·98½	177	160·18½
18	16·29	58	52·49	98	88·69	138	124·89	178	161·09
19	17·19½	59	53·39½	99	89·59½	139	125·79½	179	161·99½
20	18·10	**60**	54·30	**100**	90·50	**140**	126·70	**180**	162·90
21	19·00½	61	55·20½	101	91·40½	141	127·60½	181	163·80½
22	19·91	62	56·11	102	92·31	142	128·51	182	164·71
23	20·81½	63	57·01½	103	93·21½	143	129·41½	183	165·61½
24	21·72	64	57·92	104	94·12	144	130·32	184	166·52
25	22·62½	65	58·82½	105	95·02½	145	131·22½	185	167·42½
26	23·53	66	59·73	106	95·93	146	132·13	186	168·33
27	24·43½	67	60·63½	107	96·83½	147	133·03½	187	169·23½
28	25·34	68	61·54	108	97·74	148	133·94	188	170·14
29	26·24½	69	62·44½	109	98·64½	149	134·84½	189	171·04½
30	27·15	**70**	63·35	**110**	99·55	**150**	135·75	**190**	171·95
31	28·05½	71	64·25½	111	100·45½	151	136·65½	191	172·85½
32	28·96	72	65·16	112	101·36	152	137·56	192	173·76
33	29·86½	73	66·06½	113	102·26½	153	138·46½	193	174·66½
34	30·77	74	66·97	114	103·17	154	139·37	194	175·57
35	31·67½	75	67·87½	115	104·07½	155	140·27½	195	176·47½
36	32·58	76	68·78	116	104·98	156	141·18	196	177·38
37	33·48½	77	69·68½	117	105·88½	157	142·08½	197	178·28½
38	34·39	78	70·59	118	106·79	158	142·99	198	179·19
39	35·29½	79	71·49½	119	107·69½	159	143·89½	199	180·09½
40	36·20	**80**	72·40	**120**	108·60	**160**	144·80	**200**	181·00

$\frac{1}{16}=0{\cdot}05^{7}$ $\frac{1}{9}=0{\cdot}10^{1}$ $\frac{1}{8}=0{\cdot}11^{3}$ $\frac{1}{6}=0{\cdot}15^{1}$ $\frac{1}{5}=0{\cdot}18^{1}$ $\frac{1}{4}=0{\cdot}22^{6}$ $\frac{1}{3}=0{\cdot}30^{2}$

$\frac{3}{8}=0{\cdot}33^{9}$ $\frac{1}{2}=0{\cdot}45^{3}$ $\frac{5}{8}=0{\cdot}56^{6}$ $\frac{3}{4}=0{\cdot}67^{9}$ $\frac{7}{8}=0{\cdot}79^{2}$ $\frac{2}{3}=0{\cdot}60^{3}$ $\frac{5}{6}=0{\cdot}75^{4}$

$\frac{3}{16}=0{\cdot}17^{0}$ $\frac{5}{16}=0{\cdot}28^{3}$ $\frac{7}{16}=0{\cdot}39^{6}$ $\frac{9}{16}=0{\cdot}50^{9}$ $\frac{11}{16}=0{\cdot}62^{2}$ $\frac{13}{16}=0{\cdot}73^{5}$ $\frac{15}{16}=0{\cdot}84^{8}$

1	0·91	41	37·31	81	73·71	121	110·11	161	146·51
2	1·82	42	38·22	82	74·62	122	111·02	162	147·42
3	2·73	43	39·13	83	75·53	123	111·93	163	148·33
4	3·64	44	40·04	84	76·44	124	112·84	164	149·24
5	4·55	45	40·95	85	77·35	125	113·75	165	150·15
6	5·46	46	41·86	86	78·26	126	114·66	166	151·06
7	6·37	47	42·77	87	79·17	127	115·57	167	151·97
8	7·28	48	43·68	88	80·08	128	116·48	168	152·88
9	8·19	49	44·59	89	80·99	129	117·39	169	153·79
10	9·10	**50**	45·50	**90**	81·90	**130**	118·30	**170**	154·70
11	10·01	51	46·41	91	82·81	131	119·21	171	155·61
12	10·92	52	47·32	92	83·72	132	120·12	172	156·52
13	11·83	53	48·23	93	84·63	133	121·03	173	157·43
14	12·74	54	49·14	94	85·54	134	121·94	174	158·34
15	13·65	55	50·05	95	86·45	135	122·85	175	159·25
16	14·56	56	50·96	96	87·36	136	123·76	176	160·16
17	15·47	57	51·87	97	88·27	137	124·67	177	161·07
18	16·38	58	52·78	98	89·18	138	125·58	178	161·98
19	17·29	59	53·69	99	90·09	139	126·49	179	162·89
20	18·20	**60**	54·60	**100**	91·00	**140**	127·40	**180**	163·80
21	19·11	61	55·51	101	91·91	141	128·31	181	164·71
22	20·02	62	56·42	102	92·82	142	129·22	182	165·62
23	20·93	63	57·33	103	93·73	143	130·13	183	166·53
24	21·84	64	58·24	104	94·64	144	131·04	184	167·44
25	22·75	65	59·15	105	95·55	145	131·95	185	168·35
26	23·66	66	60·06	106	96·46	146	132·86	186	169·26
27	24·57	67	60·97	107	97·37	147	133·77	187	170·17
28	25·48	68	61·88	108	98·28	148	134·68	188	171·08
29	26·39	69	62·79	109	99·19	149	135·59	189	171·99
30	27·30	**70**	63·70	**110**	100·10	**150**	136·50	**190**	172·90
31	28·21	71	64·61	111	101·01	151	137·41	191	173·81
32	29·12	72	65·52	112	101·92	152	138·32	192	174·72
33	30·03	73	66·43	113	102·83	153	139·23	193	175·63
34	30·94	74	67·34	114	103·74	154	140·14	194	176·54
35	31·85	75	68·25	115	104·65	155	141·05	195	177·45
36	32·76	76	69·16	116	105·56	156	141·96	196	178·36
37	33·67	77	70·07	117	106·47	157	142·87	197	179·27
38	34·58	78	70·98	118	107·38	158	143·78	198	180·18
39	35·49	79	71·89	119	108·29	159	144·69	199	181·09
40	36·40	**80**	72·80	**120**	109·20	**160**	145·60	**200**	182·00

$\frac{1}{16}=0{\cdot}05^7$ $\frac{1}{9}=0{\cdot}10^1$ $\frac{1}{8}=0{\cdot}11^4$ $\frac{1}{6}=0{\cdot}15^2$ $\frac{1}{5}=0{\cdot}18^2$ $\frac{1}{4}=0{\cdot}22^8$ $\frac{1}{3}=0{\cdot}30^3$

$\frac{3}{8}=0{\cdot}34^1$ $\frac{1}{2}=0{\cdot}45^5$ $\frac{5}{8}=0{\cdot}56^9$ $\frac{3}{4}=0{\cdot}68^3$ $\frac{7}{8}=0{\cdot}79^6$ $\frac{2}{3}=0{\cdot}60^7$ $\frac{5}{6}=0{\cdot}75^8$

$\frac{3}{16}=0{\cdot}17^1$ $\frac{5}{16}=0{\cdot}28^4$ $\frac{7}{16}=0{\cdot}39^8$ $\frac{9}{16}=0{\cdot}51^2$ $\frac{11}{16}=0{\cdot}62^6$ $\frac{13}{16}=0{\cdot}73^9$ $\frac{15}{16}=0{\cdot}85^3$

(1.093=1£) **£0.91½**

1	0·91½	41	37·51½	81	74·11½	121	110·71½	161	147·31½
2	1·83	42	38·43	82	75·03	122	111·63	162	148·23
3	2·74½	43	39·34½	83	75·94½	123	112·54½	163	149·14½
4	3·66	44	40·26	84	76·86	124	113·46	164	150·06
5	4·57½	45	41·17½	85	77·77½	125	114·37½	165	150·97½
6	5·49	46	42·09	86	78·69	126	115·29	166	151·89
7	6·40½	47	43·00½	87	79·60½	127	116·20½	167	152·80½
8	7·32	48	43·92	88	80·52	128	117·12	168	153·72
9	8·23½	49	44·83½	89	81·43½	129	118·03½	169	154·63½
10	9·15	**50**	45·75	**90**	82·35	**130**	118·95	**170**	155·55
11	10·06½	51	46·66½	91	83·26½	131	119·86½	171	156·46½
12	10·98	52	47·58	92	84·18	132	120·78	172	157·38
13	11·89½	53	48·49½	93	85·09½	133	121·69½	173	158·29½
14	12·81	54	49·41	94	86·01	134	122·61	174	159·21
15	13·72½	55	50·32½	95	86·92½	135	123·52½	175	160·12½
16	14·64	56	51·24	96	87·84	136	124·44	176	161·04
17	15·55½	57	52·15½	97	88·75½	137	125·35½	177	161·95½
18	16·47	58	53·07	98	89·67	138	126·27	178	162·87
19	17·38½	59	53·98½	99	90·58½	139	127·18½	179	163·78½
20	18·30	**60**	54·90	**100**	91·50	**140**	128·10	**180**	164·70
21	19·21½	61	55·81½	101	92·41½	141	129·01½	181	165·61½
22	20·13	62	56·73	102	93·33	142	129·93	182	166·53
23	21·04½	63	57·64½	103	94·24½	143	130·84½	183	167·44½
24	21·96	64	58·56	104	95·16	144	131·76	184	168·36
25	22·87½	65	59·47½	105	96·07½	145	132·67½	185	169·27½
26	23·79	66	60·39	106	96·99	146	133·59	186	170·19
27	24·70½	67	61·30½	107	97·90½	147	134·50½	187	171·10½
28	25·62	68	62·22	108	98·82	148	135·42	188	172·02
29	26·53½	69	63·13½	109	99·73½	149	136·33½	189	172·93½
30	27·45	**70**	64·05	**110**	100·65	**150**	137·25	**190**	173·85
31	28·36½	71	64·96½	111	101·56½	151	138·16½	191	174·76½
32	29·28	72	65·88	112	102·48	152	139·08	192	175·68
33	30·19½	73	66·79½	113	103·39½	153	139·99½	193	176·59½
34	31·11	74	67·71	114	104·31	154	140·91	194	177·51
35	32·02½	75	68·62½	115	105·22½	155	141·82½	195	178·42½
36	32·94	76	69·54	116	106·14	156	142·74	196	179·34
37	33·85½	77	70·45½	117	107·05½	157	143·65½	197	180·25½
38	34·77	78	71·37	118	107·97	158	144·57	198	181·17
39	35·68½	79	72·28½	119	108·88½	159	145·48½	199	182·08½
40	36·60	**80**	73·20	**120**	109·80	**160**	146·40	**200**	183·00

$\frac{1}{16}$=0·05^{7} $\frac{1}{9}$=0·10^{2} $\frac{1}{8}$=0·11^{4} $\frac{1}{6}$=0·15^{3} $\frac{1}{5}$=0·18^{3} $\frac{1}{4}$=0·22^{9} $\frac{1}{3}$=0·30^{5}

$\frac{3}{8}$=0·34^{3} $\frac{1}{2}$=0·45^{8} $\frac{5}{8}$=0·57^{2} $\frac{3}{4}$=0·68^{6} $\frac{7}{8}$=0·80^{1} $\frac{2}{3}$=0·61^{0} $\frac{5}{6}$=0·76^{3}

$\frac{3}{16}$=0·17^{2} $\frac{5}{16}$=0·28^{6} $\frac{7}{16}$=0·40^{0} $\frac{9}{16}$=0·51^{5} $\frac{11}{16}$=0·62^{9} $\frac{13}{16}$=0·74^{3} $\frac{15}{16}$=0·85^{8}

1 yard=0·9144 metres

(1.087=1£) **£0.92**

1	0·92	41	37·72	81	74·52	121	111·32	161	148·12
2	1·84	42	38·64	82	75·44	122	112·24	162	149·04
3	2·76	43	39·56	83	76·36	123	113·16	163	149·96
4	3·68	44	40·48	84	77·28	124	114·08	164	150·88
5	4·60	45	41·40	85	78·20	125	115·00	165	151·80
6	5·52	46	42·32	86	79·12	126	115·92	166	152·72
7	6·44	47	43·24	87	80·04	127	116·84	167	153·64
8	7·36	48	44·16	88	80·96	128	117·76	168	154·56
9	8·28	49	45·08	89	81·88	129	118·68	169	155·48
10	9·20	**50**	46·00	**90**	82·80	**130**	119·60	**170**	156·40
11	10·12	51	46·92	91	83·72	131	120·52	171	157·32
12	11·04	52	47·84	92	84·64	132	121·44	172	158·24
13	11·96	53	48·76	93	85·56	133	122·36	173	159·16
14	12·88	54	49·68	94	86·48	134	123·28	174	160·08
15	13·80	55	50·60	95	87·40	135	124·20	175	161·00
16	14·72	56	51·52	96	88·32	136	125·12	176	161·92
17	15·64	57	52·44	97	89·24	137	126·04	177	162·84
18	16·56	58	53·36	98	90·16	138	126·96	178	163·76
19	17·48	59	54·28	99	91·08	139	127·88	179	164·68
20	18·40	**60**	55·20	**100**	92·00	**140**	128·80	**180**	165·60
21	19·32	61	56·12	101	92·92	141	129·72	181	166·52
22	20·24	62	57·04	102	93·84	142	130·64	182	167·44
23	21·16	63	57·96	103	94·76	143	131·56	183	168·36
24	22·08	64	58·88	104	95·68	144	132·48	184	169·28
25	23·00	65	59·80	105	96·60	145	133·40	185	170·20
26	23·92	66	60·72	106	97·52	146	134·32	186	171·12
27	24·84	67	61·64	107	98·44	147	135·24	187	172·04
28	25·76	68	62·56	108	99·36	148	136·16	188	172·96
29	26·68	69	63·48	109	100·28	149	137·08	189	173·88
30	27·60	**70**	64·40	**110**	101·20	**150**	138·00	**190**	174·80
31	28·52	71	65·32	111	102·12	151	138·92	191	175·72
32	29·44	72	66·24	112	103·04	152	139·84	192	176·64
33	30·36	73	67·16	113	103·96	153	140·76	193	177·56
34	31·28	74	68·08	114	104·88	154	141·68	194	178·48
35	32·20	75	69·00	115	105·80	155	142·60	195	179·40
36	33·12	76	69·92	116	106·72	156	143·52	196	180·32
37	34·04	77	70·84	117	107·64	157	144·44	197	181·24
38	34·96	78	71·76	118	108·56	158	145·36	198	182·16
39	35·88	79	72·68	119	109·48	159	146·28	199	183·08
40	36·80	**80**	73·60	**120**	110·40	**160**	147·20	**200**	184·00

$\frac{1}{16}=0{\cdot}05^{8}$ $\frac{1}{9}=0{\cdot}10^{2}$ $\frac{1}{8}=0{\cdot}11^{5}$ $\frac{1}{6}=0{\cdot}15^{3}$ $\frac{1}{5}=0{\cdot}18^{4}$ $\frac{1}{4}=0{\cdot}23^{0}$ $\frac{1}{3}=0{\cdot}30^{7}$

$\frac{3}{8}=0{\cdot}34^{5}$ $\frac{1}{2}=0{\cdot}46^{0}$ $\frac{5}{8}=0{\cdot}57^{5}$ $\frac{3}{4}=0{\cdot}69^{0}$ $\frac{7}{8}=0{\cdot}80^{5}$ $\frac{2}{3}=0{\cdot}61^{3}$ $\frac{5}{6}=0{\cdot}76^{7}$

$\frac{3}{16}=0{\cdot}17^{3}$ $\frac{5}{16}=0{\cdot}28^{8}$ $\frac{7}{16}=0{\cdot}40^{3}$ $\frac{9}{16}=0{\cdot}51^{8}$ $\frac{11}{16}=0{\cdot}63^{3}$ $\frac{13}{16}=0{\cdot}74^{8}$ $\frac{15}{16}=0{\cdot}86^{3}$

(1.081=1£) **£0.92½**

1	0·92½	41	37·92½	81	74·92½	121	111·92½	161	148·92½
2	1·85	42	38·85	82	75·85	122	112·85	162	149·85
3	2·77½	43	39·77½	83	76·77½	123	113·77½	163	150·77½
4	3·70	44	40·70	84	77·70	124	114·70	164	151·70
5	4·62½	45	41·62½	85	78·62½	125	115·62½	165	152·62½
6	5·55	46	42·55	86	79·55	126	116·55	166	153·55
7	6·47½	47	43·47½	87	80·47½	127	117·47½	167	154·47½
8	7·40	48	44·40	88	81·40	128	118·40	168	155·40
9	8·32½	49	45·32½	89	82·32½	129	119·32½	169	156·32½
10	9·25	**50**	46·25	**90**	83·25	**130**	120·25	**170**	157·25
11	10·17½	51	47·17½	91	84·17½	131	121·17½	171	158·17½
12	11·10	52	48·10	92	85·10	132	122·10	172	159·10
13	12·02½	53	49·02½	93	86·02½	133	123·02½	173	160·02½
14	12·95	54	49·95	94	86·95	134	123·95	174	160·95
15	13·87½	55	50·87½	95	87·87½	135	124·87½	175	161·87½
16	14·80	56	51·80	96	88·80	136	125·80	176	162·80
17	15·72½	57	52·72½	97	89·72½	137	126·72½	177	163·72½
18	16·65	58	53·65	98	90·65	138	127·65	178	164·65
19	17·57½	59	54·57½	99	91·57½	139	128·57½	179	165·57½
20	18·50	**60**	55·50	**100**	92·50	**140**	129·50	**180**	166·50
21	19·42½	61	56·42½	101	93·42½	141	130·42½	181	167·42½
22	20·35	62	57·35	102	94·35	142	131·35	182	168·35
23	21·27½	63	58·27½	103	95·27½	143	132·27½	183	169·27½
24	22·20	64	59·20	104	96·20	144	133·20	184	170·20
25	23·12½	65	60·12½	105	97·12½	145	134·12½	185	171·12½
26	24·05	66	61·05	106	98·05	146	135·05	186	172·05
27	24·97½	67	61·97½	107	98·97½	147	135·97½	187	172·97½
28	25·90	68	62·90	108	99·90	148	136·90	188	173·90
29	26·82½	69	63·82½	109	100·82½	149	137·82½	189	174·82½
30	27·75	**70**	64·75	**110**	101·75	**150**	138·75	**190**	175·75
31	28·67½	71	65·67½	111	102·67½	151	139·67½	191	176·67½
32	29·60	72	66·60	112	103·60	152	140·60	192	177·60
33	30·52½	73	67·52½	113	104·52½	153	141·52½	193	178·52½
34	31·45	74	68·45	114	105·45	154	142·45	194	179·45
35	32·37½	75	69·37½	115	106·37½	155	143·37½	195	180·37½
36	33·30	76	70·30	116	107·30	156	144·30	196	181·30
37	34·22½	77	71·22½	117	108·22½	157	145·22½	197	182·22½
38	35·15	78	72·15	118	109·15	158	146·15	198	183·15
39	36·07½	79	73·07½	119	110·07½	159	147·07½	199	184·07½
40	37·00	**80**	74·00	**120**	111·00	**160**	148·00	**200**	185·00

$\frac{1}{16}=0{\cdot}05^{8}$ $\frac{1}{9}=0{\cdot}10^{3}$ $\frac{1}{8}=0{\cdot}11^{6}$ $\frac{1}{6}=0{\cdot}15^{4}$ $\frac{1}{5}=0{\cdot}18^{5}$ $\frac{1}{4}=0{\cdot}23^{1}$ $\frac{1}{3}=0{\cdot}30^{8}$

$\frac{3}{8}=0{\cdot}34^{7}$ $\frac{1}{2}=0{\cdot}46^{3}$ $\frac{5}{8}=0{\cdot}57^{8}$ $\frac{3}{4}=0{\cdot}69^{4}$ $\frac{7}{8}=0{\cdot}80^{9}$ $\frac{2}{3}=0{\cdot}61^{7}$ $\frac{5}{6}=0{\cdot}77^{1}$

$\frac{3}{16}=0{\cdot}17^{3}$ $\frac{5}{16}=0{\cdot}28^{9}$ $\frac{7}{16}=0{\cdot}40^{5}$ $\frac{9}{16}=0{\cdot}52^{0}$ $\frac{11}{16}=0{\cdot}63^{6}$ $\frac{13}{16}=0{\cdot}75^{2}$ $\frac{15}{16}=0{\cdot}86^{7}$

(1.075=1£) **£0.93**

1	0·93	41	38·13	81	75·33	121	112·53	161	149·73
2	1·86	42	39·06	82	76·26	122	113·46	162	150·66
3	2·79	43	39·99	83	77·19	123	114·39	163	151·59
4	3·72	44	40·92	84	78·12	124	115·32	164	152·52
5	4·65	45	41·85	85	79·05	125	116·25	165	153·45
6	5·58	46	42·78	86	79·98	126	117·18	166	154·38
7	6·51	47	43·71	87	80·91	127	118·11	167	155·31
8	7·44	48	44·64	88	81·84	128	119·04	168	156·24
9	8·37	49	45·57	89	82·77	129	119·97	169	157·17
10	9·30	**50**	46·50	**90**	83·70	**130**	120·90	**170**	158·10
11	10·23	51	47·43	91	84·63	131	121·83	171	159·03
12	11·16	52	48·36	92	85·56	132	122·76	172	159·96
13	12·09	53	49·29	93	86·49	133	123·69	173	160·89
14	13·02	54	50·22	94	87·42	134	124·62	174	161·82
15	13·95	55	51·15	95	88·35	135	125·55	175	162·75
16	14·88	56	52·08	96	89·28	136	126·48	176	163·68
17	15·81	57	53·01	97	90·21	137	127·41	177	164·61
18	16·74	58	53·94	98	91·14	138	128·34	178	165·54
19	17·67	59	54·87	99	92·07	139	129·27	179	166·47
20	18·60	**60**	55·80	**100**	93·00	**140**	130·20	**180**	167·40
21	19·53	61	56·73	101	93·93	141	131·13	181	168·33
22	20·46	62	57·66	102	94·86	142	132·06	182	169·26
23	21·39	63	58·59	103	95·79	143	132·99	183	170·19
24	22·32	64	59·52	104	96·72	144	133·92	184	171·12
25	23·25	65	60·45	105	97·65	145	134·85	185	172·05
26	24·18	66	61·38	106	98·58	146	135·78	186	172·98
27	25·11	67	62·31	107	99·51	147	136·71	187	173·91
28	26·04	68	63·24	108	100·44	148	137·64	188	174·84
29	26·97	69	64·17	109	101·37	149	138·57	189	175·77
30	27·90	**70**	65·10	**110**	102·30	**150**	139·50	**190**	176·70
31	28·83	71	66·03	111	103·23	151	140·43	191	177·63
32	29·76	72	66·96	112	104·16	152	141·36	192	178·56
33	30·69	73	67·89	113	105·09	153	142·29	193	179·49
34	31·62	74	68·82	114	106·02	154	143·22	194	180·42
35	32·55	75	69·75	115	106·95	155	144·15	195	181·35
36	33·48	76	70·68	116	107·88	156	145·08	196	182·28
37	34·41	77	71·61	117	108·81	157	146·01	197	183·21
38	35·34	78	72·54	118	109·74	158	146·94	198	184·14
39	36·27	79	73·47	119	110·67	159	147·87	199	185·07
40	37·20	**80**	74·40	**120**	111·60	**160**	148·80	**200**	186·00

$\frac{1}{16}$=0·05^{8} $\frac{1}{9}$=0·10^{3} $\frac{1}{8}$=0·11^{6} $\frac{1}{6}$=0·15^{5} $\frac{1}{5}$=0·18^{6} $\frac{1}{4}$=0·23^{3} $\frac{1}{3}$=0·31^{0}

$\frac{3}{8}$=0·34^{9} $\frac{1}{2}$=0·46^{5} $\frac{5}{8}$=0·58^{1} $\frac{3}{4}$=0·69^{8} $\frac{7}{8}$=0·81^{4} $\frac{2}{3}$=0·62^{0} $\frac{5}{6}$=0·77^{5}

$\frac{3}{16}$=0·17^{4} $\frac{5}{16}$=0·29^{1} $\frac{7}{16}$=0·40^{7} $\frac{9}{16}$=0·52^{3} $\frac{11}{16}$=0·63^{9} $\frac{13}{16}$=0·75^{6} $\frac{15}{16}$=0·87^{2}

1 square foot=929 square centimetres

(1.070=1£) £0.93½

1	0·93½	41	38·33½	81	75·73½	121	113·13½	161	150·53½
2	1·87	42	39·27	82	76·67	122	114·07	162	151·47
3	2·80½	43	40·20½	83	77·60½	123	115·00½	163	152·40½
4	3·74	44	41·14	84	78·54	124	115·94	164	153·34
5	4·67½	45	42·07½	85	79·47½	125	116·87½	165	154·27½
6	5·61	46	43·01	86	80·41	126	117·81	166	155·21
7	6·54½	47	43·94½	87	81·34½	127	118·74½	167	156·14½
8	7·48	48	44·88	88	82·28	128	119·68	168	157·08
9	8·41½	49	45·81½	89	83·21½	129	120·61½	169	158·01½
10	9·35	**50**	46·75	**90**	84·15	**130**	121·55	**170**	158·95
11	10·28½	51	47·68½	91	85·08½	131	122·48½	171	159·88½
12	11·22	52	48·62	92	86·02	132	123·42	172	160·82
13	12·15½	53	49·55½	93	86·95½	133	124·35½	173	161·75½
14	13·09	54	50·49	94	87·89	134	125·29	174	162·69
15	14·02½	55	51·42½	95	88·82½	135	126·22½	175	163·62½
16	14·96	56	52·36	96	89·76	136	127·16	176	164·56
17	15·89½	57	53·29½	97	90·69½	137	128·09½	177	165·49½
18	16·83	58	54·23	98	91·63	138	129·03	178	166·43
19	17·76½	59	55·16½	99	92·56½	139	129·96½	179	167·36½
20	18·70	**60**	56·10	**100**	93·50	**140**	130·90	**180**	168·30
21	19·63½	61	57·03½	101	94·43½	141	131·83½	181	169·23½
22	20·57	62	57·97	102	95·37	142	132·77	182	170·17
23	21·50½	63	58·90½	103	96·30½	143	133·70½	183	171·10½
24	22·44	64	59·84	104	97·24	144	134·64	184	172·04
25	23·37½	65	60·77½	105	98·17½	145	135·57½	185	172·97½
26	24·31	66	61·71	106	99·11	146	136·51	186	173·91
27	25·24½	67	62·64½	107	100·04½	147	137·44½	187	174·84½
28	26·18	68	63·58	108	100·98	148	138·38	188	175·78
29	27·11½	69	64·51½	109	101·91½	149	139·31½	189	176·71½
30	28·05	**70**	65·45	**110**	102·85	**150**	140·25	**190**	177·65
31	28·98½	71	66·38½	111	103·78½	151	141·18½	191	178·58½
32	29·92	72	67·32	112	104·72	152	142·12	192	179·52
33	30·85½	73	68·25½	113	105·65½	153	143·05½	193	180·45½
34	31·79	74	69·19	114	106·59	154	143·99	194	181·39
35	32·72½	75	70·12½	115	107·52½	155	144·92½	195	182·32½
36	33·66	76	71·06	116	108·46	156	145·86	196	183·26
37	34·59½	77	71·99½	117	109·39½	157	146·79½	197	184·19½
38	35·53	78	72·93	118	110·33	158	147·73	198	185·13
39	36·46½	79	73·86½	119	111·26½	159	148·66½	199	186·06½
40	37·40	**80**	74·80	**120**	112·20	**160**	149·60	**200**	187·00

$\frac{1}{16}=0{\cdot}05^{8}$ $\frac{1}{9}=0{\cdot}10^{4}$ $\frac{1}{8}=0{\cdot}11^{7}$ $\frac{1}{6}=0{\cdot}15^{6}$ $\frac{1}{5}=0{\cdot}18^{7}$ $\frac{1}{4}=0{\cdot}23^{4}$ $\frac{1}{3}=0{\cdot}31^{2}$

$\frac{3}{8}=0{\cdot}35^{1}$ $\frac{1}{2}=0{\cdot}46^{8}$ $\frac{5}{8}=0{\cdot}58^{4}$ $\frac{3}{4}=0{\cdot}70^{1}$ $\frac{7}{8}=0{\cdot}81^{8}$ $\frac{2}{3}=0{\cdot}62^{3}$ $\frac{5}{6}=0{\cdot}77^{9}$

$\frac{3}{16}=0{\cdot}17^{5}$ $\frac{5}{16}=0{\cdot}29^{2}$ $\frac{7}{16}=0{\cdot}40^{9}$ $\frac{9}{16}=0{\cdot}52^{6}$ $\frac{11}{16}=0{\cdot}64^{3}$ $\frac{13}{16}=0{\cdot}76^{0}$ $\frac{15}{16}=0{\cdot}87^{7}$

(1.064=1£) **£0.94**

1	0·94	41	38·54	81	76·14	121	113·74	161	151·34
2	1·88	42	39·48	82	77·08	122	114·68	162	152·28
3	2·82	43	40·42	83	78·02	123	115·62	163	153·22
4	3·76	44	41·36	84	78·96	124	116·56	164	154·16
5	4·70	45	42·30	85	79·90	125	117·50	165	155·10
6	5·64	46	43·24	86	80·84	126	118·44	166	156·04
7	6·58	47	44·18	87	81·78	127	119·38	167	156·98
8	7·52	48	45·12	88	82·72	128	120·32	168	157·92
9	8·46	49	46·06	89	83·66	129	121·26	169	158·86
10	9·40	**50**	47·00	**90**	84·60	**130**	122·20	**170**	159·80
11	10·34	51	47·94	91	85·54	131	123·14	171	160·74
12	11·28	52	48·88	92	86·48	132	124·08	172	161·68
13	12·22	53	49·82	93	87·42	133	125·02	173	162·62
14	13·16	54	50·76	94	88·36	134	125·96	174	163·56
15	14·10	55	51·70	95	89·30	135	126·90	175	164·50
16	15·04	56	52·64	96	90·24	136	127·84	176	165·44
17	15·98	57	53·58	97	91·18	137	128·78	177	166·38
18	16·92	58	54·52	98	92·12	138	129·72	178	167·32
19	17·86	59	55·46	99	93·06	139	130·66	179	168·26
20	18·80	**60**	56·40	**100**	94·00	**140**	131·60	**180**	169·20
21	19·74	61	57·34	101	94·94	141	132·54	181	170·14
22	20·68	62	58·28	102	95·88	142	133·48	182	171·08
23	21·62	63	59·22	103	96·82	143	134·42	183	172·02
24	22·56	64	60·16	104	97·76	144	135·36	184	172·96
25	23·50	65	61·10	105	98·70	145	136·30	185	173·90
26	24·44	66	62·04	106	99·64	146	137·24	186	174·84
27	25·38	67	62·98	107	100·58	147	138·18	187	175·78
28	26·32	68	63·92	108	101·52	148	139·12	188	176·72
29	27·26	69	64·86	109	102·46	149	140·06	189	177·66
30	28·20	**70**	65·80	**110**	103·40	**150**	141·00	**190**	178·60
31	29·14	71	66·74	111	104·34	151	141·94	191	179·54
32	30·08	72	67·68	112	105·28	152	142·88	192	180·48
33	31·02	73	68·62	113	106·22	153	143·82	193	181·42
34	31·96	74	69·56	114	107·16	154	144·76	194	182·36
35	32·90	75	70·50	115	108·10	155	145·70	195	183·30
36	33·84	76	71·44	116	109·04	156	146·64	196	184·24
37	34·78	77	72·38	117	109·98	157	147·58	197	185·18
38	35·72	78	73·32	118	110·92	158	148·52	198	186·12
39	36·66	79	74·26	119	111·86	159	149·46	199	187·06
40	37·60	**80**	75·20	**120**	112·80	**160**	150·40	**200**	188·00

$\frac{1}{16}=0{\cdot}05^{9}$ $\frac{1}{9}=0{\cdot}10^{4}$ $\frac{1}{8}=0{\cdot}11^{8}$ $\frac{1}{6}=0{\cdot}15^{7}$ $\frac{1}{5}=0{\cdot}18^{8}$ $\frac{1}{4}=0{\cdot}23^{5}$ $\frac{1}{3}=0{\cdot}31^{3}$

$\frac{3}{8}=0{\cdot}35^{3}$ $\frac{1}{2}=0{\cdot}47^{0}$ $\frac{5}{8}=0{\cdot}58^{8}$ $\frac{3}{4}=0{\cdot}70^{5}$ $\frac{7}{8}=0{\cdot}82^{3}$ $\frac{2}{3}=0{\cdot}62^{7}$ $\frac{5}{6}=0{\cdot}78^{3}$

$\frac{3}{16}=0{\cdot}17^{6}$ $\frac{5}{16}=0{\cdot}29^{4}$ $\frac{7}{16}=0{\cdot}41^{1}$ $\frac{9}{16}=0{\cdot}52^{9}$ $\frac{11}{16}=0{\cdot}64^{6}$ $\frac{13}{16}=0{\cdot}76^{4}$ $\frac{15}{16}=0{\cdot}88^{1}$

(1.058=1£) **£0.94½**

1	0·94½	41	38·74½	81	76·54½	121	114·34½	161	152·14½
2	1·89	42	39·69	82	77·49	122	115·29	162	153·09
3	2·83½	43	40·63½	83	78·43½	123	116·23½	163	154·03½
4	3·78	44	41·58	84	79·38	124	117·18	164	154·98
5	4·72½	45	42·52½	85	80·32½	125	118·12½	165	155·92½
6	5·67	46	43·47	86	81·27	126	119·07	166	156·87
7	6·61½	47	44·41½	87	82·21½	127	120·01½	167	157·81½
8	7·56	48	45·36	88	83·16	128	120·96	168	158·76
9	8·50½	49	46·30½	89	84·10½	129	121·90½	169	159·70½
10	9·45	**50**	47·25	**90**	85·05	**130**	122·85	**170**	160·65
11	10·39½	51	48·19½	91	85·99½	131	123·79½	171	161·59½
12	11·34	52	49·14	92	86·94	132	124·74	172	162·54
13	12·28½	53	50·08½	93	87·88½	133	125·68½	173	163·48½
14	13·23	54	51·03	94	88·83	134	126·63	174	164·43
15	14·17½	55	51·97½	95	89·77½	135	127·57½	175	165·37½
16	15·12	56	52·92	96	90·72	136	128·52	176	166·32
17	16·06½	57	53·86½	97	91·66½	137	129·46½	177	167·26½
18	17·01	58	54·81	98	92·61	138	130·41	178	168·21
19	17·95½	59	55·75½	99	93·55½	139	131·35½	179	169·15½
20	18·90	**60**	56·70	**100**	94·50	**140**	132·30	**180**	170·10
21	19·84½	61	57·64½	101	95·44½	141	133·24½	181	171·04½
22	20·79	62	58·59	102	96·39	142	134·19	182	171·99
23	21·73½	63	59·53½	103	97·33½	143	135·13½	183	172·93½
24	22·68	64	60·48	104	98·28	144	136·08	184	173·88
25	23·62½	65	61·42½	105	99·22½	145	137·02½	185	174·82½
26	24·57	66	62·37	106	100·17	146	137·97	186	175·77
27	25·51½	67	63·31½	107	101·11½	147	138·91½	187	176·71½
28	26·46	68	64·26	108	102·06	148	139·86	188	177·66
29	27·40½	69	65·20½	109	103·00½	149	140·80½	189	178·60½
30	28·35	**70**	66·15	**110**	103·95	**150**	141·75	**190**	179·55
31	29·29½	71	67·09½	111	104·89½	151	142·69½	191	180·49½
32	30·24	72	68·04	112	105·84	152	143·64	192	181·44
33	31·18½	73	68·98½	113	106·78½	153	144·58½	193	182·38½
34	32·13	74	69·93	114	107·73	154	145·53	194	183·33
35	33·07½	75	70·87½	115	108·67½	155	146·47½	195	184·27½
36	34·02	76	71·82	116	109·62	156	147·42	196	185·22
37	34·96½	77	72·76½	117	110·56½	157	148·36½	197	186·16½
38	35·91	78	73·71	118	111·51	158	149·31	198	187·11
39	36·85½	79	74·65½	119	112·45½	159	150·25½	199	188·05½
40	37·80	**80**	75·60	**120**	113·40	**160**	151·20	**200**	189·00

$\frac{1}{16}=0{\cdot}05^{9}$ $\frac{1}{9}=0{\cdot}10^{5}$ $\frac{1}{8}=0{\cdot}11^{8}$ $\frac{1}{6}=0{\cdot}15^{8}$ $\frac{1}{5}=0{\cdot}18^{9}$ $\frac{1}{4}=0{\cdot}23^{6}$ $\frac{1}{3}=0{\cdot}31^{5}$

$\frac{3}{8}=0{\cdot}35^{4}$ $\frac{1}{2}=0{\cdot}47^{3}$ $\frac{5}{8}=0{\cdot}59^{1}$ $\frac{3}{4}=0{\cdot}70^{9}$ $\frac{7}{8}=0{\cdot}82^{7}$ $\frac{2}{3}=0{\cdot}63^{0}$ $\frac{5}{6}=0{\cdot}78^{8}$

$\frac{3}{16}=0{\cdot}17^{7}$ $\frac{5}{16}=0{\cdot}29^{5}$ $\frac{7}{16}=0{\cdot}41^{3}$ $\frac{9}{16}=0{\cdot}53^{2}$ $\frac{11}{16}=0{\cdot}65^{0}$ $\frac{13}{16}=0{\cdot}76^{8}$ $\frac{15}{16}=0{\cdot}88^{6}$

1 Kilo joule=0·947 British Thermal Units

(1.053=1£) **£0.95**

1	0·95	41	38·95	81	76·95	121	114·95	161	152·95
2	1·90	42	39·90	82	77·90	122	115·90	162	153·90
3	2·85	43	40·85	83	78·85	123	116·85	163	154·85
4	3·80	44	41·80	84	79·80	124	117·80	164	155·80
5	4·75	45	42·75	85	80·75	125	118·75	165	156·75
6	5·70	46	43·70	86	81·70	126	119·70	166	157·70
7	6·65	47	44·65	87	82·65	127	120·65	167	158·65
8	7·60	48	45·60	88	83·60	128	121·60	168	159·60
9	8·55	49	46·55	89	84·55	129	122·55	169	160·55
10	9·50	**50**	47·50	**90**	85·50	**130**	123·50	**170**	161·50
11	10·45	51	48·45	91	86·45	131	124·45	171	162·45
12	11·40	52	49·40	92	87·40	132	125·40	172	163·40
13	12·35	53	50·35	93	88·35	133	126·35	173	164·35
14	13·30	54	51·30	94	89·30	134	127·30	174	165·30
15	14·25	55	52·25	95	90·25	135	128·25	175	166·25
16	15·20	56	53·20	96	91·20	136	129·20	176	167·20
17	16·15	57	54·15	97	92·15	137	130·15	177	168·15
18	17·10	58	55·10	98	93·10	138	131·10	178	169·10
19	18·05	59	56·05	99	94·05	139	132·05	179	170·05
20	19·00	**60**	57·00	**100**	95·00	**140**	133·00	**180**	171·00
21	19·95	61	57·95	101	95·95	141	133·95	181	171·95
22	20·90	62	58·90	102	96·90	142	134·90	182	172·90
23	21·85	63	59·85	103	97·85	143	135·85	183	173·85
24	22·80	64	60·80	104	98·80	144	136·80	184	174·80
25	23·75	65	61·75	105	99·75	145	137·75	185	175·75
26	24·70	66	62·70	106	100·70	146	138·70	186	176·70
27	25·65	67	63·65	107	101·65	147	139·65	187	177·65
28	26·60	68	64·60	108	102·60	148	140·60	188	178·60
29	27·55	69	65·55	109	103·55	149	141·55	189	179·55
30	28·50	**70**	66·50	**110**	104·50	**150**	142·50	**190**	180·50
31	29·45	71	67·45	111	105·45	151	143·45	191	181·45
32	30·40	72	68·40	112	106·40	152	144·40	192	182·40
33	31·35	73	69·35	113	107·35	153	145·35	193	183·35
34	32·30	74	70·30	114	108·30	154	146·30	194	184·30
35	33·25	75	71·25	115	109·25	155	147·25	195	185·25
36	34·20	76	72·20	116	110·20	156	148·20	196	186·20
37	35·15	77	73·15	117	111·15	157	149·15	197	187·15
38	36·10	78	74·10	118	112·10	158	150·10	198	188·10
39	37·05	79	75·05	119	113·05	159	151·05	199	189·05
40	38·00	**80**	76·00	**120**	114·00	**160**	152·00	**200**	190·00

$\frac{1}{16}=0{\cdot}05^{9}$ $\frac{1}{9}=0{\cdot}10^{6}$ $\frac{1}{8}=0{\cdot}11^{9}$ $\frac{1}{6}=0{\cdot}15^{8}$ $\frac{1}{5}=0{\cdot}19^{0}$ $\frac{1}{4}=0{\cdot}23^{8}$ $\frac{1}{3}=0{\cdot}31^{7}$

$\frac{3}{8}=0{\cdot}35^{6}$ $\frac{1}{2}=0{\cdot}47^{5}$ $\frac{5}{8}=0{\cdot}59^{4}$ $\frac{3}{4}=0{\cdot}71^{3}$ $\frac{7}{8}=0{\cdot}83^{1}$ $\frac{2}{3}=0{\cdot}63^{3}$ $\frac{5}{6}=0{\cdot}79^{2}$

$\frac{3}{16}=0{\cdot}17^{8}$ $\frac{5}{16}=0{\cdot}29^{7}$ $\frac{7}{16}=0{\cdot}41^{6}$ $\frac{9}{16}=0{\cdot}53^{4}$ $\frac{11}{16}=0{\cdot}65^{3}$ $\frac{13}{16}=0{\cdot}77^{2}$ $\frac{15}{16}=0{\cdot}89^{1}$

(1.047=1£) **£0.95½**

1	0·95½	41	39·15½	81	77·35½	121	115·55½	161	153·75½
2	1·91	42	40·11	82	78·31	122	116·51	162	154·71
3	2·86½	43	41·06½	83	79·26½	123	117·46½	163	155·66½
4	3·82	44	42·02	84	80·22	124	118·42	164	156·62
5	4·77½	45	42·97½	85	81·17½	125	119·37½	165	157·57½
6	5·73	46	43·93	86	82·13	126	120·33	166	158·53
7	6·68½	47	44·88½	87	83·08½	127	121·28½	167	159·48½
8	7·64	48	45·84	88	84·04	128	122·24	168	160·44
9	8·59½	49	46·79½	89	84·99½	129	123·19½	169	161·39½
10	9·55	**50**	47·75	**90**	85·95	**130**	124·15	**170**	162·35
11	10·50½	51	48·70½	91	86·90½	131	125·10½	171	163·30½
12	11·46	52	49·66	92	87·86	132	126·06	172	164·26
13	12·41½	53	50·61½	93	88·81½	133	127·01½	173	165·21½
14	13·37	54	51·57	94	89·77	134	127·97	174	166·17
15	14·32½	55	52·52½	95	90·72½	135	128·92½	175	167·12½
16	15·28	56	53·48	96	91·68	136	129·88	176	168·08
17	16·23½	57	54·43½	97	92·63½	137	130·83½	177	169·03½
18	17·19	58	55·39	98	93·59	138	131·79	178	169·99
19	18·14½	59	56·34½	99	94·54½	139	132·74½	179	170·94½
20	19·10	**60**	57·30	**100**	95·50	**140**	133·70	**180**	171·90
21	20·05½	61	58·25½	101	96·45½	141	134·65½	181	172·85½
22	21·01	62	59·21	102	97·41	142	135·61	182	173·81
23	21·96½	63	60·16½	103	98·36½	143	136·56½	183	174·76½
24	22·92	64	61·12	104	99·32	144	137·52	184	175·72
25	23·87½	65	62·07½	105	100·27½	145	138·47½	185	176·67½
26	24·83	66	63·03	106	101·23	146	139·43	186	177·63
27	25·78½	67	63·98½	107	102·18½	147	140·38½	187	178·58½
28	26·74	68	64·94	108	103·14	148	141·34	188	179·54
29	27·69½	69	65·89½	109	104·09½	149	142·29½	189	180·49½
30	28·65	**70**	66·85	**110**	105·05	**150**	143·25	**190**	181·45
31	29·60½	71	67·80½	111	106·00½	151	144·20½	191	182·40½
32	30·56	72	68·76	112	106·96	152	145·16	192	183·36
33	31·51½	73	69·71½	113	107·91½	153	146·11½	193	184·31½
34	32·47	74	70·67	114	108·87	154	147·07	194	185·27
35	33·42½	75	71·62½	115	109·82½	155	148·02½	195	186·22½
36	34·38	76	72·58	116	110·78	156	148·98	196	187·18
37	35·33½	77	73·53½	117	111·73½	157	149·93½	197	188·13½
38	36·29	78	74·49	118	112·69	158	150·89	198	189·09
39	37·24½	79	75·44½	119	113·64½	159	151·84½	199	190·04½
40	38·20	**80**	76·40	**120**	114·60	**160**	152·80	**200**	191·00

$\frac{1}{16}=0{\cdot}06^{0}$ $\frac{1}{9}=0{\cdot}10^{6}$ $\frac{1}{8}=0{\cdot}11^{9}$ $\frac{1}{6}=0{\cdot}15^{9}$ $\frac{1}{5}=0{\cdot}19^{1}$ $\frac{1}{4}=0{\cdot}23^{9}$ $\frac{1}{3}=0{\cdot}31^{8}$

$\frac{3}{8}=0{\cdot}35^{8}$ $\frac{1}{2}=0{\cdot}47^{8}$ $\frac{5}{8}=0{\cdot}59^{7}$ $\frac{3}{4}=0{\cdot}71^{6}$ $\frac{7}{8}=0{\cdot}83^{6}$ $\frac{2}{3}=0{\cdot}63^{7}$ $\frac{5}{6}=0{\cdot}79^{6}$

$\frac{3}{16}=0{\cdot}17^{9}$ $\frac{5}{16}=0{\cdot}29^{8}$ $\frac{7}{16}=0{\cdot}41^{8}$ $\frac{9}{16}=0{\cdot}53^{7}$ $\frac{11}{16}=0{\cdot}65^{7}$ $\frac{13}{16}=0{\cdot}77^{6}$ $\frac{15}{16}=0{\cdot}89^{5}$

(1.042=1£) **£0.96**

1	0·96	41	39·36	81	77·76	121	116·16	161	154·56
2	1·92	42	40·32	82	78·72	122	117·12	162	155·52
3	2·88	43	41·28	83	79·68	123	118·08	163	156·48
4	3·84	44	42·24	84	80·64	124	119·04	164	157·44
5	4·80	45	43·20	85	81·60	125	120·00	165	158·40
6	5·76	46	44·16	86	82·56	126	120·96	166	159·36
7	6·72	47	45·12	87	83·52	127	121·92	167	160·32
8	7·68	48	46·08	88	84·48	128	122·88	168	161·28
9	8·64	49	47·04	89	85·44	129	123·84	169	162·24
10	9·60	**50**	48·00	**90**	86·40	**130**	124·80	**170**	163·20
11	10·56	51	48·96	91	87·36	131	125·76	171	164·16
12	11·52	52	49·92	92	88·32	132	126·72	172	165·12
13	12·48	53	50·88	93	89·28	133	127·68	173	166·08
14	13·44	54	51·84	94	90·24	134	128·64	174	167·04
15	14·40	55	52·80	95	91·20	135	129·60	175	168·00
16	15·36	56	53·76	96	92·16	136	130·56	176	168·96
17	16·32	57	54·72	97	93·12	137	131·52	177	169·92
18	17·28	58	55·68	98	94·08	138	132·48	178	170·88
19	18·24	59	56·64	99	95·04	139	133·44	179	171·84
20	19·20	**60**	57·60	**100**	96·00	**140**	134·40	**180**	172·80
21	20·16	61	58·56	101	96·96	141	135·36	181	173·76
22	21·12	62	59·52	102	97·92	142	136·32	182	174·72
23	22·08	63	60·48	103	98·88	143	137·28	183	175·68
24	23·04	64	61·44	104	99·84	144	138·24	184	176·64
25	24·00	65	62·40	105	100·80	145	139·20	185	177·60
26	24·96	66	63·36	106	101·76	146	140·16	186	178·56
27	25·92	67	64·32	107	102·72	147	141·12	187	179·52
28	26·88	68	65·28	108	103·68	148	142·08	188	180·48
29	27·84	69	66·24	109	104·64	149	143·04	189	181·44
30	28·80	**70**	67·20	**110**	105·60	**150**	144·00	**190**	182·40
31	29·76	71	68·16	111	106·56	151	144·96	191	183·36
32	30·72	72	69·12	112	107·52	152	145·92	192	184·32
33	31·68	73	70·08	113	108·48	153	146·88	193	185·28
34	32·64	74	71·04	114	109·44	154	147·84	194	186·24
35	33·60	75	72·00	115	110·40	155	148·80	195	187·20
36	34·56	76	72·96	116	111·36	156	149·76	196	188·16
37	35·52	77	73·92	117	112·32	157	150·72	197	189·12
38	36·48	78	74·88	118	113·28	158	151·68	198	190·08
39	37·44	79	75·84	119	114·24	159	152·64	199	191·04
40	38·40	**80**	76·80	**120**	115·20	**160**	153·60	**200**	192·00

$\frac{1}{16}$=0·06^{0} $\frac{1}{9}$=0·10^{7} $\frac{1}{8}$=0·12^{0} $\frac{1}{6}$=0·16^{0} $\frac{1}{5}$=0·19^{2} $\frac{1}{4}$=0·24^{0} $\frac{1}{3}$=0·32^{0}

$\frac{3}{8}$=0·36^{0} $\frac{1}{2}$=0·48^{0} $\frac{5}{8}$=0·60^{0} $\frac{3}{4}$=0·72^{0} $\frac{7}{8}$=0·84^{0} $\frac{2}{3}$=0·64^{0} $\frac{5}{6}$=0·80^{0}

$\frac{3}{16}$=0·18^{0} $\frac{5}{16}$=0·30^{0} $\frac{7}{16}$=0·42^{0} $\frac{9}{16}$=0·54^{0} $\frac{11}{16}$=0·66^{0} $\frac{13}{16}$=0·78^{0} $\frac{15}{16}$=0·90^{0}

1	0·96½	41	39·56½	81	78·16½	121	116·76½	161	155·36½
2	1·93	42	40·53	82	79·13	122	117·73	162	156·33
3	2·89½	43	41·49½	83	80·09½	123	118·69½	163	157·29½
4	3·86	44	42·46	84	81·06	124	119·66	164	158·26
5	4·82½	45	43·42½	85	82·02½	125	120·62½	165	159·22½
6	5·79	46	44·39	86	82·99	126	121·59	166	160·19
7	6·75½	47	45·35½	87	83·95½	127	122·55½	167	161·15½
8	7·72	48	46·32	88	84·92	128	123·52	168	162·12
9	8·68½	49	47·28½	89	85·88½	129	124·48½	169	163·08½
10	9·65	**50**	48·25	**90**	86·85	**130**	125·45	**170**	164·05
11	10·61½	51	49·21½	91	87·81½	131	126·41½	171	165·01½
12	11·58	52	50·18	92	88·78	132	127·38	172	165·98
13	12·54½	53	51·14½	93	89·74½	133	128·34½	173	166·94½
14	13·51	54	52·11	94	90·71	134	129·31	174	167·91
15	14·47½	55	53·07½	95	91·67½	135	130·27½	175	168·87½
16	15·44	56	54·04	96	92·64	136	131·24	176	169·84
17	16·40½	57	55·00½	97	93·60½	137	132·20½	177	170·80½
18	17·37	58	55·97	98	94·57	138	133·17	178	171·77
19	18·33½	59	56·93½	99	95·53½	139	134·13½	179	172·73½
20	19·30	**60**	57·90	**100**	96·50	**140**	135·10	**180**	173·70
21	20·26½	61	58·86½	101	97·46½	141	136·06½	181	174·66½
22	21·23	62	59·83	102	98·43	142	137·03	182	175·63
23	22·19½	63	60·79½	103	99·39½	143	137·99½	183	176·59½
24	23·16	64	61·76	104	100·36	144	138·96	184	177·56
25	24·12½	65	62·72½	105	101·32½	145	139·92½	185	178·52½
26	25·09	66	63·69	106	102·29	146	140·89	186	179·49
27	26·05½	67	64·65½	107	103·25½	147	141·85½	187	180·45½
28	27·02	68	65·62	108	104·22	148	142·82	188	181·42
29	27·98½	69	66·58½	109	105·18½	149	143·78½	189	182·38½
30	28·95	**70**	67·55	**110**	106·15	**150**	144·75	**190**	183·35
31	29·91½	71	68·51½	111	107·11½	151	145·71½	191	184·31½
32	30·88	72	69·48	112	108·08	152	146·68	192	185·28
33	31·84½	73	70·44½	113	109·04½	153	147·64½	193	186·24½
34	32·81	74	71·41	114	110·01	154	148·61	194	187·21
35	33·77½	75	72·37½	115	110·97½	155	149·57½	195	188·17½
36	34·74	76	73·34	116	111·94	156	150·54	196	189·14
37	35·70½	77	74·30½	117	112·90½	157	151·50½	197	190·10½
38	36·67	78	75·27	118	113·87	158	152·47	198	191·07
39	37·63½	79	76·23½	119	114·83½	159	153·43½	199	192·03½
40	38·60	**80**	77·20	**120**	115·80	**160**	154·40	**200**	193·00

$\frac{1}{16}$=0·06^{0} $\frac{1}{9}$=0·10^{7} $\frac{1}{8}$=0·12^{1} $\frac{1}{6}$=0·16^{1} $\frac{1}{5}$=0·19^{3} $\frac{1}{4}$=0·24^{1} $\frac{1}{3}$=0·32^{2}

$\frac{3}{8}$=0·36^{2} $\frac{1}{2}$=0·48^{3} $\frac{5}{8}$=0·60^{3} $\frac{3}{4}$=0·72^{4} $\frac{7}{8}$=0·84^{4} $\frac{2}{3}$=0·64^{3} $\frac{5}{6}$=0·80^{4}

$\frac{3}{16}$=0·18^{1} $\frac{5}{16}$=0·30^{2} $\frac{7}{16}$=0·42^{2} $\frac{9}{16}$=0·54^{3} $\frac{11}{16}$=0·66^{3} $\frac{13}{16}$=0·78^{4} $\frac{15}{16}$=0·90^{5}

(1.031=1£) **£0.97**

1	0·97	41	39·77	81	78·57	121	117·37	161	156·17
2	1·94	42	40·74	82	79·54	122	118·34	162	157·14
3	2·91	43	41·71	83	80·51	123	119·31	163	158·11
4	3·88	44	42·68	84	81·48	124	120·28	164	159·08
5	4·85	45	43·65	85	82·45	125	121·25	165	160·05
6	5·82	46	44·62	86	83·42	126	122·22	166	161·02
7	6·79	47	45·59	87	84·39	127	123·19	167	161·99
8	7·76	48	46·56	88	85·36	128	124·16	168	162·96
9	8·73	49	47·53	89	86·33	129	125·13	169	163·93
10	9·70	**50**	48·50	**90**	87·30	**130**	126·10	**170**	164·90
11	10·67	51	49·47	91	88·27	131	127·07	171	165·87
12	11·64	52	50·44	92	89·24	132	128·04	172	166·84
13	12·61	53	51·41	93	90·21	133	129·01	173	167·81
14	13·58	54	52·38	94	91·18	134	129·98	174	168·78
15	14·55	55	53·35	95	92·15	135	130·95	175	169·75
16	15·52	56	54·32	96	93·12	136	131·92	176	170·72
17	16·49	57	55·29	97	94·09	137	132·89	177	171·69
18	17·46	58	56·26	98	95·06	138	133·86	178	172·66
19	18·43	59	57·23	99	96·03	139	134·83	179	173·63
20	19·40	**60**	58·20	**100**	97·00	**140**	135·80	**180**	174·60
21	20·37	61	59·17	101	97·97	141	136·77	181	175·57
22	21·34	62	60·14	102	98·94	142	137·74	182	176·54
23	22·31	63	61·11	103	99·91	143	138·71	183	177·51
24	23·28	64	62·08	104	100·88	144	139·68	184	178·48
25	24·25	65	63·05	105	101·85	145	140·65	185	179·45
26	25·22	66	64·02	106	102·82	146	141·62	186	180·42
27	26·19	67	64·99	107	103·79	147	142·59	187	181·39
28	27·16	68	65·96	108	104·76	148	143·56	188	182·36
29	28·13	69	66·93	109	105·73	149	144·53	189	183·33
30	29·10	**70**	67·90	**110**	106·70	**150**	145·50	**190**	184·30
31	30·07	71	68·87	111	107·67	151	146·47	191	185·27
32	31·04	72	69·84	112	108·64	152	147·44	192	186·24
33	32·01	73	70·81	113	109·61	153	148·41	193	187·21
34	32·98	74	71·78	114	110·58	154	149·38	194	188·18
35	33·95	75	72·75	115	111·55	155	150·35	195	189·15
36	34·92	76	73·72	116	112·52	156	151·32	196	190·12
37	35·89	77	74·69	117	113·49	157	152·29	197	191·09
38	36·86	78	75·66	118	114·46	158	153·26	198	192·06
39	37·83	79	76·63	119	115·43	159	154·23	199	193·03
40	38·80	**80**	77·60	**120**	116·40	**160**	155·20	**200**	194·00

$\frac{1}{16}=0{\cdot}06^{1}$ $\frac{1}{9}=0{\cdot}10^{8}$ $\frac{1}{8}=0{\cdot}12^{1}$ $\frac{1}{6}=0{\cdot}16^{2}$ $\frac{1}{5}=0{\cdot}19^{4}$ $\frac{1}{4}=0{\cdot}24^{3}$ $\frac{1}{3}=0{\cdot}32^{3}$

$\frac{3}{8}=0{\cdot}36^{4}$ $\frac{1}{2}=0{\cdot}48^{5}$ $\frac{5}{8}=0{\cdot}60^{6}$ $\frac{3}{4}=0{\cdot}72^{8}$ $\frac{7}{8}=0{\cdot}84^{9}$ $\frac{2}{3}=0{\cdot}64^{7}$ $\frac{5}{6}=0{\cdot}80^{8}$

$\frac{3}{16}=0{\cdot}18^{2}$ $\frac{5}{16}=0{\cdot}30^{3}$ $\frac{7}{16}=0{\cdot}42^{4}$ $\frac{9}{16}=0{\cdot}54^{6}$ $\frac{11}{16}=0{\cdot}66^{7}$ $\frac{13}{16}=0{\cdot}78^{8}$ $\frac{15}{16}=0{\cdot}90^{9}$

(1.026=1£) **£0.97½**

1	0·97½	41	39·97½	81	78·97½	121	117·97½	161	156·97½
2	1·95	42	40·95	82	79·95	122	118·95	162	157·95
3	2·92½	43	41·92½	83	80·92½	123	119·92½	163	158·92½
4	3·90	44	42·90	84	81·90	124	120·90	164	159·90
5	4·87½	45	43·87½	85	82·87½	125	121·87½	165	160·87½
6	5·85	46	44·85	86	83·85	126	122·85	166	161·85
7	6·82½	47	45·82½	87	84·82½	127	123·82½	167	162·82½
8	7·80	48	46·80	88	85·80	128	124·80	168	163·80
9	8·77½	49	47·77½	89	86·77½	129	125·77½	169	164·77½
10	9·75	**50**	48·75	**90**	87·75	**130**	126·75	**170**	165·75
11	10·72½	51	49·72½	91	88·72½	131	127·72½	171	166·72½
12	11·70	52	50·70	92	89·70	132	128·70	172	167·70
13	12·67½	53	51·67½	93	90·67½	133	129·67½	173	168·67½
14	13·65	54	52·65	94	91·65	134	130·65	174	169·65
15	14·62½	55	53·62½	95	92·62½	135	131·62½	175	170·62½
16	15·60	56	54·60	96	93·60	136	132·60	176	171·60
17	16·57½	57	55·57½	97	94·57½	137	133·57½	177	172·57½
18	17·55	58	56·55	98	95·55	138	134·55	178	173·55
19	18·52½	59	57·52½	99	96·52½	139	135·52½	179	174·52½
20	19·50	**60**	58·50	**100**	97·50	**140**	136·50	**180**	175·50
21	20·47½	61	59·47½	101	98·47½	141	137·47½	181	176·47½
22	21·45	62	60·45	102	99·45	142	138·45	182	177·45
23	22·42½	63	61·42½	103	100·42½	143	139·42½	183	178·42½
24	23·40	64	62·40	104	101·40	144	140·40	184	179·40
25	24·37½	65	63·37½	105	102·37½	145	141·37½	185	180·37½
26	25·35	66	64·35	106	103·35	146	142·35	186	181·35
27	26·32½	67	65·32½	107	104·32½	147	143·32½	187	182·32½
28	27·30	68	66·30	108	105·30	148	144·30	188	183·30
29	28·27½	69	67·27½	109	106·27½	149	145·27½	189	184·27½
30	29·25	**70**	68·25	**110**	107·25	**150**	146·25	**190**	185·25
31	30·22½	71	69·22½	111	108·22½	151	147·22½	191	186·22½
32	31·20	72	70·20	112	109·20	152	148·20	192	187·20
33	32·17½	73	71·17½	113	110·17½	153	149·17½	193	188·17½
34	33·15	74	72·15	114	111·15	154	150·15	194	189·15
35	34·12½	75	73·12½	115	112·12½	155	151·12½	195	190·12½
36	35·10	76	74·10	116	113·10	156	152·10	196	191·10
37	36·07½	77	75·07½	117	114·07½	157	153·07½	197	192·07½
38	37·05	78	76·05	118	115·05	158	154·05	198	193·05
39	38·02½	79	77·02½	119	116·02½	159	155·02½	199	194·02½
40	39·00	**80**	78·00	**120**	117·00	**160**	156·00	**200**	195·00

$\frac{1}{16}$=0·06^{1} $\frac{1}{9}$=0·10^{8} $\frac{1}{8}$=0·12^{2} $\frac{1}{6}$=0·16^{3} $\frac{1}{5}$=0·19^{5} $\frac{1}{4}$=0·24^{4} $\frac{1}{3}$=0·32^{5}

$\frac{3}{8}$=0·36^{6} $\frac{1}{2}$=0·48^{8} $\frac{5}{8}$=0·60^{9} $\frac{3}{4}$=0·73^{1} $\frac{7}{8}$=0·85^{3} $\frac{2}{3}$=0·65^{0} $\frac{5}{6}$=0·81^{3}

$\frac{3}{16}$=0·18^{3} $\frac{5}{16}$=0·30^{5} $\frac{7}{16}$=0·42^{7} $\frac{9}{16}$=0·54^{8} $\frac{11}{16}$=0·67^{0} $\frac{13}{16}$=0·79^{2} $\frac{15}{16}$=0·91^{4}

(1.020=1£) **£0.98**

1	0·98	41	40·18	81	79·38	121	118·58	161	157·78
2	1·96	42	41·16	82	80·36	122	119·56	162	158·76
3	2·94	43	42·14	83	81·34	123	120·54	163	159·74
4	3·92	44	43·12	84	82·32	124	121·52	164	160·72
5	4·90	45	44·10	85	83·30	125	122·50	165	161·70
6	5·88	46	45·08	86	84·28	126	123·48	166	162·68
7	6·86	47	46·06	87	85·26	127	124·46	167	163·66
8	7·84	48	47·04	88	86·24	128	125·44	168	164·64
9	8·82	49	48·02	89	87·22	129	126·42	169	165·62
10	9·80	**50**	49·00	**90**	88·20	**130**	127·40	**170**	166·60
11	10·78	51	49·98	91	89·18	131	128·38	171	167·58
12	11·76	52	50·96	92	90·16	132	129·36	172	168·56
13	12·74	53	51·94	93	91·14	133	130·34	173	169·54
14	13·72	54	52·92	94	92·12	134	131·32	174	170·52
15	14·70	55	53·90	95	93·10	135	132·30	175	171·50
16	15·68	56	54·88	96	94·08	136	133·28	176	172·48
17	16·66	57	55·86	97	95·06	137	134·26	177	173·46
18	17·64	58	56·84	98	96·04	138	135·24	178	174·44
19	18·62	59	57·82	99	97·02	139	136·22	179	175·42
20	19·60	**60**	58·80	**100**	98·00	**140**	137·20	**180**	176·40
21	20·58	61	59·78	101	98·98	141	138·18	181	177·38
22	21·56	62	60·76	102	99·96	142	139·16	182	178·36
23	22·54	63	61·74	103	100·94	143	140·14	183	179·34
24	23·52	64	62·72	104	101·92	144	141·12	184	180·32
25	24·50	65	63·70	105	102·90	145	142·10	185	181·30
26	25·48	66	64·68	106	103·88	146	143·08	186	182·28
27	26·46	67	65·66	107	104·86	147	144·06	187	183·26
28	27·44	68	66·64	108	105·84	148	145·04	188	184·24
29	28·42	69	67·62	109	106·82	149	146·02	189	185·22
30	29·40	**70**	68·60	**110**	107·80	**150**	147·00	**190**	186·20
31	30·38	71	69·58	111	108·78	151	147·98	191	187·18
32	31·36	72	70·56	112	109·76	152	148·96	192	188·16
33	32·34	73	71·54	113	110·74	153	149·94	193	189·14
34	33·32	74	72·52	114	111·72	154	150·92	194	190·12
35	34·30	75	73·50	115	112·70	155	151·90	195	191·10
36	35·28	76	74·48	116	113·68	156	152·88	196	192·08
37	36·26	77	75·46	117	114·66	157	153·86	197	193·06
38	37·24	78	76·44	118	115·64	158	154·84	198	194·04
39	38·22	79	77·42	119	116·62	159	155·82	199	195·02
40	39·20	**80**	78·40	**120**	117·60	**160**	156·80	**200**	196·00

$\frac{1}{16}$=0·06^{1} $\frac{1}{9}$=0·10^{9} $\frac{1}{8}$=0·12^{3} $\frac{1}{6}$=0·16^{3} $\frac{1}{5}$=0·19^{6} $\frac{1}{4}$=0·24^{5} $\frac{1}{3}$=0·32^{7}

$\frac{3}{8}$=0·36^{8} $\frac{1}{2}$=0·49^{0} $\frac{5}{8}$=0·61^{3} $\frac{3}{4}$=0·73^{5} $\frac{7}{8}$=0·85^{8} $\frac{2}{3}$=0·65^{3} $\frac{5}{6}$=0·81^{7}

$\frac{3}{16}$=0·18^{4} $\frac{5}{16}$=0·30^{6} $\frac{7}{16}$=0·42^{9} $\frac{9}{16}$=0·55^{1} $\frac{11}{16}$=0·67^{4} $\frac{13}{16}$=0·79^{6} $\frac{15}{16}$=0·91^{9}

1	0·98½	41	40·38½	81	79·78½	121	119·18½	161	158·58½
2	1·97	42	41·37	82	80·77	122	120·17	162	159·57
3	2·95½	43	42·35½	83	81·75½	123	121·15½	163	160·55½
4	3·94	44	43·34	84	82·74	124	122·14	164	161·54
5	4·92½	45	44·32½	85	83·72½	125	123·12½	165	162·52½
6	5·91	46	45·31	86	84·71	126	124·11	166	163·51
7	6·89½	47	46·29½	87	85·69½	127	125·09½	167	164·49½
8	7·88	48	47·28	88	86·68	128	126·08	168	165·48
9	8·86½	49	48·26½	89	87·66½	129	127·06½	169	166·46½
10	9·85	**50**	49·25	**90**	88·65	**130**	128·05	**170**	167·45
11	10·83½	51	50·23½	91	89·63½	131	129·03½	171	168·43½
12	11·82	52	51·22	92	90·62	132	130·02	172	169·42
13	12·80½	53	52·20½	93	91·60½	133	131·00½	173	170·40½
14	13·79	54	53·19	94	92·59	134	131·99	174	171·39
15	14·77½	55	54·17½	95	93·57½	135	132·97½	175	172·37½
16	15·76	56	55·16	96	94·56	136	133·96	176	173·36
17	16·74½	57	56·14½	97	95·54½	137	134·94½	177	174·34½
18	17·73	58	57·13	98	96·53	138	135·93	178	175·33
19	18·71½	59	58·11½	99	97·51½	139	136·91½	179	176·31½
20	19·70	**60**	59·10	**100**	98·50	**140**	137·90	**180**	177·30
21	20·68½	61	60·08½	101	99·48½	141	138·88½	181	178·28½
22	21·67	62	61·07	102	100·47	142	139·87	182	179·27
23	22·65½	63	62·05½	103	101·45½	143	140·85½	183	180·25½
24	23·64	64	63·04	104	102·44	144	141·84	184	181·24
25	24·62½	65	64·02½	105	103·42½	145	142·82½	185	182·22½
26	25·61	66	65·01	106	104·41	146	143·81	186	183·21
27	26·59½	67	65·99½	107	105·39½	147	144·79½	187	184·19½
28	27·58	68	66·98	108	106·38	148	145·78	188	185·18
29	28·56½	69	67·96½	109	107·36½	149	146·76½	189	186·16½
30	29·55	**70**	68·95	**110**	108·35	**150**	147·75	**190**	187·15
31	30·53½	71	69·93½	111	109·33½	151	148·73½	191	188·13½
32	31·52	72	70·92	112	110·32	152	149·72	192	189·12
33	32·50½	73	71·90½	113	111·30½	153	150·70½	193	190·10½
34	33·49	74	72·89	114	112·29	154	151·69	194	191·09
35	34·47½	75	73·87½	115	113·27½	155	152·67½	195	192·07½
36	35·46	76	74·86	116	114·26	156	153·66	196	193·06
37	36·44½	77	75·84½	117	115·24½	157	154·64½	197	194·04½
38	37·43	78	76·83	118	116·23	158	155·63	198	195·03
39	38·41½	79	77·81½	119	117·21½	159	156·61½	199	196·01½
40	39·40	**80**	78·80	**120**	118·20	**160**	157·60	**200**	197·00

$\frac{1}{16}$=0·06^{2} $\frac{1}{9}$=0·10^{9} $\frac{1}{8}$=0·12^{3} $\frac{1}{6}$=0·16^{4} $\frac{1}{5}$=0·19^{7} $\frac{1}{4}$=0·24^{6} $\frac{1}{3}$=0·32^{8}

$\frac{3}{8}$=0·36^{9} $\frac{1}{2}$=0·49^{3} $\frac{5}{8}$=0·61^{6} $\frac{3}{4}$=0·73^{9} $\frac{7}{8}$=0·86^{2} $\frac{2}{3}$=0·65^{7} $\frac{5}{6}$=0·82^{1}

$\frac{3}{16}$=0·18^{5} $\frac{5}{16}$=0·30^{8} $\frac{7}{16}$=0·43^{1} $\frac{9}{16}$=0·55^{4} $\frac{11}{16}$=0·67^{7} $\frac{13}{16}$=0·80^{0} $\frac{15}{16}$=0·92^{3}

(1.010=1£) **£0.99**

1	0·99	41	40·59	81	80·19	121	119·79	161	159·39
2	1·98	42	41·58	82	81·18	122	120·78	162	160·38
3	2·97	43	42·57	83	82·17	123	121·77	163	161·37
4	3·96	44	43·56	84	83·16	124	122·76	164	162·36
5	4·95	45	44·55	85	84·15	125	123·75	165	163·35
6	5·94	46	45·54	86	85·14	126	124·74	166	164·34
7	6·93	47	46·53	87	86·13	127	125·73	167	165·33
8	7·92	48	47·52	88	87·12	128	126·72	168	166·32
9	8·91	49	48·51	89	88·11	129	127·71	169	167·31
10	9·90	**50**	49·50	**90**	89·10	**130**	128·70	**170**	168·30
11	10·89	51	50·49	91	90·09	131	129·69	171	169·29
12	11·88	52	51·48	92	91·08	132	130·68	172	170·28
13	12·87	53	52·47	93	92·07	133	131·67	173	171·27
14	13·86	54	53·46	94	93·06	134	132·66	174	172·26
15	14·85	55	54·45	95	94·05	135	133·65	175	173·25
16	15·84	56	55·44	96	95·04	136	134·64	176	174·24
17	16·83	57	56·43	97	96·03	137	135·63	177	175·23
18	17·82	58	57·42	98	97·02	138	136·62	178	176·22
19	18·81	59	58·41	99	98·01	139	137·61	179	177·21
20	19·80	**60**	59·40	**100**	99·00	**140**	138·60	**180**	178·20
21	20·79	61	60·39	101	99·99	141	139·59	181	179·19
22	21·78	62	61·38	102	100·98	142	140·58	182	180·18
23	22·77	63	62·37	103	101·97	143	141·57	183	181·17
24	23·76	64	63·36	104	102·96	144	142·56	184	182·16
25	24·75	65	64·35	105	103·95	145	143·55	185	183·15
26	25·74	66	65·34	106	104·94	146	144·54	186	184·14
27	26·73	67	66·33	107	105·93	147	145·53	187	185·13
28	27·72	68	67·32	108	106·92	148	146·52	188	186·12
29	28·71	69	68·31	109	107·91	149	147·51	189	187·11
30	29·70	**70**	69·30	**110**	108·90	**150**	148·50	**190**	188·10
31	30·69	71	70·29	111	109·89	151	149·49	191	189·09
32	31·68	72	71·28	112	110·88	152	150·48	192	190·08
33	32·67	73	72·27	113	111·87	153	151·47	193	191·07
34	33·66	74	73·26	114	112·86	154	152·46	194	192·06
35	34·65	75	74·25	115	113·85	155	153·45	195	193·05
36	35·64	76	75·24	116	114·84	156	154·44	196	194·04
37	36·63	77	76·23	117	115·83	157	155·43	197	195·03
38	37·62	78	77·22	118	116·82	158	156·42	198	196·02
39	38·61	79	78·21	119	117·81	159	157·41	199	197·01
40	39·60	**80**	79·20	**120**	118·80	**160**	158·40	**200**	198·00

$\frac{1}{16}=0{\cdot}06^{2}$ $\frac{1}{9}=0{\cdot}11^{0}$ $\frac{1}{8}=0{\cdot}12^{4}$ $\frac{1}{6}=0{\cdot}16^{5}$ $\frac{1}{5}=0{\cdot}19^{8}$ $\frac{1}{4}=0{\cdot}24^{8}$ $\frac{1}{3}=0{\cdot}33^{0}$

$\frac{3}{8}=0{\cdot}37^{1}$ $\frac{1}{2}=0{\cdot}49^{5}$ $\frac{5}{8}=0{\cdot}61^{9}$ $\frac{3}{4}=0{\cdot}74^{3}$ $\frac{7}{8}=0{\cdot}86^{6}$ $\frac{2}{3}=0{\cdot}66^{0}$ $\frac{5}{6}=0{\cdot}82^{5}$

$\frac{3}{16}=0{\cdot}18^{6}$ $\frac{5}{16}=0{\cdot}30^{9}$ $\frac{7}{16}=0{\cdot}43^{3}$ $\frac{9}{16}=0{\cdot}55^{7}$ $\frac{11}{16}=0{\cdot}68^{1}$ $\frac{13}{16}=0{\cdot}80^{4}$ $\frac{15}{16}=0{\cdot}92^{8}$

(1.005=1£) **£0.99½**

1	0·99½	41	40·79½	81	80·59½	121	120·39½	161	160·19½
2	1·99	42	41·79	82	81·59	122	121·39	162	161·19
3	2·98½	43	42·78½	83	82·58½	123	122·38½	163	162·18½
4	3·98	44	43·78	84	83·58	124	123·38	164	163·18
5	4·97½	45	44·77½	85	84·57½	125	124·37½	165	164·17½
6	5·97	46	45·77	86	85·57	126	125·37	166	165·17
7	6·96½	47	46·76½	87	86·56½	127	126·36½	167	166·16½
8	7·96	48	47·76	88	87·56	128	127·36	168	167·16
9	8·95½	49	48·75½	89	88·55½	129	128·35½	169	168·15½
10	9·95	**50**	49·75	**90**	89·55	**130**	129·35	**170**	169·15
11	10·94½	51	50·74½	91	90·54½	131	130·34½	171	170·14½
12	11·94	52	51·74	92	91·54	132	131·34	172	171·14
13	12·93½	53	52·73½	93	92·53½	133	132·33½	173	172·13½
14	13·93	54	53·73	94	93·53	134	133·33	174	173·13
15	14·92½	55	54·72½	95	94·52½	135	134·32½	175	174·12½
16	15·92	56	55·72	96	95·52	136	135·32	176	175·12
17	16·91½	57	56·71½	97	96·51½	137	136·31½	177	176·11½
18	17·91	58	57·71	98	97·51	138	137·31	178	177·11
19	18·90½	59	58·70½	99	98·50½	139	138·30½	179	178·10½
20	19·90	**60**	59·70	**100**	99·50	**140**	139·30	**180**	179·10
21	20·89½	61	60·69½	101	100·49½	141	140·29½	181	180·09½
22	21·89	62	61·69	102	101·49	142	141·29	182	181·09
23	22·88½	63	62·68½	103	102·48½	143	142·28½	183	182·08½
24	23·88	64	63·68	104	103·48	144	143·28	184	183·08
25	24·87½	65	64·67½	105	104·47½	145	144·27½	185	184·07½
26	25·87	66	65·67	106	105·47	146	145·27	186	185·07
27	26·86½	67	66·66½	107	106·46½	147	146·26½	187	186·06½
28	27·86	68	67·66	108	107·46	148	147·26	188	187·06
29	28·85½	69	68·65½	109	108·45½	149	148·25½	189	188·05½
30	29·85	**70**	69·65	**110**	109·45	**150**	149·25	**190**	189·05
31	30·84½	71	70·64½	111	110·44½	151	150·24½	191	190·04½
32	31·84	72	71·64	112	111·44	152	151·24	192	191·04
33	32·83½	73	72·63½	113	112·43½	153	152·23½	193	192·03½
34	33·83	74	73·63	114	113·43	154	153·23	194	193·03
35	34·82½	75	74·62½	115	114·42½	155	154·22½	195	194·02½
36	35·82	76	75·62	116	115·42	156	155·22	196	195·02
37	36·81½	77	76·61½	117	116·41½	157	156·21½	197	196·01½
38	37·81	78	77·61	118	117·41	158	157·21	198	197·01
39	38·80½	79	78·60½	119	118·40½	159	158·20½	199	198·00½
40	39·80	**80**	79·60	**120**	119·40	**160**	159·20	**200**	199·00

$\frac{1}{16}=0{\cdot}06^{2}$ $\frac{1}{9}=0{\cdot}11^{1}$ $\frac{1}{8}=0{\cdot}12^{4}$ $\frac{1}{6}=0{\cdot}16^{6}$ $\frac{1}{5}=0{\cdot}19^{9}$ $\frac{1}{4}=0{\cdot}24^{9}$ $\frac{1}{3}=0{\cdot}33^{2}$

$\frac{3}{8}=0{\cdot}37^{3}$ $\frac{1}{2}=0{\cdot}49^{8}$ $\frac{5}{8}=0{\cdot}62^{2}$ $\frac{3}{4}=0{\cdot}74^{6}$ $\frac{7}{8}=0{\cdot}87^{1}$ $\frac{2}{3}=0{\cdot}66^{3}$ $\frac{5}{6}=0{\cdot}82^{9}$

$\frac{3}{16}=0{\cdot}18^{7}$ $\frac{5}{16}=0{\cdot}31^{1}$ $\frac{7}{16}=0{\cdot}43^{5}$ $\frac{9}{16}=0{\cdot}56^{0}$ $\frac{11}{16}=0{\cdot}68^{4}$ $\frac{13}{16}=0{\cdot}80^{8}$ $\frac{15}{16}=0{\cdot}93^{3}$

1	1·00	41	41·00	81	81·00	121	121·00	161	161·00
2	2·00	42	42·00	82	82·00	122	122·00	162	162·00
3	3·00	43	43·00	83	83·00	123	123·00	163	163·00
4	4·00	44	44·00	84	84·00	124	124·00	164	164·00
5	5·00	45	45·00	85	85·00	125	125·00	165	165·00
6	6·00	46	46·00	86	86·00	126	126·00	166	166·00
7	7·00	47	47·00	87	87·00	127	127·00	167	167·00
8	8·00	48	48·00	88	88·00	128	128·00	168	168·00
9	9·00	49	49·00	89	89·00	129	129·00	169	169·00
10	10·00	**50**	50·00	**90**	90·00	**130**	130·00	**170**	170·00
11	11·00	51	51·00	91	91·00	131	131·00	171	171·00
12	12·00	52	52·00	92	92·00	132	132·00	172	172·00
13	13·00	53	53·00	93	93·00	133	133·00	173	173·00
14	14·00	54	54·00	94	94·00	134	134·00	174	174·00
15	15·00	55	55·00	95	95·00	135	135·00	175	175·00
16	16·00	56	56·00	96	96·00	136	136·00	176	176·00
17	17·00	57	57·00	97	97·00	137	137·00	177	177·00
18	18·00	58	58·00	98	98·00	138	138·00	178	178·00
19	19·00	59	59·00	99	99·00	139	139·00	179	179·00
20	20·00	**60**	60·00	**100**	100·00	**140**	140·00	**180**	180·00
21	21·00	61	61·00	101	101·00	141	141·00	181	181·00
22	22·00	62	62·00	102	102·00	142	142·00	182	182·00
23	23·00	63	63·00	103	103·00	143	143·00	183	183·00
24	24·00	64	64·00	104	104·00	144	144·00	184	184·00
25	25·00	65	65·00	105	105·00	145	145·00	185	185·00
26	26·00	66	66·00	106	106·00	146	146·00	186	186·00
27	27·00	67	67·00	107	107·00	147	147·00	187	187·00
28	28·00	68	68·00	108	108·00	148	148·00	188	188·00
29	29·00	69	69·00	109	109·00	149	149·00	189	189·00
30	30·00	**70**	70·00	**110**	110·00	**150**	150·00	**190**	190·00
31	31·00	71	71·00	111	111·00	151	151·00	191	191·00
32	32·00	72	72·00	112	112·00	152	152·00	192	192·00
33	33·00	73	73·00	113	113·00	153	153·00	193	193·00
34	34·00	74	74·00	114	114·00	154	154·00	194	194·00
35	35·00	75	75·00	115	115·00	155	155·00	195	195·00
36	36·00	76	76·00	116	116·00	156	156·00	196	196·00
37	37·00	77	77·00	117	117·00	157	157·00	197	197·00
38	38·00	78	78·00	118	118·00	158	158·00	198	198·00
39	39·00	79	79·00	119	119·00	159	159·00	199	199·00
40	40·00	**80**	80·00	**120**	120·00	**160**	160·00	**200**	200·00

$\frac{1}{16}=0{\cdot}06^3$ $\frac{1}{9}=0{\cdot}11^1$ $\frac{1}{8}=0{\cdot}12^5$ $\frac{1}{6}=0{\cdot}16^7$ $\frac{1}{5}=0{\cdot}20^0$ $\frac{1}{4}=0{\cdot}25^0$ $\frac{1}{3}=0{\cdot}33^3$

$\frac{3}{8}=0{\cdot}37^5$ $\frac{1}{2}=0{\cdot}50^0$ $\frac{5}{8}=0{\cdot}62^5$ $\frac{3}{4}=0{\cdot}75^0$ $\frac{7}{8}=0{\cdot}87^5$ $\frac{2}{3}=0{\cdot}66^7$ $\frac{5}{6}=0{\cdot}83^3$

$\frac{3}{16}=0{\cdot}18^8$ $\frac{5}{16}=0{\cdot}31^3$ $\frac{7}{16}=0{\cdot}43^8$ $\frac{9}{16}=0{\cdot}56^3$ $\frac{11}{16}=0{\cdot}68^8$ $\frac{13}{16}=0{\cdot}81^3$ $\frac{15}{16}=0{\cdot}93^8$

(0.995=1£) **£1.00½**

1	1·00½	41	41·20½	81	81·40½	121	121·60½	161	161·80½
2	2·01	42	42·21	82	82·41	122	122·61	162	162·81
3	3·01½	43	43·21½	83	83·41½	123	123·61½	163	163·81½
4	4·02	44	44·22	84	84·42	124	124·62	164	164·82
5	5·02½	45	45·22½	85	85·42½	125	125·62½	165	165·82½
6	6·03	46	46·23	86	86·43	126	126·63	166	166·83
7	7·03½	47	47·23½	87	87·43½	127	127·63½	167	167·83½
8	8·04	48	48·24	88	88·44	128	128·64	168	168·84
9	9·04½	49	49·24½	89	89·44½	129	129·64½	169	169·84½
10	10·05	**50**	50·25	**90**	90·45	**130**	130·65	**170**	170·85
11	11·05½	51	51·25½	91	91·45½	131	131·65½	171	171·85½
12	12·06	52	52·26	92	92·46	132	132·66	172	172·86
13	13·06½	53	53·26½	93	93·46½	133	133·66½	173	173·86½
14	14·07	54	54·27	94	94·47	134	134·67	174	174·87
15	15·07½	55	55·27½	95	95·47½	135	135·67½	175	175·87½
16	16·08	56	56·28	96	96·48	136	136·68	176	176·88
17	17·08½	57	57·28½	97	97·48½	137	137·68½	177	177·88½
18	18·09	58	58·29	98	98·49	138	138·69	178	178·89
19	19·09½	59	59·29½	99	99·49½	139	139·69½	179	179·89½
20	20·10	**60**	60·30	**100**	100·50	**140**	140·70	**180**	180·90
21	21·10½	61	61·30½	101	101·50½	141	141·70½	181	181·90½
22	22·11	62	62·31	102	102·51	142	142·71	182	182·91
23	23·11½	63	63·31½	103	103·51½	143	143·71½	183	183·91½
24	24·12	64	64·32	104	104·52	144	144·72	184	184·92
25	25·12½	65	65·32½	105	105·52½	145	145·72½	185	185·92½
26	26·13	66	66·33	106	106·53	146	146·73	186	186·93
27	27·13½	67	67·33½	107	107·53½	147	147·73½	187	187·93½
28	28·14	68	68·34	108	108·54	148	148·74	188	188·94
29	29·14½	69	69·34½	109	109·54½	149	149·74½	189	189·94½
30	30·15	**70**	70·35	**110**	110·55	**150**	150·75	**190**	190·95
31	31·15½	71	71·35½	111	111·55½	151	151·75½	191	191·95½
32	32·16	72	72·36	112	112·56	152	152·76	192	192·96
33	33·16½	73	73·36½	113	113·56½	153	153·76½	193	193·96½
34	34·17	74	74·37	114	114·57	154	154·77	194	194·97
35	35·17½	75	75·37½	115	115·57½	155	155·77½	195	195·97½
36	36·18	76	76·38	116	116·58	156	156·78	196	196·98
37	37·18½	77	77·38½	117	117·58½	157	157·78½	197	197·98½
38	38·19	78	78·39	118	118·59	158	158·79	198	198·99
39	39·19½	79	79·39½	119	119·59½	159	159·79½	199	199·99½
40	40·20	**80**	80·40	**120**	120·60	**160**	160·80	**200**	201·00

$\frac{1}{16}=0{\cdot}06^{3}$	$\frac{1}{9}=0{\cdot}11^{2}$	$\frac{1}{8}=0{\cdot}12^{6}$	$\frac{1}{6}=0{\cdot}16^{8}$	$\frac{1}{5}=0{\cdot}20^{1}$	$\frac{1}{4}=0{\cdot}25^{1}$	$\frac{1}{3}=0{\cdot}33^{5}$
$\frac{3}{8}=0{\cdot}37^{7}$	$\frac{1}{2}=0{\cdot}50^{3}$	$\frac{5}{8}=0{\cdot}62^{8}$	$\frac{3}{4}=0{\cdot}75^{4}$	$\frac{7}{8}=0{\cdot}87^{9}$	$\frac{2}{3}=0{\cdot}67^{0}$	$\frac{5}{6}=0{\cdot}83^{8}$
$\frac{3}{16}=0{\cdot}18^{8}$	$\frac{5}{16}=0{\cdot}31^{4}$	$\frac{7}{16}=0{\cdot}44^{0}$	$\frac{9}{16}=0{\cdot}56^{5}$	$\frac{11}{16}=0{\cdot}69^{1}$	$\frac{13}{16}=0{\cdot}81^{7}$	$\frac{15}{16}=0{\cdot}94^{2}$

(0.990=1£) £1.01

1	1·01	41	41·41	81	81·81	121	122·21	161	162·61
2	2·02	42	42·42	82	82·82	122	123·22	162	163·62
3	3·03	43	43·43	83	83·83	123	124·23	163	164·63
4	4·04	44	44·44	84	84·84	124	125·24	164	165·64
5	5·05	45	45·45	85	85·85	125	126·25	165	166·65
6	6·06	46	46·46	86	86·86	126	127·26	166	167·66
7	7·07	47	47·47	87	87·87	127	128·27	167	168·67
8	8·08	48	48·48	88	88·88	128	129·28	168	169·68
9	9·09	49	49·49	89	89·89	129	130·29	169	170·69
10	10·10	**50**	50·50	**90**	90·90	**130**	131·30	**170**	171·70
11	11·11	51	51·51	91	91·91	131	132·31	171	172·71
12	12·12	52	52·52	92	92·92	132	133·32	172	173·72
13	13·13	53	53·53	93	93·93	133	134·33	173	174·73
14	14·14	54	54·54	94	94·94	134	135·34	174	175·74
15	15·15	55	55·55	95	95·95	135	136·35	175	176·75
16	16·16	56	56·56	96	96·96	136	137·36	176	177·76
17	17·17	57	57·57	97	97·97	137	138·37	177	178·77
18	18·18	58	58·58	98	98·98	138	139·38	178	179·78
19	19·19	59	59·59	99	99·99	139	140·39	179	180·79
20	20·20	**60**	60·60	**100**	101·00	**140**	141·40	**180**	181·80
21	21·21	61	61·61	101	102·01	141	142·41	181	182·81
22	22·22	62	62·62	102	103·02	142	143·42	182	183·82
23	23·23	63	63·63	103	104·03	143	144·43	183	184·83
24	24·24	64	64·64	104	105·04	144	145·44	184	185·84
25	25·25	65	65·65	105	106·05	145	146·45	185	186·85
26	26·26	66	66·66	106	107·06	146	147·46	186	187·86
27	27·27	67	67·67	107	108·07	147	148·47	187	188·87
28	28·28	68	68·68	108	109·08	148	149·48	188	189·88
29	29·29	69	69·69	109	110·09	149	150·49	189	190·89
30	30·30	**70**	70·70	**110**	111·10	**150**	151·50	**190**	191·90
31	31·31	71	71·71	111	112·11	151	152·51	191	192·91
32	32·32	72	72·72	112	113·12	152	153·52	192	193·92
33	33·33	73	73·73	113	114·13	153	154·53	193	194·93
34	34·34	74	74·74	114	115·14	154	155·54	194	195·94
35	35·35	75	75·75	115	116·15	155	156·55	195	196·95
36	36·36	76	76·76	116	117·16	156	157·56	196	197·96
37	37·37	77	77·77	117	118·17	157	158·57	197	198·97
38	38·38	78	78·78	118	119·18	158	159·58	198	199·98
39	39·39	79	79·79	119	120·19	159	160·59	199	200·99
40	40·40	**80**	80·80	**120**	121·20	**160**	161·60	**200**	202·00

$\frac{1}{16}$=0·06^{3} $\frac{1}{9}$=0·11^{2} $\frac{1}{8}$=0·12^{6} $\frac{1}{6}$=0·16^{8} $\frac{1}{5}$=0·20^{2} $\frac{1}{4}$=0·25^{3} $\frac{1}{3}$=0·33^{7}

$\frac{3}{8}$=0·37^{9} $\frac{1}{2}$=0·50^{5} $\frac{5}{8}$=0·63^{1} $\frac{3}{4}$=0·75^{8} $\frac{7}{8}$=0·88^{4} $\frac{2}{3}$=0·67^{3} $\frac{5}{6}$=0·84^{2}

$\frac{3}{16}$=0·18^{9} $\frac{5}{16}$=0·31^{6} $\frac{7}{16}$=0·44^{2} $\frac{9}{16}$=0·56^{8} $\frac{11}{16}$=0·69^{4} $\frac{13}{16}$=0·82^{1} $\frac{15}{16}$=0·94^{7}

1	1·01½	41	41·61½	81	82·21½	121	122·81½	161	163·41½
2	2·03	42	42·63	82	83·23	122	123·83	162	164·43
3	3·04½	43	43·64½	83	84·24½	123	124·84½	163	165·44½
4	4·06	44	44·66	84	85·26	124	125·86	164	166·46
5	5·07½	45	45·67½	85	86·27½	125	126·87½	165	167·47½
6	6·09	46	46·69	86	87·29	126	127·89	166	168·49
7	7·10½	47	47·70½	87	88·30½	127	128·90½	167	169·50½
8	8·12	48	48·72	88	89·32	128	129·92	168	170·52
9	9·13½	49	49·73½	89	90·33½	129	130·93½	169	171·53½
10	10·15	**50**	50·75	**90**	91·35	**130**	131·95	**170**	172·55
11	11·16½	51	51·76½	91	92·36½	131	132·96½	171	173·56½
12	12·18	52	52·78	92	93·38	132	133·98	172	174·58
13	13·19½	53	53·79½	93	94·39½	133	134·99½	173	175·59½
14	14·21	54	54·81	94	95·41	134	136·01	174	176·61
15	15·22½	55	55·82½	95	96·42½	135	137·02½	175	177·62½
16	16·24	56	56·84	96	97·44	136	138·04	176	178·64
17	17·25½	57	57·85½	97	98·45½	137	139·05½	177	179·65½
18	18·27	58	58·87	98	99·47	138	140·07	178	180·67
19	19·28½	59	59·88½	99	100·48½	139	141·08½	179	181·68½
20	20·30	**60**	60·90	**100**	101·50	**140**	142·10	**180**	182·70
21	21·31½	61	61·91½	101	102·51½	141	143·11½	181	183·71½
22	22·33	62	62·93	102	103·53	142	144·13	182	184·73
23	23·34½	63	63·94½	103	104·54½	143	145·14½	183	185·74½
24	24·36	64	64·96	104	105·56	144	146·16	184	186·76
25	25·37½	65	65·97½	105	106·57½	145	147·17½	185	187·77½
26	26·39	66	66·99	106	107·59	146	148·19	186	188·79
27	27·40½	67	68·00½	107	108·60½	147	149·20½	187	189·80½
28	28·42	68	69·02	108	109·62	148	150·22	188	190·82
29	29·43½	69	70·03½	109	110·63½	149	151·23½	189	191·83½
30	30·45	**70**	71·05	**110**	111·65	**150**	152·25	**190**	192·85
31	31·46½	71	72·06½	111	112·66½	151	153·26½	191	193·86½
32	32·48	72	73·08	112	113·68	152	154·28	192	194·88
33	33·49½	73	74·09½	113	114·69½	153	155·29½	193	195·89½
34	34·51	74	75·11	114	115·71	154	156·31	194	196·91
35	35·52½	75	76·12½	115	116·72½	155	157·32½	195	197·92½
36	36·54	76	77·14	116	117·74	156	158·34	196	198·94
37	37·55½	77	78·15½	117	118·75½	157	159·35½	197	199·95½
38	38·57	78	79·17	118	119·77	158	160·37	198	200·97
39	39·58½	79	80·18½	119	120·78½	159	161·38½	199	201·98½
40	40·60	**80**	81·20	**120**	121·80	**160**	162·40	**200**	203·00

$\frac{1}{16}=0{\cdot}06^{3}$ $\frac{1}{9}=0{\cdot}11^{3}$ $\frac{1}{8}=0{\cdot}12^{7}$ $\frac{1}{6}=0{\cdot}16^{9}$ $\frac{1}{5}=0{\cdot}20^{3}$ $\frac{1}{4}=0{\cdot}25^{4}$ $\frac{1}{3}=0{\cdot}33^{8}$

$\frac{3}{8}=0{\cdot}38^{1}$ $\frac{1}{2}=0{\cdot}50^{8}$ $\frac{5}{8}=0{\cdot}63^{4}$ $\frac{3}{4}=0{\cdot}76^{1}$ $\frac{7}{8}=0{\cdot}88^{8}$ $\frac{2}{3}=0{\cdot}67^{7}$ $\frac{5}{6}=0{\cdot}$[illegible]4^{6}

$\frac{3}{16}=0{\cdot}19^{0}$ $\frac{5}{16}=0{\cdot}31^{7}$ $\frac{7}{16}=0{\cdot}44^{4}$ $\frac{9}{16}=0{\cdot}57^{1}$ $\frac{11}{16}=0{\cdot}69^{8}$ $\frac{13}{16}=0{\cdot}82^{5}$ $\frac{15}{16}=0{\cdot}95^{2}$

1	1·02	41	41·82	81	82·62	121	123·42	161	164·22
2	2·04	42	42·84	82	83·64	122	124·44	162	165·24
3	3·06	43	43·86	83	84·66	123	125·46	163	166·26
4	4·08	44	44·88	84	85·68	124	126·48	164	167·28
5	5·10	45	45·90	85	86·70	125	127·50	165	168·30
6	6·12	46	46·92	86	87·72	126	128·52	166	169·32
7	7·14	47	47·94	87	88·74	127	129·54	167	170·34
8	8·16	48	48·96	88	89·76	128	130·56	168	171·36
9	9·18	49	49·98	89	90·78	129	131·58	169	172·38
10	10·20	**50**	51·00	**90**	91·80	**130**	132·60	**170**	173·40
11	11·22	51	52·02	91	92·82	131	133·62	171	174·42
12	12·24	52	53·04	92	93·84	132	134·64	172	175·44
13	13·26	53	54·06	93	94·86	133	135·66	173	176·46
14	14·28	54	55·08	94	95·88	134	136·68	174	177·48
15	15·30	55	56·10	95	96·90	135	137·70	175	178·50
16	16·32	56	57·12	96	97·92	136	138·72	176	179·52
17	17·34	57	58·14	97	98·94	137	139·74	177	180·54
18	18·36	58	59·16	98	99·96	138	140·76	178	181·56
19	19·38	59	60·18	99	100·98	139	141·78	179	182·58
20	20·40	**60**	61·20	**100**	102·00	**140**	142·80	**180**	183·60
21	21·42	61	62·22	101	103·02	141	143·82	181	184·62
22	22·44	62	63·24	102	104·04	142	144·84	182	185·64
23	23·46	63	64·26	103	105·06	143	145·86	183	186·66
24	24·48	64	65·28	104	106·08	144	146·88	184	187·68
25	25·50	65	66·30	105	107·10	145	147·90	185	188·70
26	26·52	66	67·32	106	108·12	146	148·92	186	189·72
27	27·54	67	68·34	107	109·14	147	149·94	187	190·74
28	28·56	68	69·36	108	110·16	148	150·96	188	191·76
29	29·58	69	70·38	109	111·18	149	151·98	189	192·78
30	30·60	**70**	71·40	**110**	112·20	**150**	153·00	**190**	193·80
31	31·62	71	72·42	111	113·22	151	154·02	191	194·82
32	32·64	72	73·44	112	114·24	152	155·04	192	195·84
33	33·66	73	74·46	113	115·26	153	156·06	193	196·86
34	34·68	74	75·48	114	116·28	154	157·08	194	197·88
35	35·70	75	76·50	115	117·30	155	158·10	195	198·90
36	36·72	76	77·52	116	118·32	156	159·12	196	199·92
37	37·74	77	78·54	117	119·34	157	160·14	197	200·94
38	38·76	78	79·56	118	120·36	158	161·16	198	201·96
39	39·78	79	80·58	119	121·38	159	162·18	199	202·98
40	40·80	**80**	81·60	**120**	122·40	**160**	163·20	**200**	204·00

$\frac{1}{16}=0{\cdot}06^{4}$ $\frac{1}{9}=0{\cdot}11^{3}$ $\frac{1}{8}=0{\cdot}12^{8}$ $\frac{1}{6}=0{\cdot}17^{0}$ $\frac{1}{5}=0{\cdot}20^{4}$ $\frac{1}{4}=0{\cdot}25^{5}$ $\frac{1}{3}=0{\cdot}34^{0}$

$\frac{3}{8}=0{\cdot}38^{3}$ $\frac{1}{2}=0{\cdot}51^{0}$ $\frac{5}{8}=0{\cdot}63^{8}$ $\frac{3}{4}=0{\cdot}76^{5}$ $\frac{7}{8}=0{\cdot}89^{3}$ $\frac{2}{3}=0{\cdot}68^{0}$ $\frac{5}{6}=0{\cdot}85^{0}$

$\frac{3}{16}=0{\cdot}19^{1}$ $\frac{5}{16}=0{\cdot}31^{9}$ $\frac{7}{16}=0{\cdot}44^{6}$ $\frac{9}{16}=0{\cdot}57^{4}$ $\frac{11}{16}=0{\cdot}70^{1}$ $\frac{13}{16}=0{\cdot}82^{9}$ $\frac{15}{16}=0{\cdot}95^{6}$

1	1·02½	41	42·02½	81	83·02½	121	124·02½	161	165·02½
2	2·05	42	43·05	82	84·05	122	125·05	162	166·05
3	3·07½	43	44·07½	83	85·07½	123	126·07½	163	167·07½
4	4·10	44	45·10	84	86·10	124	127·10	164	168·10
5	5·12½	45	46·12½	85	87·12½	125	128·12½	165	169·12½
6	6·15	46	47·15	86	88·15	126	129·15	166	170·15
7	7·17½	47	48·17½	87	89·17½	127	130·17½	167	171·17½
8	8·20	48	49·20	88	90·20	128	131·20	168	172·20
9	9·22½	49	50·22½	89	91·22½	129	132·22½	169	173·22½
10	10·25	**50**	51·25	**90**	92·25	**130**	133·25	**170**	174·25
11	11·27½	51	52·27½	91	93·27½	131	134·27½	171	175·27½
12	12·30	52	53·30	92	94·30	132	135·30	172	176·30
13	13·32½	53	54·32½	93	95·32½	133	136·32½	173	177·32½
14	14·35	54	55·35	94	96·35	134	137·35	174	178·35
15	15·37½	55	56·37½	95	97·37½	135	138·37½	175	179·37½
16	16·40	56	57·40	96	98·40	136	139·40	176	180·40
17	17·42½	57	58·42½	97	99·42½	137	140·42½	177	181·42½
18	18·45	58	59·45	98	100·45	138	141·45	178	182·45
19	19·47½	59	60·47½	99	101·47½	139	142·47½	179	183·47½
20	20·50	**60**	61·50	**100**	102·50	**140**	143·50	**180**	184·50
21	21·52½	61	62·52½	101	103·52½	141	144·52½	181	185·52½
22	22·55	62	63·55	102	104·55	142	145·55	182	186·55
23	23·57½	63	64·57½	103	105·57½	143	146·57½	183	187·57½
24	24·60	64	65·60	104	106·60	144	147·60	184	188·60
25	25·62½	65	66·62½	105	107·62½	145	148·62½	185	189·62½
26	26·65	66	67·65	106	108·65	146	149·65	186	190·65
27	27·67½	67	68·67½	107	109·67½	147	150·67½	187	191·67½
28	28·70	68	69·70	108	110·70	148	151·70	188	192·70
29	29·72½	69	70·72½	109	111·72½	149	152·72½	189	193·72½
30	30·75	**70**	71·75	**110**	112·75	**150**	153·75	**190**	194·75
31	31·77½	71	72·77½	111	113·77½	151	154·77½	191	195·77½
32	32·80	72	73·80	112	114·80	152	155·80	192	196·80
33	33·82½	73	74·82½	113	115·82½	153	156·82½	193	197·82½
34	34·85	74	75·85	114	116·85	154	157·85	194	198·85
35	35·87½	75	76·87½	115	117·87½	155	158·87½	195	199·87½
36	36·90	76	77·90	116	118·90	156	159·90	196	200·90
37	37·92½	77	78·92½	117	119·92½	157	160·92½	197	201·92½
38	38·95	78	79·95	118	120·95	158	161·95	198	202·95
39	39·97½	79	80·97½	119	121·97½	159	162·97½	199	203·97½
40	41·00	**80**	82·00	**120**	123·00	**160**	164·00	**200**	205·00

$\frac{1}{16}$=0·06^{4} $\frac{1}{9}$=0·11^{4} $\frac{1}{8}$=0·12^{8} $\frac{1}{6}$=0·17^{1} $\frac{1}{5}$=0·20^{5} $\frac{1}{4}$=0·25^{6} $\frac{1}{3}$=0·34^{2}

$\frac{3}{8}$=0·38^{4} $\frac{1}{2}$=0·51^{3} $\frac{5}{8}$=0·64^{1} $\frac{3}{4}$=0·76^{9} $\frac{7}{8}$=0·89^{7} $\frac{2}{3}$=0·68^{3} $\frac{5}{6}$=0·85^{4}

$\frac{3}{16}$=0·19^{2} $\frac{5}{16}$=0·32^{0} $\frac{7}{16}$=0·44^{8} $\frac{9}{16}$=0·57^{7} $\frac{11}{16}$=0·70^{5} $\frac{13}{16}$=0·83^{3} $\frac{15}{16}$=0·96^{1}

(0.971=1£) **£1.03**

1	1·03	41	42·23	81	83·43	121	124·63	161	165·83
2	2·06	42	43·26	82	84·46	122	125·66	162	166·86
3	3·09	43	44·29	83	85·49	123	126·69	163	167·89
4	4·12	44	45·32	84	86·52	124	127·72	164	168·92
5	5·15	45	46·35	85	87·55	125	128·75	165	169·95
6	6·18	46	47·38	86	88·58	126	129·78	166	170·98
7	7·21	47	48·41	87	89·61	127	130·81	167	172·01
8	8·24	48	49·44	88	90·64	128	131·84	168	173·04
9	9·27	49	50·47	89	91·67	129	132·87	169	174·07
10	10·30	**50**	51·50	**90**	92·70	**130**	133·90	**170**	175·10
11	11·33	51	52·53	91	93·73	131	134·93	171	176·13
12	12·36	52	53·56	92	94·76	132	135·96	172	177·16
13	13·39	53	54·59	93	95·79	133	136·99	173	178·19
14	14·42	54	55·62	94	96·82	134	138·02	174	179·22
15	15·45	55	56·65	95	97·85	135	139·05	175	180·25
16	16·48	56	57·68	96	98·88	136	140·08	176	181·28
17	17·51	57	58·71	97	99·91	137	141·11	177	182·31
18	18·54	58	59·74	98	100·94	138	142·14	178	183·34
19	19·57	59	60·77	99	101·97	139	143·17	179	184·37
20	20·60	**60**	61·80	**100**	103·00	**140**	144·20	**180**	185·40
21	21·63	61	62·83	101	104·03	141	145·23	181	186·43
22	22·66	62	63·86	102	105·06	142	146·26	182	187·46
23	23·69	63	64·89	103	106·09	143	147·29	183	188·49
24	24·72	64	65·92	104	107·12	144	148·32	184	189·52
25	25·75	65	66·95	105	108·15	145	149·35	185	190·55
26	26·78	66	67·98	106	109·18	146	150·38	186	191·58
27	27·81	67	69·01	107	110·21	147	151·41	187	192·61
28	28·84	68	70·04	108	111·24	148	152·44	188	193·64
29	29·87	69	71·07	109	112·27	149	153·47	189	194·67
30	30·90	**70**	72·10	**110**	113·30	**150**	154·50	**190**	195·70
31	31·93	71	73·13	111	114·33	151	155·53	191	196·73
32	32·96	72	74·16	112	115·36	152	156·56	192	197·76
33	33·99	73	75·19	113	116·39	153	157·59	193	198·79
34	35·02	74	76·22	114	117·42	154	158·62	194	199·82
35	36·05	75	77·25	115	118·45	155	159·65	195	200·85
36	37·08	76	78·28	116	119·48	156	160·68	196	201·88
37	38·11	77	79·31	117	120·51	157	161·71	197	202·91
38	39·14	78	80·34	118	121·54	158	162·74	198	203·94
39	40·17	79	81·37	119	122·57	159	163·77	199	204·97
40	41·20	**80**	82·40	**120**	123·60	**160**	164·80	**200**	206·00

$\frac{1}{16}=0{\cdot}06^4$ $\frac{1}{9}=0{\cdot}11^4$ $\frac{1}{8}=0{\cdot}12^9$ $\frac{1}{6}=0{\cdot}17^2$ $\frac{1}{5}=0{\cdot}20^6$ $\frac{1}{4}=0{\cdot}25^8$ $\frac{1}{3}=0{\cdot}34^3$

$\frac{3}{8}=0{\cdot}38^6$ $\frac{1}{2}=0{\cdot}51^5$ $\frac{5}{8}=0{\cdot}64^4$ $\frac{3}{4}=0{\cdot}77^3$ $\frac{7}{8}=0{\cdot}90^1$ $\frac{2}{3}=0{\cdot}68^7$ $\frac{5}{6}=0{\cdot}85^8$

$\frac{3}{16}=0{\cdot}19^3$ $\frac{5}{16}=0{\cdot}32^2$ $\frac{7}{16}=0{\cdot}45^1$ $\frac{9}{16}=0{\cdot}57^9$ $\frac{11}{16}=0{\cdot}70^8$ $\frac{13}{16}=0{\cdot}83^7$ $\frac{15}{16}=0{\cdot}96^6$

(0.966=1£) **£1.03½**

1	1·03½	41	42·43½	81	83·83½	121	125·23½	161	166·63½
2	2·07	42	43·47	82	84·87	122	126·27	162	167·67
3	3·10½	43	44·50½	83	85·90½	123	127·30½	163	168·70½
4	4·14	44	45·54	84	86·94	124	128·34	164	169·74
5	5·17½	45	46·57½	85	87·97½	125	129·37½	165	170·77½
6	6·21	46	47·61	86	89·01	126	130·41	166	171·81
7	7·24½	47	48·64½	87	90·04½	127	131·44½	167	172·84½
8	8·28	48	49·68	88	91·08	128	132·48	168	173·88
9	9·31½	49	50·71½	89	92·11½	129	133·51½	169	174·91½
10	10·35	**50**	51·75	**90**	93·15	**130**	134·55	**170**	175·95
11	11·38½	51	52·78½	91	94·18½	131	135·58½	171	176·98½
12	12·42	52	53·82	92	95·22	132	136·62	172	178·02
13	13·45½	53	54·85½	93	96·25½	133	137·65½	173	179·05½
14	14·49	54	55·89	94	97·29	134	138·69	174	180·09
15	15·52½	55	56·92½	95	98·32½	135	139·72½	175	181·12½
16	16·56	56	57·96	96	99·36	136	140·76	176	182·16
17	17·59½	57	58·99½	97	100·39½	137	141·79½	177	183·19½
18	18·63	58	60·03	98	101·43	138	142·83	178	184·23
19	19·66½	59	61·06½	99	102·46½	139	143·86½	179	185·26½
20	20·70	**60**	62·10	**100**	103·50	**140**	144·90	**180**	186·30
21	21·73½	61	63·13½	101	104·53½	141	145·93½	181	187·33½
22	22·77	62	64·17	102	105·57	142	146·97	182	188·37
23	23·80½	63	65·20½	103	106·60½	143	148·00½	183	189·40½
24	24·84	64	66·24	104	107·64	144	149·04	184	190·44
25	25·87½	65	67·27½	105	108·67½	145	150·07½	185	191·47½
26	26·91	66	68·31	106	109·71	146	151·11	186	192·51
27	27·94½	67	69·34½	107	110·74½	147	152·14½	187	193·54½
28	28·98	68	70·38	108	111·78	148	153·18	188	194·58
29	30·01½	69	71·41½	109	112·81½	149	154·21½	189	195·61½
30	31·05	**70**	72·45	**110**	113·85	**150**	155·25	**190**	196·65
31	32·08½	71	73·48½	111	114·88½	151	156·28½	191	197·68½
32	33·12	72	74·52	112	115·92	152	157·32	192	198·72
33	34·15½	73	75·55½	113	116·95½	153	158·35½	193	199·75½
34	35·19	74	76·59	114	117·99	154	159·39	194	200·79
35	36·22½	75	77·62½	115	119·02½	155	160·42½	195	201·82½
36	37·26	76	78·66	116	120·06	156	161·46	196	202·86
37	38·29½	77	79·69½	117	121·09½	157	162·49½	197	203·89½
38	39·33	78	80·73	118	122·13	158	163·53	198	204·93
39	40·36½	79	81·76½	119	123·16½	159	164·56½	199	205·96½
40	41·40	**80**	82·80	**120**	124·20	**160**	165·60	**200**	207·00

$\frac{1}{16}=0{\cdot}06^{5}$ $\frac{1}{9}=0{\cdot}11^{5}$ $\frac{1}{8}=0{\cdot}12^{9}$ $\frac{1}{6}=0{\cdot}17^{3}$ $\frac{1}{5}=0{\cdot}20^{7}$ $\frac{1}{4}=0{\cdot}25^{9}$ $\frac{1}{3}=0{\cdot}34^{5}$

$\frac{3}{8}=0{\cdot}38^{8}$ $\frac{1}{2}=0{\cdot}51^{8}$ $\frac{5}{8}=0{\cdot}64^{7}$ $\frac{3}{4}=0{\cdot}77^{6}$ $\frac{7}{8}=0{\cdot}90^{6}$ $\frac{2}{3}=0{\cdot}69^{0}$ $\frac{5}{6}=0{\cdot}86^{3}$

$\frac{3}{16}=0{\cdot}19^{4}$ $\frac{5}{16}=0{\cdot}32^{3}$ $\frac{7}{16}=0{\cdot}45^{3}$ $\frac{9}{16}=0{\cdot}58^{2}$ $\frac{11}{16}=0{\cdot}71^{2}$ $\frac{13}{16}=0{\cdot}84^{1}$ $\frac{15}{16}=0{\cdot}97^{0}$

(0.962=1£) **£1.04**

1	1·04	41	42·64	81	84·24	121	125·84	161	167·44
2	2·08	42	43·68	82	85·28	122	126·88	162	168·48
3	3·12	43	44·72	83	86·32	123	127·92	163	169·52
4	4·16	44	45·76	84	87·36	124	128·96	164	170·56
5	5·20	45	46·80	85	88·40	125	130·00	165	171·60
6	6·24	46	47·84	86	89·44	126	131·04	166	172·64
7	7·28	47	48·88	87	90·48	127	132·08	167	173·68
8	8·32	48	49·92	88	91·52	128	133·12	168	174·72
9	9·36	49	50·96	89	92·56	129	134·16	169	175·76
10	10·40	**50**	52·00	**90**	93·60	**130**	135·20	**170**	176·80
11	11·44	51	53·04	91	94·64	131	136·24	171	177·84
12	12·48	52	54·08	92	95·68	132	137·28	172	178·88
13	13·52	53	55·12	93	96·72	133	138·32	173	179·92
14	14·56	54	56·16	94	97·76	134	139·36	174	180·96
15	15·60	55	57·20	95	98·80	135	140·40	175	182·00
16	16·64	56	58·24	96	99·84	136	141·44	176	183·04
17	17·68	57	59·28	97	100·88	137	142·48	177	184·08
18	18·72	58	60·32	98	101·92	138	143·52	178	185·12
19	19·76	59	61·36	99	102·96	139	144·56	179	186·16
20	20·80	**60**	62·40	**100**	104·00	**140**	145·60	**180**	187·20
21	21·84	61	63·44	101	105·04	141	146·64	181	188·24
22	22·88	62	64·48	102	106·08	142	147·68	182	189·28
23	23·92	63	65·52	103	107·12	143	148·72	183	190·32
24	24·96	64	66·56	104	108·16	144	149·76	184	191·36
25	26·00	65	67·60	105	109·20	145	150·80	185	192·40
26	27·04	66	68·64	106	110·24	146	151·84	186	193·44
27	28·08	67	69·68	107	111·28	147	152·88	187	194·48
28	29·12	68	70·72	108	112·32	148	153·92	188	195·52
29	30·16	69	71·76	109	113·36	149	154·96	189	196·56
30	31·20	**70**	72·80	**110**	114·40	**150**	156·00	**190**	197·60
31	32·24	71	73·84	111	115·44	151	157·04	191	198·64
32	33·28	72	74·88	112	116·48	152	158·08	192	199·68
33	34·32	73	75·92	113	117·52	153	159·12	193	200·72
34	35·36	74	76·96	114	118·56	154	160·16	194	201·76
35	36·40	75	78·00	115	119·60	155	161·20	195	202·80
36	37·44	76	79·04	116	120·64	156	162·24	196	203·84
37	38·48	77	80·08	117	121·68	157	163·28	197	204·88
38	39·52	78	81·12	118	122·72	158	164·32	198	205·92
39	40·56	79	82·16	119	123·76	159	165·36	199	206·96
40	41·60	**80**	83·20	**120**	124·80	**160**	166·40	**200**	208·00

$\frac{1}{16}=0{\cdot}06^{5}$ $\frac{1}{9}=0{\cdot}11^{6}$ $\frac{1}{8}=0{\cdot}13^{0}$ $\frac{1}{6}=0{\cdot}17^{3}$ $\frac{1}{5}=0{\cdot}20^{8}$ $\frac{1}{4}=0{\cdot}26^{0}$ $\frac{1}{3}=0{\cdot}34^{7}$

$\frac{3}{8}=0{\cdot}39^{0}$ $\frac{1}{2}=0{\cdot}52^{0}$ $\frac{5}{8}=0{\cdot}65^{0}$ $\frac{3}{4}=0{\cdot}78^{0}$ $\frac{7}{8}=0{\cdot}91^{0}$ $\frac{2}{3}=0{\cdot}69^{3}$ $\frac{5}{6}=0{\cdot}86^{7}$

$\frac{3}{16}=0{\cdot}19^{5}$ $\frac{5}{16}=0{\cdot}32^{5}$ $\frac{7}{16}=0{\cdot}45^{5}$ $\frac{9}{16}=0{\cdot}58^{5}$ $\frac{11}{16}=0{\cdot}71^{5}$ $\frac{13}{16}=0{\cdot}84^{5}$ $\frac{15}{16}=0{\cdot}97^{5}$

1	1·04½	41	42·84½	81	84·64½	121	126·44½	161	168·24½
2	2·09	42	43·89	82	85·69	122	127·49	162	169·29
3	3·13½	43	44·93½	83	86·73½	123	128·53½	163	170·33½
4	4·18	44	45·98	84	87·78	124	129·58	164	171·38
5	5·22½	45	47·02½	85	88·82½	125	130·62½	165	172·42½
6	6·27	46	48·07	86	89·87	126	131·67	166	173·47
7	7·31½	47	49·11½	87	90·91½	127	132·71½	167	174·51½
8	8·36	48	50·16	88	91·96	128	133·76	168	175·56
9	9·40½	49	51·20½	89	93·00½	129	134·80½	169	176·60½
10	10·45	**50**	52·25	**90**	94·05	**130**	135·85	**170**	177·65
11	11·49½	51	53·29½	91	95·09½	131	136·89½	171	178·69½
12	12·54	52	54·34	92	96·14	132	137·94	172	179·74
13	13·58½	53	55·38½	93	97·18½	133	138·98½	173	180·78½
14	14·63	54	56·43	94	98·23	134	140·03	174	181·83
15	15·67½	55	57·47½	95	99·27½	135	141·07½	175	182·87½
16	16·72	56	58·52	96	100·32	136	142·12	176	183·92
17	17·76½	57	59·56½	97	101·36½	137	143·16½	177	184·96½
18	18·81	58	60·61	98	102·41	138	144·21	178	186·01
19	19·85½	59	61·65½	99	103·45½	139	145·25½	179	187·05½
20	20·90	**60**	62·70	**100**	104·50	**140**	146·30	**180**	188·10
21	21·94½	61	63·74½	101	105·54½	141	147·34½	181	189·14½
22	22·99	62	64·79	102	106·59	142	148·39	182	190·19
23	24·03½	63	65·83½	103	107·63½	143	149·43½	183	191·23½
24	25·08	64	66·88	104	108·68	144	150·48	184	192·28
25	26·12½	65	67·92½	105	109·72½	145	151·52½	185	193·32½
26	27·17	66	68·97	106	110·77	146	152·57	186	194·37
27	28·21½	67	70·01½	107	111·81½	147	153·61½	187	195·41½
28	29·26	68	71·06	108	112·86	148	154·66	188	196·46
29	30·30½	69	72·10½	109	113·90½	149	155·70½	189	197·50½
30	31·35	**70**	73·15	**110**	114·95	**150**	156·75	**190**	198·55
31	32·39½	71	74·19½	111	115·99½	151	157·79½	191	199·59½
32	33·44	72	75·24	112	117·04	152	158·84	192	200·64
33	34·48½	73	76·28½	113	118·08½	153	159·88½	193	201·68½
34	35·53	74	77·33	114	119·13	154	160·93	194	202·73
35	36·57½	75	78·37½	115	120·17½	155	161·97½	195	203·77½
36	37·62	76	79·42	116	121·22	156	163·02	196	204·82
37	38·66½	77	80·46½	117	122·26½	157	164·06½	197	205·86½
38	39·71	78	81·51	118	123·31	158	165·11	198	206·91
39	40·75½	79	82·55½	119	124·35½	159	166·15½	199	207·95½
40	41·80	**80**	83·60	**120**	125·40	**160**	167·20	**200**	209·00

$\frac{1}{16}=0{\cdot}06^{5}$ $\frac{1}{9}=0{\cdot}11^{6}$ $\frac{1}{8}=0{\cdot}13^{1}$ $\frac{1}{6}=0{\cdot}17^{4}$ $\frac{1}{5}=0{\cdot}20^{9}$ $\frac{1}{4}=0{\cdot}26^{1}$ $\frac{1}{3}=0{\cdot}34^{8}$

$\frac{3}{8}=0{\cdot}39^{2}$ $\frac{1}{2}=0{\cdot}52^{3}$ $\frac{5}{8}=0{\cdot}65^{3}$ $\frac{3}{4}=0{\cdot}78^{4}$ $\frac{7}{8}=0{\cdot}91^{4}$ $\frac{2}{3}=0{\cdot}69^{7}$ $\frac{5}{6}=0{\cdot}87^{1}$

$\frac{3}{16}=0{\cdot}19^{6}$ $\frac{5}{16}=0{\cdot}32^{7}$ $\frac{7}{16}=0{\cdot}45^{7}$ $\frac{9}{16}=0{\cdot}58^{8}$ $\frac{11}{16}=0{\cdot}71^{8}$ $\frac{13}{16}=0{\cdot}84^{9}$ $\frac{15}{16}=0{\cdot}98^{0}$

(0.952=1£) **£1.05**

1	1·05	41	43·05	81	85·05	121	127·05	161	169·05
2	2·10	42	44·10	82	86·10	122	128·10	162	170·10
3	3·15	43	45·15	83	87·15	123	129·15	163	171·15
4	4·20	44	46·20	84	88·20	124	130·20	164	172·20
5	5·25	45	47·25	85	89·25	125	131·25	165	173·25
6	6·30	46	48·30	86	90·30	126	132·30	166	174·30
7	7·35	47	49 35	87	91·35	127	133·35	167	175·35
8	8·40	48	50·40	88	92·40	128	134·40	168	176·40
9	9·45	49	51·45	89	93·45	129	135·45	169	177·45
10	10·50	**50**	52·50	**90**	94·50	**130**	136·50	**170**	178·50
11	11·55	51	53·55	91	95·55	131	137·55	171	179·55
12	12·60	52	54·60	92	96·60	132	138·60	172	180·60
13	13·65	53	55·65	93	97·65	133	139·65	173	181·65
14	14·70	54	56·70	94	98·70	134	140·70	174	182·70
15	15·75	55	57·75	95	99·75	135	141·75	175	183·75
16	16·80	56	58·80	96	100·80	136	142·80	176	184·80
17	17·85	57	59·85	97	101·85	137	143·85	177	185·85
18	18·90	58	60·90	98	102·90	138	144·90	178	186·90
19	19·95	59	61·95	99	103·95	139	145·95	179	187·95
20	21·00	**60**	63·00	**100**	105·00	**140**	147·00	**180**	189·00
21	22·05	61	64·05	101	106·05	141	148·05	181	190·05
22	23·10	62	65·10	102	107·10	142	149·10	182	191·10
23	24·15	63	66·15	103	108·15	143	150·15	183	192·15
24	25·20	64	67·20	104	109·20	144	151·20	184	193·20
25	26·25	65	68·25	105	110·25	145	152·25	185	194·25
26	27·30	66	69·30	106	111·30	146	153·30	186	195·30
27	28·35	67	70·35	107	112·35	147	154·35	187	196·35
28	29·40	68	71·40	108	113·40	148	155·40	188	197·40
29	30·45	69	72·45	109	114·45	149	156·45	189	198·45
30	31·50	**70**	73·50	**110**	115·50	**150**	157·50	**190**	199·50
31	32·55	71	74·55	111	116·55	151	158·55	191	200·55
32	33·60	72	75·60	112	117·60	152	159·60	192	201·60
33	34·65	73	76·65	113	118·65	153	160·65	193	202·65
34	35·70	74	77·70	114	119·70	154	161·70	194	203·70
35	36·75	75	78·75	115	120·75	155	162·75	195	204·75
36	37·80	76	79·80	116	121·80	156	163·80	196	205·80
37	38·85	77	80·85	117	122·85	157	164·85	197	206·85
38	39·90	78	81·90	118	123·90	158	165·90	198	207·90
39	40·95	79	82·95	119	124·95	159	166·95	199	208·95
40	42·00	**80**	84·00	**120**	126·00	**160**	168·00	**200**	210·00

$\frac{1}{16}$=0·06^{6} $\frac{1}{9}$=0·11^{7} $\frac{1}{8}$=0·13^{1} $\frac{1}{6}$=0·17^{5} $\frac{1}{5}$=0·21^{0} $\frac{1}{4}$=0·26^{3} $\frac{1}{3}$=0·35^{0}

$\frac{3}{8}$=0·39^{4} $\frac{1}{2}$=0·52^{5} $\frac{5}{8}$=0·65^{6} $\frac{3}{4}$=0·78^{8} $\frac{7}{8}$=0·91^{9} $\frac{2}{3}$=0·70^{0} $\frac{5}{6}$=0·87^{5}

$\frac{3}{16}$=0·19^{7} $\frac{5}{16}$=0·32^{8} $\frac{7}{16}$=0·45^{9} $\frac{9}{16}$=0·59^{1} $\frac{11}{16}$=0·72^{2} $\frac{13}{16}$=0·85^{3} $\frac{15}{16}$=0·98^{4}

1	1·05½	41	43·25½	81	85·45½	121	127·65½	161	169·85½
2	2·11	42	44·31	82	86·51	122	128·71	162	170·91
3	3·16½	43	45·36½	83	87·56½	123	129·76½	163	171·96½
4	4·22	44	46·42	84	88·62	124	130·82	164	173·02
5	5·27½	45	47·47½	85	89·67½	125	131·87½	165	174·07½
6	6·33	46	48·53	86	90·73	126	132·93	166	175·13
7	7·38½	47	49·58½	87	91·78½	127	133·98½	167	176·18½
8	8·44	48	50·64	88	92·84	128	135·04	168	177·24
9	9·49½	49	51·69½	89	93·89½	129	136·09½	169	178·29½
10	10·55	**50**	52·75	**90**	94·95	**130**	137·15	**170**	179·35
11	11·60½	51	53·80½	91	96·00½	131	138·20½	171	180·40½
12	12·66	52	54·86	92	97·06	132	139·26	172	181·46
13	13·71½	53	55·91½	93	98·11½	133	140·31½	173	182·51½
14	14·77	54	56·97	94	99·17	134	141·37	174	183·57
15	15·82½	55	58·02½	95	100·22½	135	142·42½	175	184·62½
16	16·88	56	59·08	96	101·28	136	143·48	176	185·68
17	17·93½	57	60·13½	97	102·33½	137	144·53½	177	186·73½
18	18·99	58	61·19	98	103·39	138	145·59	178	187·79
19	20·04½	59	62·24½	99	104·44½	139	146·64½	179	188·84½
20	21·10	**60**	63·30	**100**	105·50	**140**	147·70	**180**	189·90
21	22·15½	61	64·35½	101	106·55½	141	148·75½	181	190·95½
22	23·21	62	65·41	102	107·61	142	149·81	182	192·01
23	24·26½	63	66·46½	103	108·66½	143	150·86½	183	193·06½
24	25·32	64	67·52	104	109·72	144	151·92	184	194·12
25	26·37½	65	68·57½	105	110·77½	145	152·97½	185	195·17½
26	27·43	66	69·63	106	111·83	146	154·03	186	196·23
27	28·48½	67	70·68½	107	112·88½	147	155·08½	187	197·28½
28	29·54	68	71·74	108	113·94	148	156·14	188	198·34
29	30·59½	69	72·79½	109	114·99½	149	157·19½	189	199·39½
30	31·65	**70**	73·85	**110**	116·05	**150**	158·25	**190**	200·45
31	32·70½	71	74·90½	111	117·10½	151	159·30½	191	201·50½
32	33·76	72	75·96	112	118·16	152	160·36	192	202·56
33	34·81½	73	77·01½	113	119·21½	153	161·41½	193	203·61½
34	35·87	74	78·07	114	120·27	154	162·47	194	204·67
35	36·92½	75	79·12½	115	121·32½	155	163·52½	195	205·72½
36	37·98	76	80·18	116	122·38	156	164·58	196	206·78
37	39·03½	77	81·23½	117	123·43½	157	165·63½	197	207·83½
38	40·09	78	82·29	118	124·49	158	166·69	198	208·89
39	41·14½	79	83·34½	119	125·54½	159	167·74½	199	209·94½
40	42·20	**80**	84·40	**120**	126·60	**160**	168·80	**200**	211·00

$\frac{1}{16}=0{\cdot}06^{6}$ $\frac{1}{9}=0{\cdot}11^{7}$ $\frac{1}{8}=0{\cdot}13^{2}$ $\frac{1}{6}=0{\cdot}17^{6}$ $\frac{1}{5}=0{\cdot}21^{1}$ $\frac{1}{4}=0{\cdot}26^{4}$ $\frac{1}{3}=0{\cdot}35^{2}$

$\frac{3}{8}=0{\cdot}39^{6}$ $\frac{1}{2}=0{\cdot}52^{8}$ $\frac{5}{8}=0{\cdot}65^{9}$ $\frac{3}{4}=0{\cdot}79^{1}$ $\frac{7}{8}=0{\cdot}92^{3}$ $\frac{2}{3}=0{\cdot}70^{3}$ $\frac{5}{6}=0{\cdot}87^{9}$

$\frac{3}{16}=0{\cdot}19^{8}$ $\frac{5}{16}=0{\cdot}33^{0}$ $\frac{7}{16}=0{\cdot}46^{2}$ $\frac{9}{16}=0{\cdot}59^{3}$ $\frac{11}{16}=0{\cdot}72^{5}$ $\frac{13}{16}=0{\cdot}85^{7}$ $\frac{15}{16}=0{\cdot}98^{9}$

1 British Thermal Unit=1·055 kilojoules

(0.943=1£) **£1.06**

1	1·06	41	43·46	81	85·86	121	128·26	161	170·66
2	2·12	42	44·52	82	86·92	122	129·32	162	171·72
3	3·18	43	45·58	83	87·98	123	130·38	163	172·78
4	4·24	44	46·64	84	89·04	124	131·44	164	173·84
5	5·30	45	47·70	85	90·10	125	132·50	165	174·90
6	6·36	46	48·76	86	91·16	126	133·56	166	175·96
7	7·42	47	49·82	87	92·22	127	134·62	167	177·02
8	8·48	48	50·88	88	93·28	128	135·68	168	178·08
9	9·54	49	51·94	89	94·34	129	136·74	169	179·14
10	10·60	**50**	53·00	**90**	95·40	**130**	137·80	**170**	180·20
11	11·66	51	54·06	91	96·46	131	138·86	171	181·26
12	12·72	52	55·12	92	97·52	132	139·92	172	182·32
13	13·78	53	56·18	93	98·58	133	140·98	173	183·38
14	14·84	54	57·24	94	99·64	134	142·04	174	184·44
15	15·90	55	58·30	95	100·70	135	143·10	175	185·50
16	16·96	56	59·36	96	101·76	136	144·16	176	186·56
17	18·02	57	60·42	97	102·82	137	145·22	177	187·62
18	19·08	58	61·48	98	103·88	138	146·28	178	188·68
19	20·14	59	62·54	99	104·94	139	147·34	179	189·74
20	21·20	**60**	63·60	**100**	106·00	**140**	148·40	**180**	190·80
21	22·26	61	64·66	101	107·06	141	149·46	181	191·86
22	23·32	62	65·72	102	108·12	142	150·52	182	192·92
23	24·38	63	66·78	103	109·18	143	151·58	183	193·98
24	25·44	64	67·84	104	110·24	144	152·64	184	195·04
25	26·50	65	68·90	105	111·30	145	153·70	185	196·10
26	27·56	66	69·96	106	112·36	146	154·76	186	197·16
27	28·62	67	71·02	107	113·42	147	155·82	187	198·22
28	29·68	68	72·08	108	114·48	148	156·88	188	199·28
29	30·74	69	73·14	109	115·54	149	157·94	189	200·34
30	31·80	**70**	74·20	**110**	116·60	**150**	159·00	**190**	201·40
31	32·86	71	75·26	111	117·66	151	160·06	191	202·46
32	33·92	72	76·32	112	118·72	152	161·12	192	203·52
33	34·98	73	77·38	113	119·78	153	162·18	193	204·58
34	36·04	74	78·44	114	120·84	154	163·24	194	205·64
35	37·10	75	79·50	115	121·90	155	164·30	195	206·70
36	38·16	76	80·56	116	122·96	156	165·36	196	207·76
37	39·22	77	81·62	117	124·02	157	166·42	197	208·82
38	40·28	78	82·68	118	125·08	158	167·48	198	209·88
39	41·34	79	83·74	119	126·14	159	168·54	199	210·94
40	42·40	**80**	84·80	**120**	127·20	**160**	169·60	**200**	212·00

$\frac{1}{16}=0{\cdot}06^{6}$ $\frac{1}{9}=0{\cdot}11^{8}$ $\frac{1}{8}=0{\cdot}13^{3}$ $\frac{1}{6}=0{\cdot}17^{7}$ $\frac{1}{5}=0{\cdot}21^{2}$ $\frac{1}{4}=0{\cdot}26^{5}$ $\frac{1}{3}=0{\cdot}35^{3}$

$\frac{3}{8}=0{\cdot}39^{8}$ $\frac{1}{2}=0{\cdot}53^{0}$ $\frac{5}{8}=0{\cdot}66^{3}$ $\frac{3}{4}=0{\cdot}79^{5}$ $\frac{7}{8}=0{\cdot}92^{8}$ $\frac{2}{3}=0{\cdot}70^{7}$ $\frac{5}{6}=0{\cdot}88^{3}$

$\frac{3}{16}=0{\cdot}19^{9}$ $\frac{5}{16}=0{\cdot}33^{1}$ $\frac{7}{16}=0{\cdot}46^{4}$ $\frac{9}{16}=0{\cdot}59^{6}$ $\frac{11}{16}=0{\cdot}72^{9}$ $\frac{13}{16}=0{\cdot}86^{1}$ $\frac{15}{16}=0{\cdot}99^{4}$

(0.939=1£) **£1.06½**

1	1·06½	41	43·66½	81	86·26½	121	128·86½	161	171·46½
2	2·13	42	44·73	82	87·33	122	129·93	162	172·53
3	3·19½	43	45·79½	83	88·39½	123	130·99½	163	173·59½
4	4·26	44	46·86	84	89·46	124	132·06	164	174·66
5	5·32½	45	47·92½	85	90·52½	125	133·12½	165	175·72½
6	6·39	46	48·99	86	91·59	126	134·19	166	176·79
7	7·45½	47	50·05½	87	92·65½	127	135·25½	167	177·85½
8	8·52	48	51·12	88	93·72	128	136·32	168	178·92
9	9·58½	49	52·18½	89	94·78½	129	137·38½	169	179·98½
10	10·65	**50**	53·25	**90**	95·85	**130**	138·45	**170**	181·05
11	11·71½	51	54·31½	91	96·91½	131	139·51½	171	182·11½
12	12·78	52	55·38	92	97·98	132	140·58	172	183·18
13	13·84½	53	56·44½	93	99·04½	133	141·64½	173	184·24½
14	14·91	54	57·51	94	100·11	134	142·71	174	185·31
15	15·97½	55	58·57½	95	101·17½	135	143·77½	175	186·37½
16	17·04	56	59·64	96	102·24	136	144·84	176	187·44
17	18·10½	57	60·70½	97	103·30½	137	145·90½	177	188·50½
18	19·17	58	61·77	98	104·37	138	146·97	178	189·57
19	20·23½	59	62·83½	99	105·43½	139	148·03½	179	190·63½
20	21·30	**60**	63·90	**100**	106·50	**140**	149·10	**180**	191·70
21	22·36½	61	64·96½	101	107·56½	141	150·16½	181	192·76½
22	23·43	62	66·03	102	108·63	142	151·23	182	193·83
23	24·49½	63	67·09½	103	109·69½	143	152·29½	183	194·89½
24	25·56	64	68·16	104	110·76	144	153·36	184	195·96
25	26·62½	65	69·22½	105	111·82½	145	154·42½	185	197·02½
26	27·69	66	70·29	106	112·89	146	155·49	186	198·09
27	28·75½	67	71·35½	107	113·95½	147	156·55½	187	199·15½
28	29·82	68	72·42	108	115·02	148	157·62	188	200·22
29	30·88½	69	73·48½	109	116·08½	149	158·68½	189	201·28½
30	31·95	**70**	74·55	**110**	117·15	**150**	159·75	**190**	202·35
31	33·01½	71	75·61½	111	118·21½	151	160·81½	191	203·41½
32	34·08	72	76·68	112	119·28	152	161·88	192	204·48
33	35·14½	73	77·74½	113	120·34½	153	162·94½	193	205·54½
34	36·21	74	78·81	114	121·41	154	164·01	194	206·61
35	37·27½	75	79·87½	115	122·47½	155	165·07½	195	207·67½
36	38·34	76	80·94	116	123·54	156	166·14	196	208·74
37	39·40½	77	82·00½	117	124·60½	157	167·20½	197	209·80½
38	40·47	78	83·07	118	125·67	158	168·27	198	210·87
39	41·53½	79	84·13½	119	126·73½	159	169·33½	199	211·93½
40	42·60	**80**	85·20	**120**	127·80	**160**	170·40	**200**	213·00

$\frac{1}{16}=0{\cdot}06^{7}$	$\frac{1}{9}=0{\cdot}11^{8}$	$\frac{1}{8}=0{\cdot}13^{3}$	$\frac{1}{6}=0{\cdot}17^{8}$	$\frac{1}{5}=0{\cdot}21^{3}$	$\frac{1}{4}=0{\cdot}26^{6}$	$\frac{1}{3}=0{\cdot}35^{5}$
$\frac{3}{8}=0{\cdot}39^{9}$	$\frac{1}{2}=0{\cdot}53^{3}$	$\frac{5}{8}=0{\cdot}66^{6}$	$\frac{3}{4}=0{\cdot}79^{9}$	$\frac{7}{8}=0{\cdot}93^{2}$	$\frac{2}{3}=0{\cdot}71^{0}$	$\frac{5}{6}=0{\cdot}88^{8}$
$\frac{3}{16}=0{\cdot}20^{0}$	$\frac{5}{16}=0{\cdot}33^{3}$	$\frac{7}{16}=0{\cdot}46^{6}$	$\frac{9}{16}=0{\cdot}59^{9}$	$\frac{11}{16}=0{\cdot}73^{2}$	$\frac{13}{16}=0{\cdot}86^{5}$	$\frac{15}{16}=0{\cdot}99^{8}$

(0.935=1£) **£1.07**

1	1·07	41	43·87	81	86·67	121	129·47	161	172·27
2	2·14	42	44·94	82	87·74	122	130·54	162	173·34
3	3·21	43	46·01	83	88·81	123	131·61	163	174·41
4	4·28	44	47·08	84	89·88	124	132·68	164	175·48
5	5·35	45	48·15	85	90·95	125	133·75	165	176·55
6	6·42	46	49·22	86	92·02	126	134·82	166	177·62
7	7·49	47	50·29	87	93·09	127	135·89	167	178·69
8	8·56	48	51·36	88	94·16	128	136·96	168	179·76
9	9·63	49	52·43	89	95·23	129	138·03	169	180·83
10	10·70	**50**	53·50	**90**	96·30	**130**	139·10	**170**	181·90
11	11·77	51	54·57	91	97·37	131	140·17	171	182·97
12	12·84	52	55·64	92	98·44	132	141·24	172	184·04
13	13·91	53	56·71	93	99·51	133	142·31	173	185·11
14	14·98	54	57·78	94	100·58	134	143·38	174	186·18
15	16·05	55	58·85	95	101·65	135	144·45	175	187·25
16	17·12	56	59·92	96	102·72	136	145·52	176	188·32
17	18·19	57	60·99	97	103·79	137	146·59	177	189·39
18	19·26	58	62·06	98	104·86	138	147·66	178	190·46
19	20·33	59	63·13	99	105·93	139	148·73	179	191·53
20	21·40	**60**	64·20	**100**	107·00	**140**	149·80	**180**	192·60
21	22·47	61	65·27	101	108·07	141	150·87	181	193·67
22	23·54	62	66·34	102	109·14	142	151·94	182	194·74
23	24·61	63	67·41	103	110·21	143	153·01	183	195·81
24	25·68	64	68·48	104	111·28	144	154·08	184	196·88
25	26·75	65	69·55	105	112·35	145	155·15	185	197·95
26	27·82	66	70·62	106	113·42	146	156·22	186	199·02
27	28·89	67	71·69	107	114·49	147	157·29	187	200·09
28	29·96	68	72·76	108	115·56	148	158·36	188	201·16
29	31·03	69	73·83	109	116·63	149	159·43	189	202·23
30	32·10	**70**	74·90	**110**	117·70	**150**	160·50	**190**	203·30
31	33·17	71	75·97	111	118·77	151	161·57	191	204·37
32	34·24	72	77·04	112	119·84	152	162·64	192	205·44
33	35·31	73	78·11	113	120·91	153	163·71	193	206·51
34	36·38	74	79·18	114	121·98	154	164·78	194	207·58
35	37·45	75	80·25	115	123·05	155	165·85	195	208·65
36	38·52	76	81·32	116	124·12	156	166·92	196	209·72
37	39·59	77	82·39	117	125·19	157	167·99	197	210·79
38	40·66	78	83·46	118	126·26	158	169·06	198	211·86
39	41·73	79	84·53	119	127·33	159	170·13	199	212·93
40	42·80	**80**	85·60	**120**	128·40	**160**	171·20	**200**	214·00

$\frac{1}{16}=0{\cdot}06^{7}$ $\frac{1}{9}=0{\cdot}11^{9}$ $\frac{1}{8}=0{\cdot}13^{4}$ $\frac{1}{6}=0{\cdot}17^{8}$ $\frac{1}{5}=0{\cdot}21^{4}$ $\frac{1}{4}=0{\cdot}26^{8}$ $\frac{1}{3}=0{\cdot}35^{7}$

$\frac{3}{8}=0{\cdot}40^{1}$ $\frac{1}{2}=0{\cdot}53^{5}$ $\frac{5}{8}=0{\cdot}66^{9}$ $\frac{3}{4}=0{\cdot}80^{3}$ $\frac{7}{8}=0{\cdot}93^{6}$ $\frac{2}{3}=0{\cdot}71^{3}$ $\frac{5}{6}=0{\cdot}89^{2}$

$\frac{3}{16}=0{\cdot}20^{1}$ $\frac{5}{16}=0{\cdot}33^{4}$ $\frac{7}{16}=0{\cdot}46^{8}$ $\frac{9}{16}=0{\cdot}60^{2}$ $\frac{11}{16}=0{\cdot}73^{6}$ $\frac{13}{16}=0{\cdot}86^{9}$ $\frac{15}{16}=1{\cdot}00^{3}$

(0.930=1£) **£1.07½**

1	1·07½	41	44·07½	81	87·07½	121	130·07½	161	173·07½
2	2·15	42	45·15	82	88·15	122	131·15	162	174·15
3	3·22½	43	46·22½	83	89·22½	123	132·22½	163	175·22½
4	4·30	44	47·30	84	90·30	124	133·30	164	176·30
5	5·37½	45	48·37½	85	91·37½	125	134·37½	165	177·37½
6	6·45	46	49·45	86	92·45	126	135·45	166	178·45
7	7·52½	47	50·52½	87	93·52½	127	136·52½	167	179·52½
8	8·60	48	51·60	88	94·60	128	137·60	168	180·60
9	9·67½	49	52·67½	89	95·67½	129	138·67½	169	181·67½
10	10·75	**50**	53·75	**90**	96·75	**130**	139·75	**170**	182·75
11	11·82½	51	54·82½	91	97·82½	131	140·82½	171	183·82½
12	12·90	52	55·90	92	98·90	132	141·90	172	184·90
13	13·97½	53	56·97½	93	99·97½	133	142·97½	173	185·97½
14	15·05	54	58·05	94	101·05	134	144·05	174	187·05
15	16·12½	55	59·12½	95	102·12½	135	145·12½	175	188·12½
16	17·20	56	60·20	96	103·20	136	146·20	176	189·20
17	18·27½	57	61·27½	97	104·27½	137	147·27½	177	190·27½
18	19·35	58	62·35	98	105·35	138	148·35	178	191·35
19	20·42½	59	63·42½	99	106·42½	139	149·42½	179	192·42½
20	21·50	**60**	64·50	**100**	107·50	**140**	150·50	**180**	193·50
21	22·57½	61	65·57½	101	108·57½	141	151·57½	181	194·57½
22	23·65	62	66·65	102	109·65	142	152·65	182	195·65
23	24·72½	63	67·72½	103	110·72½	143	153·72½	183	196·72½
24	25·80	64	68·80	104	111·80	144	154·80	184	197·80
25	26·87½	65	69·87½	105	112·87½	145	155·87½	185	198·87½
26	27·95	66	70·95	106	113·95	146	156·95	186	199·95
27	29·02½	67	72·02½	107	115·02½	147	158·02½	187	201·02½
28	30·10	68	73·10	108	116·10	148	159·10	188	202·10
29	31·17½	69	74·17½	109	117·17½	149	160·17½	189	203·17½
30	32·25	**70**	75·25	**110**	118·25	**150**	161·25	**190**	204·25
31	33·32½	71	76·32½	111	119·32½	151	162·32½	191	205·32½
32	34·40	72	77·40	112	120·40	152	163·40	192	206·40
33	35·47½	73	78·47½	113	121·47½	153	164·47½	193	207·47½
34	36·55	74	79·55	114	122·55	154	165·55	194	208·55
35	37·62½	75	80·62½	115	123·62½	155	166·62½	195	209·62½
36	38·70	76	81·70	116	124·70	156	167·70	196	210·70
37	39·77½	77	82·77½	117	125·77½	157	168·77½	197	211·77½
38	40·85	78	83·85	118	126·85	158	169·85	198	212·85
39	41·92½	79	84·92½	119	127·92½	159	170·92½	199	213·92½
40	43·00	**80**	86·00	**120**	129·00	**160**	172·00	**200**	215·00

$\frac{1}{16}=0{\cdot}06^{7}$ $\frac{1}{9}=0{\cdot}11^{9}$ $\frac{1}{8}=0{\cdot}13^{4}$ $\frac{1}{6}=0{\cdot}17^{9}$ $\frac{1}{5}=0{\cdot}21^{5}$ $\frac{1}{4}=0{\cdot}26^{9}$ $\frac{1}{3}=0{\cdot}35^{8}$

$\frac{3}{8}=0{\cdot}40^{3}$ $\frac{1}{2}=0{\cdot}53^{8}$ $\frac{5}{8}=0{\cdot}67^{2}$ $\frac{3}{4}=0{\cdot}80^{6}$ $\frac{7}{8}=0{\cdot}94^{1}$ $\frac{2}{3}=0{\cdot}71^{7}$ $\frac{5}{6}=0{\cdot}89^{6}$

$\frac{3}{16}=0{\cdot}20^{2}$ $\frac{5}{16}=0{\cdot}33^{6}$ $\frac{7}{16}=0{\cdot}47^{0}$ $\frac{9}{16}=0{\cdot}60^{5}$ $\frac{11}{16}=0{\cdot}73^{9}$ $\frac{13}{16}=0{\cdot}87^{3}$ $\frac{15}{16}=1{\cdot}00^{8}$

(0.926=1£) **£1.08**

1	1·08	41	44·28	81	87·48	121	130·68	161	173·88
2	2·16	42	45·36	82	88·56	122	131·76	162	174·96
3	3·24	43	46·44	83	89·64	123	132·84	163	176·04
4	4·32	44	47·52	84	90·72	124	133·92	164	177·12
5	5·40	45	48·60	85	91·80	125	135·00	165	178·20
6	6·48	46	49·68	86	92·88	126	136·08	166	179·28
7	7·56	47	50·76	87	93·96	127	137·16	167	180·36
8	8·64	48	51·84	88	95·04	128	138·24	168	181·44
9	9·72	49	52·92	89	96·12	129	139·32	169	182·52
10	10·80	**50**	54·00	**90**	97·20	**130**	140·40	**170**	183·60
11	11·88	51	55·08	91	98·28	131	141·48	171	184·68
12	12·96	52	56·16	92	99·36	132	142·56	172	185·76
13	14·04	53	57·24	93	100·44	133	143·64	173	186·84
14	15·12	54	58·32	94	101·52	134	144·72	174	187·92
15	16·20	55	59·40	95	102·60	135	145·80	175	189·00
16	17·28	56	60·48	96	103·68	136	146·88	176	190·08
17	18·36	57	61·56	97	104·76	137	147·96	177	191·16
18	19·44	58	62·64	98	105·84	138	149·04	178	192·24
19	20·52	59	63·72	99	106·92	139	150·12	179	193·32
20	21·60	**60**	64·80	**100**	108·00	**140**	151·20	**180**	194·40
21	22·68	61	65·88	101	109·08	141	152·28	181	195·48
22	23·76	62	66·96	102	110·16	142	153·36	182	196·56
23	24·84	63	68·04	103	111·24	143	154·44	183	197·64
24	25·92	64	69·12	104	112·32	144	155·52	184	198·72
25	27·00	65	70·20	105	113·40	145	156·60	185	199·80
26	28·08	66	71·28	106	114·48	146	157·68	186	200·88
27	29·16	67	72·36	107	115·56	147	158·76	187	201·96
28	30·24	68	73·44	108	116·64	148	159·84	188	203·04
29	31·32	69	74·52	109	117·72	149	160·92	189	204·12
30	32·40	**70**	75·60	**110**	118·80	**150**	162·00	**190**	205·20
31	33·48	71	76·68	111	119·88	151	163·08	191	206·28
32	34·56	72	77·76	112	120·96	152	164·16	192	207·36
33	35·64	73	78·84	113	122·04	153	165·24	193	208·44
34	36·72	74	79·92	114	123·12	154	166·32	194	209·52
35	37·80	75	81·00	115	124·20	155	167·40	195	210·60
36	38·88	76	82·08	116	125·28	156	168·48	196	211·68
37	39·96	77	83·16	117	126·36	157	169·56	197	212·76
38	41·04	78	84·24	118	127·44	158	170·64	198	213·84
39	42·12	79	85·32	119	128·52	159	171·72	199	214·92
40	43·20	**80**	86·40	**120**	129·60	**160**	172·80	**200**	216·00

$\frac{1}{16}=0{\cdot}06^{8}$ $\frac{1}{9}=0{\cdot}12^{0}$ $\frac{1}{8}=0{\cdot}13^{5}$ $\frac{1}{6}=0{\cdot}18^{0}$ $\frac{1}{5}=0{\cdot}21^{6}$ $\frac{1}{4}=0{\cdot}27^{0}$ $\frac{1}{3}=0{\cdot}36^{0}$

$\frac{3}{8}=0{\cdot}40^{5}$ $\frac{1}{2}=0{\cdot}54^{0}$ $\frac{5}{8}=0{\cdot}67^{5}$ $\frac{3}{4}=0{\cdot}81^{0}$ $\frac{7}{8}=0{\cdot}94^{5}$ $\frac{2}{3}=0{\cdot}72^{0}$ $\frac{5}{6}=0{\cdot}90^{0}$

$\frac{3}{16}=0{\cdot}20^{3}$ $\frac{5}{16}=0{\cdot}33^{8}$ $\frac{7}{16}=0{\cdot}47^{3}$ $\frac{9}{16}=0{\cdot}60^{8}$ $\frac{11}{16}=0{\cdot}74^{3}$ $\frac{13}{16}=0{\cdot}87^{8}$ $\frac{15}{16}=1{\cdot}01^{3}$

(0.922=1£) **£1.08½**

1	1·08½	41	44·48½	81	87·88½	121	131·28½	161	174·68½
2	2·17	42	45·57	82	88·97	122	132·37	162	175·77
3	3·25½	43	46·65½	83	90·05½	123	133·45½	163	176·85½
4	4·34	44	47·74	84	91·14	124	134·54	164	177·94
5	5·42½	45	48·82½	85	92·22½	125	135·62½	165	179·02½
6	6·51	46	49·91	86	93·31	126	136·71	166	180·11
7	7·59½	47	50·99½	87	94·39½	127	137·79½	167	181·19½
8	8·68	48	52·08	88	95·48	128	138·88	168	182·28
9	9·76½	49	53·16½	89	96·56½	129	139·96½	169	183·36½
10	10·85	**50**	54·25	**90**	97·65	**130**	141·05	**170**	184·45
11	11·93½	51	55·33½	91	98·73½	131	142·13½	171	185·53½
12	13·02	52	56·42	92	99·82	132	143·22	172	186·62
13	14·10½	53	57·50½	93	100·90½	133	144·30½	173	187·70½
14	15·19	54	58·59	94	101·99	134	145·39	174	188·79
15	16·27½	55	59·67½	95	103·07½	135	146·47½	175	189·87½
16	17·36	56	60·76	96	104·16	136	147·56	176	190·96
17	18·44½	57	61·84½	97	105·24½	137	148·64½	177	192·04½
18	19·53	58	62·93	98	106·33	138	149·73	178	193·13
19	20·61½	59	64·01½	99	107·41½	139	150·81½	179	194·21½
20	21·70	**60**	65·10	**100**	108·50	**140**	151·90	**180**	195·30
21	22·78½	61	66·18½	101	109·58½	141	152·98½	181	196·38½
22	23·87	62	67·27	102	110·67	142	154·07	182	197·47
23	24·95½	63	68·35½	103	111·75½	143	155·15½	183	198·55½
24	26·04	64	69·44	104	112·84	144	156·24	184	199·64
25	27·12½	65	70·52½	105	113·92½	145	157·32½	185	200·72½
26	28·21	66	71·61	106	115·01	146	158·41	186	201·81
27	29·29½	67	72·69½	107	116·09½	147	159·49½	187	202·89½
28	30·38	68	73·78	108	117·18	148	160·58	188	203·98
29	31·46½	69	74·86½	109	118·26½	149	161·66½	189	205·06½
30	32·55	**70**	75·95	**110**	119·35	**150**	162·75	**190**	206·15
31	33·63½	71	77·03½	111	120·43½	151	163·83½	191	207·23½
32	34·72	72	78·12	112	121·52	152	164·92	192	208·32
33	35·80½	73	79·20½	113	122·60½	153	166·00½	193	209·40½
34	36·89	74	80·29	114	123·69	154	167·09	194	210·49
35	37·97½	75	81·37½	115	124·77½	155	168·17½	195	211·57½
36	39·06	76	82·46	116	125·86	156	169·26	196	212·66
37	40·14½	77	83·54½	117	126·94½	157	170·34½	197	213·74½
38	41·23	78	84·63	118	128·03	158	171·43	198	214·83
39	42·31½	79	85·71½	119	129·11½	159	172·51½	199	215·91½
40	43·40	**80**	86·80	**120**	130·20	**160**	173·60	**200**	217·00

$\frac{1}{16}=0{\cdot}06^{8}$ $\frac{1}{9}=0{\cdot}12^{1}$ $\frac{1}{8}=0{\cdot}13^{6}$ $\frac{1}{6}=0{\cdot}18^{1}$ $\frac{1}{5}=0{\cdot}21^{7}$ $\frac{1}{4}=0{\cdot}27^{1}$ $\frac{1}{3}=0{\cdot}36^{2}$

$\frac{3}{8}=0{\cdot}40^{7}$ $\frac{1}{2}=0{\cdot}54^{3}$ $\frac{5}{8}=0{\cdot}67^{8}$ $\frac{3}{4}=0{\cdot}81^{4}$ $\frac{7}{8}=0{\cdot}94^{9}$ $\frac{2}{3}=0{\cdot}72^{3}$ $\frac{5}{6}=0{\cdot}90^{4}$

(0.917=1£) **£1.09**

1	1·09	41	44·69	81	88·29	121	131·89	161	175·49
2	2·18	42	45·78	82	89·38	122	132·98	162	176·58
3	3·27	43	46·87	83	90·47	123	134·07	163	177·67
4	4·36	44	47·96	84	91·56	124	135·16	164	178·76
5	5·45	45	49·05	85	92·65	125	136·25	165	179·85
6	6·54	46	50·14	86	93·74	126	137·34	166	180·94
7	7·63	47	51·23	87	94·83	127	138·43	167	182·03
8	8·72	48	52·32	88	95·92	128	139·52	168	183·12
9	9·81	49	53·41	89	97·01	129	140·61	169	184·21
10	10·90	**50**	54·50	**90**	98·10	**130**	141·70	**170**	185·30
11	11·99	51	55·59	91	99·19	131	142·79	171	186·39
12	13·08	52	56·68	92	100·28	132	143·88	172	187·48
13	14·17	53	57·77	93	101·37	133	144·97	173	188·57
14	15·26	54	58·86	94	102·46	134	146·06	174	189·66
15	16·35	55	59·95	95	103·55	135	147·15	175	190·75
16	17·44	56	61·04	96	104·64	136	148·24	176	191·84
17	18·53	57	62·13	97	105·73	137	149·33	177	192·93
18	19·62	58	63·22	98	106·82	138	150·42	178	194·02
19	20·71	59	64·31	99	107·91	139	151·51	179	195·11
20	21·80	**60**	65·40	**100**	109·00	**140**	152·60	**180**	196·20
21	22·89	61	66·49	101	110·09	141	153·69	181	197·29
22	23·98	62	67·58	102	111·18	142	154·78	182	198·38
23	25·07	63	68·67	103	112·27	143	155·87	183	199·47
24	26·16	64	69·76	104	113·36	144	156·96	184	200·56
25	27·25	65	70·85	105	114·45	145	158·05	185	201·65
26	28·34	66	71·94	106	115·54	146	159·14	186	202·74
27	29·43	67	73·03	107	116·63	147	160·23	187	203·83
28	30·52	68	74·12	108	117·72	148	161·32	188	204·92
29	31·61	69	75·21	109	118·81	149	162·41	189	206·01
30	32·70	**70**	76·30	**110**	119·90	**150**	163·50	**190**	207·10
31	33·79	71	77·39	111	120·99	151	164·59	191	208·19
32	34·88	72	78·48	112	122·08	152	165·68	192	209·28
33	35·97	73	79·57	113	123·17	153	166·77	193	210·37
34	37·06	74	80·66	114	124·26	154	167·86	194	211·46
35	38·15	75	81·75	115	125·35	155	168·95	195	212·55
36	39·24	76	82·84	116	126·44	156	170·04	196	213·64
37	40·33	77	83·93	117	127·53	157	171·13	197	214·73
38	41·42	78	85·02	118	128·62	158	172·22	198	215·82
39	42·51	79	86·11	119	129·71	159	173·31	199	216·91
40	43·60	**80**	87·20	**120**	130·80	**160**	174·40	**200**	218·00

$\frac{1}{16}=0{\cdot}06^{8}$ $\frac{1}{9}=0{\cdot}12^{1}$ $\frac{1}{8}=0{\cdot}13^{6}$ $\frac{1}{6}=0{\cdot}18^{2}$ $\frac{1}{5}=0{\cdot}21^{8}$ $\frac{1}{4}=0{\cdot}27^{3}$ $\frac{1}{3}=0{\cdot}36^{3}$

$\frac{3}{8}=0{\cdot}40^{9}$ $\frac{1}{2}=0{\cdot}54^{5}$ $\frac{5}{8}=0{\cdot}68^{1}$ $\frac{3}{4}=0{\cdot}81^{8}$ $\frac{7}{8}=0{\cdot}95^{4}$ $\frac{2}{3}=0{\cdot}72^{7}$ $\frac{5}{6}=0{\cdot}90^{8}$

$\frac{3}{16}=0{\cdot}20^{4}$ $\frac{5}{16}=0{\cdot}34^{1}$ $\frac{7}{16}=0{\cdot}47^{7}$ $\frac{9}{16}=0{\cdot}61^{3}$ $\frac{11}{16}=0{\cdot}74^{9}$ $\frac{13}{16}=0{\cdot}88^{6}$ $\frac{15}{16}=1{\cdot}02^{2}$

(0.913=1£) **£1.09½**

1	1·09½	41	44·89½	81	88·69½	121	132·49½	161	176·29½
2	2·19	42	45·99	82	89·79	122	133·59	162	177·39
3	3·28½	43	47·08½	83	90·88½	123	134·68½	163	178·48½
4	4·38	44	48·18	84	91·98	124	135·78	164	179·58
5	5·47½	45	49·27½	85	93·07½	125	136·87½	165	180·67½
6	6·57	46	50·37	86	94·17	126	137·97	166	181·77
7	7·66½	47	51·46½	87	95·26½	127	139·06½	167	182·86½
8	8·76	48	52·56	88	96·36	128	140·16	168	183·96
9	9·85½	49	53·65½	89	97·45½	129	141·25½	169	185·05½
10	10·95	**50**	54·75	**90**	98·55	**130**	142·35	**170**	186·15
11	12·04½	51	55·84½	91	99·64½	131	143·44½	171	187·24½
12	13·14	52	56·94	92	100·74	132	144·54	172	188·34
13	14·23½	53	58·03½	93	101·83½	133	145·63½	173	189·43½
14	15·33	54	59·13	94	102·93	134	146·73	174	190·53
15	16·42½	55	60·22½	95	104·02½	135	147·82½	175	191·62½
16	17·52	56	61·32	96	105·12	136	148·92	176	192·72
17	18·61½	57	62·41½	97	106·21½	137	150·01½	177	193·81½
18	19·71	58	63·51	98	107·31	138	151·11	178	194·91
19	20·80½	59	64·60½	99	108·40½	139	152·20½	179	196·00½
20	21·90	**60**	65·70	**100**	109·50	**140**	153·30	**180**	197·10
21	22·99½	61	66·79½	101	110·59½	141	154·39½	181	198·19½
22	24·09	62	67·89	102	111·69	142	155·49	182	199·29
23	25·18½	63	68·98½	103	112·78½	143	156·58½	183	200·38½
24	26·28	64	70·08	104	113·88	144	157·68	184	201·48
25	27·37½	65	71·17½	105	114·97½	145	158·77½	185	202·57½
26	28·47	66	72·27	106	116·07	146	159·87	186	203·67
27	29·56½	67	73·36½	107	117·16½	147	160·96½	187	204·76½
28	30·66	68	74·46	108	118·26	148	162·06	188	205·86
29	31·75½	69	75·55½	109	119·35½	149	163·15½	189	206·95½
30	32·85	**70**	76·65	**110**	120·45	**150**	164·25	**190**	208·05
31	33·94½	71	77·74½	111	121·54½	151	165·34½	191	209·14½
32	35·04	72	78·84	112	122·64	152	166·44	192	210·24
33	36·13½	73	79·93½	113	123·73½	153	167·53½	193	211·33½
34	37·23	74	81·03	114	124·83	154	168·63	194	212·43
35	38·32½	75	82·12½	115	125·92½	155	169·72½	195	213·52½
36	39·42	76	83·22	116	127·02	156	170·82	196	214·62
37	40·51½	77	84·31½	117	128·11½	157	171·91½	197	215·71½
38	41·61	78	85·41	118	129·21	158	173·01	198	216·81
39	42·70½	79	86·50½	119	130·30½	159	174·10½	199	217·90½
40	43·80	**80**	87·60	**120**	131·40	**160**	175·20	**200**	219·00

$\frac{1}{16}$=0·06^{8} $\frac{1}{9}$=0·12^{2} $\frac{1}{8}$=0·13^{7} $\frac{1}{6}$=0·18^{3} $\frac{1}{5}$=0·21^{9} $\frac{1}{4}$=0·27^{4} $\frac{1}{3}$=0·36^{5}

$\frac{3}{8}$=0·41^{1} $\frac{1}{2}$=0·54^{8} $\frac{5}{8}$=0·68^{4} $\frac{3}{4}$=0·82^{1} $\frac{7}{8}$=0·95^{8} $\frac{2}{3}$=0·73^{0} $\frac{5}{6}$=0·91^{3}

$\frac{3}{16}$=0·20^{5} $\frac{5}{16}$=0·34^{2} $\frac{7}{16}$=0·47^{9} $\frac{9}{16}$=0·61^{6} $\frac{11}{16}$=0·75^{3} $\frac{13}{16}$=0·89^{0} $\frac{15}{16}$=1·02^{7}

1 metre=1·0936 yards

1	1·10	41	45·10	81	89·10	121	133·10	161	177·10
2	2·20	42	46·20	82	90·20	122	134·20	162	178·20
3	3·30	43	47·30	83	91·30	123	135·30	163	179·30
4	4·40	44	48·40	84	92·40	124	136·40	164	180·40
5	5·50	45	49·50	85	93·50	125	137·50	165	181·50
6	6·60	46	50·60	86	94·60	126	138·60	166	182·60
7	7·70	47	51·70	87	95·70	127	139·70	167	183·70
8	8·80	48	52·80	88	96·80	128	140·80	168	184·80
9	9·90	49	53·90	89	97·90	129	141·90	169	185·90
10	11·00	**50**	55·00	**90**	99·00	**130**	143·00	**170**	187·00
11	12·10	51	56·10	91	100·10	131	144·10	171	188·10
12	13·20	52	57·20	92	101·20	132	145·20	172	189·20
13	14·30	53	58·30	93	102·30	133	146·30	173	190·30
14	15·40	54	59·40	94	103·40	134	147·40	174	191·40
15	16·50	55	60·50	95	104·50	135	148·50	175	192·50
16	17·60	56	61·60	96	105·60	136	149·60	176	193·60
17	18·70	57	62·70	97	106·70	137	150·70	177	194·70
18	19·80	58	63·80	98	107·80	138	151·80	178	195·80
19	20·90	59	64·90	99	108·90	139	152·90	179	196·90
20	22·00	**60**	66·00	**100**	110·00	**140**	154·00	**180**	198·00
21	23·10	61	67·10	101	111·10	141	155·10	181	199·10
22	24·20	62	68·20	102	112·20	142	156·20	182	200·20
23	25·30	63	69·30	103	113·30	143	157·30	183	201·30
24	26·40	64	70·40	104	114·40	144	158·40	184	202·40
25	27·50	65	71·50	105	115·50	145	159·50	185	203·50
26	28·60	66	72·60	106	116·60	146	160·60	186	204·60
27	29·70	67	73·70	107	117·70	147	161·70	187	205·70
28	30·80	68	74·80	108	118·80	148	162·80	188	206·80
29	31·90	69	75·90	109	119·90	149	163·90	189	207·90
30	33·00	**70**	77·00	**110**	121·00	**150**	165·00	**190**	209·00
31	34·10	71	78·10	111	122·10	151	166·10	191	210·10
32	35·20	72	79·20	112	123·20	152	167·20	192	211·20
33	36·30	73	80·30	113	124·30	153	168·30	193	212·30
34	37·40	74	81·40	114	125·40	154	169·40	194	213·40
35	38·50	75	82·50	115	126·50	155	170·50	195	214·50
36	39·60	76	83·60	116	127·60	156	171·60	196	215·60
37	40·70	77	84·70	117	128·70	157	172·70	197	216·70
38	41·80	78	85·80	118	129·80	158	173·80	198	217·80
39	42·90	79	86·90	119	130·90	159	174·90	199	218·90
40	44·00	**80**	88·00	**120**	132·00	**160**	176·00	**200**	220·00

$\frac{1}{16}$=0·06^{9} $\frac{1}{9}$=0·12^{2} $\frac{1}{8}$=0·13^{8} $\frac{1}{6}$=0·18^{3} $\frac{1}{5}$=0·22^{0} $\frac{1}{4}$=0·27^{5} $\frac{1}{3}$=0·36^{7}

$\frac{3}{8}$=0·41^{3} $\frac{1}{2}$=0·55^{0} $\frac{5}{8}$=0·68^{8} $\frac{3}{4}$=0·82^{5} $\frac{7}{8}$=0·96^{3} $\frac{2}{3}$=0·73^{3} $\frac{5}{6}$=0·91^{7}

$\frac{3}{16}$=0·20^{6} $\frac{5}{16}$=0·34^{4} $\frac{7}{16}$=0·48^{1} $\frac{9}{16}$=0·61^{9} $\frac{11}{16}$=0·75^{6} $\frac{13}{16}$=0·89^{4} $\frac{15}{16}$=1·03^{1}

(0.905=1£) £1.10½

1	1·10½	41	45·30½	81	89·50½	121	133·70½	161	177·90½
2	2·21	42	46·41	82	90·61	122	134·81	162	179·01
3	3·31½	43	47·51½	83	91·71½	123	135·91½	163	180·11½
4	4·42	44	48·62	84	92·82	124	137·02	164	181·22
5	5·52½	45	49·72½	85	93·92½	125	138·12½	165	182·32½
6	6·63	46	50·83	86	95·03	126	139·23	166	183·43
7	7·73½	47	51·93½	87	96·13½	127	140·33½	167	184·53½
8	8·84	48	53·04	88	97·24	128	141·44	168	185·64
9	9·94½	49	54·14½	89	98·34½	129	142·54½	169	186·74½
10	11·05	**50**	55·25	**90**	99·45	**130**	143·65	**170**	187·85
11	12·15½	51	56·35½	91	100·55½	131	144·75½	171	188·95½
12	13·26	52	57·46	92	101·66	132	145·86	172	190·06
13	14·36½	53	58·56½	93	102·76½	133	146·96½	173	191·16½
14	15·47	54	59·67	94	103·87	134	148·07	174	192·27
15	16·57½	55	60·77½	95	104·97½	135	149·17½	175	193·37½
16	17·68	56	61·88	96	106·08	136	150·28	176	194·48
17	18·78½	57	62·98½	97	107·18½	137	151·38½	177	195·58½
18	19·89	58	64·09	98	108·29	138	152·49	178	196·69
19	20·99½	59	65·19½	99	109·39½	139	153·59½	179	197·79½
20	22·10	**60**	66·30	**100**	110·50	**140**	154·70	**180**	198·90
21	23·20½	61	67·40½	101	111·60½	141	155·80½	181	200·00½
22	24·31	62	68·51	102	112·71	142	156·91	182	201·11
23	25·41½	63	69·61½	103	113·81½	143	158·01½	183	202·21½
24	26·52	64	70·72	104	114·92	144	159·12	184	203·32
25	27·62½	65	71·82½	105	116·02½	145	160·22½	185	204·42½
26	28·73	66	72·93	106	117·13	146	161·33	186	205·53
27	29·83½	67	74·03½	107	118·23½	147	162·43½	187	206·63½
28	30·94	68	75·14	108	119·34	148	163·54	188	207·74
29	32·04½	69	76·24½	109	120·44½	149	164·64½	189	208·84½
30	33·15	**70**	77·35	**110**	121·55	**150**	165·75	**190**	209·95
31	34·25½	71	78·45½	111	122·65½	151	166·85½	191	211·05½
32	35·36	72	79·56	112	123·76	152	167·96	192	212·16
33	36·46½	73	80·66½	113	124·86½	153	169·06½	193	213·26½
34	37·57	74	81·77	114	125·97	154	170·17	194	214·37
35	38·67½	75	82·87½	115	127·07½	155	171·27½	195	215·47½
36	39·78	76	83·98	116	128·18	156	172·38	196	216·58
37	40·88½	77	85·08½	117	129·28½	157	173·48½	197	217·68½
38	41·99	78	86·19	118	130·39	158	174·59	198	218·79
39	43·09½	79	87·29½	119	131·49½	159	175·69½	199	219·89½
40	44·20	**80**	88·40	**120**	132·60	**160**	176·80	**200**	221·00

$\frac{1}{16}=0{\cdot}06^{9}$ $\frac{1}{9}=0{\cdot}12^{3}$ $\frac{1}{8}=0{\cdot}13^{8}$ $\frac{1}{6}=0{\cdot}18^{4}$ $\frac{1}{5}=0{\cdot}22^{1}$ $\frac{1}{4}=0{\cdot}27^{6}$ $\frac{1}{3}=0{\cdot}36^{8}$

$\frac{3}{8}=0{\cdot}41^{4}$ $\frac{1}{2}=0{\cdot}55^{3}$ $\frac{5}{8}=0{\cdot}69^{1}$ $\frac{3}{4}=0{\cdot}82^{9}$ $\frac{7}{8}=0{\cdot}96^{7}$ $\frac{2}{3}=0{\cdot}73^{7}$ $\frac{5}{6}=0{\cdot}92^{1}$

$\frac{3}{16}=0{\cdot}20^{7}$ $\frac{5}{16}=0{\cdot}34^{5}$ $\frac{7}{16}=0{\cdot}48^{3}$ $\frac{9}{16}=0{\cdot}62^{2}$ $\frac{11}{16}=0{\cdot}76^{0}$ $\frac{13}{16}=0{\cdot}89^{8}$ $\frac{15}{16}=1{\cdot}03^{6}$

(0.901=1£) £1.11

1	1·11	41	45·51	81	89·91	121	134·31	161	178·71
2	2·22	42	46·62	82	91·02	122	135·42	162	179·82
3	3·33	43	47·73	83	92·13	123	136·53	163	180·93
4	4·44	44	48·84	84	93·24	124	137·64	164	182·04
5	5·55	45	49·95	85	94·35	125	138·75	165	183·15
6	6·66	46	51·06	86	95·46	126	139·86	166	184·26
7	7·77	47	52·17	87	96·57	127	140·97	167	185·37
8	8·88	48	53·28	88	97·68	128	142·08	168	186·48
9	9·99	49	54·39	89	98·79	129	143·19	169	187·59
10	11·10	**50**	55·50	**90**	99·90	**130**	144·30	**170**	188·70
11	12·21	51	56·61	91	101·01	131	145·41	171	189·81
12	13·32	52	57·72	92	102·12	132	146·52	172	190·92
13	14·43	53	58·83	93	103·23	133	147·63	173	192·03
14	15·54	54	59·94	94	104·34	134	148·74	174	193·14
15	16·65	55	61·05	95	105·45	135	149·85	175	194·25
16	17·76	56	62·16	96	106·56	136	150·96	176	195·36
17	18·87	57	63·27	97	107·67	137	152·07	177	196·47
18	19·98	58	64·38	98	108·78	138	153·18	178	197·58
19	21·09	59	65·49	99	109·89	139	154·29	179	198·69
20	22·20	**60**	66·60	**100**	111·00	**140**	155·40	**180**	199·80
21	23·31	61	67·71	101	112·11	141	156·51	181	200·91
22	24·42	62	68·82	102	113·22	142	157·62	182	202·02
23	25·53	63	69·93	103	114·33	143	158·73	183	203·13
24	26·64	64	71·04	104	115·44	144	159·84	184	204·24
25	27·75	65	72·15	105	116·55	145	160·95	185	205·35
26	28·86	66	73·26	106	117·66	146	162·06	186	206·46
27	29·97	67	74·37	107	118·77	147	163·17	187	207·57
28	31·08	68	75·48	108	119·88	148	164·28	188	208·68
29	32·19	69	76·59	109	120·99	149	165·39	189	209·79
30	33·30	**70**	77·70	**110**	122·10	**150**	166·50	**190**	210·90
31	34·41	71	78·81	111	123·21	151	167·61	191	212·01
32	35·52	72	79·92	112	124·32	152	168·72	192	213·12
33	36·63	73	81·03	113	125·43	153	169·83	193	214·23
34	37·74	74	82·14	114	126·54	154	170·94	194	215·34
35	38·85	75	83·25	115	127·65	155	172·05	195	216·45
36	39·96	76	84·36	116	128·76	156	173·16	196	217·56
37	41·07	77	85·47	117	129·87	157	174·27	197	218·67
38	42·18	78	86·58	118	130·98	158	175·38	198	219·78
39	43·29	79	87·69	119	132·09	159	176·49	199	220·89
40	44·40	**80**	88·80	**120**	133·20	**160**	177·60	**200**	222·00

$\frac{1}{16}=0{\cdot}06^{9}$ $\frac{1}{9}=0{\cdot}12^{3}$ $\frac{1}{8}=0{\cdot}13^{9}$ $\frac{1}{6}=0{\cdot}18^{5}$ $\frac{1}{5}=0{\cdot}22^{2}$ $\frac{1}{4}=0{\cdot}27^{8}$ $\frac{1}{3}=0{\cdot}37^{0}$

$\frac{3}{8}=0{\cdot}41^{6}$ $\frac{1}{2}=0{\cdot}55^{5}$ $\frac{5}{8}=0{\cdot}69^{4}$ $\frac{3}{4}=0{\cdot}83^{3}$ $\frac{7}{8}=0{\cdot}97^{1}$ $\frac{2}{3}=0{\cdot}74^{0}$ $\frac{5}{6}=0{\cdot}92^{5}$

$\frac{3}{16}=0{\cdot}20^{8}$ $\frac{5}{16}=0{\cdot}34^{7}$ $\frac{7}{16}=0{\cdot}48^{6}$ $\frac{9}{16}=0{\cdot}62^{4}$ $\frac{11}{16}=0{\cdot}76^{3}$ $\frac{13}{16}=0{\cdot}90^{2}$ $\frac{15}{16}=1{\cdot}04^{1}$

1	1·11½	41	45·71½	81	90·31½	121	134·91½	161	179·51½
2	2·23	42	46·83	82	91·43	122	136·03	162	180·63
3	3·34½	43	47·94½	83	92·54½	123	137·14½	163	181·74½
4	4·46	44	49·06	84	93·66	124	138·26	164	182·86
5	5·57½	45	50·17½	85	94·77½	125	139·37½	165	183·97½
6	6·69	46	51·29	86	95·89	126	140·49	166	185·09
7	7·80½	47	52·40½	87	97·00½	127	141·60½	167	186·20½
8	8·92	48	53·52	88	98·12	128	142·72	168	187·32
9	10·03½	49	54·63½	89	99·23½	129	143·83½	169	188·43½
10	11·15	**50**	55·75	**90**	100·35	**130**	144·95	**170**	189·55
11	12·26½	51	56·86½	91	101·46½	131	146·06½	171	190·66½
12	13·38	52	57·98	92	102·58	132	147·18	172	191·78
13	14·49½	53	59·09½	93	103·69½	133	148·29½	173	192·89½
14	15·61	54	60·21	94	104·81	134	149·41	174	194·01
15	16·72½	55	61·32½	95	105·92½	135	150·52½	175	195·12½
16	17·84	56	62·44	96	107·04	136	151·64	176	196·24
17	18·95½	57	63·55½	97	108·15½	137	152·75½	177	197·35½
18	20·07	58	64·67	98	109·27	138	153·87	178	198·47
19	21·18½	59	65·78½	99	110·38½	139	154·98½	179	199·58½
20	22·30	**60**	66·90	**100**	111·50	**140**	156·10	**180**	200·70
21	23·41½	61	68·01½	101	112·61½	141	157·21½	181	201·81½
22	24·53	62	69·13	102	113·73	142	158·33	182	202·93
23	25·64½	63	70·24½	103	114·84½	143	159·44½	183	204·04½
24	26·76	64	71·36	104	115·96	144	160·56	184	205·16
25	27·87½	65	72·47½	105	117·07½	145	161·67½	185	206·27½
26	28·99	66	73·59	106	118·19	146	162·79	186	207·39
27	30·10½	67	74·70½	107	119·30½	147	163·90½	187	208·50½
28	31·22	68	75·82	108	120·42	148	165·02	188	209·62
29	32·33½	69	76·93½	109	121·53½	149	166·13½	189	210·73½
30	33·45	**70**	78·05	**110**	122·65	**150**	167·25	**190**	211·85
31	34·56½	71	79·16½	111	123·76½	151	168·36½	191	212·96½
32	35·68	72	80·28	112	124·88	152	169·48	192	214·08
33	36·79½	73	81·39½	113	125·99½	153	170·59½	193	215·19½
34	37·91	74	82·51	114	127·11	154	171·71	194	216·31
35	39·02½	75	83·62½	115	128·22½	155	172·82½	195	217·42½
36	40·14	76	84·74	116	129·34	156	173·94	196	218·54
37	41·25½	77	85·85½	117	130·45½	157	175·05½	197	219·65½
38	42·37	78	86·97	118	131·57	158	176·17	198	220·77
39	43·48½	79	88·08½	119	132·68½	159	177·28½	199	221·88½
40	44·60	**80**	89·20	**120**	133·80	**160**	178·40	**200**	223·00

$\frac{1}{16}=0{\cdot}07^{0}$ $\frac{1}{9}=0{\cdot}12^{4}$ $\frac{1}{8}=0{\cdot}13^{9}$ $\frac{1}{6}=0{\cdot}18^{6}$ $\frac{1}{5}=0{\cdot}22^{3}$ $\frac{1}{4}=0{\cdot}27^{9}$ $\frac{1}{3}=0{\cdot}37^{2}$

$\frac{3}{8}=0{\cdot}41^{8}$ $\frac{1}{2}=0{\cdot}55^{8}$ $\frac{5}{8}=0{\cdot}69^{7}$ $\frac{3}{4}=0{\cdot}83^{6}$ $\frac{7}{8}=0{\cdot}97^{6}$ $\frac{2}{3}=0{\cdot}74^{3}$ $\frac{5}{6}=0{\cdot}92^{9}$

$\frac{3}{16}=0{\cdot}20^{9}$ $\frac{5}{16}=0{\cdot}34^{8}$ $\frac{7}{16}=0{\cdot}48^{8}$ $\frac{9}{16}=0{\cdot}62^{7}$ $\frac{11}{16}=0{\cdot}76^{7}$ $\frac{13}{16}=0{\cdot}90^{6}$ $\frac{15}{16}=1{\cdot}04^{5}$

1	1·12	41	45·92	81	90·72	121	135·52	161	180·32
2	2·24	42	47·04	82	91·84	122	136·64	162	181·44
3	3·36	43	48·16	83	92·96	123	137·76	163	182·56
4	4·48	44	49·28	84	94·08	124	138·88	164	183·68
5	5·60	45	50·40	85	95·20	125	140·00	165	184·80
6	6·72	46	51·52	86	96·32	126	141·12	166	185·92
7	7·84	47	52·64	87	97·44	127	142·24	167	187·04
8	8·96	48	53·76	88	98·56	128	143·36	168	188·16
9	10·08	49	54·88	89	99·68	129	144·48	169	189·28
10	11·20	**50**	56·00	**90**	100·80	**130**	145·60	**170**	190·40
11	12·32	51	57·12	91	101·92	131	146·72	171	191·52
12	13·44	52	58·24	92	103·04	132	147·84	172	192·64
13	14·56	53	59·36	93	104·16	133	148·96	173	193·76
14	15·68	54	60·48	94	105·28	134	150·08	174	194·88
15	16·80	55	61·60	95	106·40	135	151·20	175	196·00
16	17·92	56	62·72	96	107·52	136	152·32	176	197·12
17	19·04	57	63·84	97	108·64	137	153·44	177	198·24
18	20·16	58	64·96	98	109·76	138	154·56	178	199·36
19	21·28	59	66·08	99	110·88	139	155·68	179	200·48
20	22·40	**60**	67·20	**100**	112·00	**140**	156·80	**180**	201·60
21	23·52	61	68·32	101	113·12	141	157·92	181	202·72
22	24·64	62	69·44	102	114·24	142	159·04	182	203·84
23	25·76	63	70·56	103	115·36	143	160·16	183	204·96
24	26·88	64	71·68	104	116·48	144	161·28	184	206·08
25	28·00	65	72·80	105	117·60	145	162·40	185	207·20
26	29·12	66	73·92	106	118·72	146	163·52	186	208·32
27	30·24	67	75·04	107	119·84	147	164·64	187	209·44
28	31·36	68	76·16	108	120·96	148	165·76	188	210·56
29	32·48	69	77·28	109	122·08	149	166·88	189	211·68
30	33·60	**70**	78·40	**110**	123·20	**150**	168·00	**190**	212·80
31	34·72	71	79·52	111	124·32	151	169·12	191	213·92
32	35·84	72	80·64	112	125·44	152	170·24	192	215·04
33	36·96	73	81·76	113	126·56	153	171·36	193	216·16
34	38·08	74	82·88	114	127·68	154	172·48	194	217·28
35	39·20	75	84·00	115	128·80	155	173·60	195	218·40
36	40·32	76	85·12	116	129·92	156	174·72	196	219·52
37	41·44	77	86·24	117	131·04	157	175·84	197	220·64
38	42·56	78	87·36	118	132·16	158	176·96	198	221·76
39	43·68	79	88·48	119	133·28	159	178·08	199	222·88
40	44·80	**80**	89·60	**120**	134·40	**160**	179·20	**200**	224·00

$\frac{1}{16}=0{\cdot}07^0$ $\frac{1}{9}=0{\cdot}12^4$ $\frac{1}{8}=0{\cdot}14^0$ $\frac{1}{6}=0{\cdot}18^7$ $\frac{1}{5}=0{\cdot}22^4$ $\frac{1}{4}=0{\cdot}28^0$ $\frac{1}{3}=0{\cdot}37^3$

$\frac{3}{8}=0{\cdot}42^0$ $\frac{1}{2}=0{\cdot}56^0$ $\frac{5}{8}=0{\cdot}70^0$ $\frac{3}{4}=0{\cdot}84^0$ $\frac{7}{8}=0{\cdot}98^0$ $\frac{2}{3}=0{\cdot}74^7$ $\frac{5}{6}=0{\cdot}93^3$

$\frac{3}{16}=0{\cdot}21^0$ $\frac{5}{16}=0{\cdot}35^0$ $\frac{7}{16}=0{\cdot}49^0$ $\frac{9}{16}=0{\cdot}63^0$ $\frac{11}{16}=0{\cdot}77^0$ $\frac{13}{16}=0{\cdot}91^0$ $\frac{15}{16}=1{\cdot}05^0$

(0.889=1£) **£1.12½**

1	1·12½	41	46·12½	81	91·12½	121	136·12½	161	181·12½
2	2·25	42	47·25	82	92·25	122	137·25	162	182·25
3	3·37½	43	48·37½	83	93·37½	123	138·37½	163	183·37½
4	4·50	44	49·50	84	94·50	124	139·50	164	184·50
5	5·62½	45	50·62½	85	95·62½	125	140·62½	165	185·62½
6	6·75	46	51·75	86	96·75	126	141·75	166	186·75
7	7·87½	47	52·87½	87	97·87½	127	142·87½	167	187·87½
8	9·00	48	54·00	88	99·00	128	144·00	168	189·00
9	10·12½	49	55·12½	89	100·12½	129	145·12½	169	190·12½
10	11·25	**50**	56·25	**90**	101·25	**130**	146·25	**170**	191·25
11	12·37½	51	57·37½	91	102·37½	131	147·37½	171	192·37½
12	13·50	52	58·50	92	103·50	132	148·50	172	193·50
13	14·62½	53	59·62½	93	104·62½	133	149·62½	173	194·62½
14	15·75	54	60·75	94	105·75	134	150·75	174	195·75
15	16·87½	55	61·87½	95	106·87½	135	151·87½	175	196·87½
16	18·00	56	63·00	96	108·00	136	153·00	176	198·00
17	19·12½	57	64·12½	97	109·12½	137	154·12½	177	199·12½
18	20·25	58	65·25	98	110·25	138	155·25	178	200·25
19	21·37½	59	66·37½	99	111·37½	139	156·37½	179	201·37½
20	22·50	**60**	67·50	**100**	112·50	**140**	157·50	**180**	202·50
21	23·62½	61	68·62½	101	113·62½	141	158·62½	181	203·62½
22	24·75	62	69·75	102	114·75	142	159·75	182	204·75
23	25·87½	63	70·87½	103	115·87½	143	160·87½	183	205·87½
24	27·00	64	72·00	104	117·00	144	162·00	184	207·00
25	28·12½	65	73·12½	105	118·12½	145	163·12½	185	208·12½
26	29·25	66	74·25	106	119·25	146	164·25	186	209·25
27	30·37½	67	75·37½	107	120·37½	147	165·37½	187	210·37½
28	31·50	68	76·50	108	121·50	148	166·50	188	211·50
29	32·62½	69	77·62½	109	122·62½	149	167·62½	189	212·62½
30	33·75	**70**	78·75	**110**	123·75	**150**	168·75	**190**	213·75
31	34·87½	71	79·87½	111	124·87½	151	169·87½	191	214·87½
32	36·00	72	81·00	112	126·00	152	171·00	192	216·00
33	37·12½	73	82·12½	113	127·12½	153	172·12½	193	217·12½
34	38·25	74	83·25	114	128·25	154	173·25	194	218·25
35	39·37½	75	84·37½	115	129·37½	155	174·37½	195	219·37½
36	40·50	76	85·50	116	130·50	156	175·50	196	220·50
37	41·62½	77	86·62½	117	131·62½	157	176·62½	197	221·62½
38	42·75	78	87·75	118	132·75	158	177·75	198	222·75
39	43·87½	79	88·87½	119	133·87½	159	178·87½	199	223·87½
40	45·00	**80**	90·00	**120**	135·00	**160**	180·00	**200**	225·00

$\frac{1}{16}=0{\cdot}07^{0}$ $\frac{1}{9}=0{\cdot}12^{5}$ $\frac{1}{8}=0{\cdot}14^{1}$ $\frac{1}{6}=0{\cdot}18^{8}$ $\frac{1}{5}=0{\cdot}22^{5}$ $\frac{1}{4}=0{\cdot}28^{1}$ $\frac{1}{3}=0{\cdot}37^{5}$

$\frac{3}{8}=0{\cdot}42^{2}$ $\frac{1}{2}=0{\cdot}56^{3}$ $\frac{5}{8}=0{\cdot}70^{3}$ $\frac{3}{4}=0{\cdot}84^{4}$ $\frac{7}{8}=0{\cdot}98^{4}$ $\frac{2}{3}=0{\cdot}75^{0}$ $\frac{5}{6}=0{\cdot}93^{8}$

$\frac{3}{16}=0{\cdot}21^{1}$ $\frac{5}{16}=0{\cdot}35^{2}$ $\frac{7}{16}=0{\cdot}49^{2}$ $\frac{9}{16}=0{\cdot}63^{3}$ $\frac{11}{16}=0{\cdot}77^{3}$ $\frac{13}{16}=0{\cdot}91^{4}$ $\frac{15}{16}=1{\cdot}05^{5}$

(0.885=1£) **£1.13**

1	1·13	41	46·33	81	91·53	121	136·73	161	181·93
2	2·26	42	47·46	82	92·66	122	137·86	162	183·06
3	3·39	43	48·59	83	93·79	123	138·99	163	184·19
4	4·52	44	49·72	84	94·92	124	140·12	164	185·32
5	5·65	45	50·85	85	96·05	125	141·25	165	186·45
6	6·78	46	51·98	86	97·18	126	142·38	166	187·58
7	7·91	47	53·11	87	98·31	127	143·51	167	188·71
8	9·04	48	54·24	88	99·44	128	144·64	168	189·84
9	10·17	49	55·37	89	100·57	129	145·77	169	190·97
10	11·30	**50**	56·50	**90**	101·70	**130**	146·90	**170**	192·10
11	12·43	51	57·63	91	102·83	131	148·03	171	193·23
12	13·56	52	58·76	92	103·96	132	149·16	172	194·36
13	14·69	53	59·89	93	105·09	133	150·29	173	195·49
14	15·82	54	61·02	94	106·22	134	151·42	174	196·62
15	16·95	55	62·15	95	107·35	135	152·55	175	197·75
16	18·08	56	63·28	96	108·48	136	153·68	176	198·88
17	19·21	57	64·41	97	109·61	137	154·81	177	200·01
18	20·34	58	65·54	98	110·74	138	155·94	178	201·14
19	21·47	59	66·67	99	111·87	139	157·07	179	202·27
20	22·60	**60**	67·80	**100**	113·00	**140**	158·20	**180**	203·40
21	23·73	61	68·93	101	114·13	141	159·33	181	204·53
22	24·86	62	70·06	102	115·26	142	160·46	182	205·66
23	25·99	63	71·19	103	116·39	143	161·59	183	206·79
24	27·12	64	72·32	104	117·52	144	162·72	184	207·92
25	28·25	65	73·45	105	118·65	145	163·85	185	209·05
26	29·38	66	74·58	106	119·78	146	164·98	186	210·18
27	30·51	67	75·71	107	120·91	147	166·11	187	211·31
28	31·64	68	76·84	108	122·04	148	167·24	188	212·44
29	32·77	69	77·97	109	123·17	149	168·37	189	213·57
30	33·90	**70**	79·10	**110**	124·30	**150**	169·50	**190**	214·70
31	35·03	71	80·23	111	125·43	151	170·63	191	215·83
32	36·16	72	81·36	112	126·56	152	171·76	192	216·96
33	37·29	73	82·49	113	127·69	153	172·89	193	218·09
34	38·42	74	83·62	114	128·82	154	174·02	194	219·22
35	39·55	75	84·75	115	129·95	155	175·15	195	220·35
36	40·68	76	85·88	116	131·08	156	176·28	196	221·48
37	41·81	77	87·01	117	132·21	157	177·41	197	222·61
38	42·94	78	88·14	118	133·34	158	178·54	198	223·74
39	44·07	79	89·27	119	134·47	159	179·67	199	224·87
40	45·20	**80**	90·40	**120**	135·60	**160**	180·80	**200**	226·00

$\frac{1}{16}=0{\cdot}07^{1}$ $\frac{1}{9}=0{\cdot}12^{6}$ $\frac{1}{8}=0{\cdot}14^{1}$ $\frac{1}{6}=0{\cdot}18^{8}$ $\frac{1}{5}=0{\cdot}22^{6}$ $\frac{1}{4}=0{\cdot}28^{3}$ $\frac{1}{3}=0{\cdot}37^{7}$

$\frac{3}{8}=0{\cdot}42^{4}$ $\frac{1}{2}=0{\cdot}56^{5}$ $\frac{5}{8}=0{\cdot}70^{6}$ $\frac{3}{4}=0{\cdot}84^{8}$ $\frac{7}{8}=0{\cdot}98^{9}$ $\frac{2}{3}=0{\cdot}75^{3}$ $\frac{5}{6}=0{\cdot}94^{2}$

$\frac{3}{16}=0{\cdot}21^{2}$ $\frac{5}{16}=0{\cdot}35^{3}$ $\frac{7}{16}=0{\cdot}49^{4}$ $\frac{9}{16}=0{\cdot}63^{6}$ $\frac{11}{16}=0{\cdot}77^{7}$ $\frac{13}{16}=0{\cdot}91^{8}$ $\frac{15}{16}=1{\cdot}05^{9}$

(0.881=1£) **£1.13½**

1	1·13½	41	46·53½	81	91·93½	121	137·33½	161	182·73½
2	2·27	42	47·67	82	93·07	122	138·47	162	183·87
3	3·40½	43	48·80½	83	94·20½	123	139·60½	163	185·00½
4	4·54	44	49·94	84	95·34	124	140·74	164	186·14
5	5·67½	45	51·07½	85	96·47½	125	141·87½	165	187·27½
6	6·81	46	52·21	86	97·61	126	143·01	166	188·41
7	7·94½	47	53·34½	87	98·74½	127	144·14½	167	189·54½
8	9·08	48	54·48	88	99·88	128	145·28	168	190·68
9	10·21½	49	55·61½	89	101·01½	129	146·41½	169	191·81½
10	11·35	**50**	56·75	**90**	102·15	**130**	147·55	**170**	192·95
11	12·48½	51	57·88½	91	103·28½	131	148·68½	171	194·08½
12	13·62	52	59·02	92	104·42	132	149·82	172	195·22
13	14·75½	53	60·15½	93	105·55½	133	150·95½	173	196·35½
14	15·89	54	61·29	94	106·69	134	152·09	174	197·49
15	17·02½	55	62·42½	95	107·82½	135	153·22½	175	198·62½
16	18·16	56	63·56	96	108·96	136	154·36	176	199·76
17	19·29½	57	64·69½	97	110·09½	137	155·49½	177	200·89½
18	20·43	58	65·83	98	111·23	138	156·63	178	202·03
19	21·56½	59	66·96½	99	112·36½	139	157·76½	179	203·16½
20	22·70	**60**	68·10	**100**	113·50	**140**	158·90	**180**	204·30
21	23·83½	61	69·23½	101	114·63½	141	160·03½	181	205·43½
22	24·97	62	70·37	102	115·77	142	161·17	182	206·57
23	26·10½	63	71·50½	103	116·90½	143	162·30½	183	207·70½
24	27·24	64	72·64	104	118·04	144	163·44	184	208·84
25	28·37½	65	73·77½	105	119·17½	145	164·57½	185	209·97½
26	29·51	66	74·91	106	120·31	146	165·71	186	211·11
27	30·64½	67	76·04½	107	121·44½	147	166·84½	187	212·24½
28	31·78	68	77·18	108	122·58	148	167·98	188	213·38
29	32·91½	69	78·31½	109	123·71½	149	169·11½	189	214·51½
30	34·05	**70**	79·45	**110**	124·85	**150**	170·25	**190**	215·65
31	35·18½	71	80·58½	111	125·98½	151	171·38½	191	216·78½
32	36·32	72	81·72	112	127·12	152	172·52	192	217·92
33	37·45½	73	82·85½	113	128·25½	153	173·65½	193	219·05½
34	38·59	74	83·99	114	129·39	154	174·79	194	220·19
35	39·72½	75	85·12½	115	130·52½	155	175·92½	195	221·32½
36	40·86	76	86·26	116	131·66	156	177·06	196	222·46
37	41·99½	77	87·39½	117	132·79½	157	178·19½	197	223·59½
38	43·13	78	88·53	118	133·93	158	179·33	198	224·73
39	44·26½	79	89·66½	119	135·06½	159	180·46½	199	225·86½
40	45·40	**80**	90·80	**120**	136·20	**160**	181·60	**200**	227·00

$\frac{1}{16}$=0·07^{1}	$\frac{1}{9}$=0·12^{6}	$\frac{1}{8}$=0·14^{2}	$\frac{1}{6}$=0·18^{9}	$\frac{1}{5}$=0·22^{7}	$\frac{1}{4}$=0·28^{4}	$\frac{1}{3}$=0·37^{8}
$\frac{3}{8}$=0·42^{6}	$\frac{1}{2}$=0·56^{8}	$\frac{5}{8}$=0·70^{9}	$\frac{3}{4}$=0·85^{1}	$\frac{7}{8}$=0·99^{3}	$\frac{2}{3}$=0·75^{7}	$\frac{5}{6}$=0·94^{6}
$\frac{3}{16}$=0·21^{3}	$\frac{5}{16}$=0·35^{5}	$\frac{7}{16}$=0·49^{7}	$\frac{9}{16}$=0·63^{8}	$\frac{11}{16}$=0·78^{0}	$\frac{13}{16}$=0·92^{2}	$\frac{15}{16}$=1·06^{4}

(0.877=1£) **£1.14**

1	1·14	41	46·74	81	92·34	121	137·94	161	183·54
2	2·28	42	47·88	82	93·48	122	139·08	162	184·68
3	3·42	43	49·02	83	94·62	123	140·22	163	185·82
4	4·56	44	50·16	84	95·76	124	141·36	164	186·96
5	5·70	45	51·30	85	96·90	125	142·50	165	188·10
6	6·84	46	52·44	86	98·04	126	143·64	166	189·24
7	7·98	47	53·58	87	99·18	127	144·78	167	190·38
8	9·12	48	54·72	88	100·32	128	145·92	168	191·52
9	10·26	49	55·86	89	101·46	129	147·06	169	192·66
10	11·40	**50**	57·00	**90**	102·60	**130**	148·20	**170**	193·80
11	12·54	51	58·14	91	103·74	131	149·34	171	194·94
12	13·68	52	59·28	92	104·88	132	150·48	172	196·08
13	14·82	53	60·42	93	106·02	133	151·62	173	197·22
14	15·96	54	61·56	94	107·16	134	152·76	174	198·36
15	17·10	55	62·70	95	108·30	135	153·90	175	199·50
16	18·24	56	63·84	96	109·44	136	155·04	176	200·64
17	19·38	57	64·98	97	110·58	137	156·18	177	201·78
18	20·52	58	66·12	98	111·72	138	157·32	178	202·92
19	21·66	59	67·26	99	112·86	139	158·46	179	204·06
20	22·80	**60**	68·40	**100**	114·00	**140**	159·60	**180**	205·20
21	23·94	61	69·54	101	115·14	141	160·74	181	206·34
22	25·08	62	70·68	102	116·28	142	161·88	182	207·48
23	26·22	63	71·82	103	117·42	143	163·02	183	208·62
24	27·36	64	72·96	104	118·56	144	164·16	184	209·76
25	28·50	65	74·10	105	119·70	145	165·30	185	210·90
26	29·64	66	75·24	106	120·84	146	166·44	186	212·04
27	30·78	67	76·38	107	121·98	147	167·58	187	213·18
28	31·92	68	77·52	108	123·12	148	168·72	188	214·32
29	33·06	69	78·66	109	124·26	149	169·86	189	215·46
30	34·20	**70**	79·80	**110**	125·40	**150**	171·00	**190**	216·60
31	35·34	71	80·94	111	126·54	151	172·14	191	217·74
32	36·48	72	82·08	112	127·68	152	173·28	192	218·88
33	37·62	73	83·22	113	128·82	153	174·42	193	220·02
34	38·76	74	84·36	114	129·96	154	175·56	194	221·16
35	39·90	75	85·50	115	131·10	155	176·70	195	222·30
36	41·04	76	86·64	116	132·24	156	177·84	196	223·44
37	42·18	77	87·78	117	133·38	157	178·98	197	224·58
38	43·32	78	88·92	118	134·52	158	180·12	198	225·72
39	44·46	79	90·06	119	135·66	159	181·26	199	226·86
40	45·60	**80**	91·20	**120**	136·80	**160**	182·40	**200**	228·00

$\frac{1}{16}=0{\cdot}07^{1}$ $\frac{1}{9}=0{\cdot}12^{7}$ $\frac{1}{8}=0{\cdot}14^{3}$ $\frac{1}{6}=0{\cdot}19^{0}$ $\frac{1}{5}=0{\cdot}22^{8}$ $\frac{1}{4}=0{\cdot}28^{5}$ $\frac{1}{3}=0{\cdot}38^{0}$

$\frac{3}{8}=0{\cdot}42^{8}$ $\frac{1}{2}=0{\cdot}57^{0}$ $\frac{5}{8}=0{\cdot}71^{3}$ $\frac{3}{4}=0{\cdot}85^{5}$ $\frac{7}{8}=0{\cdot}99^{8}$ $\frac{2}{3}=0{\cdot}76^{0}$ $\frac{5}{6}=0{\cdot}95^{0}$

$\frac{3}{16}=0{\cdot}21^{4}$ $\frac{5}{16}=0{\cdot}35^{6}$ $\frac{7}{16}=0{\cdot}49^{9}$ $\frac{9}{16}=0{\cdot}64^{1}$ $\frac{11}{16}=0{\cdot}78^{4}$ $\frac{13}{16}=0{\cdot}92^{6}$ $\frac{15}{16}=1{\cdot}06^{9}$

(0.873=1£) **£1.14½**

1	1·14½	41	46·94½	81	92·74½	121	138·54½	161	184·34½
2	2·29	42	48·09	82	93·89	122	139·69	162	185·49
3	3·43½	43	49·23½	83	95·03½	123	140·83½	163	186·63½
4	4·58	44	50·38	84	96·18	124	141·98	164	187·78
5	5·72½	45	51·52½	85	97·32½	125	143·12½	165	188·92½
6	6·87	46	52·67	86	98·47	126	144·27	166	190·07
7	8·01½	47	53·81½	87	99·61½	127	145·41½	167	191·21½
8	9·16	48	54·96	88	100·76	128	146·56	168	192·36
9	10·30½	49	56·10½	89	101·90½	129	147·70½	169	193·50½
10	11·45	**50**	57·25	**90**	103·05	**130**	148·85	**170**	194·65
11	12·59½	51	58·39½	91	104·19½	131	149·99½	171	195·79½
12	13·74	52	59·54	92	105·34	132	151·14	172	196·94
13	14·88½	53	60·68½	93	106·48½	133	152·28½	173	198·08½
14	16·03	54	61·83	94	107·63	134	153·43	174	199·23
15	17·17½	55	62·97½	95	108·77½	135	154·57½	175	200·37½
16	18·32	56	64·12	96	109·92	136	155·72	176	201·52
17	19·46½	57	65·26½	97	111·06½	137	156·86½	177	202·66½
18	20·61	58	66·41	98	112·21	138	158·01	178	203·81
19	21·75½	59	67·55½	99	113·35½	139	159·15½	179	204·95½
20	22·90	**60**	68·70	**100**	114·50	**140**	160·30	**180**	206·10
21	24·04½	61	69·84½	101	115·64½	141	161·44½	181	207·24½
22	25·19	62	70·99	102	116·79	142	162·59	182	208·39
23	26·33½	63	72·13½	103	117·93½	143	163·73½	183	209·53½
24	27·48	64	73·28	104	119·08	144	164·88	184	210·68
25	28·62½	65	74·42½	105	120·22½	145	166·02½	185	211·82½
26	29·77	66	75·57	106	121·37	146	167·17	186	212·97
27	30·91½	67	76·71½	107	122·51½	147	168·31½	187	214·11½
28	32·06	68	77·86	108	123·66	148	169·46	188	215·26
29	33·20½	69	79·00½	109	124·80½	149	170·60½	189	216·40½
30	34·35	**70**	80·15	**110**	125·95	**150**	171·75	**190**	217·55
31	35·49½	71	81·29½	111	127·09½	151	172·89½	191	218·69½
32	36·64	72	82·44	112	128·24	152	174·04	192	219·84
33	37·78½	73	83·58½	113	129·38½	153	175·18½	193	220·98½
34	38·93	74	84·73	114	130·53	154	176·33	194	222·13
35	40·07½	75	85·87½	115	131·67½	155	177·47½	195	223·27½
36	41·22	76	87·02	116	132·82	156	178·62	196	224·42
37	42·36½	77	88·16½	117	133·96½	157	179·76½	197	225·56½
38	43·51	78	89·31	118	135·11	158	180·91	198	226·71
39	44·65½	79	90·45½	119	136·25½	159	182·05½	199	227·85½
40	45·80	**80**	91·60	**120**	137·40	**160**	183·20	**200**	229·00

$\frac{1}{16}$=0·07^{2} $\frac{1}{9}$=0·12^{7} $\frac{1}{8}$=0·14^{3} $\frac{1}{6}$=0·19^{1} $\frac{1}{5}$=0·22^{9} $\frac{1}{4}$=0·28^{6} $\frac{1}{3}$=0·38^{2}

$\frac{3}{8}$=0·42^{9} $\frac{1}{2}$=0·57^{3} $\frac{5}{8}$=0·71^{6} $\frac{3}{4}$=0·85^{9} $\frac{7}{8}$=1·00^{2} $\frac{2}{3}$=0·76^{3} $\frac{5}{6}$=0·95^{4}

$\frac{3}{16}$=0·21^{5} $\frac{5}{16}$=0·35^{8} $\frac{7}{16}$=0·50^{1} $\frac{9}{16}$=0·64^{4} $\frac{11}{16}$=0·78^{7} $\frac{13}{16}$=0·93^{0} $\frac{15}{16}$=1·07^{3}

(0.870=1£) **£1.15**

1	1·15	41	47·15	81	93·15	121	139·15	161	185·15
2	2·30	42	48·30	82	94·30	122	140·30	162	186·30
3	3·45	43	49·45	83	95·45	123	141·45	163	187·45
4	4·60	44	50·60	84	96·60	124	142·60	164	188·60
5	5·75	45	51·75	85	97·75	125	143·75	165	189·75
6	6·90	46	52·90	86	98·90	126	144·90	166	190·90
7	8·05	47	54·05	87	100·05	127	146·05	167	192·05
8	9·20	48	55·20	88	101·20	128	147·20	168	193·20
9	10·35	49	56·35	89	102·35	129	148·35	169	194·35
10	11·50	**50**	57·50	**90**	103·50	**130**	149·50	**170**	195·50
11	12·65	51	58·65	91	104·65	131	150·65	171	196·65
12	13·80	52	59·80	92	105·80	132	151·80	172	197·80
13	14·95	53	60·95	93	106·95	133	152·95	173	198·95
14	16·10	54	62·10	94	108·10	134	154·10	174	200·10
15	17·25	55	63·25	95	109·25	135	155·25	175	201·25
16	18·40	56	64·40	96	110·40	136	156·40	176	202·40
17	19·55	57	65·55	97	111·55	137	157·55	177	203·55
18	20·70	58	66·70	98	112·70	138	158·70	178	204·70
19	21·85	59	67·85	99	113·85	139	159·85	179	205·85
20	23·00	**60**	69·00	**100**	115·00	**140**	161·00	**180**	207·00
21	24·15	61	70·15	101	116·15	141	162·15	181	208·15
22	25·30	62	71·30	102	117·30	142	163·30	182	209·30
23	26·45	63	72·45	103	118·45	143	164·45	183	210·45
24	27·60	64	73·60	104	119·60	144	165·60	184	211·60
25	28·75	65	74·75	105	120·75	145	166·75	185	212·75
26	29·90	66	75·90	106	121·90	146	167·90	186	213·90
27	31·05	67	77·05	107	123·05	147	169·05	187	215·05
28	32·20	68	78·20	108	124·20	148	170·20	188	216·20
29	33·35	69	79·35	109	125·35	149	171·35	189	217·35
30	34·50	**70**	80·50	**110**	126·50	**150**	172·50	**190**	218·50
31	35·65	71	81·65	111	127·65	151	173·65	191	219·65
32	36·80	72	82·80	112	128·80	152	174·80	192	220·80
33	37·95	73	83·95	113	129·95	153	175·95	193	221·95
34	39·10	74	85·10	114	131·10	154	177·10	194	223·10
35	40·25	75	86·25	115	132·25	155	178·25	195	224·25
36	41·40	76	87·40	116	133·40	156	179·40	196	225·40
37	42·55	77	88·55	117	134·55	157	180·55	197	226·55
38	43·70	78	89·70	118	135·70	158	181·70	198	227·70
39	44·85	79	90·85	119	136·85	159	182·85	199	228·85
40	46·00	**80**	92·00	**120**	138·00	**160**	184·00	**200**	230·00

$\frac{1}{16}$=0·07^{2} $\frac{1}{9}$=0·12^{8} $\frac{1}{8}$=0·14^{4} $\frac{1}{6}$=0·19^{2} $\frac{1}{5}$=0·23^{0} $\frac{1}{4}$=0·28^{8} $\frac{1}{3}$=0·38^{3}

$\frac{3}{8}$=0·43^{1} $\frac{1}{2}$=0·57^{5} $\frac{5}{8}$=0·71^{9} $\frac{3}{4}$=0·86^{3} $\frac{7}{8}$=1·00^{6} $\frac{2}{3}$=0·76^{7} $\frac{5}{6}$=0·95^{8}

$\frac{3}{16}$=0·21^{6} $\frac{5}{16}$=0·35^{9} $\frac{7}{16}$=0·50^{3} $\frac{9}{16}$=0·64^{7} $\frac{11}{16}$=0·79^{1} $\frac{13}{16}$=0·93^{4} $\frac{15}{16}$=1·07^{8}

(0.866=1£) £1.15½

1	1·15½	41	47·35½	81	93·55½	121	139·75½	161	185·95½
2	2·31	42	48·51	82	94·71	122	140·91	162	187·11
3	3·46½	43	49·66½	83	95·86½	123	142·06½	163	188·26½
4	4·62	44	50·82	84	97·02	124	143·22	164	189·42
5	5·77½	45	51·97½	85	98·17½	125	144·37½	165	190·57½
6	6·93	46	53·13	86	99·33	126	145·53	166	191·73
7	8·08½	47	54·28½	87	100·48½	127	146·68½	167	192·88½
8	9·24	48	55·44	88	101·64	128	147·84	168	194·04
9	10·39½	49	56·59½	89	102·79½	129	148·99½	169	195·19½
10	11·55	**50**	57·75	**90**	103·95	**130**	150·15	**170**	196·35
11	12·70½	51	58·90½	91	105·10½	131	151·30½	171	197·50½
12	13·86	52	60·06	92	106·26	132	152·46	172	198·66
13	15·01½	53	61·21½	93	107·41½	133	153·61½	173	199·81½
14	16·17	54	62·37	94	108·57	134	154·77	174	200·97
15	17·32½	55	63·52½	95	109·72½	135	155·92½	175	202·12½
16	18·48	56	64·68	96	110·88	136	157·08	176	203·28
17	19·63½	57	65·83½	97	112·03½	137	158·23½	177	204·43½
18	20·79	58	66·99	98	113·19	138	159·39	178	205·59
19	21·94½	59	68·14½	99	114·34½	139	160·54½	179	206·74½
20	23·10	**60**	69·30	**100**	115·50	**140**	161·70	**180**	207·90
21	24·25½	61	70·45½	101	116·65½	141	162·85½	181	209·05½
22	25·41	62	71·61	102	117·81	142	164·01	182	210·21
23	26·56½	63	72·76½	103	118·96½	143	165·16½	183	211·36½
24	27·72	64	73·92	104	120·12	144	166·32	184	212·52
25	28·87½	65	75·07½	105	121·27½	145	167·47½	185	213·67½
26	30·03	66	76·23	106	122·43	146	168·63	186	214·83
27	31·18½	67	77·38½	107	123·58½	147	169·78½	187	215·98½
28	32·34	68	78·54	108	124·74	148	170·94	188	217·14
29	33·49½	69	79·69½	109	125·89½	149	172·09½	189	218·29½
30	34·65	**70**	80·85	**110**	127·05	**150**	173·25	**190**	219·45
31	35·80½	71	82·00½	111	128·20½	151	174·40½	191	220·60½
32	36·96	72	83·16	112	129·36	152	175·56	192	221·76
33	38·11½	73	84·31½	113	130·51½	153	176·71½	193	222·91½
34	39·27	74	85·47	114	131·67	154	177·87	194	224·07
35	40·42½	75	86·62½	115	132·82½	155	179·02½	195	225·22½
36	41·58	76	87·78	116	133·98	156	180·18	196	226·38
37	42·73½	77	88·93½	117	135·13½	157	181·33½	197	227·53½
38	43·89	78	90·09	118	136·29	158	182·49	198	228·69
39	45·04½	79	91·24½	119	137·44½	159	183·64½	199	229·84½
40	46·20	**80**	92·40	**120**	138·60	**160**	184·80	**200**	231·00

$\frac{1}{16}=0{\cdot}07^{2}$ $\frac{1}{9}=0{\cdot}12^{8}$ $\frac{1}{8}=0{\cdot}14^{4}$ $\frac{1}{6}=0{\cdot}19^{3}$ $\frac{1}{5}=0{\cdot}23^{1}$ $\frac{1}{4}=0{\cdot}28^{9}$ $\frac{1}{3}=0{\cdot}38^{5}$

$\frac{3}{8}=0{\cdot}43^{3}$ $\frac{1}{2}=0{\cdot}57^{8}$ $\frac{5}{8}=0{\cdot}72^{2}$ $\frac{3}{4}=0{\cdot}86^{6}$ $\frac{7}{8}=1{\cdot}01^{1}$ $\frac{2}{3}=0{\cdot}77^{0}$ $\frac{5}{6}=0{\cdot}96^{3}$

$\frac{3}{16}=0{\cdot}21^{7}$ $\frac{5}{16}=0{\cdot}36^{1}$ $\frac{7}{16}=0{\cdot}50^{5}$ $\frac{9}{16}=0{\cdot}65^{0}$ $\frac{11}{16}=0{\cdot}79^{4}$ $\frac{13}{16}=0{\cdot}93^{8}$ $\frac{15}{16}=1{\cdot}08^{3}$

(0.862=1£) **£1.16**

1	1·16	41	47·56	81	93·96	121	140·36	161	186·76
2	2·32	42	48·72	82	95·12	122	141·52	162	187·92
3	3·48	43	49·88	83	96·28	123	142·68	163	189·08
4	4·64	44	51·04	84	97·44	124	143·84	164	190·24
5	5·80	45	52·20	85	98·60	125	145·00	165	191·40
6	6·96	46	53·36	86	99·76	126	146·16	166	192·56
7	8·12	47	54·52	87	100·92	127	147·32	167	193·72
8	9·28	48	55·68	88	102·08	128	148·48	168	194·88
9	10·44	49	56·84	89	103·24	129	149·64	169	196·04
10	11·60	**50**	58·00	**90**	104·40	**130**	150·80	**170**	197·20
11	12·76	51	59·16	91	105·56	131	151·96	171	198·36
12	13·92	52	60·32	92	106·72	132	153·12	172	199·52
13	15·08	53	61·48	93	107·88	133	154·28	173	200·68
14	16·24	54	62·64	94	109·04	134	155·44	174	201·84
15	17·40	55	63·80	95	110·20	135	156·60	175	203·00
16	18·56	56	64·96	96	111·36	136	157·76	176	204·16
17	19·72	57	66·12	97	112·52	137	158·92	177	205·32
18	20·88	58	67·28	98	113·68	138	160·08	178	206·48
19	22·04	59	68·44	99	114·84	139	161·24	179	207·64
20	23·20	**60**	69·60	**100**	116·00	**140**	162·40	**180**	208·80
21	24·36	61	70·76	101	117·16	141	163·56	181	209·96
22	25·52	62	71·92	102	118·32	142	164·72	182	211·12
23	26·68	63	73·08	103	119·48	143	165·88	183	212·28
24	27·84	64	74·24	104	120·64	144	167·04	184	213·44
25	29·00	65	75·40	105	121·80	145	168·20	185	214·60
26	30·16	66	76·56	106	122·96	146	169·36	186	215·76
27	31·32	67	77·72	107	124·12	147	170·52	187	216·92
28	32·48	68	78·88	108	125·28	148	171·68	188	218·08
29	33·64	69	80·04	109	126·44	149	172·84	189	219·24
30	34·80	**70**	81·20	**110**	127·60	**150**	174·00	**190**	220·40
31	35·96	71	82·36	111	128·76	151	175·16	191	221·56
32	37·12	72	83·52	112	129·92	152	176·32	192	222·72
33	38·28	73	84·68	113	131·08	153	177·48	193	223·88
34	39·44	74	85·84	114	132·24	154	178·64	194	225·04
35	40·60	75	87·00	115	133·40	155	179·80	195	226·20
36	41·76	76	88·16	116	134·56	156	180·96	196	227·36
37	42·92	77	89·32	117	135·72	157	182·12	197	228·52
38	44·08	78	90·48	118	136·88	158	183·28	198	229·68
39	45·24	79	91·64	119	138·04	159	184·44	199	230·84
40	46·40	**80**	92·80	**120**	139·20	**160**	185·60	**200**	232·00

$\frac{1}{16}=0{\cdot}07^{3}$ $\frac{1}{9}=0{\cdot}12^{9}$ $\frac{1}{8}=0{\cdot}14^{5}$ $\frac{1}{6}=0{\cdot}19^{3}$ $\frac{1}{5}=0{\cdot}23^{2}$ $\frac{1}{4}=0{\cdot}29^{0}$ $\frac{1}{3}=0{\cdot}38^{7}$

$\frac{3}{8}=0{\cdot}43^{5}$ $\frac{1}{2}=0{\cdot}58^{0}$ $\frac{5}{8}=0{\cdot}72^{5}$ $\frac{3}{4}=0{\cdot}87^{0}$ $\frac{7}{8}=1{\cdot}01^{5}$ $\frac{2}{3}=0{\cdot}77^{3}$ $\frac{5}{6}=0{\cdot}96^{7}$

$\frac{3}{16}=0{\cdot}21^{8}$ $\frac{5}{16}=0{\cdot}36^{3}$ $\frac{7}{16}=0{\cdot}50^{8}$ $\frac{9}{16}=0{\cdot}65^{3}$ $\frac{11}{16}=0{\cdot}79^{8}$ $\frac{13}{16}=0{\cdot}94^{3}$ $\frac{15}{16}=1{\cdot}08^{8}$

(0.858=1£) **£1.16½**

1	1·16½	41	47·76½	81	94·36½	121	140·96½	161	187·56½
2	2·33	42	48·93	82	95·53	122	142·13	162	188·73
3	3·49½	43	50·09½	83	96·69½	123	143·29½	163	189·89½
4	4·66	44	51·26	84	97·86	124	144·46	164	191·06
5	5·82½	45	52·42½	85	99·02½	125	145·62½	165	192·22½
6	6·99	46	53·59	86	100·19	126	146·79	166	193·39
7	8·15½	47	54·75½	87	101·35½	127	147·95½	167	194·55½
8	9·32	48	55·92	88	102·52	128	149·12	168	195·72
9	10·48½	49	57·08½	89	103·68½	129	150·28½	169	196·88½
10	11·65	**50**	58·25	**90**	104·85	**130**	151·45	**170**	198·05
11	12·81½	51	59·41½	91	106·01½	131	152·61½	171	199·21½
12	13·98	52	60·58	92	107·18	132	153·78	172	200·38
13	15·14½	53	61·74½	93	108·34½	133	154·94½	173	201·54½
14	16·31	54	62·91	94	109·51	134	156·11	174	202·71
15	17·47½	55	64·07½	95	110·67½	135	157·27½	175	203·87½
16	18·64	56	65·24	96	111·84	136	158·44	176	205·04
17	19·80½	57	66·40½	97	113·00½	137	159·60½	177	206·20½
18	20·97	58	67·57	98	114·17	138	160·77	178	207·37
19	22·13½	59	68·73½	99	115·33½	139	161·93½	179	208·53½
20	23·30	**60**	69·90	**100**	116·50	**140**	163·10	**180**	209·70
21	24·46½	61	71·06½	101	117·66½	141	164·26½	181	210·86½
22	25·63	62	72·23	102	118·83	142	165·43	182	212·03
23	26·79½	63	73·39½	103	119·99½	143	166·59½	183	213·19½
24	27·96	64	74·56	104	121·16	144	167·76	184	214·36
25	29·12½	65	75·72½	105	122·32½	145	168·92½	185	215·52½
26	30·29	66	76·89	106	123·49	146	170·09	186	216·69
27	31·45½	67	78·05½	107	124·65½	147	171·25½	187	217·85½
28	32·62	68	79·22	108	125·82	148	172·42	188	219·02
29	33·78½	69	80·38½	109	126·98½	149	173·58½	189	220·18½
30	34·95	**70**	81·55	**110**	128·15	**150**	174·75	**190**	221·35
31	36·11½	71	82·71½	111	129·31½	151	175·91½	191	222·51½
32	37·28	72	83·88	112	130·48	152	177·08	192	223·68
33	38·44½	73	85·04½	113	131·64½	153	178·24½	193	224·84½
34	39·61	74	86·21	114	132·81	154	179·41	194	226·01
35	40·77½	75	87·37½	115	133·97½	155	180·57½	195	227·17½
36	41·94	76	88·54	116	135·14	156	181·74	196	228·34
37	43·10½	77	89·70½	117	136·30½	157	182·90½	197	229·50½
38	44·27	78	90·87	118	137·47	158	184·07	198	230·67
39	45·43½	79	92·03½	119	138·63½	159	185·23½	199	231·83½
40	46·60	**80**	93·20	**120**	139·80	**160**	186·40	**200**	233·00

$\frac{1}{16}=0{\cdot}07^{3}$	$\frac{1}{9}=0{\cdot}12^{9}$	$\frac{1}{8}=0{\cdot}14^{6}$	$\frac{1}{6}=0{\cdot}19^{4}$	$\frac{1}{5}=0{\cdot}23^{3}$	$\frac{1}{4}=0{\cdot}29^{1}$	$\frac{1}{3}=0{\cdot}38^{8}$
$\frac{3}{8}=0{\cdot}43^{7}$	$\frac{1}{2}=0{\cdot}58^{3}$	$\frac{5}{8}=0{\cdot}72^{8}$	$\frac{3}{4}=0{\cdot}87^{4}$	$\frac{7}{8}=1{\cdot}01^{9}$	$\frac{2}{3}=0{\cdot}77^{7}$	$\frac{5}{6}=0{\cdot}97^{1}$
$\frac{3}{16}=0{\cdot}21^{8}$	$\frac{5}{16}=0{\cdot}36^{4}$	$\frac{7}{16}=0{\cdot}51^{0}$	$\frac{9}{16}=0{\cdot}65^{5}$	$\frac{11}{16}=0{\cdot}80^{1}$	$\frac{13}{16}=0{\cdot}94^{7}$	$\frac{15}{16}=1{\cdot}09^{2}$

(0.855=1£) **£1.17**

1	1·17	41	47·97	81	94·77	121	141·57	161	188·37
2	2·34	42	49·14	82	95·94	122	142·74	162	189·54
3	3·51	43	50·31	83	97·11	123	143·91	163	190·71
4	4·68	44	51·48	84	98·28	124	145·08	164	191·88
5	5·85	45	52·65	85	99·45	125	146·25	165	193·05
6	7·02	46	53·82	86	100·62	126	147·42	166	194·22
7	8·19	47	54·99	87	101·79	127	148·59	167	195·39
8	9·36	48	56·16	88	102·96	128	149·76	168	196·56
9	10·53	49	57·33	89	104·13	129	150·93	169	197·73
10	11·70	**50**	58·50	**90**	105·30	**130**	152·10	**170**	198·90
11	12·87	51	59·67	91	106·47	131	153·27	171	200·07
12	14·04	52	60·84	92	107·64	132	154·44	172	201·24
13	15·21	53	62·01	93	108·81	133	155·61	173	202·41
14	16·38	54	63·18	94	109·98	134	156·78	174	203·58
15	17·55	55	64·35	95	111·15	135	157·95	175	204·75
16	18·72	56	65·52	96	112·32	136	159·12	176	205·92
17	19·89	57	66·69	97	113·49	137	160·29	177	207·09
18	21·06	58	67·86	98	114·66	138	161·46	178	208·26
19	22·23	59	69·03	99	115·83	139	162·63	179	209·43
20	23·40	**60**	70·20	**100**	117·00	**140**	163·80	**180**	210·60
21	24·57	61	71·37	101	118·17	141	164·97	181	211·77
22	25·74	62	72·54	102	119·34	142	166·14	182	212·94
23	26·91	63	73·71	103	120·51	143	167·31	183	214·11
24	28·08	64	74·88	104	121·68	144	168·48	184	215·28
25	29·25	65	76·05	105	122·85	145	169·65	185	216·45
26	30·42	66	77·22	106	124·02	146	170·82	186	217·62
27	31·59	67	78·39	107	125·19	147	171·99	187	218·79
28	32·76	68	79·56	108	126·36	148	173·16	188	219·96
29	33·93	69	80·73	109	127·53	149	174·33	189	221·13
30	35·10	**70**	81·90	**110**	128·70	**150**	175·50	**190**	222·30
31	36·27	71	83·07	111	129·87	151	176·67	191	223·47
32	37·44	72	84·24	112	131·04	152	177·84	192	224·64
33	38·61	73	85·41	113	132·21	153	179·01	193	225·81
34	39·78	74	86·58	114	133·38	154	180·18	194	226·98
35	40·95	75	87·75	115	134·55	155	181·35	195	228·15
36	42·12	76	88·92	116	135·72	156	182·52	196	229·32
37	43·29	77	90·09	117	136·89	157	183·69	197	230·49
38	44·46	78	91·26	118	138·06	158	184·86	198	231·66
39	45·63	79	92·43	119	139·23	159	186·03	199	232·83
40	46·80	**80**	93·60	**120**	140·40	**160**	187·20	**200**	234·00

$\frac{1}{16}=0{\cdot}07^{3}$	$\frac{1}{9}=0{\cdot}13^{0}$	$\frac{1}{8}=0{\cdot}14^{6}$	$\frac{1}{6}=0{\cdot}19^{5}$	$\frac{1}{5}=0{\cdot}23^{4}$	$\frac{1}{4}=0{\cdot}29^{3}$	$\frac{1}{3}=0{\cdot}39^{0}$
$\frac{3}{8}=0{\cdot}43^{9}$	$\frac{1}{2}=0{\cdot}58^{5}$	$\frac{5}{8}=0{\cdot}73^{1}$	$\frac{3}{4}=0{\cdot}87^{8}$	$\frac{7}{8}=1{\cdot}02^{4}$	$\frac{2}{3}=0{\cdot}78^{0}$	$\frac{5}{6}=0{\cdot}97^{5}$
$\frac{3}{16}=0{\cdot}21^{9}$	$\frac{5}{16}=0{\cdot}36^{6}$	$\frac{7}{16}=0{\cdot}51^{2}$	$\frac{9}{16}=0{\cdot}65^{8}$	$\frac{11}{16}=0{\cdot}80^{4}$	$\frac{13}{16}=0{\cdot}95^{1}$	$\frac{15}{16}=1{\cdot}09^{7}$

(0.851=1£) **£1.17½**

1	1·17½	41	48·17½	81	95·17½	121	142·17½	161	189·17½
2	2·35	42	49·35	82	96·35	122	143·35	162	190·35
3	3·52½	43	50·52½	83	97·52½	123	144·52½	163	191·52½
4	4·70	44	51·70	84	98·70	124	145·70	164	192·70
5	5·87½	45	52·87½	85	99·87½	125	146·87½	165	193·87½
6	7·05	46	54·05	86	101·05	126	148·05	166	195·05
7	8·22½	47	55·22½	87	102·22½	127	149·22½	167	196·22½
8	9·40	48	56·40	88	103·40	128	150·40	168	197·40
9	10·57½	49	57·57½	89	104·57½	129	151·57½	169	198·57½
10	11·75	**50**	58·75	**90**	105·75	**130**	152·75	**170**	199·75
11	12·92½	51	59·92½	91	106·92½	131	153·92½	171	200·92½
12	14·10	52	61·10	92	108·10	132	155·10	172	202·10
13	15·27½	53	62·27½	93	109·27½	133	156·27½	173	203·27½
14	16·45	54	63·45	94	110·45	134	157·45	174	204·45
15	17·62½	55	64·62½	95	111·62½	135	158·62½	175	205·62½
16	18·80	56	65·80	96	112·80	136	159·80	176	206·80
17	19·97½	57	66·97½	97	113·97½	137	160·97½	177	207·97½
18	21·15	58	68·15	98	115·15	138	162·15	178	209·15
19	22·32½	59	69·32½	99	116·32½	139	163·32½	179	210·32½
20	23·50	**60**	70·50	**100**	117·50	**140**	164·50	**180**	211·50
21	24·67½	61	71·67½	101	118·67½	141	165·67½	181	212·67½
22	25·85	62	72·85	102	119·85	142	166·85	182	213·85
23	27·02½	63	74·02½	103	121·02½	143	168·02½	183	215·02½
24	28·20	64	75·20	104	122·20	144	169·20	184	216·20
25	29·37½	65	76·37½	105	123·37½	145	170·37½	185	217·37½
26	30·55	66	77·55	106	124·55	146	171·55	186	218·55
27	31·72½	67	78·72½	107	125·72½	147	172·72½	187	219·72½
28	32·90	68	79·90	108	126·90	148	173·90	188	220·90
29	34·07½	69	81·07½	109	128·07½	149	175·07½	189	222·07½
30	35·25	**70**	82·25	**110**	129·25	**150**	176·25	**190**	223·25
31	36·42½	71	83·42½	111	130·42½	151	177·42½	191	224·42½
32	37·60	72	84·60	112	131·60	152	178·60	192	225·60
33	38·77½	73	85·77½	113	132·77½	153	179·77½	193	226·77½
34	39·95	74	86·95	114	133·95	154	180·95	194	227·95
35	41·12½	75	88·12½	115	135·12½	155	182·12½	195	229·12½
36	42·30	76	89·30	116	136·30	156	183·30	196	230·30
37	43·47½	77	90·47½	117	137·47½	157	184·47½	197	231·47½
38	44·65	78	91·65	118	138·65	158	185·65	198	232·65
39	45·82½	79	92·82½	119	139·82½	159	186·82½	199	233·82½
40	47·00	**80**	94·00	**120**	141·00	**160**	188·00	**200**	235·00

$\frac{1}{16}=0{\cdot}07^{3}$ $\frac{1}{9}=0{\cdot}13^{1}$ $\frac{1}{8}=0{\cdot}14^{7}$ $\frac{1}{6}=0{\cdot}19^{6}$ $\frac{1}{5}=0{\cdot}23^{5}$ $\frac{1}{4}=0{\cdot}29^{4}$ $\frac{1}{3}=0{\cdot}39^{2}$

$\frac{3}{8}=0{\cdot}44^{1}$ $\frac{1}{2}=0{\cdot}58^{8}$ $\frac{5}{8}=0{\cdot}73^{4}$ $\frac{3}{4}=0{\cdot}88^{1}$ $\frac{7}{8}=1{\cdot}02^{8}$ $\frac{2}{3}=0{\cdot}78^{3}$ $\frac{5}{6}=0{\cdot}97^{9}$

$\frac{3}{16}=0{\cdot}22^{0}$ $\frac{5}{16}=0{\cdot}36^{7}$ $\frac{7}{16}=0{\cdot}51^{4}$ $\frac{9}{16}=0{\cdot}66^{1}$ $\frac{11}{16}=0{\cdot}80^{8}$ $\frac{13}{16}=0{\cdot}95^{5}$ $\frac{15}{16}=1{\cdot}10^{2}$

(0.847=1£) **£1.18**

1	1·18	41	48·38	81	95·58	121	142·78	161	189·98
2	2·36	42	49·56	82	96·76	122	143·96	162	191·16
3	3·54	43	50·74	83	97·94	123	145·14	163	192·34
4	4·72	44	51·92	84	99·12	124	146·32	164	193·52
5	5·90	45	53·10	85	100·30	125	147·50	165	194·70
6	7·08	46	54·28	86	101·48	126	148·68	166	195·88
7	8·26	47	55·46	87	102·66	127	149·86	167	197·06
8	9·44	48	56·64	88	103·84	128	151·04	168	198·24
9	10·62	49	57·82	89	105·02	129	152·22	169	199·42
10	11·80	**50**	59·00	**90**	106·20	**130**	153·40	**170**	200·60
11	12·98	51	60·18	91	107·38	131	154·58	171	201·78
12	14·16	52	61·36	92	108·56	132	155·76	172	202·96
13	15·34	53	62·54	93	109·74	133	156·94	173	204·14
14	16·52	54	63·72	94	110·92	134	158·12	174	205·32
15	17·70	55	64·90	95	112·10	135	159·30	175	206·50
16	18·88	56	66·08	96	113·28	136	160·48	176	207·68
17	20·06	57	67·26	97	114·46	137	161·66	177	208·86
18	21·24	58	68·44	98	115·64	138	162·84	178	210·04
19	22·42	59	69·62	99	116·82	139	164·02	179	211·22
20	23·60	**60**	70·80	**100**	118·00	**140**	165·20	**180**	212·40
21	24·78	61	71·98	101	119·18	141	166·38	181	213·58
22	25·96	62	73·16	102	120·36	142	167·56	182	214·76
23	27·14	63	74·34	103	121·54	143	168·74	183	215·94
24	28·32	64	75·52	104	122·72	144	169·92	184	217·12
25	29·50	65	76·70	105	123·90	145	171·10	185	218·30
26	30·68	66	77·88	106	125·08	146	172·28	186	219·48
27	31·86	67	79·06	107	126·26	147	173·46	187	220·66
28	33·04	68	80·24	108	127·44	148	174·64	188	221·84
29	34·22	69	81·42	109	128·62	149	175·82	189	223·02
30	35·40	**70**	82·60	**110**	129·80	**150**	177·00	**190**	224·20
31	36·58	71	83·78	111	130·98	151	178·18	191	225·38
32	37·76	72	84·96	112	132·16	152	179·36	192	226·56
33	38·94	73	86·14	113	133·34	153	180·54	193	227·74
34	40·12	74	87·32	114	134·52	154	181·72	194	228·92
35	41·30	75	88·50	115	135·70	155	182·90	195	230·10
36	42·48	76	89·68	116	136·88	156	184·08	196	231·28
37	43·66	77	90·86	117	138·06	157	185·26	197	232·46
38	44·84	78	92·04	118	139·24	158	186·44	198	233·64
39	46·02	79	93·22	119	140·42	159	187·62	199	234·82
40	47·20	**80**	94·40	**120**	141·60	**160**	188·80	**200**	236·00

$\frac{1}{16}=0{\cdot}07^4$ $\frac{1}{9}=0{\cdot}13^1$ $\frac{1}{8}=0{\cdot}14^8$ $\frac{1}{6}=0{\cdot}19^7$ $\frac{1}{5}=0{\cdot}23^6$ $\frac{1}{4}=0{\cdot}29^5$ $\frac{1}{3}=0{\cdot}39^3$

$\frac{3}{8}=0{\cdot}44^3$ $\frac{1}{2}=0{\cdot}59^0$ $\frac{5}{8}=0{\cdot}73^8$ $\frac{3}{4}=0{\cdot}88^5$ $\frac{7}{8}=1{\cdot}03^3$ $\frac{2}{3}=0{\cdot}78^7$ $\frac{5}{6}=0{\cdot}98^3$

$\frac{3}{16}=0{\cdot}22^1$ $\frac{5}{16}=0{\cdot}36^9$ $\frac{7}{16}=0{\cdot}51^6$ $\frac{9}{16}=0{\cdot}66^4$ $\frac{11}{16}=0{\cdot}81^1$ $\frac{13}{16}=0{\cdot}95^9$ $\frac{15}{16}=1{\cdot}10^6$

1	1·18½	41	48·58½	81	95·98½	121	143·38½	161	190·78½
2	2·37	42	49·77	82	97·17	122	144·57	162	191·97
3	3·55½	43	50·95½	83	98·35½	123	145·75½	163	193·15½
4	4·74	44	52·14	84	99·54	124	146·94	164	194·34
5	5·92½	45	53·32½	85	100·72½	125	148·12½	165	195·52½
6	7·11	46	54·51	86	101·91	126	149·31	166	196·71
7	8·29½	47	55·69½	87	103·09½	127	150·49½	167	197·89½
8	9·48	48	56·88	88	104·28	128	151·68	168	199·08
9	10·66½	49	58·06½	89	105·46½	129	152·86½	169	200·26½
10	11·85	**50**	59·25	**90**	106·65	**130**	154·05	**170**	201·45
11	13·03½	51	60·43½	91	107·83½	131	155·23½	171	202·63½
12	14·22	52	61·62	92	109·02	132	156·42	172	203·82
13	15·40½	53	62·80½	93	110·20½	133	157·60½	173	205·00½
14	16·59	54	63·99	94	111·39	134	158·79	174	206·19
15	17·77½	55	65·17½	95	112·57½	135	159·97½	175	207·37½
16	18·96	56	66·36	96	113·76	136	161·16	176	208·56
17	20·14½	57	67·54½	97	114·94½	137	162·34½	177	209·74½
18	21·33	58	68·73	98	116·13	138	163·53	178	210·93
19	22·51½	59	69·91½	99	117·31½	139	164·71½	179	212·11½
20	23·70	**60**	71·10	**100**	118·50	**140**	165·90	**180**	213·30
21	24·88½	61	72·28½	101	119·68½	141	167·08½	181	214·48½
22	26·07	62	73·47	102	120·87	142	168·27	182	215·67
23	27·25½	63	74·65½	103	122·05½	143	169·45½	183	216·85½
24	28·44	64	75·84	104	123·24	144	170·64	184	218·04
25	29·62½	65	77·02½	105	124·42½	145	171·82½	185	219·22½
26	30·81	66	78·21	106	125·61	146	173·01	186	220·41
27	31·99½	67	79·39½	107	126·79½	147	174·19½	187	221·59½
28	33·18	68	80·58	108	127·98	148	175·38	188	222·78
29	34·36½	69	81·76½	109	129·16½	149	176·56½	189	223·96½
30	35·55	**70**	82·95	**110**	130·35	**150**	177·75	**190**	225·15
31	36·73½	71	84·13½	111	131·53½	151	178·93½	191	226·33½
32	37·92	72	85·32	112	132·72	152	180·12	192	227·52
33	39·10½	73	86·50½	113	133·90½	153	181·30½	193	228·70½
34	40·29	74	87·69	114	135·09	154	182·49	194	229·89
35	41·47½	75	88·87½	115	136·27½	155	183·67½	195	231·07½
36	42·66	76	90·06	116	137·46	156	184·86	196	232·26
37	43·84½	77	91·24½	117	138·64½	157	186·04½	197	233·44½
38	45·03	78	92·43	118	139·83	158	187·23	198	234·63
39	46·21½	79	93·61½	119	141·01½	159	188·41½	199	235·81½
40	47·40	**80**	94·80	**120**	142·20	**160**	189·60	**200**	237·00

$\frac{1}{16}$=0·07^{4} $\frac{1}{9}$=0·13^{2} $\frac{1}{8}$=0·14^{8} $\frac{1}{6}$=0·19^{8} $\frac{1}{5}$=0·23^{7} $\frac{1}{4}$=0·29^{6} $\frac{1}{3}$=0·39^{5}

$\frac{3}{8}$=0·44^{4} $\frac{1}{2}$=0·59^{3} $\frac{5}{8}$=0·74^{1} $\frac{3}{4}$=0·88^{9} $\frac{7}{8}$=1·03^{7} $\frac{2}{3}$=0·79^{0} $\frac{5}{6}$=0·98^{8}

$\frac{3}{16}$=0·22^{2} $\frac{5}{16}$=0·37^{0} $\frac{7}{16}$=0·51^{8} $\frac{9}{16}$=0·66^{7} $\frac{11}{16}$=0·81^{5} $\frac{13}{16}$=0·96^{3} $\frac{15}{16}$=1·11^{1}

(0.840=1£) **£1.19**

1	1·19	41	48·79	81	96·39	121	143·99	161	191·59
2	2·38	42	49·98	82	97·58	122	145·18	162	192·78
3	3·57	43	51·17	83	98·77	123	146·37	163	193·97
4	4·76	44	52·36	84	99·96	124	147·56	164	195·16
5	5·95	45	53·55	85	101·15	125	148·75	165	196·35
6	7·14	46	54·74	86	102·34	126	149·94	166	197·54
7	8·33	47	55·93	87	103·53	127	151·13	167	198·73
8	9·52	48	57·12	88	104·72	128	152·32	168	199·92
9	10·71	49	58·31	89	105·91	129	153·51	169	201·11
10	11·90	**50**	59·50	**90**	107·10	**130**	154·70	**170**	202·30
11	13·09	51	60·69	91	108·29	131	155·89	171	203·49
12	14·28	52	61·88	92	109·48	132	157·08	172	204·68
13	15·47	53	63·07	93	110·67	133	158·27	173	205·87
14	16·66	54	64·26	94	111·86	134	159·46	174	207·06
15	17·85	55	65·45	95	113·05	135	160·65	175	208·25
16	19·04	56	66·64	96	114·24	136	161·84	176	209·44
17	20·23	57	67·83	97	115·43	137	163·03	177	210·63
18	21·42	58	69·02	98	116·62	138	164·22	178	211·82
19	22·61	59	70·21	99	117·81	139	165·41	179	213·01
20	23·80	**60**	71·40	**100**	119·00	**140**	166·60	**180**	214·20
21	24·99	61	72·59	101	120·19	141	167·79	181	215·39
22	26·18	62	73·78	102	121·38	142	168·98	182	216·58
23	27·37	63	74·97	103	122·57	143	170·17	183	217·77
24	28·56	64	76·16	104	123·76	144	171·36	184	218·96
25	29·75	65	77·35	105	124·95	145	172·55	185	220·15
26	30·94	66	78·54	106	126·14	146	173·74	186	221·34
27	32·13	67	79·73	107	127·33	147	174·93	187	222·53
28	33·32	68	80·92	108	128·52	148	176·12	188	223·72
29	34·51	69	82·11	109	129·71	149	177·31	189	224·91
30	35·70	**70**	83·30	**110**	130·90	**150**	178·50	**190**	226·10
31	36·89	71	84·49	111	132·09	151	179·69	191	227·29
32	38·08	72	85·68	112	133·28	152	180·88	192	228·48
33	39·27	73	86·87	113	134·47	153	182·07	193	229·67
34	40·46	74	88·06	114	135·66	154	183·26	194	230·86
35	41·65	75	89·25	115	136·85	155	184·45	195	232·05
36	42·84	76	90·44	116	138·04	156	185·64	196	233·24
37	44·03	77	91·63	117	139·23	157	186·83	197	234·43
38	45·22	78	92·82	118	140·42	158	188·02	198	235·62
39	46·41	79	94·01	119	141·61	159	189·21	199	236·81
40	47·60	**80**	95·20	**120**	142·80	**160**	190·40	**200**	238·00

$\frac{1}{16}=0{\cdot}07^{4}$ $\frac{1}{9}=0{\cdot}13^{2}$ $\frac{1}{8}=0{\cdot}14^{9}$ $\frac{1}{6}=0{\cdot}19^{8}$ $\frac{1}{5}=0{\cdot}23^{8}$ $\frac{1}{4}=0{\cdot}29^{8}$ $\frac{1}{3}=0{\cdot}39^{7}$

$\frac{3}{8}=0{\cdot}44^{6}$ $\frac{1}{2}=0{\cdot}59^{5}$ $\frac{5}{8}=0{\cdot}74^{4}$ $\frac{3}{4}=0{\cdot}89^{3}$ $\frac{7}{8}=1{\cdot}04^{1}$ $\frac{2}{3}=0{\cdot}79^{3}$ $\frac{5}{6}=0{\cdot}99^{2}$

$\frac{3}{16}=0{\cdot}22^{3}$ $\frac{5}{16}=0{\cdot}37^{2}$ $\frac{7}{16}=0{\cdot}52^{1}$ $\frac{9}{16}=0{\cdot}66^{9}$ $\frac{11}{16}=0{\cdot}81^{8}$ $\frac{13}{16}=0{\cdot}96^{7}$ $\frac{15}{16}=1{\cdot}11^{6}$

(0.837=1£) **£1.19½**

1	1·19½	41	48·99½	81	96·79½	121	144·59½	161	192·39½
2	2·39	42	50·19	82	97·99	122	145·79	162	193·59
3	3·58½	43	51·38½	83	99·18½	123	146·98½	163	194·78½
4	4·78	44	52·58	84	100·38	124	148·18	164	195·98
5	5·97½	45	53·77½	85	101·57½	125	149·37½	165	197·17½
6	7·17	46	54·97	86	102·77	126	150·57	166	198·37
7	8·36½	47	56·16½	87	103·96½	127	151·76½	167	199·56½
8	9·56	48	57·36	88	105·16	128	152·96	168	200·76
9	10·75½	49	58·55½	89	106·35½	129	154·15½	169	201·95½
10	11·95	**50**	59·75	**90**	107·55	**130**	155·35	**170**	203·15
11	13·14½	51	60·94½	91	108·74½	131	156·54½	171	204·34½
12	14·34	52	62·14	92	109·94	132	157·74	172	205·54
13	15·53½	53	63·33½	93	111·13½	133	158·93½	173	206·73½
14	16·73	54	64·53	94	112·33	134	160·13	174	207·93
15	17·92½	55	65·72½	95	113·52½	135	161·32½	175	209·12½
16	19·12	56	66·92	96	114·72	136	162·52	176	210·32
17	20·31½	57	68·11½	97	115·91½	137	163·71½	177	211·51½
18	21·51	58	69·31	98	117·11	138	164·91	178	212·71
19	22·70½	59	70·50½	99	118·30½	139	166·10½	179	213·90½
20	23·90	**60**	71·70	**100**	119·50	**140**	167·30	**180**	215·10
21	25·09½	61	72·89½	101	120·69½	141	168·49½	181	216·29½
22	26·29	62	74·09	102	121·89	142	169·69	182	217·49
23	27·48½	63	75·28½	103	123·08½	143	170·88½	183	218·68½
24	28·68	64	76·48	104	124·28	144	172·08	184	219·88
25	29·87½	65	77·67½	105	125·47½	145	173·27½	185	221·07½
26	31·07	66	78·87	106	126·67	146	174·47	186	222·27
27	32·26½	67	80·06½	107	127·86½	147	175·66½	187	223·46½
28	33·46	68	81·26	108	129·06	148	176·86	188	224·66
29	34·65½	69	82·45½	109	130·25½	149	178·05½	189	225·85½
30	35·85	**70**	83·65	**110**	131·45	**150**	179·25	**190**	227·05
31	37·04½	71	84·84½	111	132·64½	151	180·44½	191	228·24½
32	38·24	72	86·04	112	133·84	152	181·64	192	229·44
33	39·43½	73	87·23½	113	135·03½	153	182·83½	193	230·63½
34	40·63	74	88·43	114	136·23	154	184·03	194	231·83
35	41·82½	75	89·62½	115	137·42½	155	185·22½	195	233·02½
36	43·02	76	90·82	116	138·62	156	186·42	196	234·22
37	44·21½	77	92·01½	117	139·81½	157	187·61½	197	235·41½
38	45·41	78	93·21	118	141·01	158	188·81	198	236·61
39	46·60½	79	94·40½	119	142·20½	159	190·00½	199	237·80½
40	47·80	**80**	95·60	**120**	143·40	**160**	191·20	**200**	239·00

$\frac{1}{16}=0{\cdot}07^5$ $\frac{1}{9}=0{\cdot}13^3$ $\frac{1}{8}=0{\cdot}14^9$ $\frac{1}{6}=0{\cdot}19^9$ $\frac{1}{5}=0{\cdot}23^9$ $\frac{1}{4}=0{\cdot}29^9$ $\frac{1}{3}=0{\cdot}39^8$

$\frac{3}{8}=0{\cdot}44^8$ $\frac{1}{2}=0{\cdot}59^8$ $\frac{5}{8}=0{\cdot}74^7$ $\frac{3}{4}=0{\cdot}89^6$ $\frac{7}{8}=1{\cdot}04^6$ $\frac{2}{3}=0{\cdot}79^7$ $\frac{5}{6}=0{\cdot}99^6$

$\frac{3}{16}=0{\cdot}22^4$ $\frac{5}{16}=0{\cdot}37^3$ $\frac{7}{16}=0{\cdot}52^3$ $\frac{9}{16}=0{\cdot}67^2$ $\frac{11}{16}=0{\cdot}82^2$ $\frac{13}{16}=0{\cdot}97^1$ $\frac{15}{16}=1{\cdot}12^0$

1 square metre=1·196 square yards

(0.833=1£) **£1.20**

1	1·20	41	49·20	81	97·20	121	145·20	161	193·20
2	2·40	42	50·40	82	98·40	122	146·40	162	194·40
3	3·60	43	51·60	83	99·60	123	147·60	163	195·60
4	4·80	44	52·80	84	100·80	124	148·80	164	196·80
5	6·00	45	54·00	85	102·00	125	150·00	165	198·00
6	7·20	46	55·20	86	103·20	126	151·20	166	199·20
7	8·40	47	56·40	87	104·40	127	152·40	167	200·40
8	9·60	48	57·60	88	105·60	128	153·60	168	201·60
9	10·80	49	58·80	89	106·80	129	154·80	169	202·80
10	12·00	**50**	60·00	**90**	108·00	**130**	156·00	**170**	204·00
11	13·20	51	61·20	91	109·20	131	157·20	171	205·20
12	14·40	52	62·40	92	110·40	132	158·40	172	206·40
13	15·60	53	63·60	93	111·60	133	159·60	173	207·60
14	16·80	54	64·80	94	112·80	134	160·80	174	208·80
15	18·00	55	66·00	95	114·00	135	162·00	175	210·00
16	19·20	56	67·20	96	115·20	136	163·20	176	211·20
17	20·40	57	68·40	97	116·40	137	164·40	177	212·40
18	21·60	58	69·60	98	117·60	138	165·60	178	213·60
19	22·80	59	70·80	99	118·80	139	166·80	179	214·80
20	24·00	**60**	72·00	**100**	120·00	**140**	168·00	**180**	216·00
21	25·20	61	73·20	101	121·20	141	169·20	181	217·20
22	26·40	62	74·40	102	122·40	142	170·40	182	218·40
23	27·60	63	75·60	103	123·60	143	171·60	183	219·60
24	28·80	64	76·80	104	124·80	144	172·80	184	220·80
25	30·00	65	78·00	105	126·00	145	174·00	185	222·00
26	31·20	66	79·20	106	127·20	146	175·20	186	223·20
27	32·40	67	80·40	107	128·40	147	176·40	187	224·40
28	33·60	68	81·60	108	129·60	148	177·60	188	225·60
29	34·80	69	82·80	109	130·80	149	178·80	189	226·80
30	36·00	**70**	84·00	**110**	132·00	**150**	180·00	**190**	228·00
31	37·20	71	85·20	111	133·20	151	181·20	191	229·20
32	38·40	72	86·40	112	134·40	152	182·40	192	230·40
33	39·60	73	87·60	113	135·60	153	183·60	193	231·60
34	40·80	74	88·80	114	136·80	154	184·80	194	232·80
35	42·00	75	90·00	115	138·00	155	186·00	195	234·00
36	43·20	76	91·20	116	139·20	156	187·20	196	235·20
37	44·40	77	92·40	117	140·40	157	188·40	197	236·40
38	45·60	78	93·60	118	141·60	158	189·60	198	237·60
39	46·80	79	94·80	119	142·80	159	190·80	199	238·80
40	48·00	**80**	96·00	**120**	144·00	**160**	192·00	**200**	240·00

$\frac{1}{16}=0{\cdot}07^5$ $\frac{1}{9}=0{\cdot}13^3$ $\frac{1}{8}=0{\cdot}15^0$ $\frac{1}{6}=0{\cdot}20^0$ $\frac{1}{5}=0{\cdot}24^0$ $\frac{1}{4}=0{\cdot}30^0$ $\frac{1}{3}=0{\cdot}40^0$

$\frac{3}{8}=0{\cdot}45^0$ $\frac{1}{2}=0{\cdot}60^0$ $\frac{5}{8}=0{\cdot}75^0$ $\frac{3}{4}=0{\cdot}90^0$ $\frac{7}{8}=1{\cdot}05^0$ $\frac{2}{3}=0{\cdot}80^0$ $\frac{5}{6}=1{\cdot}00^0$

$\frac{3}{16}=0{\cdot}22^5$ $\frac{5}{16}=0{\cdot}37^5$ $\frac{7}{16}=0{\cdot}52^5$ $\frac{9}{16}=0{\cdot}67^5$ $\frac{11}{16}=0{\cdot}82^5$ $\frac{13}{16}=0{\cdot}97^5$ $\frac{15}{16}=1{\cdot}12^5$

1	1·20½	41	49·40½	81	97·60½	121	145·80½	161	194·00½
2	2·41	42	50·61	82	98·81	122	147·01	162	195·21
3	3·61½	43	51·81½	83	100·01½	123	148·21½	163	196·41½
4	4·82	44	53·02	84	101·22	124	149·42	164	197·62
5	6·02½	45	54·22½	85	102·42½	125	150·62½	165	198·82½
6	7·23	46	55·43	86	103·63	126	151·83	166	200·03
7	8·43½	47	56·63½	87	104·83½	127	153·03½	167	201·23½
8	9·64	48	57·84	88	106·04	128	154·24	168	202·44
9	10·84½	49	59·04½	89	107·24½	129	155·44½	169	203·64½
10	12·05	**50**	60·25	**90**	108·45	**130**	156·65	**170**	204·85
11	13·25½	51	61·45½	91	109·65½	131	157·85½	171	206·05½
12	14·46	52	62·66	92	110·86	132	159·06	172	207·26
13	15·66½	53	63·86½	93	112·06½	133	160·26½	173	208·46½
14	16·87	54	65·07	94	113·27	134	161·47	174	209·67
15	18·07½	55	66·27½	95	114·47½	135	162·67½	175	210·87½
16	19·28	56	67·48	96	115·68	136	163·88	176	212·08
17	20·48½	57	68·68½	97	116·88½	137	165·08½	177	213·28½
18	21·69	58	69·89	98	118·09	138	166·29	178	214·49
19	22·89½	59	71·09½	99	119·29½	139	167·49½	179	215·69½
20	24·10	**60**	72·30	**100**	120·50	**140**	168·70	**180**	216·90
21	25·30½	61	73·50½	101	121·70½	141	169·90½	181	218·10½
22	26·51	62	74·71	102	122·91	142	171·11	182	219·31
23	27·71½	63	75·91½	103	124·11½	143	172·31½	183	220·51½
24	28·92	64	77·12	104	125·32	144	173·52	184	221·72
25	30·12½	65	78·32½	105	126·52½	145	174·72½	185	222·92½
26	31·33	66	79·53	106	127·73	146	175·93	186	224·13
27	32·53½	67	80·73½	107	128·93½	147	177·13½	187	225·33½
28	33·74	68	81·94	108	130·14	148	178·34	188	226·54
29	34·94½	69	83·14½	109	131·34½	149	179·54½	189	227·74½
30	36·15	**70**	84·35	**110**	132·55	**150**	180·75	**190**	228·95
31	37·35½	71	85·55½	111	133·75½	151	181·95½	191	230·15½
32	38·56	72	86·76	112	134·96	152	183·16	192	231·36
33	39·76½	73	87·96½	113	136·16½	153	184·36½	193	232·56½
34	40·97	74	89·17	114	137·37	154	185·57	194	233·77
35	42·17½	75	90·37½	115	138·57½	155	186·77½	195	234·97½
36	43·38	76	91·58	116	139·78	156	187·98	196	236·18
37	44·58½	77	92·78½	117	140·98½	157	189·18½	197	237·38½
38	45·79	78	93·99	118	142·19	158	190·39	198	238·59
39	46·99½	79	95·19½	119	143·39½	159	191·59½	199	239·79½
40	48·20	**80**	96·40	**120**	144·60	**160**	192·80	**200**	241·00

$\frac{1}{16}=0{\cdot}07^{5}$ $\frac{1}{9}=0{\cdot}13^{4}$ $\frac{1}{8}=0{\cdot}15^{1}$ $\frac{1}{6}=0{\cdot}20^{1}$ $\frac{1}{5}=0{\cdot}24^{1}$ $\frac{1}{4}=0{\cdot}30^{1}$ $\frac{1}{3}=0{\cdot}40^{2}$

$\frac{3}{8}=0{\cdot}45^{2}$ $\frac{1}{2}=0{\cdot}60^{3}$ $\frac{5}{8}=0{\cdot}75^{3}$ $\frac{3}{4}=0{\cdot}90^{4}$ $\frac{7}{8}=1{\cdot}05^{4}$ $\frac{2}{3}=0{\cdot}80^{3}$ $\frac{5}{6}=1{\cdot}00^{4}$

$\frac{3}{16}=0{\cdot}22^{6}$ $\frac{5}{16}=0{\cdot}37^{7}$ $\frac{7}{16}=0{\cdot}52^{7}$ $\frac{9}{16}=0{\cdot}67^{8}$ $\frac{11}{16}=0{\cdot}82^{8}$ $\frac{13}{16}=0{\cdot}97^{9}$ $\frac{15}{16}=1{\cdot}13^{0}$

(0.826=1£) **£1.21**

1	1·21	41	49·61	81	98·01	121	146·41	161	194·81
2	2·42	42	50·82	82	99·22	122	147·62	162	196·02
3	3·63	43	52·03	83	100·43	123	148·83	163	197·23
4	4·84	44	53·24	84	101·64	124	150·04	164	198·44
5	6·05	45	54·45	85	102·85	125	151·25	165	199·65
6	7·26	46	55·66	86	104·06	126	152·46	166	200·86
7	8·47	47	56·87	87	105·27	127	153·67	167	202·07
8	9·68	48	58·08	88	106·48	128	154·88	168	203·28
9	10·89	49	59·29	89	107·69	129	156·09	169	204·49
10	12·10	**50**	60·50	**90**	108·90	**130**	157·30	**170**	205·70
11	13·31	51	61·71	91	110·11	131	158·51	171	206·91
12	14·52	52	62·92	92	111·32	132	159·72	172	208·12
13	15·73	53	64·13	93	112·53	133	160·93	173	209·33
14	16·94	54	65·34	94	113·74	134	162·14	174	210·54
15	18·15	55	66·55	95	114·95	135	163·35	175	211·75
16	19·36	56	67·76	96	116·16	136	164·56	176	212·96
17	20·57	57	68·97	97	117·37	137	165·77	177	214·17
18	21·78	58	70·18	98	118·58	138	166·98	178	215·38
19	22·99	59	71·39	99	119·79	139	168·19	179	216·59
20	24·20	**60**	72·60	**100**	121·00	**140**	169·40	**180**	217·80
21	25·41	61	73·81	101	122·21	141	170·61	181	219·01
22	26·62	62	75·02	102	123·42	142	171·82	182	220·22
23	27·83	63	76·23	103	124·63	143	173·03	183	221·43
24	29·04	64	77·44	104	125·84	144	174·24	184	222·64
25	30·25	65	78·65	105	127·05	145	175·45	185	223·85
26	31·46	66	79·86	106	128·26	146	176·66	186	225·06
27	32·67	67	81·07	107	129·47	147	177·87	187	226·27
28	33·88	68	82·28	108	130·68	148	179·08	188	227·48
29	35·09	69	83·49	109	131·89	149	180·29	189	228·69
30	36·30	**70**	84·70	**110**	133·10	**150**	181·50	**190**	229·90
31	37·51	71	85·91	111	134·31	151	182·71	191	231·11
32	38·72	72	87·12	112	135·52	152	183·92	192	232·32
33	39·93	73	88·33	113	136·73	153	185·13	193	233·53
34	41·14	74	89·54	114	137·94	154	186·34	194	234·74
35	42·35	75	90·75	115	139·15	155	187·55	195	235·95
36	43·56	76	91·96	116	140·36	156	188·76	196	237·16
37	44·77	77	93·17	117	141·57	157	189·97	197	238·37
38	45·98	78	94·38	118	142·78	158	191·18	198	239·58
39	47·19	79	95·59	119	143·99	159	192·39	199	240·79
40	48·40	**80**	96·80	**120**	145·20	**160**	193·60	**200**	242·00

$\frac{1}{16}=0{\cdot}07^{6}$ $\frac{1}{9}=0{\cdot}13^{4}$ $\frac{1}{8}=0{\cdot}15^{1}$ $\frac{1}{6}=0{\cdot}20^{2}$ $\frac{1}{5}=0{\cdot}24^{2}$ $\frac{1}{4}=0{\cdot}30^{3}$ $\frac{1}{3}=0{\cdot}40^{3}$

$\frac{3}{8}=0{\cdot}45^{4}$ $\frac{1}{2}=0{\cdot}60^{5}$ $\frac{5}{8}=0{\cdot}75^{6}$ $\frac{3}{4}=0{\cdot}90^{8}$ $\frac{7}{8}=1{\cdot}05^{9}$ $\frac{2}{3}=0{\cdot}80^{7}$ $\frac{5}{6}=1{\cdot}00^{8}$

$\frac{3}{16}=0{\cdot}22^{7}$ $\frac{5}{16}=0{\cdot}37^{8}$ $\frac{7}{16}=0{\cdot}52^{9}$ $\frac{9}{16}=0{\cdot}68^{1}$ $\frac{11}{16}=0{\cdot}83^{2}$ $\frac{13}{16}=0{\cdot}98^{3}$ $\frac{15}{16}=1{\cdot}13^{4}$

(0.823=1£) **£1.21½**

1	1·21½	41	49·81½	81	98·41½	121	147·01½	161	195·61½
2	2·43	42	51·03	82	99·63	122	148·23	162	196·83
3	3·64½	43	52·24½	83	100·84½	123	149·44½	163	198·04½
4	4·86	44	53·46	84	102·06	124	150·66	164	199·26
5	6·07½	45	54·67½	85	103·27½	125	151·87½	165	200·47½
6	7·29	46	55·89	86	104·49	126	153·09	166	201·69
7	8·50½	47	57·10½	87	105·70½	127	154·30½	167	202·90½
8	9·72	48	58·32	88	106·92	128	155·52	168	204·12
9	10·93½	49	59·53½	89	108·13½	129	156·73½	169	205·33½
10	12·15	**50**	60·75	**90**	109·35	**130**	157·95	**170**	206·55
11	13·36½	51	61·96½	91	110·56½	131	159·16½	171	207·76½
12	14·58	52	63·18	92	111·78	132	160·38	172	208·98
13	15·79½	53	64·39½	93	112·99½	133	161·59½	173	210·19½
14	17·01	54	65·61	94	114·21	134	162·81	174	211·41
15	18·22½	55	66·82½	95	115·42½	135	164·02½	175	212·62½
16	19·44	56	68·04	96	116·64	136	165·24	176	213·84
17	20·65½	57	69·25½	97	117·85½	137	166·45½	177	215·05½
18	21·87	58	70·47	98	119·07	138	167·67	178	216·27
19	23·08½	59	71·68½	99	120·28½	139	168·88½	179	217·48½
20	24·30	**60**	72·90	**100**	121·50	**140**	170·10	**180**	218·70
21	25·51½	61	74·11½	101	122·71½	141	171·31½	181	219·91½
22	26·73	62	75·33	102	123·93	142	172·53	182	221·13
23	27·94½	63	76·54½	103	125·14½	143	173·74½	183	222·34½
24	29·16	64	77·76	104	126·36	144	174·96	184	223·56
25	30·37½	65	78·97½	105	127·57½	145	176·17½	185	224·77½
26	31·59	66	80·19	106	128·79	146	177·39	186	225·99
27	32·80½	67	81·40½	107	130·00½	147	178·60½	187	227·20½
28	34·02	68	82·62	108	131·22	148	179·82	188	228·42
29	35·23½	69	83·83½	109	132·43½	149	181·03½	189	229·63½
30	36·45	**70**	85·05	**110**	133·65	**150**	182·25	**190**	230·85
31	37·66½	71	86·26½	111	134·86½	151	183·46½	191	232·06½
32	38·88	72	87·48	112	136·08	152	184·68	192	233·28
33	40·09½	73	88·69½	113	137·29½	153	185·89½	193	234·49½
34	41·31	74	89·91	114	138·51	154	187·11	194	235·71
35	42·52½	75	91·12½	115	139·72½	155	188·32½	195	236·92½
36	43·74	76	92·34	116	140·94	156	189·54	196	238·14
37	44·95½	77	93·55½	117	142·15½	157	190·75½	197	239·35½
38	46·17	78	94·77	118	143·37	158	191·97	198	240·57
39	47·38½	79	95·98½	119	144·58½	159	193·18½	199	241·78½
40	48·60	**80**	97·20	**120**	145·80	**160**	194·40	**200**	243·00

$\frac{1}{16}=0{\cdot}07^{6}$ $\frac{1}{9}=0{\cdot}13^{5}$ $\frac{1}{8}=0{\cdot}15^{2}$ $\frac{1}{6}=0{\cdot}20^{3}$ $\frac{1}{5}=0{\cdot}24^{3}$ $\frac{1}{4}=0{\cdot}30^{4}$ $\frac{1}{3}=0{\cdot}40^{5}$

$\frac{3}{8}=0{\cdot}45^{6}$ $\frac{1}{2}=0{\cdot}60^{8}$ $\frac{5}{8}=0{\cdot}75^{9}$ $\frac{3}{4}=0{\cdot}91^{1}$ $\frac{7}{8}=1{\cdot}06^{3}$ $\frac{2}{3}=0{\cdot}81^{0}$ $\frac{5}{6}=1{\cdot}01^{3}$

$\frac{3}{16}=0{\cdot}22^{8}$ $\frac{5}{16}=0{\cdot}38^{0}$ $\frac{7}{16}=0{\cdot}53^{2}$ $\frac{9}{16}=0{\cdot}68^{3}$ $\frac{11}{16}=0{\cdot}83^{5}$ $\frac{13}{16}=0{\cdot}98^{7}$ $\frac{15}{16}=1{\cdot}13^{9}$

(0.820=1£) **£1.22**

1	1·22	41	50·02	81	98·82	121	147·62	161	196·42
2	2·44	42	51·24	82	100·04	122	148·84	162	197·64
3	3·66	43	52·46	83	101·26	123	150·06	163	198·86
4	4·88	44	53·68	84	102·48	124	151·28	164	200·08
5	6·10	45	54·90	85	103·70	125	152·50	165	201·30
6	7·32	46	56·12	86	104·92	126	153·72	166	202·52
7	8·54	47	57·34	87	106·14	127	154·94	167	203·74
8	9·76	48	58·56	88	107·36	128	156·16	168	204·96
9	10·98	49	59·78	89	108·58	129	157·38	169	206·18
10	12·20	**50**	61·00	**90**	109·80	**130**	158·60	**170**	207·40
11	13·42	51	62·22	91	111·02	131	159·82	171	208·62
12	14·64	52	63·44	92	112·24	132	161·04	172	209·84
13	15·86	53	64·66	93	113·46	133	162·26	173	211·06
14	17·08	54	65·88	94	114·68	134	163·48	174	212·28
15	18·30	55	67·10	95	115·90	135	164·70	175	213·50
16	19·52	56	68·32	96	117·12	136	165·92	176	214·72
17	20·74	57	69·54	97	118·34	137	167·14	177	215·94
18	21·96	58	70·76	98	119·56	138	168·36	178	217·16
19	23·18	59	71·98	99	120·78	139	169·58	179	218·38
20	24·40	**60**	73·20	**100**	122·00	**140**	170·80	**180**	219·60
21	25·62	61	74·42	101	123·22	141	172·02	181	220·82
22	26·84	62	75·64	102	124·44	142	173·24	182	222·04
23	28·06	63	76·86	103	125·66	143	174·46	183	223·26
24	29·28	64	78·08	104	126·88	144	175·68	184	224·48
25	30·50	65	79·30	105	128·10	145	176·90	185	225·70
26	31·72	66	80·52	106	129·32	146	178·12	186	226·92
27	32·94	67	81·74	107	130·54	147	179·34	187	228·14
28	34·16	68	82·96	108	131·76	148	180·56	188	229·36
29	35·38	69	84·18	109	132·98	149	181·78	189	230·58
30	36·60	**70**	85·40	**110**	134·20	**150**	183·00	**190**	231·80
31	37·82	71	86·62	111	135·42	151	184·22	191	233·02
32	39·04	72	87·84	112	136·64	152	185·44	192	234·24
33	40·26	73	89·06	113	137·86	153	186·66	193	235·46
34	41·48	74	90·28	114	139·08	154	187·88	194	236·68
35	42·70	75	91·50	115	140·30	155	189·10	195	237·90
36	43·92	76	92·72	116	141·52	156	190·32	196	239·12
37	45·14	77	93·94	117	142·74	157	191·54	197	240·34
38	46·36	78	95·16	118	143·96	158	192·76	198	241·56
39	47·58	79	96·38	119	145·18	159	193·98	199	242·78
40	48·80	**80**	97·60	**120**	146·40	**160**	195·20	**200**	244·00

$\frac{1}{16}=0{\cdot}07^{6}$ $\frac{1}{9}=0{\cdot}13^{6}$ $\frac{1}{8}=0{\cdot}15^{3}$ $\frac{1}{6}=0{\cdot}20^{3}$ $\frac{1}{5}=0{\cdot}24^{4}$ $\frac{1}{4}=0{\cdot}30^{5}$ $\frac{1}{3}=0{\cdot}40^{7}$

$\frac{3}{8}=0{\cdot}45^{8}$ $\frac{1}{2}=0{\cdot}61^{0}$ $\frac{5}{8}=0{\cdot}76^{3}$ $\frac{3}{4}=0{\cdot}91^{5}$ $\frac{7}{8}=1{\cdot}06^{8}$ $\frac{2}{3}=0{\cdot}81^{3}$ $\frac{5}{6}=1{\cdot}01^{7}$

$\frac{3}{16}=0{\cdot}22^{9}$ $\frac{5}{16}=0{\cdot}38^{1}$ $\frac{7}{16}=0{\cdot}53^{4}$ $\frac{9}{16}=0{\cdot}68^{6}$ $\frac{11}{16}=0{\cdot}83^{9}$ $\frac{13}{16}=0{\cdot}99^{1}$ $\frac{15}{16}=1{\cdot}14^{4}$

1	1·22½	41	50·22½	81	99·22½	121	148·22½	161	197·22½
2	2·45	42	51·45	82	100·45	122	149·45	162	198·45
3	3·67½	43	52·67½	83	101·67½	123	150·67½	163	199·67½
4	4·90	44	53·90	84	102·90	124	151·90	164	200·90
5	6·12½	45	55·12½	85	104·12½	125	153·12½	165	202·12½
6	7·35	46	56·35	86	105·35	126	154·35	166	203·35
7	8·57½	47	57·57½	87	106·57½	127	155·57½	167	204·57½
8	9·80	48	58·80	88	107·80	128	156·80	168	205·80
9	11·02½	49	60·02½	89	109·02½	129	158·02½	169	207·02½
10	12·25	**50**	61·25	**90**	110·25	**130**	159·25	**170**	208·25
11	13·47½	51	62·47½	91	111·47½	131	160·47½	171	209·47½
12	14·70	52	63·70	92	112·70	132	161·70	172	210·70
13	15·92½	53	64·92½	93	113·92½	133	162·92½	173	211·92½
14	17·15	54	66·15	94	115·15	134	164·15	174	213·15
15	18·37½	55	67·37½	95	116·37½	135	165·37½	175	214·37½
16	19·60	56	68·60	96	117·60	136	166·60	176	215·60
17	20·82½	57	69·82½	97	118·82½	137	167·82½	177	216·82½
18	22·05	58	71·05	98	120·05	138	169·05	178	218·05
19	23·27½	59	72·27½	99	121·27½	139	170·27½	179	219·27½
20	24·50	**60**	73·50	**100**	122·50	**140**	171·50	**180**	220·50
21	25·72½	61	74·72½	101	123·72½	141	172·72½	181	221·72½
22	26·95	62	75·95	102	124·95	142	173·95	182	222·95
23	28·17½	63	77·17½	103	126·17½	143	175·17½	183	224·17½
24	29·40	64	78·40	104	127·40	144	176·40	184	225·40
25	30·62½	65	79·62½	105	128·62½	145	177·62½	185	226·62½
26	31·85	66	80·85	106	129·85	146	178·85	186	227·85
27	33·07½	67	82·07½	107	131·07½	147	180·07½	187	229·07½
28	34·30	68	83·30	108	132·30	148	181·30	188	230·30
29	35·52½	69	84·52½	109	133·52½	149	182·52½	189	231·52½
30	36·75	**70**	85·75	**110**	134·75	**150**	183·75	**190**	232·75
31	37·97½	71	86·97½	111	135·97½	151	184·97½	191	233·97½
32	39·20	72	88·20	112	137·20	152	186·20	192	235·20
33	40·42½	73	89·42½	113	138·42½	153	187·42½	193	236·42½
34	41·65	74	90·65	114	139·65	154	188·65	194	237·65
35	42·87½	75	91·87½	115	140·87½	155	189·87½	195	238·87½
36	44·10	76	93·10	116	142·10	156	191·10	196	240·10
37	45·32½	77	94·32½	117	143·32½	157	192·32½	197	241·32½
38	46·55	78	95·55	118	144·55	158	193·55	198	242·55
39	47·77½	79	96·77½	119	145·77½	159	194·77½	199	243·77½
40	49·00	**80**	98·00	**120**	147·00	**160**	196·00	**200**	245·00

$\frac{1}{16}$=0·07^{7} $\frac{1}{9}$=0·13^{6} $\frac{1}{8}$=0·15^{3} $\frac{1}{6}$=0·20^{4} $\frac{1}{5}$=0·24^{5} $\frac{1}{4}$=0·30^{6} $\frac{1}{3}$=0·40^{8}

$\frac{3}{8}$=0·45^{9} $\frac{1}{2}$=0·61^{3} $\frac{5}{8}$=0·76^{6} $\frac{3}{4}$=0·91^{9} $\frac{7}{8}$=1·07^{2} $\frac{2}{3}$=0·81^{7} $\frac{5}{6}$=1·02^{1}

$\frac{3}{16}$=0·23^{0} $\frac{5}{16}$=0·38^{3} $\frac{7}{16}$=0·53^{6} $\frac{9}{16}$=0·68^{9} $\frac{11}{16}$=0·84^{2} $\frac{13}{16}$=0·99^{5} $\frac{15}{16}$=1·14^{8}

(0.813=1£) **£1.23**

1	1·23	41	50·43	81	99·63	121	148·83	161	198·03
2	2·46	42	51·66	82	100·86	122	150·06	162	199·26
3	3·69	43	52·89	83	102·09	123	151·29	163	200·49
4	4·92	44	54·12	84	103·32	124	152·52	164	201·72
5	6·15	45	55·35	85	104·55	125	153·75	165	202·95
6	7·38	46	56·58	86	105·78	126	154·98	166	204·18
7	8·61	47	57·81	87	107·01	127	156·21	167	205·41
8	9·84	48	59·04	88	108·24	128	157·44	168	206·64
9	11·07	49	60·27	89	109·47	129	158·67	169	207·87
10	12·30	**50**	61·50	**90**	110·70	**130**	159·90	**170**	209·10
11	13·53	51	62·73	91	111·93	131	161·13	171	210·33
12	14·76	52	63·96	92	113·16	132	162·36	172	211·56
13	15·99	53	65·19	93	114·39	133	163·59	173	212·79
14	17·22	54	66·42	94	115·62	134	164·82	174	214·02
15	18·45	55	67·65	95	116·85	135	166·05	175	215·25
16	19·68	56	68·88	96	118·08	136	167·28	176	216·48
17	20·91	57	70·11	97	119·31	137	168·51	177	217·71
18	22·14	58	71·34	98	120·54	138	169·74	178	218·94
19	23·37	59	72·57	99	121·77	139	170·97	179	220·17
20	24·60	**60**	73·80	**100**	123·00	**140**	172·20	**180**	221·40
21	25·83	61	75·03	101	124·23	141	173·43	181	222·63
22	27·06	62	76·26	102	125·46	142	174·66	182	223·86
23	28·29	63	77·49	103	126·69	143	175·89	183	225·09
24	29·52	64	78·72	104	127·92	144	177·12	184	226·32
25	30·75	65	79·95	105	129·15	145	178·35	185	227·55
26	31·98	66	81·18	106	130·38	146	179·58	186	228·78
27	33·21	67	82·41	107	131·61	147	180·81	187	230·01
28	34·44	68	83·64	108	132·84	148	182·04	188	231·24
29	35·67	69	84·87	109	134·07	149	183·27	189	232·47
30	36·90	**70**	86·10	**110**	135·30	**150**	184·50	**190**	233·70
31	38·13	71	87·33	111	136·53	151	185·73	191	234·93
32	39·36	72	88·56	112	137·76	152	186·96	192	236·16
33	40·59	73	89·79	113	138·99	153	188·19	193	237·39
34	41·82	74	91·02	114	140·22	154	189·42	194	238·62
35	43·05	75	92·25	115	141·45	155	190·65	195	239·85
36	44·28	76	93·48	116	142·68	156	191·88	196	241·08
37	45·51	77	94·71	117	143·91	157	193·11	197	242·31
38	46·74	78	95·94	118	145·14	158	194·34	198	243·54
39	47·97	79	97·17	119	146·37	159	195·57	199	244·77
40	49·20	**80**	98·40	**120**	147·60	**160**	196·80	**200**	246·00

$\frac{1}{16}=0{\cdot}07^{7}$	$\frac{1}{9}=0{\cdot}13^{7}$	$\frac{1}{8}=0{\cdot}15^{4}$	$\frac{1}{6}=0{\cdot}20^{5}$	$\frac{1}{5}=0{\cdot}24^{6}$	$\frac{1}{4}=0{\cdot}30^{8}$	$\frac{1}{3}=0{\cdot}41^{0}$
$\frac{3}{8}=0{\cdot}46^{1}$	$\frac{1}{2}=0{\cdot}61^{5}$	$\frac{5}{8}=0{\cdot}76^{9}$	$\frac{3}{4}=0{\cdot}92^{3}$	$\frac{7}{8}=1{\cdot}07^{6}$	$\frac{2}{3}=0{\cdot}82^{0}$	$\frac{5}{6}=1{\cdot}02^{5}$
$\frac{3}{16}=0{\cdot}23^{1}$	$\frac{5}{16}=0{\cdot}38^{4}$	$\frac{7}{16}=0{\cdot}53^{8}$	$\frac{9}{16}=0{\cdot}69^{2}$	$\frac{11}{16}=0{\cdot}84^{6}$	$\frac{13}{16}=0{\cdot}99^{9}$	$\frac{15}{16}=1{\cdot}15^{3}$

(0.810=1£) **£1.23½**

1	1·23½	41	50·63½	81	100·03½	121	149·43½	161	198·83½
2	2·47	42	51·87	82	101·27	122	150·67	162	200·07
3	3·70½	43	53·10½	83	102·50½	123	151·90½	163	201·30½
4	4·94	44	54·34	84	103·74	124	153·14	164	202·54
5	6·17½	45	55·57½	85	104·97½	125	154·37½	165	203·77½
6	7·41	46	56·81	86	106·21	126	155·61	166	205·01
7	8·64½	47	58·04½	87	107·44½	127	156·84½	167	206·24½
8	9·88	48	59·28	88	108·68	128	158·08	168	207·48
9	11·11½	49	60·51½	89	109·91½	129	159·31½	169	208·71½
10	12·35	**50**	61·75	**90**	111·15	**130**	160·55	**170**	209·95
11	13·58½	51	62·98½	91	112·38½	131	161·78½	171	211·18½
12	14·82	52	64·22	92	113·62	132	163·02	172	212·42
13	16·05½	53	65·45½	93	114·85½	133	164·25½	173	213·65½
14	17·29	54	66·69	94	116·09	134	165·49	174	214·89
15	18·52½	55	67·92½	95	117·32½	135	166·72½	175	216·12½
16	19·76	56	69·16	96	118·56	136	167·96	176	217·36
17	20·99½	57	70·39½	97	119·79½	137	169·19½	177	218·59½
18	22·23	58	71·63	98	121·03	138	170·43	178	219·83
19	23·46½	59	72·86½	99	122·26½	139	171·66½	179	221·06½
20	24·70	**60**	74·10	**100**	123·50	**140**	172·90	**180**	222·30
21	25·93½	61	75·33½	101	124·73½	141	174·13½	181	223·53½
22	27·17	62	76·57	102	125·97	142	175·37	182	224·77
23	28·40½	63	77·80½	103	127·20½	143	176·60½	183	226·00½
24	29·64	64	79·04	104	128·44	144	177·84	184	227·24
25	30·87½	65	80·27½	105	129·67½	145	179·07½	185	228·47½
26	32·11	66	81·51	106	130·91	146	180·31	186	229·71
27	33·34½	67	82·74½	107	132·14½	147	181·54½	187	230·94½
28	34·58	68	83·98	108	133·38	148	182·78	188	232·18
29	35·81½	69	85·21½	109	134·61½	149	184·01½	189	233·41½
30	37·05	**70**	86·45	**110**	135·85	**150**	185·25	**190**	234·65
31	38·28½	71	87·68½	111	137·08½	151	186·48½	191	235·88½
32	39·52	72	88·92	112	138·32	152	187·72	192	237·12
33	40·75½	73	90·15½	113	139·55½	153	188·95½	193	238·35½
34	41·99	74	91·39	114	140·79	154	190·19	194	239·59
35	43·22½	75	92·62½	115	142·02½	155	191·42½	195	240·82½
36	44·46	76	93·86	116	143·26	156	192·66	196	242·06
37	45·69½	77	95·09½	117	144·49½	157	193·89½	197	243·29½
38	46·93	78	96·33	118	145·73	158	195·13	198	244·53
39	48·16½	79	97·56½	119	146·96½	159	196·36½	199	245·76½
40	49·40	**80**	98·80	**120**	148·20	**160**	197·60	**200**	247·00

$\frac{1}{16}=0{\cdot}07^{7}$ $\frac{1}{9}=0{\cdot}13^{7}$ $\frac{1}{8}=0{\cdot}15^{4}$ $\frac{1}{6}=0{\cdot}20^{6}$ $\frac{1}{5}=0{\cdot}24^{7}$ $\frac{1}{4}=0{\cdot}30^{9}$ $\frac{1}{3}=0{\cdot}41^{2}$

$\frac{3}{8}=0{\cdot}46^{3}$ $\frac{1}{2}=0{\cdot}61^{8}$ $\frac{5}{8}=0{\cdot}77^{2}$ $\frac{3}{4}=0{\cdot}92^{6}$ $\frac{7}{8}=1{\cdot}08^{1}$ $\frac{2}{3}=0{\cdot}82^{3}$ $\frac{5}{6}=1{\cdot}02^{9}$

$\frac{3}{16}=0{\cdot}23^{2}$ $\frac{5}{16}=0{\cdot}38^{6}$ $\frac{7}{16}=0{\cdot}54^{0}$ $\frac{9}{16}=0{\cdot}69^{5}$ $\frac{11}{16}=0{\cdot}84^{9}$ $\frac{13}{16}=1{\cdot}00^{3}$ $\frac{15}{16}=1{\cdot}15^{8}$

1	1·24	41	50·84	81	100·44	121	150·04	161	199·64
2	2·48	42	52·08	82	101·68	122	151·28	162	200·88
3	3·72	43	53·32	83	102·92	123	152·52	163	202·12
4	4·96	44	54·56	84	104·16	124	153·76	164	203·36
5	6·20	45	55·80	85	105·40	125	155·00	165	204·60
6	7·44	46	57·04	86	106·64	126	156·24	166	205·84
7	8·68	47	58·28	87	107·88	127	157·48	167	207·08
8	9·92	48	59·52	88	109·12	128	158·72	168	208·32
9	11·16	49	60·76	89	110·36	129	159·96	169	209·56
10	12·40	**50**	62·00	**90**	111·60	**130**	161·20	**170**	210·80
11	13·64	51	63·24	91	112·84	131	162·44	171	212·04
12	14·88	52	64·48	92	114·08	132	163·68	172	213·28
13	16·12	53	65·72	93	115·32	133	164·92	173	214·52
14	17·36	54	66·96	94	116·56	134	166·16	174	215·76
15	18·60	55	68·20	95	117·80	135	167·40	175	217·00
16	19·84	56	69·44	96	119·04	136	168·64	176	218·24
17	21·08	57	70·68	97	120·28	137	169·88	177	219·48
18	22·32	58	71·92	98	121·52	138	171·12	178	220·72
19	23·56	59	73·16	99	122·76	139	172·36	179	221·96
20	24·80	**60**	74·40	**100**	124·00	**140**	173·60	**180**	223·20
21	26·04	61	75·64	101	125·24	141	174·84	181	224·44
22	27·28	62	76·88	102	126·48	142	176·08	182	225·68
23	28·52	63	78·12	103	127·72	143	177·32	183	226·92
24	29·76	64	79·36	104	128·96	144	178·56	184	228·16
25	31·00	65	80·60	105	130·20	145	179·80	185	229·40
26	32·24	66	81·84	106	131·44	146	181·04	186	230·64
27	33·48	67	83·08	107	132·68	147	182·28	187	231·88
28	34·72	68	84·32	108	133·92	148	183·52	188	233·12
29	35·96	69	85·56	109	135·16	149	184·76	189	234·36
30	37·20	**70**	86·80	**110**	136·40	**150**	186·00	**190**	235·60
31	38·44	71	88·04	111	137·64	151	187·24	191	236·84
32	39·68	72	89·28	112	138·88	152	188·48	192	238·08
33	40·92	73	90·52	113	140·12	153	189·72	193	239·32
34	42·16	74	91·76	114	141·36	154	190·96	194	240·56
35	43·40	75	93·00	115	142·60	155	192·20	195	241·80
36	44·64	76	94·24	116	143·84	156	193·44	196	243·04
37	45·88	77	95·48	117	145·08	157	194·68	197	244·28
38	47·12	78	96·72	118	146·32	158	195·92	198	245·52
39	48·36	79	97·96	119	147·56	159	197·16	199	246·76
40	49·60	**80**	99·20	**120**	148·80	**160**	198·40	**200**	248·00

$\frac{1}{16}=0{\cdot}07^{8}$ $\frac{1}{9}=0{\cdot}13^{8}$ $\frac{1}{8}=0{\cdot}15^{5}$ $\frac{1}{6}=0{\cdot}20^{7}$ $\frac{1}{5}=0{\cdot}24^{8}$ $\frac{1}{4}=0{\cdot}31^{0}$ $\frac{1}{3}=0{\cdot}41^{3}$

$\frac{3}{8}=0{\cdot}46^{5}$ $\frac{1}{2}=0{\cdot}62^{0}$ $\frac{5}{8}=0{\cdot}77^{5}$ $\frac{3}{4}=0{\cdot}93^{0}$ $\frac{7}{8}=1{\cdot}08^{5}$ $\frac{2}{3}=0{\cdot}82^{7}$ $\frac{5}{6}=1{\cdot}03^{3}$

$\frac{3}{16}=0{\cdot}23^{3}$ $\frac{5}{16}=0{\cdot}38^{8}$ $\frac{7}{16}=0{\cdot}54^{3}$ $\frac{9}{16}=0{\cdot}69^{8}$ $\frac{11}{16}=0{\cdot}85^{3}$ $\frac{13}{16}=1{\cdot}00^{8}$ $\frac{15}{16}=1{\cdot}16^{3}$

1	1·24½	41	51·04½	81	100·84½	121	150·64½	161	200·44½
2	2·49	42	52·29	82	102·09	122	151·89	162	201·69
3	3·73½	43	53·53½	83	103·33½	123	153·13½	163	202·93½
4	4·98	44	54·78	84	104·58	124	154·38	164	204·18
5	6·22½	45	56·02½	85	105·82½	125	155·62½	165	205·42½
6	7·47	46	57·27	86	107·07	126	156·87	166	206·67
7	8·71½	47	58·51½	87	108·31½	127	158·11½	167	207·91½
8	9·96	48	59·76	88	109·56	128	159·36	168	209·16
9	11·20½	49	61·00½	89	110·80½	129	160·60½	169	210·40½
10	12·45	**50**	62·25	**90**	112·05	**130**	161·85	**170**	211·65
11	13·69½	51	63·49½	91	113·29½	131	163·09½	171	212·89½
12	14·94	52	64·74	92	114·54	132	164·34	172	214·14
13	16·18½	53	65·98½	93	115·78½	133	165·58½	173	215·38½
14	17·43	54	67·23	94	117·03	134	166·83	174	216·63
15	18·67½	55	68·47½	95	118·27½	135	168·07½	175	217·87½
16	19·92	56	69·72	96	119·52	136	169·32	176	219·12
17	21·16½	57	70·96½	97	120·76½	137	170·56½	177	220·36½
18	22·41	58	72·21	98	122·01	138	171·81	178	221·61
19	23·65½	59	73·45½	99	123·25½	139	173·05½	179	222·85½
20	24·90	**60**	74·70	**100**	124·50	**140**	174·30	**180**	224·10
21	26·14½	61	75·94½	101	125·74½	141	175·54½	181	225·34½
22	27·39	62	77·19	102	126·99	142	176·79	182	226·59
23	28·63½	63	78·43½	103	128·23½	143	178·03½	183	227·83½
24	29·88	64	79·68	104	129·48	144	179·28	184	229·08
25	31·12½	65	80·92½	105	130·72½	145	180·52½	185	230·32½
26	32·37	66	82·17	106	131·97	146	181·77	186	231·57
27	33·61½	67	83·41½	107	133·21½	147	183·01½	187	232·81½
28	34·86	68	84·66	108	134·46	148	184·26	188	234·06
29	36·10½	69	85·90½	109	135·70½	149	185·50½	189	235·30½
30	37·35	**70**	87·15	**110**	136·95	**150**	186·75	**190**	236·55
31	38·59½	71	88·39½	111	138·19½	151	187·99½	191	237·79½
32	39·84	72	89·64	112	139·44	152	189·24	192	239·04
33	41·08½	73	90·88½	113	140·68½	153	190·48½	193	240·28½
34	42·33	74	92·13	114	141·93	154	191·73	194	241·53
35	43·57½	75	93·37½	115	143·17½	155	192·97½	195	242·77½
36	44·82	76	94·62	116	144·42	156	194·22	196	244·02
37	46·06½	77	95·86½	117	145·66½	157	195·46½	197	245·26½
38	47·31	78	97·11	118	146·91	158	196·71	198	246·51
39	48·55½	79	98·35½	119	148·15½	159	197·95½	199	247·75½
40	49·80	**80**	99·60	**120**	149·40	**160**	199·20	**200**	249·00

$\frac{1}{16}=0{\cdot}07^{8}$ $\frac{1}{9}=0{\cdot}13^{8}$ $\frac{1}{8}=0{\cdot}15^{6}$ $\frac{1}{6}=0{\cdot}20^{8}$ $\frac{1}{5}=0{\cdot}24^{9}$ $\frac{1}{4}=0{\cdot}31^{1}$ $\frac{1}{3}=0{\cdot}41^{5}$

$\frac{3}{8}=0{\cdot}46^{7}$ $\frac{1}{2}=0{\cdot}62^{3}$ $\frac{5}{8}=0{\cdot}77^{8}$ $\frac{3}{4}=0{\cdot}93^{4}$ $\frac{7}{8}=1{\cdot}08^{9}$ $\frac{2}{3}=0{\cdot}83^{0}$ $\frac{5}{6}=1{\cdot}03^{8}$

$\frac{3}{16}=0{\cdot}23^{3}$ $\frac{5}{16}=0{\cdot}38^{9}$ $\frac{7}{16}=0{\cdot}54^{5}$ $\frac{9}{16}=0{\cdot}70^{0}$ $\frac{11}{16}=0{\cdot}85^{6}$ $\frac{13}{16}=1{\cdot}01^{2}$ $\frac{15}{16}=1{\cdot}16^{7}$

1	1·25	41	51·25	81	101·25	121	151·25	161	201·25
2	2·50	42	52·50	82	102·50	122	152·50	162	202·50
3	3·75	43	53·75	83	103·75	123	153·75	163	203·75
4	5·00	44	55·00	84	105·00	124	155·00	164	205·00
5	6·25	45	56·25	85	106·25	125	156·25	165	206·25
6	7·50	46	57·50	86	107·50	126	157·50	166	207·50
7	8·75	47	58·75	87	108·75	127	158·75	167	208·75
8	10·00	48	60·00	88	110·00	128	160·00	168	210·00
9	11·25	49	61·25	89	111·25	129	161·25	169	211·25
10	12·50	**50**	62·50	**90**	112·50	**130**	162·50	**170**	212·50
11	13·75	51	63·75	91	113·75	131	163·75	171	213·75
12	15·00	52	65·00	92	115·00	132	165·00	172	215·00
13	16·25	53	66·25	93	116·25	133	166·25	173	216·25
14	17·50	54	67·50	94	117·50	134	167·50	174	217·50
15	18·75	55	68·75	95	118·75	135	168·75	175	218·75
16	20·00	56	70·00	96	120·00	136	170·00	176	220·00
17	21·25	57	71·25	97	121·25	137	171·25	177	221·25
18	22·50	58	72·50	98	122·50	138	172·50	178	222·50
19	23·75	59	73·75	99	123·75	139	173·75	179	223·75
20	25·00	**60**	75·00	**100**	125·00	**140**	175·00	**180**	225·00
21	26·25	61	76·25	101	126·25	141	176·25	181	226·25
22	27·50	62	77·50	102	127·50	142	177·50	182	227·50
23	28·75	63	78·75	103	128·75	143	178·75	183	228·75
24	30·00	64	80·00	104	130·00	144	180·00	184	230·00
25	31·25	65	81·25	105	131·25	145	181·25	185	231·25
26	32·50	66	82·50	106	132·50	146	182·50	186	232·50
27	33·75	67	83·75	107	133·75	147	183·75	187	233·75
28	35·00	68	85·00	108	135·00	148	185·00	188	235·00
29	36·25	69	86·25	109	136·25	149	186·25	189	236·25
30	37·50	**70**	87·50	**110**	137·50	**150**	187·50	**190**	237·50
31	38·75	71	88·75	111	138·75	151	188·75	191	238·75
32	40·00	72	90·00	112	140·00	152	190·00	192	240·00
33	41·25	73	91·25	113	141·25	153	191·25	193	241·25
34	42·50	74	92·50	114	142·50	154	192·50	194	242·50
35	43·75	75	93·75	115	143·75	155	193·75	195	243·75
36	45·00	76	95·00	116	145·00	156	195·00	196	245·00
37	46·25	77	96·25	117	146·25	157	196·25	197	246·25
38	47·50	78	97·50	118	147·50	158	197·50	198	247·50
39	48·75	79	98·75	119	148·75	159	198·75	199	248·75
40	50·00	**80**	100·00	**120**	150·00	**160**	200·00	**200**	250·00

$\frac{1}{16}=0{\cdot}07^{8}$ $\frac{1}{9}=0{\cdot}13^{9}$ $\frac{1}{8}=0{\cdot}15^{6}$ $\frac{1}{6}=0{\cdot}20^{8}$ $\frac{1}{5}=0{\cdot}25^{0}$ $\frac{1}{4}=0{\cdot}31^{3}$ $\frac{1}{3}=0{\cdot}41^{7}$

$\frac{3}{8}=0{\cdot}46^{9}$ $\frac{1}{2}=0{\cdot}62^{5}$ $\frac{5}{8}=0{\cdot}78^{1}$ $\frac{3}{4}=0{\cdot}93^{8}$ $\frac{7}{8}=1{\cdot}09^{4}$ $\frac{2}{3}=0{\cdot}83^{3}$ $\frac{5}{6}=1{\cdot}04^{2}$

$\frac{3}{16}=0{\cdot}23^{4}$ $\frac{5}{16}=0{\cdot}39^{1}$ $\frac{7}{16}=0{\cdot}54^{7}$ $\frac{9}{16}=0{\cdot}70^{3}$ $\frac{11}{16}=0{\cdot}85^{9}$ $\frac{13}{16}=1{\cdot}01^{6}$ $\frac{15}{16}=1{\cdot}17^{2}$

1	1·25½	41	51·45½	81	101·65½	121	151·85½	161	202·05½
2	2·51	42	52·71	82	102·91	122	153·11	162	203·31
3	3·76½	43	53·96½	83	104·16½	123	154·36½	163	204·56½
4	5·02	44	55·22	84	105·42	124	155·62	164	205·82
5	6·27½	45	56·47½	85	106·67½	125	156·87½	165	207·07½
6	7·53	46	57·73	86	107·93	126	158·13	166	208·33
7	8·78½	47	58·98½	87	109·18½	127	159·38½	167	209·58½
8	10·04	48	60·24	88	110·44	128	160·64	168	210·84
9	11·29½	49	61·49½	89	111·69½	129	161·89½	169	212·09½
10	12·55	**50**	62·75	**90**	112·95	**130**	163·15	**170**	213·35
11	13·80½	51	64·00½	91	114·20½	131	164·40½	171	214·60½
12	15·06	52	65·26	92	115·46	132	165·66	172	215·86
13	16·31½	53	66·51½	93	116·71½	133	166·91½	173	217·11½
14	17·57	54	67·77	94	117·97	134	168·17	174	218·37
15	18·82½	55	69·02½	95	119·22½	135	169·42½	175	219·62½
16	20·08	56	70·28	96	120·48	136	170·68	176	220·88
17	21·33½	57	71·53½	97	121·73½	137	171·93½	177	222·13½
18	22·59	58	72·79	98	122·99	138	173·19	178	223·39
19	23·84½	59	74·04½	99	124·24½	139	174·44½	179	224·64½
20	25·10	**60**	75·30	**100**	125·50	**140**	175·70	**180**	225·90
21	26·35½	61	76·55½	101	126·75½	141	176·95½	181	227·15½
22	27·61	62	77·81	102	128·01	142	178·21	182	228·41
23	28·86½	63	79·06½	103	129·26½	143	179·46½	183	229·66½
24	30·12	64	80·32	104	130·52	144	180·72	184	230·92
25	31·37½	65	81·57½	105	131·77½	145	181·97½	185	232·17½
26	32·63	66	82·83	106	133·03	146	183·23	186	233·43
27	33·88½	67	84·08½	107	134·28½	147	184·48½	187	234·68½
28	35·14	68	85·34	108	135·54	148	185·74	188	235·94
29	36·39½	69	86·59½	109	136·79½	149	186·99½	189	237·19½
30	37·65	**70**	87·85	**110**	138·05	**150**	188·25	**190**	238·45
31	38·90½	71	89·10½	111	139·30½	151	189·50½	191	239·70½
32	40·16	72	90·36	112	140·56	152	190·76	192	240·96
33	41·41½	73	91·61½	113	141·81½	153	192·01½	193	242·21½
34	42·67	74	92·87	114	143·07	154	193·27	194	243·47
35	43·92½	75	94·12½	115	144·32½	155	194·52½	195	244·72½
36	45·18	76	95·38	116	145·58	156	195·78	196	245·98
37	46·43½	77	96·63½	117	146·83½	157	197·03½	197	247·23½
38	47·69	78	97·89	118	148·09	158	198·29	198	248·49
39	48·94½	79	99·14½	119	149·34½	159	199·54½	199	249·74½
40	50·20	**80**	100·40	**120**	150·60	**160**	200·80	**200**	251·00

$\frac{1}{16}=0{\cdot}07^8$ $\frac{1}{9}=0{\cdot}13^9$ $\frac{1}{8}=0{\cdot}15^7$ $\frac{1}{6}=0{\cdot}20^9$ $\frac{1}{5}=0{\cdot}25^1$ $\frac{1}{4}=0{\cdot}31^4$ $\frac{1}{3}=0{\cdot}41^8$

$\frac{3}{8}=0{\cdot}47^1$ $\frac{1}{2}=0{\cdot}62^8$ $\frac{5}{8}=0{\cdot}78^4$ $\frac{3}{4}=0{\cdot}94^1$ $\frac{7}{8}=1{\cdot}09^8$ $\frac{2}{3}=0{\cdot}83^7$ $\frac{5}{6}=1{\cdot}04^6$

$\frac{3}{16}=0{\cdot}23^5$ $\frac{5}{16}=0{\cdot}39^2$ $\frac{7}{16}=0{\cdot}54^9$ $\frac{9}{16}=0{\cdot}70^6$ $\frac{11}{16}=0{\cdot}86^3$ $\frac{13}{16}=1{\cdot}02^0$ $\frac{15}{16}=1{\cdot}17^7$

1 cwt per acre=125·535 Kg per hectare

(0.794=1£) £1.26

1	1·26	41	51·66	81	102·06	121	152·46	161	202·86
2	2·52	42	52·92	82	103·32	122	153·72	162	204·12
3	3·78	43	54·18	83	104·58	123	154·98	163	205·38
4	5·04	44	55·44	84	105·84	124	156·24	164	206·64
5	6·30	45	56·70	85	107·10	125	157·50	165	207·90
6	7·56	46	57·96	86	108·36	126	158·76	166	209·16
7	8·82	47	59·22	87	109·62	127	160·02	167	210·42
8	10·08	48	60·48	88	110·88	128	161·28	168	211·68
9	11·34	49	61·74	89	112·14	129	162·54	169	212·94
10	12·60	**50**	63·00	**90**	113·40	**130**	163·80	**170**	214·20
11	13·86	51	64·26	91	114·66	131	165·06	171	215·46
12	15·12	52	65·52	92	115·92	132	166·32	172	216·72
13	16·38	53	66·78	93	117·18	133	167·58	173	217·98
14	17·64	54	68·04	94	118·44	134	168·84	174	219·24
15	18·90	55	69·30	95	119·70	135	170·10	175	220·50
16	20·16	56	70·56	96	120·96	136	171·36	176	221·76
17	21·42	57	71·82	97	122·22	137	172·62	177	223·02
18	22·68	58	73·08	98	123·48	138	173·88	178	224·28
19	23·94	59	74·34	99	124·74	139	175·14	179	225·54
20	25·20	**60**	75·60	**100**	126·00	**140**	176·40	**180**	226·80
21	26·46	61	76·86	101	127·26	141	177·66	181	228·06
22	27·72	62	78·12	102	128·52	142	178·92	182	229·32
23	28·98	63	79·38	103	129·78	143	180·18	183	230·58
24	30·24	64	80·64	104	131·04	144	181·44	184	231·84
25	31·50	65	81·90	105	132·30	145	182·70	185	233·10
26	32·76	66	83·16	106	133·56	146	183·96	186	234·36
27	34·02	67	84·42	107	134·82	147	185·22	187	235·62
28	35·28	68	85·68	108	136·08	148	186·48	188	236·88
29	36·54	69	86·94	109	137·34	149	187·74	189	238·14
30	37·80	**70**	88·20	**110**	138·60	**150**	189·00	**190**	239·40
31	39·06	71	89·46	111	139·86	151	190·26	191	240·66
32	40·32	72	90·72	112	141·12	152	191·52	192	241·92
33	41·58	73	91·98	113	142·38	153	192·78	193	243·18
34	42·84	74	93·24	114	143·64	154	194·04	194	244·44
35	44·10	75	94·50	115	144·90	155	195·30	195	245·70
36	45·36	76	95·76	116	146·16	156	196·56	196	246·96
37	46·62	77	97·02	117	147·42	157	197·82	197	248·22
38	47·88	78	98·28	118	148·68	158	199·08	198	249·48
39	49·14	79	99·54	119	149·94	159	200·34	199	250·74
40	50·40	**80**	100·80	**120**	151·20	**160**	201·60	**200**	252·00

$\frac{1}{16}=0{\cdot}07^{9}$	$\frac{1}{9}=0{\cdot}14^{0}$	$\frac{1}{8}=0{\cdot}15^{8}$	$\frac{1}{6}=0{\cdot}21^{0}$	$\frac{1}{5}=0{\cdot}25^{2}$	$\frac{1}{4}=0{\cdot}31^{5}$	$\frac{1}{3}=0{\cdot}42^{0}$
$\frac{3}{8}=0{\cdot}47^{3}$	$\frac{1}{2}=0{\cdot}63^{0}$	$\frac{5}{8}=0{\cdot}78^{8}$	$\frac{3}{4}=0{\cdot}94^{5}$	$\frac{7}{8}=1{\cdot}10^{3}$	$\frac{2}{3}=0{\cdot}84^{0}$	$\frac{5}{6}=1{\cdot}05^{0}$
$\frac{3}{16}=0{\cdot}23^{6}$	$\frac{5}{16}=0{\cdot}39^{4}$	$\frac{7}{16}=0{\cdot}55^{1}$	$\frac{9}{16}=0{\cdot}70^{9}$	$\frac{11}{16}=0{\cdot}86^{6}$	$\frac{13}{16}=1{\cdot}02^{4}$	$\frac{15}{16}=1{\cdot}18^{1}$

(0.791=1£) **£1.26½**

1	1·26½	41	51·86½	81	102·46½	121	153·06½	161	203·66½
2	2·53	42	53·13	82	103·73	122	154·33	162	204·93
3	3·79½	43	54·39½	83	104·99½	123	155·59½	163	206·19½
4	5·06	44	55·66	84	106·26	124	156·86	164	207·46
5	6·32½	45	56·92½	85	107·52½	125	158·12½	165	208·72½
6	7·59	46	58·19	86	108·79	126	159·39	166	209·99
7	8·85½	47	59·45½	87	110·05½	127	160·65½	167	211·25½
8	10·12	48	60·72	88	111·32	128	161·92	168	212·52
9	11·38½	49	61·98½	89	112·58½	129	163·18½	169	213·78½
10	12·65	**50**	63·25	**90**	113·85	**130**	164·45	**170**	215·05
11	13·91½	51	64·51½	91	115·11½	131	165·71½	171	216·31½
12	15·18	52	65·78	92	116·38	132	166·98	172	217·58
13	16·44½	53	67·04½	93	117·64½	133	168·24½	173	218·84½
14	17·71	54	68·31	94	118·91	134	169·51	174	220·11
15	18·97½	55	69·57½	95	120·17½	135	170·77½	175	221·37½
16	20·24	56	70·84	96	121·44	136	172·04	176	222·64
17	21·50½	57	72·10½	97	122·70½	137	173·30½	177	223·90½
18	22·77	58	73·37	98	123·97	138	174·57	178	225·17
19	24·03½	59	74·63½	99	125·23½	139	175·83½	179	226·43½
20	25·30	**60**	75·90	**100**	126·50	**140**	177·10	**180**	227·70
21	26·56½	61	77·16½	101	127·76½	141	178·36½	181	228·96½
22	27·83	62	78·43	102	129·03	142	179·63	182	230·23
23	29·09½	63	79·69½	103	130·29½	143	180·89½	183	231·49½
24	30·36	64	80·96	104	131·56	144	182·16	184	232·76
25	31·62½	65	82·22½	105	132·82½	145	183·42½	185	234·02½
26	32·89	66	83·49	106	134·09	146	184·69	186	235·29
27	34·15½	67	84·75½	107	135·35½	147	185·95½	187	236·55½
28	35·42	68	86·02	108	136·62	148	187·22	188	237·82
29	36·68½	69	87·28½	109	137·88½	149	188·48½	189	239·08½
30	37·95	**70**	88·55	**110**	139·15	**150**	189·75	**190**	240·35
31	39·21½	71	89·81½	111	140·41½	151	191·01½	191	241·61½
32	40·48	72	91·08	112	141·68	152	192·28	192	242·88
33	41·74½	73	92·34½	113	142·94½	153	193·54½	193	244·14½
34	43·01	74	93·61	114	144·21	154	194·81	194	245·41
35	44·27½	75	94·87½	115	145·47½	155	196·07½	195	246·67½
36	45·54	76	96·14	116	146·74	156	197·34	196	247·94
37	46·80½	77	97·40½	117	148·00½	157	198·60½	197	249·20½
38	48·07	78	98·67	118	149·27	158	199·87	198	250·47
39	49·33½	79	99·93½	119	150·53½	159	201·13½	199	251·73½
40	50·60	**80**	101·20	**120**	151·80	**160**	202·40	**200**	253·00

$\frac{1}{16}=0{\cdot}07^{9}$ $\frac{1}{9}=0{\cdot}14^{1}$ $\frac{1}{8}=0{\cdot}15^{8}$ $\frac{1}{6}=0{\cdot}21^{1}$ $\frac{1}{5}=0{\cdot}25^{3}$ $\frac{1}{4}=0{\cdot}31^{6}$ $\frac{1}{3}=0{\cdot}42^{2}$

$\frac{3}{8}=0{\cdot}47^{4}$ $\frac{1}{2}=0{\cdot}63^{3}$ $\frac{5}{8}=0{\cdot}79^{1}$ $\frac{3}{4}=0{\cdot}94^{9}$ $\frac{7}{8}=1{\cdot}10^{7}$ $\frac{2}{3}=0{\cdot}84^{3}$ $\frac{5}{6}=1{\cdot}05^{4}$

$\frac{3}{16}=0{\cdot}23^{7}$ $\frac{5}{16}=0{\cdot}39^{5}$ $\frac{7}{16}=0{\cdot}55^{3}$ $\frac{9}{16}=0{\cdot}71^{2}$ $\frac{11}{16}=0{\cdot}87^{0}$ $\frac{13}{16}=1{\cdot}02^{8}$ $\frac{15}{16}=1{\cdot}18^{6}$

(0.787=1£) **£1.27**

1	1·27	41	52·07	81	102·87	121	153·67	161	204·47
2	2·54	42	53·34	82	104·14	122	154·94	162	205·74
3	3·81	43	54·61	83	105·41	123	156·21	163	207·01
4	5·08	44	55·88	84	106·68	124	157·48	164	208·28
5	6·35	45	57·15	85	107·95	125	158·75	165	209·55
6	7·62	46	58·42	86	109·22	126	160·02	166	210·82
7	8·89	47	59·69	87	110·49	127	161·29	167	212·09
8	10·16	48	60·96	88	111·76	128	162·56	168	213·36
9	11·43	49	62·23	89	113·03	129	163·83	169	214·63
10	12·70	**50**	63·50	**90**	114·30	**130**	165·10	**170**	215·90
11	13·97	51	64·77	91	115·57	131	166·37	171	217·17
12	15·24	52	66·04	92	116·84	132	167·64	172	218·44
13	16·51	53	67·31	93	118·11	133	168·91	173	219·71
14	17·78	54	68·58	94	119·38	134	170·18	174	220·98
15	19·05	55	69·85	95	120·65	135	171·45	175	222·25
16	20·32	56	71·12	96	121·92	136	172·72	176	223·52
17	21·59	57	72·39	97	123·19	137	173·99	177	224·79
18	22·86	58	73·66	98	124·46	138	175·26	178	226·06
19	24·13	59	74·93	99	125·73	139	176·53	179	227·33
20	25·40	**60**	76·20	**100**	127·00	**140**	177·80	**180**	228·60
21	26·67	61	77·47	101	128·27	141	179·07	181	229·87
22	27·94	62	78·74	102	129·54	142	180·34	182	231·14
23	29·21	63	80·01	103	130·81	143	181·61	183	232·41
24	30·48	64	81·28	104	132·08	144	182·88	184	233·68
25	31·75	65	82·55	105	133·35	145	184·15	185	234·95
26	33·02	66	83·82	106	134·62	146	185·42	186	236·22
27	34·29	67	85·09	107	135·89	147	186·69	187	237·49
28	35·56	68	86·36	108	137·16	148	187·96	188	238·76
29	36·83	69	87·63	109	138·43	149	189·23	189	240·03
30	38·10	**70**	88·90	**110**	139·70	**150**	190·50	**190**	241·30
31	39·37	71	90·17	111	140·97	151	191·77	191	242·57
32	40·64	72	91·44	112	142·24	152	193·04	192	243·84
33	41·91	73	92·71	113	143·51	153	194·31	193	245·11
34	43·18	74	93·98	114	144·78	154	195·58	194	246·38
35	44·45	75	95·25	115	146·05	155	196·85	195	247·65
36	45·72	76	96·52	116	147·32	156	198·12	196	248·92
37	46·99	77	97·79	117	148·59	157	199·39	197	250·19
38	48·26	78	99·06	118	149·86	158	200·66	198	251·46
39	49·53	79	100·33	119	151·13	159	201·93	199	252·73
40	50·80	**80**	101·60	**120**	152·40	**160**	203·20	**200**	254·00

$\frac{1}{16}=0{\cdot}07^{9}$ $\frac{1}{9}=0{\cdot}14^{1}$ $\frac{1}{8}=0{\cdot}15^{9}$ $\frac{1}{6}=0{\cdot}21^{2}$ $\frac{1}{5}=0{\cdot}25^{4}$ $\frac{1}{4}=0{\cdot}31^{8}$ $\frac{1}{3}=0{\cdot}42^{3}$

$\frac{3}{8}=0{\cdot}47^{6}$ $\frac{1}{2}=0{\cdot}63^{5}$ $\frac{5}{8}=0{\cdot}79^{4}$ $\frac{3}{4}=0{\cdot}95^{3}$ $\frac{7}{8}=1{\cdot}11^{1}$ $\frac{2}{3}=0{\cdot}84^{7}$ $\frac{5}{6}=1{\cdot}05^{8}$

$\frac{3}{16}=0{\cdot}23^{8}$ $\frac{5}{16}=0{\cdot}39^{7}$ $\frac{7}{16}=0{\cdot}55^{6}$ $\frac{9}{16}=0{\cdot}71^{4}$ $\frac{11}{16}=0{\cdot}87^{3}$ $\frac{13}{16}=1{\cdot}03^{2}$ $\frac{15}{16}=1{\cdot}19^{1}$

(0.784=1£) £1.27½

1	1·27½	41	52·27½	81	103·27½	121	154·27½	161	205·27½
2	2·55	42	53·55	82	104·55	122	155·55	162	206·55
3	3·82½	43	54·82½	83	105·82½	123	156·82½	163	207·82½
4	5·10	44	56·10	84	107·10	124	158·10	164	209·10
5	6·37½	45	57·37½	85	108·37½	125	159·37½	165	210·37½
6	7·65	46	58·65	86	109·65	126	160·65	166	211·65
7	8·92½	47	59·92½	87	110·92½	127	161·92½	167	212·92½
8	10·20	48	61·20	88	112·20	128	163·20	168	214·20
9	11·47½	49	62·47½	89	113·47½	129	164·47½	169	215·47½
10	12·75	**50**	63·75	**90**	114·75	**130**	165·75	**170**	216·75
11	14·02½	51	65·02½	91	116·02½	131	167·02½	171	218·02½
12	15·30	52	66·30	92	117·30	132	168·30	172	219·30
13	16·57½	53	67·57½	93	118·57½	133	169·57½	173	220·57½
14	17·85	54	68·85	94	119·85	134	170·85	174	221·85
15	19·12½	55	70·12½	95	121·12½	135	172·12½	175	223·12½
16	20·40	56	71·40	96	122·40	136	173·40	176	224·40
17	21·67½	57	72·67½	97	123·67½	137	174·67½	177	225·67½
18	22·95	58	73·95	98	124·95	138	175·95	178	226·95
19	24·22½	59	75·22½	99	126·22½	139	177·22½	179	228·22½
20	25·50	**60**	76·50	**100**	127·50	**140**	178·50	**180**	229·50
21	26·77½	61	77·77½	101	128·77½	141	179·77½	181	230·77½
22	28·05	62	79·05	102	130·05	142	181·05	182	232·05
23	29·32½	63	80·32½	103	131·32½	143	182·32½	183	233·32½
24	30·60	64	81·60	104	132·60	144	183·60	184	234·60
25	31·87½	65	82·87½	105	133·87½	145	184·87½	185	235·87½
26	33·15	66	84·15	106	135·15	146	186·15	186	237·15
27	34·42½	67	85·42½	107	136·42½	147	187·42½	187	238·42½
28	35·70	68	86·70	108	137·70	148	188·70	188	239·70
29	36·97½	69	87·97½	109	138·97½	149	189·97½	189	240·97½
30	38·25	**70**	89·25	**110**	140·25	**150**	191·25	**190**	242·25
31	39·52½	71	90·52½	111	141·52½	151	192·52½	191	243·52½
32	40·80	72	91·80	112	142·80	152	193·80	192	244·80
33	42·07½	73	93·07½	113	144·07½	153	195·07½	193	246·07½
34	43·35	74	94·35	114	145·35	154	196·35	194	247·35
35	44·62½	75	95·62½	115	146·62½	155	197·62½	195	248·62½
36	45·90	76	96·90	116	147·90	156	198·90	196	249·90
37	47·17½	77	98·17½	117	149·17½	157	200·17½	197	251·17½
38	48·45	78	99·45	118	150·45	158	201·45	198	252·45
39	49·72½	79	100·72½	119	151·72½	159	202·72½	199	253·72½
40	51·00	**80**	102·00	**120**	153·00	**160**	204·00	**200**	255·00

$\frac{1}{16}=0{\cdot}08^{0}$ $\frac{1}{9}=0{\cdot}14^{2}$ $\frac{1}{8}=0{\cdot}15^{9}$ $\frac{1}{6}=0{\cdot}21^{3}$ $\frac{1}{5}=0{\cdot}25^{5}$ $\frac{1}{4}=0{\cdot}31^{9}$ $\frac{1}{3}=0{\cdot}42^{5}$

$\frac{3}{8}=0{\cdot}47^{8}$ $\frac{1}{2}=0{\cdot}63^{8}$ $\frac{5}{8}=0{\cdot}79^{7}$ $\frac{3}{4}=0{\cdot}95^{6}$ $\frac{7}{8}=1{\cdot}11^{6}$ $\frac{2}{3}=0{\cdot}85^{0}$ $\frac{5}{6}=1{\cdot}06^{3}$

$\frac{3}{16}=0{\cdot}23^{9}$ $\frac{5}{16}=0{\cdot}39^{8}$ $\frac{7}{16}=0{\cdot}55^{8}$ $\frac{9}{16}=0{\cdot}71^{7}$ $\frac{11}{16}=0{\cdot}87^{7}$ $\frac{13}{16}=1{\cdot}03^{6}$ $\frac{15}{16}=1{\cdot}19^{5}$

1	1·28	41	52·48	81	103·68	121	154·88	161	206·08
2	2·56	42	53·76	82	104·96	122	156·16	162	207·36
3	3·84	43	55·04	83	106·24	123	157·44	163	208·64
4	5·12	44	56·32	84	107·52	124	158·72	164	209·92
5	6·40	45	57·60	85	108·80	125	160·00	165	211·20
6	7·68	46	58·88	86	110·08	126	161·28	166	212·48
7	8·96	47	60·16	87	111·36	127	162·56	167	213·76
8	10·24	48	61·44	88	112·64	128	163·84	168	215·04
9	11·52	49	62·72	89	113·92	129	165·12	169	216·32
10	12·80	**50**	64·00	**90**	115·20	**130**	166·40	**170**	217·60
11	14·08	51	65·28	91	116·48	131	167·68	171	218·88
12	15·36	52	66·56	92	117·76	132	168·96	172	220·16
13	16·64	53	67·84	93	119·04	133	170·24	173	221·44
14	17·92	54	69·12	94	120·32	134	171·52	174	222·72
15	19·20	55	70·40	95	121·60	135	172·80	175	224·00
16	20·48	56	71·68	96	122·88	136	174·08	176	225·28
17	21·76	57	72·96	97	124·16	137	175·36	177	226·56
18	23·04	58	74·24	98	125·44	138	176·64	178	227·84
19	24·32	59	75·52	99	126·72	139	177·92	179	229·12
20	25·60	**60**	76·80	**100**	128·00	**140**	179·20	**180**	230·40
21	26·88	61	78·08	101	129·28	141	180·48	181	231·68
22	28·16	62	79·36	102	130·56	142	181·76	182	232·96
23	29·44	63	80·64	103	131·84	143	183·04	183	234·24
24	30·72	64	81·92	104	133·12	144	184·32	184	235·52
25	32·00	65	83·20	105	134·40	145	185·60	185	236·80
26	33·28	66	84·48	106	135·68	146	186·88	186	238·08
27	34·56	67	85·76	107	136·96	147	188·16	187	239·36
28	35·84	68	87·04	108	138·24	148	189·44	188	240·64
29	37·12	69	88·32	109	139·52	149	190·72	189	241·92
30	38·40	**70**	89·60	**110**	140·80	**150**	192·00	**190**	243·20
31	39·68	71	90·88	111	142·08	151	193·28	191	244·48
32	40·96	72	92·16	112	143·36	152	194·56	192	245·76
33	42·24	73	93·44	113	144·64	153	195·84	193	247·04
34	43·52	74	94·72	114	145·92	154	197·12	194	248·32
35	44·80	75	96·00	115	147·20	155	198·40	195	249·60
36	46·08	76	97·28	116	148·48	156	199·68	196	250·88
37	47·36	77	98·56	117	149·76	157	200·96	197	252·16
38	48·64	78	99·84	118	151·04	158	202·24	198	253·44
39	49·92	79	101·12	119	152·32	159	203·52	199	254·72
40	51·20	**80**	102·40	**120**	153·60	**160**	204·80	**200**	256·00

$\frac{1}{16}=0{\cdot}08^0$ $\frac{1}{9}=0{\cdot}14^2$ $\frac{1}{8}=0{\cdot}16^0$ $\frac{1}{6}=0{\cdot}21^3$ $\frac{1}{5}=0{\cdot}25^6$ $\frac{1}{4}=0{\cdot}32^0$ $\frac{1}{3}=0{\cdot}42^7$

$\frac{3}{8}=0{\cdot}48^0$ $\frac{1}{2}=0{\cdot}64^0$ $\frac{5}{8}=0{\cdot}80^0$ $\frac{3}{4}=0{\cdot}96^0$ $\frac{7}{8}=1{\cdot}12^0$ $\frac{2}{3}=0{\cdot}85^3$ $\frac{5}{6}=1{\cdot}06^7$

$\frac{3}{16}=0{\cdot}24^0$ $\frac{5}{16}=0{\cdot}40^0$ $\frac{7}{16}=0{\cdot}56^0$ $\frac{9}{16}=0{\cdot}72^0$ $\frac{11}{16}=0{\cdot}88^0$ $\frac{13}{16}=1{\cdot}04^0$ $\frac{15}{16}=1{\cdot}20^0$

1	1·28½	41	52·68½	81	104·08½	121	155·48½	161	206·88½
2	2·57	42	53·97	82	105·37	122	156·77	162	208·17
3	3·85½	43	55·25½	83	106·65½	123	158·05½	163	209·45½
4	5·14	44	56·54	84	107·94	124	159·34	164	210·74
5	6·42½	45	57·82½	85	109·22½	125	160·62½	165	212·02½
6	7·71	46	59·11	86	110·51	126	161·91	166	213·31
7	8·99½	47	60·39½	87	111·79½	127	163·19½	167	214·59½
8	10·28	48	61·68	88	113·08	128	164·48	168	215·88
9	11·56½	49	62·96½	89	114·36½	129	165·76½	169	217·16½
10	12·85	**50**	64·25	**90**	115·65	**130**	167·05	**170**	218·45
11	14·13½	51	65·53½	91	116·93½	131	168·33½	171	219·73½
12	15·42	52	66·82	92	118·22	132	169·62	172	221·02
13	16·70½	53	68·10½	93	119·50½	133	170·90½	173	222·30½
14	17·99	54	69·39	94	120·79	134	172·19	174	223·59
15	19·27½	55	70·67½	95	122·07½	135	173·47½	175	224·87½
16	20·56	56	71·96	96	123·36	136	174·76	176	226·16
17	21·84½	57	73·24½	97	124·64½	137	176·04½	177	227·44½
18	23·13	58	74·53	98	125·93	138	177·33	178	228·73
19	24·41½	59	75·81½	99	127·21½	139	178·61½	179	230·01½
20	25·70	**60**	77·10	**100**	128·50	**140**	179·90	**180**	231·30
21	26·98½	61	78·38½	101	129·78½	141	181·18½	181	232·58½
22	28·27	62	79·67	102	131·07	142	182·47	182	233·87
23	29·55½	63	80·95½	103	132·35½	143	183·75½	183	235·15½
24	30·84	64	82·24	104	133·64	144	185·04	184	236·44
25	32·12½	65	83·52½	105	134·92½	145	186·32½	185	237·72½
26	33·41	66	84·81	106	136·21	146	187·61	186	239·01
27	34·69½	67	86·09½	107	137·49½	147	188·89½	187	240·29½
28	35·98	68	87·38	108	138·78	148	190·18	188	241·58
29	37·26½	69	88·66½	109	140·06½	149	191·46½	189	242·86½
30	38·55	**70**	89·95	**110**	141·35	**150**	192·75	**190**	244·15
31	39·83½	71	91·23½	111	142·63½	151	194·03½	191	245·43½
32	41·12	72	92·52	112	143·92	152	195·32	192	246·72
33	42·40½	73	93·80½	113	145·20½	153	196·60½	193	248·00½
34	43·69	74	95·09	114	146·49	154	197·89	194	249·29
35	44·97½	75	96·37½	115	147·77½	155	199·17½	195	250·57½
36	46·26	76	97·66	116	149·06	156	200·46	196	251·86
37	47·54½	77	98·94½	117	150·34½	157	201·74½	197	253·14½
38	48·83	78	100·23	118	151·63	158	203·03	198	254·43
39	50·11½	79	101·51½	119	152·91½	159	204·31½	199	255·71½
40	51·40	**80**	102·80	**120**	154·20	**160**	205·60	**200**	257·00

$\frac{1}{16}$=0·08⁰	$\frac{1}{9}$=0·14³	$\frac{1}{8}$=0·16¹	$\frac{1}{6}$=0·21⁴	$\frac{1}{5}$=0·25⁷	$\frac{1}{4}$=0·32¹	$\frac{1}{3}$=0·42⁸
$\frac{3}{8}$=0·48²	$\frac{1}{2}$=0·64³	$\frac{5}{8}$=0·80³	$\frac{3}{4}$=0·96⁴	$\frac{7}{8}$=1·12⁴	$\frac{2}{3}$=0·85⁷	$\frac{5}{6}$=1·07¹
$\frac{3}{16}$=0·24¹	$\frac{5}{16}$=0·40²	$\frac{7}{16}$=0·56²	$\frac{9}{16}$=0·72³	$\frac{11}{16}$=0·88³	$\frac{13}{16}$=1·04⁴	$\frac{15}{16}$=1·20⁵

1	1·29	41	52·89	81	104·49	121	156·09	161	207·69
2	2·58	42	54·18	82	105·78	122	157·38	162	208·98
3	3·87	43	55·47	83	107·07	123	158·67	163	210·27
4	5·16	44	56·76	84	108·36	124	159·96	164	211·56
5	6·45	45	58·05	85	109·65	125	161·25	165	212·85
6	7·74	46	59·34	86	110·94	126	162·54	166	214·14
7	9·03	47	60·63	87	112·23	127	163·83	167	215·43
8	10·32	48	61·92	88	113·52	128	165·12	168	216·72
9	11·61	49	63·21	89	114·81	129	166·41	169	218·01
10	12·90	**50**	64·50	**90**	116·10	**130**	167·70	**170**	219·30
11	14·19	51	65·79	91	117·39	131	168·99	171	220·59
12	15·48	52	67·08	92	118·68	132	170·28	172	221·88
13	16·77	53	68·37	93	119·97	133	171·57	173	223·17
14	18·06	54	69·66	94	121·26	134	172·86	174	224·46
15	19·35	55	70·95	95	122·55	135	174·15	175	225·75
16	20·64	56	72·24	96	123·84	136	175·44	176	227·04
17	21·93	57	73·53	97	125·13	137	176·73	177	228·33
18	23·22	58	74·82	98	126·42	138	178·02	178	229·62
19	24·51	59	76·11	99	127·71	139	179·31	179	230·91
20	25·80	**60**	77·40	**100**	129·00	**140**	180·60	**180**	232·20
21	27·09	61	78·69	101	130·29	141	181·89	181	233·49
22	28·38	62	79·98	102	131·58	142	183·18	182	234·78
23	29·67	63	81·27	103	132·87	143	184·47	183	236·07
24	30·96	64	82·56	104	134·16	144	185·76	184	237·36
25	32·25	65	83·85	105	135·45	145	187·05	185	238·65
26	33·54	66	85·14	106	136·74	146	188·34	186	239·94
27	34·83	67	86·43	107	138·03	147	189·63	187	241·23
28	36·12	68	87·72	108	139·32	148	190·92	188	242·52
29	37·41	69	89·01	109	140·61	149	192·21	189	243·81
30	38·70	**70**	90·30	**110**	141·90	**150**	193·50	**190**	245·10
31	39·99	71	91·59	111	143·19	151	194·79	191	246·39
32	41·28	72	92·88	112	144·48	152	196·08	192	247·68
33	42·57	73	94·17	113	145·77	153	197·37	193	248·97
34	43·86	74	95·46	114	147·06	154	198·66	194	250·26
35	45·15	75	96·75	115	148·35	155	199·95	195	251·55
36	46·44	76	98·04	116	149·64	156	201·24	196	252·84
37	47·73	77	99·33	117	150·93	157	202·53	197	254·13
38	49·02	78	100·62	118	152·22	158	203·82	198	255·42
39	50·31	79	101·91	119	153·51	159	205·11	199	256·71
40	51·60	**80**	103·20	**120**	154·80	**160**	206·40	**200**	258·00

$\frac{1}{16}$=0·08^{1} $\frac{1}{9}$=0·14^{3} $\frac{1}{8}$=0·16^{1} $\frac{1}{6}$=0·21^{5} $\frac{1}{5}$=0·25^{8} $\frac{1}{4}$=0·32^{3} $\frac{1}{3}$=0·43^{0}

$\frac{3}{8}$=0·48^{4} $\frac{1}{2}$=0·64^{5} $\frac{5}{8}$=0·80^{6} $\frac{3}{4}$=0·96^{8} $\frac{7}{8}$=1·12^{9} $\frac{2}{3}$=0·86^{0} $\frac{5}{6}$=1·07^{5}

$\frac{3}{16}$=0·24^{2} $\frac{5}{16}$=0·40^{3} $\frac{7}{16}$=0·56^{4} $\frac{9}{16}$=0·72^{6} $\frac{11}{16}$=0·88^{7} $\frac{13}{16}$=1·04^{8} $\frac{15}{16}$=1·20^{9}

1	1·29½	41	53·09½	81	104·89½	121	156·69½	161	208·49½
2	2·59	42	54·39	82	106·19	122	157·99	162	209·79
3	3·88½	43	55·68½	83	107·48½	123	159·28½	163	211·08½
4	5·18	44	56·98	84	108·78	124	160·58	164	212·38
5	6·47½	45	58·27½	85	110·07½	125	161·87½	165	213·67½
6	7·77	46	59·57	86	111·37	126	163·17	166	214·97
7	9·06½	47	60·86½	87	112·66½	127	164·46½	167	216·26½
8	10·36	48	62·16	88	113·96	128	165·76	168	217·56
9	11·65½	49	63·45½	89	115·25½	129	167·05½	169	218·85½
10	12·95	**50**	64·75	**90**	116·55	**130**	168·35	**170**	220·15
11	14·24½	51	66·04½	91	117·84½	131	169·64½	171	221·44½
12	15·54	52	67·34	92	119·14	132	170·94	172	222·74
13	16·83½	53	68·63½	93	120·43½	133	172·23½	173	224·03½
14	18·13	54	69·93	94	121·73	134	173·53	174	225·33
15	19·42½	55	71·22½	95	123·02½	135	174·82½	175	226·62½
16	20·72	56	72·52	96	124·32	136	176·12	176	227·92
17	22·01½	57	73·81½	97	125·61½	137	177·41½	177	229·21½
18	23·31	58	75·11	98	126·91	138	178·71	178	230·51
19	24·60½	59	76·40½	99	128·20½	139	180·00½	179	231·80½
20	25·90	**60**	77·70	**100**	129·50	**140**	181·30	**180**	233·10
21	27·19½	61	78·99½	101	130·79½	141	182·59½	181	234·39½
22	28·49	62	80·29	102	132·09	142	183·89	182	235·69
23	29·78½	63	81·58½	103	133·38½	143	185·18½	183	236·98½
24	31·08	64	82·88	104	134·68	144	186·48	184	238·28
25	32·37½	65	84·17½	105	135·97½	145	187·77½	185	239·57½
26	33·67	66	85·47	106	137·27	146	189·07	186	240·87
27	34·96½	67	86·76½	107	138·56½	147	190·36½	187	242·16½
28	36·26	68	88·06	108	139·86	148	191·66	188	243·46
29	37·55½	69	89·35½	109	141·15½	149	192·95½	189	244·75½
30	38·85	**70**	90·65	**110**	142·45	**150**	194·25	**190**	246·05
31	40·14½	71	91·94½	111	143·74½	151	195·54½	191	247·34½
32	41·44	72	93·24	112	145·04	152	196·84	192	248·64
33	42·73½	73	94·53½	113	146·33½	153	198·13½	193	249·93½
34	44·03	74	95·83	114	147·63	154	199·43	194	251·23
35	45·32½	75	97·12½	115	148·92½	155	200·72½	195	252·52½
36	46·62	76	98·42	116	150·22	156	202·02	196	253·82
37	47·91½	77	99·71½	117	151·51½	157	203·31½	197	255·11½
38	49·21	78	101·01	118	152·81	158	204·61	198	256·41
39	50·50½	79	102·30½	119	154·10½	159	205·90½	199	257·70½
40	51·80	**80**	103·60	**120**	155·40	**160**	207·20	**200**	259·00

$\frac{1}{16}=0{\cdot}08^1$	$\frac{1}{9}=0{\cdot}14^4$	$\frac{1}{8}=0{\cdot}16^2$	$\frac{1}{6}=0{\cdot}21^6$	$\frac{1}{5}=0{\cdot}25^9$	$\frac{1}{4}=0{\cdot}32^4$	$\frac{1}{3}=0{\cdot}43^2$
$\frac{3}{8}=0{\cdot}48^6$	$\frac{1}{2}=0{\cdot}64^8$	$\frac{5}{8}=0{\cdot}80^9$	$\frac{3}{4}=0{\cdot}97^1$	$\frac{7}{8}=1{\cdot}13^3$	$\frac{2}{3}=0{\cdot}86^3$	$\frac{5}{6}=1{\cdot}07^9$
$\frac{3}{16}=0{\cdot}24^3$	$\frac{5}{16}=0{\cdot}40^5$	$\frac{7}{16}=0{\cdot}56^7$	$\frac{9}{16}=0{\cdot}72^8$	$\frac{11}{16}=0{\cdot}89^0$	$\frac{13}{16}=1{\cdot}05^2$	$\frac{15}{16}=1{\cdot}21^4$

(0.769=1£) **£1.30**

1	1·30	41	53·30	81	105·30	121	157·30	161	209·30
2	2·60	42	54·60	82	106·60	122	158·60	162	210·60
3	3·90	43	55·90	83	107·90	123	159·90	163	211·90
4	5·20	44	57·20	84	109·20	124	161·20	164	213·20
5	6·50	45	58·50	85	110·50	125	162·50	165	214·50
6	7·80	46	59·80	86	111·80	126	163·80	166	215·80
7	9·10	47	61·10	87	113·10	127	165·10	167	217·10
8	10·40	48	62·40	88	114·40	128	166·40	168	218·40
9	11·70	49	63·70	89	115·70	129	167·70	169	219·70
10	13·00	**50**	65·00	**90**	117·00	**130**	169·00	**170**	221·00
11	14·30	51	66·30	91	118·30	131	170·30	171	222·30
12	15·60	52	67·60	92	119·60	132	171·60	172	223·60
13	16·90	53	68·90	93	120·90	133	172·90	173	224·90
14	18·20	54	70·20	94	122·20	134	174·20	174	226·20
15	19·50	55	71·50	95	123·50	135	175·50	175	227·50
16	20·80	56	72·80	96	124·80	136	176·80	176	228·80
17	22·10	57	74·10	97	126·10	137	178·10	177	230·10
18	23·40	58	75·40	98	127·40	138	179·40	178	231·40
19	24·70	59	76·70	99	128·70	139	180·70	179	232·70
20	26·00	**60**	78·00	**100**	130·00	**140**	182·00	**180**	234·00
21	27·30	61	79·30	101	131·30	141	183·30	181	235·30
22	28·60	62	80·60	102	132·60	142	184·60	182	236·60
23	29·90	63	81·90	103	133·90	143	185·90	183	237·90
24	31·20	64	83·20	104	135·20	144	187·20	184	239·20
25	32·50	65	84·50	105	136·50	145	188·50	185	240·50
26	33·80	66	85·80	106	137·80	146	189·80	186	241·80
27	35·10	67	87·10	107	139·10	147	191·10	187	243·10
28	36·40	68	88·40	108	140·40	148	192·40	188	244·40
29	37·70	69	89·70	109	141·70	149	193·70	189	245·70
30	39·00	**70**	91·00	**110**	143·00	**150**	195·00	**190**	247·00
31	40·30	71	92·30	111	144·30	151	196·30	191	248·30
32	41·60	72	93·60	112	145·60	152	197·60	192	249·60
33	42·90	73	94·90	113	146·90	153	198·90	193	250·90
34	44·20	74	96·20	114	148·20	154	200·20	194	252·20
35	45·50	75	97·50	115	149·50	155	201·50	195	253·50
36	46·80	76	98·80	116	150·80	156	202·80	196	254·80
37	48·10	77	100·10	117	152·10	157	204·10	197	256·10
38	49·40	78	101·40	118	153·40	158	205·40	198	257·40
39	50·70	79	102·70	119	154·70	159	206·70	199	258·70
40	52·00	**80**	104·00	**120**	156·00	**160**	208·00	**200**	260·00

$\frac{1}{16}$=0·08^{1} $\frac{1}{9}$=0·14^{4} $\frac{1}{8}$=0·16^{3} $\frac{1}{6}$=0·21^{7} $\frac{1}{5}$=0·26^{0} $\frac{1}{4}$=0·32^{5} $\frac{1}{3}$=0·43^{3}

$\frac{3}{8}$=0·48^{8} $\frac{1}{2}$=0·65^{0} $\frac{5}{8}$=0·81^{3} $\frac{3}{4}$=0·97^{5} $\frac{7}{8}$=1·13^{8} $\frac{2}{3}$=0·86^{7} $\frac{5}{6}$=1·08^{3}

$\frac{3}{16}$=0·24^{4} $\frac{5}{16}$=0·40^{6} $\frac{7}{16}$=0·56^{9} $\frac{9}{16}$=0·73^{1} $\frac{11}{16}$=0·89^{4} $\frac{13}{16}$=1·05^{6} $\frac{15}{16}$=1·21^{9}

(0.766=1£) **£1.30½**

1	1·30½	41	53·50½	81	105·70½	121	157·90½	161	210·10½
2	2·61	42	54·81	82	107·01	122	159·21	162	211·41
3	3·91½	43	56·11½	83	108·31½	123	160·51½	163	212·71½
4	5·22	44	57·42	84	109·62	124	161·82	164	214·02
5	6·52½	45	58·72½	85	110·92½	125	163·12½	165	215·32½
6	7·83	46	60·03	86	112·23	126	164·43	166	216·63
7	9·13½	47	61·33½	87	113·53½	127	165·73½	167	217·93½
8	10·44	48	62·64	88	114·84	128	167·04	168	219·24
9	11·74½	49	63·94½	89	116·14½	129	168·34½	169	220·54½
10	13·05	**50**	65·25	**90**	117·45	**130**	169·65	**170**	221·85
11	14·35½	51	66·55½	91	118·75½	131	170·95½	171	223·15½
12	15·66	52	67·86	92	120·06	132	172·26	172	224·46
13	16·96½	53	69·16½	93	121·36½	133	173·56½	173	225·76½
14	18·27	54	70·47	94	122·67	134	174·87	174	227·07
15	19·57½	55	71·77½	95	123·97½	135	176·17½	175	228·37½
16	20·88	56	73·08	96	125·28	136	177·48	176	229·68
17	22·18½	57	74·38½	97	126·58½	137	178·78½	177	230·98½
18	23·49	58	75·69	98	127·89	138	180·09	178	232·29
19	24·79½	59	76·99½	99	129·19½	139	181·39½	179	233·59½
20	26·10	**60**	78·30	**100**	130·50	**140**	182·70	**180**	234·90
21	27·40½	61	79·60½	101	131·80½	141	184·00½	181	236·20½
22	28·71	62	80·91	102	133·11	142	185·31	182	237·51
23	30·01½	63	82·21½	103	134·41½	143	186·61½	183	238·81½
24	31·32	64	83·52	104	135·72	144	187·92	184	240·12
25	32·62½	65	84·82½	105	137·02½	145	189·22½	185	241·42½
26	33·93	66	86·13	106	138·33	146	190·53	186	242·73
27	35·23½	67	87·43½	107	139·63½	147	191·83½	187	244·03½
28	36·54	68	88·74	108	140·94	148	193·14	188	245·34
29	37·84½	69	90·04½	109	142·24½	149	194·44½	189	246·64½
30	39·15	**70**	91·35	**110**	143·55	**150**	195·75	**190**	247·95
31	40·45½	71	92·65½	111	144·85½	151	197·05½	191	249·25½
32	41·76	72	93·96	112	146·16	152	198·36	192	250·56
33	43·06½	73	95·26½	113	147·46½	153	199·66½	193	251·86½
34	44·37	74	96·57	114	148·77	154	200·97	194	253·17
35	45·67½	75	97·87½	115	150·07½	155	202·27½	195	254·47½
36	46·98	76	99·18	116	151·38	156	203·58	196	255·78
37	48·28½	77	100·48½	117	152·68½	157	204·88½	197	257·08½
38	49·59	78	101·79	118	153·99	158	206·19	198	258·39
39	50·89½	79	103·09½	119	155·29½	159	207·49½	199	259·69½
40	52·20	**80**	104·40	**120**	156·60	**160**	208·80	**200**	261·00

$\frac{1}{16}=0{\cdot}08^{2}$ $\frac{1}{9}=0{\cdot}14^{5}$ $\frac{1}{8}=0{\cdot}16^{3}$ $\frac{1}{6}=0{\cdot}21^{8}$ $\frac{1}{5}=0{\cdot}26^{1}$ $\frac{1}{4}=0{\cdot}32^{6}$ $\frac{1}{3}=0{\cdot}43^{5}$

$\frac{3}{8}=0{\cdot}48^{9}$ $\frac{1}{2}=0{\cdot}65^{3}$ $\frac{5}{8}=0{\cdot}81^{6}$ $\frac{3}{4}=0{\cdot}97^{9}$ $\frac{7}{8}=1{\cdot}14^{2}$ $\frac{2}{3}=0{\cdot}87^{0}$ $\frac{5}{6}=1{\cdot}08^{8}$

$\frac{3}{16}=0{\cdot}24^{5}$ $\frac{5}{16}=0{\cdot}40^{8}$ $\frac{7}{16}=0{\cdot}57^{1}$ $\frac{9}{16}=0{\cdot}73^{4}$ $\frac{11}{16}=0{\cdot}89^{7}$ $\frac{13}{16}=1{\cdot}06^{0}$ $\frac{15}{16}=1{\cdot}22^{3}$

(0.763=1£) **£1.31**

1	1·31	41	53·71	81	106·11	121	158·51	161	210·91
2	2·62	42	55·02	82	107·42	122	159·82	162	212·22
3	3·93	43	56·33	83	108·73	123	161·13	163	213·53
4	5·24	44	57·64	84	110·04	124	162·44	164	214·84
5	6·55	45	58·95	85	111·35	125	163·75	165	216·15
6	7·86	46	60·26	86	112·66	126	165·06	166	217·46
7	9·17	47	61·57	87	113·97	127	166·37	167	218·77
8	10·48	48	62·88	88	115·28	128	167·68	168	220·08
9	11·79	49	64·19	89	116·59	129	168·99	169	221·39
10	13·10	**50**	65·50	**90**	117·90	**130**	170·30	**170**	222·70
11	14·41	51	66·81	91	119·21	131	171·61	171	224·01
12	15·72	52	68·12	92	120·52	132	172·92	172	225·32
13	17·03	53	69·43	93	121·83	133	174·23	173	226·63
14	18·34	54	70·74	94	123·14	134	175·54	174	227·94
15	19·65	55	72·05	95	124·45	135	176·85	175	229·25
16	20·96	56	73·36	96	125·76	136	178·16	176	230·56
17	22·27	57	74·67	97	127·07	137	179·47	177	231·87
18	23·58	58	75·98	98	128·38	138	180·78	178	233·18
19	24·89	59	77·29	99	129·69	139	182·09	179	234·49
20	26·20	**60**	78·60	**100**	131·00	**140**	183·40	**180**	235·80
21	27·51	61	79·91	101	132·31	141	184·71	181	237·11
22	28·82	62	81·22	102	133·62	142	186·02	182	238·42
23	30·13	63	82·53	103	134·93	143	187·33	183	239·73
24	31·44	64	83·84	104	136·24	144	188·64	184	241·04
25	32·75	65	85·15	105	137·55	145	189·95	185	242·35
26	34·06	66	86·46	106	138·86	146	191·26	186	243·66
27	35·37	67	87·77	107	140·17	147	192·57	187	244·97
28	36·68	68	89·08	108	141·48	148	193·88	188	246·28
29	37·99	69	90·39	109	142·79	149	195·19	189	247·59
30	39·30	**70**	91·70	**110**	144·10	**150**	196·50	**190**	248·90
31	40·61	71	93·01	111	145·41	151	197·81	191	250·21
32	41·92	72	94·32	112	146·72	152	199·12	192	251·52
33	43·23	73	95·63	113	148·03	153	200·43	193	252·83
34	44·54	74	96·94	114	149·34	154	201·74	194	254·14
35	45·85	75	98·25	115	150·65	155	203·05	195	255·45
36	47·16	76	99·56	116	151·96	156	204·36	196	256·76
37	48·47	77	100·87	117	153·27	157	205·67	197	258·07
38	49·78	78	102·18	118	154·58	158	206·98	198	259·38
39	51·09	79	103·49	119	155·89	159	208·29	199	260·69
40	52·40	**80**	104·80	**120**	157·20	**160**	209·60	**200**	262·00

$\frac{1}{16}$=0·08^{2} $\frac{1}{9}$=0·14^{6} $\frac{1}{8}$=0·16^{4} $\frac{1}{6}$=0·21^{8} $\frac{1}{5}$=0·26^{2} $\frac{1}{4}$=0·32^{8} $\frac{1}{3}$=0·43^{7}

$\frac{3}{8}$=0·49^{1} $\frac{1}{2}$=0·65^{5} $\frac{5}{8}$=0·81^{9} $\frac{3}{4}$=0·98^{3} $\frac{7}{8}$=1·14^{6} $\frac{2}{3}$=0·87^{3} $\frac{5}{6}$=1·09^{2}

$\frac{3}{16}$=0·24^{6} $\frac{5}{16}$=0·40^{9} $\frac{7}{16}$=0·57^{3} $\frac{9}{16}$=0·73^{7} $\frac{11}{16}$=0·90^{1} $\frac{13}{16}$=1·06^{4} $\frac{15}{16}$=1·22^{8}

1 cubic metre=1·308 cubic yards

(0.760=1£) **£1.31½**

1	1·31½	41	53·91½	81	106·51½	121	159·11½	161	211·71½
2	2·63	42	55·23	82	107·83	122	160·43	162	213·03
3	3·94½	43	56·54½	83	109·14½	123	161·74½	163	214·34½
4	5·26	44	57·86	84	110·46	124	163·06	164	215·66
5	6·57½	45	59·17½	85	111·77½	125	164·37½	165	216·97½
6	7·89	46	60·49	86	113·09	126	165·69	166	218·29
7	9·20½	47	61·80½	87	114·40½	127	167·00½	167	219·60½
8	10·52	48	63·12	88	115·72	128	168·32	168	220·92
9	11·83½	49	64·43½	89	117·03½	129	169·63½	169	222·23½
10	13·15	**50**	65·75	**90**	118·35	**130**	170·95	**170**	223·55
11	14·46½	51	67·06½	91	119·66½	131	172·26½	171	224·86½
12	15·78	52	68·38	92	120·98	132	173·58	172	226·18
13	17·09½	53	69·69½	93	122·29½	133	174·89½	173	227·49½
14	18·41	54	71·01	94	123·61	134	176·21	174	228·81
15	19·72½	55	72·32½	95	124·92½	135	177·52½	175	230·12½
16	21·04	56	73·64	96	126·24	136	178·84	176	231·44
17	22·35½	57	74·95½	97	127·55½	137	180·15½	177	232·75½
18	23·67	58	76·27	98	128·87	138	181·47	178	234·07
19	24·98½	59	77·58½	99	130·18½	139	182·78½	179	235·38½
20	26·30	**60**	78·90	**100**	131·50	**140**	184·10	**180**	236·70
21	27·61½	61	80·21½	101	132·81½	141	185·41½	181	238·01½
22	28·93	62	81·53	102	134·13	142	186·73	182	239·33
23	30·24½	63	82·84½	103	135·44½	143	188·04½	183	240·64½
24	31·56	64	84·16	104	136·76	144	189·36	184	241·96
25	32·87½	65	85·47½	105	138·07½	145	190·67½	185	243·27½
26	34·19	66	86·79	106	139·39	146	191·99	186	244·59
27	35·50½	67	88·10½	107	140·70½	147	193·30½	187	245·90½
28	36·82	68	89·42	108	142·02	148	194·62	188	247·22
29	38·13½	69	90·73½	109	143·33½	149	195·93½	189	248·53½
30	39·45	**70**	92·05	**110**	144·65	**150**	197·25	**190**	249·85
31	40·76½	71	93·36½	111	145·96½	151	198·56½	191	251·16½
32	42·08	72	94·68	112	147·28	152	199·88	192	252·48
33	43·39½	73	95·99½	113	148·59½	153	201·19½	193	253·79½
34	44·71	74	97·31	114	149·91	154	202·51	194	255·11
35	46·02½	75	98·62½	115	151·22½	155	203·82½	195	256·42½
36	47·34	76	99·94	116	152·54	156	205·14	196	257·74
37	48·65½	77	101·25½	117	153·85½	157	206·45½	197	259·05½
38	49·97	78	102·57	118	155·17	158	207·77	198	260·37
39	51·28½	79	103·88½	119	156·48½	159	209·08½	199	261·68½
40	52·60	**80**	105·20	**120**	157·80	**160**	210·40	**200**	263·00

$\frac{1}{16}=0{\cdot}08^{2}$ $\frac{1}{9}=0{\cdot}14^{6}$ $\frac{1}{8}=0{\cdot}16^{4}$ $\frac{1}{6}=0{\cdot}21^{9}$ $\frac{1}{5}=0{\cdot}26^{3}$ $\frac{1}{4}=0{\cdot}32^{9}$ $\frac{1}{3}=0{\cdot}43^{8}$

$\frac{3}{8}=0{\cdot}49^{3}$ $\frac{1}{2}=0{\cdot}65^{8}$ $\frac{5}{8}=0{\cdot}82^{2}$ $\frac{3}{4}=0{\cdot}98^{6}$ $\frac{7}{8}=1{\cdot}15^{1}$ $\frac{2}{3}=0{\cdot}87^{7}$ $\frac{5}{6}=1{\cdot}09^{6}$

$\frac{3}{16}=0{\cdot}24^{7}$ $\frac{5}{16}=0{\cdot}41^{1}$ $\frac{7}{16}=0{\cdot}57^{5}$ $\frac{9}{16}=0{\cdot}74^{0}$ $\frac{11}{16}=0{\cdot}90^{4}$ $\frac{13}{16}=1{\cdot}06^{8}$ $\frac{15}{16}=1{\cdot}23^{3}$

(0.758=1£) £1.32

1	1·32	41	54·12	81	106·92	121	159·72	161	212·52
2	2·64	42	55·44	82	108·24	122	161·04	162	213·84
3	3·96	43	56·76	83	109·56	123	162·36	163	215·16
4	5·28	44	58·08	84	110·88	124	163·68	164	216·48
5	6·60	45	59·40	85	112·20	125	165·00	165	217·80
6	7·92	46	60·72	86	113·52	126	166·32	166	219·12
7	9·24	47	62·04	87	114·84	127	167·64	167	220·44
8	10·56	48	63·36	88	116·16	128	168·96	168	221·76
9	11·88	49	64·68	89	117·48	129	170·28	169	223·08
10	13·20	**50**	66·00	**90**	118·80	**130**	171·60	**170**	224·40
11	14·52	51	67·32	91	120·12	131	172·92	171	225·72
12	15·84	52	68·64	92	121·44	132	174·24	172	227·04
13	17·16	53	69·96	93	122·76	133	175·56	173	228·36
14	18·48	54	71·28	94	124·08	134	176·88	174	229·68
15	19·80	55	72·60	95	125·40	135	178·20	175	231·00
16	21·12	56	73·92	96	126·72	136	179·52	176	232·32
17	22·44	57	75·24	97	128·04	137	180·84	177	233·64
18	23·76	58	76·56	98	129·36	138	182·16	178	234·96
19	25·08	59	77·88	99	130·68	139	183·48	179	236·28
20	26·40	**60**	79·20	**100**	132·00	**140**	184·80	**180**	237·60
21	27·72	61	80·52	101	133·32	141	186·12	181	238·92
22	29·04	62	81·84	102	134·64	142	187·44	182	240·24
23	30·36	63	83·16	103	135·96	143	188·76	183	241·56
24	31·68	64	84·48	104	137·28	144	190·08	184	242·88
25	33·00	65	85·80	105	138·60	145	191·40	185	244·20
26	34·32	66	87·12	106	139·92	146	192·72	186	245·52
27	35·64	67	88·44	107	141·24	147	194·04	187	246·84
28	36·96	68	89·76	108	142·56	148	195·36	188	248·16
29	38·28	69	91·08	109	143·88	149	196·68	189	249·48
30	39·60	**70**	92·40	**110**	145·20	**150**	198·00	**190**	250·80
31	40·92	71	93·72	111	146·52	151	199·32	191	252·12
32	42·24	72	95·04	112	147·84	152	200·64	192	253·44
33	43·56	73	96·36	113	149·16	153	201·96	193	254·76
34	44·88	74	97·68	114	150·48	154	203·28	194	256·08
35	46·20	75	99·00	115	151·80	155	204·60	195	257·40
36	47·52	76	100·32	116	153·12	156	205·92	196	258·72
37	48·84	77	101·64	117	154·44	157	207·24	197	260·04
38	50·16	78	102·96	118	155·76	158	208·56	198	261·36
39	51·48	79	104·28	119	157·08	159	209·88	199	262·68
40	52·80	**80**	105·60	**120**	158·40	**160**	211·20	**200**	264·00

$\frac{1}{16}=0{\cdot}08^{3}$ $\frac{1}{9}=0{\cdot}14^{7}$ $\frac{1}{8}=0{\cdot}16^{5}$ $\frac{1}{6}=0{\cdot}22^{0}$ $\frac{1}{5}=0{\cdot}26^{4}$ $\frac{1}{4}=0{\cdot}33^{0}$ $\frac{1}{3}=0{\cdot}44^{0}$

$\frac{3}{8}=0{\cdot}49^{5}$ $\frac{1}{2}=0{\cdot}66^{0}$ $\frac{5}{8}=0{\cdot}82^{5}$ $\frac{3}{4}=0{\cdot}99^{0}$ $\frac{7}{8}=1{\cdot}15^{5}$ $\frac{2}{3}=0{\cdot}88^{0}$ $\frac{5}{6}=1{\cdot}10^{0}$

$\frac{3}{16}=0{\cdot}24^{8}$ $\frac{5}{16}=0{\cdot}41^{3}$ $\frac{7}{16}=0{\cdot}57^{8}$ $\frac{9}{16}=0{\cdot}74^{3}$ $\frac{11}{16}=0{\cdot}90^{8}$ $\frac{13}{16}=1{\cdot}07^{3}$ $\frac{15}{16}=1{\cdot}23^{8}$

(0.755=1£) **£1.32½**

1	1·32½	41	54·32½	81	107·32½	121	160·32½	161	213·32½
2	2·65	42	55·65	82	108·65	122	161·65	162	214·65
3	3·97½	43	56·97½	83	109·97½	123	162·97½	163	215·97½
4	5·30	44	58·30	84	111·30	124	164·30	164	217·30
5	6·62½	45	59·62½	85	112·62½	125	165·62½	165	218·62½
6	7·95	46	60·95	86	113·95	126	166·95	166	219·95
7	9·27½	47	62·27½	87	115·27½	127	168·27½	167	221·27½
8	10·60	48	63·60	88	116·60	128	169·60	168	222·60
9	11·92½	49	64·92½	89	117·92½	129	170·92½	169	223·92½
10	13·25	**50**	66·25	**90**	119·25	**130**	172·25	**170**	225·25
11	14·57½	51	67·57½	91	120·57½	131	173·57½	171	226·57½
12	15·90	52	68·90	92	121·90	132	174·90	172	227·90
13	17·22½	53	70·22½	93	123·22½	133	176·22½	173	229·22½
14	18·55	54	71·55	94	124·55	134	177·55	174	230·55
15	19·87½	55	72·87½	95	125·87½	135	178·87½	175	231·87½
16	21·20	56	74·20	96	127·20	136	180·20	176	233·20
17	22·52½	57	75·52½	97	128·52½	137	181·52½	177	234·52½
18	23·85	58	76·85	98	129·85	138	182·85	178	235·85
19	25·17½	59	78·17½	99	131·17½	139	184·17½	179	237·17½
20	26·50	**60**	79·50	**100**	132·50	**140**	185·50	**180**	238·50
21	27·82½	61	80·82½	101	133·82½	141	186·82½	181	239·82½
22	29·15	62	82·15	102	135·15	142	188·15	182	241·15
23	30·47½	63	83·47½	103	136·47½	143	189·47½	183	242·47½
24	31·80	64	84·80	104	137·80	144	190·80	184	243·80
25	33·12½	65	86·12½	105	139·12½	145	192·12½	185	245·12½
26	34·45	66	87·45	106	140·45	146	193·45	186	246·45
27	35·77½	67	88·77½	107	141·77½	147	194·77½	187	247·77½
28	37·10	68	90·10	108	143·10	148	196·10	188	249·10
29	38·42½	69	91·42½	109	144·42½	149	197·42½	189	250·42½
30	39·75	**70**	92·75	**110**	145·75	**150**	198·75	**190**	251·75
31	41·07½	71	94·07½	111	147·07½	151	200·07½	191	253·07½
32	42·40	72	95·40	112	148·40	152	201·40	192	254·40
33	43·72½	73	96·72½	113	149·72½	153	202·72½	193	255·72½
34	45·05	74	98·05	114	151·05	154	204·05	194	257·05
35	46·37½	75	99·37½	115	152·37½	155	205·37½	195	258·37½
36	47·70	76	100·70	116	153·70	156	206·70	196	259·70
37	49·02½	77	102·02½	117	155·02½	157	208·02½	197	261·02½
38	50·35	78	103·35	118	156·35	158	209·35	198	262·35
39	51·67½	79	104·67½	119	157·67½	159	210·67½	199	263·67½
40	53·00	**80**	106·00	**120**	159·00	**160**	212·00	**200**	265·00

$\frac{1}{16}=0{\cdot}08^{3}$ $\frac{1}{9}=0{\cdot}14^{7}$ $\frac{1}{8}=0{\cdot}16^{6}$ $\frac{1}{6}=0{\cdot}22^{1}$ $\frac{1}{5}=0{\cdot}26^{5}$ $\frac{1}{4}=0{\cdot}33^{1}$ $\frac{1}{3}=0{\cdot}44^{2}$

$\frac{3}{8}=0{\cdot}49^{7}$ $\frac{1}{2}=0{\cdot}66^{3}$ $\frac{5}{8}=0{\cdot}82^{8}$ $\frac{3}{4}=0{\cdot}99^{4}$ $\frac{7}{8}=1{\cdot}15^{9}$ $\frac{2}{3}=0{\cdot}88^{3}$ $\frac{5}{6}=1{\cdot}10^{4}$

$\frac{3}{16}=0{\cdot}24^{8}$ $\frac{5}{16}=0{\cdot}41^{4}$ $\frac{7}{16}=0{\cdot}58^{0}$ $\frac{9}{16}=0{\cdot}74^{5}$ $\frac{11}{16}=0{\cdot}91^{1}$ $\frac{13}{16}=1{\cdot}07^{7}$ $\frac{15}{16}=1{\cdot}24^{2}$

(0.752=1£) **£1.33**

1	1·33	41	54·53	81	107·73	121	160·93	161	214·13
2	2·66	42	55·86	82	109·06	122	162·26	162	215·46
3	3·99	43	57·19	83	110·39	123	163·59	163	216·79
4	5·32	44	58·52	84	111·72	124	164·92	164	218·12
5	6·65	45	59·85	85	113·05	125	166·25	165	219·45
6	7·98	46	61·18	86	114·38	126	167·58	166	220·78
7	9·31	47	62·51	87	115·71	127	168·91	167	222·11
8	10·64	48	63·84	88	117·04	128	170·24	168	223·44
9	11·97	49	65·17	89	118·37	129	171·57	169	224·77
10	13·30	**50**	66·50	**90**	119·70	**130**	172·90	**170**	226·10
11	14·63	51	67·83	91	121·03	131	174·23	171	227·43
12	15·96	52	69·16	92	122·36	132	175·56	172	228·76
13	17·29	53	70·49	93	123·69	133	176·89	173	230·09
14	18·62	54	71·82	94	125·02	134	178·22	174	231·42
15	19·95	55	73·15	95	126·35	135	179·55	175	232·75
16	21·28	56	74·48	96	127·68	136	180·88	176	234·08
17	22·61	57	75·81	97	129·01	137	182·21	177	235·41
18	23·94	58	77·14	98	130·34	138	183·54	178	236·74
19	25·27	59	78·47	99	131·67	139	184·87	179	238·07
20	26·60	**60**	79·80	**100**	133·00	**140**	186·20	**180**	239·40
21	27·93	61	81·13	101	134·33	141	187·53	181	240·73
22	29·26	62	82·46	102	135·66	142	188·86	182	242·06
23	30·59	63	83·79	103	136·99	143	190·19	183	243·39
24	31·92	64	85·12	104	138·32	144	191·52	184	244·72
25	33·25	65	86·45	105	139·65	145	192·85	185	246·05
26	34·58	66	87·78	106	140·98	146	194·18	186	247·38
27	35·91	67	89·11	107	142·31	147	195·51	187	248·71
28	37·24	68	90·44	108	143·64	148	196·84	188	250·04
29	38·57	69	91·77	109	144·97	149	198·17	189	251·37
30	39·90	**70**	93·10	**110**	146·30	**150**	199·50	**190**	252·70
31	41·23	71	94·43	111	147·63	151	200·83	191	254·03
32	42·56	72	95·76	112	148·96	152	202·16	192	255·36
33	43·89	73	97·09	113	150·29	153	203·49	193	256·69
34	45·22	74	98·42	114	151·62	154	204·82	194	258·02
35	46·55	75	99·75	115	152·95	155	206·15	195	259·35
36	47·88	76	101·08	116	154·28	156	207·48	196	260·68
37	49·21	77	102·41	117	155·61	157	208·81	197	262·01
38	50·54	78	103·74	118	156·94	158	210·14	198	263·34
39	51·87	79	105·07	119	158·27	159	211·47	199	264·67
40	53·20	**80**	106·40	**120**	159·60	**160**	212·80	**200**	266·00

$\frac{1}{16}=0{\cdot}08^{3}$ $\frac{1}{9}=0{\cdot}14^{8}$ $\frac{1}{8}=0{\cdot}16^{6}$ $\frac{1}{6}=0{\cdot}22^{2}$ $\frac{1}{5}=0{\cdot}26^{6}$ $\frac{1}{4}=0{\cdot}33^{3}$ $\frac{1}{3}=0{\cdot}44^{3}$

$\frac{3}{8}=0{\cdot}49^{9}$ $\frac{1}{2}=0{\cdot}66^{5}$ $\frac{5}{8}=0{\cdot}83^{1}$ $\frac{3}{4}=0{\cdot}99^{8}$ $\frac{7}{8}=1{\cdot}16^{4}$ $\frac{2}{3}=0{\cdot}88^{7}$ $\frac{5}{6}=1{\cdot}10^{8}$

$\frac{3}{16}=0{\cdot}24^{9}$ $\frac{5}{16}=0{\cdot}41^{6}$ $\frac{7}{16}=0{\cdot}58^{2}$ $\frac{9}{16}=0{\cdot}74^{8}$ $\frac{11}{16}=0{\cdot}91^{4}$ $\frac{13}{16}=1{\cdot}08^{1}$ $\frac{15}{16}=1{\cdot}24^{7}$

(0.749=1£) **£1.33½**

1	1·33½	41	54·73½	81	108·13½	121	161·53½	161	214·93½
2	2·67	42	56·07	82	109·47	122	162·87	162	216·27
3	4·00½	43	57·40½	83	110·80½	123	164·20½	163	217·60½
4	5·34	44	58·74	84	112·14	124	165·54	164	218·94
5	6·67½	45	60·07½	85	113·47½	125	166·87½	165	220·27½
6	8·01	46	61·41	86	114·81	126	168·21	166	221·61
7	9·34½	47	62·74½	87	116·14½	127	169·54½	167	222·94½
8	10·68	48	64·08	88	117·48	128	170·88	168	224·28
9	12·01½	49	65·41½	89	118·81½	129	172·21½	169	225·61½
10	13·35	**50**	66·75	**90**	120·15	**130**	173·55	**170**	226·95
11	14·68½	51	68·08½	91	121·48½	131	174·88½	171	228·28½
12	16·02	52	69·42	92	122·82	132	176·22	172	229·62
13	17·35½	53	70·75½	93	124·15½	133	177·55½	173	230·95½
14	18·69	54	72·09	94	125·49	134	178·89	174	232·29
15	20·02½	55	73·42½	95	126·82½	135	180·22½	175	233·62½
16	21·36	56	74·76	96	128·16	136	181·56	176	234·96
17	22·69½	57	76·09½	97	129·49½	137	182·89½	177	236·29½
18	24·03	58	77·43	98	130·83	138	184·23	178	237·63
19	25·36½	59	78·76½	99	132·16½	139	185·56½	179	238·96½
20	26·70	**60**	80·10	**100**	133·50	**140**	186·90	**180**	240·30
21	28·03½	61	81·43½	101	134·83½	141	188·23½	181	241·63½
22	29·37	62	82·77	102	136·17	142	189·57	182	242·97
23	30·70½	63	84·10½	103	137·50½	143	190·90½	183	244·30½
24	32·04	64	85·44	104	138·84	144	192·24	184	245·64
25	33·37½	65	86·77½	105	140·17½	145	193·57½	185	246·97½
26	34·71	66	88·11	106	141·51	146	194·91	186	248·31
27	36·04½	67	89·44½	107	142·84½	147	196·24½	187	249·64½
28	37·38	68	90·78	108	144·18	148	197·58	188	250·98
29	38·71½	69	92·11½	109	145·51½	149	198·91½	189	252·31½
30	40·05	**70**	93·45	**110**	146·85	**150**	200·25	**190**	253·65
31	41·38½	71	94·78½	111	148·18½	151	201·58½	191	254·98½
32	42·72	72	96·12	112	149·52	152	202·92	192	256·32
33	44·05½	73	97·45½	113	150·85½	153	204·25½	193	257·65½
34	45·39	74	98·79	114	152·19	154	205·59	194	258·99
35	46·72½	75	100·12½	115	153·52½	155	206·92½	195	260·32½
36	48·06	76	101·46	116	154·86	156	208·26	196	261·66
37	49·39½	77	102·79½	117	156·19½	157	209·59½	197	262·99½
38	50·73	78	104·13	118	157·53	158	210·93	198	264·33
39	52·06½	79	105·46½	119	158·86½	159	212·26½	199	265·66½
40	53·40	**80**	106·80	**120**	160·20	**160**	213·60	**200**	267·00

$\frac{1}{16}=0{\cdot}08^{3}$	$\frac{1}{9}=0{\cdot}14^{8}$	$\frac{1}{8}=0{\cdot}16^{7}$	$\frac{1}{6}=0{\cdot}22^{3}$	$\frac{1}{5}=0{\cdot}26^{7}$	$\frac{1}{4}=0{\cdot}33^{4}$	$\frac{1}{3}=0{\cdot}44^{5}$
$\frac{3}{8}=0{\cdot}50^{1}$	$\frac{1}{2}=0{\cdot}66^{8}$	$\frac{5}{8}=0{\cdot}83^{4}$	$\frac{3}{4}=1{\cdot}00^{1}$	$\frac{7}{8}=1{\cdot}16^{8}$	$\frac{2}{3}=0{\cdot}89^{0}$	$\frac{5}{6}=1{\cdot}11^{3}$
$\frac{3}{16}=0{\cdot}25^{0}$	$\frac{5}{16}=0{\cdot}41^{7}$	$\frac{7}{16}=0{\cdot}58^{4}$	$\frac{9}{16}=0{\cdot}75^{1}$	$\frac{11}{16}=0{\cdot}91^{8}$	$\frac{13}{16}=1{\cdot}08^{5}$	$\frac{15}{16}=1{\cdot}25^{2}$

(0.746=1£) **£1.34**

1	1·34	41	54·94	81	108·54	121	162·14	161	215·74
2	2·68	42	56·28	82	109·88	122	163·48	162	217·08
3	4·02	43	57·62	83	111·22	123	164·82	163	218·42
4	5·36	44	58·96	84	112·56	124	166·16	164	219·76
5	6·70	45	60·30	85	113·90	125	167·50	165	221·10
6	8·04	46	61·64	86	115·24	126	168·84	166	222·44
7	9·38	47	62·98	87	116·58	127	170·18	167	223·78
8	10·72	48	64·32	88	117·92	128	171·52	168	225·12
9	12·06	49	65·66	89	119·26	129	172·86	169	226·46
10	13·40	**50**	67·00	**90**	120·60	**130**	174·20	**170**	227·80
11	14·74	51	68·34	91	121·94	131	175·54	171	229·14
12	16·08	52	69·68	92	123·28	132	176·88	172	230·48
13	17·42	53	71·02	93	124·62	133	178·22	173	231·82
14	18·76	54	72·36	94	125·96	134	179·56	174	233·16
15	20·10	55	73·70	95	127·30	135	180·90	175	234·50
16	21·44	56	75·04	96	128·64	136	182·24	176	235·84
17	22·78	57	76·38	97	129·98	137	183·58	177	237·18
18	24·12	58	77·72	98	131·32	138	184·92	178	238·52
19	25·46	59	79·06	99	132·66	139	186·26	179	239·86
20	26·80	**60**	80·40	**100**	134·00	**140**	187·60	**180**	241·20
21	28·14	61	81·74	101	135·34	141	188·94	181	242·54
22	29·48	62	83·08	102	136·68	142	190·28	182	243·88
23	30·82	63	84·42	103	138·02	143	191·62	183	245·22
24	32·16	64	85·76	104	139·36	144	192·96	184	246·56
25	33·50	65	87·10	105	140·70	145	194·30	185	247·90
26	34·84	66	88·44	106	142·04	146	195·64	186	249·24
27	36·18	67	89·78	107	143·38	147	196·98	187	250·58
28	37·52	68	91·12	108	144·72	148	198·32	188	251·92
29	38·86	69	92·46	109	146·06	149	199·66	189	253·26
30	40·20	**70**	93·80	**110**	147·40	**150**	201·00	**190**	254·60
31	41·54	71	95·14	111	148·74	151	202·34	191	255·94
32	42·88	72	96·48	112	150·08	152	203·68	192	257·28
33	44·22	73	97·82	113	151·42	153	205·02	193	258·62
34	45·56	74	99·16	114	152·76	154	206·36	194	259·96
35	46·90	75	100·50	115	154·10	155	207·70	195	261·30
36	48·24	76	101·84	116	155·44	156	209·04	196	262·64
37	49·58	77	103·18	117	156·78	157	210·38	197	263·98
38	50·92	78	104·52	118	158·12	158	211·72	198	265·32
39	52·26	79	105·86	119	159·46	159	213·06	199	266·66
40	53·60	**80**	107·20	**120**	160·80	**160**	214·40	**200**	268·00

$\frac{1}{16}=0{\cdot}08^{4}$ $\frac{1}{9}=0{\cdot}14^{9}$ $\frac{1}{8}=0{\cdot}16^{8}$ $\frac{1}{6}=0{\cdot}22^{3}$ $\frac{1}{5}=0{\cdot}26^{8}$ $\frac{1}{4}=0{\cdot}33^{5}$ $\frac{1}{3}=0{\cdot}44^{7}$

$\frac{3}{8}=0{\cdot}50^{3}$ $\frac{1}{2}=0{\cdot}67^{0}$ $\frac{5}{8}=0{\cdot}83^{8}$ $\frac{3}{4}=1{\cdot}00^{5}$ $\frac{7}{8}=1{\cdot}17^{3}$ $\frac{2}{3}=0{\cdot}89^{3}$ $\frac{5}{6}=1{\cdot}11^{7}$

$\frac{3}{16}=0{\cdot}25^{1}$ $\frac{5}{16}=0{\cdot}41^{9}$ $\frac{7}{16}=0{\cdot}58^{6}$ $\frac{9}{16}=0{\cdot}75^{4}$ $\frac{11}{16}=0{\cdot}92^{1}$ $\frac{13}{16}=1{\cdot}08^{9}$ $\frac{15}{16}=1{\cdot}25^{6}$

(0.743=1£) **£1.34½**

1	1·34½	41	55·14½	81	108·94½	121	162·74½	161	216·54½
2	2·69	42	56·49	82	110·29	122	164·09	162	217·89
3	4·03½	43	57·83½	83	111·63½	123	165·43½	163	219·23½
4	5·38	44	59·18	84	112·98	124	166·78	164	220·58
5	6·72½	45	60·52½	85	114·32½	125	168·12½	165	221·92½
6	8·07	46	61·87	86	115·67	126	169·47	166	223·27
7	9·41½	47	63·21½	87	117·01½	127	170·81½	167	224·61½
8	10·76	48	64·56	88	118·36	128	172·16	168	225·96
9	12·10½	49	65·90½	89	119·70½	129	173·50½	169	227·30½
10	13·45	**50**	67·25	**90**	121·05	**130**	174·85	**170**	228·65
11	14·79½	51	68·59½	91	122·39½	131	176·19½	171	229·99½
12	16·14	52	69·94	92	123·74	132	177·54	172	231·34
13	17·48½	53	71·28½	93	125·08½	133	178·88½	173	232·68½
14	18·83	54	72·63	94	126·43	134	180·23	174	234·03
15	20·17½	55	73·97½	95	127·77½	135	181·57½	175	235·37½
16	21·52	56	75·32	96	129·12	136	182·92	176	236·72
17	22·86½	57	76·66½	97	130·46½	137	184·26½	177	238·06½
18	24·21	58	78·01	98	131·81	138	185·61	178	239·41
19	25·55½	59	79·35½	99	133·15½	139	186·95½	179	240·75½
20	26·90	**60**	80·70	**100**	134·50	**140**	188·30	**180**	242·10
21	28·24½	61	82·04½	101	135·84½	141	189·64½	181	243·44½
22	29·59	62	83·39	102	137·19	142	190·99	182	244·79
23	30·93½	63	84·73½	103	138·53½	143	192·33½	183	246·13½
24	32·28	64	86·08	104	139·88	144	193·68	184	247·48
25	33·62½	65	87·42½	105	141·22½	145	195·02½	185	248·82½
26	34·97	66	88·77	106	142·57	146	196·37	186	250·17
27	36·31½	67	90·11½	107	143·91½	147	197·71½	187	251·51½
28	37·66	68	91·46	108	145·26	148	199·06	188	252·86
29	39·00½	69	92·80½	109	146·60½	149	200·40½	189	254·20½
30	40·35	**70**	94·15	**110**	147·95	**150**	201·75	**190**	255·55
31	41·69½	71	95·49½	111	149·29½	151	203·09½	191	256·89½
32	43·04	72	96·84	112	150·64	152	204·44	192	258·24
33	44·38½	73	98·18½	113	151·98½	153	205·78½	193	259·58½
34	45·73	74	99·53	114	153·33	154	207·13	194	260·93
35	47·07½	75	100·87½	115	154·67½	155	208·47½	195	262·27½
36	48·42	76	102·22	116	156·02	156	209·82	196	263·62
37	49·76½	77	103·56½	117	157·36½	157	211·16½	197	264·96½
38	51·11	78	104·91	118	158·71	158	212·51	198	266·31
39	52·45½	79	106·25½	119	160·05½	159	213·85½	199	267·65½
40	53·80	**80**	107·60	**120**	161·40	**160**	215·20	**200**	269·00

$\frac{1}{16}=0{\cdot}08^{4}$ $\frac{1}{9}=0{\cdot}14^{9}$ $\frac{1}{8}=0{\cdot}16^{8}$ $\frac{1}{6}=0{\cdot}22^{4}$ $\frac{1}{5}=0{\cdot}26^{9}$ $\frac{1}{4}=0{\cdot}33^{6}$ $\frac{1}{3}=0{\cdot}44^{8}$

$\frac{3}{8}=0{\cdot}50^{4}$ $\frac{1}{2}=0{\cdot}67^{3}$ $\frac{5}{8}=0{\cdot}84^{1}$ $\frac{3}{4}=1{\cdot}00^{9}$ $\frac{7}{8}=1{\cdot}17^{7}$ $\frac{2}{3}=0{\cdot}89^{7}$ $\frac{5}{6}=1{\cdot}12^{1}$

$\frac{3}{16}=0{\cdot}25^{2}$ $\frac{5}{16}=0{\cdot}42^{0}$ $\frac{7}{16}=0{\cdot}58^{8}$ $\frac{9}{16}=0{\cdot}75^{7}$ $\frac{11}{16}=0{\cdot}92^{5}$ $\frac{13}{16}=1{\cdot}09^{3}$ $\frac{15}{16}=1{\cdot}26^{1}$

(0.741=1£) **£1.35**

1	1·35	41	55·35	81	109·35	121	163·35	161	217·35
2	2·70	42	56·70	82	110·70	122	164·70	162	218·70
3	4·05	43	58·05	83	112·05	123	166·05	163	220·05
4	5·40	44	59·40	84	113·40	124	167·40	164	221·40
5	6·75	45	60·75	85	114·75	125	168·75	165	222·75
6	8·10	46	62·10	86	116·10	126	170·10	166	224·10
7	9·45	47	63·45	87	117·45	127	171·45	167	225·45
8	10·80	48	64·80	88	118·80	128	172·80	168	226·80
9	12·15	49	66·15	89	120·15	129	174·15	169	228·15
10	13·50	**50**	67·50	**90**	121·50	**130**	175·50	**170**	229·50
11	14·85	51	68·85	91	122·85	131	176·85	171	230·85
12	16·20	52	70·20	92	124·20	132	178·20	172	232·20
13	17·55	53	71·55	93	125·55	133	179·55	173	233·55
14	18·90	54	72·90	94	126·90	134	180·90	174	234·90
15	20·25	55	74·25	95	128·25	135	182·25	175	236·25
16	21·60	56	75·60	96	129·60	136	183·60	176	237·60
17	22·95	57	76·95	97	130·95	137	184·95	177	238·95
18	24·30	58	78·30	98	132·30	138	186·30	178	240·30
19	25·65	59	79·65	99	133·65	139	187·65	179	241·65
20	27·00	**60**	81·00	**100**	135·00	**140**	189·00	**180**	243·00
21	28·35	61	82·35	101	136·35	141	190·35	181	244·35
22	29·70	62	83·70	102	137·70	142	191·70	182	245·70
23	31·05	63	85·05	103	139·05	143	193·05	183	247·05
24	32·40	64	86·40	104	140·40	144	194·40	184	248·40
25	33·75	65	87·75	105	141·75	145	195·75	185	249·75
26	35·10	66	89·10	106	143·10	146	197·10	186	251·10
27	36·45	67	90·45	107	144·45	147	198·45	187	252·45
28	37·80	68	91·80	108	145·80	148	199·80	188	253·80
29	39·15	69	93·15	109	147·15	149	201·15	189	255·15
30	40·50	**70**	94·50	**110**	148·50	**150**	202·50	**190**	256·50
31	41·85	71	95·85	111	149·85	151	203·85	191	257·85
32	43·20	72	97·20	112	151·20	152	205·20	192	259·20
33	44·55	73	98·55	113	152·55	153	206·55	193	260·55
34	45·90	74	99·90	114	153·90	154	207·90	194	261·90
35	47·25	75	101·25	115	155·25	155	209·25	195	263·25
36	48·60	76	102·60	116	156·60	156	210·60	196	264·60
37	49·95	77	103·95	117	157·95	157	211·95	197	265·95
38	51·30	78	105·30	118	159·30	158	213·30	198	267·30
39	52·65	79	106·65	119	160·65	159	214·65	199	268·65
40	54·00	**80**	108·00	**120**	162·00	**160**	216·00	**200**	270·00

$\frac{1}{16}=0{\cdot}08^4$ $\frac{1}{9}=0{\cdot}15^0$ $\frac{1}{8}=0{\cdot}16^9$ $\frac{1}{6}=0{\cdot}22^5$ $\frac{1}{5}=0{\cdot}27^0$ $\frac{1}{4}=0{\cdot}33^8$ $\frac{1}{3}=0{\cdot}45^0$

$\frac{3}{8}=0{\cdot}50^6$ $\frac{1}{2}=0{\cdot}67^5$ $\frac{5}{8}=0{\cdot}84^4$ $\frac{3}{4}=1{\cdot}01^3$ $\frac{7}{8}=1{\cdot}18^1$ $\frac{2}{3}=0{\cdot}90^0$ $\frac{5}{6}=1{\cdot}12^5$

$\frac{3}{16}=0{\cdot}25^3$ $\frac{5}{16}=0{\cdot}42^2$ $\frac{7}{16}=0{\cdot}59^1$ $\frac{9}{16}=0{\cdot}75^9$ $\frac{11}{16}=0{\cdot}92^8$ $\frac{13}{16}=1{\cdot}09^7$ $\frac{15}{16}=1{\cdot}26^6$

(0.738=1£) **£1.35½**

1	1·35½	41	55·55½	81	109·75½	121	163·95½	161	218·15½
2	2·71	42	56·91	82	111·11	122	165·31	162	219·51
3	4·06½	43	58·26½	83	112·46½	123	166·66½	163	220·86½
4	5·42	44	59·62	84	113·82	124	168·02	164	222·22
5	6·77½	45	60·97½	85	115·17½	125	169·37½	165	223·57½
6	8·13	46	62·33	86	116·53	126	170·73	166	224·93
7	9·48½	47	63·68½	87	117·88½	127	172·08½	167	226·28½
8	10·84	48	65·04	88	119·24	128	173·44	168	227·64
9	12·19½	49	66·39½	89	120·59½	129	174·79½	169	228·99½
10	13·55	**50**	67·75	**90**	121·95	**130**	176·15	**170**	230·35
11	14·90½	51	69·10½	91	123·30½	131	177·50½	171	231·70½
12	16·26	52	70·46	92	124·66	132	178·86	172	233·06
13	17·61½	53	71·81½	93	126·01½	133	180·21½	173	234·41½
14	18·97	54	73·17	94	127·37	134	181·57	174	235·77
15	20·32½	55	74·52½	95	128·72½	135	182·92½	175	237·12½
16	21·68	56	75·88	96	130·08	136	184·28	176	238·48
17	23·03½	57	77·23½	97	131·43½	137	185·63½	177	239·83½
18	24·39	58	78·59	98	132·79	138	186·99	178	241·19
19	25·74½	59	79·94½	99	134·14½	139	188·34½	179	242·54½
20	27·10	**60**	81·30	**100**	135·50	**140**	189·70	**180**	243·90
21	28·45½	61	82·65½	101	136·85½	141	191·05½	181	245·25½
22	29·81	62	84·01	102	138·21	142	192·41	182	246·61
23	31·16½	63	85·36½	103	139·56½	143	193·76½	183	247·96½
24	32·52	64	86·72	104	140·92	144	195·12	184	249·32
25	33·87½	65	88·07½	105	142·27½	145	196·47½	185	250·67½
26	35·23	66	89·43	106	143·63	146	197·83	186	252·03
27	36·58½	67	90·78½	107	144·98½	147	199·18½	187	253·38½
28	37·94	68	92·14	108	146·34	148	200·54	188	254·74
29	39·29½	69	93·49½	109	147·69½	149	201·89½	189	256·09½
30	40·65	**70**	94·85	**110**	149·05	**150**	203·25	**190**	257·45
31	42·00½	71	96·20½	111	150·40½	151	204·60½	191	258·80½
32	43·36	72	97·56	112	151·76	152	205·96	192	260·16
33	44·71½	73	98·91½	113	153·11½	153	207·31½	193	261·51½
34	46·07	74	100·27	114	154·47	154	208·67	194	262·87
35	47·42½	75	101·62½	115	155·82½	155	210·02½	195	264·22½
36	48·78	76	102·98	116	157·18	156	211·38	196	265·58
37	50·13½	77	104·33½	117	158·53½	157	212·73½	197	266·93½
38	51·49	78	105·69	118	159·89	158	214·09	198	268·29
39	52·84½	79	107·04½	119	161·24½	159	215·44½	199	269·64½
40	54·20	**80**	108·40	**120**	162·60	**160**	216·80	**200**	271·00

$\frac{1}{16}=0{\cdot}08^5$ $\frac{1}{9}=0{\cdot}15^1$ $\frac{1}{8}=0{\cdot}16^9$ $\frac{1}{6}=0{\cdot}22^6$ $\frac{1}{5}=0{\cdot}27^1$ $\frac{1}{4}=0{\cdot}33^9$ $\frac{1}{3}=0{\cdot}45^2$

$\frac{3}{8}=0{\cdot}50^8$ $\frac{1}{2}=0{\cdot}67^8$ $\frac{5}{8}=0{\cdot}84^7$ $\frac{3}{4}=1{\cdot}01^6$ $\frac{7}{8}=1{\cdot}18^6$ $\frac{2}{3}=0{\cdot}90^3$ $\frac{5}{6}=1{\cdot}12^9$

$\frac{3}{16}=0{\cdot}25^4$ $\frac{5}{16}=0{\cdot}42^3$ $\frac{7}{16}=0{\cdot}59^3$ $\frac{9}{16}=0{\cdot}76^2$ $\frac{11}{16}=0{\cdot}93^2$ $\frac{13}{16}=1{\cdot}10^1$ $\frac{15}{16}=1{\cdot}27^0$

1	1·36	41	55·76	81	110·16	121	164·56	161	218·96
2	2·72	42	57·12	82	111·52	122	165·92	162	220·32
3	4·08	43	58·48	83	112·88	123	167·28	163	221·68
4	5·44	44	59·84	84	114·24	124	168·64	164	223·04
5	6·80	45	61·20	85	115·60	125	170·00	165	224·40
6	8·16	46	62·56	86	116·96	126	171·36	166	225·76
7	9·52	47	63·92	87	118·32	127	172·72	167	227·12
8	10·88	48	65·28	88	119·68	128	174·08	168	228·48
9	12·24	49	66·64	89	121·04	129	175·44	169	229·84
10	13·60	**50**	68·00	**90**	122·40	**130**	176·80	**170**	231·20
11	14·96	51	69·36	91	123·76	131	178·16	171	232·56
12	16·32	52	70·72	92	125·12	132	179·52	172	233·92
13	17·68	53	72·08	93	126·48	133	180·88	173	235·28
14	19·04	54	73·44	94	127·84	134	182·24	174	236·64
15	20·40	55	74·80	95	129·20	135	183·60	175	238·00
16	21·76	56	76·16	96	130·56	136	184·96	176	239·36
17	23·12	57	77·52	97	131·92	137	186·32	177	240·72
18	24·48	58	78·88	98	133·28	138	187·68	178	242·08
19	25·84	59	80·24	99	134·64	139	189·04	179	243·44
20	27·20	**60**	81·60	**100**	136·00	**140**	190·40	**180**	244·80
21	28·56	61	82·96	101	137·36	141	191·76	181	246·16
22	29·92	62	84·32	102	138·72	142	193·12	182	247·52
23	31·28	63	85·68	103	140·08	143	194·48	183	248·88
24	32·64	64	87·04	104	141·44	144	195·84	184	250·24
25	34·00	65	88·40	105	142·80	145	197·20	185	251·60
26	35·36	66	89·76	106	144·16	146	198·56	186	252·96
27	36·72	67	91·12	107	145·52	147	199·92	187	254·32
28	38·08	68	92·48	108	146·88	148	201·28	188	255·68
29	39·44	69	93·84	109	148·24	149	202·64	189	257·04
30	40·80	**70**	95·20	**110**	149·60	**150**	204·00	**190**	258·40
31	42·16	71	96·56	111	150·96	151	205·36	191	259·76
32	43·52	72	97·92	112	152·32	152	206·72	192	261·12
33	44·88	73	99·28	113	153·68	153	208·08	193	262·48
34	46·24	74	100·64	114	155·04	154	209·44	194	263·84
35	47·60	75	102·00	115	156·40	155	210·80	195	265·20
36	48·96	76	103·36	116	157·76	156	212·16	196	266·56
37	50·32	77	104·72	117	159·12	157	213·52	197	267·92
38	51·68	78	106·08	118	160·48	158	214·88	198	269·28
39	53·04	79	107·44	119	161·84	159	216·24	199	270·64
40	54·40	**80**	108·80	**120**	163·20	**160**	217·60	**200**	272·00

$\frac{1}{16}$=0·08^{5} $\frac{1}{9}$=0·15^{1} $\frac{1}{8}$=0·17^{0} $\frac{1}{6}$=0·22^{7} $\frac{1}{5}$=0·27^{2} $\frac{1}{4}$=0·34^{0} $\frac{1}{3}$=0·45^{3}

$\frac{3}{8}$=0·51^{0} $\frac{1}{2}$=0·68^{0} $\frac{5}{8}$=0·85^{0} $\frac{3}{4}$=1·02^{0} $\frac{7}{8}$=1·19^{0} $\frac{2}{3}$=0·90^{7} $\frac{5}{6}$=1·13^{3}

$\frac{3}{16}$=0·25^{5} $\frac{5}{16}$=0·42^{5} $\frac{7}{16}$=0·59^{5} $\frac{9}{16}$=0·76^{5} $\frac{11}{16}$=0·93^{5} $\frac{13}{16}$=1·10^{5} $\frac{15}{16}$=1·27^{5}

(0.733=1£) £1.36½

1	1·36½	41	55·96½	81	110·56½	121	165·16½	161	219·76½
2	2·73	42	57·33	82	111·93	122	166·53	162	221·13
3	4·09½	43	58·69½	83	113·29½	123	167·89½	163	222·49½
4	5·46	44	60·06	84	114·66	124	169·26	164	223·86
5	6·82½	45	61·42½	85	116·02½	125	170·62½	165	225·22½
6	8·19	46	62·79	86	117·39	126	171·99	166	226·59
7	9·55½	47	64·15½	87	118·75½	127	173·35½	167	227·95½
8	10·92	48	65·52	88	120·12	128	174·72	168	229·32
9	12·28½	49	66·88½	89	121·48½	129	176·08½	169	230·68½
10	13·65	**50**	68·25	**90**	122·85	**130**	177·45	**170**	232·05
11	15·01½	51	69·61½	91	124·21½	131	178·81½	171	233·41½
12	16·38	52	70·98	92	125·58	132	180·18	172	234·78
13	17·74½	53	72·34½	93	126·94½	133	181·54½	173	236·14½
14	19·11	54	73·71	94	128·31	134	182·91	174	237·51
15	20·47½	55	75·07½	95	129·67½	135	184·27½	175	238·87½
16	21·84	56	76·44	96	131·04	136	185·64	176	240·24
17	23·20½	57	77·80½	97	132·40½	137	187·00½	177	241·60½
18	24·57	58	79·17	98	133·77	138	188·37	178	242·97
19	25·93½	59	80·53½	99	135·13½	139	189·73½	179	244·33½
20	27·30	**60**	81·90	**100**	136·50	**140**	191·10	**180**	245·70
21	28·66½	61	83·26½	101	137·86½	141	192·46½	181	247·06½
22	30·03	62	84·63	102	139·23	142	193·83	182	248·43
23	31·39½	63	85·99½	103	140·59½	143	195·19½	183	249·79½
24	32·76	64	87·36	104	141·96	144	196·56	184	251·16
25	34·12½	65	88·72½	105	143·32½	145	197·92½	185	252·52½
26	35·49	66	90·09	106	144·69	146	199·29	186	253·89
27	36·85½	67	91·45½	107	146·05½	147	200·65½	187	255·25½
28	38·22	68	92·82	108	147·42	148	202·02	188	256·62
29	39·58½	69	94·18½	109	148·78½	149	203·38½	189	257·98½
30	40·95	**70**	95·55	**110**	150·15	**150**	204·75	**190**	259·35
31	42·31½	71	96·91½	111	151·51½	151	206·11½	191	260·71½
32	43·68	72	98·28	112	152·88	152	207·48	192	262·08
33	45·04½	73	99·64½	113	154·24½	153	208·84½	193	263·44½
34	46·41	74	101·01	114	155·61	154	210·21	194	264·81
35	47·77½	75	102·37½	115	156·97½	155	211·57½	195	266·17½
36	49·14	76	103·74	116	158·34	156	212·94	196	267·54
37	50·50½	77	105·10½	117	159·70½	157	214·30½	197	268·90½
38	51·87	78	106·47	118	161·07	158	215·67	198	270·27
39	53·23½	79	107·83½	119	162·43½	159	217·03½	199	271·63½
40	54·60	**80**	109·20	**120**	163·80	**160**	218·40	**200**	273·00

$\frac{1}{16}=0{\cdot}08^{5}$	$\frac{1}{9}=0{\cdot}15^{2}$	$\frac{1}{8}=0{\cdot}17^{1}$	$\frac{1}{6}=0{\cdot}22^{8}$	$\frac{1}{5}=0{\cdot}27^{3}$	$\frac{1}{4}=0{\cdot}34^{1}$	$\frac{1}{3}=0{\cdot}45^{5}$
$\frac{3}{8}=0{\cdot}51^{2}$	$\frac{1}{2}=0{\cdot}68^{3}$	$\frac{5}{8}=0{\cdot}85^{3}$	$\frac{3}{4}=1{\cdot}02^{4}$	$\frac{7}{8}=1{\cdot}19^{4}$	$\frac{2}{3}=0{\cdot}91^{0}$	$\frac{5}{6}=1{\cdot}13^{8}$
$\frac{3}{16}=0{\cdot}25^{6}$	$\frac{5}{16}=0{\cdot}42^{7}$	$\frac{7}{16}=0{\cdot}59^{7}$	$\frac{9}{16}=0{\cdot}76^{8}$	$\frac{11}{16}=0{\cdot}93^{8}$	$\frac{13}{16}=1{\cdot}10^{9}$	$\frac{15}{16}=1{\cdot}28^{0}$

1	1·37	41	56·17	81	110·97	121	165·77	161	220·57
2	2·74	42	57·54	82	112·34	122	167·14	162	221·94
3	4·11	43	58·91	83	113·71	123	168·51	163	223·31
4	5·48	44	60·28	84	115·08	124	169·88	164	224·68
5	6·85	45	61·65	85	116·45	125	171·25	165	226·05
6	8·22	46	63·02	86	117·82	126	172·62	166	227·42
7	9·59	47	64·39	87	119·19	127	173·99	167	228·79
8	10·96	48	65·76	88	120·56	128	175·36	168	230·16
9	12·33	49	67·13	89	121·93	129	176·73	169	231·53
10	13·70	**50**	68·50	**90**	123·30	**130**	178·10	**170**	232·90
11	15·07	51	69·87	91	124·67	131	179·47	171	234·27
12	16·44	52	71·24	92	126·04	132	180·84	172	235·64
13	17·81	53	72·61	93	127·41	133	182·21	173	237·01
14	19·18	54	73·98	94	128·78	134	183·58	174	238·38
15	20·55	55	75·35	95	130·15	135	184·95	175	239·75
16	21·92	56	76·72	96	131·52	136	186·32	176	241·12
17	23·29	57	78·09	97	132·89	137	187·69	177	242·49
18	24·66	58	79·46	98	134·26	138	189·06	178	243·86
19	26·03	59	80·83	99	135·63	139	190·43	179	245·23
20	27·40	**60**	82·20	**100**	137·00	**140**	191·80	**180**	246·60
21	28·77	61	83·57	101	138·37	141	193·17	181	247·97
22	30·14	62	84·94	102	139·74	142	194·54	182	249·34
23	31·51	63	86·31	103	141·11	143	195·91	183	250·71
24	32·88	64	87·68	104	142·48	144	197·28	184	252·08
25	34·25	65	89·05	105	143·85	145	198·65	185	253·45
26	35·62	66	90·42	106	145·22	146	200·02	186	254·82
27	36·99	67	91·79	107	146·59	147	201·39	187	256·19
28	38·36	68	93·16	108	147·96	148	202·76	188	257·56
29	39·73	69	94·53	109	149·33	149	204·13	189	258·93
30	41·10	**70**	95·90	**110**	150·70	**150**	205·50	**190**	260·30
31	42·47	71	97·27	111	152·07	151	206·87	191	261·67
32	43·84	72	98·64	112	153·44	152	208·24	192	263·04
33	45·21	73	100·01	113	154·81	153	209·61	193	264·41
34	46·58	74	101·38	114	156·18	154	210·98	194	265·78
35	47·95	75	102·75	115	157·55	155	212·35	195	267·15
36	49·32	76	104·12	116	158·92	156	213·72	196	268·52
37	50·69	77	105·49	117	160·29	157	215·09	197	269·89
38	52·06	78	106·86	118	161·66	158	216·46	198	271·26
39	53·43	79	108·23	119	163·03	159	217·83	199	272·63
40	54·80	**80**	109·60	**120**	164·40	**160**	219·20	**200**	274·00

$\frac{1}{16}=0{\cdot}08^6$ $\frac{1}{9}=0{\cdot}15^2$ $\frac{1}{8}=0{\cdot}17^1$ $\frac{1}{6}=0{\cdot}22^8$ $\frac{1}{5}=0{\cdot}27^4$ $\frac{1}{4}=0{\cdot}34^3$ $\frac{1}{3}=0{\cdot}45^7$

$\frac{3}{8}=0{\cdot}51^4$ $\frac{1}{2}=0{\cdot}68^5$ $\frac{5}{8}=0{\cdot}85^6$ $\frac{3}{4}=1{\cdot}02^8$ $\frac{7}{8}=1{\cdot}19^9$ $\frac{2}{3}=0{\cdot}91^3$ $\frac{5}{6}=1{\cdot}14^2$

$\frac{3}{16}=0{\cdot}25^7$ $\frac{5}{16}=0{\cdot}42^8$ $\frac{7}{16}=0{\cdot}59^9$ $\frac{9}{16}=0{\cdot}77^1$ $\frac{11}{16}=0{\cdot}94^2$ $\frac{13}{16}=1{\cdot}11^3$ $\frac{15}{16}=1{\cdot}28^4$

(0.727=1£) **£1.37½**

1	1·37½	41	56·37½	81	111·37½	121	166·37½	161	221·37½
2	2·75	42	57·75	82	112·75	122	167·75	162	222·75
3	4·12½	43	59·12½	83	114·12½	123	169·12½	163	224·12½
4	5·50	44	60·50	84	115·50	124	170·50	164	225·50
5	6·87½	45	61·87½	85	116·87½	125	171·87½	165	226·87½
6	8·25	46	63·25	86	118·25	126	173·25	166	228·25
7	9·62½	47	64·62½	87	119·62½	127	174·62½	167	229·62½
8	11·00	48	66·00	88	121·00	128	176·00	168	231·00
9	12·37½	49	67·37½	89	122·37½	129	177·37½	169	232·37½
10	13·75	**50**	68·75	**90**	123·75	**130**	178·75	**170**	233·75
11	15·12½	51	70·12½	91	125·12½	131	180·12½	171	235·12½
12	16·50	52	71·50	92	126·50	132	181·50	172	236·50
13	17·87½	53	72·87½	93	127·87½	133	182·87½	173	237·87½
14	19·25	54	74·25	94	129·25	134	184·25	174	239·25
15	20·62½	55	75·62½	95	130·62½	135	185·62½	175	240·62½
16	22·00	56	77·00	96	132·00	136	187·00	176	242·00
17	23·37½	57	78·37½	97	133·37½	137	188·37½	177	243·37½
18	24·75	58	79·75	98	134·75	138	189·75	178	244·75
19	26·12½	59	81·12½	99	136·12½	139	191·12½	179	246·12½
20	27·50	**60**	82·50	**100**	137·50	**140**	192·50	**180**	247·50
21	28·87½	61	83·87½	101	138·87½	141	193·87½	181	248·87½
22	30·25	62	85·25	102	140·25	142	195·25	182	250·25
23	31·62½	63	86·62½	103	141·62½	143	196·62½	183	251·62½
24	33·00	64	88·00	104	143·00	144	198·00	184	253·00
25	34·37½	65	89·37½	105	144·37½	145	199·37½	185	254·37½
26	35·75	66	90·75	106	145·75	146	200·75	186	255·75
27	37·12½	67	92·12½	107	147·12½	147	202·12½	187	257·12½
28	38·50	68	93·50	108	148·50	148	203·50	188	258·50
29	39·87½	69	94·87½	109	149·87½	149	204·87½	189	259·87½
30	41·25	**70**	96·25	**110**	151·25	**150**	206·25	**190**	261·25
31	42·62½	71	97·62½	111	152·62½	151	207·62½	191	262·62½
32	44·00	72	99·00	112	154·00	152	209·00	192	264·00
33	45·37½	73	100·37½	113	155·37½	153	210·37½	193	265·37½
34	46·75	74	101·75	114	156·75	154	211·75	194	266·75
35	48·12½	75	103·12½	115	158·12½	155	213·12½	195	268·12½
36	49·50	76	104·50	116	159·50	156	214·50	196	269·50
37	50·87½	77	105·87½	117	160·87½	157	215·87½	197	270·87½
38	52·25	78	107·25	118	162·25	158	217·25	198	272·25
39	53·62½	79	108·62½	119	163·62½	159	218·62½	199	273·62½
40	55·00	**80**	110·00	**120**	165·00	**160**	220·00	**200**	275·00

$\frac{1}{16}=0\cdot08^{6}$	$\frac{1}{9}=0\cdot15^{3}$	$\frac{1}{8}=0\cdot17^{2}$	$\frac{1}{6}=0\cdot22^{9}$	$\frac{1}{5}=0\cdot27^{5}$	$\frac{1}{4}=0\cdot34^{4}$	$\frac{1}{3}=0\cdot45^{8}$
$\frac{3}{8}=0\cdot51^{6}$	$\frac{1}{2}=0\cdot68^{8}$	$\frac{5}{8}=0\cdot85^{9}$	$\frac{3}{4}=1\cdot03^{1}$	$\frac{7}{8}=1\cdot20^{3}$	$\frac{2}{3}=0\cdot91^{7}$	$\frac{5}{6}=1\cdot14^{6}$
$\frac{3}{16}=0\cdot25^{8}$	$\frac{5}{16}=0\cdot43^{0}$	$\frac{7}{16}=0\cdot60^{2}$	$\frac{9}{16}=0\cdot77^{3}$	$\frac{11}{16}=0\cdot94^{5}$	$\frac{13}{16}=1\cdot11^{7}$	$\frac{15}{16}=1\cdot28^{9}$

(0.725=1£) **£1.38**

1	1·38	41	56·58	81	111·78	121	166·98	161	222·18
2	2·76	42	57·96	82	113·16	122	168·36	162	223·56
3	4·14	43	59·34	83	114·54	123	169·74	163	224·94
4	5·52	44	60·72	84	115·92	124	171·12	164	226·32
5	6·90	45	62·10	85	117·30	125	172·50	165	227·70
6	8·28	46	63·48	86	118·68	126	173·88	166	229·08
7	9·66	47	64·86	87	120·06	127	175·26	167	230·46
8	11·04	48	66·24	88	121·44	128	176·64	168	231·84
9	12·42	49	67·62	89	122·82	129	178·02	169	233·22
10	13·80	**50**	69·00	**90**	124·20	**130**	179·40	**170**	234·60
11	15·18	51	70·38	91	125·58	131	180·78	171	235·98
12	16·56	52	71·76	92	126·96	132	182·16	172	237·36
13	17·94	53	73·14	93	128·34	133	183·54	173	238·74
14	19·32	54	74·52	94	129·72	134	184·92	174	240·12
15	20·70	55	75·90	95	131·10	135	186·30	175	241·50
16	22·08	56	77·28	96	132·48	136	187·68	176	242·88
17	23·46	57	78·66	97	133·86	137	189·06	177	244·26
18	24·84	58	80·04	98	135·24	138	190·44	178	245·64
19	26·22	59	81·42	99	136·62	139	191·82	179	247·02
20	27·60	**60**	82·80	**100**	138·00	**140**	193·20	**180**	248·40
21	28·98	61	84·18	101	139·38	141	194·58	181	249·78
22	30·36	62	85·56	102	140·76	142	195·96	182	251·16
23	31·74	63	86·94	103	142·14	143	197·34	183	252·54
24	33·12	64	88·32	104	143·52	144	198·72	184	253·92
25	34·50	65	89·70	105	144·90	145	200·10	185	255·30
26	35·88	66	91·08	106	146·28	146	201·48	186	256·68
27	37·26	67	92·46	107	147·66	147	202·86	187	258·06
28	38·64	68	93·84	108	149·04	148	204·24	188	259·44
29	40·02	69	95·22	109	150·42	149	205·62	189	260·82
30	41·40	**70**	96·60	**110**	151·80	**150**	207·00	**190**	262·20
31	42·78	71	97·98	111	153·18	151	208·38	191	263·58
32	44·16	72	99·36	112	154·56	152	209·76	192	264·96
33	45·54	73	100·74	113	155·94	153	211·14	193	266·34
34	46·92	74	102·12	114	157·32	154	212·52	194	267·72
35	48·30	75	103·50	115	158·70	155	213·90	195	269·10
36	49·68	76	104·88	116	160·08	156	215·28	196	270·48
37	51·06	77	106·26	117	161·46	157	216·66	197	271·86
38	52·44	78	107·64	118	162·84	158	218·04	198	273·24
39	53·82	79	109·02	119	164·22	159	219·42	199	274·62
40	55·20	**80**	110·40	**120**	165·60	**160**	220·80	**200**	276·00

$\frac{1}{16}$=0·08^{6} $\frac{1}{9}$=0·15^{3} $\frac{1}{8}$=0·17^{3} $\frac{1}{6}$=0·23^{0} $\frac{1}{5}$=0·27^{6} $\frac{1}{4}$=0·34^{5} $\frac{1}{3}$=0·46^{0}

$\frac{3}{8}$=0·51^{8} $\frac{1}{2}$=0·69^{0} $\frac{5}{8}$=0·86^{3} $\frac{3}{4}$=1·03^{5} $\frac{7}{8}$=1·20^{8} $\frac{2}{3}$=0·92^{0} $\frac{5}{6}$=1·15^{0}

$\frac{3}{16}$=0·25^{9} $\frac{5}{16}$=0·43^{1} $\frac{7}{16}$=0·60^{4} $\frac{9}{16}$=0·77^{6} $\frac{11}{16}$=0·94^{9} $\frac{13}{16}$=1·12^{1} $\frac{15}{16}$=1·29^{4}

(0.722=1£) **£1.38½**

1	1·38½	41	56·78½	81	112·18½	121	167·58½	161	222·98½
2	2·77	42	58·17	82	113·57	122	168·97	162	224·37
3	4·15½	43	59·55½	83	114·95½	123	170·35½	163	225·75½
4	5·54	44	60·94	84	116·34	124	171·74	164	227·14
5	6·92½	45	62·32½	85	117·72½	125	173·12½	165	228·52½
6	8·31	46	63·71	86	119·11	126	174·51	166	229·91
7	9·69½	47	65·09½	87	120·49½	127	175·89½	167	231·29½
8	11·08	48	66·48	88	121·88	128	177·28	168	232·68
9	12·46½	49	67·86½	89	123·26½	129	178·66½	169	234·06½
10	13·85	**50**	69·25	**90**	124·65	**130**	180·05	**170**	235·45
11	15·23½	51	70·63½	91	126·03½	131	181·43½	171	236·83½
12	16·62	52	72·02	92	127·42	132	182·82	172	238·22
13	18·00½	53	73·40½	93	128·80½	133	184·20½	173	239·60½
14	19·39	54	74·79	94	130·19	134	185·59	174	240·99
15	20·77½	55	76·17½	95	131·57½	135	186·97½	175	242·37½
16	22·16	56	77·56	96	132·96	136	188·36	176	243·76
17	23·54½	57	78·94½	97	134·34½	137	189·74½	177	245·14½
18	24·93	58	80·33	98	135·73	138	191·13	178	246·53
19	26·31½	59	81·71½	99	137·11½	139	192·51½	179	247·91½
20	27·70	**60**	83·10	**100**	138·50	**140**	193·90	**180**	249·30
21	29·08½	61	84·48½	101	139·88½	141	195·28½	181	250·68½
22	30·47	62	85·87	102	141·27	142	196·67	182	252·07
23	31·85½	63	87·25½	103	142·65½	143	198·05½	183	253·45½
24	33·24	64	88·64	104	144·04	144	199·44	184	254·84
25	34·62½	65	90·02½	105	145·42½	145	200·82½	185	256·22½
26	36·01	66	91·41	106	146·81	146	202·21	186	257·61
27	37·39½	67	92·79½	107	148·19½	147	203·59½	187	258·99½
28	38·78	68	94·18	108	149·58	148	204·98	188	260·38
29	40·16½	69	95·56½	109	150·96½	149	206·36½	189	261·76½
30	41·55	**70**	96·95	**110**	152·35	**150**	207·75	**190**	263·15
31	42·93½	71	98·33½	111	153·73½	151	209·13½	191	264·53½
32	44·32	72	99·72	112	155·12	152	210·52	192	265·92
33	45·70½	73	101·10½	113	156·50½	153	211·90½	193	267·30½
34	47·09	74	102·49	114	157·89	154	213·29	194	268·69
35	48·47½	75	103·87½	115	159·27½	155	214·67½	195	270·07½
36	49·86	76	105·26	116	160·66	156	216·06	196	271·46
37	51·24½	77	106·64½	117	162·04½	157	217·44½	197	272·84½
38	52·63	78	108·03	118	163·43	158	218·83	198	274·23
39	54·01½	79	109·41½	119	164·81½	159	220·21½	199	275·61½
40	55·40	**80**	110·80	**120**	166·20	**160**	221·60	**200**	277·00

$\frac{1}{16}=0{\cdot}08^{7}$ $\frac{1}{9}=0{\cdot}15^{4}$ $\frac{1}{8}=0{\cdot}17^{3}$ $\frac{1}{6}=0{\cdot}23^{1}$ $\frac{1}{5}=0{\cdot}27^{7}$ $\frac{1}{4}=0{\cdot}34^{6}$ $\frac{1}{3}=0{\cdot}46^{2}$

$\frac{3}{8}=0{\cdot}51^{9}$ $\frac{1}{2}=0{\cdot}69^{3}$ $\frac{5}{8}=0{\cdot}86^{6}$ $\frac{3}{4}=1{\cdot}03^{9}$ $\frac{7}{8}=1{\cdot}21^{2}$ $\frac{2}{3}=0{\cdot}92^{3}$ $\frac{5}{6}=1{\cdot}15^{4}$

$\frac{3}{16}=0{\cdot}26^{0}$ $\frac{5}{16}=0{\cdot}43^{3}$ $\frac{7}{16}=0{\cdot}60^{6}$ $\frac{9}{16}=0{\cdot}77^{9}$ $\frac{11}{16}=0{\cdot}95^{2}$ $\frac{13}{16}=1{\cdot}12^{5}$ $\frac{15}{16}=1{\cdot}29^{8}$

1	1·39	41	56·99	81	112·59	121	168·19	161	223·79
2	2·78	42	58·38	82	113·98	122	169·58	162	225·18
3	4·17	43	59·77	83	115·37	123	170·97	163	226·57
4	5·56	44	61·16	84	116·76	124	172·36	164	227·96
5	6·95	45	62·55	85	118·15	125	173·75	165	229·35
6	8·34	46	63·94	86	119·54	126	175·14	166	230·74
7	9·73	47	65·33	87	120·93	127	176·53	167	232·13
8	11·12	48	66·72	88	122·32	128	177·92	168	233·52
9	12·51	49	68·11	89	123·71	129	179·31	169	234·91
10	13·90	**50**	69·50	**90**	125·10	**130**	180·70	**170**	236·30
11	15·29	51	70·89	91	126·49	131	182·09	171	237·69
12	16·68	52	72·28	92	127·88	132	183·48	172	239·08
13	18·07	53	73·67	93	129·27	133	184·87	173	240·47
14	19·46	54	75·06	94	130·66	134	186·26	174	241·86
15	20·85	55	76·45	95	132·05	135	187·65	175	243·25
16	22·24	56	77·84	96	133·44	136	189·04	176	244·64
17	23·63	57	79·23	97	134·83	137	190·43	177	246·03
18	25·02	58	80·62	98	136·22	138	191·82	178	247·42
19	26·41	59	82·01	99	137·61	139	193·21	179	248·81
20	27·80	**60**	83·40	**100**	139·00	**140**	194·60	**180**	250·20
21	29·19	61	84·79	101	140·39	141	195·99	181	251·59
22	30·58	62	86·18	102	141·78	142	197·38	182	252·98
23	31·97	63	87·57	103	143·17	143	198·77	183	254·37
24	33·36	64	88·96	104	144·56	144	200·16	184	255·76
25	34·75	65	90·35	105	145·95	145	201·55	185	257·15
26	36·14	66	91·74	106	147·34	146	202·94	186	258·54
27	37·53	67	93·13	107	148·73	147	204·33	187	259·93
28	38·92	68	94·52	108	150·12	148	205·72	188	261·32
29	40·31	69	95·91	109	151·51	149	207·11	189	262·71
30	41·70	**70**	97·30	**110**	152·90	**150**	208·50	**190**	264·10
31	43·09	71	98·69	111	154·29	151	209·89	191	265·49
32	44·48	72	100·08	112	155·68	152	211·28	192	266·88
33	45·87	73	101·47	113	157·07	153	212·67	193	268·27
34	47·26	74	102·86	114	158·46	154	214·06	194	269·66
35	48·65	75	104·25	115	159·85	155	215·45	195	271·05
36	50·04	76	105·64	116	161·24	156	216·84	196	272·44
37	51·43	77	107·03	117	162·63	157	218·23	197	273·83
38	52·82	78	108·42	118	164·02	158	219·62	198	275·22
39	54·21	79	109·81	119	165·41	159	221·01	199	276·61
40	55·60	**80**	111·20	**120**	166·80	**160**	222·40	**200**	278·00

$\frac{1}{16}=0{\cdot}08^{7}$ $\frac{1}{9}=0{\cdot}15^{4}$ $\frac{1}{8}=0{\cdot}17^{4}$ $\frac{1}{6}=0{\cdot}23^{2}$ $\frac{1}{5}=0{\cdot}27^{8}$ $\frac{1}{4}=0{\cdot}34^{8}$ $\frac{1}{3}=0{\cdot}46^{3}$

$\frac{3}{8}=0{\cdot}52^{1}$ $\frac{1}{2}=0{\cdot}69^{5}$ $\frac{5}{8}=0{\cdot}86^{9}$ $\frac{3}{4}=1{\cdot}04^{3}$ $\frac{7}{8}=1{\cdot}21^{6}$ $\frac{2}{3}=0{\cdot}92^{7}$ $\frac{5}{6}=1{\cdot}15^{8}$

$\frac{3}{16}=0{\cdot}26^{1}$ $\frac{5}{16}=0{\cdot}43^{4}$ $\frac{7}{16}=0{\cdot}60^{8}$ $\frac{9}{16}=0{\cdot}78^{2}$ $\frac{11}{16}=0{\cdot}95^{6}$ $\frac{13}{16}=1{\cdot}12^{9}$ $\frac{15}{16}=1{\cdot}30^{3}$

1	1·39½	41	57·19½	81	112·99½	121	168·79½	161	224·59½
2	2·79	42	58·59	82	114·39	122	170·19	162	225·99
3	4·18½	43	59·98½	83	115·78½	123	171·58½	163	227·38½
4	5·58	44	61·38	84	117·18	124	172·98	164	228·78
5	6·97½	45	62·77½	85	118·57½	125	174·37½	165	230·17½
6	8·37	46	64·17	86	119·97	126	175·77	166	231·57
7	9·76½	47	65·56½	87	121·36½	127	177·16½	167	232·96½
8	11·16	48	66·96	88	122·76	128	178·56	168	234·36
9	12·55½	49	68·35½	89	124·15½	129	179·95½	169	235·75½
10	13·95	**50**	69·75	**90**	125·55	**130**	181·35	**170**	237·15
11	15·34½	51	71·14½	91	126·94½	131	182·74½	171	238·54½
12	16·74	52	72·54	92	128·34	132	184·14	172	239·94
13	18·13½	53	73·93½	93	129·73½	133	185·53½	173	241·33½
14	19·53	54	75·33	94	131·13	134	186·93	174	242·73
15	20·92½	55	76·72½	95	132·52½	135	188·32½	175	244·12½
16	22·32	56	78·12	96	133·92	136	189·72	176	245·52
17	23·71½	57	79·51½	97	135·31½	137	191·11½	177	246·91½
18	25·11	58	80·91	98	136·71	138	192·51	178	248·31
19	26·50½	59	82·30½	99	138·10½	139	193·90½	179	249·70½
20	27·90	**60**	83·70	**100**	139·50	**140**	195·30	**180**	251·10
21	29·29½	61	85·09½	101	140·89½	141	196·69½	181	252·49½
22	30·69	62	86·49	102	142·29	142	198·09	182	253·89
23	32·08½	63	87·88½	103	143·68½	143	199·48½	183	255·28½
24	33·48	64	89·28	104	145·08	144	200·88	184	256·68
25	34·87½	65	90·67½	105	146·47½	145	202·27½	185	258·07½
26	36·27	66	92·07	106	147·87	146	203·67	186	259·47
27	37·66½	67	93·46½	107	149·26½	147	205·06½	187	260·86½
28	39·06	68	94·86	108	150·66	148	206·46	188	262·26
29	40·45½	69	96·25½	109	152·05½	149	207·85½	189	263·65½
30	41·85	**70**	97·65	**110**	153·45	**150**	209·25	**190**	265·05
31	43·24½	71	99·04½	111	154·84½	151	210·64½	191	266·44½
32	44·64	72	100·44	112	156·24	152	212·04	192	267·84
33	46·03½	73	101·83½	113	157·63½	153	213·43½	193	269·23½
34	47·43	74	103·23	114	159·03	154	214·83	194	270·63
35	48·82½	75	104·62½	115	160·42½	155	216·22½	195	272·02½
36	50·22	76	106·02	116	161·82	156	217·62	196	273·42
37	51·61½	77	107·41½	117	163·21½	157	219·01½	197	274·81½
38	53·01	78	108·81	118	164·61	158	220·41	198	276·21
39	54·40½	79	110·20½	119	166·00½	159	221·80½	199	277·60½
40	55·80	**80**	111·60	**120**	167·40	**160**	223·20	**200**	279·00

$\frac{1}{16}=0{\cdot}08^{7}$ $\frac{1}{9}=0{\cdot}15^{5}$ $\frac{1}{8}=0{\cdot}17^{4}$ $\frac{1}{6}=0{\cdot}23^{3}$ $\frac{1}{5}=0{\cdot}27^{9}$ $\frac{1}{4}=0{\cdot}34^{9}$ $\frac{1}{3}=0{\cdot}46^{5}$

$\frac{3}{8}=0{\cdot}52^{3}$ $\frac{1}{2}=0{\cdot}69^{8}$ $\frac{5}{8}=0{\cdot}87^{2}$ $\frac{3}{4}=1{\cdot}04^{6}$ $\frac{7}{8}=1{\cdot}22^{1}$ $\frac{2}{3}=0{\cdot}93^{0}$ $\frac{5}{6}=1{\cdot}16^{3}$

$\frac{3}{16}=0{\cdot}26^{2}$ $\frac{5}{16}=0{\cdot}43^{6}$ $\frac{7}{16}=0{\cdot}61^{0}$ $\frac{9}{16}=0{\cdot}78^{5}$ $\frac{11}{16}=0{\cdot}95^{9}$ $\frac{13}{16}=1{\cdot}13^{3}$ $\frac{15}{16}=1{\cdot}30^{8}$

1	1·40	41	57·40	81	113·40	121	169·40	161	225·40
2	2·80	42	58·80	82	114·80	122	170·80	162	226·80
3	4·20	43	60·20	83	116·20	123	172·20	163	228·20
4	5·60	44	61·60	84	117·60	124	173·60	164	229·60
5	7·00	45	63·00	85	119·00	125	175·00	165	231·00
6	8·40	46	64·40	86	120·40	126	176·40	166	232·40
7	9·80	47	65·80	87	121·80	127	177·80	167	233·80
8	11·20	48	67·20	88	123·20	128	179·20	168	235·20
9	12·60	49	68·60	89	124·60	129	180·60	169	236·60
10	14·00	**50**	70·00	**90**	126·00	**130**	182·00	**170**	238·00
11	15·40	51	71·40	91	127·40	131	183·40	171	239·40
12	16·80	52	72·80	92	128·80	132	184·80	172	240·80
13	18·20	53	74·20	93	130·20	133	186·20	173	242·20
14	19·60	54	75·60	94	131·60	134	187·60	174	243·60
15	21·00	55	77·00	95	133·00	135	189·00	175	245·00
16	22·40	56	78·40	96	134·40	136	190·40	176	246·40
17	23·80	57	79·80	97	135·80	137	191·80	177	247·80
18	25·20	58	81·20	98	137·20	138	193·20	178	249·20
19	26·60	59	82·60	99	138·60	139	194·60	179	250·60
20	28·00	**60**	84·00	**100**	140·00	**140**	196·00	**180**	252·00
21	29·40	61	85·40	101	141·40	141	197·40	181	253·40
22	30·80	62	86·80	102	142·80	142	198·80	182	254·80
23	32·20	63	88·20	103	144·20	143	200·20	183	256·20
24	33·60	64	89·60	104	145·60	144	201·60	184	257·60
25	35·00	65	91·00	105	147·00	145	203·00	185	259·00
26	36·40	66	92·40	106	148·40	146	204·40	186	260·40
27	37·80	67	93·80	107	149·80	147	205·80	187	261·80
28	39·20	68	95·20	108	151·20	148	207·20	188	263·20
29	40·60	69	96·60	109	152·60	149	208·60	189	264·60
30	42·00	**70**	98·00	**110**	154·00	**150**	210·00	**190**	266·00
31	43·40	71	99·40	111	155·40	151	211·40	191	267·40
32	44·80	72	100·80	112	156·80	152	212·80	192	268·80
33	46·20	73	102·20	113	158·20	153	214·20	193	270·20
34	47·60	74	103·60	114	159·60	154	215·60	194	271·60
35	49·00	75	105·00	115	161·00	155	217·00	195	273·00
36	50·40	76	106·40	116	162·40	156	218·40	196	274·40
37	51·80	77	107·80	117	163·80	157	219·80	197	275·80
38	53·20	78	109·20	118	165·20	158	221·20	198	277·20
39	54·60	79	110·60	119	166·60	159	222·60	199	278·60
40	56·00	**80**	112·00	**120**	168·00	**160**	224·00	**200**	280·00

$\frac{1}{16}=0{\cdot}08^{8}$ $\frac{1}{9}=0{\cdot}15^{6}$ $\frac{1}{8}=0{\cdot}17^{5}$ $\frac{1}{6}=0{\cdot}23^{3}$ $\frac{1}{5}=0{\cdot}28^{0}$ $\frac{1}{4}=0{\cdot}35^{0}$ $\frac{1}{3}=0{\cdot}46^{7}$

$\frac{3}{8}=0{\cdot}52^{5}$ $\frac{1}{2}=0{\cdot}70^{0}$ $\frac{5}{8}=0{\cdot}87^{5}$ $\frac{3}{4}=1{\cdot}05^{0}$ $\frac{7}{8}=1{\cdot}22^{5}$ $\frac{2}{3}=0{\cdot}93^{3}$ $\frac{5}{6}=1{\cdot}16^{7}$

$\frac{3}{16}=0{\cdot}26^{3}$ $\frac{5}{16}=0{\cdot}43^{8}$ $\frac{7}{16}=0{\cdot}61^{3}$ $\frac{9}{16}=0{\cdot}78^{8}$ $\frac{11}{16}=0{\cdot}96^{3}$ $\frac{13}{16}=1{\cdot}13^{8}$ $\frac{15}{16}=1{\cdot}31^{3}$

(0.712=1£) **£1.40½**

1	1·40½	41	57·60½	81	113·80½	121	170·00½	161	226·20½
2	2·81	42	59·01	82	115·21	122	171·41	162	227·61
3	4·21½	43	60·41½	83	116·61½	123	172·81½	163	229·01½
4	5·62	44	61·82	84	118·02	124	174·22	164	230·42
5	7·02½	45	63·22½	85	119·42½	125	175·62½	165	231·82½
6	8·43	46	64·63	86	120·83	126	177·03	166	233·23
7	9·83½	47	66·03½	87	122·23½	127	178·43½	167	234·63½
8	11·24	48	67·44	88	123·64	128	179·84	168	236·04
9	12·64½	49	68·84½	89	125·04½	129	181·24½	169	237·44½
10	14·05	**50**	70·25	**90**	126·45	**130**	182·65	**170**	238·85
11	15·45½	51	71·65½	91	127·85½	131	184·05½	171	240·25½
12	16·86	52	73·06	92	129·26	132	185·46	172	241·66
13	18·26½	53	74·46½	93	130·66½	133	186·86½	173	243·06½
14	19·67	54	75·87	94	132·07	134	188·27	174	244·47
15	21·07½	55	77·27½	95	133·47½	135	189·67½	175	245·87½
16	22·48	56	78·68	96	134·88	136	191·08	176	247·28
17	23·88½	57	80·08½	97	136·28½	137	192·48½	177	248·68½
18	25·29	58	81·49	98	137·69	138	193·89	178	250·09
19	26·69½	59	82·89½	99	139·09½	139	195·29½	179	251·49½
20	28·10	**60**	84·30	**100**	140·50	**140**	196·70	**180**	252·90
21	29·50½	61	85·70½	101	141·90½	141	198·10½	181	254·30½
22	30·91	62	87·11	102	143·31	142	199·51	182	255·71
23	32·31½	63	88·51½	103	144·71½	143	200·91½	183	257·11½
24	33·72	64	89·92	104	146·12	144	202·32	184	258·52
25	35·12½	65	91·32½	105	147·52½	145	203·72½	185	259·92½
26	36·53	66	92·73	106	148·93	146	205·13	186	261·33
27	37·93½	67	94·13½	107	150·33½	147	206·53½	187	262·73½
28	39·34	68	95·54	108	151·74	148	207·94	188	264·14
29	40·74½	69	96·94½	109	153·14½	149	209·34½	189	265·54½
30	42·15	**70**	98·35	**110**	154·55	**150**	210·75	**190**	266·95
31	43·55½	71	99·75½	111	155·95½	151	212·15½	191	268·35½
32	44·96	72	101·16	112	157·36	152	213·56	192	269·76
33	46·36½	73	102·56½	113	158·76½	153	214·96½	193	271·16½
34	47·77	74	103·97	114	160·17	154	216·37	194	272·57
35	49·17½	75	105·37½	115	161·57½	155	217·77½	195	273·97½
36	50·58	76	106·78	116	162·98	156	219·18	196	275·38
37	51·98½	77	108·18½	117	164·38½	157	220·58½	197	276·78½
38	53·39	78	109·59	118	165·79	158	221·99	198	278·19
39	54·79½	79	110·99½	119	167·19½	159	223·39½	199	279·59½
40	56·20	**80**	112·40	**120**	168·60	**160**	224·80	**200**	281·00

$\frac{1}{16}=0{\cdot}08^{8}$ $\frac{1}{9}=0{\cdot}15^{6}$ $\frac{1}{8}=0{\cdot}17^{6}$ $\frac{1}{6}=0{\cdot}23^{4}$ $\frac{1}{5}=0{\cdot}28^{1}$ $\frac{1}{4}=0{\cdot}35^{1}$ $\frac{1}{3}=0{\cdot}46^{8}$

$\frac{3}{8}=0{\cdot}52^{7}$ $\frac{1}{2}=0{\cdot}70^{3}$ $\frac{5}{8}=0{\cdot}87^{8}$ $\frac{3}{4}=1{\cdot}05^{4}$ $\frac{7}{8}=1{\cdot}22^{9}$ $\frac{2}{3}=0{\cdot}93^{7}$ $\frac{5}{6}=1{\cdot}17^{1}$

$\frac{3}{16}=0{\cdot}26^{3}$ $\frac{5}{16}=0{\cdot}43^{9}$ $\frac{7}{16}=0{\cdot}61^{5}$ $\frac{9}{16}=0{\cdot}79^{0}$ $\frac{11}{16}=0{\cdot}96^{6}$ $\frac{13}{16}=1{\cdot}14^{2}$ $\frac{15}{16}=1{\cdot}31^{7}$

1	1·41	41	57·81	81	114·21	121	170·61	161	227·01
2	2·82	42	59·22	82	115·62	122	172·02	162	228·42
3	4·23	43	60·63	83	117·03	123	173·43	163	229·83
4	5·64	44	62·04	84	118·44	124	174·84	164	231·24
5	7·05	45	63·45	85	119·85	125	176·25	165	232·65
6	8·46	46	64·86	86	121·26	126	177·66	166	234·06
7	9·87	47	66·27	87	122·67	127	179·07	167	235·47
8	11·28	48	67·68	88	124·08	128	180·48	168	236·88
9	12·69	49	69·09	89	125·49	129	181·89	169	238·29
10	14·10	**50**	70·50	**90**	126·90	**130**	183·30	**170**	239·70
11	15·51	51	71·91	91	128·31	131	184·71	171	241·11
12	16·92	52	73·32	92	129·72	132	186·12	172	242·52
13	18·33	53	74·73	93	131·13	133	187·53	173	243·93
14	19·74	54	76·14	94	132·54	134	188·94	174	245·34
15	21·15	55	77·55	95	133·95	135	190·35	175	246·75
16	22·56	56	78·96	96	135·36	136	191·76	176	248·16
17	23·97	57	80·37	97	136·77	137	193·17	177	249·57
18	25·38	58	81·78	98	138·18	138	194·58	178	250·98
19	26·79	59	83·19	99	139·59	139	195·99	179	252·39
20	28·20	**60**	84·60	**100**	141·00	**140**	197·40	**180**	253·80
21	29·61	61	86·01	101	142·41	141	198·81	181	255·21
22	31·02	62	87·42	102	143·82	142	200·22	182	256·62
23	32·43	63	88·83	103	145·23	143	201·63	183	258·03
24	33·84	64	90·24	104	146·64	144	203·04	184	259·44
25	35·25	65	91·65	105	148·05	145	204·45	185	260·85
26	36·66	66	93·06	106	149·46	146	205·86	186	262·26
27	38·07	67	94·47	107	150·87	147	207·27	187	263·67
28	39·48	68	95·88	108	152·28	148	208·68	188	265·08
29	40·89	69	97·29	109	153·69	149	210·09	189	266·49
30	42·30	**70**	98·70	**110**	155·10	**150**	211·50	**190**	267·90
31	43·71	71	100·11	111	156·51	151	212·91	191	269·31
32	45·12	72	101·52	112	157·92	152	214·32	192	270·72
33	46·53	73	102·93	113	159·33	153	215·73	193	272·13
34	47·94	74	104·34	114	160·74	154	217·14	194	273·54
35	49·35	75	105·75	115	162·15	155	218·55	195	274·95
36	50·76	76	107·16	116	163·56	156	219·96	196	276·36
37	52·17	77	108·57	117	164·97	157	221·37	197	277·77
38	53·58	78	109·98	118	166·38	158	222·78	198	279·18
39	54·99	79	111·39	119	167·79	159	224·19	199	280·59
40	56·40	**80**	112·80	**120**	169·20	**160**	225·60	**200**	282·00

$\frac{1}{16}$=0·08^{8} $\frac{1}{9}$=0·15^{7} $\frac{1}{8}$=0·17^{6} $\frac{1}{6}$=0·23^{5} $\frac{1}{5}$=0·28^{2} $\frac{1}{4}$=0·35^{3} $\frac{1}{3}$=0·47^{0}

$\frac{3}{8}$=0·52^{9} $\frac{1}{2}$=0·70^{5} $\frac{5}{8}$=0·88^{1} $\frac{3}{4}$=1·05^{8} $\frac{7}{8}$=1·23^{4} $\frac{2}{3}$=0·94^{0} $\frac{5}{6}$=1·17^{5}

$\frac{3}{16}$=0·26^{4} $\frac{5}{16}$=0·44^{1} $\frac{7}{16}$=0·61^{7} $\frac{9}{16}$=0·79^{3} $\frac{11}{16}$=0·96^{9} $\frac{13}{16}$=1·14^{6} $\frac{15}{16}$=1·32^{2}

1	1·41½	41	58·01½	81	114·61½	121	171·21½	161	227·81½
2	2·83	42	59·43	82	116·03	122	172·63	162	229·23
3	4·24½	43	60·84½	83	117·44½	123	174·04½	163	230·64½
4	5·66	44	62·26	84	118·86	124	175·46	164	232·06
5	7·07½	45	63·67½	85	120·27½	125	176·87½	165	233·47½
6	8·49	46	65·09	86	121·69	126	178·29	166	234·89
7	9·90½	47	66·50½	87	123·10½	127	179·70½	167	236·30½
8	11·32	48	67·92	88	124·52	128	181·12	168	237·72
9	12·73½	49	69·33½	89	125·93½	129	182·53½	169	239·13½
10	14·15	**50**	70·75	**90**	127·35	**130**	183·95	**170**	240·55
11	15·56½	51	72·16½	91	128·76½	131	185·36½	171	241·96½
12	16·98	52	73·58	92	130·18	132	186·78	172	243·38
13	18·39½	53	74·99½	93	131·59½	133	188·19½	173	244·79½
14	19·81	54	76·41	94	133·01	134	189·61	174	246·21
15	21·22½	55	77·82½	95	134·42½	135	191·02½	175	247·62½
16	22·64	56	79·24	96	135·84	136	192·44	176	249·04
17	24·05½	57	80·65½	97	137·25½	137	193·85½	177	250·45½
18	25·47	58	82·07	98	138·67	138	195·27	178	251·87
19	26·88½	59	83·48½	99	140·08½	139	196·68½	179	253·28½
20	28·30	**60**	84·90	**100**	141·50	**140**	198·10	**180**	254·70
21	29·71½	61	86·31½	101	142·91½	141	199·51½	181	256·11½
22	31·13	62	87·73	102	144·33	142	200·93	182	257·53
23	32·54½	63	89·14½	103	145·74½	143	202·34½	183	258·94½
24	33·96	64	90·56	104	147·16	144	203·76	184	260·36
25	35·37½	65	91·97½	105	148·57½	145	205·17½	185	261·77½
26	36·79	66	93·39	106	149·99	146	206·59	186	263·19
27	38·20½	67	94·80½	107	151·40½	147	208·00½	187	264·60½
28	39·62	68	96·22	108	152·82	148	209·42	188	266·02
29	41·03½	69	97·63½	109	154·23½	149	210·83½	189	267·43½
30	42·45	**70**	99·05	**110**	155·65	**150**	212·25	**190**	268·85
31	43·86½	71	100·46½	111	157·06½	151	213·66½	191	270·26½
32	45·28	72	101·88	112	158·48	152	215·08	192	271·68
33	46·69½	73	103·29½	113	159·89½	153	216·49½	193	273·09½
34	48·11	74	104·71	114	161·31	154	217·91	194	274·51
35	49·52½	75	106·12½	115	162·72½	155	219·32½	195	275·92½
36	50·94	76	107·54	116	164·14	156	220·74	196	277·34
37	52·35½	77	108·95½	117	165·55½	157	222·15½	197	278·75½
38	53·77	78	110·37	118	166·97	158	223·57	198	280·17
39	55·18½	79	111·78½	119	168·38½	159	224·98½	199	281·58½
40	56·60	**80**	113·20	**120**	169·80	**160**	226·40	**200**	283·00

$\frac{1}{16}=0{\cdot}08^{8}$ $\frac{1}{9}=0{\cdot}15^{7}$ $\frac{1}{8}=0{\cdot}17^{7}$ $\frac{1}{6}=0{\cdot}23^{6}$ $\frac{1}{5}=0{\cdot}28^{3}$ $\frac{1}{4}=0{\cdot}35^{4}$ $\frac{1}{3}=0{\cdot}47^{2}$

$\frac{3}{8}=0{\cdot}53^{1}$ $\frac{1}{2}=0{\cdot}70^{8}$ $\frac{5}{8}=0{\cdot}88^{4}$ $\frac{3}{4}=1{\cdot}06^{1}$ $\frac{7}{8}=1{\cdot}23^{8}$ $\frac{2}{3}=0{\cdot}94^{3}$ $\frac{5}{6}=1{\cdot}17^{9}$

$\frac{3}{16}=0{\cdot}26^{5}$ $\frac{5}{16}=0{\cdot}44^{2}$ $\frac{7}{16}=0{\cdot}61^{9}$ $\frac{9}{16}=0{\cdot}79^{6}$ $\frac{11}{16}=0{\cdot}97^{3}$ $\frac{13}{16}=1{\cdot}15^{0}$ $\frac{15}{16}=1{\cdot}32^{7}$

1·4142 is the square root of 2 1 British Thermal Unit per second=1·415 Horse power

(0.704=1£) **£1.42**

1	1·42	41	58·22	81	115·02	121	171·82	161	228·62
2	2·84	42	59·64	82	116·44	122	173·24	162	230·04
3	4·26	43	61·06	83	117·86	123	174·66	163	231·46
4	5·68	44	62·48	84	119·28	124	176·08	164	232·88
5	7·10	45	63·90	85	120·70	125	177·50	165	234·30
6	8·52	46	65·32	86	122·12	126	178·92	166	235·72
7	9·94	47	66·74	87	123·54	127	180·34	167	237·14
8	11·36	48	68·16	88	124·96	128	181·76	168	238·56
9	12·78	49	69·58	89	126·38	129	183·18	169	239·98
10	14·20	**50**	71·00	**90**	127·80	**130**	184·60	**170**	241·40
11	15·62	51	72·42	91	129·22	131	186·02	171	242·82
12	17·04	52	73·84	92	130·64	132	187·44	172	244·24
13	18·46	53	75·26	93	132·06	133	188·86	173	245·66
14	19·88	54	76·68	94	133·48	134	190·28	174	247·08
15	21·30	55	78·10	95	134·90	135	191·70	175	248·50
16	22·72	56	79·52	96	136·32	136	193·12	176	249·92
17	24·14	57	80·94	97	137·74	137	194·54	177	251·34
18	25·56	58	82·36	98	139·16	138	195·96	178	252·76
19	26·98	59	83·78	99	140·58	139	197·38	179	254·18
20	28·40	**60**	85·20	**100**	142·00	**140**	198·80	**180**	255·60
21	29·82	61	86·62	101	143·42	141	200·22	181	257·02
22	31·24	62	88·04	102	144·84	142	201·64	182	258·44
23	32·66	63	89·46	103	146·26	143	203·06	183	259·86
24	34·08	64	90·88	104	147·68	144	204·48	184	261·28
25	35·50	65	92·30	105	149·10	145	205·90	185	262·70
26	36·92	66	93·72	106	150·52	146	207·32	186	264·12
27	38·34	67	95·14	107	151·94	147	208·74	187	265·54
28	39·76	68	96·56	108	153·36	148	210·16	188	266·96
29	41·18	69	97·98	109	154·78	149	211·58	189	268·38
30	42·60	**70**	99·40	**110**	156·20	**150**	213·00	**190**	269·80
31	44·02	71	100·82	111	157·62	151	214·42	191	271·22
32	45·44	72	102·24	112	159·04	152	215·84	192	272·64
33	46·86	73	103·66	113	160·46	153	217·26	193	274·06
34	48·28	74	105·08	114	161·88	154	218·68	194	275·48
35	49·70	75	106·50	115	163·30	155	220·10	195	276·90
36	51·12	76	107·92	116	164·72	156	221·52	196	278·32
37	52·54	77	109·34	117	166·14	157	222·94	197	279·74
38	53·96	78	110·76	118	167·56	158	224·36	198	281·16
39	55·38	79	112·18	119	168·98	159	225·78	199	282·58
40	56·80	**80**	113·60	**120**	170·40	**160**	227·20	**200**	284·00

$\frac{1}{16}=0{\cdot}08^{9}$ $\frac{1}{9}=0{\cdot}15^{8}$ $\frac{1}{8}=0{\cdot}17^{8}$ $\frac{1}{6}=0{\cdot}23^{7}$ $\frac{1}{5}=0{\cdot}28^{4}$ $\frac{1}{4}=0{\cdot}35^{5}$ $\frac{1}{3}=0{\cdot}47^{3}$

$\frac{3}{8}=0{\cdot}53^{3}$ $\frac{1}{2}=0{\cdot}71^{0}$ $\frac{5}{8}=0{\cdot}88^{8}$ $\frac{3}{4}=1{\cdot}06^{5}$ $\frac{7}{8}=1{\cdot}24^{3}$ $\frac{2}{3}=0{\cdot}94^{7}$ $\frac{5}{6}=1{\cdot}18^{3}$

$\frac{3}{16}=0{\cdot}26^{6}$ $\frac{5}{16}=0{\cdot}44^{4}$ $\frac{7}{16}=0{\cdot}62^{1}$ $\frac{9}{16}=0{\cdot}79^{9}$ $\frac{11}{16}=0{\cdot}97^{6}$ $\frac{13}{16}=1{\cdot}15^{4}$ $\frac{15}{16}=1{\cdot}33^{1}$

1 kilogramme per sq. cm=14·22 lbs per square inch

(0.702=1£) £1.42½

1	1·42½	41	58·42½	81	115·42½	121	172·42½	161	229·42½
2	2·85	42	59·85	82	116·85	122	173·85	162	230·85
3	4·27½	43	61·27½	83	118·27½	123	175·27½	163	232·27½
4	5·70	44	62·70	84	119·70	124	176·70	164	233·70
5	7·12½	45	64·12½	85	121·12½	125	178·12½	165	235·12½
6	8·55	46	65·55	86	122·55	126	179·55	166	236·55
7	9·97½	47	66·97½	87	123·97½	127	180·97½	167	237·97½
8	11·40	48	68·40	88	125·40	128	182·40	168	239·40
9	12·82½	49	69·82½	89	126·82½	129	183·82½	169	240·82½
10	14·25	**50**	71·25	**90**	128·25	**130**	185·25	**170**	242·25
11	15·67½	51	72·67½	91	129·67½	131	186·67½	171	243·67½
12	17·10	52	74·10	92	131·10	132	188·10	172	245·10
13	18·52½	53	75·52½	93	132·52½	133	189·52½	173	246·52½
14	19·95	54	76·95	94	133·95	134	190·95	174	247·95
15	21·37½	55	78·37½	95	135·37½	135	192·37½	175	249·37½
16	22·80	56	79·80	96	136·80	136	193·80	176	250·80
17	24·22½	57	81·22½	97	138·22½	137	195·22½	177	252·22½
18	25·65	58	82·65	98	139·65	138	196·65	178	253·65
19	27·07½	59	84·07½	99	141·07½	139	198·07½	179	255·07½
20	28·50	**60**	85·50	**100**	142·50	**140**	199·50	**180**	256·50
21	29·92½	61	86·92½	101	143·92½	141	200·92½	181	257·92½
22	31·35	62	88·35	102	145·35	142	202·35	182	259·35
23	32·77½	63	89·77½	103	146·77½	143	203·77½	183	260·77½
24	34·20	64	91·20	104	148·20	144	205·20	184	262·20
25	35·62½	65	92·62½	105	149·62½	145	206·62½	185	263·62½
26	37·05	66	94·05	106	151·05	146	208·05	186	265·05
27	38·47½	67	95·47½	107	152·47½	147	209·47½	187	266·47½
28	39·90	68	96·90	108	153·90	148	210·90	188	267·90
29	41·32½	69	98·32½	109	155·32½	149	212·32½	189	269·32½
30	42·75	**70**	99·75	**110**	156·75	**150**	213·75	**190**	270·75
31	44·17½	71	101·17½	111	158·17½	151	215·17½	191	272·17½
32	45·60	72	102·60	112	159·60	152	216·60	192	273·60
33	47·02½	73	104·02½	113	161·02½	153	218·02½	193	275·02½
34	48·45	74	105·45	114	162·45	154	219·45	194	276·45
35	49·87½	75	106·87½	115	163·87½	155	220·87½	195	277·87½
36	51·30	76	108·30	116	165·30	156	222·30	196	279·30
37	52·72½	77	109·72½	117	166·72½	157	223·72½	197	280·72½
38	54·15	78	111·15	118	168·15	158	225·15	198	282·15
39	55·57½	79	112·57½	119	169·57½	159	226·57½	199	283·57½
40	57·00	**80**	114·00	**120**	171·00	**160**	228·00	**200**	285·00

$\frac{1}{16}=0{\cdot}08^{9}$ $\frac{1}{9}=0{\cdot}15^{8}$ $\frac{1}{8}=0{\cdot}17^{8}$ $\frac{1}{6}=0{\cdot}23^{8}$ $\frac{1}{5}=0{\cdot}28^{5}$ $\frac{1}{4}=0{\cdot}35^{6}$ $\frac{1}{3}=0{\cdot}47^{5}$

$\frac{3}{8}=0{\cdot}53^{4}$ $\frac{1}{2}=0{\cdot}71^{3}$ $\frac{5}{8}=0{\cdot}89^{1}$ $\frac{3}{4}=1{\cdot}06^{9}$ $\frac{7}{8}=1{\cdot}24^{7}$ $\frac{2}{3}=0{\cdot}95^{0}$ $\frac{5}{6}=1{\cdot}18^{8}$

$\frac{3}{16}=0{\cdot}26^{7}$ $\frac{5}{16}=0{\cdot}44^{5}$ $\frac{7}{16}=0{\cdot}62^{3}$ $\frac{9}{16}=0{\cdot}80^{2}$ $\frac{11}{16}=0{\cdot}98^{0}$ $\frac{13}{16}=1{\cdot}15^{8}$ $\frac{15}{16}=1{\cdot}33^{6}$

(0.699=1£) **£1.43**

1	1·43	41	58·63	81	115·83	121	173·03	161	230·23
2	2·86	42	60·06	82	117·26	122	174·46	162	231·66
3	4·29	43	61·49	83	118·69	123	175·89	163	233·09
4	5·72	44	62·92	84	120·12	124	177·32	164	234·52
5	7·15	45	64·35	85	121·55	125	178·75	165	235·95
6	8·58	46	65·78	86	122·98	126	180·18	166	237·38
7	10·01	47	67·21	87	124·41	127	181·61	167	238·81
8	11·44	48	68·64	88	125·84	128	183·04	168	240·24
9	12·87	49	70·07	89	127·27	129	184·47	169	241·67
10	14·30	**50**	71·50	**90**	128·70	**130**	185·90	**170**	243·10
11	15·73	51	72·93	91	130·13	131	187·33	171	244·53
12	17·16	52	74·36	92	131·56	132	188·76	172	245·96
13	18·59	53	75·79	93	132·99	133	190·19	173	247·39
14	20·02	54	77·22	94	134·42	134	191·62	174	248·82
15	21·45	55	78·65	95	135·85	135	193·05	175	250·25
16	22·88	56	80·08	96	137·28	136	194·48	176	251·68
17	24·31	57	81·51	97	138·71	137	195·91	177	253·11
18	25·74	58	82·94	98	140·14	138	197·34	178	254·54
19	27·17	59	84·37	99	141·57	139	198·77	179	255·97
20	28·60	**60**	85·80	**100**	143·00	**140**	200·20	**180**	257·40
21	30·03	61	87·23	101	144·43	141	201·63	181	258·83
22	31·46	62	88·66	102	145·86	142	203·06	182	260·26
23	32·89	63	90·09	103	147·29	143	204·49	183	261·69
24	34·32	64	91·52	104	148·72	144	205·92	184	263·12
25	35·75	65	92·95	105	150·15	145	207·35	185	264·55
26	37·18	66	94·38	106	151·58	146	208·78	186	265·98
27	38·61	67	95·81	107	153·01	147	210·21	187	267·41
28	40·04	68	97·24	108	154·44	148	211·64	188	268·84
29	41·47	69	98·67	109	155·87	149	213·07	189	270·27
30	42·90	**70**	100·10	**110**	157·30	**150**	214·50	**190**	271·70
31	44·33	71	101·53	111	158·73	151	215·93	191	273·13
32	45·76	72	102·96	112	160·16	152	217·36	192	274·56
33	47·19	73	104·39	113	161·59	153	218·79	193	275·99
34	48·62	74	105·82	114	163·02	154	220·22	194	277·42
35	50·05	75	107·25	115	164·45	155	221·65	195	278·85
36	51·48	76	108·68	116	165·88	156	223·08	196	280·28
37	52·91	77	110·11	117	167·31	157	224·51	197	281·71
38	54·34	78	111·54	118	168·74	158	225·94	198	283·14
39	55·77	79	112·97	119	170·17	159	227·37	199	284·57
40	57·20	**80**	114·40	**120**	171·60	**160**	228·80	**200**	286·00

$\frac{1}{16}=0{\cdot}08^{9}$ $\frac{1}{9}=0{\cdot}15^{9}$ $\frac{1}{8}=0{\cdot}17^{9}$ $\frac{1}{6}=0{\cdot}23^{8}$ $\frac{1}{5}=0{\cdot}28^{6}$ $\frac{1}{4}=0{\cdot}35^{8}$ $\frac{1}{3}=0{\cdot}47^{7}$

$\frac{3}{8}=0{\cdot}53^{6}$ $\frac{1}{2}=0{\cdot}71^{5}$ $\frac{5}{8}=0{\cdot}89^{4}$ $\frac{3}{4}=1{\cdot}07^{3}$ $\frac{7}{8}=1{\cdot}25^{1}$ $\frac{2}{3}=0{\cdot}95^{3}$ $\frac{5}{6}=1{\cdot}19^{2}$

$\frac{3}{16}=0{\cdot}26^{8}$ $\frac{5}{16}=0{\cdot}44^{7}$ $\frac{7}{16}=0{\cdot}62^{6}$ $\frac{9}{16}=0{\cdot}80^{4}$ $\frac{11}{16}=0{\cdot}98^{3}$ $\frac{13}{16}=1{\cdot}16^{2}$ $\frac{15}{16}=1{\cdot}34^{1}$

(0.697=1£) **£1.43½**

1	1·43½	41	58·83½	81	116·23½	121	173·63½	161	231·03½
2	2·87	42	60·27	82	117·67	122	175·07	162	232·47
3	4·30½	43	61·70½	83	119·10½	123	176·50½	163	233·90½
4	5·74	44	63·14	84	120·54	124	177·94	164	235·34
5	7·17½	45	64·57½	85	121·97½	125	179·37½	165	236·77½
6	8·61	46	66·01	86	123·41	126	180·81	166	238·21
7	10·04½	47	67·44½	87	124·84½	127	182·24½	167	239·64½
8	11·48	48	68·88	88	126·28	128	183·68	168	241·08
9	12·91½	49	70·31½	89	127·71½	129	185·11½	169	242·51½
10	14·35	**50**	71·75	**90**	129·15	**130**	186·55	**170**	243·95
11	15·78½	51	73·18½	91	130·58½	131	187·98½	171	245·38½
12	17·22	52	74·62	92	132·02	132	189·42	172	246·82
13	18·65½	53	76·05½	93	133·45½	133	190·85½	173	248·25½
14	20·09	54	77·49	94	134·89	134	192·29	174	249·69
15	21·52½	55	78·92½	95	136·32½	135	193·72½	175	251·12½
16	22·96	56	80·36	96	137·76	136	195·16	176	252·56
17	24·39½	57	81·79½	97	139·19½	137	196·59½	177	253·99½
18	25·83	58	83·23	98	140·63	138	198·03	178	255·43
19	27·26½	59	84·66½	99	142·06½	139	199·46½	179	256·86½
20	28·70	**60**	86·10	**100**	143·50	**140**	200·90	**180**	258·30
21	30·13½	61	87·53½	101	144·93½	141	202·33½	181	259·73½
22	31·57	62	88·97	102	146·37	142	203·77	182	261·17
23	33·00½	63	90·40½	103	147·80½	143	205·20½	183	262·60½
24	34·44	64	91·84	104	149·24	144	206·64	184	264·04
25	35·87½	65	93·27½	105	150·67½	145	208·07½	185	265·47½
26	37·31	66	94·71	106	152·11	146	209·51	186	266·91
27	38·74½	67	96·14½	107	153·54½	147	210·94½	187	268·34½
28	40·18	68	97·58	108	154·98	148	212·38	188	269·78
29	41·61½	69	99·01½	109	156·41½	149	213·81½	189	271·21½
30	43·05	**70**	100·45	**110**	157·85	**150**	215·25	**190**	272·65
31	44·48½	71	101·88½	111	159·28½	151	216·68½	191	274·08½
32	45·92	72	103·32	112	160·72	152	218·12	192	275·52
33	47·35½	73	104·75½	113	162·15½	153	219·55½	193	276·95½
34	48·79	74	106·19	114	163·59	154	220·99	194	278·39
35	50·22½	75	107·62½	115	165·02½	155	222·42½	195	279·82½
36	51·66	76	109·06	116	166·46	156	223·86	196	281·26
37	53·09½	77	110·49½	117	167·89½	157	225·29½	197	282·69½
38	54·53	78	111·93	118	169·33	158	226·73	198	284·13
39	55·96½	79	113·36½	119	170·76½	159	228·16½	199	285·56½
40	57·40	**80**	114·80	**120**	172·20	**160**	229·60	**200**	287·00

$\frac{1}{16}=0{\cdot}09^{0}$ $\frac{1}{9}=0{\cdot}15^{9}$ $\frac{1}{8}=0{\cdot}17^{9}$ $\frac{1}{6}=0{\cdot}23^{9}$ $\frac{1}{5}=0{\cdot}28^{7}$ $\frac{1}{4}=0{\cdot}35^{9}$ $\frac{1}{3}=0{\cdot}47^{8}$

$\frac{3}{8}=0{\cdot}53^{8}$ $\frac{1}{2}=0{\cdot}71^{8}$ $\frac{5}{8}=0{\cdot}89^{7}$ $\frac{3}{4}=1{\cdot}07^{6}$ $\frac{7}{8}=1{\cdot}25^{6}$ $\frac{2}{3}=0{\cdot}95^{7}$ $\frac{5}{6}=1{\cdot}19^{6}$

$\frac{3}{16}=0{\cdot}26^{9}$ $\frac{5}{16}=0{\cdot}44^{8}$ $\frac{7}{16}=0{\cdot}62^{8}$ $\frac{9}{16}=0{\cdot}80^{7}$ $\frac{11}{16}=0{\cdot}98^{7}$ $\frac{13}{16}=1{\cdot}16^{6}$ $\frac{15}{16}=1{\cdot}34^{5}$

(0.694=1£) **£1.44**

1	1·44	41	59·04	81	116·64	121	174·24	161	231·84
2	2·88	42	60·48	82	118·08	122	175·68	162	233·28
3	4·32	43	61·92	83	119·52	123	177·12	163	234·72
4	5·76	44	63·36	84	120·96	124	178·56	164	236·16
5	7·20	45	64·80	85	122·40	125	180·00	165	237·60
6	8·64	46	66·24	86	123·84	126	181·44	166	239·04
7	10·08	47	67·68	87	125·28	127	182·88	167	240·48
8	11·52	48	69·12	88	126·72	128	184·32	168	241·92
9	12·96	49	70·56	89	128·16	129	185·76	169	243·36
10	14·40	**50**	72·00	**90**	129·60	**130**	187·20	**170**	244·80
11	15·84	51	73·44	91	131·04	131	188·64	171	246·24
12	17·28	52	74·88	92	132·48	132	190·08	172	247·68
13	18·72	53	76·32	93	133·92	133	191·52	173	249·12
14	20·16	54	77·76	94	135·36	134	192·96	174	250·56
15	21·60	55	79·20	95	136·80	135	194·40	175	252·00
16	23·04	56	80·64	96	138·24	136	195·84	176	253·44
17	24·48	57	82·08	97	139·68	137	197·28	177	254·88
18	25·92	58	83·52	98	141·12	138	198·72	178	256·32
19	27·36	59	84·96	99	142·56	139	200·16	179	257·76
20	28·80	**60**	86·40	**100**	144·00	**140**	201·60	**180**	259·20
21	30·24	61	87·84	101	145·44	141	203·04	181	260·64
22	31·68	62	89·28	102	146·88	142	204·48	182	262·08
23	33·12	63	90·72	103	148·32	143	205·92	183	263·52
24	34·56	64	92·16	104	149·76	144	207·36	184	264·96
25	36·00	65	93·60	105	151·20	145	208·80	185	266·40
26	37·44	66	95·04	106	152·64	146	210·24	186	267·84
27	38·88	67	96·48	107	154·08	147	211·68	187	269·28
28	40·32	68	97·92	108	155·52	148	213·12	188	270·72
29	41·76	69	99·36	109	156·96	149	214·56	189	272·16
30	43·20	**70**	100·80	**110**	158·40	**150**	216·00	**190**	273·60
31	44·64	71	102·24	111	159·84	151	217·44	191	275·04
32	46·08	72	103·68	112	161·28	152	218·88	192	276·48
33	47·52	73	105·12	113	162·72	153	220·32	193	277·92
34	48·96	74	106·56	114	164·16	154	221·76	194	279·36
35	50·40	75	108·00	115	165·60	155	223·20	195	280·80
36	51·84	76	109·44	116	167·04	156	224·64	196	282·24
37	53·28	77	110·88	117	168·48	157	226·08	197	283·68
38	54·72	78	112·32	118	169·92	158	227·52	198	285·12
39	56·16	79	113·76	119	171·36	159	228·96	199	286·56
40	57·60	**80**	115·20	**120**	172·80	**160**	230·40	**200**	288·00

$\frac{1}{16}$=0·09^{0} $\frac{1}{9}$=0·16^{0} $\frac{1}{8}$=0·18^{0} $\frac{1}{6}$=0·24^{0} $\frac{1}{5}$=0·28^{8} $\frac{1}{4}$=0·36^{0} $\frac{1}{3}$=0·48^{0}

$\frac{3}{8}$=0·54^{0} $\frac{1}{2}$=0·72^{0} $\frac{5}{8}$=0·90^{0} $\frac{3}{4}$=1·08^{0} $\frac{7}{8}$=1·26^{0} $\frac{2}{3}$=0·96^{0} $\frac{5}{6}$=1·20^{0}

$\frac{3}{16}$=0·27^{0} $\frac{5}{16}$=0·45^{0} $\frac{7}{16}$=0·63^{0} $\frac{9}{16}$=0·81^{0} $\frac{11}{16}$=0·99^{0} $\frac{13}{16}$=1·17^{0} $\frac{15}{16}$=1·35^{0}

(0.692=1£) **£1.44½**

1	1·44½	41	59·24½	81	117·04½	121	174·84½	161	232·64½
2	2·89	42	60·69	82	118·49	122	176·29	162	234·09
3	4·33½	43	62·13½	83	119·93½	123	177·73½	163	235·53½
4	5·78	44	63·58	84	121·38	124	179·18	164	236·98
5	7·22½	45	65·02½	85	122·82½	125	180·62½	165	238·42½
6	8·67	46	66·47	86	124·27	126	182·07	166	239·87
7	10·11½	47	67·91½	87	125·71½	127	183·51½	167	241·31½
8	11·56	48	69·36	88	127·16	128	184·96	168	242·76
9	13·00½	49	70·80½	89	128·60½	129	186·40½	169	244·20½
10	14·45	**50**	72·25	**90**	130·05	**130**	187·85	**170**	245·65
11	15·89½	51	73·69½	91	131·49½	131	189·29½	171	247·09½
12	17·34	52	75·14	92	132·94	132	190·74	172	248·54
13	18·78½	53	76·58½	93	134·38½	133	192·18½	173	249·98½
14	20·23	54	78·03	94	135·83	134	193·63	174	251·43
15	21·67½	55	79·47½	95	137·27½	135	195·07½	175	252·87½
16	23·12	56	80·92	96	138·72	136	196·52	176	254·32
17	24·56½	57	82·36½	97	140·16½	137	197·96½	177	255·76½
18	26·01	58	83·81	98	141·61	138	199·41	178	257·21
19	27·45½	59	85·25½	99	143·05½	139	200·85½	179	258·65½
20	28·90	**60**	86·70	**100**	144·50	**140**	202·30	**180**	260·10
21	30·34½	61	88·14½	101	145·94½	141	203·74½	181	261·54½
22	31·79	62	89·59	102	147·39	142	205·19	182	262·99
23	33·23½	63	91·03½	103	148·83½	143	206·63½	183	264·43½
24	34·68	64	92·48	104	150·28	144	208·08	184	265·88
25	36·12½	65	93·92½	105	151·72½	145	209·52½	185	267·32½
26	37·57	66	95·37	106	153·17	146	210·97	186	268·77
27	39·01½	67	96·81½	107	154·61½	147	212·41½	187	270·21½
28	40·46	68	98·26	108	156·06	148	213·86	188	271·66
29	41·90½	69	99·70½	109	157·50½	149	215·30½	189	273·10½
30	43·35	**70**	101·15	**110**	158·95	**150**	216·75	**190**	274·55
31	44·79½	71	102·59½	111	160·39½	151	218·19½	191	275·99½
32	46·24	72	104·04	112	161·84	152	219·64	192	277·44
33	47·68½	73	105·48½	113	163·28½	153	221·08½	193	278·88½
34	49·13	74	106·93	114	164·73	154	222·53	194	280·33
35	50·57½	75	108·37½	115	166·17½	155	223·97½	195	281·77½
36	52·02	76	109·82	116	167·62	156	225·42	196	283·22
37	53·46½	77	111·26½	117	169·06½	157	226·86½	197	284·66½
38	54·91	78	112·71	118	170·51	158	228·31	198	286·11
39	56·35½	79	114·15½	119	171·95½	159	229·75½	199	287·55½
40	57·80	**80**	115·60	**120**	173·40	**160**	231·20	**200**	289·00

$\frac{1}{16}$=0·09^{0} $\frac{1}{9}$=0·16^{1} $\frac{1}{8}$=0·18^{1} $\frac{1}{6}$=0·24^{1} $\frac{1}{5}$=0·28^{9} $\frac{1}{4}$=0·36^{1} $\frac{1}{3}$=0·48^{2}

$\frac{3}{8}$=0·54^{2} $\frac{1}{2}$=0·72^{3} $\frac{5}{8}$=0·90^{3} $\frac{3}{4}$=1·08^{4} $\frac{7}{8}$=1·26^{4} $\frac{2}{3}$=0·96^{3} $\frac{5}{6}$=1·20^{4}

$\frac{3}{16}$=0·27^{1} $\frac{5}{16}$=0·45^{2} $\frac{7}{16}$=0·63^{2} $\frac{9}{16}$=0·81^{3} $\frac{11}{16}$=0·99^{3} $\frac{13}{16}$=1·17^{4} $\frac{15}{16}$=1·35^{5}

1	1·45	41	59·45	81	117·45	121	175·45	161	233·45
2	2·90	42	60·90	82	118·90	122	176·90	162	234·90
3	4·35	43	62·35	83	120·35	123	178·35	163	236·35
4	5·80	44	63·80	84	121·80	124	179·80	164	237·80
5	7·25	45	65·25	85	123·25	125	181·25	165	239·25
6	8·70	46	66·70	86	124·70	126	182·70	166	240·70
7	10·15	47	68·15	87	126·15	127	184·15	167	242·15
8	11·60	48	69·60	88	127·60	128	185·60	168	243·60
9	13·05	49	71·05	89	129·05	129	187·05	169	245·05
10	14·50	**50**	72·50	**90**	130·50	**130**	188·50	**170**	246·50
11	15·95	51	73·95	91	131·95	131	189·95	171	247·95
12	17·40	52	75·40	92	133·40	132	191·40	172	249·40
13	18·85	53	76·85	93	134·85	133	192·85	173	250·85
14	20·30	54	78·30	94	136·30	134	194·30	174	252·30
15	21·75	55	79·75	95	137·75	135	195·75	175	253·75
16	23·20	56	81·20	96	139·20	136	197·20	176	255·20
17	24·65	57	82·65	97	140·65	137	198·65	177	256·65
18	26·10	58	84·10	98	142·10	138	200·10	178	258·10
19	27·55	59	85·55	99	143·55	139	201·55	179	259·55
20	29·00	**60**	87·00	**100**	145·00	**140**	203·00	**180**	261·00
21	30·45	61	88·45	101	146·45	141	204·45	181	262·45
22	31·90	62	89·90	102	147·90	142	205·90	182	263·90
23	33·35	63	91·35	103	149·35	143	207·35	183	265·35
24	34·80	64	92·80	104	150·80	144	208·80	184	266·80
25	36·25	65	94·25	105	152·25	145	210·25	185	268·25
26	37·70	66	95·70	106	153·70	146	211·70	186	269·70
27	39·15	67	97·15	107	155·15	147	213·15	187	271·15
28	40·60	68	98·60	108	156·60	148	214·60	188	272·60
29	42·05	69	100·05	109	158·05	149	216·05	189	274·05
30	43·50	**70**	101·50	**110**	159·50	**150**	217·50	**190**	275·50
31	44·95	71	102·95	111	160·95	151	218·95	191	276·95
32	46·40	72	104·40	112	162·40	152	220·40	192	278·40
33	47·85	73	105·85	113	163·85	153	221·85	193	279·85
34	49·30	74	107·30	114	165·30	154	223·30	194	281·30
35	50·75	75	108·75	115	166·75	155	224·75	195	282·75
36	52·20	76	110·20	116	168·20	156	226·20	196	284·20
37	53·65	77	111·65	117	169·65	157	227·65	197	285·65
38	55·10	78	113·10	118	171·10	158	229·10	198	287·10
39	56·55	79	114·55	119	172·55	159	230·55	199	288·55
40	58·00	**80**	116·00	**120**	174·00	**160**	232·00	**200**	290·00

$\frac{1}{16}=0{\cdot}09^{1}$ $\frac{1}{9}=0{\cdot}16^{1}$ $\frac{1}{8}=0{\cdot}18^{1}$ $\frac{1}{6}=0{\cdot}24^{2}$ $\frac{1}{5}=0{\cdot}29^{0}$ $\frac{1}{4}=0{\cdot}36^{3}$ $\frac{1}{3}=0{\cdot}48^{3}$

$\frac{3}{8}=0{\cdot}54^{4}$ $\frac{1}{2}=0{\cdot}72^{5}$ $\frac{5}{8}=0{\cdot}90^{6}$ $\frac{3}{4}=1{\cdot}08^{8}$ $\frac{7}{8}=1{\cdot}26^{9}$ $\frac{2}{3}=0{\cdot}96^{7}$ $\frac{5}{6}=1{\cdot}20^{8}$

$\frac{3}{16}=0{\cdot}27^{2}$ $\frac{5}{16}=0{\cdot}45^{3}$ $\frac{7}{16}=0{\cdot}63^{4}$ $\frac{9}{16}=0{\cdot}81^{6}$ $\frac{11}{16}=0{\cdot}99^{7}$ $\frac{13}{16}=1{\cdot}17^{8}$ $\frac{15}{16}=1{\cdot}35^{9}$

(0.687=1£) **£1.45½**

1	1·45½	41	59·65½	81	117·85½	121	176·05½	161	234·25½
2	2·91	42	61·11	82	119·31	122	177·51	162	235·71
3	4·36½	43	62·56½	83	120·76½	123	178·96½	163	237·16½
4	5·82	44	64·02	84	122·22	124	180·42	164	238·62
5	7·27½	45	65·47½	85	123·67½	125	181·87½	165	240·07½
6	8·73	46	66·93	86	125·13	126	183·33	166	241·53
7	10·18½	47	68·38½	87	126·58½	127	184·78½	167	242·98½
8	11·64	48	69·84	88	128·04	128	186·24	168	244·44
9	13·09½	49	71·29½	89	129·49½	129	187·69½	169	245·89½
10	14·55	**50**	72·75	**90**	130·95	**130**	189·15	**170**	247·35
11	16·00½	51	74·20½	91	132·40½	131	190·60½	171	248·80½
12	17·46	52	75·66	92	133·86	132	192·06	172	250·26
13	18·91½	53	77·11½	93	135·31½	133	193·51½	173	251·71½
14	20·37	54	78·57	94	136·77	134	194·97	174	253·17
15	21·82½	55	80·02½	95	138·22½	135	196·42½	175	254·62½
16	23·28	56	81·48	96	139·68	136	197·88	176	256·08
17	24·73½	57	82·93½	97	141·13½	137	199·33½	177	257·53½
18	26·19	58	84·39	98	142·59	138	200·79	178	258·99
19	27·64½	59	85·84½	99	144·04½	139	202·24½	179	260·44½
20	29·10	**60**	87·30	**100**	145·50	**140**	203·70	**180**	261·90
21	30·55½	61	88·75½	101	146·95½	141	205·15½	181	263·35½
22	32·01	62	90·21	102	148·41	142	206·61	182	264·81
23	33·46½	63	91·66½	103	149·86½	143	208·06½	183	266·26½
24	34·92	64	93·12	104	151·32	144	209·52	184	267·72
25	36·37½	65	94·57½	105	152·77½	145	210·97½	185	269·17½
26	37·83	66	96·03	106	154·23	146	212·43	186	270·63
27	39·28½	67	97·48½	107	155·68½	147	213·88½	187	272·08½
28	40·74	68	98·94	108	157·14	148	215·34	188	273·54
29	42·19½	69	100·39½	109	158·59½	149	216·79½	189	274·99½
30	43·65	**70**	101·85	**110**	160·05	**150**	218·25	**190**	276·45
31	45·10½	71	103·30½	111	161·50½	151	219·70½	191	277·90½
32	46·56	72	104·76	112	162·96	152	221·16	192	279·36
33	48·01½	73	106·21½	113	164·41½	153	222·61½	193	280·81½
34	49·47	74	107·67	114	165·87	154	224·07	194	282·27
35	50·92½	75	109·12½	115	167·32½	155	225·52½	195	283·72½
36	52·38	76	110·58	116	168·78	156	226·98	196	285·18
37	53·83½	77	112·03½	117	170·23½	157	228·43½	197	286·63½
38	55·29	78	113·49	118	171·69	158	229·89	198	288·09
39	56·74½	79	114·94½	119	173·14½	159	231·34½	199	289·54½
40	58·20	**80**	116·40	**120**	174·60	**160**	232·80	**200**	291·00

$\frac{1}{16}=0{\cdot}09^{1}$ $\frac{1}{9}=0{\cdot}16^{2}$ $\frac{1}{8}=0{\cdot}18^{2}$ $\frac{1}{6}=0{\cdot}24^{3}$ $\frac{1}{5}=0{\cdot}29^{1}$ $\frac{1}{4}=0{\cdot}36^{4}$ $\frac{1}{3}=0{\cdot}48^{5}$

$\frac{3}{8}=0{\cdot}54^{6}$ $\frac{1}{2}=0{\cdot}72^{8}$ $\frac{5}{8}=0{\cdot}90^{9}$ $\frac{3}{4}=1{\cdot}09^{1}$ $\frac{7}{8}=1{\cdot}27^{3}$ $\frac{2}{3}=0{\cdot}97^{0}$ $\frac{5}{6}=1{\cdot}21^{3}$

$\frac{3}{16}=0{\cdot}27^{3}$ $\frac{5}{16}=0{\cdot}45^{5}$ $\frac{7}{16}=0{\cdot}63^{7}$ $\frac{9}{16}=0{\cdot}81^{8}$ $\frac{11}{16}=1{\cdot}00^{0}$ $\frac{13}{16}=1{\cdot}18^{2}$ $\frac{15}{16}=1{\cdot}36^{4}$

(0.685=1£) **£1.46**

1	1·46	41	59·86	81	118·26	121	176·66	161	235·06
2	2·92	42	61·32	82	119·72	122	178·12	162	236·52
3	4·38	43	62·78	83	121·18	123	179·58	163	237·98
4	5·84	44	64·24	84	122·64	124	181·04	164	239·44
5	7·30	45	65·70	85	124·10	125	182·50	165	240·90
6	8·76	46	67·16	86	125·56	126	183·96	166	242·36
7	10·22	47	68·62	87	127·02	127	185·42	167	243·82
8	11·68	48	70·08	88	128·48	128	186·88	168	245·28
9	13·14	49	71·54	89	129·94	129	188·34	169	246·74
10	14·60	**50**	73·00	**90**	131·40	**130**	189·80	**170**	248·20
11	16·06	51	74·46	91	132·86	131	191·26	171	249·66
12	17·52	52	75·92	92	134·32	132	192·72	172	251·12
13	18·98	53	77·38	93	135·78	133	194·18	173	252·58
14	20·44	54	78·84	94	137·24	134	195·64	174	254·04
15	21·90	55	80·30	95	138·70	135	197·10	175	255·50
16	23·36	56	81·76	96	140·16	136	198·56	176	256·96
17	24·82	57	83·22	97	141·62	137	200·02	177	258·42
18	26·28	58	84·68	98	143·08	138	201·48	178	259·88
19	27·74	59	86·14	99	144·54	139	202·94	179	261·34
20	29·20	**60**	87·60	**100**	146·00	**140**	204·40	**180**	262·80
21	30·66	61	89·06	101	147·46	141	205·86	181	264·26
22	32·12	62	90·52	102	148·92	142	207·32	182	265·72
23	33·58	63	91·98	103	150·38	143	208·78	183	267·18
24	35·04	64	93·44	104	151·84	144	210·24	184	268·64
25	36·50	65	94·90	105	153·30	145	211·70	185	270·10
26	37·96	66	96·36	106	154·76	146	213·16	186	271·56
27	39·42	67	97·82	107	156·22	147	214·62	187	273·02
28	40·88	68	99·28	108	157·68	148	216·08	188	274·48
29	42·34	69	100·74	109	159·14	149	217·54	189	275·94
30	43·80	**70**	102·20	**110**	160·60	**150**	219·00	**190**	277·40
31	45·26	71	103·66	111	162·06	151	220·46	191	278·86
32	46·72	72	105·12	112	163·52	152	221·92	192	280·32
33	48·18	73	106·58	113	164·98	153	223·38	193	281·78
34	49·64	74	108·04	114	166·44	154	224·84	194	283·24
35	51·10	75	109·50	115	167·90	155	226·30	195	284·70
36	52·56	76	110·96	116	169·36	156	227·76	196	286·16
37	54·02	77	112·42	117	170·82	157	229·22	197	287·62
38	55·48	78	113·88	118	172·28	158	230·68	198	289·08
39	56·94	79	115·34	119	173·74	159	232·14	199	290·54
40	58·40	**80**	116·80	**120**	175·20	**160**	233·60	**200**	292·00

$\frac{1}{16}=0{\cdot}09^{1}$ $\frac{1}{9}=0{\cdot}16^{2}$ $\frac{1}{8}=0{\cdot}18^{3}$ $\frac{1}{6}=0{\cdot}24^{3}$ $\frac{1}{5}=0{\cdot}29^{2}$ $\frac{1}{4}=0{\cdot}36^{5}$ $\frac{1}{3}=0{\cdot}48^{7}$

$\frac{3}{8}=0{\cdot}54^{8}$ $\frac{1}{2}=0{\cdot}73^{0}$ $\frac{5}{8}=0{\cdot}91^{3}$ $\frac{3}{4}=1{\cdot}09^{5}$ $\frac{7}{8}=1{\cdot}27^{8}$ $\frac{2}{3}=0{\cdot}97^{3}$ $\frac{5}{6}=1{\cdot}21^{7}$

$\frac{3}{16}=0{\cdot}27^{4}$ $\frac{5}{16}=0{\cdot}45^{6}$ $\frac{7}{16}=0{\cdot}63^{9}$ $\frac{9}{16}=0{\cdot}82^{1}$ $\frac{11}{16}=1{\cdot}00^{4}$ $\frac{13}{16}=1{\cdot}18^{6}$ $\frac{15}{16}=1{\cdot}36^{9}$

(0.683=1£) **£1.46½**

1	1·46½	41	60·06½	81	118·66½	121	177·26½	161	235·86½
2	2·93	42	61·53	82	120·13	122	178·73	162	237·33
3	4·39½	43	62·99½	83	121·59½	123	180·19½	163	238·79½
4	5·86	44	64·46	84	123·06	124	181·66	164	240·26
5	7·32½	45	65·92½	85	124·52½	125	183·12½	165	241·72½
6	8·79	46	67·39	86	125·99	126	184·59	166	243·19
7	10·25½	47	68·85½	87	127·45½	127	186·05½	167	244·65½
8	11·72	48	70·32	88	128·92	128	187·52	168	246·12
9	13·18½	49	71·78½	89	130·38½	129	188·98½	169	247·58½
10	14·65	**50**	73·25	**90**	131·85	**130**	190·45	**170**	249·05
11	16·11½	51	74·71½	91	133·31½	131	191·91½	171	250·51½
12	17·58	52	76·18	92	134·78	132	193·38	172	251·98
13	19·04½	53	77·64½	93	136·24½	133	194·84½	173	253·44½
14	20·51	54	79·11	94	137·71	134	196·31	174	254·91
15	21·97½	55	80·57½	95	139·17½	135	197·77½	175	256·37½
16	23·44	56	82·04	96	140·64	136	199·24	176	257·84
17	24·90½	57	83·50½	97	142·10½	137	200·70½	177	259·30½
18	26·37	58	84·97	98	143·57	138	202·17	178	260·77
19	27·83½	59	86·43½	99	145·03½	139	203·63½	179	262·23½
20	29·30	**60**	87·90	**100**	146·50	**140**	205·10	**180**	263·70
21	30·76½	61	89·36½	101	147·96½	141	206·56½	181	265·16½
22	32·23	62	90·83	102	149·43	142	208·03	182	266·63
23	33·69½	63	92·29½	103	150·89½	143	209·49½	183	268·09½
24	35·16	64	93·76	104	152·36	144	210·96	184	269·56
25	36·62½	65	95·22½	105	153·82½	145	212·42½	185	271·02½
26	38·09	66	96·69	106	155·29	146	213·89	186	272·49
27	39·55½	67	98·15½	107	156·75½	147	215·35½	187	273·95½
28	41·02	68	99·62	108	158·22	148	216·82	188	275·42
29	42·48½	69	101·08½	109	159·68½	149	218·28½	189	276·88½
30	43·95	**70**	102·55	**110**	161·15	**150**	219·75	**190**	278·35
31	45·41½	71	104·01½	111	162·61½	151	221·21½	191	279·81½
32	46·88	72	105·48	112	164·08	152	222·68	192	281·28
33	48·34½	73	106·94½	113	165·54½	153	224·14½	193	282·74½
34	49·81	74	108·41	114	167·01	154	225·61	194	284·21
35	51·27½	75	109·87½	115	168·47½	155	227·07½	195	285·67½
36	52·74	76	111·34	116	169·94	156	228·54	196	287·14
37	54·20½	77	112·80½	117	171·40½	157	230·00½	197	288·60½
38	55·67	78	114·27	118	172·87	158	231·47	198	290·07
39	57·13½	79	115·73½	119	174·33½	159	232·93½	199	291·53½
40	58·60	**80**	117·20	**120**	175·80	**160**	234·40	**200**	293·00

$\frac{1}{16}=0{\cdot}09^{2}$ $\frac{1}{9}=0{\cdot}16^{3}$ $\frac{1}{8}=0{\cdot}18^{3}$ $\frac{1}{6}=0{\cdot}24^{4}$ $\frac{1}{5}=0{\cdot}29^{3}$ $\frac{1}{4}=0{\cdot}36^{6}$ $\frac{1}{3}=0{\cdot}48^{8}$

$\frac{3}{8}=0{\cdot}54^{9}$ $\frac{1}{2}=0{\cdot}73^{3}$ $\frac{5}{8}=0{\cdot}91^{6}$ $\frac{3}{4}=1{\cdot}09^{9}$ $\frac{7}{8}=1{\cdot}28^{2}$ $\frac{2}{3}=0{\cdot}97^{7}$ $\frac{5}{6}=1{\cdot}22^{1}$

$\frac{3}{16}=0{\cdot}27^{5}$ $\frac{5}{16}=0{\cdot}45^{8}$ $\frac{7}{16}=0{\cdot}64^{1}$ $\frac{9}{16}=0{\cdot}82^{4}$ $\frac{11}{16}=1{\cdot}00^{7}$ $\frac{13}{16}=1{\cdot}19^{0}$ $\frac{15}{16}=1{\cdot}37^{3}$

(0.671=1£) **£1.49**

1	1·49	41	61·09	81	120·69	121	180·29	161	239·89
2	2·98	42	62·58	82	122·18	122	181·78	162	241·38
3	4·47	43	64·07	83	123·67	123	183·27	163	242·87
4	5·96	44	65·56	84	125·16	124	184·76	164	244·36
5	7·45	45	67·05	85	126·65	125	186·25	165	245·85
6	8·94	46	68·54	86	128·14	126	187·74	166	247·34
7	10·43	47	70·03	87	129·63	127	189·23	167	248·83
8	11·92	48	71·52	88	131·12	128	190·72	168	250·32
9	13·41	49	73·01	89	132·61	129	192·21	169	251·81
10	14·90	**50**	74·50	**90**	134·10	**130**	193·70	**170**	253·30
11	16·39	51	75·99	91	135·59	131	195·19	171	254·79
12	17·88	52	77·48	92	137·08	132	196·68	172	256·28
13	19·37	53	78·97	93	138·57	133	198·17	173	257·77
14	20·86	54	80·46	94	140·06	134	199·66	174	259·26
15	22·35	55	81·95	95	141·55	135	201·15	175	260·75
16	23·84	56	83·44	96	143·04	136	202·64	176	262·24
17	25·33	57	84·93	97	144·53	137	204·13	177	263·73
18	26·82	58	86·42	98	146·02	138	205·62	178	265·22
19	28·31	59	87·91	99	147·51	139	207·11	179	266·71
20	29·80	**60**	89·40	**100**	149·00	**140**	208·60	**180**	268·20
21	31·29	61	90·89	101	150·49	141	210·09	181	269·69
22	32·78	62	92·38	102	151·98	142	211·58	182	271·18
23	34·27	63	93·87	103	153·47	143	213·07	183	272·67
24	35·76	64	95·36	104	154·96	144	214·56	184	274·16
25	37·25	65	96·85	105	156·45	145	216·05	185	275·65
26	38·74	66	98·34	106	157·94	146	217·54	186	277·14
27	40·23	67	99·83	107	159·43	147	219·03	187	278·63
28	41·72	68	101·32	108	160·92	148	220·52	188	280·12
29	43·21	69	102·81	109	162·41	149	222·01	189	281·61
30	44·70	**70**	104·30	**110**	163·90	**150**	223·50	**190**	283·10
31	46·19	71	105·79	111	165·39	151	224·99	191	284·59
32	47·68	72	107·28	112	166·88	152	226·48	192	286·08
33	49·17	73	108·77	113	168·37	153	227·97	193	287·57
34	50·66	74	110·26	114	169·86	154	229·46	194	289·06
35	52·15	75	111·75	115	171·35	155	230·95	195	290·55
36	53·64	76	113·24	116	172·84	156	232·44	196	292·04
37	55·13	77	114·73	117	174·33	157	233·93	197	293·53
38	56·62	78	116·22	118	175·82	158	235·42	198	295·02
39	58·11	79	117·71	119	177·31	159	236·91	199	296·51
40	59·60	**80**	119·20	**120**	178·80	**160**	238·40	**200**	298·00

$\frac{1}{16}=0{\cdot}09^{3}$ $\frac{1}{9}=0{\cdot}16^{6}$ $\frac{1}{8}=0{\cdot}18^{6}$ $\frac{1}{6}=0{\cdot}24^{8}$ $\frac{1}{5}=0{\cdot}29^{8}$ $\frac{1}{4}=0{\cdot}37^{3}$ $\frac{1}{3}=0{\cdot}49^{7}$

$\frac{3}{8}=0{\cdot}55^{9}$ $\frac{1}{2}=0{\cdot}74^{5}$ $\frac{5}{8}=0{\cdot}93^{1}$ $\frac{3}{4}=1{\cdot}11^{8}$ $\frac{7}{8}=1{\cdot}30^{4}$ $\frac{2}{3}=0{\cdot}99^{3}$ $\frac{5}{6}=1{\cdot}24^{2}$

$\frac{3}{16}=0{\cdot}27^{9}$ $\frac{5}{16}=0{\cdot}46^{6}$ $\frac{7}{16}=0{\cdot}65^{2}$ $\frac{9}{16}=0{\cdot}83^{8}$ $\frac{11}{16}=1{\cdot}02^{4}$ $\frac{13}{16}=1{\cdot}21^{1}$ $\frac{15}{16}=1{\cdot}39^{7}$

1	1·49½	41	61·29½	81	121·09½	121	180·89½	161	240·69½
2	2·99	42	62·79	82	122·59	122	182·39	162	242·19
3	4·48½	43	64·28½	83	124·08½	123	183·88½	163	243·68½
4	5·98	44	65·78	84	125·58	124	185·38	164	245·18
5	7·47½	45	67·27½	85	127·07½	125	186·87½	165	246·67½
6	8·97	46	68·77	86	128·57	126	188·37	166	248·17
7	10·46½	47	70·26½	87	130·06½	127	189·86½	167	249·66½
8	11·96	48	71·76	88	131·56	128	191·36	168	251·16
9	13·45½	49	73·25½	89	133·05½	129	192·85½	169	252·65½
10	14·95	**50**	74·75	**90**	134·55	**130**	194·35	**170**	254·15
11	16·44½	51	76·24½	91	136·04½	131	195·84½	171	255·64½
12	17·94	52	77·74	92	137·54	132	197·34	172	257·14
13	19·43½	53	79·23½	93	139·03½	133	198·83½	173	258·63½
14	20·93	54	80·73	94	140·53	134	200·33	174	260·13
15	22·42½	55	82·22½	95	142·02½	135	201·82½	175	261·62½
16	23·92	56	83·72	96	143·52	136	203·32	176	263·12
17	25·41½	57	85·21½	97	145·01½	137	204·81½	177	264·61½
18	26·91	58	86·71	98	146·51	138	206·31	178	266·11
19	28·40½	59	88·20½	99	148·00½	139	207·80½	179	267·60½
20	29·90	**60**	89·70	**100**	149·50	**140**	209·30	**180**	269·10
21	31·39½	61	91·19½	101	150·99½	141	210·79½	181	270·59½
22	32·89	62	92·69	102	152·49	142	212·29	182	272·09
23	34·38½	63	94·18½	103	153·98½	143	213·78½	183	273·58½
24	35·88	64	95·68	104	155·48	144	215·28	184	275·08
25	37·37½	65	97·17½	105	156·97½	145	216·77½	185	276·57½
26	38·87	66	98·67	106	158·47	146	218·27	186	278·07
27	40·36½	67	100·16½	107	159·96½	147	219·76½	187	279·56½
28	41·86	68	101·66	108	161·46	148	221·26	188	281·06
29	43·35½	69	103·15½	109	162·95½	149	222·75½	189	282·55½
30	44·85	**70**	104·65	**110**	164·45	**150**	224·25	**190**	284·05
31	46·34½	71	106·14½	111	165·94½	151	225·74½	191	285·54½
32	47·84	72	107·64	112	167·44	152	227·24	192	287·04
33	49·33½	73	109·13½	113	168·93½	153	228·73½	193	288·53½
34	50·83	74	110·63	114	170·43	154	230·23	194	290·03
35	52·32½	75	112·12½	115	171·92½	155	231·72½	195	291·52½
36	53·82	76	113·62	116	173·42	156	233·22	196	293·02
37	55·31½	77	115·11½	117	174·91½	157	234·71½	197	294·51½
38	56·81	78	116·61	118	176·41	158	236·21	198	296·01
39	58·30½	79	118·10½	119	177·90½	159	237·70½	199	297·50½
40	59·80	**80**	119·60	**120**	179·40	**160**	239·20	**200**	299·00

$\frac{1}{16}$=0·09^3	$\frac{1}{9}$=0·16^6	$\frac{1}{8}$=0·18^7	$\frac{1}{6}$=0·24^9	$\frac{1}{5}$=0·29^9	$\frac{1}{4}$=0·37^4	$\frac{1}{3}$=0·49^8
$\frac{3}{8}$=0·56^1	$\frac{1}{2}$=0·74^8	$\frac{5}{8}$=0·93^4	$\frac{3}{4}$=1·12^1	$\frac{7}{8}$=1·30^8	$\frac{2}{3}$=0·99^7	$\frac{5}{6}$=1·24^6
$\frac{3}{16}$=0·28^0	$\frac{5}{16}$=0·46^7	$\frac{7}{16}$=0·65^4	$\frac{9}{16}$=0·84^1	$\frac{11}{16}$=1·02^8	$\frac{13}{16}$=1·21^5	$\frac{15}{16}$=1·40^2

(0.667=1£) **£1.50**

1	1·50	41	61·50	81	121·50	121	181·50	161	241·50
2	3·00	42	63·00	82	123·00	122	183·00	162	243·00
3	4·50	43	64·50	83	124·50	123	184·50	163	244·50
4	6·00	44	66·00	84	126·00	124	186·00	164	246·00
5	7·50	45	67·50	85	127·50	125	187·50	165	247·50
6	9·00	46	69·00	86	129·00	126	189·00	166	249·00
7	10·50	47	70·50	87	130·50	127	190·50	167	250·50
8	12·00	48	72·00	88	132·00	128	192·00	168	252·00
9	13·50	49	73·50	89	133·50	129	193·50	169	253·50
10	15·00	**50**	75·00	**90**	135·00	**130**	195·00	**170**	255·00
11	16·50	51	76·50	91	136·50	131	196·50	171	256·50
12	18·00	52	78·00	92	138·00	132	198·00	172	258·00
13	19·50	53	79·50	93	139·50	133	199·50	173	259·50
14	21·00	54	81·00	94	141·00	134	201·00	174	261·00
15	22·50	55	82·50	95	142·50	135	202·50	175	262·50
16	24·00	56	84·00	96	144·00	136	204·00	176	264·00
17	25·50	57	85·50	97	145·50	137	205·50	177	265·50
18	27·00	58	87·00	98	147·00	138	207·00	178	267·00
19	28·50	59	88·50	99	148·50	139	208·50	179	268·50
20	30·00	**60**	90·00	**100**	150·00	**140**	210·00	**180**	270·00
21	31·50	61	91·50	101	151·50	141	211·50	181	271·50
22	33·00	62	93·00	102	153·00	142	213·00	182	273·00
23	34·50	63	94·50	103	154·50	143	214·50	183	274·50
24	36·00	64	96·00	104	156·00	144	216·00	184	276·00
25	37·50	65	97·50	105	157·50	145	217·50	185	277·50
26	39·00	66	99·00	106	159·00	146	219·00	186	279·00
27	40·50	67	100·50	107	160·50	147	220·50	187	280·50
28	42·00	68	102·00	108	162·00	148	222·00	188	282·00
29	43·50	69	103·50	109	163·50	149	223·50	189	283·50
30	45·00	**70**	105·00	**110**	165·00	**150**	225·00	**190**	285·00
31	46·50	71	106·50	111	166·50	151	226·50	191	286·50
32	48·00	72	108·00	112	168·00	152	228·00	192	288·00
33	49·50	73	109·50	113	169·50	153	229·50	193	289·50
34	51·00	74	111·00	114	171·00	154	231·00	194	291·00
35	52·50	75	112·50	115	172·50	155	232·50	195	292·50
36	54·00	76	114·00	116	174·00	156	234·00	196	294·00
37	55·50	77	115·50	117	175·50	157	235·50	197	295·50
38	57·00	78	117·00	118	177·00	158	237·00	198	297·00
39	58·50	79	118·50	119	178·50	159	238·50	199	298·50
40	60·00	**80**	120·00	**120**	180·00	**160**	240·00	**200**	300·00

$\frac{1}{16}=0{\cdot}09^{4}$ $\frac{1}{9}=0{\cdot}16^{7}$ $\frac{1}{8}=0{\cdot}18^{8}$ $\frac{1}{6}=0{\cdot}25^{0}$ $\frac{1}{5}=0{\cdot}30^{0}$ $\frac{1}{4}=0{\cdot}37^{5}$ $\frac{1}{3}=0{\cdot}50^{0}$

$\frac{3}{8}=0{\cdot}56^{3}$ $\frac{1}{2}=0{\cdot}75^{0}$ $\frac{5}{8}=0{\cdot}93^{8}$ $\frac{3}{4}=1{\cdot}12^{5}$ $\frac{7}{8}=1{\cdot}31^{3}$ $\frac{2}{3}=1{\cdot}00^{0}$ $\frac{5}{6}=1{\cdot}25^{0}$

$\frac{3}{16}=0{\cdot}28^{1}$ $\frac{5}{16}=0{\cdot}46^{9}$ $\frac{7}{16}=0{\cdot}65^{6}$ $\frac{9}{16}=0{\cdot}84^{4}$ $\frac{11}{16}=1{\cdot}03^{1}$ $\frac{13}{16}=1{\cdot}21^{9}$ $\frac{15}{16}=1{\cdot}40^{6}$

1	1·50½	41	61·70½	81	121·90½	121	182·10½	161	242·30½
2	3·01	42	63·21	82	123·41	122	183·61	162	243·81
3	4·51½	43	64·71½	83	124·91½	123	185·11½	163	245·31½
4	6·02	44	66·22	84	126·42	124	186·62	164	246·82
5	7·52½	45	67·72½	85	127·92½	125	188·12½	165	248·32½
6	9·03	46	69·23	86	129·43	126	189·63	166	249·83
7	10·53½	47	70·73½	87	130·93½	127	191·13½	167	251·33½
8	12·04	48	72·24	88	132·44	128	192·64	168	252·84
9	13·54½	49	73·74½	89	133·94½	129	194·14½	169	254·34½
10	15·05	**50**	75·25	**90**	135·45	**130**	195·65	**170**	255·85
11	16·55½	51	76·75½	91	136·95½	131	197·15½	171	257·35½
12	18·06	52	78·26	92	138·46	132	198·66	172	258·86
13	19·56½	53	79·76½	93	139·96½	133	200·16½	173	260·36½
14	21·07	54	81·27	94	141·47	134	201·67	174	261·87
15	22·57½	55	82·77½	95	142·97½	135	203·17½	175	263·37½
16	24·08	56	84·28	96	144·48	136	204·68	176	264·88
17	25·58½	57	85·78½	97	145·98½	137	206·18½	177	266·38½
18	27·09	58	87·29	98	147·49	138	207·69	178	267·89
19	28·59½	59	88·79½	99	148·99½	139	209·19½	179	269·39½
20	30·10	**60**	90·30	**100**	150·50	**140**	210·70	**180**	270·90
21	31·60½	61	91·80½	101	152·00½	141	212·20½	181	272·40½
22	33·11	62	93·31	102	153·51	142	213·71	182	273·91
23	34·61½	63	94·81½	103	155·01½	143	215·21½	183	275·41½
24	36·12	64	96·32	104	156·52	144	216·72	184	276·92
25	37·62½	65	97·82½	105	158·02½	145	218·22½	185	278·42½
26	39·13	66	99·33	106	159·53	146	219·73	186	279·93
27	40·63½	67	100·83½	107	161·03½	147	221·23½	187	281·43½
28	42·14	68	102·34	108	162·54	148	222·74	188	282·94
29	43·64½	69	103·84½	109	164·04½	149	224·24½	189	284·44½
30	45·15	**70**	105·35	**110**	165·55	**150**	225·75	**190**	285·95
31	46·65½	71	106·85½	111	167·05½	151	227·25½	191	287·45½
32	48·16	72	108·36	112	168·56	152	228·76	192	288·96
33	49·66½	73	109·86½	113	170·06½	153	230·26½	193	290·46½
34	51·17	74	111·37	114	171·57	154	231·77	194	291·97
35	52·67½	75	112·87½	115	173·07½	155	233·27½	195	293·47½
36	54·18	76	114·38	116	174·58	156	234·78	196	294·98
37	55·68½	77	115·88½	117	176·08½	157	236·28½	197	296·48½
38	57·19	78	117·39	118	177·59	158	237·79	198	297·99
39	58·69½	79	118·89½	119	179·09½	159	239·29½	199	299·49½
40	60·20	**80**	120·40	**120**	180·60	**160**	240·80	**200**	301·00

$\frac{1}{16}$=0·09^{4} $\frac{1}{9}$=0·16^{7} $\frac{1}{8}$=0·18^{8} $\frac{1}{6}$=0·25^{1} $\frac{1}{5}$=0·30^{1} $\frac{1}{4}$=0·37^{6} $\frac{1}{3}$=0·50^{2}

$\frac{3}{8}$=0·56^{4} $\frac{1}{2}$=0·75^{3} $\frac{5}{8}$=0·94^{1} $\frac{3}{4}$=1·12^{9} $\frac{7}{8}$=1 31^{7} $\frac{2}{3}$=1·00^{3} $\frac{5}{6}$=1·25^{4}

$\frac{3}{16}$=0·28^{2} $\frac{5}{16}$=0·47^{0} $\frac{7}{16}$=0·65^{8} $\frac{9}{16}$=0·84^{7} $\frac{11}{16}$=1·03^{5} $\frac{13}{16}$=1·22^{3} $\frac{15}{16}$=1·41^{1}

(0.662=1£) **£1.51**

1	1·51	41	61·91	81	122·31	121	182·71	161	243·11
2	3·02	42	63·42	82	123·82	122	184·22	162	244·62
3	4·53	43	64·93	83	125·33	123	185·73	163	246·13
4	6·04	44	66·44	84	126·84	124	187·24	164	247·64
5	7·55	45	67·95	85	128·35	125	188·75	165	249·15
6	9·06	46	69·46	86	129·86	126	190·26	166	250·66
7	10·57	47	70·97	87	131·37	127	191·77	167	252·17
8	12·08	48	72·48	88	132·88	128	193·28	168	253·68
9	13·59	49	73·99	89	134·39	129	194·79	169	255·19
10	15·10	**50**	75·50	**90**	135·90	**130**	196·30	**170**	256·70
11	16·61	51	77·01	91	137·41	131	197·81	171	258·21
12	18·12	52	78·52	92	138·92	132	199·32	172	259·72
13	19·63	53	80·03	93	140·43	133	200·83	173	261·23
14	21·14	54	81·54	94	141·94	134	202·34	174	262·74
15	22·65	55	83·05	95	143·45	135	203·85	175	264·25
16	24·16	56	84·56	96	144·96	136	205·36	176	265·76
17	25·67	57	86·07	97	146·47	137	206·87	177	267·27
18	27·18	58	87·58	98	147·98	138	208·38	178	268·78
19	28·69	59	89·09	99	149·49	139	209·89	179	270·29
20	30·20	**60**	90·60	**100**	151·00	**140**	211·40	**180**	271·80
21	31·71	61	92·11	101	152·51	141	212·91	181	273·31
22	33·22	62	93·62	102	154·02	142	214·42	182	274·82
23	34·73	63	95·13	103	155·53	143	215·93	183	276·33
24	36·24	64	96·64	104	157·04	144	217·44	184	277·84
25	37·75	65	98·15	105	158·55	145	218·95	185	279·35
26	39·26	66	99·66	106	160·06	146	220·46	186	280·86
27	40·77	67	101·17	107	161·57	147	221·97	187	282·37
28	42·28	68	102·68	108	163·08	148	223·48	188	283·88
29	43·79	69	104·19	109	164·59	149	224·99	189	285·39
30	45·30	**70**	105·70	**110**	166·10	**150**	226·50	**190**	286·90
31	46·81	71	107·21	111	167·61	151	228·01	191	288·41
32	48·32	72	108·72	112	169·12	152	229·52	192	289·92
33	49·83	73	110·23	113	170·63	153	231·03	193	291·43
34	51·34	74	111·74	114	172·14	154	232·54	194	292·94
35	52·85	75	113·25	115	173·65	155	234·05	195	294·45
36	54·36	76	114·76	116	175·16	156	235·56	196	295·96
37	55·87	77	116·27	117	176·67	157	237·07	197	297·47
38	57·38	78	117·78	118	178·18	158	238·58	198	298·98
39	58·89	79	119·29	119	179·69	159	240·09	199	300·49
40	60·40	**80**	120·80	**120**	181·20	**160**	241·60	**200**	302·00

$\frac{1}{16}=0{\cdot}09^{4}$ $\frac{1}{9}=0{\cdot}16^{8}$ $\frac{1}{8}=0{\cdot}18^{9}$ $\frac{1}{6}=0{\cdot}25^{2}$ $\frac{1}{5}=0{\cdot}30^{2}$ $\frac{1}{4}=0{\cdot}37^{8}$ $\frac{1}{3}=0{\cdot}50^{3}$

$\frac{3}{8}=0{\cdot}56^{6}$ $\frac{1}{2}=0{\cdot}75^{5}$ $\frac{5}{8}=0{\cdot}94^{4}$ $\frac{3}{4}=1{\cdot}13^{3}$ $\frac{7}{8}=1{\cdot}32^{1}$ $\frac{2}{3}=1{\cdot}00^{7}$ $\frac{5}{6}=1{\cdot}25^{8}$

$\frac{3}{16}=0{\cdot}28^{3}$ $\frac{5}{16}=0{\cdot}47^{2}$ $\frac{7}{16}=0{\cdot}66^{1}$ $\frac{9}{16}=0{\cdot}84^{9}$ $\frac{11}{16}=1{\cdot}03^{8}$ $\frac{13}{16}=1{\cdot}22^{7}$ $\frac{15}{16}=1{\cdot}41^{6}$

(0.660=1£) £1.51½

1	1·51½	41	62·11½	81	122·71½	121	183·31½	161	243·91½
2	3·03	42	63·63	82	124·23	122	184·83	162	245·43
3	4·54½	43	65·14½	83	125·74½	123	186·34½	163	246·94½
4	6·06	44	66·66	84	127·26	124	187·86	164	248·46
5	7·57½	45	68·17½	85	128·77½	125	189·37½	165	249·97½
6	9·09	46	69·69	86	130·29	126	190·89	166	251·49
7	10·60½	47	71·20½	87	131·80½	127	192·40½	167	253·00½
8	12·12	48	72·72	88	133·32	128	193·92	168	254·52
9	13·63½	49	74·23½	89	134·83½	129	195·43½	169	256·03½
10	15·15	**50**	75·75	**90**	136·35	**130**	196·95	**170**	257·55
11	16·66½	51	77·26½	91	137·86½	131	198·46½	171	259·06½
12	18·18	52	78·78	92	139·38	132	199·98	172	260·58
13	19·69½	53	80·29½	93	140·89½	133	201·49½	173	262·09½
14	21·21	54	81·81	94	142·41	134	203·01	174	263·61
15	22·72½	55	83·32½	95	143·92½	135	204·52½	175	265·12½
16	24·24	56	84·84	96	145·44	136	206·04	176	266·64
17	25·75½	57	86·35½	97	146·95½	137	207·55½	177	268·15½
18	27·27	58	87·87	98	148·47	138	209·07	178	269·67
19	28·78½	59	89·38½	99	149·98½	139	210·58½	179	271·18½
20	30·30	**60**	90·90	**100**	151·50	**140**	212·10	**180**	272·70
21	31·81½	61	92·41½	101	153·01½	141	213·61½	181	274·21½
22	33·33	62	93·93	102	154·53	142	215·13	182	275·73
23	34·84½	63	95·44½	103	156·04½	143	216·64½	183	277·24½
24	36·36	64	96·96	104	157·56	144	218·16	184	278·76
25	37·87½	65	98·47½	105	159·07½	145	219·67½	185	280·27½
26	39·39	66	99·99	106	160·59	146	221·19	186	281·79
27	40·90½	67	101·50½	107	162·10½	147	222·70½	187	283·30½
28	42·42	68	103·02	108	163·62	148	224·22	188	284·82
29	43·93½	69	104·53½	109	165·13½	149	225·73½	189	286·33½
30	45·45	**70**	106·05	**110**	166·65	**150**	227·25	**190**	287·85
31	46·96½	71	107·56½	111	168·16½	151	228·76½	191	289·36½
32	48·48	72	109·08	112	169·68	152	230·28	192	290·88
33	49·99½	73	110·59½	113	171·19½	153	231·79½	193	292·39½
34	51·51	74	112·11	114	172·71	154	233·31	194	293·91
35	53·02½	75	113·62½	115	174·22½	155	234·82½	195	295·42½
36	54·54	76	115·14	116	175·74	156	236·34	196	296·94
37	56·05½	77	116·65½	117	177·25½	157	237·85½	197	298·45½
38	57·57	78	118·17	118	178·77	158	239·37	198	299·97
39	59·08½	79	119·68½	119	180·28½	159	240·88½	199	301·48½
40	60·60	**80**	121·20	**120**	181·80	**160**	242·40	**200**	303·00

$\frac{1}{16}=0{\cdot}09^{5}$	$\frac{1}{9}=0{\cdot}16^{8}$	$\frac{1}{8}=0{\cdot}18^{9}$	$\frac{1}{6}=0{\cdot}25^{3}$	$\frac{1}{5}=0{\cdot}30^{3}$	$\frac{1}{4}=0{\cdot}37^{9}$	$\frac{1}{3}=0{\cdot}50^{5}$
$\frac{3}{8}=0{\cdot}56^{8}$	$\frac{1}{2}=0{\cdot}75^{8}$	$\frac{5}{8}=0{\cdot}94^{7}$	$\frac{3}{4}=1{\cdot}13^{6}$	$\frac{7}{8}=1{\cdot}32^{6}$	$\frac{2}{3}=1{\cdot}01^{0}$	$\frac{5}{6}=1{\cdot}26^{3}$
$\frac{3}{16}=0{\cdot}28^{4}$	$\frac{5}{16}=0{\cdot}47^{3}$	$\frac{7}{16}=0{\cdot}66^{3}$	$\frac{9}{16}=0{\cdot}85^{2}$	$\frac{11}{16}=1{\cdot}04^{2}$	$\frac{13}{16}=1{\cdot}23^{1}$	$\frac{15}{16}=1{\cdot}42^{0}$

1	1·52	41	62·32	81	123·12	121	183·92	161	244·72
2	3·04	42	63·84	82	124·64	122	185·44	162	246·24
3	4·56	43	65·36	83	126·16	123	186·96	163	247·76
4	6·08	44	66·88	84	127·68	124	188·48	164	249·28
5	7·60	45	68·40	85	129·20	125	190·00	165	250·80
6	9·12	46	69·92	86	130·72	126	191·52	166	252·32
7	10·64	47	71·44	87	132·24	127	193·04	167	253·84
8	12·16	48	72·96	88	133·76	128	194·56	168	255·36
9	13·68	49	74·48	89	135·28	129	196·08	169	256·88
10	15·20	**50**	76·00	**90**	136·80	**130**	197·60	**170**	258·40
11	16·72	51	77·52	91	138·32	131	199·12	171	259·92
12	18·24	52	79·04	92	139·84	132	200·64	172	261·44
13	19·76	53	80·56	93	141·36	133	202·16	173	262·96
14	21·28	54	82·08	94	142·88	134	203·68	174	264·48
15	22·80	55	83·60	95	144·40	135	205·20	175	266·00
16	24·32	56	85·12	96	145·92	136	206·72	176	267·52
17	25·84	57	86·64	97	147·44	137	208·24	177	269·04
18	27·36	58	88·16	98	148·96	138	209·76	178	270·56
19	28·88	59	89·68	99	150·48	139	211·28	179	272·08
20	30·40	**60**	91·20	**100**	152·00	**140**	212·80	**180**	273·60
21	31·92	61	92·72	101	153·52	141	214·32	181	275·12
22	33·44	62	94·24	102	155·04	142	215·84	182	276·64
23	34·96	63	95·76	103	156·56	143	217·36	183	278·16
24	36·48	64	97·28	104	158·08	144	218·88	184	279·68
25	38·00	65	98·80	105	159·60	145	220·40	185	281·20
26	39·52	66	100·32	106	161·12	146	221·92	186	282·72
27	41·04	67	101·84	107	162·64	147	223·44	187	284·24
28	42·56	68	103·36	108	164·16	148	224·96	188	285·76
29	44·08	69	104·88	109	165·68	149	226·48	189	287·28
30	45·60	**70**	106·40	**110**	167·20	**150**	228·00	**190**	288·80
31	47·12	71	107·92	111	168·72	151	229·52	191	290·32
32	48·64	72	109·44	112	170·24	152	231·04	192	291·84
33	50·16	73	110·96	113	171·76	153	232·56	193	293·36
34	51·68	74	112·48	114	173·28	154	234·08	194	294·88
35	53·20	75	114·00	115	174·80	155	235·60	195	296·40
36	54·72	76	115·52	116	176·32	156	237·12	196	297·92
37	56·24	77	117·04	117	177·84	157	238·64	197	299·44
38	57·76	78	118·56	118	179·36	158	240·16	198	300·96
39	59·28	79	120·08	119	180·88	159	241·68	199	302·48
40	60·80	**80**	121·60	**120**	182·40	**160**	243·20	**200**	304·00

$\frac{1}{16}$=0·09^{5} $\frac{1}{9}$=0·16^{9} $\frac{1}{8}$=0·19^{0} $\frac{1}{6}$=0·25^{3} $\frac{1}{5}$=0·30^{4} $\frac{1}{4}$=0·38^{0} $\frac{1}{3}$=0·50^{7}

$\frac{3}{8}$=0·57^{0} $\frac{1}{2}$=0·76^{0} $\frac{5}{8}$=0·95^{0} $\frac{3}{4}$=1·14^{0} $\frac{7}{8}$=1·33^{0} $\frac{2}{3}$=1·01^{3} $\frac{5}{6}$=1·26^{7}

$\frac{3}{16}$=0·28^{5} $\frac{5}{16}$=0·47^{5} $\frac{7}{16}$=0·66^{5} $\frac{9}{16}$=0·85^{5} $\frac{11}{16}$=1·04^{5} $\frac{13}{16}$=1·23^{5} $\frac{15}{16}$=1·42^{5}

(0.656=1£) £1.52½

1	1·52½	41	62·52½	81	123·52½	121	184·52½	161	245·52½
2	3·05	42	64·05	82	125·05	122	186·05	162	247·05
3	4·57½	43	65·57½	83	126·57½	123	187·57½	163	248·57½
4	6·10	44	67·10	84	128·10	124	189·10	164	250·10
5	7·62½	45	68·62½	85	129·62½	125	190·62½	165	251·62½
6	9·15	46	70·15	86	131·15	126	192·15	166	253·15
7	10·67½	47	71·67½	87	132·67½	127	193·67½	167	254·67½
8	12·20	48	73·20	88	134·20	128	195·20	168	256·20
9	13·72½	49	74·72½	89	135·72½	129	196·72½	169	257·72½
10	15·25	**50**	76·25	**90**	137·25	**130**	198·25	**170**	259·25
11	16·77½	51	77·77½	91	138·77½	131	199·77½	171	260·77½
12	18·30	52	79·30	92	140·30	132	201·30	172	262·30
13	19·82½	53	80·82½	93	141·82½	133	202·82½	173	263·82½
14	21·35	54	82·35	94	143·35	134	204·35	174	265·35
15	22·87½	55	83·87½	95	144·87½	135	205·87½	175	266·87½
16	24·40	56	85·40	96	146·40	136	207·40	176	268·40
17	25·92½	57	86·92½	97	147·92½	137	208·92½	177	269·92½
18	27·45	58	88·45	98	149·45	138	210·45	178	271·45
19	28·97½	59	89·97½	99	150·97½	139	211·97½	179	272·97½
20	30·50	**60**	91·50	**100**	152·50	**140**	213·50	**180**	274·50
21	32·02½	61	93·02½	101	154·02½	141	215·02½	181	276·02½
22	33·55	62	94·55	102	155·55	142	216·55	182	277·55
23	35·07½	63	96·07½	103	157·07½	143	218·07½	183	279·07½
24	36·60	64	97·60	104	158·60	144	219·60	184	280·60
25	38·12½	65	99·12½	105	160·12½	145	221·12½	185	282·12½
26	39·65	66	100·65	106	161·65	146	222·65	186	283·65
27	41·17½	67	102·17½	107	163·17½	147	224·17½	187	285·17½
28	42·70	68	103·70	108	164·70	148	225·70	188	286·70
29	44·22½	69	105·22½	109	166·22½	149	227·22½	189	288·22½
30	45·75	**70**	106·75	**110**	167·75	**150**	228·75	**190**	289·75
31	47·27½	71	108·27½	111	169·27½	151	230·27½	191	291·27½
32	48·80	72	109·80	112	170·80	152	231·80	192	292·80
33	50·32½	73	111·32½	113	172·32½	153	233·32½	193	294·32½
34	51·85	74	112·85	114	173·85	154	234·85	194	295·85
35	53·37½	75	114·37½	115	175·37½	155	236·37½	195	297·37½
36	54·90	76	115·90	116	176·90	156	237·90	196	298·90
37	56·42½	77	117·42½	117	178·42½	157	239·42½	197	300·42½
38	57·95	78	118·95	118	179·95	158	240·95	198	301·95
39	59·47½	79	120·47½	119	181·47½	159	242·47½	199	303·47½
40	61·00	**80**	122·00	**120**	183·00	**160**	244·00	**200**	305·00

$\frac{1}{16}=0{\cdot}09^{5}$ $\frac{1}{9}=0{\cdot}16^{9}$ $\frac{1}{8}=0{\cdot}19^{1}$ $\frac{1}{6}=0{\cdot}25^{4}$ $\frac{1}{5}=0{\cdot}30^{5}$ $\frac{1}{4}=0{\cdot}38^{1}$ $\frac{1}{3}=0{\cdot}50^{8}$

$\frac{3}{8}=0{\cdot}57^{2}$ $\frac{1}{2}=0{\cdot}76^{3}$ $\frac{5}{8}=0{\cdot}95^{3}$ $\frac{3}{4}=1{\cdot}14^{4}$ $\frac{7}{8}=1{\cdot}33^{4}$ $\frac{2}{3}=1{\cdot}01^{7}$ $\frac{5}{6}=1{\cdot}27^{1}$

$\frac{3}{16}=0{\cdot}28^{6}$ $\frac{5}{16}=0{\cdot}47^{7}$ $\frac{7}{16}=0{\cdot}66^{7}$ $\frac{9}{16}=0{\cdot}85^{8}$ $\frac{11}{16}=1{\cdot}04^{8}$ $\frac{13}{16}=1{\cdot}23^{9}$ $\frac{15}{16}=1{\cdot}43^{0}$

1	1·53	41	62·73	81	123·93	121	185·13	161	246·33
2	3·06	42	64·26	82	125·46	122	186·66	162	247·86
3	4·59	43	65·79	83	126·99	123	188·19	163	249·39
4	6·12	44	67·32	84	128·52	124	189·72	164	250·92
5	7·65	45	68·85	85	130·05	125	191·25	165	252·45
6	9·18	46	70·38	86	131·58	126	192·78	166	253·98
7	10·71	47	71·91	87	133·11	127	194·31	167	255·51
8	12·24	48	73·44	88	134·64	128	195·84	168	257·04
9	13·77	49	74·97	89	136·17	129	197·37	169	258·57
10	15·30	**50**	76·50	**90**	137·70	**130**	198·90	**170**	260·10
11	16·83	51	78·03	91	139·23	131	200·43	171	261·63
12	18·36	52	79·56	92	140·76	132	201·96	172	263·16
13	19·89	53	81·09	93	142·29	133	203·49	173	264·69
14	21·42	54	82·62	94	143·82	134	205·02	174	266·22
15	22·95	55	84·15	95	145·35	135	206·55	175	267·75
16	24·48	56	85·68	96	146·88	136	208·08	176	269·28
17	26·01	57	87·21	97	148·41	137	209·61	177	270·81
18	27·54	58	88·74	98	149·94	138	211·14	178	272·34
19	29·07	59	90·27	99	151·47	139	212·67	179	273·87
20	30·60	**60**	91·80	**100**	153·00	**140**	214·20	**180**	275·40
21	32·13	61	93·33	101	154·53	141	215·73	181	276·93
22	33·66	62	94·86	102	156·06	142	217·26	182	278·46
23	35·19	63	96·39	103	157·59	143	218·79	183	279·99
24	36·72	64	97·92	104	159·12	144	220·32	184	281·52
25	38·25	65	99·45	105	160·65	145	221·85	185	283·05
26	39·78	66	100·98	106	162·18	146	223·38	186	284·58
27	41·31	67	102·51	107	163·71	147	224·91	187	286·11
28	42·84	68	104·04	108	165·24	148	226·44	188	287·64
29	44·37	69	105·57	109	166·77	149	227·97	189	289·17
30	45·90	**70**	107·10	**110**	168·30	**150**	229·50	**190**	290·70
31	47·43	71	108·63	111	169·83	151	231·03	191	292·23
32	48·96	72	110·16	112	171·36	152	232·56	192	293·76
33	50·49	73	111·69	113	172·89	153	234·09	193	295·29
34	52·02	74	113·22	114	174·42	154	235·62	194	296·82
35	53·55	75	114·75	115	175·95	155	237·15	195	298·35
36	55·08	76	116·28	116	177·48	156	238·68	196	299·88
37	56·61	77	117·81	117	179·01	157	240·21	197	301·41
38	58·14	78	119·34	118	180·54	158	241·74	198	302·94
39	59·67	79	120·87	119	182·07	159	243·27	199	304·47
40	61·20	**80**	122·40	**120**	183·60	**160**	244·80	**200**	306·00

$\frac{1}{16}=0{\cdot}09^6$ $\frac{1}{9}=0{\cdot}17^0$ $\frac{1}{8}=0{\cdot}19^1$ $\frac{1}{6}=0{\cdot}25^5$ $\frac{1}{5}=0{\cdot}30^6$ $\frac{1}{4}=0{\cdot}38^3$ $\frac{1}{3}=0{\cdot}51^0$

$\frac{3}{8}=0{\cdot}57^4$ $\frac{1}{2}=0{\cdot}76^5$ $\frac{5}{8}=0{\cdot}95^6$ $\frac{3}{4}=1{\cdot}14^8$ $\frac{7}{8}=1{\cdot}33^9$ $\frac{2}{3}=1{\cdot}02^0$ $\frac{5}{6}=1{\cdot}27^5$

$\frac{3}{16}=0{\cdot}28^7$ $\frac{5}{16}=0{\cdot}47^8$ $\frac{7}{16}=0{\cdot}66^9$ $\frac{9}{16}=0{\cdot}86^1$ $\frac{11}{16}=1{\cdot}05^2$ $\frac{13}{16}=1{\cdot}24^3$ $\frac{15}{16}=1{\cdot}43^4$

(0.651=1£) **£1.53½**

1	1·53½	41	62·93½	81	124·33½	121	185·73½	161	247·13½
2	3·07	42	64·47	82	125·87	122	187·27	162	248·67
3	4·60½	43	66·00½	83	127·40½	123	188·80½	163	250·20½
4	6·14	44	67·54	84	128·94	124	190·34	164	251·74
5	7·67½	45	69·07½	85	130·47½	125	191·87½	165	253·27½
6	9·21	46	70·61	86	132·01	126	193·41	166	254·81
7	10·74½	47	72·14½	87	133·54½	127	194·94½	167	256·34½
8	12·28	48	73·68	88	135·08	128	196·48	168	257·88
9	13·81½	49	75·21½	89	136·61½	129	198·01½	169	259·41½
10	15·35	**50**	76·75	**90**	138·15	**130**	199·55	**170**	260·95
11	16·88½	51	78·28½	91	139·68½	131	201·08½	171	262·48½
12	18·42	52	79·82	92	141·22	132	202·62	172	264·02
13	19·95½	53	81·35½	93	142·75½	133	204·15½	173	265·55½
14	21·49	54	82·89	94	144·29	134	205·69	174	267·09
15	23·02½	55	84·42½	95	145·82½	135	207·22½	175	268·62½
16	24·56	56	85·96	96	147·36	136	208·76	176	270·16
17	26·09½	57	87·49½	97	148·89½	137	210·29½	177	271·69½
18	27·63	58	89·03	98	150·43	138	211·83	178	273·23
19	29·16½	59	90·56½	99	151·96½	139	213·36½	179	274·76½
20	30·70	**60**	92·10	**100**	153·50	**140**	214·90	**180**	276·30
21	32·23½	61	93·63½	101	155·03½	141	216·43½	181	277·83½
22	33·77	62	95·17	102	156·57	142	217·97	182	279·37
23	35·30½	63	96·70½	103	158·10½	143	219·50½	183	280·90½
24	36·84	64	98·24	104	159·64	144	221·04	184	282·44
25	38·37½	65	99·77½	105	161·17½	145	222·57½	185	283·97½
26	39·91	66	101·31	106	162·71	146	224·11	186	285·51
27	41·44½	67	102·84½	107	164·24½	147	225·64½	187	287·04½
28	42·98	68	104·38	108	165·78	148	227·18	188	288·58
29	44·51½	69	105·91½	109	167·31½	149	228·71½	189	290·11½
30	46·05	**70**	107·45	**110**	168·85	**150**	230·25	**190**	291·65
31	47·58½	71	108·98½	111	170·38½	151	231·78½	191	293·18½
32	49·12	72	110·52	112	171·92	152	233·32	192	294·72
33	50·65½	73	112·05½	113	173·45½	153	234·85½	193	296·25½
34	52·19	74	113·59	114	174·99	154	236·39	194	297·79
35	53·72½	75	115·12½	115	176·52½	155	237·92½	195	299·32½
36	55·26	76	116·66	116	178·06	156	239·46	196	300·86
37	56·79½	77	118·19½	117	179·59½	157	240·99½	197	302·39½
38	58·33	78	119·73	118	181·13	158	242·53	198	303·93
39	59·86½	79	121·26½	119	182·66½	159	244·06½	199	305·46½
40	61·40	**80**	122·80	**120**	184·20	**160**	245·60	**200**	307·00

$\frac{1}{16}=0{\cdot}09^{6}$ $\frac{1}{9}=0{\cdot}17^{1}$ $\frac{1}{8}=0{\cdot}19^{2}$ $\frac{1}{6}=0{\cdot}25^{6}$ $\frac{1}{5}=0{\cdot}30^{7}$ $\frac{1}{4}=0{\cdot}38^{4}$ $\frac{1}{3}=0{\cdot}51^{2}$

$\frac{3}{8}=0{\cdot}57^{6}$ $\frac{1}{2}=0{\cdot}76^{8}$ $\frac{5}{8}=0{\cdot}95^{9}$ $\frac{3}{4}=1{\cdot}15^{1}$ $\frac{7}{8}=1{\cdot}34^{3}$ $\frac{2}{3}=1{\cdot}02^{3}$ $\frac{5}{6}=1{\cdot}27^{9}$

$\frac{3}{16}=0{\cdot}28^{8}$ $\frac{5}{16}=0{\cdot}48^{0}$ $\frac{7}{16}=0{\cdot}67^{2}$ $\frac{9}{16}=0{\cdot}86^{3}$ $\frac{11}{16}=1{\cdot}05^{5}$ $\frac{13}{16}=1{\cdot}24^{7}$ $\frac{15}{16}=1{\cdot}43^{9}$

(0.649=1£) **£1.54**

1	1·54	41	63·14	81	124·74	121	186·34	161	247·94
2	3·08	42	64·68	82	126·28	122	187·88	162	249·48
3	4·62	43	66·22	83	127·82	123	189·42	163	251·02
4	6·16	44	67·76	84	129·36	124	190·96	164	252·56
5	7·70	45	69·30	85	130·90	125	192·50	165	254·10
6	9·24	46	70·84	86	132·44	126	194·04	166	255·64
7	10·78	47	72·38	87	133·98	127	195·58	167	257·18
8	12·32	48	73·92	88	135·52	128	197·12	168	258·72
9	13·86	49	75·46	89	137·06	129	198·66	169	260·26
10	15·40	**50**	77·00	**90**	138·60	**130**	200·20	**170**	261·80
11	16·94	51	78·54	91	140·14	131	201·74	171	263·34
12	18·48	52	80·08	92	141·68	132	203·28	172	264·88
13	20·02	53	81·62	93	143·22	133	204·82	173	266·42
14	21·56	54	83·16	94	144·76	134	206·36	174	267·96
15	23·10	55	84·70	95	146·30	135	207·90	175	269·50
16	24·64	56	86·24	96	147·84	136	209·44	176	271·04
17	26·18	57	87·78	97	149·38	137	210·98	177	272·58
18	27·72	58	89·32	98	150·92	138	212·52	178	274·12
19	29·26	59	90·86	99	152·46	139	214·06	179	275·66
20	30·80	**60**	92·40	**100**	154·00	**140**	215·60	**180**	277·20
21	32·34	61	93·94	101	155·54	141	217·14	181	278·74
22	33·88	62	95·48	102	157·08	142	218·68	182	280·28
23	35·42	63	97·02	103	158·62	143	220·22	183	281·82
24	36·96	64	98·56	104	160·16	144	221·76	184	283·36
25	38·50	65	100·10	105	161·70	145	223·30	185	284·90
26	40·04	66	101·64	106	163·24	146	224·84	186	286·44
27	41·58	67	103·18	107	164·78	147	226·38	187	287·98
28	43·12	68	104·72	108	166·32	148	227·92	188	289·52
29	44·66	69	106·26	109	167·86	149	229·46	189	291·06
30	46·20	**70**	107·80	**110**	169·40	**150**	231·00	**190**	292·60
31	47·74	71	109·34	111	170·94	151	232·54	191	294·14
32	49·28	72	110·88	112	172·48	152	234·08	192	295·68
33	50·82	73	112·42	113	174·02	153	235·62	193	297·22
34	52·36	74	113·96	114	175·56	154	237·16	194	298·76
35	53·90	75	115·50	115	177·10	155	238·70	195	300·30
36	55·44	76	117·04	116	178·64	156	240·24	196	301·84
37	56·98	77	118·58	117	180·18	157	241·78	197	303·38
38	58·52	78	120·12	118	181·72	158	243·32	198	304·92
39	60·06	79	121·66	119	183·26	159	244·86	199	306·46
40	61·60	**80**	123·20	**120**	184·80	**160**	246·40	**200**	308·00

$\frac{1}{16}=0{\cdot}09^{6}$ $\frac{1}{9}=0{\cdot}17^{1}$ $\frac{1}{8}=0{\cdot}19^{3}$ $\frac{1}{6}=0{\cdot}25^{7}$ $\frac{1}{5}=0{\cdot}30^{8}$ $\frac{1}{4}=0{\cdot}38^{5}$ $\frac{1}{3}=0{\cdot}51^{3}$

$\frac{3}{8}=0{\cdot}57^{8}$ $\frac{1}{2}=0{\cdot}77^{0}$ $\frac{5}{8}=0{\cdot}96^{3}$ $\frac{3}{4}=1{\cdot}15^{5}$ $\frac{7}{8}=1{\cdot}34^{8}$ $\frac{2}{3}=1{\cdot}02^{7}$ $\frac{5}{6}=1{\cdot}28^{3}$

$\frac{3}{16}=0{\cdot}28^{9}$ $\frac{5}{16}=0{\cdot}48^{1}$ $\frac{7}{16}=0{\cdot}67^{4}$ $\frac{9}{16}=0{\cdot}86^{6}$ $\frac{11}{16}=1{\cdot}05^{9}$ $\frac{13}{16}=1{\cdot}25^{1}$ $\frac{15}{16}=1{\cdot}44^{4}$

(0.647=1£) **£1.54½**

1	1·54½	41	63·34½	81	125·14½	121	186·94½	161	248·74½
2	3·09	42	64·89	82	126·69	122	188·49	162	250·29
3	4·63½	43	66·43½	83	128·23½	123	190·03½	163	251·83½
4	6·18	44	67·98	84	129·78	124	191·58	164	253·38
5	7·72½	45	69·52½	85	131·32½	125	193·12½	165	254·92½
6	9·27	46	71·07	86	132·87	126	194·67	166	256·47
7	10·81½	47	72·61½	87	134·41½	127	196·21½	167	258·01½
8	12·36	48	74·16	88	135·96	128	197·76	168	259·56
9	13·90½	49	75·70½	89	137·50½	129	199·30½	169	261·10½
10	15·45	**50**	77·25	**90**	139·05	**130**	200·85	**170**	262·65
11	16·99½	51	78·79½	91	140·59½	131	202·39½	171	264·19½
12	18·54	52	80·34	92	142·14	132	203·94	172	265·74
13	20·08½	53	81·88½	93	143·68½	133	205·48½	173	267·28½
14	21·63	54	83·43	94	145·23	134	207·03	174	268·83
15	23·17½	55	84·97½	95	146·77½	135	208·57½	175	270·37½
16	24·72	56	86·52	96	148·32	136	210·12	176	271·92
17	26·26½	57	88·06½	97	149·86½	137	211·66½	177	273·46½
18	27·81	58	89·61	98	151·41	138	213·21	178	275·01
19	29·35½	59	91·15½	99	152·95½	139	214·75½	179	276·55½
20	30·90	**60**	92·70	**100**	154·50	**140**	216·30	**180**	278·10
21	32·44½	61	94·24½	101	156·04½	141	217·84½	181	279·64½
22	33·99	62	95·79	102	157·59	142	219·39	182	281·19
23	35·53½	63	97·33½	103	159·13½	143	220·93½	183	282·73½
24	37·08	64	98·88	104	160·68	144	222·48	184	284·28
25	38·62½	65	100·42½	105	162·22½	145	224·02½	185	285·82½
26	40·17	66	101·97	106	163·77	146	225·57	186	287·37
27	41·71½	67	103·51½	107	165·31½	147	227·11½	187	288·91½
28	43·26	68	105·06	108	166·86	148	228·66	188	290·46
29	44·80½	69	106·60½	109	168·40½	149	230·20½	189	292·00½
30	46·35	**70**	108·15	**110**	169·95	**150**	231·75	**190**	293·55
31	47·89½	71	109·69½	111	171·49½	151	233·29½	191	295·09½
32	49·44	72	111·24	112	173·04	152	234·84	192	296·64
33	50·98½	73	112·78½	113	174·58½	153	236·38½	193	298·18½
34	52·53	74	114·33	114	176·13	154	237·93	194	299·73
35	54·07½	75	115·87½	115	177·67½	155	239·47½	195	301·27½
36	55·62	76	117·42	116	179·22	156	241·02	196	302·82
37	57·16½	77	118·96½	117	180·76½	157	242·56½	197	304·36½
38	58·71	78	120·51	118	182·31	158	244·11	198	305·91
39	60·25½	79	122·05½	119	183·85½	159	245·65½	199	307·45½
40	61·80	**80**	123·60	**120**	185·40	**160**	247·20	**200**	309·00

$\frac{1}{16}=0{\cdot}09^{7}$	$\frac{1}{9}=0{\cdot}17^{2}$	$\frac{1}{8}=0{\cdot}19^{3}$	$\frac{1}{6}=0{\cdot}25^{8}$	$\frac{1}{5}=0{\cdot}30^{9}$	$\frac{1}{4}=0{\cdot}38^{6}$	$\frac{1}{3}=0{\cdot}51^{5}$
$\frac{3}{8}=0{\cdot}57^{9}$	$\frac{1}{2}=0{\cdot}77^{3}$	$\frac{5}{8}=0{\cdot}96^{6}$	$\frac{3}{4}=1{\cdot}15^{9}$	$\frac{7}{8}=1{\cdot}35^{2}$	$\frac{2}{3}=1{\cdot}03^{0}$	$\frac{5}{6}=1{\cdot}28^{8}$
$\frac{3}{16}=0{\cdot}29^{0}$	$\frac{5}{16}=0{\cdot}48^{3}$	$\frac{7}{16}=0{\cdot}67^{6}$	$\frac{9}{16}=0{\cdot}86^{9}$	$\frac{11}{16}=1{\cdot}06^{2}$	$\frac{13}{16}=1{\cdot}25^{5}$	$\frac{15}{16}=1{\cdot}44^{8}$

(0.645=1£) £1.55

1	1·55	41	63·55	81	125·55	121	187·55	161	249·55
2	3·10	42	65·10	82	127·10	122	189·10	162	251·10
3	4·65	43	66·65	83	128·65	123	190·65	163	252·65
4	6·20	44	68·20	84	130·20	124	192·20	164	254·20
5	7·75	45	69·75	85	131·75	125	193·75	165	255·75
6	9·30	46	71·30	86	133·30	126	195·30	166	257·30
7	10·85	47	72·85	87	134·85	127	196·85	167	258·85
8	12·40	48	74·40	88	136·40	128	198·40	168	260·40
9	13·95	49	75·95	89	137·95	129	199·95	169	261·95
10	15·50	**50**	77·50	**90**	139·50	**130**	201·50	**170**	263·50
11	17·05	51	79·05	91	141·05	131	203·05	171	265·05
12	18·60	52	80·60	92	142·60	132	204·60	172	266·60
13	20·15	53	82·15	93	144·15	133	206·15	173	268·15
14	21·70	54	83·70	94	145·70	134	207·70	174	269·70
15	23·25	55	85·25	95	147·25	135	209·25	175	271·25
16	24·80	56	86·80	96	148·80	136	210·80	176	272·80
17	26·35	57	88·35	97	150·35	137	212·35	177	274·35
18	27·90	58	89·90	98	151·90	138	213·90	178	275·90
19	29·45	59	91·45	99	153·45	139	215·45	179	277·45
20	31·00	**60**	93·00	**100**	155·00	**140**	217·00	**180**	279·00
21	32·55	61	94·55	101	156·55	141	218·55	181	280·55
22	34·10	62	96·10	102	158·10	142	220·10	182	282·10
23	35·65	63	97·65	103	159·65	143	221·65	183	283·65
24	37·20	64	99·20	104	161·20	144	223·20	184	285·20
25	38·75	65	100·75	105	162·75	145	224·75	185	286·75
26	40·30	66	102·30	106	164·30	146	226·30	186	288·30
27	41·85	67	103·85	107	165·85	147	227·85	187	289·85
28	43·40	68	105·40	108	167·40	148	229·40	188	291·40
29	44·95	69	106·95	109	168·95	149	230·95	189	292·95
30	46·50	**70**	108·50	**110**	170·50	**150**	232·50	**190**	294·50
31	48·05	71	110·05	111	172·05	151	234·05	191	296·05
32	49·60	72	111·60	112	173·60	152	235·60	192	297·60
33	51·15	73	113·15	113	175·15	153	237·15	193	299·15
34	52·70	74	114·70	114	176·70	154	238·70	194	300·70
35	54·25	75	116·25	115	178·25	155	240·25	195	302·25
36	55·80	76	117·80	116	179·80	156	241·80	196	303·80
37	57·35	77	119·35	117	181·35	157	243·35	197	305·35
38	58·90	78	120·90	118	182·90	158	244·90	198	306·90
39	60·45	79	122·45	119	184·45	159	246·45	199	308·45
40	62·00	**80**	124·00	**120**	186·00	**160**	248·00	**200**	310·00

$\frac{1}{16}=0{\cdot}09^{7}$ $\frac{1}{9}=0{\cdot}17^{2}$ $\frac{1}{8}=0{\cdot}19^{4}$ $\frac{1}{6}=0{\cdot}25^{8}$ $\frac{1}{5}=0{\cdot}31^{0}$ $\frac{1}{4}=0{\cdot}38^{8}$ $\frac{1}{3}=0{\cdot}51^{7}$

$\frac{3}{8}=0{\cdot}58^{1}$ $\frac{1}{2}=0{\cdot}77^{5}$ $\frac{5}{8}=0{\cdot}96^{9}$ $\frac{3}{4}=1{\cdot}16^{3}$ $\frac{7}{8}=1{\cdot}35^{6}$ $\frac{2}{3}=1{\cdot}03^{3}$ $\frac{5}{6}=1{\cdot}29^{2}$

$\frac{3}{16}=0{\cdot}29^{1}$ $\frac{5}{16}=0{\cdot}48^{4}$ $\frac{7}{16}=0{\cdot}67^{8}$ $\frac{9}{16}=0{\cdot}87^{2}$ $\frac{11}{16}=1{\cdot}06^{6}$ $\frac{13}{16}=1{\cdot}25^{9}$ $\frac{15}{16}=1{\cdot}45^{3}$

1 square centimetre=0·155 square inches

(0.643=1£) **£1.55½**

1	1·55½	41	63·75½	81	125·95½	121	188·15½	161	250·35½
2	3·11	42	65·31	82	127·51	122	189·71	162	251·91
3	4·66½	43	66·86½	83	129·06½	123	191·26½	163	253·46½
4	6·22	44	68·42	84	130·62	124	192·82	164	255·02
5	7·77½	45	69·97½	85	132·17½	125	194·37½	165	256·57½
6	9·33	46	71·53	86	133·73	126	195·93	166	258·13
7	10·88½	47	73·08½	87	135·28½	127	197·48½	167	259·68½
8	12·44	48	74·64	88	136·84	128	199·04	168	261·24
9	13·99½	49	76·19½	89	138·39½	129	200·59½	169	262·79½
10	15·55	**50**	77·75	**90**	139·95	**130**	202·15	**170**	264·35
11	17·10½	51	79·30½	91	141·50½	131	203·70½	171	265·90½
12	18·66	52	80·86	92	143·06	132	205·26	172	267·46
13	20·21½	53	82·41½	93	144·61½	133	206·81½	173	269·01½
14	21·77	54	83·97	94	146·17	134	208·37	174	270·57
15	23·32½	55	85·52½	95	147·72½	135	209·92½	175	272·12½
16	24·88	56	87·08	96	149·28	136	211·48	176	273·68
17	26·43½	57	88·63½	97	150·83½	137	213·03½	177	275·23½
18	27·99	58	90·19	98	152·39	138	214·59	178	276·79
19	29·54½	59	91·74½	99	153·94½	139	216·14½	179	278·34½
20	31·10	**60**	93·30	**100**	155·50	**140**	217·70	**180**	279·90
21	32·65½	61	94·85½	101	157·05½	141	219·25½	181	281·45½
22	34·21	62	96·41	102	158·61	142	220·81	182	283·01
23	35·76½	63	97·96½	103	160·16½	143	222·36½	183	284·56½
24	37·32	64	99·52	104	161·72	144	223·92	184	286·12
25	38·87½	65	101·07½	105	163·27½	145	225·47½	185	287·67½
26	40·43	66	102·63	106	164·83	146	227·03	186	289·23
27	41·98½	67	104·18½	107	166·38½	147	228·58½	187	290·78½
28	43·54	68	105·74	108	167·94	148	230·14	188	292·34
29	45·09½	69	107·29½	109	169·49½	149	231·69½	189	293·89½
30	46·65	**70**	108·85	**110**	171·05	**150**	233·25	**190**	295·45
31	48·20½	71	110·40½	111	172·60½	151	234·80½	191	297·00½
32	49·76	72	111·96	112	174·16	152	236·36	192	298·56
33	51·31½	73	113·51½	113	175·71½	153	237·91½	193	300·11½
34	52·87	74	115·07	114	177·27	154	239·47	194	301·67
35	54·42½	75	116·62½	115	178·82½	155	241·02½	195	303·22½
36	55·98	76	118·18	116	180·38	156	242·58	196	304·78
37	57·53½	77	119·73½	117	181·93½	157	244·13½	197	306·33½
38	59·09	78	121·29	118	183·49	158	245·69	198	307·89
39	60·64½	79	122·84½	119	185·04½	159	247·24½	199	309·44½
40	62·20	**80**	124·40	**120**	186·60	**160**	248·80	**200**	311·00

$\frac{1}{16}=0{\cdot}09^7$ $\frac{1}{9}=0{\cdot}17^3$ $\frac{1}{8}=0{\cdot}19^4$ $\frac{1}{6}=0{\cdot}25^9$ $\frac{1}{5}=0{\cdot}31^1$ $\frac{1}{4}=0{\cdot}38^9$ $\frac{1}{3}=0{\cdot}51^8$

$\frac{3}{8}=0{\cdot}58^3$ $\frac{1}{2}=0{\cdot}77^8$ $\frac{5}{8}=0{\cdot}97^2$ $\frac{3}{4}=1{\cdot}16^6$ $\frac{7}{8}=1{\cdot}36^1$ $\frac{2}{3}=1{\cdot}03^7$ $\frac{5}{6}=1{\cdot}29^6$

$\frac{3}{16}=0{\cdot}29^2$ $\frac{5}{16}=0{\cdot}48^6$ $\frac{7}{16}=0{\cdot}68^0$ $\frac{9}{16}=0{\cdot}87^5$ $\frac{11}{16}=1{\cdot}06^9$ $\frac{13}{16}=1{\cdot}26^3$ $\frac{15}{16}=1{\cdot}45^8$

1	1·56	41	63·96	81	126·36	121	188·76	161	251·16
2	3·12	42	65·52	82	127·92	122	190·32	162	252·72
3	4·68	43	67·08	83	129·48	123	191·88	163	254·28
4	6·24	44	68·64	84	131·04	124	193·44	164	255·84
5	7·80	45	70·20	85	132·60	125	195·00	165	257·40
6	9·36	46	71·76	86	134·16	126	196·56	166	258·96
7	10·92	47	73·32	87	135·72	127	198·12	167	260·52
8	12·48	48	74·88	88	137·28	128	199·68	168	262·08
9	14·04	49	76·44	89	138·84	129	201·24	169	263·64
10	15·60	**50**	78·00	**90**	140·40	**130**	202·80	**170**	265·20
11	17·16	51	79·56	91	141·96	131	204·36	171	266·76
12	18·72	52	81·12	92	143·52	132	205·92	172	268·32
13	20·28	53	82·68	93	145·08	133	207·48	173	269·88
14	21·84	54	84·24	94	146·64	134	209·04	174	271·44
15	23·40	55	85·80	95	148·20	135	210·60	175	273·00
16	24·96	56	87·36	96	149·76	136	212·16	176	274·56
17	26·52	57	88·92	97	151·32	137	213·72	177	276·12
18	28·08	58	90·48	98	152·88	138	215·28	178	277·68
19	29·64	59	92·04	99	154·44	139	216·84	179	279·24
20	31·20	**60**	93·60	**100**	156·00	**140**	218·40	**180**	280·80
21	32·76	61	95·16	101	157·56	141	219·96	181	282·36
22	34·32	62	96·72	102	159·12	142	221·52	182	283·92
23	35·88	63	98·28	103	160·68	143	223·08	183	285·48
24	37·44	64	99·84	104	162·24	144	224·64	184	287·04
25	39·00	65	101·40	105	163·80	145	226·20	185	288·60
26	40·56	66	102·96	106	165·36	146	227·76	186	290·16
27	42·12	67	104·52	107	166·92	147	229·32	187	291·72
28	43·68	68	106·08	108	168·48	148	230·88	188	293·28
29	45·24	69	107·64	109	170·04	149	232·44	189	294·84
30	46·80	**70**	109·20	**110**	171·60	**150**	234·00	**190**	296·40
31	48·36	71	110·76	111	173·16	151	235·56	191	297·96
32	49·92	72	112·32	112	174·72	152	237·12	192	299·52
33	51·48	73	113·88	113	176·28	153	238·68	193	301·08
34	53·04	74	115·44	114	177·84	154	240·24	194	302·64
35	54·60	75	117·00	115	179·40	155	241·80	195	304·20
36	56·16	76	118·56	116	180·96	156	243·36	196	305·76
37	57·72	77	120·12	117	182·52	157	244·92	197	307·32
38	59·28	78	121·68	118	184·08	158	246·48	198	308·88
39	60·84	79	123·24	119	185·64	159	248·04	199	310·44
40	62·40	**80**	124·80	**120**	187·20	**160**	249·60	**200**	312·00

$\frac{1}{16}=0{\cdot}09^{8}$ $\frac{1}{9}=0{\cdot}17^{3}$ $\frac{1}{8}=0{\cdot}19^{5}$ $\frac{1}{6}=0{\cdot}26^{0}$ $\frac{1}{5}=0{\cdot}31^{2}$ $\frac{1}{4}=0{\cdot}39^{0}$ $\frac{1}{3}=0{\cdot}52^{0}$

$\frac{3}{8}=0{\cdot}58^{5}$ $\frac{1}{2}=0{\cdot}78^{0}$ $\frac{5}{8}=0{\cdot}97^{5}$ $\frac{3}{4}=1{\cdot}17^{0}$ $\frac{7}{8}=1{\cdot}36^{5}$ $\frac{2}{3}=1{\cdot}04^{0}$ $\frac{5}{6}=1{\cdot}30^{0}$

$\frac{3}{16}=0{\cdot}29^{3}$ $\frac{5}{16}=0{\cdot}48^{8}$ $\frac{7}{16}=0{\cdot}68^{3}$ $\frac{9}{16}=0{\cdot}87^{8}$ $\frac{11}{16}=1{\cdot}07^{3}$ $\frac{13}{16}=1{\cdot}26^{8}$ $\frac{15}{16}=1{\cdot}46^{3}$

(0.639=1£) **£1.56½**

1	1·56½	41	64·16½	81	126·76½	121	189·36½	161	251·96½
2	3·13	42	65·73	82	128·33	122	190·93	162	253·53
3	4·69½	43	67·29½	83	129·89½	123	192·49½	163	255·09½
4	6·26	44	68·86	84	131·46	124	194·06	164	256·66
5	7·82½	45	70·42½	85	133·02½	125	195·62½	165	258·22½
6	9·39	46	71·99	86	134·59	126	197·19	166	259·79
7	10·95½	47	73·55½	87	136·15½	127	198·75½	167	261·35½
8	12·52	48	75·12	88	137·72	128	200·32	168	262·92
9	14·08½	49	76·68½	89	139·28½	129	201·88½	169	264·48½
10	15·65	**50**	78·25	**90**	140·85	**130**	203·45	**170**	266·05
11	17·21½	51	79·81½	91	142·41½	131	205·01½	171	267·61½
12	18·78	52	81·38	92	143·98	132	206·58	172	269·18
13	20·34½	53	82·94½	93	145·54½	133	208·14½	173	270·74½
14	21·91	54	84·51	94	147·11	134	209·71	174	272·31
15	23·47½	55	86·07½	95	148·67½	135	211·27½	175	273·87½
16	25·04	56	87·64	96	150·24	136	212·84	176	275·44
17	26·60½	57	89·20½	97	151·80½	137	214·40½	177	277·00½
18	28·17	58	90·77	98	153·37	138	215·97	178	278·57
19	29·73½	59	92·33½	99	154·93½	139	217·53½	179	280·13½
20	31·30	**60**	93·90	**100**	156·50	**140**	219·10	**180**	281·70
21	32·86½	61	95·46½	101	158·06½	141	220·66½	181	283·26½
22	34·43	62	97·03	102	159·63	142	222·23	182	284·83
23	35·99½	63	98·59½	103	161·19½	143	223·79½	183	286·39½
24	37·56	64	100·16	104	162·76	144	225·36	184	287·96
25	39·12½	65	101·72½	105	164·32½	145	226·92½	185	289·52½
26	40·69	66	103·29	106	165·89	146	228·49	186	291·09
27	42·25½	67	104·85½	107	167·45½	147	230·05½	187	292·65½
28	43·82	68	106·42	108	169·02	148	231·62	188	294·22
29	45·38½	69	107·98½	109	170·58½	149	233·18½	189	295·78½
30	46·95	**70**	109·55	**110**	172·15	**150**	234·75	**190**	297·35
31	48·51½	71	111·11½	111	173·71½	151	236·31½	191	298·91½
32	50·08	72	112·68	112	175·28	152	237·88	192	300·48
33	51·64½	73	114·24½	113	176·84½	153	239·44½	193	302·04½
34	53·21	74	115·81	114	178·41	154	241·01	194	303·61
35	54·77½	75	117·37½	115	179·97½	155	242·57½	195	305·17½
36	56·34	76	118·94	116	181·54	156	244·14	196	306·74
37	57·90½	77	120·50½	117	183·10½	157	245·70½	197	308·30½
38	59·47	78	122·07	118	184·67	158	247·27	198	309·87
39	61·03½	79	123·63½	119	186·23½	159	248·83½	199	311·43½
40	62·60	**80**	125·20	**120**	187·80	**160**	250·40	**200**	313·00

$\frac{1}{16}=0{\cdot}09^{8}$ $\frac{1}{9}=0{\cdot}17^{4}$ $\frac{1}{8}=0{\cdot}19^{6}$ $\frac{1}{6}=0{\cdot}26^{1}$ $\frac{1}{5}=0{\cdot}31^{3}$ $\frac{1}{4}=0{\cdot}39^{1}$ $\frac{1}{3}=0{\cdot}52^{2}$

$\frac{3}{8}=0{\cdot}58^{7}$ $\frac{1}{2}=0{\cdot}78^{3}$ $\frac{5}{8}=0{\cdot}97^{8}$ $\frac{3}{4}=1{\cdot}17^{4}$ $\frac{7}{8}=1{\cdot}36^{9}$ $\frac{2}{3}=1{\cdot}04^{3}$ $\frac{5}{6}=1{\cdot}30^{4}$

$\frac{3}{16}=0{\cdot}29^{3}$ $\frac{5}{16}=0{\cdot}48^{9}$ $\frac{7}{16}=0{\cdot}68^{5}$ $\frac{9}{16}=0{\cdot}88^{0}$ $\frac{11}{16}=1{\cdot}07^{6}$ $\frac{13}{16}=1{\cdot}27^{2}$ $\frac{15}{16}=1{\cdot}46^{7}$

(0.637=1£) **£1.57**

1	1·57	41	64·37	81	127·17	121	189·97	161	252·77
2	3·14	42	65·94	82	128·74	122	191·54	162	254·34
3	4·71	43	67·51	83	130·31	123	193·11	163	255·91
4	6·28	44	69·08	84	131·88	124	194·68	164	257·48
5	7·85	45	70·65	85	133·45	125	196·25	165	259·05
6	9·42	46	72·22	86	135·02	126	197·82	166	260·62
7	10·99	47	73·79	87	136·59	127	199·39	167	262·19
8	12·56	48	75·36	88	138·16	128	200·96	168	263·76
9	14·13	49	76·93	89	139·73	129	202·53	169	265·33
10	15·70	**50**	78·50	**90**	141·30	**130**	204·10	**170**	266·90
11	17·27	51	80·07	91	142·87	131	205·67	171	268·47
12	18·84	52	81·64	92	144·44	132	207·24	172	270·04
13	20·41	53	83·21	93	146·01	133	208·81	173	271·61
14	21·98	54	84·78	94	147·58	134	210·38	174	273·18
15	23·55	55	86·35	95	149·15	135	211·95	175	274·75
16	25·12	56	87·92	96	150·72	136	213·52	176	276·32
17	26·69	57	89·49	97	152·29	137	215·09	177	277·89
18	28·26	58	91·06	98	153·86	138	216·66	178	279·46
19	29·83	59	92·63	99	155·43	139	218·23	179	281·03
20	31·40	**60**	94·20	**100**	157·00	**140**	219·80	**180**	282·60
21	32·97	61	95·77	101	158·57	141	221·37	181	284·17
22	34·54	62	97·34	102	160·14	142	222·94	182	285·74
23	36·11	63	98·91	103	161·71	143	224·51	183	287·31
24	37·68	64	100·48	104	163·28	144	226·08	184	288·88
25	39·25	65	102·05	105	164·85	145	227·65	185	290·45
26	40·82	66	103·62	106	166·42	146	229·22	186	292·02
27	42·39	67	105·19	107	167·99	147	230·79	187	293·59
28	43·96	68	106·76	108	169·56	148	232·36	188	295·16
29	45·53	69	108·33	109	171·13	149	233·93	189	296·73
30	47·10	**70**	109·90	**110**	172·70	**150**	235·50	**190**	298·30
31	48·67	71	111·47	111	174·27	151	237·07	191	299·87
32	50·24	72	113·04	112	175·84	152	238·64	192	301·44
33	51·81	73	114·61	113	177·41	153	240·21	193	303·01
34	53·38	74	116·18	114	178·98	154	241·78	194	304·58
35	54·95	75	117·75	115	180·55	155	243·35	195	306·15
36	56·52	76	119·32	116	182·12	156	244·92	196	307·72
37	58·09	77	120·89	117	183·69	157	246·49	197	309·29
38	59·66	78	122·46	118	185·26	158	248·06	198	310·86
39	61·23	79	124·03	119	186·83	159	249·63	199	312·43
40	62·80	**80**	125·60	**120**	188·40	**160**	251·20	**200**	314·00

$\frac{1}{16}=0{\cdot}09^{8}$	$\frac{1}{9}=0{\cdot}17^{4}$	$\frac{1}{8}=0{\cdot}19^{6}$	$\frac{1}{6}=0{\cdot}26^{2}$	$\frac{1}{5}=0{\cdot}31^{4}$	$\frac{1}{4}=0{\cdot}39^{3}$	$\frac{1}{3}=0{\cdot}52^{3}$
$\frac{3}{8}=0{\cdot}58^{9}$	$\frac{1}{2}=0{\cdot}78^{5}$	$\frac{5}{8}=0{\cdot}98^{1}$	$\frac{3}{4}=1{\cdot}17^{8}$	$\frac{7}{8}=1{\cdot}37^{4}$	$\frac{2}{3}=1{\cdot}04^{7}$	$\frac{5}{6}=1{\cdot}30^{8}$
$\frac{3}{16}=0{\cdot}29^{4}$	$\frac{5}{16}=0{\cdot}49^{1}$	$\frac{7}{16}=0{\cdot}68^{7}$	$\frac{9}{16}=0{\cdot}88^{3}$	$\frac{11}{16}=1{\cdot}07^{9}$	$\frac{13}{16}=1{\cdot}27^{6}$	$\frac{15}{16}=1{\cdot}47^{2}$

(0.635=1£) **£1.57½**

1	1·57½	41	64·57½	81	127·57½	121	190·57½	161	253·57½
2	3·15	42	66·15	82	129·15	122	192·15	162	255·15
3	4·72½	43	67·72½	83	130·72½	123	193·72½	163	256·72½
4	6·30	44	69·30	84	132·30	124	195·30	164	258·30
5	7·87½	45	70·87½	85	133·87½	125	196·87½	165	259·87½
6	9·45	46	72·45	86	135·45	126	198·45	166	261·45
7	11·02½	47	74·02½	87	137·02½	127	200·02½	167	263·02½
8	12·60	48	75·60	88	138·60	128	201·60	168	264·60
9	14·17½	49	77·17½	89	140·17½	129	203·17½	169	266·17½
10	15·75	**50**	78·75	**90**	141·75	**130**	204·75	**170**	267·75
11	17·32½	51	80·32½	91	143·32½	131	206·32½	171	269·32½
12	18·90	52	81·90	92	144·90	132	207·90	172	270·90
13	20·47½	53	83·47½	93	146·47½	133	209·47½	173	272·47½
14	22·05	54	85·05	94	148·05	134	211·05	174	274·05
15	23·62½	55	86·62½	95	149·62½	135	212·62½	175	275·62½
16	25·20	56	88·20	96	151·20	136	214·20	176	277·20
17	26·77½	57	89·77½	97	152·77½	137	215·77½	177	278·77½
18	28·35	58	91·35	98	154·35	138	217·35	178	280·35
19	29·92½	59	92·92½	99	155·92½	139	218·92½	179	281·92½
20	31·50	**60**	94·50	**100**	157·50	**140**	220·50	**180**	283·50
21	33·07½	61	96·07½	101	159·07½	141	222·07½	181	285·07½
22	34·65	62	97·65	102	160·65	142	223·65	182	286·65
23	36·22½	63	99·22½	103	162·22½	143	225·22½	183	288·22½
24	37·80	64	100·80	104	163·80	144	226·80	184	289·80
25	39·37½	65	102·37½	105	165·37½	145	228·37½	185	291·37½
26	40·95	66	103·95	106	166·95	146	229·95	186	292·95
27	42·52½	67	105·52½	107	168·52½	147	231·52½	187	294·52½
28	44·10	68	107·10	108	170·10	148	233·10	188	296·10
29	45·67½	69	108·67½	109	171·67½	149	234·67½	189	297·67½
30	47·25	**70**	110·25	**110**	173·25	**150**	236·25	**190**	299·25
31	48·82½	71	111·82½	111	174·82½	151	237·82½	191	300·82½
32	50·40	72	113·40	112	176·40	152	239·40	192	302·40
33	51·97½	73	114·97½	113	177·97½	153	240·97½	193	303·97½
34	53·55	74	116·55	114	179·55	154	242·55	194	305·55
35	55·12½	75	118·12½	115	181·12½	155	244·12½	195	307·12½
36	56·70	76	119·70	116	182·70	156	245·70	196	308·70
37	58·27½	77	121·27½	117	184·27½	157	247·27½	197	310·27½
38	59·85	78	122·85	118	185·85	158	248·85	198	311·85
39	61·42½	79	124·42½	119	187·42½	159	250·42½	199	313·42½
40	63·00	**80**	126·00	**120**	189·00	**160**	252·00	**200**	315·00

$\frac{1}{16}=0{\cdot}09^{8}$ $\frac{1}{9}=0{\cdot}17^{5}$ $\frac{1}{8}=0{\cdot}19^{7}$ $\frac{1}{6}=0{\cdot}26^{3}$ $\frac{1}{5}=0{\cdot}31^{5}$ $\frac{1}{4}=0{\cdot}39^{4}$ $\frac{1}{3}=0{\cdot}52^{5}$

$\frac{3}{8}=0{\cdot}59^{1}$ $\frac{1}{2}=0{\cdot}78^{8}$ $\frac{5}{8}=0{\cdot}98^{4}$ $\frac{3}{4}=1{\cdot}18^{1}$ $\frac{7}{8}=1{\cdot}37^{8}$ $\frac{2}{3}=1{\cdot}05^{0}$ $\frac{5}{6}=1{\cdot}31^{3}$

$\frac{3}{16}=0{\cdot}29^{5}$ $\frac{5}{16}=0{\cdot}49^{2}$ $\frac{7}{16}=0{\cdot}68^{9}$ $\frac{9}{16}=0{\cdot}88^{6}$ $\frac{11}{16}=1{\cdot}08^{3}$ $\frac{13}{16}=1{\cdot}28^{0}$ $\frac{15}{16}=1{\cdot}47^{7}$

(0.633=1£) **£1.58**

1	1·58	41	64·78	81	127·98	121	191·18	161	254·38
2	3·16	42	66·36	82	129·56	122	192·76	162	255·96
3	4·74	43	67·94	83	131·14	123	194·34	163	257·54
4	6·32	44	69·52	84	132·72	124	195·92	164	259·12
5	7·90	45	71·10	85	134·30	125	197·50	165	260·70
6	9·48	46	72·68	86	135·88	126	199·08	166	262·28
7	11·06	47	74·26	87	137·46	127	200·66	167	263·86
8	12·64	48	75·84	88	139·04	128	202·24	168	265·44
9	14·22	49	77·42	89	140·62	129	203·82	169	267·02
10	15·80	**50**	79·00	**90**	142·20	**130**	205·40	**170**	268·60
11	17·38	51	80·58	91	143·78	131	206·98	171	270·18
12	18·96	52	82·16	92	145·36	132	208·56	172	271·76
13	20·54	53	83·74	93	146·94	133	210·14	173	273·34
14	22·12	54	85·32	94	148·52	134	211·72	174	274·92
15	23·70	55	86·90	95	150·10	135	213·30	175	276·50
16	25·28	56	88·48	96	151·68	136	214·88	176	278·08
17	26·86	57	90·06	97	153·26	137	216·46	177	279·66
18	28·44	58	91·64	98	154·84	138	218·04	178	281·24
19	30·02	59	93·22	99	156·42	139	219·62	179	282·82
20	31·60	**60**	94·80	**100**	158·00	**140**	221·20	**180**	284·40
21	33·18	61	96·38	101	159·58	141	222·78	181	285·98
22	34·76	62	97·96	102	161·16	142	224·36	182	287·56
23	36·34	63	99·54	103	162·74	143	225·94	183	289·14
24	37·92	64	101·12	104	164·32	144	227·52	184	290·72
25	39·50	65	102·70	105	165·90	145	229·10	185	292·30
26	41·08	66	104·28	106	167·48	146	230·68	186	293·88
27	42·66	67	105·86	107	169·06	147	232·26	187	295·46
28	44·24	68	107·44	108	170·64	148	233·84	188	297·04
29	45·82	69	109·02	109	172·22	149	235·42	189	298·62
30	47·40	**70**	110·60	**110**	173·80	**150**	237·00	**190**	300·20
31	48·98	71	112·18	111	175·38	151	238·58	191	301·78
32	50·56	72	113·76	112	176·96	152	240·16	192	303·36
33	52·14	73	115·34	113	178·54	153	241·74	193	304·94
34	53·72	74	116·92	114	180·12	154	243·32	194	306·52
35	55·30	75	118·50	115	181·70	155	244·90	195	308·10
36	56·88	76	120·08	116	183·28	156	246·48	196	309·68
37	58·46	77	121·66	117	184·86	157	248·06	197	311·26
38	60·04	78	123·24	118	186·44	158	249·64	198	312·84
39	61·62	79	124·82	119	188·02	159	251·22	199	314·42
40	63·20	**80**	126·40	**120**	189·60	**160**	252·80	**200**	316·00

$\frac{1}{16}$=0·09^{9} $\frac{1}{9}$=0·17^{6} $\frac{1}{8}$=0·19^{8} $\frac{1}{6}$=0·26^{3} $\frac{1}{5}$=0·31^{6} $\frac{1}{4}$=0·39^{5} $\frac{1}{3}$=0·52^{7}

$\frac{3}{8}$=0·59^{3} $\frac{1}{2}$=0·79^{0} $\frac{5}{8}$=0·98^{8} $\frac{3}{4}$=1·18^{5} $\frac{7}{8}$=1·38^{3} $\frac{2}{3}$=1·05^{3} $\frac{5}{6}$=1·31^{7}

$\frac{3}{16}$=0·29^{6} $\frac{5}{16}$=0·49^{4} $\frac{7}{16}$=0·69^{1} $\frac{9}{16}$=0·88^{9} $\frac{11}{16}$=1·08^{6} $\frac{13}{16}$=1·28^{4} $\frac{15}{16}$=1·48^{1}

1	1·58½	41	64·98½	81	128·38½	121	191·78½	161	255·18½
2	3·17	42	66·57	82	129·97	122	193·37	162	256·77
3	4·75½	43	68·15½	83	131·55½	123	194·95½	163	258·35½
4	6·34	44	69·74	84	133·14	124	196·54	164	259·94
5	7·92½	45	71·32½	85	134·72½	125	198·12½	165	261·52½
6	9·51	46	72·91	86	136·31	126	199·71	166	263·11
7	11·09½	47	74·49½	87	137·89½	127	201·29½	167	264·69½
8	12·68	48	76·08	88	139·48	128	202·88	168	266·28
9	14·26½	49	77·66½	89	141·06½	129	204·46½	169	267·86½
10	15·85	**50**	79·25	**90**	142·65	**130**	206·05	**170**	269·45
11	17·43½	51	80·83½	91	144·23½	131	207·63½	171	271·03½
12	19·02	52	82·42	92	145·82	132	209·22	172	272·62
13	20·60½	53	84·00½	93	147·40½	133	210·80½	173	274·20½
14	22·19	54	85·59	94	148·99	134	212·39	174	275·79
15	23·77½	55	87·17½	95	150·57½	135	213·97½	175	277·37½
16	25·36	56	88·76	96	152·16	136	215·56	176	278·96
17	26·94½	57	90·34½	97	153·74½	137	217·14½	177	280·54½
18	28·53	58	91·93	98	155·33	138	218·73	178	282·13
19	30·11½	59	93·51½	99	156·91½	139	220·31½	179	283·71½
20	31·70	**60**	95·10	**100**	158·50	**140**	221·90	**180**	285·30
21	33·28½	61	96·68½	101	160·08½	141	223·48½	181	286·88½
22	34·87	62	98·27	102	161·67	142	225·07	182	288·47
23	36·45½	63	99·85½	103	163·25½	143	226·65½	183	290·05½
24	38·04	64	101·44	104	164·84	144	228·24	184	291·64
25	39·62½	65	103·02½	105	166·42½	145	229·82½	185	293·22½
26	41·21	66	104·61	106	168·01	146	231·41	186	294·81
27	42·79½	67	106·19½	107	169·59½	147	232·99½	187	296·39½
28	44·38	68	107·78	108	171·18	148	234·58	188	297·98
29	45·96½	69	109·36½	109	172·76½	149	236·16½	189	299·56½
30	47·55	**70**	110·95	**110**	174·35	**150**	237·75	**190**	301·15
31	49·13½	71	112·53½	111	175·93½	151	239·33½	191	302·73½
32	50·72	72	114·12	112	177·52	152	240·92	192	304·32
33	52·30½	73	115·70½	113	179·10½	153	242·50½	193	305·90½
34	53·89	74	117·29	114	180·69	154	244·09	194	307·49
35	55·47½	75	118·87½	115	182·27½	155	245·67½	195	309·07½
36	57·06	76	120·46	116	183·86	156	247·26	196	310·66
37	58·64½	77	122·04½	117	185·44½	157	248·84½	197	312·24½
38	60·23	78	123·63	118	187·03	158	250·43	198	313·83
39	61·81½	79	125·21½	119	188·61½	159	252·01½	199	315·41½
40	63·40	**80**	126·80	**120**	190·20	**160**	253·60	**200**	317·00

$\frac{1}{16}=0{\cdot}09^{9}$ $\frac{1}{9}=0{\cdot}17^{6}$ $\frac{1}{8}=0{\cdot}19^{8}$ $\frac{1}{6}=0{\cdot}26^{4}$ $\frac{1}{5}=0{\cdot}31^{7}$ $\frac{1}{4}=0{\cdot}39^{6}$ $\frac{1}{3}=0{\cdot}52^{8}$

$\frac{3}{8}=0{\cdot}59^{4}$ $\frac{1}{2}=0{\cdot}79^{3}$ $\frac{5}{8}=0{\cdot}99^{1}$ $\frac{3}{4}=1{\cdot}18^{9}$ $\frac{7}{8}=1{\cdot}38^{7}$ $\frac{2}{3}=1{\cdot}05^{7}$ $\frac{5}{6}=1{\cdot}32^{1}$

$\frac{3}{16}=0{\cdot}29^{7}$ $\frac{5}{16}=0{\cdot}49^{5}$ $\frac{7}{16}=0{\cdot}69^{3}$ $\frac{9}{16}=0{\cdot}89^{2}$ $\frac{11}{16}=1{\cdot}09^{0}$ $\frac{13}{16}=1{\cdot}28^{8}$ $\frac{15}{16}=1{\cdot}48^{6}$

(0.629=1£) **£1.59**

1	1·59	41	65·19	81	128·79	121	192·39	161	255·99
2	3·18	42	66·78	82	130·38	122	193·98	162	257·58
3	4·77	43	68·37	83	131·97	123	195·57	163	259·17
4	6·36	44	69·96	84	133·56	124	197·16	164	260·76
5	7·95	45	71·55	85	135·15	125	198·75	165	262·35
6	9·54	46	73·14	86	136·74	126	200·34	166	263·94
7	11·13	47	74·73	87	138·33	127	201·93	167	265·53
8	12·72	48	76·32	88	139·92	128	203·52	168	267·12
9	14·31	49	77·91	89	141·51	129	205·11	169	268·71
10	15·90	**50**	79·50	**90**	143·10	**130**	206·70	**170**	270·30
11	17·49	51	81·09	91	144·69	131	208·29	171	271·89
12	19·08	52	82·68	92	146·28	132	209·88	172	273·48
13	20·67	53	84·27	93	147·87	133	211·47	173	275·07
14	22·26	54	85·86	94	149·46	134	213·06	174	276·66
15	23·85	55	87·45	95	151·05	135	214·65	175	278·25
16	25·44	56	89·04	96	152·64	136	216·24	176	279·84
17	27·03	57	90·63	97	154·23	137	217·83	177	281·43
18	28·62	58	92·22	98	155·82	138	219·42	178	283·02
19	30·21	59	93·81	99	157·41	139	221·01	179	284·61
20	31·80	**60**	95·40	**100**	159·00	**140**	222·60	**180**	286·20
21	33·39	61	96·99	101	160·59	141	224·19	181	287·79
22	34·98	62	98·58	102	162·18	142	225·78	182	289·38
23	36·57	63	100·17	103	163·77	143	227·37	183	290·97
24	38·16	64	101·76	104	165·36	144	228·96	184	292·56
25	39·75	65	103·35	105	166·95	145	230·55	185	294·15
26	41·34	66	104·94	106	168·54	146	232·14	186	295·74
27	42·93	67	106·53	107	170·13	147	233·73	187	297·33
28	44·52	68	108·12	108	171·72	148	235·32	188	298·92
29	46·11	69	109·71	109	173·31	149	236·91	189	300·51
30	47·70	**70**	111·30	**110**	174·90	**150**	238·50	**190**	302·10
31	49·29	71	112·89	111	176·49	151	240·09	191	303·69
32	50·88	72	114·48	112	178·08	152	241·68	192	305·28
33	52·47	73	116·07	113	179·67	153	243·27	193	306·87
34	54·06	74	117·66	114	181·26	154	244·86	194	308·46
35	55·65	75	119·25	115	182·85	155	246·45	195	310·05
36	57·24	76	120·84	116	184·44	156	248·04	196	311·64
37	58·83	77	122·43	117	186·03	157	249·63	197	313·23
38	60·42	78	124·02	118	187·62	158	251·22	198	314·82
39	62·01	79	125·61	119	189·21	159	252·81	199	316·41
40	63·60	**80**	127·20	**120**	190·80	**160**	254·40	**200**	318·00

$\frac{1}{16}=0{\cdot}09^{9}$ $\frac{1}{9}=0{\cdot}17^{7}$ $\frac{1}{8}=0{\cdot}19^{9}$ $\frac{1}{6}=0{\cdot}26^{5}$ $\frac{1}{5}=0{\cdot}31^{8}$ $\frac{1}{4}=0{\cdot}39^{8}$ $\frac{1}{3}=0{\cdot}53^{0}$

$\frac{3}{8}=0{\cdot}59^{6}$ $\frac{1}{2}=0{\cdot}79^{5}$ $\frac{5}{8}=0{\cdot}99^{4}$ $\frac{3}{4}=1{\cdot}19^{3}$ $\frac{7}{8}=1{\cdot}39^{1}$ $\frac{2}{3}=1{\cdot}06^{0}$ $\frac{5}{6}=1{\cdot}32^{5}$

$\frac{3}{16}=0{\cdot}29^{8}$ $\frac{5}{16}=0{\cdot}49^{7}$ $\frac{7}{16}=0{\cdot}69^{6}$ $\frac{9}{16}=0{\cdot}89^{4}$ $\frac{11}{16}=1{\cdot}09^{3}$ $\frac{13}{16}=1{\cdot}29^{2}$ $\frac{15}{16}=1{\cdot}49^{1}$

(0.627=1£) **£1.59½**

1	1·59½	41	65·39½	81	129·19½	121	192·99½	161	256·79½
2	3·19	42	66·99	82	130·79	122	194·59	162	258·39
3	4·78½	43	68·58½	83	132·38½	123	196·18½	163	259·98½
4	6·38	44	70·18	84	133·98	124	197·78	164	261·58
5	7·97½	45	71·77½	85	135·57½	125	199·37½	165	263·17½
6	9·57	46	73·37	86	137·17	126	200·97	166	264·77
7	11·16½	47	74·96½	87	138·76½	127	202·56½	167	266·36½
8	12·76	48	76·56	88	140·36	128	204·16	168	267·96
9	14·35½	49	78·15½	89	141·95½	129	205·75½	169	269·55½
10	15·95	**50**	79·75	**90**	143·55	**130**	207·35	**170**	271·15
11	17·54½	51	81·34½	91	145·14½	131	208·94½	171	272·74½
12	19·14	52	82·94	92	146·74	132	210·54	172	274·34
13	20·73½	53	84·53½	93	148·33½	133	212·13½	173	275·93½
14	22·33	54	86·13	94	149·93	134	213·73	174	277·53
15	23·92½	55	87·72½	95	151·52½	135	215·32½	175	279·12½
16	25·52	56	89·32	96	153·12	136	216·92	176	280·72
17	27·11½	57	90·91½	97	154·71½	137	218·51½	177	282·31½
18	28·71	58	92·51	98	156·31	138	220·11	178	283·91
19	30·30½	59	94·10½	99	157·90½	139	221·70½	179	285·50½
20	31·90	**60**	95·70	**100**	159·50	**140**	223·30	**180**	287·10
21	33·49½	61	97·29½	101	161·09½	141	224·89½	181	288·69½
22	35·09	62	98·89	102	162·69	142	226·49	182	290·29
23	36·68½	63	100·48½	103	164·28½	143	228·08½	183	291·88½
24	38·28	64	102·08	104	165·88	144	229·68	184	293·48
25	39·87½	65	103·67½	105	167·47½	145	231·27½	185	295·07½
26	41·47	66	105·27	106	169·07	146	232·87	186	296·67
27	43·06½	67	106·86½	107	170·66½	147	234·46½	187	298·26½
28	44·66	68	108·46	108	172·26	148	236·06	188	299·86
29	46·25½	69	110·05½	109	173·85½	149	237·65½	189	301·45½
30	47·85	**70**	111·65	**110**	175·45	**150**	239·25	**190**	303·05
31	49·44½	71	113·24½	111	177·04½	151	240·84½	191	304·64½
32	51·04	72	114·84	112	178·64	152	242·44	192	306·24
33	52·63½	73	116·43½	113	180·23½	153	244·03½	193	307·83½
34	54·23	74	118·03	114	181·83	154	245·63	194	309·43
35	55·82½	75	119·62½	115	183·42½	155	247·22½	195	311·02½
36	57·42	76	121·22	116	185·02	156	248·82	196	312·62
37	59·01½	77	122·81½	117	186·61½	157	250·41½	197	314·21½
38	60·61	78	124·41	118	188·21	158	252·01	198	315·81
39	62·20½	79	126·00½	119	189·80½	159	253·60½	199	317·40½
40	63·80	**80**	127·60	**120**	191·40	**160**	255·20	**200**	319·00

$\frac{1}{16}=0{\cdot}10^{0}$ $\frac{1}{9}=0{\cdot}17^{7}$ $\frac{1}{8}=0{\cdot}19^{9}$ $\frac{1}{6}=0{\cdot}26^{6}$ $\frac{1}{5}=0{\cdot}31^{9}$ $\frac{1}{4}=0{\cdot}39^{9}$ $\frac{1}{3}=0{\cdot}53^{2}$

$\frac{3}{8}=0{\cdot}59^{8}$ $\frac{1}{2}=0{\cdot}79^{8}$ $\frac{5}{8}=0{\cdot}99^{7}$ $\frac{3}{4}=1{\cdot}19^{6}$ $\frac{7}{8}=1{\cdot}39^{6}$ $\frac{2}{3}=1{\cdot}06^{3}$ $\frac{5}{6}=1{\cdot}32^{9}$

$\frac{3}{16}=0{\cdot}29^{9}$ $\frac{5}{16}=0{\cdot}49^{8}$ $\frac{7}{16}=0{\cdot}69^{8}$ $\frac{9}{16}=0{\cdot}89^{7}$ $\frac{11}{16}=1{\cdot}09^{7}$ $\frac{13}{16}=1{\cdot}29^{6}$ $\frac{15}{16}=1{\cdot}49^{5}$

(0.625=1£) **£1.60**

1	1·60	41	65·60	81	129·60	121	193·60	161	257·60
2	3·20	42	67·20	82	131·20	122	195·20	162	259·20
3	4·80	43	68·80	83	132·80	123	196·80	163	260·80
4	6·40	44	70·40	84	134·40	124	198·40	164	262·40
5	8·00	45	72·00	85	136·00	125	200·00	165	264·00
6	9·60	46	73·60	86	137·60	126	201·60	166	265·60
7	11·20	47	75·20	87	139·20	127	203·20	167	267·20
8	12·80	48	76·80	88	140·80	128	204·80	168	268·80
9	14·40	49	78·40	89	142·40	129	206·40	169	270·40
10	16·00	**50**	80·00	**90**	144·00	**130**	208·00	**170**	272·00
11	17·60	51	81·60	91	145·60	131	209·60	171	273·60
12	19·20	52	83·20	92	147·20	132	211·20	172	275·20
13	20·80	53	84·80	93	148·80	133	212·80	173	276·80
14	22·40	54	86·40	94	150·40	134	214·40	174	278·40
15	24·00	55	88·00	95	152·00	135	216·00	175	280·00
16	25·60	56	89·60	96	153·60	136	217·60	176	281·60
17	27·20	57	91·20	97	155·20	137	219·20	177	283·20
18	28·80	58	92·80	98	156·80	138	220·80	178	284·80
19	30·40	59	94·40	99	158·40	139	222·40	179	286·40
20	32·00	**60**	96·00	**100**	160·00	**140**	224·00	**180**	288·00
21	33·60	61	97·60	101	161·60	141	225·60	181	289·60
22	35·20	62	99·20	102	163·20	142	227·20	182	291·20
23	36·80	63	100·80	103	164·80	143	228·80	183	292·80
24	38·40	64	102·40	104	166·40	144	230·40	184	294·40
25	40·00	65	104·00	105	168·00	145	232·00	185	296·00
26	41·60	66	105·60	106	169·60	146	233·60	186	297·60
27	43·20	67	107·20	107	171·20	147	235·20	187	299·20
28	44·80	68	108·80	108	172·80	148	236·80	188	300·80
29	46·40	69	110·40	109	174·40	149	238·40	189	302·40
30	48·00	**70**	112·00	**110**	176·00	**150**	240·00	**190**	304·00
31	49·60	71	113·60	111	177·60	151	241·60	191	305·60
32	51·20	72	115·20	112	179·20	152	243·20	192	307·20
33	52·80	73	116·80	113	180·80	153	244·80	193	308·80
34	54·40	74	118·40	114	182·40	154	246·40	194	310·40
35	56·00	75	120·00	115	184·00	155	248·00	195	312·00
36	57·60	76	121·60	116	185·60	156	249·60	196	313·60
37	59·20	77	123·20	117	187·20	157	251·20	197	315·20
38	60·80	78	124·80	118	188·80	158	252·80	198	316·80
39	62·40	79	126·40	119	190·40	159	254·40	199	318·40
40	64·00	**80**	128·00	**120**	192·00	**160**	256·00	**200**	320·00

$\frac{1}{16}=0{\cdot}10^{0}$ $\frac{1}{9}=0{\cdot}17^{8}$ $\frac{1}{8}=0{\cdot}20^{0}$ $\frac{1}{6}=0{\cdot}26^{7}$ $\frac{1}{5}=0{\cdot}32^{0}$ $\frac{1}{4}=0{\cdot}40^{0}$ $\frac{1}{3}=0{\cdot}53^{3}$

$\frac{3}{8}=0{\cdot}60^{0}$ $\frac{1}{2}=0{\cdot}80^{0}$ $\frac{5}{8}=1{\cdot}00^{0}$ $\frac{3}{4}=1{\cdot}20^{0}$ $\frac{7}{8}=1{\cdot}40^{0}$ $\frac{2}{3}=1{\cdot}06^{7}$ $\frac{5}{6}=1{\cdot}33^{3}$

$\frac{3}{16}=0{\cdot}30^{0}$ $\frac{5}{16}=0{\cdot}50^{0}$ $\frac{7}{16}=0{\cdot}70^{0}$ $\frac{9}{16}=0{\cdot}90^{0}$ $\frac{11}{16}=1{\cdot}10^{0}$ $\frac{13}{16}=1{\cdot}30^{0}$ $\frac{15}{16}=1{\cdot}50^{0}$

1	1·60½	41	65·80½	81	130·00½	121	194·20½	161	258·40½
2	3·21	42	67·41	82	131·61	122	195·81	162	260·01
3	4·81½	43	69·01½	83	133·21½	123	197·41½	163	261·61½
4	6·42	44	70·62	84	134·82	124	199·02	164	263·22
5	8·02½	45	72·22½	85	136·42½	125	200·62½	165	264·82½
6	9·63	46	73·83	86	138·03	126	202·23	166	266·43
7	11·23½	47	75·43½	87	139·63½	127	203·83½	167	268·03½
8	12·84	48	77·04	88	141·24	128	205·44	168	269·64
9	14·44½	49	78·64½	89	142·84½	129	207·04½	169	271·24½
10	16·05	**50**	80·25	**90**	144·45	**130**	208·65	**170**	272·85
11	17·65½	51	81·85½	91	146·05½	131	210·25½	171	274·45½
12	19·26	52	83·46	92	147·66	132	211·86	172	276·06
13	20·86½	53	85·06½	93	149·26½	133	213·46½	173	277·66½
14	22·47	54	86·67	94	150·87	134	215·07	174	279·27
15	24·07½	55	88·27½	95	152·47½	135	216·67½	175	280·87½
16	25·68	56	89·88	96	154·08	136	218·28	176	282·48
17	27·28½	57	91·48½	97	155·68½	137	219·88½	177	284·08½
18	28·89	58	93·09	98	157·29	138	221·49	178	285·69
19	30·49½	59	94·69½	99	158·89½	139	223·09½	179	287·29½
20	32·10	**60**	96·30	**100**	160·50	**140**	224·70	**180**	288·90
21	33·70½	61	97·90½	101	162·10½	141	226·30½	181	290·50½
22	35·31	62	99·51	102	163·71	142	227·91	182	292·11
23	36·91½	63	101·11½	103	165·31½	143	229·51½	183	293·71½
24	38·52	64	102·72	104	166·92	144	231·12	184	295·32
25	40·12½	65	104·32½	105	168·52½	145	232·72½	185	296·92½
26	41·73	66	105·93	106	170·13	146	234·33	186	298·53
27	43·33½	67	107·53½	107	171·73½	147	235·93½	187	300·13½
28	44·94	68	109·14	108	173·34	148	237·54	188	301·74
29	46·54½	69	110·74½	109	174·94½	149	239·14½	189	303·34½
30	48·15	**70**	112·35	**110**	176·55	**150**	240·75	**190**	304·95
31	49·75½	71	113·95½	111	178·15½	151	242·35½	191	306·55½
32	51·36	72	115·56	112	179·76	152	243·96	192	308·16
33	52·96½	73	117·16½	113	181·36½	153	245·56½	193	309·76½
34	54·57	74	118·77	114	182·97	154	247·17	194	311·37
35	56·17½	75	120·37½	115	184·57½	155	248·77½	195	312·97½
36	57·78	76	121·98	116	186·18	156	250·38	196	314·58
37	59·38½	77	123·58½	117	187·78½	157	251·98½	197	316·18½
38	60·99	78	125·19	118	189·39	158	253·59	198	317·79
39	62·59½	79	126·79½	119	190·99½	159	255·19½	199	319·39½
40	64·20	**80**	128·40	**120**	192·60	**160**	256·80	**200**	321·00

$\frac{1}{16}=0{\cdot}10^{0}$	$\frac{1}{9}=0{\cdot}17^{8}$	$\frac{1}{8}=0{\cdot}20^{1}$	$\frac{1}{6}=0{\cdot}26^{8}$	$\frac{1}{5}=0{\cdot}32^{1}$	$\frac{1}{4}=0{\cdot}40^{1}$	$\frac{1}{3}=0{\cdot}53^{5}$
$\frac{3}{8}=0{\cdot}60^{2}$	$\frac{1}{2}=0{\cdot}80^{3}$	$\frac{5}{8}=1{\cdot}00^{3}$	$\frac{3}{4}=1{\cdot}20^{4}$	$\frac{7}{8}=1{\cdot}40^{4}$	$\frac{2}{3}=1{\cdot}07^{0}$	$\frac{5}{6}=1{\cdot}33^{8}$
$\frac{3}{16}=0{\cdot}30^{1}$	$\frac{5}{16}=0{\cdot}50^{2}$	$\frac{7}{16}=0{\cdot}70^{2}$	$\frac{9}{16}=0{\cdot}90^{3}$	$\frac{11}{16}=1{\cdot}10^{3}$	$\frac{13}{16}=1{\cdot}30^{4}$	$\frac{15}{16}=1{\cdot}50^{5}$

(0.621=1£) **£1.61**

1	1·61	41	66·01	81	130·41	121	194·81	161	259·21
2	3·22	42	67·62	82	132·02	122	196·42	162	260·82
3	4·83	43	69·23	83	133·63	123	198·03	163	262·43
4	6·44	44	70·84	84	135·24	124	199·64	164	264·04
5	8·05	45	72·45	85	136·85	125	201·25	165	265·65
6	9·66	46	74·06	86	138·46	126	202·86	166	267·26
7	11·27	47	75·67	87	140·07	127	204·47	167	268·87
8	12·88	48	77·28	88	141·68	128	206·08	168	270·48
9	14·49	49	78·89	89	143·29	129	207·69	169	272·09
10	16·10	**50**	80·50	**90**	144·90	**130**	209·30	**170**	273·70
11	17·71	51	82·11	91	146·51	131	210·91	171	275·31
12	19·32	52	83·72	92	148·12	132	212·52	172	276·92
13	20·93	53	85·33	93	149·73	133	214·13	173	278·53
14	22·54	54	86·94	94	151·34	134	215·74	174	280·14
15	24·15	55	88·55	95	152·95	135	217·35	175	281·75
16	25·76	56	90·16	96	154·56	136	218·96	176	283·36
17	27·37	57	91·77	97	156·17	137	220·57	177	284·97
18	28·98	58	93·38	98	157·78	138	222·18	178	286·58
19	30·59	59	94·99	99	159·39	139	223·79	179	288·19
20	32·20	**60**	96·60	**100**	161·00	**140**	225·40	**180**	289·80
21	33·81	61	98·21	101	162·61	141	227·01	181	291·41
22	35·42	62	99·82	102	164·22	142	228·62	182	293·02
23	37·03	63	101·43	103	165·83	143	230·23	183	294·63
24	38·64	64	103·04	104	167·44	144	231·84	184	296·24
25	40·25	65	104·65	105	169·05	145	233·45	185	297·85
26	41·86	66	106·26	106	170·66	146	235·06	186	299·46
27	43·47	67	107·87	107	172·27	147	236·67	187	301·07
28	45·08	68	109·48	108	173·88	148	238·28	188	302·68
29	46·69	69	111·09	109	175·49	149	239·89	189	304·29
30	48·30	**70**	112·70	**110**	177·10	**150**	241·50	**190**	305·90
31	49·91	71	114·31	111	178·71	151	243·11	191	307·51
32	51·52	72	115·92	112	180·32	152	244·72	192	309·12
33	53·13	73	117·53	113	181·93	153	246·33	193	310·73
34	54·74	74	119·14	114	183·54	154	247·94	194	312·34
35	56·35	75	120·75	115	185·15	155	249·55	195	313·95
36	57·96	76	122·36	116	186·76	156	251·16	196	315·56
37	59·57	77	123·97	117	188·37	157	252·77	197	317·17
38	61·18	78	125·58	118	189·98	158	254·38	198	318·78
39	62·79	79	127·19	119	191·59	159	255·99	199	320·39
40	64·40	**80**	128·80	**120**	193·20	**160**	257·60	**200**	322·00

$\frac{1}{16}=0{\cdot}10^{1}$ $\frac{1}{9}=0{\cdot}17^{9}$ $\frac{1}{8}=0{\cdot}20^{1}$ $\frac{1}{6}=0{\cdot}26^{8}$ $\frac{1}{5}=0{\cdot}32^{2}$ $\frac{1}{4}=0{\cdot}40^{3}$ $\frac{1}{3}=0{\cdot}53^{7}$

$\frac{3}{8}=0{\cdot}60^{4}$ $\frac{1}{2}=0{\cdot}80^{5}$ $\frac{5}{8}=1{\cdot}00^{6}$ $\frac{3}{4}=1{\cdot}20^{8}$ $\frac{7}{8}=1{\cdot}40^{9}$ $\frac{2}{3}=1{\cdot}07^{3}$ $\frac{5}{6}=1{\cdot}34^{2}$

$\frac{3}{16}=0{\cdot}30^{2}$ $\frac{5}{16}=0{\cdot}50^{3}$ $\frac{7}{16}=0{\cdot}70^{4}$ $\frac{9}{16}=0{\cdot}90^{6}$ $\frac{11}{16}=1{\cdot}10^{7}$ $\frac{13}{16}=1{\cdot}30^{8}$ $\frac{15}{16}=1{\cdot}50^{9}$

1 mile=1·609 kilometres

1	1·61½	41	66·21½	81	130·81½	121	195·41½	161	260·01½
2	3·23	42	67·83	82	132·43	122	197·03	162	261·63
3	4·84½	43	69·44½	83	134·04½	123	198·64½	163	263·24½
4	6·46	44	71·06	84	135·66	124	200·26	164	264·86
5	8·07½	45	72·67½	85	137·27½	125	201·87½	165	266·47½
6	9·69	46	74·29	86	138·89	126	203·49	166	268·09
7	11·30½	47	75·90½	87	140·50½	127	205·10½	167	269·70½
8	12·92	48	77·52	88	142·12	128	206·72	168	271·32
9	14·53½	49	79·13½	89	143·73½	129	208·33½	169	272·93½
10	16·15	**50**	80·75	**90**	145·35	**130**	209·95	**170**	274·55
11	17·76½	51	82·36½	91	146·96½	131	211·56½	171	276·16½
12	19·38	52	83·98	92	148·58	132	213·18	172	277·78
13	20·99½	53	85·59½	93	150·19½	133	214·79½	173	279·39½
14	22·61	54	87·21	94	151·81	134	216·41	174	281·01
15	24·22½	55	88·82½	95	153·42½	135	218·02½	175	282·62½
16	25·84	56	90·44	96	155·04	136	219·64	176	284·24
17	27·45½	57	92·05½	97	156·65½	137	221·25½	177	285·85½
18	29·07	58	93·67	98	158·27	138	222·87	178	287·47
19	30·68½	59	95·28½	99	159·88½	139	224·48½	179	289·08½
20	32·30	**60**	96·90	**100**	161·50	**140**	226·10	**180**	290·70
21	33·91½	61	98·51½	101	163·11½	141	227·71½	181	292·31½
22	35·53	62	100·13	102	164·73	142	229·33	182	293·93
23	37·14½	63	101·74½	103	166·34½	143	230·94½	183	295·54½
24	38·76	64	103·36	104	167·96	144	232·56	184	297·16
25	40·37½	65	104·97½	105	169·57½	145	234·17½	185	298·77½
26	41·99	66	106·59	106	171·19	146	235·79	186	300·39
27	43·60½	67	108·20½	107	172·80½	147	237·40½	187	302·00½
28	45·22	68	109·82	108	174·42	148	239·02	188	303·62
29	46·83½	69	111·43½	109	176·03½	149	240·63½	189	305·23½
30	48·45	**70**	113·05	**110**	177·65	**150**	242·25	**190**	306·85
31	50·06½	71	114·66½	111	179·26½	151	243·86½	191	308·46½
32	51·68	72	116·28	112	180·88	152	245·48	192	310·08
33	53·29½	73	117·89½	113	182·49½	153	247·09½	193	311·69½
34	54·91	74	119·51	114	184·11	154	248·71	194	313·31
35	56·52½	75	121·12½	115	185·72½	155	250·32½	195	314·92½
36	58·14	76	122·74	116	187·34	156	251·94	196	316·54
37	59·75½	77	124·35½	117	188·95½	157	253·55½	197	318·15½
38	61·37	78	125·97	118	190·57	158	255·17	198	319·77
39	62·98½	79	127·58½	119	192·18½	159	256·78½	199	321·38½
40	64·60	**80**	129·20	**120**	193·80	**160**	258·40	**200**	323·00

$\frac{1}{16}=0{\cdot}10^{1}$	$\frac{1}{9}=0{\cdot}17^{9}$	$\frac{1}{8}=0{\cdot}20^{2}$	$\frac{1}{6}=0{\cdot}26^{9}$	$\frac{1}{5}=0{\cdot}32^{3}$	$\frac{1}{4}=0{\cdot}40^{4}$	$\frac{1}{3}=0{\cdot}53^{8}$
$\frac{3}{8}=0{\cdot}60^{6}$	$\frac{1}{2}=0{\cdot}80^{8}$	$\frac{5}{8}=1{\cdot}00^{9}$	$\frac{3}{4}=1{\cdot}21^{1}$	$\frac{7}{8}=1{\cdot}41^{3}$	$\frac{2}{3}=1{\cdot}07^{7}$	$\frac{5}{6}=1{\cdot}34^{6}$
$\frac{3}{16}=0{\cdot}30^{3}$	$\frac{5}{16}=0{\cdot}50^{5}$	$\frac{7}{16}=0{\cdot}70^{7}$	$\frac{9}{16}=0{\cdot}90^{8}$	$\frac{11}{16}=1{\cdot}11^{0}$	$\frac{13}{16}=1{\cdot}31^{2}$	$\frac{15}{16}=1{\cdot}51^{4}$

1	1·62	41	66·42	81	131·22	121	196·02	161	260·82
2	3·24	42	68·04	82	132·84	122	197·64	162	262·44
3	4·86	43	69·66	83	134·46	123	199·26	163	264·06
4	6·48	44	71·28	84	136·08	124	200·88	164	265·68
5	8·10	45	72·90	85	137·70	125	202·50	165	267·30
6	9·72	46	74·52	86	139·32	126	204·12	166	268·92
7	11·34	47	76·14	87	140·94	127	205·74	167	270·54
8	12·96	48	77·76	88	142·56	128	207·36	168	272·16
9	14·58	49	79·38	89	144·18	129	208·98	169	273·78
10	16·20	**50**	81·00	**90**	145·80	**130**	210·60	**170**	275·40
11	17·82	51	82·62	91	147·42	131	212·22	171	277·02
12	19·44	52	84·24	92	149·04	132	213·84	172	278·64
13	21·06	53	85·86	93	150·66	133	215·46	173	280·26
14	22·68	54	87·48	94	152·28	134	217·08	174	281·88
15	24·30	55	89·10	95	153·90	135	218·70	175	283·50
16	25·92	56	90·72	96	155·52	136	220·32	176	285·12
17	27·54	57	92·34	97	157·14	137	221·94	177	286·74
18	29·16	58	93·96	98	158·76	138	223·56	178	288·36
19	30·78	59	95·58	99	160·38	139	225·18	179	289·98
20	32·40	**60**	97·20	**100**	162·00	**140**	226·80	**180**	291·60
21	34·02	61	98·82	101	163·62	141	228·42	181	293·22
22	35·64	62	100·44	102	165·24	142	230·04	182	294·84
23	37·26	63	102·06	103	166·86	143	231·66	183	296·46
24	38·88	64	103·68	104	168·48	144	233·28	184	298·08
25	40·50	65	105·30	105	170·10	145	234·90	185	299·70
26	42·12	66	106·92	106	171·72	146	236·52	186	301·32
27	43·74	67	108·54	107	173·34	147	238·14	187	302·94
28	45·36	68	110·16	108	174·96	148	239·76	188	304·56
29	46·98	69	111·78	109	176·58	149	241·38	189	306·18
30	48·60	**70**	113·40	**110**	178·20	**150**	243·00	**190**	307·80
31	50·22	71	115·02	111	179·82	151	244·62	191	309·42
32	51·84	72	116·64	112	181·44	152	246·24	192	311·04
33	53·46	73	118·26	113	183·06	153	247·86	193	312·66
34	55·08	74	119·88	114	184·68	154	249·48	194	314·28
35	56·70	75	121·50	115	186·30	155	251·10	195	315·90
36	58·32	76	123·12	116	187·92	156	252·72	196	317·52
37	59·94	77	124·74	117	189·54	157	254·34	197	319·14
38	61·56	78	126·36	118	191·16	158	255·96	198	320·76
39	63·18	79	127·98	119	192·78	159	257·58	199	322·38
40	64·80	**80**	129·60	**120**	194·40	**160**	259·20	**200**	324·00

$\frac{1}{16}=0{\cdot}10^1$	$\frac{1}{9}=0{\cdot}18^0$	$\frac{1}{8}=0{\cdot}20^3$	$\frac{1}{6}=0{\cdot}27^0$	$\frac{1}{5}=0{\cdot}32^4$	$\frac{1}{4}=0{\cdot}40^5$	$\frac{1}{3}=0{\cdot}54^0$
$\frac{3}{8}=0{\cdot}60^8$	$\frac{1}{2}=0{\cdot}81^0$	$\frac{5}{8}=1{\cdot}01^3$	$\frac{3}{4}=1{\cdot}21^5$	$\frac{7}{8}=1{\cdot}41^8$	$\frac{2}{3}=1{\cdot}08^0$	$\frac{5}{6}=1{\cdot}35^0$
$\frac{3}{16}=0{\cdot}30^4$	$\frac{5}{16}=0{\cdot}50^6$	$\frac{7}{16}=0{\cdot}70^9$	$\frac{9}{16}=0{\cdot}91^1$	$\frac{11}{16}=1{\cdot}11^4$	$\frac{13}{16}=1{\cdot}31^6$	$\frac{15}{16}=1{\cdot}51^9$

(0.615=1£) **£1.62½**

1	1·62½	41	66·62½	81	131·62½	121	196·62½	161	261·62½
2	3·25	42	68·25	82	133·25	122	198·25	162	263·25
3	4·87½	43	69·87½	83	134·87½	123	199·87½	163	264·87½
4	6·50	44	71·50	84	136·50	124	201·50	164	266·50
5	8·12½	45	73·12½	85	138·12½	125	203·12½	165	268·12½
6	9·75	46	74·75	86	139·75	126	204·75	166	269·75
7	11·37½	47	76·37½	87	141·37½	127	206·37½	167	271·37½
8	13·00	48	78·00	88	143·00	128	208·00	168	273·00
9	14·62½	49	79·62½	89	144·62½	129	209·62½	169	274·62½
10	16·25	**50**	81·25	**90**	146·25	**130**	211·25	**170**	276·25
11	17·87½	51	82·87½	91	147·87½	131	212·87½	171	277·87½
12	19·50	52	84·50	92	149·50	132	214·50	172	279·50
13	21·12½	53	86·12½	93	151·12½	133	216·12½	173	281·12½
14	22·75	54	87·75	94	152·75	134	217·75	174	282·75
15	24·37½	55	89·37½	95	154·37½	135	219·37½	175	284·37½
16	26·00	56	91·00	96	156·00	136	221·00	176	286·00
17	27·62½	57	92·62½	97	157·62½	137	222·62½	177	287·62½
18	29·25	58	94·25	98	159·25	138	224·25	178	289·25
19	30·87½	59	95·87½	99	160·87½	139	225·87½	179	290·87½
20	32·50	**60**	97·50	**100**	162·50	**140**	227·50	**180**	292·50
21	34·12½	61	99·12½	101	164·12½	141	229·12½	181	294·12½
22	35·75	62	100·75	102	165·75	142	230·75	182	295·75
23	37·37½	63	102·37½	103	167·37½	143	232·37½	183	297·37½
24	39·00	64	104·00	104	169·00	144	234·00	184	299·00
25	40·62½	65	105·62½	105	170·62½	145	235·62½	185	300·62½
26	42·25	66	107·25	106	172·25	146	237·25	186	302·25
27	43·87½	67	108·87½	107	173·87½	147	238·87½	187	303·87½
28	45·50	68	110·50	108	175·50	148	240·50	188	305·50
29	47·12½	69	112·12½	109	177·12½	149	242·12½	189	307·12½
30	48·75	**70**	113·75	**110**	178·75	**150**	243·75	**190**	308·75
31	50·37½	71	115·37½	111	180·37½	151	245·37½	191	310·37½
32	52·00	72	117·00	112	182·00	152	247·00	192	312·00
33	53·62½	73	118·62½	113	183·62½	153	248·62½	193	313·62½
34	55·25	74	120·25	114	185·25	154	250·25	194	315·25
35	56·87½	75	121·87½	115	186·87½	155	251·87½	195	316·87½
36	58·50	76	123·50	116	188·50	156	253·50	196	318·50
37	60·12½	77	125·12½	117	190·12½	157	255·12½	197	320·12½
38	61·75	78	126·75	118	191·75	158	256·75	198	321·75
39	63·37½	79	128·37½	119	193·37½	159	258·37½	199	323·37½
40	65·00	**80**	130·00	**120**	195·00	**160**	260·00	**200**	325·00

$\frac{1}{16}=0{\cdot}10^{2}$ $\frac{1}{9}=0{\cdot}18^{1}$ $\frac{1}{8}=0{\cdot}20^{3}$ $\frac{1}{6}=0{\cdot}27^{1}$ $\frac{1}{5}=0{\cdot}32^{5}$ $\frac{1}{4}=0{\cdot}40^{6}$ $\frac{1}{3}=0{\cdot}54^{2}$

$\frac{3}{8}=0{\cdot}60^{9}$ $\frac{1}{2}=0{\cdot}81^{3}$ $\frac{5}{8}=1{\cdot}01^{6}$ $\frac{3}{4}=1{\cdot}21^{9}$ $\frac{7}{8}=1{\cdot}42^{2}$ $\frac{2}{3}=1{\cdot}08^{3}$ $\frac{5}{6}=1{\cdot}35^{4}$

$\frac{3}{16}=0{\cdot}30^{5}$ $\frac{5}{16}=0{\cdot}50^{8}$ $\frac{7}{16}=0{\cdot}71^{1}$ $\frac{9}{16}=0{\cdot}91^{4}$ $\frac{11}{16}=1{\cdot}11^{7}$ $\frac{13}{16}=1{\cdot}32^{0}$ $\frac{15}{16}=1{\cdot}52^{3}$

(0.613=1£) **£1.63**

1	1·63	41	66·83	81	132·03	121	197·23	161	262·43
2	3·26	42	68·46	82	133·66	122	198·86	162	264·06
3	4·89	43	70·09	83	135·29	123	200·49	163	265·69
4	6·52	44	71·72	84	136·92	124	202·12	164	267·32
5	8·15	45	73·35	85	138·55	125	203·75	165	268·95
6	9·78	46	74·98	86	140·18	126	205·38	166	270·58
7	11·41	47	76·61	87	141·81	127	207·01	167	272·21
8	13·04	48	78·24	88	143·44	128	208·64	168	273·84
9	14·67	49	79·87	89	145·07	129	210·27	169	275·47
10	16·30	**50**	81·50	**90**	146·70	**130**	211·90	**170**	277·10
11	17·93	51	83·13	91	148·33	131	213·53	171	278·73
12	19·56	52	84·76	92	149·96	132	215·16	172	280·36
13	21·19	53	86·39	93	151·59	133	216·79	173	281·99
14	22·82	54	88·02	94	153·22	134	218·42	174	283·62
15	24·45	55	89·65	95	154·85	135	220·05	175	285·25
16	26·08	56	91·28	96	156·48	136	221·68	176	286·88
17	27·71	57	92·91	97	158·11	137	223·31	177	288·51
18	29·34	58	94·54	98	159·74	138	224·94	178	290·14
19	30·97	59	96·17	99	161·37	139	226·57	179	291·77
20	32·60	**60**	97·80	**100**	163·00	**140**	228·20	**180**	293·40
21	34·23	61	99·43	101	164·63	141	229·83	181	295·03
22	35·86	62	101·06	102	166·26	142	231·46	182	296·66
23	37·49	63	102·69	103	167·89	143	233·09	183	298·29
24	39·12	64	104·32	104	169·52	144	234·72	184	299·92
25	40·75	65	105·95	105	171·15	145	236·35	185	301·55
26	42·38	66	107·58	106	172·78	146	237·98	186	303·18
27	44·01	67	109·21	107	174·41	147	239·61	187	304·81
28	45·64	68	110·84	108	176·04	148	241·24	188	306·44
29	47·27	69	112·47	109	177·67	149	242·87	189	308·07
30	48·90	**70**	114·10	**110**	179·30	**150**	244·50	**190**	309·70
31	50·53	71	115·73	111	180·93	151	246·13	191	311·33
32	52·16	72	117·36	112	182·56	152	247·76	192	312·96
33	53·79	73	118·99	113	184·19	153	249·39	193	314·59
34	55·42	74	120·62	114	185·82	154	251·02	194	316·22
35	57·05	75	122·25	115	187·45	155	252·65	195	317·85
36	58·68	76	123·88	116	189·08	156	254·28	196	319·48
37	60·31	77	125·51	117	190·71	157	255·91	197	321·11
38	61·94	78	127·14	118	192·34	158	257·54	198	322·74
39	63·57	79	128·77	119	193·97	159	259·17	199	324·37
40	65·20	**80**	130·40	**120**	195·60	**160**	260·80	**200**	326·00

$\frac{1}{16}=0\cdot10^{2}$	$\frac{1}{9}=0\cdot18^{1}$	$\frac{1}{8}=0\cdot20^{4}$	$\frac{1}{6}=0\cdot27^{2}$	$\frac{1}{5}=0\cdot32^{6}$	$\frac{1}{4}=0\cdot40^{8}$	$\frac{1}{3}=0\cdot54^{3}$
$\frac{3}{8}=0\cdot61^{1}$	$\frac{1}{2}=0\cdot81^{5}$	$\frac{5}{8}=1\cdot01^{9}$	$\frac{3}{4}=1\cdot22^{3}$	$\frac{7}{8}=1\cdot42^{6}$	$\frac{2}{3}=1\cdot08^{7}$	$\frac{5}{6}=1\cdot35^{8}$
$\frac{3}{16}=0\cdot30^{6}$	$\frac{5}{16}=0\cdot50^{9}$	$\frac{7}{16}=0\cdot71^{3}$	$\frac{9}{16}=0\cdot91^{7}$	$\frac{11}{16}=1\cdot12^{1}$	$\frac{13}{16}=1\cdot32^{4}$	$\frac{15}{16}=1\cdot52^{8}$

(0.612=1£) **£1.63½**

1	1·63½	41	67·03½	81	132·43½	121	197·83½	161	263·23½
2	3·27	42	68·67	82	134·07	122	199·47	162	264·87
3	4·90½	43	70·30½	83	135·70½	123	201·10½	163	266·50½
4	6·54	44	71·94	84	137·34	124	202·74	164	268·14
5	8·17½	45	73·57½	85	138·97½	125	204·37½	165	269·77½
6	9·81	46	75·21	86	140·61	126	206·01	166	271·41
7	11·44½	47	76·84½	87	142·24½	127	207·64½	167	273·04½
8	13·08	48	78·48	88	143·88	128	209·28	168	274·68
9	14·71½	49	80·11½	89	145·51½	129	210·91½	169	276·31½
10	16·35	**50**	81·75	**90**	147·15	**130**	212·55	**170**	277·95
11	17·98½	51	83·38½	91	148·78½	131	214·18½	171	279·58½
12	19·62	52	85·02	92	150·42	132	215·82	172	281·22
13	21·25½	53	86·65½	93	152·05½	133	217·45½	173	282·85½
14	22·89	54	88·29	94	153·69	134	219·09	174	284·49
15	24·52½	55	89·92½	95	155·32½	135	220·72½	175	286·12½
16	26·16	56	91·56	96	156·96	136	222·36	176	287·76
17	27·79½	57	93·19½	97	158·59½	137	223·99½	177	289·39½
18	29·43	58	94·83	98	160·23	138	225·63	178	291·03
19	31·06½	59	96·46½	99	161·86½	139	227·26½	179	292·66½
20	32·70	**60**	98·10	**100**	163·50	**140**	228·90	**180**	294·30
21	34·33½	61	99·73½	101	165·13½	141	230·53½	181	295·93½
22	35·97	62	101·37	102	166·77	142	232·17	182	297·57
23	37·60½	63	103·00½	103	168·40½	143	233·80½	183	299·20½
24	39·24	64	104·64	104	170·04	144	235·44	184	300·84
25	40·87½	65	106·27½	105	171·67½	145	237·07½	185	302·47½
26	42·51	66	107·91	106	173·31	146	238·71	186	304·11
27	44·14½	67	109·54½	107	174·94½	147	240·34½	187	305·74½
28	45·78	68	111·18	108	176·58	148	241·98	188	307·38
29	47·41½	69	112·81½	109	178·21½	149	243·61½	189	309·01½
30	49·05	**70**	114·45	**110**	179·85	**150**	245·25	**190**	310·65
31	50·68½	71	116·08½	111	181·48½	151	246·88½	191	312·28½
32	52·32	72	117·72	112	183·12	152	248·52	192	313·92
33	53·95½	73	119·35½	113	184·75½	153	250·15½	193	315·55½
34	55·59	74	120·99	114	186·39	154	251·79	194	317·19
35	57·22½	75	122·62½	115	188·02½	155	253·42½	195	318·82½
36	58·86	76	124·26	116	189·66	156	255·06	196	320·46
37	60·49½	77	125·89½	117	191·29½	157	256·69½	197	322·09½
38	62·13	78	127·53	118	192·93	158	258·33	198	323·73
39	63·76½	79	129·16½	119	194·56½	159	259·96½	199	325·36½
40	65·40	**80**	130·80	**120**	196·20	**160**	261·60	**200**	327·00

$\frac{1}{16}=0{\cdot}10^{2}$ $\frac{1}{9}=0{\cdot}1\overline{8}^{2}$ $\frac{1}{8}=0{\cdot}20^{4}$ $\frac{1}{6}=0{\cdot}27^{3}$ $\frac{1}{5}=0{\cdot}32^{7}$ $\frac{1}{4}=0{\cdot}40^{9}$ $\frac{1}{3}=0{\cdot}54^{5}$

$\frac{3}{8}=0{\cdot}61^{3}$ $\frac{1}{2}=0{\cdot}81^{8}$ $\frac{5}{8}=1{\cdot}02^{2}$ $\frac{3}{4}=1{\cdot}22^{6}$ $\frac{7}{8}=1{\cdot}43^{1}$ $\frac{2}{3}=1{\cdot}09^{0}$ $\frac{5}{6}=1{\cdot}36^{3}$

$\frac{3}{16}=0{\cdot}30^{7}$ $\frac{5}{16}=0{\cdot}51^{1}$ $\frac{7}{16}=0{\cdot}71^{5}$ $\frac{9}{16}=0{\cdot}92^{0}$ $\frac{11}{16}=1{\cdot}12^{4}$ $\frac{13}{16}=1{\cdot}32^{8}$ $\frac{15}{16}=1{\cdot}53^{3}$

(0.610=1£) **£1.64**

1	1·64	41	67·24	81	132·84	121	198·44	161	264·04
2	3·28	42	68·88	82	134·48	122	200·08	162	265·68
3	4·92	43	70·52	83	136·12	123	201·72	163	267·32
4	6·56	44	72·16	84	137·76	124	203·36	164	268·96
5	8·20	45	73·80	85	139·40	125	205·00	165	270·60
6	9·84	46	75·44	86	141·04	126	206·64	166	272·24
7	11·48	47	77·08	87	142·68	127	208·28	167	273·88
8	13·12	48	78·72	88	144·32	128	209·92	168	275·52
9	14·76	49	80·36	89	145·96	129	211·56	169	277·16
10	16·40	**50**	82·00	**90**	147·60	**130**	213·20	**170**	278·80
11	18·04	51	83·64	91	149·24	131	214·84	171	280·44
12	19·68	52	85·28	92	150·88	132	216·48	172	282·08
13	21·32	53	86·92	93	152·52	133	218·12	173	283·72
14	22·96	54	88·56	94	154·16	134	219·76	174	285·36
15	24·60	55	90·20	95	155·80	135	221·40	175	287·00
16	26·24	56	91·84	96	157·44	136	223·04	176	288·64
17	27·88	57	93·48	97	159·08	137	224·68	177	290·28
18	29·52	58	95·12	98	160·72	138	226·32	178	291·92
19	31·16	59	96·76	99	162·36	139	227·96	179	293·56
20	32·80	**60**	98·40	**100**	164·00	**140**	229·60	**180**	295·20
21	34·44	61	100·04	101	165·64	141	231·24	181	296·84
22	36·08	62	101·68	102	167·28	142	232·88	182	298·48
23	37·72	63	103·32	103	168·92	143	234·52	183	300·12
24	39·36	64	104·96	104	170·56	144	236·16	184	301·76
25	41·00	65	106·60	105	172·20	145	237·80	185	303·40
26	42·64	66	108·24	106	173·84	146	239·44	186	305·04
27	44·28	67	109·88	107	175·48	147	241·08	187	306·68
28	45·92	68	111·52	108	177·12	148	242·72	188	308·32
29	47·56	69	113·16	109	178·76	149	244·36	189	309·96
30	49·20	**70**	114·80	**110**	180·40	**150**	246·00	**190**	311·60
31	50·84	71	116·44	111	182·04	151	247·64	191	313·24
32	52·48	72	118·08	112	183·68	152	249·28	192	314·88
33	54·12	73	119·72	113	185·32	153	250·92	193	316·52
34	55·76	74	121·36	114	186·96	154	252·56	194	318·16
35	57·40	75	123·00	115	188·60	155	254·20	195	319·80
36	59·04	76	124·64	116	190·24	156	255·84	196	321·44
37	60·68	77	126·28	117	191·88	157	257·48	197	323·08
38	62·32	78	127·92	118	193·52	158	259·12	198	324·72
39	63·96	79	129·56	119	195·16	159	260·76	199	326·36
40	65·60	**80**	131·20	**120**	196·80	**160**	262·40	**200**	328·00

$\frac{1}{16}$=0·10^{3} $\frac{1}{9}$=0·18^{2} $\frac{1}{8}$=0·20^{5} $\frac{1}{6}$=0·27^{3} $\frac{1}{5}$=0·32^{8} $\frac{1}{4}$=0·41^{0} $\frac{1}{3}$=0·54^{7}

$\frac{3}{8}$=0·61^{5} $\frac{1}{2}$=0·82^{0} $\frac{5}{8}$=1·02^{5} $\frac{3}{4}$=1·23^{0} $\frac{7}{8}$=1·43^{5} $\frac{2}{3}$=1·09^{3} $\frac{5}{6}$=1·36^{7}

$\frac{3}{16}$=0·30^{8} $\frac{5}{16}$=0·51^{3} $\frac{7}{16}$=0·71^{8} $\frac{9}{16}$=0·92^{3} $\frac{11}{16}$=1·12^{8} $\frac{13}{16}$=1·33^{3} $\frac{15}{16}$=1·53^{8}

1 cubic inch=16·387 cubic centimetres

(0.608=1£) **£1.64½**

1	1·64½	41	67·44½	81	133·24½	121	199·04½	161	264·84½
2	3·29	42	69·09	82	134·89	122	200·69	162	266·49
3	4·93½	43	70·73½	83	136·53½	123	202·33½	163	268·13½
4	6·58	44	72·38	84	138·18	124	203·98	164	269·78
5	8·22½	45	74·02½	85	139·82½	125	205·62½	165	271·42½
6	9·87	46	75·67	86	141·47	126	207·27	166	273·07
7	11·51½	47	77·31½	87	143·11½	127	208·91½	167	274·71½
8	13·16	48	78·96	88	144·76	128	210·56	168	276·36
9	14·80½	49	80·60½	89	146·40½	129	212·20½	169	278·00½
10	16·45	**50**	82·25	**90**	148·05	**130**	213·85	**170**	279·65
11	18·09½	51	83·89½	91	149·69½	131	215·49½	171	281·29½
12	19·74	52	85·54	92	151·34	132	217·14	172	282·94
13	21·38½	53	87·18½	93	152·98½	133	218·78½	173	284·58½
14	23·03	54	88·83	94	154·63	134	220·43	174	286·23
15	24·67½	55	90·47½	95	156·27½	135	222·07½	175	287·87½
16	26·32	56	92·12	96	157·92	136	223·72	176	289·52
17	27·96½	57	93·76½	97	159·56½	137	225·36½	177	291·16½
18	29·61	58	95·41	98	161·21	138	227·01	178	292·81
19	31·25½	59	97·05½	99	162·85½	139	228·65½	179	294·45½
20	32·90	**60**	98·70	**100**	164·50	**140**	230·30	**180**	296·10
21	34·54½	61	100·34½	101	166·14½	141	231·94½	181	297·74½
22	36·19	62	101·99	102	167·79	142	233·59	182	299·39
23	37·83½	63	103·63½	103	169·43½	143	235·23½	183	301·03½
24	39·48	64	105·28	104	171·08	144	236·88	184	302·68
25	41·12½	65	106·92½	105	172·72½	145	238·52½	185	304·32½
26	42·77	66	108·57	106	174·37	146	240·17	186	305·97
27	44·41½	67	110·21½	107	176·01½	147	241·81½	187	307·61½
28	46·06	68	111·86	108	177·66	148	243·46	188	309·26
29	47·70½	69	113·50½	109	179·30½	149	245·10½	189	310·90½
30	49·35	**70**	115·15	**110**	180·95	**150**	246·75	**190**	312·55
31	50·99½	71	116·79½	111	182·59½	151	248·39½	191	314·19½
32	52·64	72	118·44	112	184·24	152	250·04	192	315·84
33	54·28½	73	120·08½	113	185·88½	153	251·68½	193	317·48½
34	55·93	74	121·73	114	187·53	154	253·33	194	319·13
35	57·57½	75	123·37½	115	189·17½	155	254·97½	195	320·77½
36	59·22	76	125·02	116	190·82	156	256·62	196	322·42
37	60·86½	77	126·66½	117	192·46½	157	258·26½	197	324·06½
38	62·51	78	128·31	118	194·11	158	259·91	198	325·71
39	64·15½	79	129·95½	119	195·75½	159	261·55½	199	327·35½
40	65·80	**80**	131·60	**120**	197·40	**160**	263·20	**200**	329·00

$\frac{1}{16}=0{\cdot}10^{3}$ $\frac{1}{9}=0{\cdot}18^{3}$ $\frac{1}{8}=0{\cdot}20^{6}$ $\frac{1}{6}=0{\cdot}27^{4}$ $\frac{1}{5}=0{\cdot}32^{9}$ $\frac{1}{4}=0{\cdot}41^{1}$ $\frac{1}{3}=0{\cdot}54^{8}$

$\frac{3}{8}=0{\cdot}61^{7}$ $\frac{1}{2}=0{\cdot}82^{3}$ $\frac{5}{8}=1{\cdot}02^{8}$ $\frac{3}{4}=1{\cdot}23^{4}$ $\frac{7}{8}=1{\cdot}43^{9}$ $\frac{2}{3}=1{\cdot}09^{7}$ $\frac{5}{6}=1{\cdot}37^{1}$

$\frac{3}{16}=0{\cdot}30^{8}$ $\frac{5}{16}=0{\cdot}51^{4}$ $\frac{7}{16}=0{\cdot}72^{0}$ $\frac{9}{16}=0{\cdot}92^{5}$ $\frac{11}{16}=1{\cdot}13^{1}$ $\frac{13}{16}=1{\cdot}33^{7}$ $\frac{15}{16}=1{\cdot}54^{2}$

(0.606=1£) **£1.65**

1	1·65	41	67·65	81	133·65	121	199·65	161	265·65
2	3·30	42	69·30	82	135·30	122	201·30	162	267·30
3	4·95	43	70·95	83	136·95	123	202·95	163	268·95
4	6·60	44	72·60	84	138·60	124	204·60	164	270·60
5	8·25	45	74·25	85	140·25	125	206·25	165	272·25
6	9·90	46	75·90	86	141·90	126	207·90	166	273·90
7	11·55	47	77·55	87	143·55	127	209·55	167	275·55
8	13·20	48	79·20	88	145·20	128	211·20	168	277·20
9	14·85	49	80·85	89	146·85	129	212·85	169	278·85
10	16·50	**50**	82·50	**90**	148·50	**130**	214·50	**170**	280·50
11	18·15	51	84·15	91	150·15	131	216·15	171	282·15
12	19·80	52	85·80	92	151·80	132	217·80	172	283·80
13	21·45	53	87·45	93	153·45	133	219·45	173	285·45
14	23·10	54	89·10	94	155·10	134	221·10	174	287·10
15	24·75	55	90·75	95	156·75	135	222·75	175	288·75
16	26·40	56	92·40	96	158·40	136	224·40	176	290·40
17	28·05	57	94·05	97	160·05	137	226·05	177	292·05
18	29·70	58	95·70	98	161·70	138	227·70	178	293·70
19	31·35	59	97·35	99	163·35	139	229·35	179	295·35
20	33·00	**60**	99·00	**100**	165·00	**140**	231·00	**180**	297·00
21	34·65	61	100·65	101	166·65	141	232·65	181	298·65
22	36·30	62	102·30	102	168·30	142	234·30	182	300·30
23	37·95	63	103·95	103	169·95	143	235·95	183	301·95
24	39·60	64	105·60	104	171·60	144	237·60	184	303·60
25	41·25	65	107·25	105	173·25	145	239·25	185	305·25
26	42·90	66	108·90	106	174·90	146	240·90	186	306·90
27	44·55	67	110·55	107	176·55	147	242·55	187	308·55
28	46·20	68	112·20	108	178·20	148	244·20	188	310·20
29	47·85	69	113·85	109	179·85	149	245·85	189	311·85
30	49·50	**70**	115·50	**110**	181·50	**150**	247·50	**190**	313·50
31	51·15	71	117·15	111	183·15	151	249·15	191	315·15
32	52·80	72	118·80	112	184·80	152	250·80	192	316·80
33	54·45	73	120·45	113	186·45	153	252·45	193	318·45
34	56·10	74	122·10	114	188·10	154	254·10	194	320·10
35	57·75	75	123·75	115	189·75	155	255·75	195	321·75
36	59·40	76	125·40	116	191·40	156	257·40	196	323·40
37	61·05	77	127·05	117	193·05	157	259·05	197	325·05
38	62·70	78	128·70	118	194·70	158	260·70	198	326·70
39	64·35	79	130·35	119	196·35	159	262·35	199	328·35
40	66·00	**80**	132·00	**120**	198·00	**160**	264·00	**200**	330·00

$\frac{1}{16}=0{\cdot}10^{3}$ $\frac{1}{9}=0{\cdot}18^{3}$ $\frac{1}{8}=0{\cdot}20^{6}$ $\frac{1}{6}=0{\cdot}27^{5}$ $\frac{1}{5}=0{\cdot}33^{0}$ $\frac{1}{4}=0{\cdot}41^{3}$ $\frac{1}{3}=0{\cdot}55^{0}$

$\frac{3}{8}=0{\cdot}61^{9}$ $\frac{1}{2}=0{\cdot}82^{5}$ $\frac{5}{8}=1{\cdot}03^{1}$ $\frac{3}{4}=1{\cdot}23^{8}$ $\frac{7}{8}=1{\cdot}44^{4}$ $\frac{2}{3}=1{\cdot}10^{0}$ $\frac{5}{6}=1{\cdot}37^{5}$

$\frac{3}{16}=0{\cdot}30^{9}$ $\frac{5}{16}=0{\cdot}51^{6}$ $\frac{7}{16}=0{\cdot}72^{2}$ $\frac{9}{16}=0{\cdot}92^{8}$ $\frac{11}{16}=1{\cdot}13^{4}$ $\frac{13}{16}=1{\cdot}34^{1}$ $\frac{15}{16}=1{\cdot}54^{7}$

1	1·65½	41	67·85½	81	134·05½	121	200·25½	161	266·45½
2	3·31	42	69·51	82	135·71	122	201·91	162	268·11
3	4·96½	43	71·16½	83	137·36½	123	203·56½	163	269·76½
4	6·62	44	72·82	84	139·02	124	205·22	164	271·42
5	8·27½	45	74·47½	85	140·67½	125	206·87½	165	273·07½
6	9·93	46	76·13	86	142·33	126	208·53	166	274·73
7	11·58½	47	77·78½	87	143·98½	127	210·18½	167	276·38½
8	13·24	48	79·44	88	145·64	128	211·84	168	278·04
9	14·89½	49	81·09½	89	147·29½	129	213·49½	169	279·69½
10	16·55	**50**	82·75	**90**	148·95	**130**	215·15	**170**	281·35
11	18·20½	51	84·40½	91	150·60½	131	216·80½	171	283·00½
12	19·86	52	86·06	92	152·26	132	218·46	172	284·66
13	21·51½	53	87·71½	93	153·91½	133	220·11½	173	286·31½
14	23·17	54	89·37	94	155·57	134	221·77	174	287·97
15	24·82½	55	91·02½	95	157·22½	135	223·42½	175	289·62½
16	26·48	56	92·68	96	158·88	136	225·08	176	291·28
17	28·13½	57	94·33½	97	160·53½	137	226·73½	177	292·93½
18	29·79	58	95·99	98	162·19	138	228·39	178	294·59
19	31·44½	59	97·64½	99	163·84½	139	230·04½	179	296·24½
20	33·10	**60**	99·30	**100**	165·50	**140**	231·70	**180**	297·90
21	34·75½	61	100·95½	101	167·15½	141	233·35½	181	299·55½
22	36·41	62	102·61	102	168·81	142	235·01	182	301·21
23	38·06½	63	104·26½	103	170·46½	143	236·66½	183	302·86½
24	39·72	64	105·92	104	172·12	144	238·32	184	304·52
25	41·37½	65	107·57½	105	173·77½	145	239·97½	185	306·17½
26	43·03	66	109·23	106	175·43	146	241·63	186	307·83
27	44·68½	67	110·88½	107	177·08½	147	243·28½	187	309·48½
28	46·34	68	112·54	108	178·74	148	244·94	188	311·14
29	47·99½	69	114·19½	109	180·39½	149	246·59½	189	312·79½
30	49·65	**70**	115·85	**110**	182·05	**150**	248·25	**190**	314·45
31	51·30½	71	117·50½	111	183·70½	151	249·90½	191	316·10½
32	52·96	72	119·16	112	185·36	152	251·56	192	317·76
33	54·61½	73	120·81½	113	187·01½	153	253·21½	193	319·41½
34	56·27	74	122·47	114	188·67	154	254·87	194	321·07
35	57·92½	75	124·12½	115	190·32½	155	256·52½	195	322·72½
36	59·58	76	125·78	116	191·98	156	258·18	196	324·38
37	61·23½	77	127·43½	117	193·63½	157	259·83½	197	326·03½
38	62·89	78	129·09	118	195·29	158	261·49	198	327·69
39	64·54½	79	130·74½	119	196·94½	159	263·14½	199	329·34½
40	66·20	**80**	132·40	**120**	198·60	**160**	264·80	**200**	331·00

$\frac{1}{16}$=0·10[3] $\frac{1}{9}$=0·18[4] $\frac{1}{8}$=0·20[7] $\frac{1}{6}$=0·27[6] $\frac{1}{5}$=0·33[1] $\frac{1}{4}$=0·41[4] $\frac{1}{3}$=0·55[2]

$\frac{3}{8}$=0·62[1] $\frac{1}{2}$=0·82[8] $\frac{5}{8}$=1·03[4] $\frac{3}{4}$=1·24[1] $\frac{7}{8}$=1·44[8] $\frac{2}{3}$=1·10[3] $\frac{5}{6}$=1·37[9]

$\frac{3}{16}$=0·31[0] $\frac{5}{16}$=0·51[7] $\frac{7}{16}$=0·72[4] $\frac{9}{16}$=0·93[1] $\frac{11}{16}$=1·13[8] $\frac{13}{16}$=1·34[5] $\frac{15}{16}$=1·55[2]

(0.602=1£) **£1.66**

1	1·66	41	68·06	81	134·46	121	200·86	161	267·26
2	3·32	42	69·72	82	136·12	122	202·52	162	268·92
3	4·98	43	71·38	83	137·78	123	204·18	163	270·58
4	6·64	44	73·04	84	139·44	124	205·84	164	272·24
5	8·30	45	74·70	85	141·10	125	207·50	165	273·90
6	9·96	46	76·36	86	142·76	126	209·16	166	275·56
7	11·62	47	78·02	87	144·42	127	210·82	167	277·22
8	13·28	48	79·68	88	146·08	128	212·48	168	278·88
9	14·94	49	81·34	89	147·74	129	214·14	169	280·54
10	16·60	**50**	83·00	**90**	149·40	**130**	215·80	**170**	282·20
11	18·26	51	84·66	91	151·06	131	217·46	171	283·86
12	19·92	52	86·32	92	152·72	132	219·12	172	285·52
13	21·58	53	87·98	93	154·38	133	220·78	173	287·18
14	23·24	54	89·64	94	156·04	134	222·44	174	288·84
15	24·90	55	91·30	95	157·70	135	224·10	175	290·50
16	26·56	56	92·96	96	159·36	136	225·76	176	292·16
17	28·22	57	94·62	97	161·02	137	227·42	177	293·82
18	29·88	58	96·28	98	162·68	138	229·08	178	295·48
19	31·54	59	97·94	99	164·34	139	230·74	179	297·14
20	33·20	**60**	99·60	**100**	166·00	**140**	232·40	**180**	298·80
21	34·86	61	101·26	101	167·66	141	234·06	181	300·46
22	36·52	62	102·92	102	169·32	142	235·72	182	302·12
23	38·18	63	104·58	103	170·98	143	237·38	183	303·78
24	39·84	64	106·24	104	172·64	144	239·04	184	305·44
25	41·50	65	107·90	105	174·30	145	240·70	185	307·10
26	43·16	66	109·56	106	175·96	146	242·36	186	308·76
27	44·82	67	111·22	107	177·62	147	244·02	187	310·42
28	46·48	68	112·88	108	179·28	148	245·68	188	312·08
29	48·14	69	114·54	109	180·94	149	247·34	189	313·74
30	49·80	**70**	116·20	**110**	182·60	**150**	249·00	**190**	315·40
31	51·46	71	117·86	111	184·26	151	250·66	191	317·06
32	53·12	72	119·52	112	185·92	152	252·32	192	318·72
33	54·78	73	121·18	113	187·58	153	253·98	193	320·38
34	56·44	74	122·84	114	189·24	154	255·64	194	322·04
35	58·10	75	124·50	115	190·90	155	257·30	195	323·70
36	59·76	76	126·16	116	192·56	156	258·96	196	325·36
37	61·42	77	127·82	117	194·22	157	260·62	197	327·02
38	63·08	78	129·48	118	195·88	158	262·28	198	328·68
39	64·74	79	131·14	119	197·54	159	263·94	199	330·34
40	66·40	**80**	132·80	**120**	199·20	**160**	265·60	**200**	332·00

$\frac{1}{16}=0{\cdot}10^{4}$ $\frac{1}{9}=0{\cdot}18^{4}$ $\frac{1}{8}=0{\cdot}20^{8}$ $\frac{1}{6}=0{\cdot}27^{7}$ $\frac{1}{5}=0{\cdot}33^{2}$ $\frac{1}{4}=0{\cdot}41^{5}$ $\frac{1}{3}=0{\cdot}55^{3}$

$\frac{3}{8}=0{\cdot}62^{3}$ $\frac{1}{2}=0{\cdot}83^{0}$ $\frac{5}{8}=1{\cdot}03^{8}$ $\frac{3}{4}=1{\cdot}24^{5}$ $\frac{7}{8}=1{\cdot}45^{3}$ $\frac{2}{3}=1{\cdot}10^{7}$ $\frac{5}{6}=1{\cdot}38^{3}$

$\frac{3}{16}=0{\cdot}31^{1}$ $\frac{5}{16}=0{\cdot}51^{9}$ $\frac{7}{16}=0{\cdot}72^{6}$ $\frac{9}{16}=0{\cdot}93^{4}$ $\frac{11}{16}=1{\cdot}14^{1}$ $\frac{13}{16}=1{\cdot}34^{9}$ $\frac{15}{16}=1{\cdot}55^{6}$

1	1·66½	41	68·26½	81	134·86½	121	201·46½	161	268·06½
2	3·33	42	69·93	82	136·53	122	203·13	162	269·73
3	4·99½	43	71·59½	83	138·19½	123	204·79½	163	271·39½
4	6·66	44	73·26	84	139·86	124	206·46	164	273·06
5	8·32½	45	74·92½	85	141·52½	125	208·12½	165	274·72½
6	9·99	46	76·59	86	143·19	126	209·79	166	276·39
7	11·65½	47	78·25½	87	144·85½	127	211·45½	167	278·05½
8	13·32	48	79·92	88	146·52	128	213·12	168	279·72
9	14·98½	49	81·58½	89	148·18½	129	214·78½	169	281·38½
10	16·65	**50**	83·25	**90**	149·85	**130**	216·45	**170**	283·05
11	18·31½	51	84·91½	91	151·51½	131	218·11½	171	284·71½
12	19·98	52	86·58	92	153·18	132	219·78	172	286·38
13	21·64½	53	88·24½	93	154·84½	133	221·44½	173	288·04½
14	23·31	54	89·91	94	156·51	134	223·11	174	289·71
15	24·97½	55	91·57½	95	158·17½	135	224·77½	175	291·37½
16	26·64	56	93·24	96	159·84	136	226·44	176	293·04
17	28·30½	57	94·90½	97	161·50½	137	228·10½	177	294·70½
18	29·97	58	96·57	98	163·17	138	229·77	178	296·37
19	31·63½	59	98·23½	99	164·83½	139	231·43½	179	298·03½
20	33·30	**60**	99·90	**100**	166·50	**140**	233·10	**180**	299·70
21	34·96½	61	101·56½	101	168·16½	141	234·76½	181	301·36½
22	36·63	62	103·23	102	169·83	142	236·43	182	303·03
23	38·29½	63	104·89½	103	171·49½	143	238·09½	183	304·69½
24	39·96	64	106·56	104	173·16	144	239·76	184	306·36
25	41·62½	65	108·22½	105	174·82½	145	241·42½	185	308·02½
26	43·29	66	109·89	106	176·49	146	243·09	186	309·69
27	44·95½	67	111·55½	107	178·15½	147	244·75½	187	311·35½
28	46·62	68	113·22	108	179·82	148	246·42	188	313·02
29	48·28½	69	114·88½	109	181·48½	149	248·08½	189	314·68½
30	49·95	**70**	116·55	**110**	183·15	**150**	249·75	**190**	316·35
31	51·61½	71	118·21½	111	184·81½	151	251·41½	191	318·01½
32	53·28	72	119·88	112	186·48	152	253·08	192	319·68
33	54·94½	73	121·54½	113	188·14½	153	254·74½	193	321·34½
34	56·61	74	123·21	114	189·81	154	256·41	194	323·01
35	58·27½	75	124·87½	115	191·47½	155	258·07½	195	324·67½
36	59·94	76	126·54	116	193·14	156	259·74	196	326·34
37	61·60½	77	128·20½	117	194·80½	157	261·40½	197	328·00½
38	63·27	78	129·87	118	196·47	158	263·07	198	329·67
39	64·93½	79	131·53½	119	198·13½	159	264·73½	199	331·33½
40	66·60	**80**	133·20	**120**	199·80	**160**	266·40	**200**	333·00

$\frac{1}{16}=0{\cdot}10^4$ $\frac{1}{9}=0{\cdot}18^5$ $\frac{1}{8}=0{\cdot}20^8$ $\frac{1}{6}=0{\cdot}27^8$ $\frac{1}{5}=0{\cdot}33^3$ $\frac{1}{4}=0{\cdot}41^6$ $\frac{1}{3}=0{\cdot}55^5$

$\frac{3}{8}=0{\cdot}62^4$ $\frac{1}{2}=0{\cdot}83^3$ $\frac{5}{8}=1{\cdot}04^1$ $\frac{3}{4}=1{\cdot}24^9$ $\frac{7}{8}=1{\cdot}45^7$ $\frac{2}{3}=1{\cdot}11^0$ $\frac{5}{6}=1{\cdot}38^8$

$\frac{3}{16}=0{\cdot}31^2$ $\frac{5}{16}=0{\cdot}52^0$ $\frac{7}{16}=0{\cdot}72^8$ $\frac{9}{16}=0{\cdot}93^7$ $\frac{11}{16}=1{\cdot}14^5$ $\frac{13}{16}=1{\cdot}35^3$ $\frac{15}{16}=1{\cdot}56^1$

(0.599=1£) **£1.67**

1	1·67	41	68·47	81	135·27	121	202·07	161	268·87
2	3·34	42	70·14	82	136·94	122	203·74	162	270·54
3	5·01	43	71·81	83	138·61	123	205·41	163	272·21
4	6·68	44	73·48	84	140·28	124	207·08	164	273·88
5	8·35	45	75·15	85	141·95	125	208·75	165	275·55
6	10·02	46	76·82	86	143·62	126	210·42	166	277·22
7	11·69	47	78·49	87	145·29	127	212·09	167	278·89
8	13·36	48	80·16	88	146·96	128	213·76	168	280·56
9	15·03	49	81·83	89	148·63	129	215·43	169	282·23
10	16·70	**50**	83·50	**90**	150·30	**130**	217·10	**170**	283·90
11	18·37	51	85·17	91	151·97	131	218·77	171	285·57
12	20·04	52	86·84	92	153·64	132	220·44	172	287·24
13	21·71	53	88·51	93	155·31	133	222·11	173	288·91
14	23·38	54	90·18	94	156·98	134	223·78	174	290·58
15	25·05	55	91·85	95	158·65	135	225·45	175	292·25
16	26·72	56	93·52	96	160·32	136	227·12	176	293·92
17	28·39	57	95·19	97	161·99	137	228·79	177	295·59
18	30·06	58	96·86	98	163·66	138	230·46	178	297·26
19	31·73	59	98·53	99	165·33	139	232·13	179	298·93
20	33·40	**60**	100·20	**100**	167·00	**140**	233·80	**180**	300·60
21	35·07	61	101·87	101	168·67	141	235·47	181	302·27
22	36·74	62	103·54	102	170·34	142	237·14	182	303·94
23	38·41	63	105·21	103	172·01	143	238·81	183	305·61
24	40·08	64	106·88	104	173·68	144	240·48	184	307·28
25	41·75	65	108·55	105	175·35	145	242·15	185	308·95
26	43·42	66	110·22	106	177·02	146	243·82	186	310·62
27	45·09	67	111·89	107	178·69	147	245·49	187	312·29
28	46·76	68	113·56	108	180·36	148	247·16	188	313·96
29	48·43	69	115·23	109	182·03	149	248·83	189	315·63
30	50·10	**70**	116·90	**110**	183·70	**150**	250·50	**190**	317·30
31	51·77	71	118·57	111	185·37	151	252·17	191	318·97
32	53·44	72	120·24	112	187·04	152	253·84	192	320·64
33	55·11	73	121·91	113	188·71	153	255·51	193	322·31
34	56·78	74	123·58	114	190·38	154	257·18	194	323·98
35	58·45	75	125·25	115	192·05	155	258·85	195	325·65
36	60·12	76	126·92	116	193·72	156	260·52	196	327·32
37	61·79	77	128·59	117	195·39	157	262·19	197	328·99
38	63·46	78	130·26	118	197·06	158	263·86	198	330·66
39	65·13	79	131·93	119	198·73	159	265·53	199	332·33
40	66·80	**80**	133·60	**120**	200·40	**160**	267·20	**200**	334·00

$\frac{1}{16}=0{\cdot}10^4$ $\frac{1}{9}=0{\cdot}18^6$ $\frac{1}{8}=0{\cdot}20^9$ $\frac{1}{6}=0{\cdot}27^8$ $\frac{1}{5}=0{\cdot}33^4$ $\frac{1}{4}=0{\cdot}41^8$ $\frac{1}{3}=0{\cdot}55^7$

$\frac{3}{8}=0{\cdot}62^6$ $\frac{1}{2}=0{\cdot}83^5$ $\frac{5}{8}=1{\cdot}04^4$ $\frac{3}{4}=1{\cdot}25^3$ $\frac{7}{8}=1{\cdot}46^1$ $\frac{2}{3}=1{\cdot}11^3$ $\frac{5}{6}=1{\cdot}39^2$

$\frac{3}{16}=0{\cdot}31^3$ $\frac{5}{16}=0{\cdot}52^2$ $\frac{7}{16}=0{\cdot}73^1$ $\frac{9}{16}=0{\cdot}93^9$ $\frac{11}{16}=1{\cdot}14^8$ $\frac{13}{16}=1{\cdot}35^7$ $\frac{15}{16}=1{\cdot}56^6$

1	1·67½	41	68·67½	81	135·67½	121	202·67½	161	269·67½
2	3·35	42	70·35	82	137·35	122	204·35	162	271·35
3	5·02½	43	72·02½	83	139·02½	123	206·02½	163	273·02½
4	6·70	44	73·70	84	140·70	124	207·70	164	274·70
5	8·37½	45	75·37½	85	142·37½	125	209·37½	165	276·37½
6	10·05	46	77·05	86	144·05	126	211·05	166	278·05
7	11·72½	47	78·72½	87	145·72½	127	212·72½	167	279·72½
8	13·40	48	80·40	88	147·40	128	214·40	168	281·40
9	15·07½	49	82·07½	89	149·07½	129	216·07½	169	283·07½
10	16·75	**50**	83·75	**90**	150·75	**130**	217·75	**170**	284·75
11	18·42½	51	85·42½	91	152·42½	131	219·42½	171	286·42½
12	20·10	52	87·10	92	154·10	132	221·10	172	288·10
13	21·77½	53	88·77½	93	155·77½	133	222·77½	173	289·77½
14	23·45	54	90·45	94	157·45	134	224·45	174	291·45
15	25·12½	55	92·12½	95	159·12½	135	226·12½	175	293·12½
16	26·80	56	93·80	96	160·80	136	227·80	176	294·80
17	28·47½	57	95·47½	97	162·47½	137	229·47½	177	296·47½
18	30·15	58	97·15	98	164·15	138	231·15	178	298·15
19	31·82½	59	98·82½	99	165·82½	139	232·82½	179	299·82½
20	33·50	**60**	100·50	**100**	167·50	**140**	234·50	**180**	301·50
21	35·17½	61	102·17½	101	169·17½	141	236·17½	181	303·17½
22	36·85	62	103·85	102	170·85	142	237·85	182	304·85
23	38·52½	63	105·52½	103	172·52½	143	239·52½	183	306·52½
24	40·20	64	107·20	104	174·20	144	241·20	184	308·20
25	41·87½	65	108·87½	105	175·87½	145	242·87½	185	309·87½
26	43·55	66	110·55	106	177·55	146	244·55	186	311·55
27	45·22½	67	112·22½	107	179·22½	147	246·22½	187	313·22½
28	46·90	68	113·90	108	180·90	148	247·90	188	314·90
29	48·57½	69	115·57½	109	182·57½	149	249·57½	189	316·57½
30	50·25	**70**	117·25	**110**	184·25	**150**	251·25	**190**	318·25
31	51·92½	71	118·92½	111	185·92½	151	252·92½	191	319·92½
32	53·60	72	120·60	112	187·60	152	254·60	192	321·60
33	55·27½	73	122·27½	113	189·27½	153	256·27½	193	323·27½
34	56·95	74	123·95	114	190·95	154	257·95	194	324·95
35	58·62½	75	125·62½	115	192·62½	155	259·62½	195	326·62½
36	60·30	76	127·30	116	194·30	156	261·30	196	328·30
37	61·97½	77	128·97½	117	195·97½	157	262·97½	197	329·97½
38	63·65	78	130·65	118	197·65	158	264·65	198	331·65
39	65·32½	79	132·32½	119	199·32½	159	266·32½	199	333·32½
40	67·00	**80**	134·00	**120**	201·00	**160**	268·00	**200**	335·00

$\frac{1}{16}=0{\cdot}10^{5}$	$\frac{1}{9}=0{\cdot}18^{6}$	$\frac{1}{8}=0{\cdot}20^{9}$	$\frac{1}{6}=0{\cdot}27^{9}$	$\frac{1}{5}=0{\cdot}33^{5}$	$\frac{1}{4}=0{\cdot}41^{9}$	$\frac{1}{3}=0{\cdot}55^{8}$
$\frac{3}{8}=0{\cdot}62^{8}$	$\frac{1}{2}=0{\cdot}83^{8}$	$\frac{5}{8}=1{\cdot}04^{7}$	$\frac{3}{4}=1{\cdot}25^{6}$	$\frac{7}{8}=1{\cdot}46^{6}$	$\frac{2}{3}=1{\cdot}11^{7}$	$\frac{5}{6}=1{\cdot}39^{6}$
$\frac{3}{16}=0{\cdot}31^{4}$	$\frac{5}{16}=0{\cdot}52^{3}$	$\frac{7}{16}=0{\cdot}73^{3}$	$\frac{9}{16}=0{\cdot}94^{2}$	$\frac{11}{16}=1{\cdot}15^{2}$	$\frac{13}{16}=1{\cdot}36^{1}$	$\frac{15}{16}=1{\cdot}57^{0}$

(0.595=1£) **£1.68**

1	1·68	41	68·88	81	136·08	121	203·28	161	270·48
2	3·36	42	70·56	82	137·76	122	204·96	162	272·16
3	5·04	43	72·24	83	139·44	123	206·64	163	273·84
4	6·72	44	73·92	84	141·12	124	208·32	164	275·52
5	8·40	45	75·60	85	142·80	125	210·00	165	277·20
6	10·08	46	77·28	86	144·48	126	211·68	166	278·88
7	11·76	47	78·96	87	146·16	127	213·36	167	280·56
8	13·44	48	80·64	88	147·84	128	215·04	168	282·24
9	15·12	49	82·32	89	149·52	129	216·72	169	283·92
10	16·80	**50**	84·00	**90**	151·20	**130**	218·40	**170**	285·60
11	18·48	51	85·68	91	152·88	131	220·08	171	287·28
12	20·16	52	87·36	92	154·56	132	221·76	172	288·96
13	21·84	53	89·04	93	156·24	133	223·44	173	290·64
14	23·52	54	90·72	94	157·92	134	225·12	174	292·32
15	25·20	55	92·40	95	159·60	135	226·80	175	294·00
16	26·88	56	94·08	96	161·28	136	228·48	176	295·68
17	28·56	57	95·76	97	162·96	137	230·16	177	297·36
18	30·24	58	97·44	98	164·64	138	231·84	178	299·04
19	31·92	59	99·12	99	166·32	139	233·52	179	300·72
20	33·60	**60**	100·80	**100**	168·00	**140**	235·20	**180**	302·40
21	35·28	61	102·48	101	169·68	141	236·88	181	304·08
22	36·96	62	104·16	102	171·36	142	238·56	182	305·76
23	38·64	63	105·84	103	173·04	143	240·24	183	307·44
24	40·32	64	107·52	104	174·72	144	241·92	184	309·12
25	42·00	65	109·20	105	176·40	145	243·60	185	310·80
26	43·68	66	110·88	106	178·08	146	245·28	186	312·48
27	45·36	67	112·56	107	179·76	147	246·96	187	314·16
28	47·04	68	114·24	108	181·44	148	248·64	188	315·84
29	48·72	69	115·92	109	183·12	149	250·32	189	317·52
30	50·40	**70**	117·60	**110**	184·80	**150**	252·00	**190**	319·20
31	52·08	71	119·28	111	186·48	151	253·68	191	320·88
32	53·76	72	120·96	112	188·16	152	255·36	192	322·56
33	55·44	73	122·64	113	189·84	153	257·04	193	324·24
34	57·12	74	124·32	114	191·52	154	258·72	194	325·92
35	58·80	75	126·00	115	193·20	155	260·40	195	327·60
36	60·48	76	127·68	116	194·88	156	262·08	196	329·28
37	62·16	77	129·36	117	196·56	157	263·76	197	330·96
38	63·84	78	131·04	118	198·24	158	265·44	198	332·64
39	65·52	79	132·72	119	199·92	159	267·12	199	334·32
40	67·20	**80**	134·40	**120**	201·60	**160**	268·80	**200**	336·00

$\frac{1}{16}$=0·10^{5} $\frac{1}{9}$=0·18^{7} $\frac{1}{8}$=0·21^{0} $\frac{1}{6}$=0·28^{0} $\frac{1}{5}$=0·33^{6} $\frac{1}{4}$=0·42^{0} $\frac{1}{3}$=0·56^{0}

$\frac{3}{8}$=0·63^{0} $\frac{1}{2}$=0·84^{0} $\frac{5}{8}$=1·05^{0} $\frac{3}{4}$=1·26^{0} $\frac{7}{8}$=1·47^{0} $\frac{2}{3}$=1·12^{0} $\frac{5}{6}$=1·40^{0}

$\frac{3}{16}$=0·31^{5} $\frac{5}{16}$=0·52^{5} $\frac{7}{16}$=0·73^{5} $\frac{9}{16}$=0·94^{5} $\frac{11}{16}$=1·15^{5} $\frac{13}{16}$=1·36^{5} $\frac{15}{16}$=1·57^{5}

(0.593=1£) **£1.68½**

1	1·68½	41	69·08½	81	136·48½	121	203·88½	161	271·28½
2	3·37	42	70·77	82	138·17	122	205·57	162	272·97
3	5·05½	43	72·45½	83	139·85½	123	207·25½	163	274·65½
4	6·74	44	74·14	84	141·54	124	208·94	164	276·34
5	8·42½	45	75·82½	85	143·22½	125	210·62½	165	278·02½
6	10·11	46	77·51	86	144·91	126	212·31	166	279·71
7	11·79½	47	79·19½	87	146·59½	127	213·99½	167	281·39½
8	13·48	48	80·88	88	148·28	128	215·68	168	283·08
9	15·16½	49	82·56½	89	149·96½	129	217·36½	169	284·76½
10	16·85	**50**	84·25	**90**	151·65	**130**	219·05	**170**	286·45
11	18·53½	51	85·93½	91	153·33½	131	220·73½	171	288·13½
12	20·22	52	87·62	92	155·02	132	222·42	172	289·82
13	21·90½	53	89·30½	93	156·70½	133	224·10½	173	291·50½
14	23·59	54	90·99	94	158·39	134	225·79	174	293·19
15	25·27½	55	92·67½	95	160·07½	135	227·47½	175	294·87½
16	26·96	56	94·36	96	161·76	136	229·16	176	296·56
17	28·64½	57	96·04½	97	163·44½	137	230·84½	177	298·24½
18	30·33	58	97·73	98	165·13	138	232·53	178	299·93
19	32·01½	59	99·41½	99	166·81½	139	234·21½	179	301·61½
20	33·70	**60**	101·10	**100**	168·50	**140**	235·90	**180**	303·30
21	35·38½	61	102·78½	101	170·18½	141	237·58½	181	304·98½
22	37·07	62	104·47	102	171·87	142	239·27	182	306·67
23	38·75½	63	106·15½	103	173·55½	143	240·95½	183	308·35½
24	40·44	64	107·84	104	175·24	144	242·64	184	310·04
25	42·12½	65	109·52½	105	176·92½	145	244·32½	185	311·72½
26	43·81	66	111·21	106	178·61	146	246·01	186	313·41
27	45·49½	67	112·89½	107	180·29½	147	247·69½	187	315·09½
28	47·18	68	114·58	108	181·98	148	249·38	188	316·78
29	48·86½	69	116·26½	109	183·66½	149	251·06½	189	318·46½
30	50·55	**70**	117·95	**110**	185·35	**150**	252·75	**190**	320·15
31	52·23½	71	119·63½	111	187·03½	151	254·43½	191	321·83½
32	53·92	72	121·32	112	188·72	152	256·12	192	323·52
33	55·60½	73	123·00½	113	190·40½	153	257·80½	193	325·20½
34	57·29	74	124·69	114	192·09	154	259·49	194	326·89
35	58·97½	75	126·37½	115	193·77½	155	261·17½	195	328·57½
36	60·66	76	128·06	116	195·46	156	262·86	196	330·26
37	62·34½	77	129·74½	117	197·14½	157	264·54½	197	331·94½
38	64·03	78	131·43	118	198·83	158	266·23	198	333·63
39	65·71½	79	133·11½	119	200·51½	159	267·91½	199	335·31½
40	67·40	**80**	134·80	**120**	202·20	**160**	269·60	**200**	337·00

$\frac{1}{16}=0{\cdot}10^5$ $\frac{1}{9}=0{\cdot}18^7$ $\frac{1}{8}=0{\cdot}21^1$ $\frac{1}{6}=0{\cdot}28^1$ $\frac{1}{5}=0{\cdot}33^7$ $\frac{1}{4}=0{\cdot}42^1$ $\frac{1}{3}=0{\cdot}56^2$

$\frac{3}{8}=0{\cdot}63^2$ $\frac{1}{2}=0{\cdot}84^3$ $\frac{5}{8}=1{\cdot}05^3$ $\frac{3}{4}=1{\cdot}26^4$ $\frac{7}{8}=1{\cdot}47^4$ $\frac{2}{3}=1{\cdot}12^3$ $\frac{5}{6}=1{\cdot}40^4$

$\frac{3}{16}=0{\cdot}31^6$ $\frac{5}{16}=0{\cdot}52^7$ $\frac{7}{16}=0{\cdot}73^7$ $\frac{9}{16}=0{\cdot}94^8$ $\frac{11}{16}=1{\cdot}15^8$ $\frac{13}{16}=1{\cdot}36^9$ $\frac{15}{16}=1{\cdot}58^0$

(0.592=1£) **£1.69**

1	1·69	41	69·29	81	136·89	121	204·49	161	272·09
2	3·38	42	70·98	82	138·58	122	206·18	162	273·78
3	5·07	43	72·67	83	140·27	123	207·87	163	275·47
4	6·76	44	74·36	84	141·96	124	209·56	164	277·16
5	8·45	45	76·05	85	143·65	125	211·25	165	278·85
6	10·14	46	77·74	86	145·34	126	212·94	166	280·54
7	11·83	47	79·43	87	147·03	127	214·63	167	282·23
8	13·52	48	81·12	88	148·72	128	216·32	168	283·92
9	15·21	49	82·81	89	150·41	129	218·01	169	285·61
10	16·90	**50**	84·50	**90**	152·10	**130**	219·70	**170**	287·30
11	18·59	51	86·19	91	153·79	131	221·39	171	288·99
12	20·28	52	87·88	92	155·48	132	223·08	172	290·68
13	21·97	53	89·57	93	157·17	133	224·77	173	292·37
14	23·66	54	91·26	94	158·86	134	226·46	174	294·06
15	25·35	55	92·95	95	160·55	135	228·15	175	295·75
16	27·04	56	94·64	96	162·24	136	229·84	176	297·44
17	28·73	57	96·33	97	163·93	137	231·53	177	299·13
18	30·42	58	98·02	98	165·62	138	233·22	178	300·82
19	32·11	59	99·71	99	167·31	139	234·91	179	302·51
20	33·80	**60**	101·40	**100**	169·00	**140**	236·60	**180**	304·20
21	35·49	61	103·09	101	170·69	141	238·29	181	305·89
22	37·18	62	104·78	102	172·38	142	239·98	182	307·58
23	38·87	63	106·47	103	174·07	143	241·67	183	309·27
24	40·56	64	108·16	104	175·76	144	243·36	184	310·96
25	42·25	65	109·85	105	177·45	145	245·05	185	312·65
26	43·94	66	111·54	106	179·14	146	246·74	186	314·34
27	45·63	67	113·23	107	180·83	147	248·43	187	316·03
28	47·32	68	114·92	108	182·52	148	250·12	188	317·72
29	49·01	69	116·61	109	184·21	149	251·81	189	319·41
30	50·70	**70**	118·30	**110**	185·90	**150**	253·50	**190**	321·10
31	52·39	71	119·99	111	187·59	151	255·19	191	322·79
32	54·08	72	121·68	112	189·28	152	256·88	192	324·48
33	55·77	73	123·37	113	190·97	153	258·57	193	326·17
34	57·46	74	125·06	114	192·66	154	260·26	194	327·86
35	59·15	75	126·75	115	194·35	155	261·95	195	329·55
36	60·84	76	128·44	116	196·04	156	263·64	196	331·24
37	62·53	77	130·13	117	197·73	157	265·33	197	332·93
38	64·22	78	131·82	118	199·42	158	267·02	198	334·62
39	65·91	79	133·51	119	201·11	159	268·71	199	336·31
40	67·60	**80**	135·20	**120**	202·80	**160**	270·40	**200**	338·00

$\frac{1}{16}=0{\cdot}10^{6}$ $\frac{1}{9}=0{\cdot}18^{8}$ $\frac{1}{8}=0{\cdot}21^{1}$ $\frac{1}{6}=0{\cdot}28^{2}$ $\frac{1}{5}=0{\cdot}33^{8}$ $\frac{1}{4}=0{\cdot}42^{3}$ $\frac{1}{3}=0{\cdot}56^{3}$

$\frac{3}{8}=0{\cdot}63^{4}$ $\frac{1}{2}=0{\cdot}84^{5}$ $\frac{5}{8}=1{\cdot}05^{6}$ $\frac{3}{4}=1{\cdot}26^{8}$ $\frac{7}{8}=1{\cdot}47^{9}$ $\frac{2}{3}=1{\cdot}12^{7}$ $\frac{5}{6}=1{\cdot}40^{8}$

$\frac{3}{16}=0{\cdot}31^{7}$ $\frac{5}{16}=0{\cdot}52^{8}$ $\frac{7}{16}=0{\cdot}73^{9}$ $\frac{9}{16}=0{\cdot}95^{1}$ $\frac{11}{16}=1{\cdot}16^{2}$ $\frac{13}{16}=1{\cdot}37^{3}$ $\frac{15}{16}=1{\cdot}58^{4}$

1	1·69½	41	69·49½	81	137·29½	121	205·09½	161	272·89½
2	3·39	42	71·19	82	138·99	122	206·79	162	274·59
3	5·08½	43	72·88½	83	140·68½	123	208·48½	163	276·28½
4	6·78	44	74·58	84	142·38	124	210·18	164	277·98
5	8·47½	45	76·27½	85	144·07½	125	211·87½	165	279·67½
6	10·17	46	77·97	86	145·77	126	213·57	166	281·37
7	11·86½	47	79·66½	87	147·46½	127	215·26½	167	283·06½
8	13·56	48	81·36	88	149·16	128	216·96	168	284·76
9	15·25½	49	83·05½	89	150·85½	129	218·65½	169	286·45½
10	16·95	**50**	84·75	**90**	152·55	**130**	220·35	**170**	288·15
11	18·64½	51	86·44½	91	154·24½	131	222·04½	171	289·84½
12	20·34	52	88·14	92	155·94	132	223·74	172	291·54
13	22·03½	53	89·83½	93	157·63½	133	225·43½	173	293·23½
14	23·73	54	91·53	94	159·33	134	227·13	174	294·93
15	25·42½	55	93·22½	95	161·02½	135	228·82½	175	296·62½
16	27·12	56	94·92	96	162·72	136	230·52	176	298·32
17	28·81½	57	96·61½	97	164·41½	137	232·21½	177	300·01½
18	30·51	58	98·31	98	166·11	138	233·91	178	301·71
19	32·20½	59	100·00½	99	167·80½	139	235·60½	179	303·40½
20	33·90	**60**	101·70	**100**	169·50	**140**	237·30	**180**	305·10
21	35·59½	61	103·39½	101	171·19½	141	238·99½	181	306·79½
22	37·29	62	105·09	102	172·89	142	240·69	182	308·49
23	38·98½	63	106·78½	103	174·58½	143	242·38½	183	310·18½
24	40·68	64	108·48	104	176·28	144	244·08	184	311·88
25	42·37½	65	110·17½	105	177·97½	145	245·77½	185	313·57½
26	44·07	66	111·87	106	179·67	146	247·47	186	315·27
27	45·76½	67	113·56½	107	181·36½	147	249·16½	187	316·96½
28	47·46	68	115·26	108	183·06	148	250·86	188	318·66
29	49·15½	69	116·95½	109	184·75½	149	252·55½	189	320·35½
30	50·85	**70**	118·65	**110**	186·45	**150**	254·25	**190**	322·05
31	52·54½	71	120·34½	111	188·14½	151	255·94½	191	323·74½
32	54·24	72	122·04	112	189·84	152	257·64	192	325·44
33	55·93½	73	123·73½	113	191·53½	153	259·33½	193	327·13½
34	57·63	74	125·43	114	193·23	154	261·03	194	328·83
35	59·32½	75	127·12½	115	194·92½	155	262·72½	195	330·52½
36	61·02	76	128·82	116	196·62	156	264·42	196	332·22
37	62·71½	77	130·51½	117	198·31½	157	266·11½	197	333·91½
38	64·41	78	132·21	118	200·01	158	267·81	198	335·61
39	66·10½	79	133·90½	119	201·70½	159	269·50½	199	337·30½
40	67·80	**80**	135·60	**120**	203·40	**160**	271·20	**200**	339·00

$\frac{1}{16}=0{\cdot}10^{6}$ $\frac{1}{9}=0{\cdot}18^{8}$ $\frac{1}{8}=0{\cdot}21^{2}$ $\frac{1}{6}=0{\cdot}28^{3}$ $\frac{1}{5}=0{\cdot}33^{9}$ $\frac{1}{4}=0{\cdot}42^{4}$ $\frac{1}{3}=0{\cdot}56^{5}$

$\frac{3}{8}=0{\cdot}63^{6}$ $\frac{1}{2}=0{\cdot}84^{8}$ $\frac{5}{8}=1{\cdot}05^{9}$ $\frac{3}{4}=1{\cdot}27^{1}$ $\frac{7}{8}=1{\cdot}48^{3}$ $\frac{2}{3}=1{\cdot}13^{0}$ $\frac{5}{6}=1{\cdot}41^{3}$

$\frac{3}{16}=0{\cdot}31^{8}$ $\frac{5}{16}=0{\cdot}53^{0}$ $\frac{7}{16}=0{\cdot}74^{2}$ $\frac{9}{16}=0{\cdot}95^{3}$ $\frac{11}{16}=1{\cdot}16^{5}$ $\frac{13}{16}=1{\cdot}37^{7}$ $\frac{15}{16}=1{\cdot}58^{9}$

(0.588=1£) **£1.70**

1	1·70	41	69·70	81	137·70	121	205·70	161	273·70
2	3·40	42	71·40	82	139·40	122	207·40	162	275·40
3	5·10	43	73·10	83	141·10	123	209·10	163	277·10
4	6·80	44	74·80	84	142·80	124	210·80	164	278·80
5	8·50	45	76·50	85	144·50	125	212·50	165	280·50
6	10·20	46	78·20	86	146·20	126	214·20	166	282·20
7	11·90	47	79·90	87	147·90	127	215·90	167	283·90
8	13·60	48	81·60	88	149·60	128	217·60	168	285·60
9	15·30	49	83·30	89	151·30	129	219·30	169	287·30
10	17·00	**50**	85·00	**90**	153·00	**130**	221·00	**170**	289·00
11	18·70	51	86·70	91	154·70	131	222·70	171	290·70
12	20·40	52	88·40	92	156·40	132	224·40	172	292·40
13	22·10	53	90·10	93	158·10	133	226·10	173	294·10
14	23·80	54	91·80	94	159·80	134	227·80	174	295·80
15	25·50	55	93·50	95	161·50	135	229·50	175	297·50
16	27·20	56	95·20	96	163·20	136	231·20	176	299·20
17	28·90	57	96·90	97	164·90	137	232·90	177	300·90
18	30·60	58	98·60	98	166·60	138	234·60	178	302·60
19	32·30	59	100·30	99	168·30	139	236·30	179	304·30
20	34·00	**60**	102·00	**100**	170·00	**140**	238·00	**180**	306·00
21	35·70	61	103·70	101	171·70	141	239·70	181	307·70
22	37·40	62	105·40	102	173·40	142	241·40	182	309·40
23	39·10	63	107·10	103	175·10	143	243·10	183	311·10
24	40·80	64	108·80	104	176·80	144	244·80	184	312·80
25	42·50	65	110·50	105	178·50	145	246·50	185	314·50
26	44·20	66	112·20	106	180·20	146	248·20	186	316·20
27	45·90	67	113·90	107	181·90	147	249·90	187	317·90
28	47·60	68	115·60	108	183·60	148	251·60	188	319·60
29	49·30	69	117·30	109	185·30	149	253·30	189	321·30
30	51·00	**70**	119·00	**110**	187·00	**150**	255·00	**190**	323·00
31	52·70	71	120·70	111	188·70	151	256·70	191	324·70
32	54·40	72	122·40	112	190·40	152	258·40	192	326·40
33	56·10	73	124·10	113	192·10	153	260·10	193	328·10
34	57·80	74	125·80	114	193·80	154	261·80	194	329·80
35	59·50	75	127·50	115	195·50	155	263·50	195	331·50
36	61·20	76	129·20	116	197·20	156	265·20	196	333·20
37	62·90	77	130·90	117	198·90	157	266·90	197	334·90
38	64·60	78	132·60	118	200·60	158	268·60	198	336·60
39	66·30	79	134·30	119	202·30	159	270·30	199	338·30
40	68·00	**80**	136·00	**120**	204·00	**160**	272·00	**200**	340·00

$\frac{1}{16}=0{\cdot}10^{6}$ $\frac{1}{9}=0{\cdot}18^{9}$ $\frac{1}{8}=0{\cdot}21^{3}$ $\frac{1}{6}=0{\cdot}28^{3}$ $\frac{1}{5}=0{\cdot}34^{0}$ $\frac{1}{4}=0{\cdot}42^{5}$ $\frac{1}{3}=0{\cdot}56^{7}$

$\frac{3}{8}=0{\cdot}63^{8}$ $\frac{1}{2}=0{\cdot}85^{0}$ $\frac{5}{8}=1{\cdot}06^{3}$ $\frac{3}{4}=1{\cdot}27^{5}$ $\frac{7}{8}=1{\cdot}48^{8}$ $\frac{2}{3}=1{\cdot}13^{3}$ $\frac{5}{6}=1{\cdot}41^{7}$

$\frac{3}{16}=0{\cdot}31^{9}$ $\frac{5}{16}=0{\cdot}53^{1}$ $\frac{7}{16}=0{\cdot}74^{4}$ $\frac{9}{16}=0{\cdot}95^{6}$ $\frac{11}{16}=1{\cdot}16^{9}$ $\frac{13}{16}=1{\cdot}38^{1}$ $\frac{15}{16}=1{\cdot}59^{4}$

1	1·70½	41	69·90½	81	138·10½	121	206·30½	161	274·50½
2	3·41	42	71·61	82	139·81	122	208·01	162	276·21
3	5·11½	43	73·31½	83	141·51½	123	209·71½	163	277·91½
4	6·82	44	75·02	84	143·22	124	211·42	164	279·62
5	8·52½	45	76·72½	85	144·92½	125	213·12½	165	281·32½
6	10·23	46	78·43	86	146·63	126	214·83	166	283·03
7	11·93½	47	80·13½	87	148·33½	127	216·53½	167	284·73½
8	13·64	48	81·84	88	150·04	128	218·24	168	286·44
9	15·34½	49	83·54½	89	151·74½	129	219·94½	169	288·14½
10	17·05	**50**	85·25	**90**	153·45	**130**	221·65	**170**	289·85
11	18·75½	51	86·95½	91	155·15½	131	223·35½	171	291·55½
12	20·46	52	88·66	92	156·86	132	225·06	172	293·26
13	22·16½	53	90·36½	93	158·56½	133	226·76½	173	294·96½
14	23·87	54	92·07	94	160·27	134	228·47	174	296·67
15	25·57½	55	93·77½	95	161·97½	135	230·17½	175	298·37½
16	27·28	56	95·48	96	163·68	136	231·88	176	300·08
17	28·98½	57	97·18½	97	165·38½	137	233·58½	177	301·78½
18	30·69	58	98·89	98	167·09	138	235·29	178	303·49
19	32·39½	59	100·59½	99	168·79½	139	236·99½	179	305·19½
20	34·10	**60**	102·30	**100**	170·50	**140**	238·70	**180**	306·90
21	35·80½	61	104·00½	101	172·20½	141	240·40½	181	308·60½
22	37·51	62	105·71	102	173·91	142	242·11	182	310·31
23	39·21½	63	107·41½	103	175·61½	143	243·81½	183	312·01½
24	40·92	64	109·12	104	177·32	144	245·52	184	313·72
25	42·62½	65	110·82½	105	179·02½	145	247·22½	185	315·42½
26	44·33	66	112·53	106	180·73	146	248·93	186	317·13
27	46·03½	67	114·23½	107	182·43½	147	250·63½	187	318·83½
28	47·74	68	115·94	108	184·14	148	252·34	188	320·54
29	49·44½	69	117·64½	109	185·84½	149	254·04½	189	322·24½
30	51·15	**70**	119·35	**110**	187·55	**150**	255·75	**190**	323·95
31	52·85½	71	121·05½	111	189·25½	151	257·45½	191	325·65½
32	54·56	72	122·76	112	190·96	152	259·16	192	327·36
33	56·26½	73	124·46½	113	192·66½	153	260·86½	193	329·06½
34	57·97	74	126·17	114	194·37	154	262·57	194	330·77
35	59·67½	75	127·87½	115	196·07½	155	264·27½	195	332·47½
36	61·38	76	129·58	116	197·78	156	265·98	196	334·18
37	63·08½	77	131·28½	117	199·48½	157	267·68½	197	335·88½
38	64·79	78	132·99	118	201·19	158	269·39	198	337·59
39	66·49½	79	134·69½	119	202·89½	159	271·09½	199	339·29½
40	68·20	**80**	136·40	**120**	204·60	**160**	272·80	**200**	341·00

$\frac{1}{16}=0{\cdot}10^{7}$ $\frac{1}{9}=0{\cdot}18^{9}$ $\frac{1}{8}=0{\cdot}21^{3}$ $\frac{1}{6}=0{\cdot}28^{4}$ $\frac{1}{5}=0{\cdot}34^{1}$ $\frac{1}{4}=0{\cdot}42^{6}$ $\frac{1}{3}=0{\cdot}56^{8}$

$\frac{3}{8}=0{\cdot}63^{9}$ $\frac{1}{2}=0{\cdot}85^{3}$ $\frac{5}{8}=1{\cdot}06^{6}$ $\frac{3}{4}=1{\cdot}27^{9}$ $\frac{7}{8}=1{\cdot}49^{2}$ $\frac{2}{3}=1{\cdot}13^{7}$ $\frac{5}{6}=1{\cdot}42^{1}$

$\frac{3}{16}=0{\cdot}32^{0}$ $\frac{5}{16}=0{\cdot}53^{3}$ $\frac{7}{16}=0{\cdot}74^{6}$ $\frac{9}{16}=0{\cdot}95^{9}$ $\frac{11}{16}=1{\cdot}17^{2}$ $\frac{13}{16}=1{\cdot}38^{5}$ $\frac{15}{16}=1{\cdot}59^{8}$

1	1·71	41	70·11	81	138·51	121	206·91	161	275·31
2	3·42	42	71·82	82	140·22	122	208·62	162	277·02
3	5·13	43	73·53	83	141·93	123	210·33	163	278·73
4	6·84	44	75·24	84	143·64	124	212·04	164	280·44
5	8·55	45	76·95	85	145·35	125	213·75	165	282·15
6	10·26	46	78·66	86	147·06	126	215·46	166	283·86
7	11·97	47	80·37	87	148·77	127	217·17	167	285·57
8	13·68	48	82·08	88	150·48	128	218·88	168	287·28
9	15·39	49	83·79	89	152·19	129	220·59	169	288·99
10	17·10	**50**	85·50	**90**	153·90	**130**	222·30	**170**	290·70
11	18·81	51	87·21	91	155·61	131	224·01	171	292·41
12	20·52	52	88·92	92	157·32	132	225·72	172	294·12
13	22·23	53	90·63	93	159·03	133	227·43	173	295·83
14	23·94	54	92·34	94	160·74	134	229·14	174	297·54
15	25·65	55	94·05	95	162·45	135	230·85	175	299·25
16	27·36	56	95·76	96	164·16	136	232·56	176	300·96
17	29·07	57	97·47	97	165·87	137	234·27	177	302·67
18	30·78	58	99·18	98	167·58	138	235·98	178	304·38
19	32·49	59	100·89	99	169·29	139	237·69	179	306·09
20	34·20	**60**	102·60	**100**	171·00	**140**	239·40	**180**	307·80
21	35·91	61	104·31	101	172·71	141	241·11	181	309·51
22	37·62	62	106·02	102	174·42	142	242·82	182	311·22
23	39·33	63	107·73	103	176·13	143	244·53	183	312·93
24	41·04	64	109·44	104	177·84	144	246·24	184	314·64
25	42·75	65	111·15	105	179·55	145	247·95	185	316·35
26	44·46	66	112·86	106	181·26	146	249·66	186	318·06
27	46·17	67	114·57	107	182·97	147	251·37	187	319·77
28	47·88	68	116·28	108	184·68	148	253·08	188	321·48
29	49·59	69	117·99	109	186·39	149	254·79	189	323·19
30	51·30	**70**	119·70	**110**	188·10	**150**	256·50	**190**	324·90
31	53·01	71	121·41	111	189·81	151	258·21	191	326·61
32	54·72	72	123·12	112	191·52	152	259·92	192	328·32
33	56·43	73	124·83	113	193·23	153	261·63	193	330·03
34	58·14	74	126·54	114	194·94	154	263·34	194	331·74
35	59·85	75	128·25	115	196·65	155	265·05	195	333·45
36	61·56	76	129·96	116	198·36	156	266·76	196	335·16
37	63·27	77	131·67	117	200·07	157	268·47	197	336·87
38	64·98	78	133·38	118	201·78	158	270·18	198	338·58
39	66·69	79	135·09	119	203·49	159	271·89	199	340·29
40	68·40	**80**	136·80	**120**	205·20	**160**	273·60	**200**	342·00

$\frac{1}{16}=0{\cdot}10^{7}$ $\frac{1}{9}=0{\cdot}19^{0}$ $\frac{1}{8}=0{\cdot}21^{4}$ $\frac{1}{6}=0{\cdot}28^{5}$ $\frac{1}{5}=0{\cdot}34^{2}$ $\frac{1}{4}=0{\cdot}42^{8}$ $\frac{1}{3}=0{\cdot}57^{0}$

$\frac{3}{8}=0{\cdot}64^{1}$ $\frac{1}{2}=0{\cdot}85^{5}$ $\frac{5}{8}=1{\cdot}06^{9}$ $\frac{3}{4}=1{\cdot}28^{3}$ $\frac{7}{8}=1{\cdot}49^{6}$ $\frac{2}{3}=1{\cdot}14^{0}$ $\frac{5}{6}=1{\cdot}42^{5}$

$\frac{3}{16}=0{\cdot}32^{1}$ $\frac{5}{16}=0{\cdot}53^{4}$ $\frac{7}{16}=0{\cdot}74^{8}$ $\frac{9}{16}=0{\cdot}96^{2}$ $\frac{11}{16}=1{\cdot}17^{6}$ $\frac{13}{16}=1{\cdot}38^{9}$ $\frac{15}{16}=1{\cdot}60^{3}$

(0.583=1£) **£1.71½**

1	1·71½	41	70·31½	81	138·91½	121	207·51½	161	276·11½
2	3·43	42	72·03	82	140·63	122	209·23	162	277·83
3	5·14½	43	73·74½	83	142·34½	123	210·94½	163	279·54½
4	6·86	44	75·46	84	144·06	124	212·66	164	281·26
5	8·57½	45	77·17½	85	145·77½	125	214·37½	165	282·97½
6	10·29	46	78·89	86	147·49	126	216·09	166	284·69
7	12·00½	47	80·60½	87	149·20½	127	217·80½	167	286·40½
8	13·72	48	82·32	88	150·92	128	219·52	168	288·12
9	15·43½	49	84·03½	89	152·63½	129	221·23½	169	289·83½
10	17·15	**50**	85·75	**90**	154·35	**130**	222·95	**170**	291·55
11	18·86½	51	87·46½	91	156·06½	131	224·66½	171	293·26½
12	20·58	52	89·18	92	157·78	132	226·38	172	294·98
13	22·29½	53	90·89½	93	159·49½	133	228·09½	173	296·69½
14	24·01	54	92·61	94	161·21	134	229·81	174	298·41
15	25·72½	55	94·32½	95	162·92½	135	231·52½	175	300·12½
16	27·44	56	96·04	96	164·64	136	233·24	176	301·84
17	29·15½	57	97·75½	97	166·35½	137	234·95½	177	303·55½
18	30·87	58	99·47	98	168·07	138	236·67	178	305·27
19	32·58½	59	101·18½	99	169·78½	139	238·38½	179	306·98½
20	34·30	**60**	102·90	**100**	171·50	**140**	240·10	**180**	308·70
21	36·01½	61	104·61½	101	173·21½	141	241·81½	181	310·41½
22	37·73	62	106·33	102	174·93	142	243·53	182	312·13
23	39·44½	63	108·04½	103	176·64½	143	245·24½	183	313·84½
24	41·16	64	109·76	104	178·36	144	246·96	184	315·56
25	42·87½	65	111·47½	105	180·07½	145	248·67½	185	317·27½
26	44·59	66	113·19	106	181·79	146	250·39	186	318·99
27	46·30½	67	114·90½	107	183·50½	147	252·10½	187	320·70½
28	48·02	68	116·62	108	185·22	148	253·82	188	322·42
29	49·73½	69	118·33½	109	186·93½	149	255·53½	189	324·13½
30	51·45	**70**	120·05	**110**	188·65	**150**	257·25	**190**	325·85
31	53·16½	71	121·76½	111	190·36½	151	258·96½	191	327·56½
32	54·88	72	123·48	112	192·08	152	260·68	192	329·28
33	56·59½	73	125·19½	113	193·79½	153	262·39½	193	330·99½
34	58·31	74	126·91	114	195·51	154	264·11	194	332·71
35	60·02½	75	128·62½	115	197·22½	155	265·82½	195	334·42½
36	61·74	76	130·34	116	198·94	156	267·54	196	336·14
37	63·45½	77	132·05½	117	200·65½	157	269·25½	197	337·85½
38	65·17	78	133·77	118	202·37	158	270·97	198	339·57
39	66·88½	79	135·48½	119	204·08½	159	272·68½	199	341·28½
40	68·60	**80**	137·20	**120**	205·80	**160**	274·40	**200**	343·00

$\frac{1}{16}$=0·10^{7} $\frac{1}{9}$=0·19^{1} $\frac{1}{8}$=0·21^{4} $\frac{1}{6}$=0·28^{6} $\frac{1}{5}$=0·34^{3} $\frac{1}{4}$=0·42^{9} $\frac{1}{3}$=0·57^{2}

$\frac{3}{8}$=0·64^{3} $\frac{1}{2}$=0·85^{8} $\frac{5}{8}$=1·07^{2} $\frac{3}{4}$=1·28^{6} $\frac{7}{8}$=1·50^{1} $\frac{2}{3}$=1·14^{3} $\frac{5}{6}$=1·42^{9}

$\frac{3}{16}$=0·32^{2} $\frac{5}{16}$=0·53^{6} $\frac{7}{16}$=0·75^{0} $\frac{9}{16}$=0·96^{5} $\frac{11}{16}$=1·17^{9} $\frac{13}{16}$=1·39^{3} $\frac{15}{16}$=1·60^{8}

1	1·72	41	70·52	81	139·32	121	208·12	161	276·92
2	3·44	42	72·24	82	141·04	122	209·84	162	278·64
3	5·16	43	73·96	83	142·76	123	211·56	163	280·36
4	6·88	44	75·68	84	144·48	124	213·28	164	282·08
5	8·60	45	77·40	85	146·20	125	215·00	165	283·80
6	10·32	46	79·12	86	147·92	126	216·72	166	285·52
7	12·04	47	80·84	87	149·64	127	218·44	167	287·24
8	13·76	48	82·56	88	151·36	128	220·16	168	288·96
9	15·48	49	84·28	89	153·08	129	221·88	169	290·68
10	17·20	**50**	86·00	**90**	154·80	**130**	223·60	**170**	292·40
11	18·92	51	87·72	91	156·52	131	225·32	171	294·12
12	20·64	52	89·44	92	158·24	132	227·04	172	295·84
13	22·36	53	91·16	93	159·96	133	228·76	173	297·56
14	24·08	54	92·88	94	161·68	134	230·48	174	299·28
15	25·80	55	94·60	95	163·40	135	232·20	175	301·00
16	27·52	56	96·32	96	165·12	136	233·92	176	302·72
17	29·24	57	98·04	97	166·84	137	235·64	177	304·44
18	30·96	58	99·76	98	168·56	138	237·36	178	306·16
19	32·68	59	101·48	99	170·28	139	239·08	179	307·88
20	34·40	**60**	103·20	**100**	172·00	**140**	240·80	**180**	309·60
21	36·12	61	104·92	101	173·72	141	242·52	181	311·32
22	37·84	62	106·64	102	175·44	142	244·24	182	313·04
23	39·56	63	108·36	103	177·16	143	245·96	183	314·76
24	41·28	64	110·08	104	178·88	144	247·68	184	316·48
25	43·00	65	111·80	105	180·60	145	249·40	185	318·20
26	44·72	66	113·52	106	182·32	146	251·12	186	319·92
27	46·44	67	115·24	107	184·04	147	252·84	187	321·64
28	48·16	68	116·96	108	185·76	148	254·56	188	323·36
29	49·88	69	118·68	109	187·48	149	256·28	189	325·08
30	51·60	**70**	120·40	**110**	189·20	**150**	258·00	**190**	326·80
31	53·32	71	122·12	111	190·92	151	259·72	191	328·52
32	55·04	72	123·84	112	192·64	152	261·44	192	330·24
33	56·76	73	125·56	113	194·36	153	263·16	193	331·96
34	58·48	74	127·28	114	196·08	154	264·88	194	333·68
35	60·20	75	129·00	115	197·80	155	266·60	195	335·40
36	61·92	76	130·72	116	199·52	156	268·32	196	337·12
37	63·64	77	132·44	117	201·24	157	270·04	197	338·84
38	65·36	78	134·16	118	202·96	158	271·76	198	340·56
39	67·08	79	135·88	119	204·68	159	273·48	199	342·28
40	68·80	**80**	137·60	**120**	206·40	**160**	275·20	**200**	344·00

$\frac{1}{16}=0{\cdot}10^{8}$ $\frac{1}{9}=0{\cdot}19^{1}$ $\frac{1}{8}=0{\cdot}21^{5}$ $\frac{1}{6}=0{\cdot}28^{7}$ $\frac{1}{5}=0{\cdot}34^{4}$ $\frac{1}{4}=0{\cdot}43^{0}$ $\frac{1}{3}=0{\cdot}57^{3}$

$\frac{3}{8}=0{\cdot}64^{5}$ $\frac{1}{2}=0{\cdot}86^{0}$ $\frac{5}{8}=1{\cdot}07^{5}$ $\frac{3}{4}=1{\cdot}29^{0}$ $\frac{7}{8}=1{\cdot}50^{5}$ $\frac{2}{3}=1{\cdot}14^{7}$ $\frac{5}{6}=1{\cdot}43^{3}$

$\frac{3}{16}=0{\cdot}32^{3}$ $\frac{5}{16}=0{\cdot}53^{8}$ $\frac{7}{16}=0{\cdot}75^{3}$ $\frac{9}{16}=0{\cdot}96^{8}$ $\frac{11}{16}=1{\cdot}18^{3}$ $\frac{13}{16}=1{\cdot}39^{8}$ $\frac{15}{16}=1{\cdot}61^{3}$

(0.580=1£) **£1.72½**

1	1·72½	41	70·72½	81	139·72½	121	208·72½	161	277·72½
2	3·45	42	72·45	82	141·45	122	210·45	162	279·45
3	5·17½	43	74·17½	83	143·17½	123	212·17½	163	281·17½
4	6·90	44	75·90	84	144·90	124	213·90	164	282·90
5	8·62½	45	77·62½	85	146·62½	125	215·62½	165	284·62½
6	10·35	46	79·35	86	148·35	126	217·35	166	286·35
7	12·07½	47	81·07½	87	150·07½	127	219·07½	167	288·07½
8	13·80	48	82·80	88	151·80	128	220·80	168	289·80
9	15·52½	49	84·52½	89	153·52½	129	222·52½	169	291·52½
10	17·25	**50**	86·25	**90**	155·25	**130**	224·25	**170**	293·25
11	18·97½	51	87·97½	91	156·97½	131	225·97½	171	294·97½
12	20·70	52	89·70	92	158·70	132	227·70	172	296·70
13	22·42½	53	91·42½	93	160·42½	133	229·42½	173	298·42½
14	24·15	54	93·15	94	162·15	134	231·15	174	300·15
15	25·87½	55	94·87½	95	163·87½	135	232·87½	175	301·87½
16	27·60	56	96·60	96	165·60	136	234·60	176	303·60
17	29·32½	57	98·32½	97	167·32½	137	236·32½	177	305·32½
18	31·05	58	100·05	98	169·05	138	238·05	178	307·05
19	32·77½	59	101·77½	99	170·77½	139	239·77½	179	308·77½
20	34·50	**60**	103·50	**100**	172·50	**140**	241·50	**180**	310·50
21	36·22½	61	105·22½	101	174·22½	141	243·22½	181	312·22½
22	37·95	62	106·95	102	175·95	142	244·95	182	313·95
23	39·67½	63	108·67½	103	177·67½	143	246·67½	183	315·67½
24	41·40	64	110·40	104	179·40	144	248·40	184	317·40
25	43·12½	65	112·12½	105	181·12½	145	250·12½	185	319·12½
26	44·85	66	113·85	106	182·85	146	251·85	186	320·85
27	46·57½	67	115·57½	107	184·57½	147	253·57½	187	322·57½
28	48·30	68	117·30	108	186·30	148	255·30	188	324·30
29	50·02½	69	119·02½	109	188·02½	149	257·02½	189	326·02½
30	51·75	**70**	120·75	**110**	189·75	**150**	258·75	**190**	327·75
31	53·47½	71	122·47½	111	191·47½	151	260·47½	191	329·47½
32	55·20	72	124·20	112	193·20	152	262·20	192	331·20
33	56·92½	73	125·92½	113	194·92½	153	263·92½	193	332·92½
34	58·65	74	127·65	114	196·65	154	265·65	194	334·65
35	60·37½	75	129·37½	115	198·37½	155	267·37½	195	336·37½
36	62·10	76	131·10	116	200·10	156	269·10	196	338·10
37	63·82½	77	132·82½	117	201·82½	157	270·82½	197	339·82½
38	65·55	78	134·55	118	203·55	158	272·55	198	341·55
39	67·27½	79	136·27½	119	205·27½	159	274·27½	199	343·27½
40	69·00	**80**	138·00	**120**	207·00	**160**	276·00	**200**	345·00

$\frac{1}{16}=0{\cdot}10^{8}$ $\frac{1}{9}=0{\cdot}19^{2}$ $\frac{1}{8}=0{\cdot}21^{6}$ $\frac{1}{6}=0{\cdot}28^{8}$ $\frac{1}{5}=0{\cdot}34^{5}$ $\frac{1}{4}=0{\cdot}43^{1}$ $\frac{1}{3}=0{\cdot}57^{5}$

$\frac{3}{8}=0{\cdot}64^{7}$ $\frac{1}{2}=0{\cdot}86^{3}$ $\frac{5}{8}=1{\cdot}07^{8}$ $\frac{3}{4}=1{\cdot}29^{4}$ $\frac{7}{8}=1{\cdot}50^{9}$ $\frac{2}{3}=1{\cdot}15^{0}$ $\frac{5}{6}=1{\cdot}43^{8}$

$\frac{3}{16}=0{\cdot}32^{3}$ $\frac{5}{16}=0{\cdot}53^{9}$ $\frac{7}{16}=0{\cdot}75^{5}$ $\frac{9}{16}=0{\cdot}97^{0}$ $\frac{11}{16}=1{\cdot}18^{6}$ $\frac{13}{16}=1{\cdot}40^{2}$ $\frac{15}{16}=1{\cdot}61^{7}$

(0.578=1£) **£1.73**

1	1·73	41	70·93	81	140·13	121	209·33	161	278·53
2	3·46	42	72·66	82	141·86	122	211·06	162	280·26
3	5·19	43	74·39	83	143·59	123	212·79	163	281·99
4	6·92	44	76·12	84	145·32	124	214·52	164	283·72
5	8·65	45	77·85	85	147·05	125	216·25	165	285·45
6	10·38	46	79·58	86	148·78	126	217·98	166	287·18
7	12·11	47	81·31	87	150·51	127	219·71	167	288·91
8	13·84	48	83·04	88	152·24	128	221·44	168	290·64
9	15·57	49	84·77	89	153·97	129	223·17	169	292·37
10	17·30	**50**	86·50	**90**	155·70	**130**	224·90	**170**	294·10
11	19·03	51	88·23	91	157·43	131	226·63	171	295·83
12	20·76	52	89·96	92	159·16	132	228·36	172	297·56
13	22·49	53	91·69	93	160·89	133	230·09	173	299·29
14	24·22	54	93·42	94	162·62	134	231·82	174	301·02
15	25·95	55	95·15	95	164·35	135	233·55	175	302·75
16	27·68	56	96·88	96	166·08	136	235·28	176	304·48
17	29·41	57	98·61	97	167·81	137	237·01	177	306·21
18	31·14	58	100·34	98	169·54	138	238·74	178	307·94
19	32·87	59	102·07	99	171·27	139	240·47	179	309·67
20	34·60	**60**	103·80	**100**	173·00	**140**	242·20	**180**	311·40
21	36·33	61	105·53	101	174·73	141	243·93	181	313·13
22	38·06	62	107·26	102	176·46	142	245·66	182	314·86
23	39·79	63	108·99	103	178·19	143	247·39	183	316·59
24	41·52	64	110·72	104	179·92	144	249·12	184	318·32
25	43·25	65	112·45	105	181·65	145	250·85	185	320·05
26	44·98	66	114·18	106	183·38	146	252·58	186	321·78
27	46·71	67	115·91	107	185·11	147	254·31	187	323·51
28	48·44	68	117·64	108	186·84	148	256·04	188	325·24
29	50·17	69	119·37	109	188·57	149	257·77	189	326·97
30	51·90	**70**	121·10	**110**	190·30	**150**	259·50	**190**	328·70
31	53·63	71	122·83	111	192·03	151	261·23	191	330·43
32	55·36	72	124·56	112	193·76	152	262·96	192	332·16
33	57·09	73	126·29	113	195·49	153	264·69	193	333·89
34	58·82	74	128·02	114	197·22	154	266·42	194	335·62
35	60·55	75	129·75	115	198·95	155	268·15	195	337·35
36	62·28	76	131·48	116	200·68	156	269·88	196	339·08
37	64·01	77	133·21	117	202·41	157	271·61	197	340·81
38	65·74	78	134·94	118	204·14	158	273·34	198	342·54
39	67·47	79	136·67	119	205·87	159	275·07	199	344·27
40	69·20	**80**	138·40	**120**	207·60	**160**	276·80	**200**	346·00

$\frac{1}{16}=0{\cdot}10^{8}$ $\frac{1}{9}=0{\cdot}19^{2}$ $\frac{1}{8}=0{\cdot}21^{6}$ $\frac{1}{6}=0{\cdot}28^{8}$ $\frac{1}{5}=0{\cdot}34^{6}$ $\frac{1}{4}=0{\cdot}43^{3}$ $\frac{1}{3}=0{\cdot}57^{7}$

$\frac{3}{8}=0{\cdot}64^{9}$ $\frac{1}{2}=0{\cdot}86^{5}$ $\frac{5}{8}=1{\cdot}08^{1}$ $\frac{3}{4}=1{\cdot}29^{8}$ $\frac{7}{8}=1{\cdot}51^{4}$ $\frac{2}{3}=1{\cdot}15^{3}$ $\frac{5}{6}=1{\cdot}44^{2}$

1·732 is the square root of 3

(0.576=1£) **£1.73½**

1	1·73½	41	71·13½	81	140·53½	121	209·93½	161	279·33½
2	3·47	42	72·87	82	142·27	122	211·67	162	281·07
3	5·20½	43	74·60½	83	144·00½	123	213·40½	163	282·80½
4	6·94	44	76·34	84	145·74	124	215·14	164	284·54
5	8·67½	45	78·07½	85	147·47½	125	216·87½	165	286·27½
6	10·41	46	79·81	86	149·21	126	218·61	166	288·01
7	12·14½	47	81·54½	87	150·94½	127	220·34½	167	289·74½
8	13·88	48	83·28	88	152·68	128	222·08	168	291·48
9	15·61½	49	85·01½	89	154·41½	129	223·81½	169	293·21½
10	17·35	**50**	86·75	**90**	156·15	**130**	225·55	**170**	294·95
11	19·08½	51	88·48½	91	157·88½	131	227·28½	171	296·68½
12	20·82	52	90·22	92	159·62	132	229·02	172	298·42
13	22·55½	53	91·95½	93	161·35½	133	230·75½	173	300·15½
14	24·29	54	93·69	94	163·09	134	232·49	174	301·89
15	26·02½	55	95·42½	95	164·82½	135	234·22½	175	303·62½
16	27·76	56	97·16	96	166·56	136	235·96	176	305·36
17	29·49½	57	98·89½	97	168·29½	137	237·69½	177	307·09½
18	31·23	58	100·63	98	170·03	138	239·43	178	308·83
19	32·96½	59	102·36½	99	171·76½	139	241·16½	179	310·56½
20	34·70	**60**	104·10	**100**	173·50	**140**	242·90	**180**	312·30
21	36·43½	61	105·83½	101	175·23½	141	244·63½	181	314·03½
22	38·17	62	107·57	102	176·97	142	246·37	182	315·77
23	39·90½	63	109·30½	103	178·70½	143	248·10½	183	317·50½
24	41·64	64	111·04	104	180·44	144	249·84	184	319·24
25	43·37½	65	112·77½	105	182·17½	145	251·57½	185	320·97½
26	45·11	66	114·51	106	183·91	146	253·31	186	322·71
27	46·84½	67	116·24½	107	185·64½	147	255·04½	187	324·44½
28	48·58	68	117·98	108	187·38	148	256·78	188	326·18
29	50·31½	69	119·71½	109	189·11½	149	258·51½	189	327·91½
30	52·05	**70**	121·45	**110**	190·85	**150**	260·25	**190**	329·65
31	53·78½	71	123·18½	111	192·58½	151	261·98½	191	331·38½
32	55·52	72	124·92	112	194·32	152	263·72	192	333·12
33	57·25½	73	126·65½	113	196·05½	153	265·45½	193	334·85½
34	58·99	74	128·39	114	197·79	154	267·19	194	336·59
35	60·72½	75	130·12½	115	199·52½	155	268·92½	195	338·32½
36	62·46	76	131·86	116	201·26	156	270·66	196	340·06
37	64·19½	77	133·59½	117	202·99½	157	272·39½	197	341·79½
38	65·93	78	135·33	118	204·73	158	274·13	198	343·53
39	67·66½	79	137·06½	119	206·46½	159	275·86½	199	345·26½
40	69·40	**80**	138·80	**120**	208·20	**160**	277·60	**200**	347·00

$\frac{1}{16}=0{\cdot}10^{8}$	$\frac{1}{9}=0{\cdot}19^{3}$	$\frac{1}{8}=0{\cdot}21^{7}$	$\frac{1}{6}=0{\cdot}28^{9}$	$\frac{1}{5}=0{\cdot}34^{7}$	$\frac{1}{4}=0{\cdot}43^{4}$	$\frac{1}{3}=0{\cdot}57^{8}$
$\frac{3}{8}=0{\cdot}65^{1}$	$\frac{1}{2}=0{\cdot}86^{8}$	$\frac{5}{8}=1{\cdot}08^{4}$	$\frac{3}{4}=1{\cdot}30^{1}$	$\frac{7}{8}=1{\cdot}51^{8}$	$\frac{2}{3}=1{\cdot}15^{7}$	$\frac{5}{6}=1{\cdot}44^{6}$
$\frac{3}{16}=0{\cdot}32^{5}$	$\frac{5}{16}=0{\cdot}54^{2}$	$\frac{7}{16}=0{\cdot}75^{9}$	$\frac{9}{16}=0{\cdot}97^{6}$	$\frac{11}{16}=1{\cdot}19^{3}$	$\frac{13}{16}=1{\cdot}41^{0}$	$\frac{15}{16}=1{\cdot}62^{7}$

1	1·74	41	71·34	81	140·94	121	210·54	161	280·14
2	3·48	42	73·08	82	142·68	122	212·28	162	281·88
3	5·22	43	74·82	83	144·42	123	214·02	163	283·62
4	6·96	44	76·56	84	146·16	124	215·76	164	285·36
5	8·70	45	78·30	85	147·90	125	217·50	165	287·10
6	10·44	46	80·04	86	149·64	126	219·24	166	288·84
7	12·18	47	81·78	87	151·38	127	220·98	167	290·58
8	13·92	48	83·52	88	153·12	128	222·72	168	292·32
9	15·66	49	85·26	89	154·86	129	224·46	169	294·06
10	17·40	**50**	87·00	**90**	156·60	**130**	226·20	**170**	295·80
11	19·14	51	88·74	91	158·34	131	227·94	171	297·54
12	20·88	52	90·48	92	160·08	132	229·68	172	299·28
13	22·62	53	92·22	93	161·82	133	231·42	173	301·02
14	24·36	54	93·96	94	163·56	134	233·16	174	302·76
15	26·10	55	95·70	95	165·30	135	234·90	175	304·50
16	27·84	56	97·44	96	167·04	136	236·64	176	306·24
17	29·58	57	99·18	97	168·78	137	238·38	177	307·98
18	31·32	58	100·92	98	170·52	138	240·12	178	309·72
19	33·06	59	102·66	99	172·26	139	241·86	179	311·46
20	34·80	**60**	104·40	**100**	174·00	**140**	243·60	**180**	313·20
21	36·54	61	106·14	101	175·74	141	245·34	181	314·94
22	38·28	62	107·88	102	177·48	142	247·08	182	316·68
23	40·02	63	109·62	103	179·22	143	248·82	183	318·42
24	41·76	64	111·36	104	180·96	144	250·56	184	320·16
25	43·50	65	113·10	105	182·70	145	252·30	185	321·90
26	45·24	66	114·84	106	184·44	146	254·04	186	323·64
27	46·98	67	116·58	107	186·18	147	255·78	187	325·38
28	48·72	68	118·32	108	187·92	148	257·52	188	327·12
29	50·46	69	120·06	109	189·66	149	259·26	189	328·86
30	52·20	**70**	121·80	**110**	191·40	**150**	261·00	**190**	330·60
31	53·94	71	123·54	111	193·14	151	262·74	191	332·34
32	55·68	72	125·28	112	194·88	152	264·48	192	334·08
33	57·42	73	127·02	113	196·62	153	266·22	193	335·82
34	59·16	74	128·76	114	198·36	154	267·96	194	337·56
35	60·90	75	130·50	115	200·10	155	269·70	195	339·30
36	62·64	76	132·24	116	201·84	156	271·44	196	341·04
37	64·38	77	133·98	117	203·58	157	273·18	197	342·78
38	66·12	78	135·72	118	205·32	158	274·92	198	344·52
39	67·86	79	137·46	119	207·06	159	276·66	199	346·26
40	69·60	**80**	139·20	**120**	208·80	**160**	278·40	**200**	348·00

$\frac{1}{16}=0{\cdot}10^{9}$	$\frac{1}{9}=0{\cdot}19^{3}$	$\frac{1}{8}=0{\cdot}21^{8}$	$\frac{1}{6}=0{\cdot}29^{0}$	$\frac{1}{5}=0{\cdot}34^{8}$	$\frac{1}{4}=0{\cdot}43^{5}$	$\frac{1}{3}=0{\cdot}58^{0}$
$\frac{3}{8}=0{\cdot}65^{3}$	$\frac{1}{2}=0{\cdot}87^{0}$	$\frac{5}{8}=1{\cdot}08^{8}$	$\frac{3}{4}=1{\cdot}30^{5}$	$\frac{7}{8}=1{\cdot}52^{3}$	$\frac{2}{3}=1{\cdot}16^{0}$	$\frac{5}{6}=1{\cdot}45^{0}$
$\frac{3}{16}=0{\cdot}32^{6}$	$\frac{5}{16}=0{\cdot}54^{4}$	$\frac{7}{16}=0{\cdot}76^{1}$	$\frac{9}{16}=0{\cdot}97^{9}$	$\frac{11}{16}=1{\cdot}19^{6}$	$\frac{13}{16}=1{\cdot}41^{4}$	$\frac{15}{16}=1{\cdot}63^{1}$

(0.573=1£) **£1.74½**

1	1·74½	41	71·54½	81	141·34½	121	211·14½	161	280·94½
2	3·49	42	73·29	82	143·09	122	212·89	162	282·69
3	5·23½	43	75·03½	83	144·83½	123	214·63½	163	284·43½
4	6·98	44	76·78	84	146·58	124	216·38	164	286·18
5	8·72½	45	78·52½	85	148·32½	125	218·12½	165	287·92½
6	10·47	46	80·27	86	150·07	126	219·87	166	289·67
7	12·21½	47	82·01½	87	151·81½	127	221·61½	167	291·41½
8	13·96	48	83·76	88	153·56	128	223·36	168	293·16
9	15·70½	49	85·50½	89	155·30½	129	225·10½	169	294·90½
10	17·45	**50**	87·25	**90**	157·05	**130**	226·85	**170**	296·65
11	19·19½	51	88·99½	91	158·79½	131	228·59½	171	298·39½
12	20·94	52	90·74	92	160·54	132	230·34	172	300·14
13	22·68½	53	92·48½	93	162·28½	133	232·08½	173	301·88½
14	24·43	54	94·23	94	164·03	134	233·83	174	303·63
15	26·17½	55	95·97½	95	165·77½	135	235·57½	175	305·37½
16	27·92	56	97·72	96	167·52	136	237·32	176	307·12
17	29·66½	57	99·46½	97	169·26½	137	239·06½	177	308·86½
18	31·41	58	101·21	98	171·01	138	240·81	178	310·61
19	33·15½	59	102·95½	99	172·75½	139	242·55½	179	312·35½
20	34·90	**60**	104·70	**100**	174·50	**140**	244·30	**180**	314·10
21	36·64½	61	106·44½	101	176·24½	141	246·04½	181	315·84½
22	38·39	62	108·19	102	177·99	142	247·79	182	317·59
23	40·13½	63	109·93½	103	179·73½	143	249·53½	183	319·33½
24	41·88	64	111·68	104	181·48	144	251·28	184	321·08
25	43·62½	65	113·42½	105	183·22½	145	253·02½	185	322·82½
26	45·37	66	115·17	106	184·97	146	254·77	186	324·57
27	47·11½	67	116·91½	107	186·71½	147	256·51½	187	326·31½
28	48·86	68	118·66	108	188·46	148	258·26	188	328·06
29	50·60½	69	120·40½	109	190·20½	149	260·00½	189	329·80½
30	52·35	**70**	122·15	**110**	191·95	**150**	261·75	**190**	331·55
31	54·09½	71	123·89½	111	193·69½	151	263·49½	191	333·29½
32	55·84	72	125·64	112	195·44	152	265·24	192	335·04
33	57·58½	73	127·38½	113	197·18½	153	266·98½	193	336·78½
34	59·33	74	129·13	114	198·93	154	268·73	194	338·53
35	61·07½	75	130·87½	115	200·67½	155	270·47½	195	340·27½
36	62·82	76	132·62	116	202·42	156	272·22	196	342·02
37	64·56½	77	134·36½	117	204·16½	157	273·96½	197	343·76½
38	66·31	78	136·11	118	205·91	158	275·71	198	345·51
39	68·05½	79	137·85½	119	207·65½	159	277·45½	199	347·25½
40	69·80	**80**	139·60	**120**	209·40	**160**	279·20	**200**	349·00

$\frac{1}{16}=0{\cdot}10^{9}$ $\frac{1}{9}=0{\cdot}19^{4}$ $\frac{1}{8}=0{\cdot}21^{8}$ $\frac{1}{6}=0{\cdot}29^{1}$ $\frac{1}{5}=0{\cdot}34^{9}$ $\frac{1}{4}=0{\cdot}43^{6}$ $\frac{1}{3}=0{\cdot}58^{2}$

$\frac{3}{8}=0{\cdot}65^{4}$ $\frac{1}{2}=0{\cdot}87^{3}$ $\frac{5}{8}=1{\cdot}09^{1}$ $\frac{3}{4}=1{\cdot}30^{9}$ $\frac{7}{8}=1{\cdot}52^{7}$ $\frac{2}{3}=1{\cdot}16^{3}$ $\frac{5}{6}=1{\cdot}45^{4}$

$\frac{3}{16}=0{\cdot}32^{7}$ $\frac{5}{16}=0{\cdot}54^{5}$ $\frac{7}{16}=0{\cdot}76^{3}$ $\frac{9}{16}=0{\cdot}98^{2}$ $\frac{11}{16}=1{\cdot}20^{0}$ $\frac{13}{16}=1{\cdot}41^{8}$ $\frac{15}{16}=1{\cdot}63^{6}$

(0.571=1£) **£1.75**

1	1·75	41	71·75	81	141·75	121	211·75	161	281·75
2	3·50	42	73·50	82	143·50	122	213·50	162	283·50
3	5·25	43	75·25	83	145·25	123	215·25	163	285·25
4	7·00	44	77·00	84	147·00	124	217·00	164	287·00
5	8·75	45	78·75	85	148·75	125	218·75	165	288·75
6	10·50	46	80·50	86	150·50	126	220·50	166	290·50
7	12·25	47	82·25	87	152·25	127	222·25	167	292·25
8	14·00	48	84·00	88	154·00	128	224·00	168	294·00
9	15·75	49	85·75	89	155·75	129	225·75	169	295·75
10	17·50	**50**	87·50	**90**	157·50	**130**	227·50	**170**	297·50
11	19·25	51	89·25	91	159·25	131	229·25	171	299·25
12	21·00	52	91·00	92	161·00	132	231·00	172	301·00
13	22·75	53	92·75	93	162·75	133	232·75	173	302·75
14	24·50	54	94·50	94	164·50	134	234·50	174	304·50
15	26·25	55	96·25	95	166·25	135	236·25	175	306·25
16	28·00	56	98·00	96	168·00	136	238·00	176	308·00
17	29·75	57	99·75	97	169·75	137	239·75	177	309·75
18	31·50	58	101·50	98	171·50	138	241·50	178	311·50
19	33·25	59	103·25	99	173·25	139	243·25	179	313·25
20	35·00	**60**	105·00	**100**	175·00	**140**	245·00	**180**	315·00
21	36·75	61	106·75	101	176·75	141	246·75	181	316·75
22	38·50	62	108·50	102	178·50	142	248·50	182	318·50
23	40·25	63	110·25	103	180·25	143	250·25	183	320·25
24	42·00	64	112·00	104	182·00	144	252·00	184	322·00
25	43·75	65	113·75	105	183·75	145	253·75	185	323·75
26	45·50	66	115·50	106	185·50	146	255·50	186	325·50
27	47·25	67	117·25	107	187·25	147	257·25	187	327·25
28	49·00	68	119·00	108	189·00	148	259·00	188	329·00
29	50·75	69	120·75	109	190·75	149	260·75	189	330·75
30	52·50	**70**	122·50	**110**	192·50	**150**	262·50	**190**	332·50
31	54·25	71	124·25	111	194·25	151	264·25	191	334·25
32	56·00	72	126·00	112	196·00	152	266·00	192	336·00
33	57·75	73	127·75	113	197·75	153	267·75	193	337·75
34	59·50	74	129·50	114	199·50	154	269·50	194	339·50
35	61·25	75	131·25	115	201·25	155	271·25	195	341·25
36	63·00	76	133·00	116	203·00	156	273·00	196	343·00
37	64·75	77	134·75	117	204·75	157	274·75	197	344·75
38	66·50	78	136·50	118	206·50	158	276·50	198	346·50
39	68·25	79	138·25	119	208·25	159	278·25	199	348·25
40	70·00	**80**	140·00	**120**	210·00	**160**	280·00	**200**	350·00

$\frac{1}{16}$=0·10^{9} $\frac{1}{9}$=0·19^{4} $\frac{1}{8}$=0·21^{9} $\frac{1}{6}$=0·29^{2} $\frac{1}{5}$=0·35^{0} $\frac{1}{4}$=0·43^{8} $\frac{1}{3}$=0·58^{3}

$\frac{3}{8}$=0·65^{6} $\frac{1}{2}$=0·87^{5} $\frac{5}{8}$=1·09^{4} $\frac{3}{4}$=1·31^{3} $\frac{7}{8}$=1·53^{1} $\frac{2}{3}$=1·16^{7} $\frac{5}{6}$=1·45^{8}

$\frac{3}{16}$=0·32^{8} $\frac{5}{16}$=0·54^{7} $\frac{7}{16}$=0·76^{6} $\frac{9}{16}$=0·98^{4} $\frac{11}{16}$=1·20^{3} $\frac{13}{16}$=1·42^{2} $\frac{15}{16}$=1·64^{1}

(0.570=1£) **£1.75½**

1	1·75½	41	71·95½	81	142·15½	121	212·35½	161	282·55½
2	3·51	42	73·71	82	143·91	122	214·11	162	284·31
3	5·26½	43	75·46½	83	145·66½	123	215·86½	163	286·06½
4	7·02	44	77·22	84	147·42	124	217·62	164	287·82
5	8·77½	45	78·97½	85	149·17½	125	219·37½	165	289·57½
6	10·53	46	80·73	86	150·93	126	221·13	166	291·33
7	12·28½	47	82·48½	87	152·68½	127	222·88½	167	293·08½
8	14·04	48	84·24	88	154·44	128	224·64	168	294·84
9	15·79½	49	85·99½	89	156·19½	129	226·39½	169	296·59½
10	17·55	**50**	87·75	**90**	157·95	**130**	228·15	**170**	298·35
11	19·30½	51	89·50½	91	159·70½	131	229·90½	171	300·10½
12	21·06	52	91·26	92	161·46	132	231·66	172	301·86
13	22·81½	53	93·01½	93	163·21½	133	233·41½	173	303·61½
14	24·57	54	94·77	94	164·97	134	235·17	174	305·37
15	26·32½	55	96·52½	95	166·72½	135	236·92½	175	307·12½
16	28·08	56	98·28	96	168·48	136	238·68	176	308·88
17	29·83½	57	100·03½	97	170·23½	137	240·43½	177	310·63½
18	31·59	58	101·79	98	171·99	138	242·19	178	312·39
19	33·34½	59	103·54½	99	173·74½	139	243·94½	179	314·14½
20	35·10	**60**	105·30	**100**	175·50	**140**	245·70	**180**	315·90
21	36·85½	61	107·05½	101	177·25½	141	247·45½	181	317·65½
22	38·61	62	108·81	102	179·01	142	249·21	182	319·41
23	40·36½	63	110·56½	103	180·76½	143	250·96½	183	321·16½
24	42·12	64	112·32	104	182·52	144	252·72	184	322·92
25	43·87½	65	114·07½	105	184·27½	145	254·47½	185	324·67½
26	45·63	66	115·83	106	186·03	146	256·23	186	326·43
27	47·38½	67	117·58½	107	187·78½	147	257·98½	187	328·18½
28	49·14	68	119·34	108	189·54	148	259·74	188	329·94
29	50·89½	69	121·09½	109	191·29½	149	261·49½	189	331·69½
30	52·65	**70**	122·85	**110**	193·05	**150**	263·25	**190**	333·45
31	54·40½	71	124·60½	111	194·80½	151	265·00½	191	335·20½
32	56·16	72	126·36	112	196·56	152	266·76	192	336·96
33	57·91½	73	128·11½	113	198·31½	153	268·51½	193	338·71½
34	59·67	74	129·87	114	200·07	154	270·27	194	340·47
35	61·42½	75	131·62½	115	201·82½	155	272·02½	195	342·22½
36	63·18	76	133·38	116	203·58	156	273·78	196	343·98
37	64·93½	77	135·13½	117	205·33½	157	275·53½	197	345·73½
38	66·69	78	136·89	118	207·09	158	277·29	198	347·49
39	68·44½	79	138·64½	119	208·84½	159	279·04½	199	349·24½
40	70·20	**80**	140·40	**120**	210·60	**160**	280·80	**200**	351·00

$\frac{1}{16}=0{\cdot}11^{0}$ $\frac{1}{9}=0{\cdot}19^{5}$ $\frac{1}{8}=0{\cdot}21^{9}$ $\frac{1}{6}=0{\cdot}29^{3}$ $\frac{1}{5}=0{\cdot}35^{1}$ $\frac{1}{4}=0{\cdot}43^{9}$ $\frac{1}{3}=0{\cdot}58^{5}$

$\frac{3}{8}=0{\cdot}65^{8}$ $\frac{1}{2}=0{\cdot}87^{8}$ $\frac{5}{8}=1{\cdot}09^{7}$ $\frac{3}{4}=1{\cdot}31^{6}$ $\frac{7}{8}=1{\cdot}53^{6}$ $\frac{2}{3}=1{\cdot}17^{0}$ $\frac{5}{6}=1{\cdot}46^{3}$

$\frac{3}{16}=0{\cdot}32^{9}$ $\frac{5}{16}=0{\cdot}54^{8}$ $\frac{7}{16}=0{\cdot}76^{8}$ $\frac{9}{16}=0{\cdot}98^{7}$ $\frac{11}{16}=1{\cdot}20^{7}$ $\frac{13}{16}=1{\cdot}42^{6}$ $\frac{15}{16}=1{\cdot}64^{5}$

(0.568=1£) **£1.76**

1	1·76	41	72·16	81	142·56	121	212·96	161	283·36
2	3·52	42	73·92	82	144·32	122	214·72	162	285·12
3	5·28	43	75·68	83	146·08	123	216·48	163	286·88
4	7·04	44	77·44	84	147·84	124	218·24	164	288·64
5	8·80	45	79·20	85	149·60	125	220·00	165	290·40
6	10·56	46	80·96	86	151·36	126	221·76	166	292·16
7	12·32	47	82·72	87	153·12	127	223·52	167	293·92
8	14·08	48	84·48	88	154·88	128	225·28	168	295·68
9	15·84	49	86·24	89	156·64	129	227·04	169	297·44
10	17·60	**50**	88·00	**90**	158·40	**130**	228·80	**170**	299·20
11	19·36	51	89·76	91	160·16	131	230·56	171	300·96
12	21·12	52	91·52	92	161·92	132	232·32	172	302·72
13	22·88	53	93·28	93	163·68	133	234·08	173	304·48
14	24·64	54	95·04	94	165·44	134	235·84	174	306·24
15	26·40	55	96·80	95	167·20	135	237·60	175	308·00
16	28·16	56	98·56	96	168·96	136	239·36	176	309·76
17	29·92	57	100·32	97	170·72	137	241·12	177	311·52
18	31·68	58	102·08	98	172·48	138	242·88	178	313·28
19	33·44	59	103·84	99	174·24	139	244·64	179	315·04
20	35·20	**60**	105·60	**100**	176·00	**140**	246·40	**180**	316·80
21	36·96	61	107·36	101	177·76	141	248·16	181	318·56
22	38·72	62	109·12	102	179·52	142	249·92	182	320·32
23	40·48	63	110·88	103	181·28	143	251·68	183	322·08
24	42·24	64	112·64	104	183·04	144	253·44	184	323·84
25	44·00	65	114·40	105	184·80	145	255·20	185	325·60
26	45·76	66	116·16	106	186·56	146	256·96	186	327·36
27	47·52	67	117·92	107	188·32	147	258·72	187	329·12
28	49·28	68	119·68	108	190·08	148	260·48	188	330·88
29	51·04	69	121·44	109	191·84	149	262·24	189	332·64
30	52·80	**70**	123·20	**110**	193·60	**150**	264·00	**190**	334·40
31	54·56	71	124·96	111	195·36	151	265·76	191	336·16
32	56·32	72	126·72	112	197·12	152	267·52	192	337·92
33	58·08	73	128·48	113	198·88	153	269·28	193	339·68
34	59·84	74	130·24	114	200·64	154	271·04	194	341·44
35	61·60	75	132·00	115	202·40	155	272·80	195	343·20
36	63·36	76	133·76	116	204·16	156	274·56	196	344·96
37	65·12	77	135·52	117	205·92	157	276·32	197	346·72
38	66·88	78	137·28	118	207·68	158	278·08	198	348·48
39	68·64	79	139·04	119	209·44	159	279·84	199	350·24
40	70·40	**80**	140·80	**120**	211·20	**160**	281·60	**200**	352·00

$\frac{1}{16}=0{\cdot}11^{0}$ $\frac{1}{9}=0{\cdot}19^{6}$ $\frac{1}{8}=0{\cdot}22^{0}$ $\frac{1}{6}=0{\cdot}29^{3}$ $\frac{1}{5}=0{\cdot}35^{2}$ $\frac{1}{4}=0{\cdot}44^{0}$ $\frac{1}{3}=0{\cdot}58^{7}$

$\frac{3}{8}=0{\cdot}66^{0}$ $\frac{1}{2}=0{\cdot}88^{0}$ $\frac{5}{8}=1{\cdot}10^{0}$ $\frac{3}{4}=1{\cdot}32^{0}$ $\frac{7}{8}=1{\cdot}54^{0}$ $\frac{2}{3}=1{\cdot}17^{3}$ $\frac{5}{6}=1{\cdot}46^{7}$

$\frac{3}{16}=0{\cdot}33^{0}$ $\frac{5}{16}=0{\cdot}55^{0}$ $\frac{7}{16}=0{\cdot}77^{0}$ $\frac{9}{16}=0{\cdot}99^{0}$ $\frac{11}{16}=1{\cdot}21^{0}$ $\frac{13}{16}=1{\cdot}43^{0}$ $\frac{15}{16}=1{\cdot}65^{0}$

1 litre=1·76 pints

(0.567=1£) **£1.76½**

1	1·76½	41	72·36½	81	142·96½	121	213·56½	161	284·16½
2	3·53	42	74·13	82	144·73	122	215·33	162	285·93
3	5·29½	43	75·89½	83	146·49½	123	217·09½	163	287·69½
4	7·06	44	77·66	84	148·26	124	218·86	164	289·46
5	8·82½	45	79·42½	85	150·02½	125	220·62½	165	291·22½
6	10·59	46	81·19	86	151·79	126	222·39	166	292·99
7	12·35½	47	82·95½	87	153·55½	127	224·15½	167	294·75½
8	14·12	48	84·72	88	155·32	128	225·92	168	296·52
9	15·88½	49	86·48½	89	157·08½	129	227·68½	169	298·28½
10	17·65	**50**	88·25	**90**	158·85	**130**	229·45	**170**	300·05
11	19·41½	51	90·01½	91	160·61½	131	231·21½	171	301·81½
12	21·18	52	91·78	92	162·38	132	232·98	172	303·58
13	22·94½	53	93·54½	93	164·14½	133	234·74½	173	305·34½
14	24·71	54	95·31	94	165·91	134	236·51	174	307·11
15	26·47½	55	97·07½	95	167·67½	135	238·27½	175	308·87½
16	28·24	56	98·84	96	169·44	136	240·04	176	310·64
17	30·00½	57	100·60½	97	171·20½	137	241·80½	177	312·40½
18	31·77	58	102·37	98	172·97	138	243·57	178	314·17
19	33·53½	59	104·13½	99	174·73½	139	245·33½	179	315·93½
20	35·30	**60**	105·90	**100**	176·50	**140**	247·10	**180**	317·70
21	37·06½	61	107·66½	101	178·26½	141	248·86½	181	319·46½
22	38·83	62	109·43	102	180·03	142	250·63	182	321·23
23	40·59½	63	111·19½	103	181·79½	143	252·39½	183	322·99½
24	42·36	64	112·96	104	183·56	144	254·16	184	324·76
25	44·12½	65	114·72½	105	185·32½	145	255·92½	185	326·52½
26	45·89	66	116·49	106	187·09	146	257·69	186	328·29
27	47·65½	67	118·25½	107	188·85½	147	259·45½	187	330·05½
28	49·42	68	120·02	108	190·62	148	261·22	188	331·82
29	51·18½	69	121·78½	109	192·38½	149	262·98½	189	333·58½
30	52·95	**70**	123·55	**110**	194·15	**150**	264·75	**190**	335·35
31	54·71½	71	125·31½	111	195·91½	151	266·51½	191	337·11½
32	56·48	72	127·08	112	197·68	152	268·28	192	338·88
33	58·24½	73	128·84½	113	199·44½	153	270·04½	193	340·64½
34	60·01	74	130·61	114	201·21	154	271·81	194	342·41
35	61·77½	75	132·37½	115	202·97½	155	273·57½	195	344·17½
36	63·54	76	134·14	116	204·74	156	275·34	196	345·94
37	65·30½	77	135·90½	117	206·50½	157	277·10½	197	347·70½
38	67·07	78	137·67	118	208·27	158	278·87	198	349·47
39	68·83½	79	139·43½	119	210·03½	159	280·63½	199	351·23½
40	70·60	**80**	141·20	**120**	211·80	**160**	282·40	**200**	353·00

$\frac{1}{16}=0{\cdot}11^{0}$ $\frac{1}{9}=0{\cdot}19^{6}$ $\frac{1}{8}=0{\cdot}22^{1}$ $\frac{1}{6}=0{\cdot}29^{4}$ $\frac{1}{5}=0{\cdot}35^{3}$ $\frac{1}{4}=0{\cdot}44^{1}$ $\frac{1}{3}=0{\cdot}58^{8}$

$\frac{3}{8}=0{\cdot}66^{2}$ $\frac{1}{2}=0{\cdot}88^{3}$ $\frac{5}{8}=1{\cdot}10^{3}$ $\frac{3}{4}=1{\cdot}32^{4}$ $\frac{7}{8}=1{\cdot}54^{4}$ $\frac{2}{3}=1{\cdot}17^{7}$ $\frac{5}{6}=1{\cdot}47^{1}$

$\frac{3}{16}=0{\cdot}33^{1}$ $\frac{5}{16}=0{\cdot}55^{2}$ $\frac{7}{16}=0{\cdot}77^{2}$ $\frac{9}{16}=0{\cdot}99^{3}$ $\frac{11}{16}=1{\cdot}21^{3}$ $\frac{13}{16}=1{\cdot}43^{4}$ $\frac{15}{16}=1{\cdot}65^{5}$

(0.565=1£) **£1.77**

1	1·77	41	72·57	81	143·37	121	214·17	161	284·97
2	3·54	42	74·34	82	145·14	122	215·94	162	286·74
3	5·31	43	76·11	83	146·91	123	217·71	163	288·51
4	7·08	44	77·88	84	148·68	124	219·48	164	290·28
5	8·85	45	79·65	85	150·45	125	221·25	165	292·05
6	10·62	46	81·42	86	152·22	126	223·02	166	293·82
7	12·39	47	83·19	87	153·99	127	224·79	167	295·59
8	14·16	48	84·96	88	155·76	128	226·56	168	297·36
9	15·93	49	86·73	89	157·53	129	228·33	169	299·13
10	17·70	**50**	88·50	**90**	159·30	**130**	230·10	**170**	300·90
11	19·47	51	90·27	91	161·07	131	231·87	171	302·67
12	21·24	52	92·04	92	162·84	132	233·64	172	304·44
13	23·01	53	93·81	93	164·61	133	235·41	173	306·21
14	24·78	54	95·58	94	166·38	134	237·18	174	307·98
15	26·55	55	97·35	95	168·15	135	238·95	175	309·75
16	28·32	56	99·12	96	169·92	136	240·72	176	311·52
17	30·09	57	100·89	97	171·69	137	242·49	177	313·29
18	31·86	58	102·66	98	173·46	138	244·26	178	315·06
19	33·63	59	104·43	99	175·23	139	246·03	179	316·83
20	35·40	**60**	106·20	**100**	177·00	**140**	247·80	**180**	318·60
21	37·17	61	107·97	101	178·77	141	249·57	181	320·37
22	38·94	62	109·74	102	180·54	142	251·34	182	322·14
23	40·71	63	111·51	103	182·31	143	253·11	183	323·91
24	42·48	64	113·28	104	184·08	144	254·88	184	325·68
25	44·25	65	115·05	105	185·85	145	256·65	185	327·45
26	46·02	66	116·82	106	187·62	146	258·42	186	329·22
27	47·79	67	118·59	107	189·39	147	260·19	187	330·99
28	49·56	68	120·36	108	191·16	148	261·96	188	332·76
29	51·33	69	122·13	109	192·93	149	263·73	189	334·53
30	53·10	**70**	123·90	**110**	194·70	**150**	265·50	**190**	336·30
31	54·87	71	125·67	111	196·47	151	267·27	191	338·07
32	56·64	72	127·44	112	198·24	152	269·04	192	339·84
33	58·41	73	129·21	113	200·01	153	270·81	193	341·61
34	60·18	74	130·98	114	201·78	154	272·58	194	343·38
35	61·95	75	132·75	115	203·55	155	274·35	195	345·15
36	63·72	76	134·52	116	205·32	156	276·12	196	346·92
37	65·49	77	136·29	117	207·09	157	277·89	197	348·69
38	67·26	78	138·06	118	208·86	158	279·66	198	350·46
39	69·03	79	139·83	119	210·63	159	281·43	199	352·23
40	70·80	**80**	141·60	**120**	212·40	**160**	283·20	**200**	354·00

$\frac{1}{16}=0{\cdot}11^{1}$ $\frac{1}{9}=0{\cdot}19^{7}$ $\frac{1}{8}=0{\cdot}22^{1}$ $\frac{1}{6}=0{\cdot}29^{5}$ $\frac{1}{5}=0{\cdot}35^{4}$ $\frac{1}{4}=0{\cdot}44^{3}$ $\frac{1}{3}=0{\cdot}59^{0}$

$\frac{3}{8}=0{\cdot}66^{4}$ $\frac{1}{2}=0{\cdot}88^{5}$ $\frac{5}{8}=1{\cdot}10^{6}$ $\frac{3}{4}=1{\cdot}32^{8}$ $\frac{7}{8}=1{\cdot}54^{9}$ $\frac{2}{3}=1{\cdot}18^{0}$ $\frac{5}{6}=1{\cdot}47^{5}$

$\frac{3}{16}=0{\cdot}33^{2}$ $\frac{5}{16}=0{\cdot}55^{3}$ $\frac{7}{16}=0{\cdot}77^{4}$ $\frac{9}{16}=0{\cdot}99^{6}$ $\frac{11}{16}=1{\cdot}21^{7}$ $\frac{13}{16}=1{\cdot}43^{8}$ $\frac{15}{16}=1{\cdot}65^{9}$

(0.563=1£) **£1.77½**

1	1·77½	41	72·77½	81	143·77½	121	214·77½	161	285·77½
2	3·55	42	74·55	82	145·55	122	216·55	162	287·55
3	5·32½	43	76·32½	83	147·32½	123	218·32½	163	289·32½
4	7·10	44	78·10	84	149·10	124	220·10	164	291·10
5	8·87½	45	79·87½	85	150·87½	125	221·87½	165	292·87½
6	10·65	46	81·65	86	152·65	126	223·65	166	294·65
7	12·42½	47	83·42½	87	154·42½	127	225·42½	167	296·42½
8	14·20	48	85·20	88	156·20	128	227·20	168	298·20
9	15·97½	49	86·97½	89	157·97½	129	228·97½	169	299·97½
10	17·75	**50**	88·75	**90**	159·75	**130**	230·75	**170**	301·75
11	19·52½	51	90·52½	91	161·52½	131	232·52½	171	303·52½
12	21·30	52	92·30	92	163·30	132	234·30	172	305·30
13	23·07½	53	94·07½	93	165·07½	133	236·07½	173	307·07½
14	24·85	54	95·85	94	166·85	134	237·85	174	308·85
15	26·62½	55	97·62½	95	168·62½	135	239·62½	175	310·62½
16	28·40	56	99·40	96	170·40	136	241·40	176	312·40
17	30·17½	57	101·17½	97	172·17½	137	243·17½	177	314·17½
18	31·95	58	102·95	98	173·95	138	244·95	178	315·95
19	33·72½	59	104·72½	99	175·72½	139	246·72½	179	317·72½
20	35·50	**60**	106·50	**100**	177·50	**140**	248·50	**180**	319·50
21	37·27½	61	108·27½	101	179·27½	141	250·27½	181	321·27½
22	39·05	62	110·05	102	181·05	142	252·05	182	323·05
23	40·82½	63	111·82½	103	182·82½	143	253·82½	183	324·82½
24	42·60	64	113·60	104	184·60	144	255·60	184	326·60
25	44·37½	65	115·37½	105	186·37½	145	257·37½	185	328·37½
26	46·15	66	117·15	106	188·15	146	259·15	186	330·15
27	47·92½	67	118·92½	107	189·92½	147	260·92½	187	331·92½
28	49·70	68	120·70	108	191·70	148	262·70	188	333·70
29	51·47½	69	122·47½	109	193·47½	149	264·47½	189	335·47½
30	53·25	**70**	124·25	**110**	195·25	**150**	266·25	**190**	337·25
31	55·02½	71	126·02½	111	197·02½	151	268·02½	191	339·02½
32	56·80	72	127·80	112	198·80	152	269·80	192	340·80
33	58·57½	73	129·57½	113	200·57½	153	271·57½	193	342·57½
34	60·35	74	131·35	114	202·35	154	273·35	194	344·35
35	62·12½	75	133·12½	115	204·12½	155	275·12½	195	346·12½
36	63·90	76	134·90	116	205·90	156	276·90	196	347·90
37	65·67½	77	136·67½	117	207·67½	157	278·67½	197	349·67½
38	67·45	78	138·45	118	209·45	158	280·45	198	351·45
39	69·22½	79	140·22½	119	211·22½	159	282·22½	199	353·22½
40	71·00	**80**	142·00	**120**	213·00	**160**	284·00	**200**	355·00

$\frac{1}{16}=0{\cdot}11^{1}$ $\frac{1}{9}=0{\cdot}19^{7}$ $\frac{1}{8}=0{\cdot}22^{2}$ $\frac{1}{6}=0{\cdot}29^{6}$ $\frac{1}{5}=0{\cdot}35^{5}$ $\frac{1}{4}=0{\cdot}44^{4}$ $\frac{1}{3}=0{\cdot}59^{2}$

$\frac{3}{8}=0{\cdot}66^{6}$ $\frac{1}{2}=0{\cdot}88^{8}$ $\frac{5}{8}=1{\cdot}10^{9}$ $\frac{3}{4}=1{\cdot}33^{1}$ $\frac{7}{8}=1{\cdot}55^{3}$ $\frac{2}{3}=1{\cdot}18^{3}$ $\frac{5}{6}=1{\cdot}47^{9}$

$\frac{3}{16}=0{\cdot}33^{3}$ $\frac{5}{16}=0{\cdot}55^{5}$ $\frac{7}{16}=0{\cdot}77^{7}$ $\frac{9}{16}=0{\cdot}99^{8}$ $\frac{11}{16}=1{\cdot}22^{0}$ $\frac{13}{16}=1{\cdot}44^{2}$ $\frac{15}{16}=1{\cdot}66^{4}$

(0.562=1£) **£1.78**

1	1·78	41	72·98	81	144·18	121	215·38	161	286·58
2	3·56	42	74·76	82	145·96	122	217·16	162	288·36
3	5·34	43	76·54	83	147·74	123	218·94	163	290·14
4	7·12	44	78·32	84	149·52	124	220·72	164	291·92
5	8·90	45	80·10	85	151·30	125	222·50	165	293·70
6	10·68	46	81·88	86	153·08	126	224·28	166	295·48
7	12·46	47	83·66	87	154·86	127	226·06	167	297·26
8	14·24	48	85·44	88	156·64	128	227·84	168	299·04
9	16·02	49	87·22	89	158·42	129	229·62	169	300·82
10	17·80	**50**	89·00	**90**	160·20	**130**	231·40	**170**	302·60
11	19·58	51	90·78	91	161·98	131	233·18	171	304·38
12	21·36	52	92·56	92	163·76	132	234·96	172	306·16
13	23·14	53	94·34	93	165·54	133	236·74	173	307·94
14	24·92	54	96·12	94	167·32	134	238·52	174	309·72
15	26·70	55	97·90	95	169·10	135	240·30	175	311·50
16	28·48	56	99·68	96	170·88	136	242·08	176	313·28
17	30·26	57	101·46	97	172·66	137	243·86	177	315·06
18	32·04	58	103·24	98	174·44	138	245·64	178	316·84
19	33·82	59	105·02	99	176·22	139	247·42	179	318·62
20	35·60	**60**	106·80	**100**	178·00	**140**	249·20	**180**	320·40
21	37·38	61	108·58	101	179·78	141	250·98	181	322·18
22	39·16	62	110·36	102	181·56	142	252·76	182	323·96
23	40·94	63	112·14	103	183·34	143	254·54	183	325·74
24	42·72	64	113·92	104	185·12	144	256·32	184	327·52
25	44·50	65	115·70	105	186·90	145	258·10	185	329·30
26	46·28	66	117·48	106	188·68	146	259·88	186	331·08
27	48·06	67	119·26	107	190·46	147	261·66	187	332·86
28	49·84	68	121·04	108	192·24	148	263·44	188	334·64
29	51·62	69	122·82	109	194·02	149	265·22	189	336·42
30	53·40	**70**	124·60	**110**	195·80	**150**	267·00	**190**	338·20
31	55·18	71	126·38	111	197·58	151	268·78	191	339·98
32	56·96	72	128·16	112	199·36	152	270·56	192	341·76
33	58·74	73	129·94	113	201·14	153	272·34	193	343·54
34	60·52	74	131·72	114	202·92	154	274·12	194	345·32
35	62·30	75	133·50	115	204·70	155	275·90	195	347·10
36	64·08	76	135·28	116	206·48	156	277·68	196	348·88
37	65·86	77	137·06	117	208·26	157	279·46	197	350·66
38	67·64	78	138·84	118	210·04	158	281·24	198	352·44
39	69·42	79	140·62	119	211·82	159	283·02	199	354·22
40	71·20	**80**	142·40	**120**	213·60	**160**	284·80	**200**	356·00

$\frac{1}{16}$=0·11^{1} $\frac{1}{9}$=0·19^{8} $\frac{1}{8}$=0·22^{3} $\frac{1}{6}$=0·29^{7} $\frac{1}{5}$=0·35^{6} $\frac{1}{4}$=0·44^{5} $\frac{1}{3}$=0·59^{3}

$\frac{3}{8}$=0·66^{8} $\frac{1}{2}$=0·89^{0} $\frac{5}{8}$=1·11^{3} $\frac{3}{4}$=1·33^{5} $\frac{7}{8}$=1·55^{8} $\frac{2}{3}$=1·18^{7} $\frac{5}{6}$=1·48^{3}

$\frac{3}{16}$=0·33^{4} $\frac{5}{16}$=0·55^{6} $\frac{7}{16}$=0·77^{9} $\frac{9}{16}$=1·00^{1} $\frac{11}{16}$=1·22^{4} $\frac{13}{16}$=1·44^{6} $\frac{15}{16}$=1·66^{9}

1	1·78½	41	73·18½	81	144·58½	121	215·98½	161	287·38½
2	3·57	42	74·97	82	146·37	122	217·77	162	289·17
3	5·35½	43	76·75½	83	148·15½	123	219·55½	163	290·95½
4	7·14	44	78·54	84	149·94	124	221·34	164	292·74
5	8·92½	45	80·32½	85	151·72½	125	223·12½	165	294·52½
6	10·71	46	82·11	86	153·51	126	224·91	166	296·31
7	12·49½	47	83·89½	87	155·29½	127	226·69½	167	298·09½
8	14·28	48	85·68	88	157·08	128	228·48	168	299·88
9	16·06½	49	87·46½	89	158·86½	129	230·26½	169	301·66½
10	17·85	**50**	89·25	**90**	160·65	**130**	232·05	**170**	303·45
11	19·63½	51	91·03½	91	162·43½	131	233·83½	171	305·23½
12	21·42	52	92·82	92	164·22	132	235·62	172	307·02
13	23·20½	53	94·60½	93	166·00½	133	237·40½	173	308·80½
14	24·99	54	96·39	94	167·79	134	239·19	174	310·59
15	26·77½	55	98·17½	95	169·57½	135	240·97½	175	312·37½
16	28·56	56	99·96	96	171·36	136	242·76	176	314·16
17	30·34½	57	101·74½	97	173·14½	137	244·54½	177	315·94½
18	32·13	58	103·53	98	174·93	138	246·33	178	317·73
19	33·91½	59	105·31½	99	176·71½	139	248·11½	179	319·51½
20	35·70	**60**	107·10	**100**	178·50	**140**	249·90	**180**	321·30
21	37·48½	61	108·88½	101	180·28½	141	251·68½	181	323·08½
22	39·27	62	110·67	102	182·07	142	253·47	182	324·87
23	41·05½	63	112·45½	103	183·85½	143	255·25½	183	326·65½
24	42·84	64	114·24	104	185·64	144	257·04	184	328·44
25	44·62½	65	116·02½	105	187·42½	145	258·82½	185	330·22½
26	46·41	66	117·81	106	189·21	146	260·61	186	332·01
27	48·19½	67	119·59½	107	190·99½	147	262·39½	187	333·79½
28	49·98	68	121·38	108	192·78	148	264·18	188	335·58
29	51·76½	69	123·16½	109	194·56½	149	265·96½	189	337·36½
30	53·55	**70**	124·95	**110**	196·35	**150**	267·75	**190**	339·15
31	55·33½	71	126·73½	111	198·13½	151	269·53½	191	340·93½
32	57·12	72	128·52	112	199·92	152	271·32	192	342·72
33	58·90½	73	130·30½	113	201·70½	153	273·10½	193	344·50½
34	60·69	74	132·09	114	203·49	154	274·89	194	346·29
35	62·47½	75	133·87½	115	205·27½	155	276·67½	195	348·07½
36	64·26	76	135·66	116	207·06	156	278·46	196	349·86
37	66·04½	77	137·44½	117	208·84½	157	280·24½	197	351·64½
38	67·83	78	139·23	118	210·63	158	282·03	198	353·43
39	69·61½	79	141·01½	119	212·41½	159	283·81½	199	355·21½
40	71·40	**80**	142·80	**120**	214·20	**160**	285·60	**200**	357·00

$\frac{1}{16}=0{\cdot}11^{2}$ $\frac{1}{9}=0{\cdot}19^{8}$ $\frac{1}{8}=0{\cdot}22^{3}$ $\frac{1}{6}=0{\cdot}29^{8}$ $\frac{1}{5}=0{\cdot}35^{7}$ $\frac{1}{4}=0{\cdot}44^{6}$ $\frac{1}{3}=0{\cdot}59^{5}$

$\frac{3}{8}=0{\cdot}66^{9}$ $\frac{1}{2}=0{\cdot}89^{3}$ $\frac{5}{8}=1{\cdot}11^{6}$ $\frac{3}{4}=1{\cdot}33^{9}$ $\frac{7}{8}=1{\cdot}56^{2}$ $\frac{2}{3}=1{\cdot}19^{0}$ $\frac{5}{6}=1{\cdot}48^{8}$

$\frac{3}{16}=0{\cdot}33^{5}$ $\frac{5}{16}=0{\cdot}55^{8}$ $\frac{7}{16}=0{\cdot}78^{1}$ $\frac{9}{16}=1{\cdot}00^{4}$ $\frac{11}{16}=1{\cdot}22^{7}$ $\frac{13}{16}=1{\cdot}45^{0}$ $\frac{15}{16}=1{\cdot}67^{3}$

(0.559=1£) **£1.79**

1	1·79	41	73·39	81	144·99	121	216·59	161	288·19
2	3·58	42	75·18	82	146·78	122	218·38	162	289·98
3	5·37	43	76·97	83	148·57	123	220·17	163	291·77
4	7·16	44	78·76	84	150·36	124	221·96	164	293·56
5	8·95	45	80·55	85	152·15	125	223·75	165	295·35
6	10·74	46	82·34	86	153·94	126	225·54	166	297·14
7	12·53	47	84·13	87	155·73	127	227·33	167	298·93
8	14·32	48	85·92	88	157·52	128	229·12	168	300·72
9	16·11	49	87·71	89	159·31	129	230·91	169	302·51
10	17·90	**50**	89·50	**90**	161·10	**130**	232·70	**170**	304·30
11	19·69	51	91·29	91	162·89	131	234·49	171	306·09
12	21·48	52	93·08	92	164·68	132	236·28	172	307·88
13	23·27	53	94·87	93	166·47	133	238·07	173	309·67
14	25·06	54	96·66	94	168·26	134	239·86	174	311·46
15	26·85	55	98·45	95	170·05	135	241·65	175	313·25
16	28·64	56	100·24	96	171·84	136	243·44	176	315·04
17	30·43	57	102·03	97	173·63	137	245·23	177	316·83
18	32·22	58	103·82	98	175·42	138	247·02	178	318·62
19	34·01	59	105·61	99	177·21	139	248·81	179	320·41
20	35·80	**60**	107·40	**100**	179·00	**140**	250·60	**180**	322·20
21	37·59	61	109·19	101	180·79	141	252·39	181	323·99
22	39·38	62	110·98	102	182·58	142	254·18	182	325·78
23	41·17	63	112·77	103	184·37	143	255·97	183	327·57
24	42·96	64	114·56	104	186·16	144	257·76	184	329·36
25	44·75	65	116·35	105	187·95	145	259·55	185	331·15
26	46·54	66	118·14	106	189·74	146	261·34	186	332·94
27	48·33	67	119·93	107	191·53	147	263·13	187	334·73
28	50·12	68	121·72	108	193·32	148	264·92	188	336·52
29	51·91	69	123·51	109	195·11	149	266·71	189	338·31
30	53·70	**70**	125·30	**110**	196·90	**150**	268·50	**190**	340·10
31	55·49	71	127·09	111	198·69	151	270·29	191	341·89
32	57·28	72	128·88	112	200·48	152	272·08	192	343·68
33	59·07	73	130·67	113	202·27	153	273·87	193	345·47
34	60·86	74	132·46	114	204·06	154	275·66	194	347·26
35	62·65	75	134·25	115	205·85	155	277·45	195	349·05
36	64·44	76	136·04	116	207·64	156	279·24	196	350·84
37	66·23	77	137·83	117	209·43	157	281·03	197	352·63
38	68·02	78	139·62	118	211·22	158	282·82	198	354·42
39	69·81	79	141·41	119	213·01	159	284·61	199	356·21
40	71·60	**80**	143·20	**120**	214·80	**160**	286·40	**200**	358·00

$\frac{1}{16}=0{\cdot}11^{2}$ $\frac{1}{9}=0{\cdot}19^{9}$ $\frac{1}{8}=0{\cdot}22^{4}$ $\frac{1}{6}=0{\cdot}29^{8}$ $\frac{1}{5}=0{\cdot}35^{8}$ $\frac{1}{4}=0{\cdot}44^{8}$ $\frac{1}{3}=0{\cdot}59^{7}$

$\frac{3}{8}=0{\cdot}67^{1}$ $\frac{1}{2}=0{\cdot}89^{5}$ $\frac{5}{8}=1{\cdot}11^{9}$ $\frac{3}{4}=1{\cdot}34^{3}$ $\frac{7}{8}=1{\cdot}56^{6}$ $\frac{2}{3}=1{\cdot}19^{3}$ $\frac{5}{6}=1{\cdot}49^{2}$

$\frac{3}{16}=0{\cdot}33^{6}$ $\frac{5}{16}=0{\cdot}55^{9}$ $\frac{7}{16}=0{\cdot}78^{3}$ $\frac{9}{16}=1{\cdot}00^{7}$ $\frac{11}{16}=1{\cdot}23^{1}$ $\frac{13}{16}=1{\cdot}45^{4}$ $\frac{15}{16}=1{\cdot}67^{8}$

1	1·79½	41	73·59½	81	145·39½	121	217·19½	161	288·99½
2	3·59	42	75·39	82	147·19	122	218·99	162	290·79
3	5·38½	43	77·18½	83	148·98½	123	220·78½	163	292·58½
4	7·18	44	78·98	84	150·78	124	222·58	164	294·38
5	8·97½	45	80·77½	85	152·57½	125	224·37½	165	296·17½
6	10·77	46	82·57	86	154·37	126	226·17	166	297·97
7	12·56½	47	84·36½	87	156·16½	127	227·96½	167	299·76½
8	14·36	48	86·16	88	157·96	128	229·76	168	301·56
9	16·15½	49	87·95½	89	159·75½	129	231·55½	169	303·35½
10	17·95	**50**	89·75	**90**	161·55	**130**	233·35	**170**	305·15
11	19·74½	51	91·54½	91	163·34½	131	235·14½	171	306·94½
12	21·54	52	93·34	92	165·14	132	236·94	172	308·74
13	23·33½	53	95·13½	93	166·93½	133	238·73½	173	310·53½
14	25·13	54	96·93	94	168·73	134	240·53	174	312·33
15	26·92½	55	98·72½	95	170·52½	135	242·32½	175	314·12½
16	28·72	56	100·52	96	172·32	136	244·12	176	315·92
17	30·51½	57	102·31½	97	174·11½	137	245·91½	177	317·71½
18	32·31	58	104·11	98	175·91	138	247·71	178	319·51
19	34·10½	59	105·90½	99	177·70½	139	249·50½	179	321·30½
20	35·90	**60**	107·70	**100**	179·50	**140**	251·30	**180**	323·10
21	37·69½	61	109·49½	101	181·29½	141	253·09½	181	324·89½
22	39·49	62	111·29	102	183·09	142	254·89	182	326·69
23	41·28½	63	113·08½	103	184·88½	143	256·68½	183	328·48½
24	43·08	64	114·88	104	186·68	144	258·48	184	330·28
25	44·87½	65	116·67½	105	188·47½	145	260·27½	185	332·07½
26	46·67	66	118·47	106	190·27	146	262·07	186	333·87
27	48·46½	67	120·26½	107	192·06½	147	263·86½	187	335·66½
28	50·26	68	122·06	108	193·86	148	265·66	188	337·46
29	52·05½	69	123·85½	109	195·65½	149	267·45½	189	339·25½
30	53·85	**70**	125·65	**110**	197·45	**150**	269·25	**190**	341·05
31	55·64½	71	127·44½	111	199·24½	151	271·04½	191	342·84½
32	57·44	72	129·24	112	201·04	152	272·84	192	344·64
33	59·23½	73	131·03½	113	202·83½	153	274·63½	193	346·43½
34	61·03	74	132·83	114	204·63	154	276·43	194	348·23
35	62·82½	75	134·62½	115	206·42½	155	278·22½	195	350·02½
36	64·62	76	136·42	116	208·22	156	280·02	196	351·82
37	66·41½	77	138·21½	117	210·01½	157	281·81½	197	353·61½
38	68·21	78	140·01	118	211·81	158	283·61	198	355·41
39	70·00½	79	141·80½	119	213·60½	159	285·40½	199	357·20½
40	71·80	**80**	143·60	**120**	215·40	**160**	287·20	**200**	359·00

$\frac{1}{16}$=0·11^{2} $\frac{1}{9}$=0·19^{9} $\frac{1}{8}$=0·22^{4} $\frac{1}{6}$=0·29^{9} $\frac{1}{5}$=0·35^{9} $\frac{1}{4}$=0·44^{9} $\frac{1}{3}$=0·59^{8}

$\frac{3}{8}$=0·67^{3} $\frac{1}{2}$=0·89^{8} $\frac{5}{8}$=1·12^{2} $\frac{3}{4}$=1·34^{6} $\frac{7}{8}$=1·57^{1} $\frac{2}{3}$=1·19^{7} $\frac{5}{6}$=1·49^{6}

$\frac{3}{16}$=0·33^{7} $\frac{5}{16}$=0·56^{1} $\frac{7}{16}$=0·78^{5} $\frac{9}{16}$=1·01^{0} $\frac{11}{16}$=1·23^{4} $\frac{13}{16}$=1·45^{8} $\frac{15}{16}$=1·68^{3}

(0.556=1£) **£1.80**

1	1·80	41	73·80	81	145·80	121	217·80	161	289·80
2	3·60	42	75·60	82	147·60	122	219·60	162	291·60
3	5·40	43	77·40	83	149·40	123	221·40	163	293·40
4	7·20	44	79·20	84	151·20	124	223·20	164	295·20
5	9·00	45	81·00	85	153·00	125	225·00	165	297·00
6	10·80	46	82·80	86	154·80	126	226·80	166	298·80
7	12·60	47	84·60	87	156·60	127	228·60	167	300·60
8	14·40	48	86·40	88	158·40	128	230·40	168	302·40
9	16·20	49	88·20	89	160·20	129	232·20	169	304·20
10	18·00	**50**	90·00	**90**	162·00	**130**	234·00	**170**	306·00
11	19·80	51	91·80	91	163·80	131	235·80	171	307·80
12	21·60	52	93·60	92	165·60	132	237·60	172	309·60
13	23·40	53	95·40	93	167·40	133	239·40	173	311·40
14	25·20	54	97·20	94	169·20	134	241·20	174	313·20
15	27·00	55	99·00	95	171·00	135	243·00	175	315·00
16	28·80	56	100·80	96	172·80	136	244·80	176	316·80
17	30·60	57	102·60	97	174·60	137	246·60	177	318·60
18	32·40	58	104·40	98	176·40	138	248·40	178	320·40
19	34·20	59	106·20	99	178·20	139	250·20	179	322·20
20	36·00	**60**	108·00	**100**	180·00	**140**	252·00	**180**	324·00
21	37·80	61	109·80	101	181·80	141	253·80	181	325·80
22	39·60	62	111·60	102	183·60	142	255·60	182	327·60
23	41·40	63	113·40	103	185·40	143	257·40	183	329·40
24	43·20	64	115·20	104	187·20	144	259·20	184	331·20
25	45·00	65	117·00	105	189·00	145	261·00	185	333·00
26	46·80	66	118·80	106	190·80	146	262·80	186	334·80
27	48·60	67	120·60	107	192·60	147	264·60	187	336·60
28	50·40	68	122·40	108	194·40	148	266·40	188	338·40
29	52·20	69	124·20	109	196·20	149	268·20	189	340·20
30	54·00	**70**	126·00	**110**	198·00	**150**	270·00	**190**	342·00
31	55·80	71	127·80	111	199·80	151	271·80	191	343·80
32	57·60	72	129·60	112	201·60	152	273·60	192	345·60
33	59·40	73	131·40	113	203·40	153	275·40	193	347·40
34	61·20	74	133·20	114	205·20	154	277·20	194	349·20
35	63·00	75	135·00	115	207·00	155	279·00	195	351·00
36	64·80	76	136·80	116	208·80	156	280·80	196	352·80
37	66·60	77	138·60	117	210·60	157	282·60	197	354·60
38	68·40	78	140·40	118	212·40	158	284·40	198	356·40
39	70·20	79	142·20	119	214·20	159	286·20	199	358·20
40	72·00	**80**	144·00	**120**	216·00	**160**	288·00	**200**	360·00

$\frac{1}{16}=0\cdot11^{3}$ $\frac{1}{9}=0\cdot20^{0}$ $\frac{1}{8}=0\cdot22^{5}$ $\frac{1}{6}=0\cdot30^{0}$ $\frac{1}{5}=0\cdot36^{0}$ $\frac{1}{4}=0\cdot45^{0}$ $\frac{1}{3}=0\cdot60^{0}$

$\frac{3}{8}=0\cdot67^{5}$ $\frac{1}{2}=0\cdot90^{0}$ $\frac{5}{8}=1\cdot12^{5}$ $\frac{3}{4}=1\cdot35^{0}$ $\frac{7}{8}=1\cdot57^{5}$ $\frac{2}{3}=1\cdot20^{0}$ $\frac{5}{6}=1\cdot50^{0}$

$\frac{3}{16}=0\cdot33^{8}$ $\frac{5}{16}=0\cdot56^{3}$ $\frac{7}{16}=0\cdot78^{8}$ $\frac{9}{16}=1\cdot01^{3}$ $\frac{11}{16}=1\cdot23^{8}$ $\frac{13}{16}=1\cdot46^{3}$ $\frac{15}{16}=1\cdot68^{8}$

(0.554=1£) **£1.80½**

1	1·80½	41	74·00½	81	146·20½	121	218·40½	161	290·60½
2	3·61	42	75·81	82	148·01	122	220·21	162	292·41
3	5·41½	43	77·61½	83	149·81½	123	222·01½	163	294·21½
4	7·22	44	79·42	84	151·62	124	223·82	164	296·02
5	9·02½	45	81·22½	85	153·42½	125	225·62½	165	297·82½
6	10·83	46	83·03	86	155·23	126	227·43	166	299·63
7	12·63½	47	84·83½	87	157·03½	127	229·23½	167	301·43½
8	14·44	48	86·64	88	158·84	128	231·04	168	303·24
9	16·24½	49	88·44½	89	160·64½	129	232·84½	169	305·04½
10	18·05	**50**	90·25	**90**	162·45	**130**	234·65	**170**	306·85
11	19·85½	51	92·05½	91	164·25½	131	236·45½	171	308·65½
12	21·66	52	93·86	92	166·06	132	238·26	172	310·46
13	23·46½	53	95·66½	93	167·86½	133	240·06½	173	312·26½
14	25·27	54	97·47	94	169·67	134	241·87	174	314·07
15	27·07½	55	99·27½	95	171·47½	135	243·67½	175	315·87½
16	28·88	56	101·08	96	173·28	136	245·48	176	317·68
17	30·68½	57	102·88½	97	175·08½	137	247·28½	177	319·48½
18	32·49	58	104·69	98	176·89	138	249·09	178	321·29
19	34·29½	59	106·49½	99	178·69½	139	250·89½	179	323·09½
20	36·10	**60**	108·30	**100**	180·50	**140**	252·70	**180**	324·90
21	37·90½	61	110·10½	101	182·30½	141	254·50½	181	326·70½
22	39·71	62	111·91	102	184·11	142	256·31	182	328·51
23	41·51½	63	113·71½	103	185·91½	143	258·11½	183	330·31½
24	43·32	64	115·52	104	187·72	144	259·92	184	332·12
25	45·12½	65	117·32½	105	189·52½	145	261·72½	185	333·92½
26	46·93	66	119·13	106	191·33	146	263·53	186	335·73
27	48·73½	67	120·93½	107	193·13½	147	265·33½	187	337·53½
28	50·54	68	122·74	108	194·94	148	267·14	188	339·34
29	52·34½	69	124·54½	109	196·74½	149	268·94½	189	341·14½
30	54·15	**70**	126·35	**110**	198·55	**150**	270·75	**190**	342·95
31	55·95½	71	128·15½	111	200·35½	151	272·55½	191	344·75½
32	57·76	72	129·96	112	202·16	152	274·36	192	346·56
33	59·56½	73	131·76½	113	203·96½	153	276·16½	193	348·36½
34	61·37	74	133·57	114	205·77	154	277·97	194	350·17
35	63·17½	75	135·37½	115	207·57½	155	279·77½	195	351·97½
36	64·98	76	137·18	116	209·38	156	281·58	196	353·78
37	66·78½	77	138·98½	117	211·18½	157	283·38½	197	355·58½
38	68·59	78	140·79	118	212·99	158	285·19	198	357·39
39	70·39½	79	142·59½	119	214·79½	159	286·99½	199	359·19½
40	72·20	**80**	144·40	**120**	216·60	**160**	288·80	**200**	361·00

$\frac{1}{16}=0{\cdot}11^{3}$ $\frac{1}{9}=0{\cdot}20^{1}$ $\frac{1}{8}=0{\cdot}22^{6}$ $\frac{1}{6}=0{\cdot}30^{1}$ $\frac{1}{5}=0{\cdot}36^{1}$ $\frac{1}{4}=0{\cdot}45^{1}$ $\frac{1}{3}=0{\cdot}60^{2}$

$\frac{3}{8}=0{\cdot}67^{7}$ $\frac{1}{2}=0{\cdot}90^{3}$ $\frac{5}{8}=1{\cdot}12^{8}$ $\frac{3}{4}=1{\cdot}35^{4}$ $\frac{7}{8}=1{\cdot}57^{9}$ $\frac{2}{3}=1{\cdot}20^{3}$ $\frac{5}{6}=1{\cdot}50^{4}$

$\frac{3}{16}=0{\cdot}33^{8}$ $\frac{5}{16}=0{\cdot}56^{4}$ $\frac{7}{16}=0{\cdot}79^{0}$ $\frac{9}{16}=1{\cdot}01^{5}$ $\frac{11}{16}=1{\cdot}24^{1}$ $\frac{13}{16}=1{\cdot}46^{7}$ $\frac{15}{16}=1{\cdot}69^{2}$

(0.552=1£) **£1.81**

1	1·81	41	74·21	81	146·61	121	219·01	161	291·41
2	3·62	42	76·02	82	148·42	122	220·82	162	293·22
3	5·43	43	77·83	83	150·23	123	222·63	163	295·03
4	7·24	44	79·64	84	152·04	124	224·44	164	296·84
5	9·05	45	81·45	85	153·85	125	226·25	165	298·65
6	10·86	46	83·26	86	155·66	126	228·06	166	300·46
7	12·67	47	85·07	87	157·47	127	229·87	167	302·27
8	14·48	48	86·88	88	159·28	128	231·68	168	304·08
9	16·29	49	88·69	89	161·09	129	233·49	169	305·89
10	18·10	**50**	90·50	**90**	162·90	**130**	235·30	**170**	307·70
11	19·91	51	92·31	91	164·71	131	237·11	171	309·51
12	21·72	52	94·12	92	166·52	132	238·92	172	311·32
13	23·53	53	95·93	93	168·33	133	240·73	173	313·13
14	25·34	54	97·74	94	170·14	134	242·54	174	314·94
15	27·15	55	99·55	95	171·95	135	244·35	175	316·75
16	28·96	56	101·36	96	173·76	136	246·16	176	318·56
17	30·77	57	103·17	97	175·57	137	247·97	177	320·37
18	32·58	58	104·98	98	177·38	138	249·78	178	322·18
19	34·39	59	106·79	99	179·19	139	251·59	179	323·99
20	36·20	**60**	108·60	**100**	181·00	**140**	253·40	**180**	325·80
21	38·01	61	110·41	101	182·81	141	255·21	181	327·61
22	39·82	62	112·22	102	184·62	142	257·02	182	329·42
23	41·63	63	114·03	103	186·43	143	258·83	183	331·23
24	43·44	64	115·84	104	188·24	144	260·64	184	333·04
25	45·25	65	117·65	105	190·05	145	262·45	185	334·85
26	47·06	66	119·46	106	191·86	146	264·26	186	336·66
27	48·87	67	121·27	107	193·67	147	266·07	187	338·47
28	50·68	68	123·08	108	195·48	148	267·88	188	340·28
29	52·49	69	124·89	109	197·29	149	269·69	189	342·09
30	54·30	**70**	126·70	**110**	199·10	**150**	271·50	**190**	343·90
31	56·11	71	128·51	111	200·91	151	273·31	191	345·71
32	57·92	72	130·32	112	202·72	152	275·12	192	347·52
33	59·73	73	132·13	113	204·53	153	276·93	193	349·33
34	61·54	74	133·94	114	206·34	154	278·74	194	351·14
35	63·35	75	135·75	115	208·15	155	280·55	195	352·95
36	65·16	76	137·56	116	209·96	156	282·36	196	354·76
37	66·97	77	139·37	117	211·77	157	284·17	197	356·57
38	68·78	78	141·18	118	213·58	158	285·98	198	358·38
39	70·59	79	142·99	119	215·39	159	287·79	199	360·19
40	72·40	**80**	144·80	**120**	217·20	**160**	289·60	**200**	362·00

$\frac{1}{16}=0{\cdot}11^{3}$ $\frac{1}{9}=0{\cdot}20^{1}$ $\frac{1}{8}=0{\cdot}22^{6}$ $\frac{1}{6}=0{\cdot}30^{2}$ $\frac{1}{5}=0{\cdot}36^{2}$ $\frac{1}{4}=0{\cdot}45^{3}$ $\frac{1}{3}=0{\cdot}60^{3}$

$\frac{3}{8}=0{\cdot}67^{9}$ $\frac{1}{2}=0{\cdot}90^{5}$ $\frac{5}{8}=1{\cdot}13^{1}$ $\frac{3}{4}=1{\cdot}35^{8}$ $\frac{7}{8}=1{\cdot}58^{4}$ $\frac{2}{3}=1{\cdot}20^{7}$ $\frac{5}{6}=1{\cdot}50^{8}$

$\frac{3}{16}=0{\cdot}33^{9}$ $\frac{5}{16}=0{\cdot}56^{6}$ $\frac{7}{16}=0{\cdot}79^{2}$ $\frac{9}{16}=1{\cdot}01^{8}$ $\frac{11}{16}=1{\cdot}24^{4}$ $\frac{13}{16}=1{\cdot}47^{1}$ $\frac{15}{16}=1{\cdot}69^{7}$

1	1·81½	41	74·41½	81	147·01½	121	219·61½	161	292·21½
2	3·63	42	76·23	82	148·83	122	221·43	162	294·03
3	5·44½	43	78·04½	83	150·64½	123	223·24½	163	295·84½
4	7·26	44	79·86	84	152·46	124	225·06	164	297·66
5	9·07½	45	81·67½	85	154·27½	125	226·87½	165	299·47½
6	10·89	46	83·49	86	156·09	126	228·69	166	301·29
7	12·70½	47	85·30½	87	157·90½	127	230·50½	167	303·10½
8	14·52	48	87·12	88	159·72	128	232·32	168	304·92
9	16·33½	49	88·93½	89	161·53½	129	234·13½	169	306·73½
10	18·15	**50**	90·75	**90**	163·35	**130**	235·95	**170**	308·55
11	19·96½	51	92·56½	91	165·16½	131	237·76½	171	310·36½
12	21·78	52	94·38	92	166·98	132	239·58	172	312·18
13	23·59½	53	96·19½	93	168·79½	133	241·39½	173	313·99½
14	25·41	54	98·01	94	170·61	134	243·21	174	315·81
15	27·22½	55	99·82½	95	172·42½	135	245·02½	175	317·62½
16	29·04	56	101·64	96	174·24	136	246·84	176	319·44
17	30·85½	57	103·45½	97	176·05½	137	248·65½	177	321·25½
18	32·67	58	105·27	98	177·87	138	250·47	178	323·07
19	34·48½	59	107·08½	99	179·68½	139	252·28½	179	324·88½
20	36·30	**60**	108·90	**100**	181·50	**140**	254·10	**180**	326·70
21	38·11½	61	110·71½	101	183·31½	141	255·91½	181	328·51½
22	39·93	62	112·53	102	185·13	142	257·73	182	330·33
23	41·74½	63	114·34½	103	186·94½	143	259·54½	183	332·14½
24	43·56	64	116·16	104	188·76	144	261·36	184	333·96
25	45·37½	65	117·97½	105	190·57½	145	263·17½	185	335·77½
26	47·19	66	119·79	106	192·39	146	264·99	186	337·59
27	49·00½	67	121·60½	107	194·20½	147	266·80½	187	339·40½
28	50·82	68	123·42	108	196·02	148	268·62	188	341·22
29	52·63½	69	125·23½	109	197·83½	149	270·43½	189	343·03½
30	54·45	**70**	127·05	**110**	199·65	**150**	272·25	**190**	344·85
31	56·26½	71	128·86½	111	201·46½	151	274·06½	191	346·66½
32	58·08	72	130·68	112	203·28	152	275·88	192	348·48
33	59·89½	73	132·49½	113	205·09½	153	277·69½	193	350·29½
34	61·71	74	134·31	114	206·91	154	279·51	194	352·11
35	63·52½	75	136·12½	115	208·72½	155	281·32½	195	353·92½
36	65·34	76	137·94	116	210·54	156	283·14	196	355·74
37	67·15½	77	139·75½	117	212·35½	157	284·95½	197	357·55½
38	68·97	78	141·57	118	214·17	158	286·77	198	359·37
39	70·78½	79	143·38½	119	215·98½	159	288·58½	199	361·18½
40	72·60	**80**	145·20	**120**	217·80	**160**	290·40	**200**	363·00

$\frac{1}{16}$=0·11^{3} $\frac{1}{9}$=0·20^{2} $\frac{1}{8}$=0·22^{7} $\frac{1}{6}$=0·30^{3} $\frac{1}{5}$=0·36^{3} $\frac{1}{4}$=0·45^{4} $\frac{1}{3}$=0·60^{5}

$\frac{3}{8}$=0·68^{1} $\frac{1}{2}$=0·90^{8} $\frac{5}{8}$=1·13^{4} $\frac{3}{4}$=1·36^{1} $\frac{7}{8}$=1·58^{8} $\frac{2}{3}$=1·21^{0} $\frac{5}{6}$=1·51^{3}

$\frac{3}{16}$=0·34^{0} $\frac{5}{16}$=0·56^{7} $\frac{7}{16}$=0·79^{4} $\frac{9}{16}$=1·02^{1} $\frac{11}{16}$=1·24^{8} $\frac{13}{16}$=1·47^{5} $\frac{15}{16}$=1·70^{2}

(0.549=1£) **£1.82**

1	1·82	41	74·62	81	147·42	121	220·22	161	293·02
2	3·64	42	76·44	82	149·24	122	222·04	162	294·84
3	5·46	43	78·26	83	151·06	123	223·86	163	296·66
4	7·28	44	80·08	84	152·88	124	225·68	164	298·48
5	9·10	45	81·90	85	154·70	125	227·50	165	300·30
6	10·92	46	83·72	86	156·52	126	229·32	166	302·12
7	12·74	47	85·54	87	158·34	127	231·14	167	303·94
8	14·56	48	87·36	88	160·16	128	232·96	168	305·76
9	16·38	49	89·18	89	161·98	129	234·78	169	307·58
10	18·20	**50**	91·00	**90**	163·80	**130**	236·60	**170**	309·40
11	20·02	51	92·82	91	165·62	131	238·42	171	311·22
12	21·84	52	94·64	92	167·44	132	240·24	172	313·04
13	23·66	53	96·46	93	169·26	133	242·06	173	314·86
14	25·48	54	98·28	94	171·08	134	243·88	174	316·68
15	27·30	55	100·10	95	172·90	135	245·70	175	318·50
16	29·12	56	101·92	96	174·72	136	247·52	176	320·32
17	30·94	57	103·74	97	176·54	137	249·34	177	322·14
18	32·76	58	105·56	98	178·36	138	251·16	178	323·96
19	34·58	59	107·38	99	180·18	139	252·98	179	325·78
20	36·40	**60**	109·20	**100**	182·00	**140**	254·80	**180**	327·60
21	38·22	61	111·02	101	183·82	141	256·62	181	329·42
22	40·04	62	112·84	102	185·64	142	258·44	182	331·24
23	41·86	63	114·66	103	187·46	143	260·26	183	333·06
24	43·68	64	116·48	104	189·28	144	262·08	184	334·88
25	45·50	65	118·30	105	191·10	145	263·90	185	336·70
26	47·32	66	120·12	106	192·92	146	265·72	186	338·52
27	49·14	67	121·94	107	194·74	147	267·54	187	340·34
28	50·96	68	123·76	108	196·56	148	269·36	188	342·16
29	52·78	69	125·58	109	198·38	149	271·18	189	343·98
30	54·60	**70**	127·40	**110**	200·20	**150**	273·00	**190**	345·80
31	56·42	71	129·22	111	202·02	151	274·82	191	347·62
32	58·24	72	131·04	112	203·84	152	276·64	192	349·44
33	60·06	73	132·86	113	205·66	153	278·46	193	351·26
34	61·88	74	134·68	114	207·48	154	280·28	194	353·08
35	63·70	75	136·50	115	209·30	155	282·10	195	354·90
36	65·52	76	138·32	116	211·12	156	283·92	196	356·72
37	67·34	77	140·14	117	212·94	157	285·74	197	358·54
38	69·16	78	141·96	118	214·76	158	287·56	198	360·36
39	70·98	79	143·78	119	216·58	159	289·38	199	362·18
40	72·80	**80**	145·60	**120**	218·40	**160**	291·20	**200**	364·00

$\frac{1}{16}=0{\cdot}11^{4}$ $\frac{1}{9}=0{\cdot}20^{2}$ $\frac{1}{8}=0{\cdot}22^{8}$ $\frac{1}{6}=0{\cdot}30^{3}$ $\frac{1}{5}=0{\cdot}36^{4}$ $\frac{1}{4}=0{\cdot}45^{5}$ $\frac{1}{3}=0{\cdot}60^{7}$

$\frac{3}{8}=0{\cdot}68^{3}$ $\frac{1}{2}=0{\cdot}91^{0}$ $\frac{5}{8}=1{\cdot}13^{8}$ $\frac{3}{4}=1{\cdot}36^{5}$ $\frac{7}{8}=1{\cdot}59^{3}$ $\frac{2}{3}=1{\cdot}21^{3}$ $\frac{5}{6}=1{\cdot}51^{7}$

$\frac{3}{16}=0{\cdot}34^{1}$ $\frac{5}{16}=0{\cdot}56^{9}$ $\frac{7}{16}=0{\cdot}79^{6}$ $\frac{9}{16}=1{\cdot}02^{4}$ $\frac{11}{16}=1{\cdot}25^{1}$ $\frac{13}{16}=1{\cdot}47^{9}$ $\frac{15}{16}=1{\cdot}70^{6}$

1	1·82½	41	74·82½	81	147·82½	121	220·82½	161	293·82½
2	3·65	42	76·65	82	149·65	122	222·65	162	295·65
3	5·47½	43	78·47½	83	151·47½	123	224·47½	163	297·47½
4	7·30	44	80·30	84	153·30	124	226·30	164	299·30
5	9·12½	45	82·12½	85	155·12½	125	228·12½	165	301·12½
6	10·95	46	83·95	86	156·95	126	229·95	166	302·95
7	12·77½	47	85·77½	87	158·77½	127	231·77½	167	304·77½
8	14·60	48	87·60	88	160·60	128	233·60	168	306·60
9	16·42½	49	89·42½	89	162·42½	129	235·42½	169	308·42½
10	18·25	**50**	91·25	**90**	164·25	**130**	237·25	**170**	310·25
11	20·07½	51	93·07½	91	166·07½	131	239·07½	171	312·07½
12	21·90	52	94·90	92	167·90	132	240·90	172	313·90
13	23·72½	53	96·72½	93	169·72½	133	242·72½	173	315·72½
14	25·55	54	98·55	94	171·55	134	244·55	174	317·55
15	27·37½	55	100·37½	95	173·37½	135	246·37½	175	319·37½
16	29·20	56	102·20	96	175·20	136	248·20	176	321·20
17	31·02½	57	104·02½	97	177·02½	137	250·02½	177	323·02½
18	32·85	58	105·85	98	178·85	138	251·85	178	324·85
19	34·67½	59	107·67½	99	180·67½	139	253·67½	179	326·67½
20	36·50	**60**	109·50	**100**	182·50	**140**	255·50	**180**	328·50
21	38·32½	61	111·32½	101	184·32½	141	257·32½	181	330·32½
22	40·15	62	113·15	102	186·15	142	259·15	182	332·15
23	41·97½	63	114·97½	103	187·97½	143	260·97½	183	333·97½
24	43·80	64	116·80	104	189·80	144	262·80	184	335·80
25	45·62½	65	118·62½	105	191·62½	145	264·62½	185	337·62½
26	47·45	66	120·45	106	193·45	146	266·45	186	339·45
27	49·27½	67	122·27½	107	195·27½	147	268·27½	187	341·27½
28	51·10	68	124·10	108	197·10	148	270·10	188	343·10
29	52·92½	69	125·92½	109	198·92½	149	271·92½	189	344·92½
30	54·75	**70**	127·75	**110**	200·75	**150**	273·75	**190**	346·75
31	56·57½	71	129·57½	111	202·57½	151	275·57½	191	348·57½
32	58·40	72	131·40	112	204·40	152	277·40	192	350·40
33	60·22½	73	133·22½	113	206·22½	153	279·22½	193	352·22½
34	62·05	74	135·05	114	208·05	154	281·05	194	354·05
35	63·87½	75	136·87½	115	209·87½	155	282·87½	195	355·87½
36	65·70	76	138·70	116	211·70	156	284·70	196	357·70
37	67·52½	77	140·52½	117	213·52½	157	286·52½	197	359·52½
38	69·35	78	142·35	118	215·35	158	288·35	198	361·35
39	71·17½	79	144·17½	119	217·17½	159	290·17½	199	363·17½
40	73·00	**80**	146·00	**120**	219·00	**160**	292·00	**200**	365·00

$\frac{1}{16}=0{\cdot}11^{4}$ $\frac{1}{9}=0{\cdot}20^{3}$ $\frac{1}{8}=0{\cdot}22^{8}$ $\frac{1}{6}=0{\cdot}30^{4}$ $\frac{1}{5}=0{\cdot}36^{5}$ $\frac{1}{4}=0{\cdot}45^{6}$ $\frac{1}{3}=0{\cdot}60^{8}$

$\frac{3}{8}=0{\cdot}68^{4}$ $\frac{1}{2}=0{\cdot}91^{3}$ $\frac{5}{8}=1{\cdot}14^{1}$ $\frac{3}{4}=1{\cdot}36^{9}$ $\frac{7}{8}=1{\cdot}59^{7}$ $\frac{2}{3}=1{\cdot}21^{7}$ $\frac{5}{6}=1{\cdot}52^{1}$

$\frac{3}{16}=0{\cdot}34^{2}$ $\frac{5}{16}=0{\cdot}57^{0}$ $\frac{7}{16}=0{\cdot}79^{8}$ $\frac{9}{16}=1{\cdot}02^{7}$ $\frac{11}{16}=1{\cdot}25^{5}$ $\frac{13}{16}=1{\cdot}48^{3}$ $\frac{15}{16}=1{\cdot}71^{1}$

(0.546=1£) **£1.83**

1	1·83	41	75·03	81	148·23	121	221·43	161	294·63
2	3·66	42	76·86	82	150·06	122	223·26	162	296·46
3	5·49	43	78·69	83	151·89	123	225·09	163	298·29
4	7·32	44	80·52	84	153·72	124	226·92	164	300·12
5	9·15	45	82·35	85	155·55	125	228·75	165	301·95
6	10·98	46	84·18	86	157·38	126	230·58	166	303·78
7	12·81	47	86·01	87	159·21	127	232·41	167	305·61
8	14·64	48	87·84	88	161·04	128	234·24	168	307·44
9	16·47	49	89·67	89	162·87	129	236·07	169	309·27
10	18·30	**50**	91·50	**90**	164·70	**130**	237·90	**170**	311·10
11	20·13	51	93·33	91	166·53	131	239·73	171	312·93
12	21·96	52	95·16	92	168·36	132	241·56	172	314·76
13	23·79	53	96·99	93	170·19	133	243·39	173	316·59
14	25·62	54	98·82	94	172·02	134	245·22	174	318·42
15	27·45	55	100·65	95	173·85	135	247·05	175	320·25
16	29·28	56	102·48	96	175·68	136	248·88	176	322·08
17	31·11	57	104·31	97	177·51	137	250·71	177	323·91
18	32·94	58	106·14	98	179·34	138	252·54	178	325·74
19	34·77	59	107·97	99	181·17	139	254·37	179	327·57
20	36·60	**60**	109·80	**100**	183·00	**140**	256·20	**180**	329·40
21	38·43	61	111·63	101	184·83	141	258·03	181	331·23
22	40·26	62	113·46	102	186·66	142	259·86	182	333·06
23	42·09	63	115·29	103	188·49	143	261·69	183	334·89
24	43·92	64	117·12	104	190·32	144	263·52	184	336·72
25	45·75	65	118·95	105	192·15	145	265·35	185	338·55
26	47·58	66	120·78	106	193·98	146	267·18	186	340·38
27	49·41	67	122·61	107	195·81	147	269·01	187	342·21
28	51·24	68	124·44	108	197·64	148	270·84	188	344·04
29	53·07	69	126·27	109	199·47	149	272·67	189	345·87
30	54·90	**70**	128·10	**110**	201·30	**150**	274·50	**190**	347·70
31	56·73	71	129·93	111	203·13	151	276·33	191	349·53
32	58·56	72	131·76	112	204·96	152	278·16	192	351·36
33	60·39	73	133·59	113	206·79	153	279·99	193	353·19
34	62·22	74	135·42	114	208·62	154	281·82	194	355·02
35	64·05	75	137·25	115	210·45	155	283·65	195	356·85
36	65·88	76	139·08	116	212·28	156	285·48	196	358·68
37	67·71	77	140·91	117	214·11	157	287·31	197	360·51
38	69·54	78	142·74	118	215·94	158	289·14	198	362·34
39	71·37	79	144·57	119	217·77	159	290·97	199	364·17
40	73·20	**80**	146·40	**120**	219·60	**160**	292·80	**200**	366·00

$\frac{1}{16}=0\cdot11^{4}$ $\frac{1}{9}=0\cdot20^{3}$ $\frac{1}{8}=0\cdot22^{9}$ $\frac{1}{6}=0\cdot30^{5}$ $\frac{1}{5}=0\cdot36^{6}$ $\frac{1}{4}=0\cdot45^{8}$ $\frac{1}{3}=0\cdot61^{0}$

$\frac{3}{8}=0\cdot68^{6}$ $\frac{1}{2}=0\cdot91^{5}$ $\frac{5}{8}=1\cdot14^{4}$ $\frac{3}{4}=1\cdot37^{3}$ $\frac{7}{8}=1\cdot60^{1}$ $\frac{2}{3}=1\cdot22^{0}$ $\frac{5}{6}=1\cdot52^{5}$

$\frac{3}{16}=0\cdot34^{3}$ $\frac{5}{16}=0\cdot57^{2}$ $\frac{7}{16}=0\cdot80^{1}$ $\frac{9}{16}=1\cdot02^{9}$ $\frac{11}{16}=1\cdot25^{8}$ $\frac{13}{16}=1\cdot48^{7}$ $\frac{15}{16}=1\cdot71^{6}$

1	1·83½	41	75·23½	81	148·63½	121	222·03½	161	295·43½
2	3·67	42	77·07	82	150·47	122	223·87	162	297·27
3	5·50½	43	78·90½	83	152·30½	123	225·70½	163	299·10½
4	7·34	44	80·74	84	154·14	124	227·54	164	300·94
5	9·17½	45	82·57½	85	155·97½	125	229·37½	165	302·77½
6	11·01	46	84·41	86	157·81	126	231·21	166	304·61
7	12·84½	47	86·24½	87	159·64½	127	233·04½	167	306·44½
8	14·68	48	88·08	88	161·48	128	234·88	168	308·28
9	16·51½	49	89·91½	89	163·31½	129	236·71½	169	310·11½
10	18·35	**50**	91·75	**90**	165·15	**130**	238·55	**170**	311·95
11	20·18½	51	93·58½	91	166·98½	131	240·38½	171	313·78½
12	22·02	52	95·42	92	168·82	132	242·22	172	315·62
13	23·85½	53	97·25½	93	170·65½	133	244·05½	173	317·45½
14	25·69	54	99·09	94	172·49	134	245·89	174	319·29
15	27·52½	55	100·92½	95	174·32½	135	247·72½	175	321·12½
16	29·36	56	102·76	96	176·16	136	249·56	176	322·96
17	31·19½	57	104·59½	97	177·99½	137	251·39½	177	324·79½
18	33·03	58	106·43	98	179·83	138	253·23	178	326·63
19	34·86½	59	108·26½	99	181·66½	139	255·06½	179	328·46½
20	36·70	**60**	110·10	**100**	183·50	**140**	256·90	**180**	330·30
21	38·53½	61	111·93½	101	185·33½	141	258·73½	181	332·13½
22	40·37	62	113·77	102	187·17	142	260·57	182	333·97
23	42·20½	63	115·60½	103	189·00½	143	262·40½	183	335·80½
24	44·04	64	117·44	104	190·84	144	264·24	184	337·64
25	45·87½	65	119·27½	105	192·67½	145	266·07½	185	339·47½
26	47·71	66	121·11	106	194·51	146	267·91	186	341·31
27	49·54½	67	122·94½	107	196·34½	147	269·74½	187	343·14½
28	51·38	68	124·78	108	198·18	148	271·58	188	344·98
29	53·21½	69	126·61½	109	200·01½	149	273·41½	189	346·81½
30	55·05	**70**	128·45	**110**	201·85	**150**	275·25	**190**	348·65
31	56·88½	71	130·28½	111	203·68½	151	277·08½	191	350·48½
32	58·72	72	132·12	112	205·52	152	278·92	192	352·32
33	60·55½	73	133·95½	113	207·35½	153	280·75½	193	354·15½
34	62·39	74	135·79	114	209·19	154	282·59	194	355·99
35	64·22½	75	137·62½	115	211·02½	155	284·42½	195	357·82½
36	66·06	76	139·46	116	212·86	156	286·26	196	359·66
37	67·89½	77	141·29½	117	214·69½	157	288·09½	197	361·49½
38	69·73	78	143·13	118	216·53	158	289·93	198	363·33
39	71·56½	79	144·96½	119	218·36½	159	291·76½	199	365·16½
40	73·40	**80**	146·80	**120**	220·20	**160**	293·60	**200**	367·00

$\frac{1}{16}=0{\cdot}11^{5}$ $\frac{1}{9}=0{\cdot}20^{4}$ $\frac{1}{8}=0{\cdot}22^{9}$ $\frac{1}{6}=0{\cdot}30^{6}$ $\frac{1}{5}=0{\cdot}36^{7}$ $\frac{1}{4}=0{\cdot}45^{9}$ $\frac{1}{3}=0{\cdot}61^{2}$

$\frac{3}{8}=0{\cdot}68^{8}$ $\frac{1}{2}=0{\cdot}91^{8}$ $\frac{5}{8}=1{\cdot}14^{7}$ $\frac{3}{4}=1{\cdot}37^{6}$ $\frac{7}{8}=1{\cdot}60^{6}$ $\frac{2}{3}=1{\cdot}22^{3}$ $\frac{5}{6}=1{\cdot}52^{9}$

$\frac{3}{16}=0{\cdot}34^{4}$ $\frac{5}{16}=0{\cdot}57^{3}$ $\frac{7}{16}=0{\cdot}80^{3}$ $\frac{9}{16}=1{\cdot}03^{2}$ $\frac{11}{16}=1{\cdot}26^{2}$ $\frac{13}{16}=1{\cdot}49^{1}$ $\frac{15}{16}=1{\cdot}72^{0}$

1	1·84	41	75·44	81	149·04	121	222·64	161	296·24
2	3·68	42	77·28	82	150·88	122	224·48	162	298·08
3	5·52	43	79·12	83	152·72	123	226·32	163	299·92
4	7·36	44	80·96	84	154·56	124	228·16	164	301·76
5	9·20	45	82·80	85	156·40	125	230·00	165	303·60
6	11·04	46	84·64	86	158·24	126	231·84	166	305·44
7	12·88	47	86·48	87	160·08	127	233·68	167	307·28
8	14·72	48	88·32	88	161·92	128	235·52	168	309·12
9	16·56	49	90·16	89	163·76	129	237·36	169	310·96
10	18·40	**50**	92·00	**90**	165·60	**130**	239·20	**170**	312·80
11	20·24	51	93·84	91	167·44	131	241·04	171	314·64
12	22·08	52	95·68	92	169·28	132	242·88	172	316·48
13	23·92	53	97·52	93	171·12	133	244·72	173	318·32
14	25·76	54	99·36	94	172·96	134	246·56	174	320·16
15	27·60	55	101·20	95	174·80	135	248·40	175	322·00
16	29·44	56	103·04	96	176·64	136	250·24	176	323·84
17	31·28	57	104·88	97	178·48	137	252·08	177	325·68
18	33·12	58	106·72	98	180·32	138	253·92	178	327·52
19	34·96	59	108·56	99	182·16	139	255·76	179	329·36
20	36·80	**60**	110·40	**100**	184·00	**140**	257·60	**180**	331·20
21	38·64	61	112·24	101	185·84	141	259·44	181	333·04
22	40·48	62	114·08	102	187·68	142	261·28	182	334·88
23	42·32	63	115·92	103	189·52	143	263·12	183	336·72
24	44·16	64	117·76	104	191·36	144	264·96	184	338·56
25	46·00	65	119·60	105	193·20	145	266·80	185	340·40
26	47·84	66	121·44	106	195·04	146	268·64	186	342·24
27	49·68	67	123·28	107	196·88	147	270·48	187	344·08
28	51·52	68	125·12	108	198·72	148	272·32	188	345·92
29	53·36	69	126·96	109	200·56	149	274·16	189	347·76
30	55·20	**70**	128·80	**110**	202·40	**150**	276·00	**190**	349·60
31	57·04	71	130·64	111	204·24	151	277·84	191	351·44
32	58·88	72	132·48	112	206·08	152	279·68	192	353·28
33	60·72	73	134·32	113	207·92	153	281·52	193	355·12
34	62·56	74	136·16	114	209·76	154	283·36	194	356·96
35	64·40	75	138·00	115	211·60	155	285·20	195	358·80
36	66·24	76	139·84	116	213·44	156	287·04	196	360·64
37	68·08	77	141·68	117	215·28	157	288·88	197	362·48
38	69·92	78	143·52	118	217·12	158	290·72	198	364·32
39	71·76	79	145·36	119	218·96	159	292·56	199	366·16
40	73·60	**80**	147·20	**120**	220·80	**160**	294·40	**200**	368·00

$\frac{1}{16}=0{\cdot}11^5$ $\frac{1}{9}=0{\cdot}20^4$ $\frac{1}{8}=0{\cdot}23^0$ $\frac{1}{6}=0{\cdot}30^7$ $\frac{1}{5}=0{\cdot}36^8$ $\frac{1}{4}=0{\cdot}46^0$ $\frac{1}{3}=0{\cdot}61^3$

$\frac{3}{8}=0{\cdot}69^0$ $\frac{1}{2}=0{\cdot}92^0$ $\frac{5}{8}=1{\cdot}15^0$ $\frac{3}{4}=1{\cdot}38^0$ $\frac{7}{8}=1{\cdot}61^0$ $\frac{2}{3}=1{\cdot}22^7$ $\frac{5}{6}=1{\cdot}53^3$

$\frac{3}{16}=0{\cdot}34^5$ $\frac{5}{16}=0{\cdot}57^5$ $\frac{7}{16}=0{\cdot}80^5$ $\frac{9}{16}=1{\cdot}03^5$ $\frac{11}{16}=1{\cdot}26^5$ $\frac{13}{16}=1{\cdot}49^5$ $\frac{15}{16}=1{\cdot}72^5$

1	1·84½	41	75·64½	81	149·44½	121	223·24½	161	297·04½
2	3·69	42	77·49	82	151·29	122	225·09	162	298·89
3	5·53½	43	79·33½	83	153·13½	123	226·93½	163	300·73½
4	7·38	44	81·18	84	154·98	124	228·78	164	302·58
5	9·22½	45	83·02½	85	156·82½	125	230·62½	165	304·42½
6	11·07	46	84·87	86	158·67	126	232·47	166	306·27
7	12·91½	47	86·71½	87	160·51½	127	234·31½	167	308·11½
8	14·76	48	88·56	88	162·36	128	236·16	168	309·96
9	16·60½	49	90·40½	89	164·20½	129	238·00½	169	311·80½
10	18·45	**50**	92·25	**90**	166·05	**130**	239·85	**170**	313·65
11	20·29½	51	94·09½	91	167·89½	131	241·69½	171	315·49½
12	22·14	52	95·94	92	169·74	132	243·54	172	317·34
13	23·98½	53	97·78½	93	171·58½	133	245·38½	173	319·18½
14	25·83	54	99·63	94	173·43	134	247·23	174	321·03
15	27·67½	55	101·47½	95	175·27½	135	249·07½	175	322·87½
16	29·52	56	103·32	96	177·12	136	250·92	176	324·72
17	31·36½	57	105·16½	97	178·96½	137	252·76½	177	326·56½
18	33·21	58	107·01	98	180·81	138	254·61	178	328·41
19	35·05½	59	108·85½	99	182·65½	139	256·45½	179	330·25½
20	36·90	**60**	110·70	**100**	184·50	**140**	258·30	**180**	332·10
21	38·74½	61	112·54½	101	186·34½	141	260·14½	181	333·94½
22	40·59	62	114·39	102	188·19	142	261·99	182	335·79
23	42·43½	63	116·23½	103	190·03½	143	263·83½	183	337·63½
24	44·28	64	118·08	104	191·88	144	265·68	184	339·48
25	46·12½	65	119·92½	105	193·72½	145	267·52½	185	341·32½
26	47·97	66	121·77	106	195·57	146	269·37	186	343·17
27	49·81½	67	123·61½	107	197·41½	147	271·21½	187	345·01½
28	51·66	68	125·46	108	199·26	148	273·06	188	346·86
29	53·50½	69	127·30½	109	201·10½	149	274·90½	189	348·70½
30	55·35	**70**	129·15	**110**	202·95	**150**	276·75	**190**	350·55
31	57·19½	71	130·99½	111	204·79½	151	278·59½	191	352·39½
32	59·04	72	132·84	112	206·64	152	280·44	192	354·24
33	60·88½	73	134·68½	113	208·48½	153	282·28½	193	356·08½
34	62·73	74	136·53	114	210·33	154	284·13	194	357·93
35	64·57½	75	138·37½	115	212·17½	155	285·97½	195	359·77½
36	66·42	76	140·22	116	214·02	156	287·82	196	361·62
37	68·26½	77	142·06½	117	215·86½	157	289·66½	197	363·46½
38	70·11	78	143·91	118	217·71	158	291·51	198	365·31
39	71·95½	79	145·75½	119	219·55½	159	293·35½	199	367·15½
40	73·80	**80**	147·60	**120**	221·40	**160**	295·20	**200**	369·00

$\frac{1}{16}=0{\cdot}11^{5}$ $\frac{1}{9}=0{\cdot}20^{5}$ $\frac{1}{8}=0{\cdot}23^{1}$ $\frac{1}{6}=0{\cdot}30^{8}$ $\frac{1}{5}=0{\cdot}36^{9}$ $\frac{1}{4}=0{\cdot}46^{1}$ $\frac{1}{3}=0{\cdot}61^{5}$

$\frac{3}{8}=0{\cdot}69^{2}$ $\frac{1}{2}=0{\cdot}92^{3}$ $\frac{5}{8}=1{\cdot}15^{3}$ $\frac{3}{4}=1{\cdot}38^{4}$ $\frac{7}{8}=1{\cdot}61^{4}$ $\frac{2}{3}=1{\cdot}23^{0}$ $\frac{5}{6}=1{\cdot}53^{8}$

$\frac{3}{16}=0{\cdot}34^{6}$ $\frac{5}{16}=0{\cdot}57^{7}$ $\frac{7}{16}=0{\cdot}80^{7}$ $\frac{9}{16}=1{\cdot}03^{8}$ $\frac{11}{16}=1{\cdot}26^{8}$ $\frac{13}{16}=1{\cdot}49^{9}$ $\frac{15}{16}=1{\cdot}73^{0}$

(0.541=1£) **£1.85**

1	1·85	41	75·85	81	149·85	121	223·85	161	297·85
2	3·70	42	77·70	82	151·70	122	225·70	162	299·70
3	5·55	43	79·55	83	153·55	123	227·55	163	301·55
4	7·40	44	81·40	84	155·40	124	229·40	164	303·40
5	9·25	45	83·25	85	157·25	125	231·25	165	305·25
6	11·10	46	85·10	86	159·10	126	233·10	166	307·10
7	12·95	47	86·95	87	160·95	127	234·95	167	308·95
8	14·80	48	88·80	88	162·80	128	236·80	168	310·80
9	16·65	49	90·65	89	164·65	129	238·65	169	312·65
10	18·50	**50**	92·50	**90**	166·50	**130**	240·50	**170**	314·50
11	20·35	51	94·35	91	168·35	131	242·35	171	316·35
12	22·20	52	96·20	92	170·20	132	244·20	172	318·20
13	24·05	53	98·05	93	172·05	133	246·05	173	320·05
14	25·90	54	99·90	94	173·90	134	247·90	174	321·90
15	27·75	55	101·75	95	175·75	135	249·75	175	323·75
16	29·60	56	103·60	96	177·60	136	251·60	176	325·60
17	31·45	57	105·45	97	179·45	137	253·45	177	327·45
18	33·30	58	107·30	98	181·30	138	255·30	178	329·30
19	35·15	59	109·15	99	183·15	139	257·15	179	331·15
20	37·00	**60**	111·00	**100**	185·00	**140**	259·00	**180**	333·00
21	38·85	61	112·85	101	186·85	141	260·85	181	334·85
22	40·70	62	114·70	102	188·70	142	262·70	182	336·70
23	42·55	63	116·55	103	190·55	143	264·55	183	338·55
24	44·40	64	118·40	104	192·40	144	266·40	184	340·40
25	46·25	65	120·25	105	194·25	145	268·25	185	342·25
26	48·10	66	122·10	106	196·10	146	270·10	186	344·10
27	49·95	67	123·95	107	197·95	147	271·95	187	345·95
28	51·80	68	125·80	108	199·80	148	273·80	188	347·80
29	53·65	69	127·65	109	201·65	149	275·65	189	349·65
30	55·50	**70**	129·50	**110**	203·50	**150**	277·50	**190**	351·50
31	57·35	71	131·35	111	205·35	151	279·35	191	353·35
32	59·20	72	133·20	112	207·20	152	281·20	192	355·20
33	61·05	73	135·05	113	209·05	153	283·05	193	357·05
34	62·90	74	136·90	114	210·90	154	284·90	194	358·90
35	64·75	75	138·75	115	212·75	155	286·75	195	360·75
36	66·60	76	140·60	116	214·60	156	288·60	196	362·60
37	68·45	77	142·45	117	216·45	157	290·45	197	364·45
38	70·30	78	144·30	118	218·30	158	292·30	198	366·30
39	72·15	79	146·15	119	220·15	159	294·15	199	368·15
40	74·00	**80**	148·00	**120**	222·00	**160**	296·00	**200**	370·00

$\frac{1}{16}$=0·11^{6} $\frac{1}{9}$=0·20^{6} $\frac{1}{8}$=0·23^{1} $\frac{1}{6}$=0·30^{8} $\frac{1}{5}$=0·37^{0} $\frac{1}{4}$=0·46^{3} $\frac{1}{3}$=0·61^{7}

$\frac{3}{8}$=0·69^{4} $\frac{1}{2}$=0·92^{5} $\frac{5}{8}$=1·15^{6} $\frac{3}{4}$=1·38^{8} $\frac{7}{8}$=1·61^{9} $\frac{2}{3}$=1·23^{3} $\frac{5}{6}$=1·54^{2}

$\frac{3}{16}$=0·34^{7} $\frac{5}{16}$=0·57^{8} $\frac{7}{16}$=0·80^{9} $\frac{9}{16}$=1·04^{1} $\frac{11}{16}$=1·27^{2} $\frac{13}{16}$=1·50^{3} $\frac{15}{16}$=1·73^{4}

1	1·85½	41	76·05½	81	150·25½	121	224·45½	161	298·65½
2	3·71	42	77·91	82	152·11	122	226·31	162	300·51
3	5·56½	43	79·76½	83	153·96½	123	228·16½	163	302·36½
4	7·42	44	81·62	84	155·82	124	230·02	164	304·22
5	9·27½	45	83·47½	85	157·67½	125	231·87½	165	306·07½
6	11·13	46	85·33	86	159·53	126	233·73	166	307·93
7	12·98½	47	87·18½	87	161·38½	127	235·58½	167	309·78½
8	14·84	48	89·04	88	163·24	128	237·44	168	311·64
9	16·69½	49	90·89½	89	165·09½	129	239·29½	169	313·49½
10	18·55	**50**	92·75	**90**	166·95	**130**	241·15	**170**	315·35
11	20·40½	51	94·60½	91	168·80½	131	243·00½	171	317·20½
12	22·26	52	96·46	92	170·66	132	244·86	172	319·06
13	24·11½	53	98·31½	93	172·51½	133	246·71½	173	320·91½
14	25·97	54	100·17	94	174·37	134	248·57	174	322·77
15	27·82½	55	102·02½	95	176·22½	135	250·42½	175	324·62½
16	29·68	56	103·88	96	178·08	136	252·28	176	326·48
17	31·53½	57	105·73½	97	179·93½	137	254·13½	177	328·33½
18	33·39	58	107·59	98	181·79	138	255·99	178	330·19
19	35·24½	59	109·44½	99	183·64½	139	257·84½	179	332·04½
20	37·10	**60**	111·30	**100**	185·50	**140**	259·70	**180**	333·90
21	38·95½	61	113·15½	101	187·35½	141	261·55½	181	335·75½
22	40·81	62	115·01	102	189·21	142	263·41	182	337·61
23	42·66½	63	116·86½	103	191·06½	143	265·26½	183	339·46½
24	44·52	64	118·72	104	192·92	144	267·12	184	341·32
25	46·37½	65	120·57½	105	194·77½	145	268·97½	185	343·17½
26	48·23	66	122·43	106	196·63	146	270·83	186	345·03
27	50·08½	67	124·28½	107	198·48½	147	272·68½	187	346·88½
28	51·94	68	126·14	108	200·34	148	274·54	188	348·74
29	53·79½	69	127·99½	109	202·19½	149	276·39½	189	350·59½
30	55·65	**70**	129·85	**110**	204·05	**150**	278·25	**190**	352·45
31	57·50½	71	131·70½	111	205·90½	151	280·10½	191	354·30½
32	59·36	72	133·56	112	207·76	152	281·96	192	356·16
33	61·21½	73	135·41½	113	209·61½	153	283·81½	193	358·01½
34	63·07	74	137·27	114	211·47	154	285·67	194	359·87
35	64·92½	75	139·12½	115	213·32½	155	287·52½	195	361·72½
36	66·78	76	140·98	116	215·18	156	289·38	196	363·58
37	68·63½	77	142·83½	117	217·03½	157	291·23½	197	365·43½
38	70·49	78	144·69	118	218·89	158	293·09	198	367·29
39	72·34½	79	146·54½	119	220·74½	159	294·94½	199	369·14½
40	74·20	**80**	148·40	**120**	222·60	**160**	296·80	**200**	371·00

$\frac{1}{16}=0{\cdot}11^{6}$ $\frac{1}{9}=0{\cdot}20^{6}$ $\frac{1}{8}=0{\cdot}23^{2}$ $\frac{1}{6}=0{\cdot}30^{9}$ $\frac{1}{5}=0{\cdot}37^{1}$ $\frac{1}{4}=0{\cdot}46^{4}$ $\frac{1}{3}=0{\cdot}61^{8}$

$\frac{3}{8}=0{\cdot}69^{6}$ $\frac{1}{2}=0{\cdot}92^{8}$ $\frac{5}{8}=1{\cdot}15^{9}$ $\frac{3}{4}=1{\cdot}39^{1}$ $\frac{7}{8}=1{\cdot}62^{3}$ $\frac{2}{3}=1{\cdot}23^{7}$ $\frac{5}{6}=1{\cdot}54^{6}$

$\frac{3}{16}=0{\cdot}34^{8}$ $\frac{5}{16}=0{\cdot}58^{0}$ $\frac{7}{16}=0{\cdot}81^{2}$ $\frac{9}{16}=1{\cdot}04^{3}$ $\frac{11}{16}=1{\cdot}27^{5}$ $\frac{13}{16}=1{\cdot}50^{7}$ $\frac{15}{16}=1{\cdot}73^{9}$

(0.538=1£) **£1.86**

1	1·86	41	76·26	81	150·66	121	225·06	161	299·46
2	3·72	42	78·12	82	152·52	122	226·92	162	301·32
3	5·58	43	79·98	83	154·38	123	228·78	163	303·18
4	7·44	44	81·84	84	156·24	124	230·64	164	305·04
5	9·30	45	83·70	85	158·10	125	232·50	165	306·90
6	11·16	46	85·56	86	159·96	126	234·36	166	308·76
7	13·02	47	87·42	87	161·82	127	236·22	167	310·62
8	14·88	48	89·28	88	163·68	128	238·08	168	312·48
9	16·74	49	91·14	89	165·54	129	239·94	169	314·34
10	18·60	**50**	93·00	**90**	167·40	**130**	241·80	**170**	316·20
11	20·46	51	94·86	91	169·26	131	243·66	171	318·06
12	22·32	52	96·72	92	171·12	132	245·52	172	319·92
13	24·18	53	98·58	93	172·98	133	247·38	173	321·78
14	26·04	54	100·44	94	174·84	134	249·24	174	323·64
15	27·90	55	102·30	95	176·70	135	251·10	175	325·50
16	29·76	56	104·16	96	178·56	136	252·96	176	327·36
17	31·62	57	106·02	97	180·42	137	254·82	177	329·22
18	33·48	58	107·88	98	182·28	138	256·68	178	331·08
19	35·34	59	109·74	99	184·14	139	258·54	179	332·94
20	37·20	**60**	111·60	**100**	186·00	**140**	260·40	**180**	334·80
21	39·06	61	113·46	101	187·86	141	262·26	181	336·66
22	40·92	62	115·32	102	189·72	142	264·12	182	338·52
23	42·78	63	117·18	103	191·58	143	265·98	183	340·38
24	44·64	64	119·04	104	193·44	144	267·84	184	342·24
25	46·50	65	120·90	105	195·30	145	269·70	185	344·10
26	48·36	66	122·76	106	197·16	146	271·56	186	345·96
27	50·22	67	124·62	107	199·02	147	273·42	187	347·82
28	52·08	68	126·48	108	200·88	148	275·28	188	349·68
29	53·94	69	128·34	109	202·74	149	277·14	189	351·54
30	55·80	**70**	130·20	**110**	204·60	**150**	279·00	**190**	353·40
31	57·66	71	132·06	111	206·46	151	280·86	191	355·26
32	59·52	72	133·92	112	208·32	152	282·72	192	357·12
33	61·38	73	135·78	113	210·18	153	284·58	193	358·98
34	63·24	74	137·64	114	212·04	154	286·44	194	360·84
35	65·10	75	139·50	115	213·90	155	288·30	195	362·70
36	66·96	76	141·36	116	215·76	156	290·16	196	364·56
37	68·82	77	143·22	117	217·62	157	292·02	197	366·42
38	70·68	78	145·08	118	219·48	158	293·88	198	368·28
39	72·54	79	146·94	119	221·34	159	295·74	199	370·14
40	74·40	**80**	148·80	**120**	223·20	**160**	297·60	**200**	372·00

$\frac{1}{16}=0{\cdot}11^{6}$ $\frac{1}{9}=0{\cdot}20^{7}$ $\frac{1}{8}=0{\cdot}23^{3}$ $\frac{1}{6}=0{\cdot}31^{0}$ $\frac{1}{5}=0{\cdot}37^{2}$ $\frac{1}{4}=0{\cdot}46^{5}$ $\frac{1}{3}=0{\cdot}62^{0}$

$\frac{3}{8}=0{\cdot}69^{8}$ $\frac{1}{2}=0{\cdot}93^{0}$ $\frac{5}{8}=1{\cdot}16^{3}$ $\frac{3}{4}=1{\cdot}39^{5}$ $\frac{7}{8}=1{\cdot}62^{8}$ $\frac{2}{3}=1{\cdot}24^{0}$ $\frac{5}{6}=1{\cdot}55^{0}$

$\frac{3}{16}=0{\cdot}34^{9}$ $\frac{5}{16}=0{\cdot}58^{1}$ $\frac{7}{16}=0{\cdot}81^{4}$ $\frac{9}{16}=1{\cdot}04^{6}$ $\frac{11}{16}=1{\cdot}27^{9}$ $\frac{13}{16}=1{\cdot}51^{1}$ $\frac{15}{16}=1{\cdot}74^{4}$

(0.536=1£) **£1.86½**

1	1·86½	41	76·46½	81	151·06½	121	225·66½	161	300·26½
2	3·73	42	78·33	82	152·93	122	227·53	162	302·13
3	5·59½	43	80·19½	83	154·79½	123	229·39½	163	303·99½
4	7·46	44	82·06	84	156·66	124	231·26	164	305·86
5	9·32½	45	83·92½	85	158·52½	125	233·12½	165	307·72½
6	11·19	46	85·79	86	160·39	126	234·99	166	309·59
7	13·05½	47	87·65½	87	162·25½	127	236·85½	167	311·45½
8	14·92	48	89·52	88	164·12	128	238·72	168	313·32
9	16·78½	49	91·38½	89	165·98½	129	240·58½	169	315·18½
10	18·65	**50**	93·25	**90**	167·85	**130**	242·45	**170**	317·05
11	20·51½	51	95·11½	91	169·71½	131	244·31½	171	318·91½
12	22·38	52	96·98	92	171·58	132	246·18	172	320·78
13	24·24½	53	98·84½	93	173·44½	133	248·04½	173	322·64½
14	26·11	54	100·71	94	175·31	134	249·91	174	324·51
15	27·97½	55	102·57½	95	177·17½	135	251·77½	175	326·37½
16	29·84	56	104·44	96	179·04	136	253·64	176	328·24
17	31·70½	57	106·30½	97	180·90½	137	255·50½	177	330·10½
18	33·57	58	108·17	98	182·77	138	257·37	178	331·97
19	35·43½	59	110·03½	99	184·63½	139	259·23½	179	333·83½
20	37·30	**60**	111·90	**100**	186·50	**140**	261·10	**180**	335·70
21	39·16½	61	113·76½	101	188·36½	141	262·96½	181	337·56½
22	41·03	62	115·63	102	190·23	142	264·83	182	339·43
23	42·89½	63	117·49½	103	192·09½	143	266·69½	183	341·29½
24	44·76	64	119·36	104	193·96	144	268·56	184	343·16
25	46·62½	65	121·22½	105	195·82½	145	270·42½	185	345·02½
26	48·49	66	123·09	106	197·69	146	272·29	186	346·89
27	50·35½	67	124·95½	107	199·55½	147	274·15½	187	348·75½
28	52·22	68	126·82	108	201·42	148	276·02	188	350·62
29	54·08½	69	128·68½	109	203·28½	149	277·88½	189	352·48½
30	55·95	**70**	130·55	**110**	205·15	**150**	279·75	**190**	354·35
31	57·81½	71	132·41½	111	207·01½	151	281·61½	191	356·21½
32	59·68	72	134·28	112	208·88	152	283·48	192	358·08
33	61·54½	73	136·14½	113	210·74½	153	285·34½	193	359·94½
34	63·41	74	138·01	114	212·61	154	287·21	194	361·81
35	65·27½	75	139·87½	115	214·47½	155	289·07½	195	363·67½
36	67·14	76	141·74	116	216·34	156	290·94	196	365·54
37	69·00½	77	143·60½	117	218·20½	157	292·80½	197	367·40½
38	70·87	78	145·47	118	220·07	158	294·67	198	369·27
39	72·73½	79	147·33½	119	221·93½	159	296·53½	199	371·13½
40	74·60	**80**	149·20	**120**	223·80	**160**	298·40	**200**	373·00

$\frac{1}{16}$=0·11^{7} $\frac{1}{9}$=0·20^{7} $\frac{1}{8}$=0·23^{3} $\frac{1}{6}$=0·31^{1} $\frac{1}{5}$=0·37^{3} $\frac{1}{4}$=0·46^{6} $\frac{1}{3}$=0·62^{2}

$\frac{3}{8}$=0·69^{9} $\frac{1}{2}$=0·93^{3} $\frac{5}{8}$=1·16^{6} $\frac{3}{4}$=1·39^{9} $\frac{7}{8}$=1·63^{2} $\frac{2}{3}$=1·24^{3} $\frac{5}{6}$=1·55^{4}

$\frac{3}{16}$=0·35^{0} $\frac{5}{16}$=0·58^{3} $\frac{7}{16}$=0·81^{6} $\frac{9}{16}$=1·04^{9} $\frac{11}{16}$=1·28^{2} $\frac{13}{16}$=1·51^{5} $\frac{15}{16}$=1·74^{8}

(0.535=1£) **£1.87**

1	1·87	41	76·67	81	151·47	121	226·27	161	301·07
2	3·74	42	78·54	82	153·34	122	228·14	162	302·94
3	5·61	43	80·41	83	155·21	123	230·01	163	304·81
4	7·48	44	82·28	84	157·08	124	231·88	164	306·68
5	9·35	45	84·15	85	158·95	125	233·75	165	308·55
6	11·22	46	86·02	86	160·82	126	235·62	166	310·42
7	13·09	47	87·89	87	162·69	127	237·49	167	312·29
8	14·96	48	89·76	88	164·56	128	239·36	168	314·16
9	16·83	49	91·63	89	166·43	129	241·23	169	316·03
10	18·70	**50**	93·50	**90**	168·30	**130**	243·10	**170**	317·90
11	20·57	51	95·37	91	170·17	131	244·97	171	319·77
12	22·44	52	97·24	92	172·04	132	246·84	172	321·64
13	24·31	53	99·11	93	173·91	133	248·71	173	323·51
14	26·18	54	100·98	94	175·78	134	250·58	174	325·38
15	28·05	55	102·85	95	177·65	135	252·45	175	327·25
16	29·92	56	104·72	96	179·52	136	254·32	176	329·12
17	31·79	57	106·59	97	181·39	137	256·19	177	330·99
18	33·66	58	108·46	98	183·26	138	258·06	178	332·86
19	35·53	59	110·33	99	185·13	139	259·93	179	334·73
20	37·40	**60**	112·20	**100**	187·00	**140**	261·80	**180**	336·60
21	39·27	61	114·07	101	188·87	141	263·67	181	338·47
22	41·14	62	115·94	102	190·74	142	265·54	182	340·34
23	43·01	63	117·81	103	192·61	143	267·41	183	342·21
24	44·88	64	119·68	104	194·48	144	269·28	184	344·08
25	46·75	65	121·55	105	196·35	145	271·15	185	345·95
26	48·62	66	123·42	106	198·22	146	273·02	186	347·82
27	50·49	67	125·29	107	200·09	147	274·89	187	349·69
28	52·36	68	127·16	108	201·96	148	276·76	188	351·56
29	54·23	69	129·03	109	203·83	149	278·63	189	353·43
30	56·10	**70**	130·90	**110**	205·70	**150**	280·50	**190**	355·30
31	57·97	71	132·77	111	207·57	151	282·37	191	357·17
32	59·84	72	134·64	112	209·44	152	284·24	192	359·04
33	61·71	73	136·51	113	211·31	153	286·11	193	360·91
34	63·58	74	138·38	114	213·18	154	287·98	194	362·78
35	65·45	75	140·25	115	215·05	155	289·85	195	364·65
36	67·32	76	142·12	116	216·92	156	291·72	196	366·52
37	69·19	77	143·99	117	218·79	157	293·59	197	368·39
38	71·06	78	145·86	118	220·66	158	295·46	198	370·26
39	72·93	79	147·73	119	222·53	159	297·33	199	372·13
40	74·80	**80**	149·60	**120**	224·40	**160**	299·20	**200**	374·00

$\frac{1}{16}=0{\cdot}11^{7}$ $\frac{1}{9}=0{\cdot}20^{8}$ $\frac{1}{8}=0{\cdot}23^{4}$ $\frac{1}{6}=0{\cdot}31^{2}$ $\frac{1}{5}=0{\cdot}37^{4}$ $\frac{1}{4}=0{\cdot}46^{8}$ $\frac{1}{3}=0{\cdot}62^{3}$

$\frac{3}{8}=0{\cdot}70^{1}$ $\frac{1}{2}=0{\cdot}93^{5}$ $\frac{5}{8}=1{\cdot}16^{9}$ $\frac{3}{4}=1{\cdot}40^{3}$ $\frac{7}{8}=1{\cdot}63^{6}$ $\frac{2}{3}=1{\cdot}24^{7}$ $\frac{5}{6}=1{\cdot}55^{8}$

$\frac{3}{16}=0{\cdot}35^{1}$ $\frac{5}{16}=0{\cdot}58^{4}$ $\frac{7}{16}=0{\cdot}81^{8}$ $\frac{9}{16}=1{\cdot}05^{2}$ $\frac{11}{16}=1{\cdot}28^{6}$ $\frac{13}{16}=1{\cdot}51^{9}$ $\frac{15}{16}=1{\cdot}75^{3}$

(0.533=1£) **£1.87½**

1	1·87½	41	76·87½	81	151·87½	121	226·87½	161	301·87½
2	3·75	42	78·75	82	153·75	122	228·75	162	303·75
3	5·62½	43	80·62½	83	155·62½	123	230·62½	163	305·62½
4	7·50	44	82·50	84	157·50	124	232·50	164	307·50
5	9·37½	45	84·37½	85	159·37½	125	234·37½	165	309·37½
6	11·25	46	86·25	86	161·25	126	236·25	166	311·25
7	13·12½	47	88·12½	87	163·12½	127	238·12½	167	313·12½
8	15·00	48	90·00	88	165·00	128	240·00	168	315·00
9	16·87½	49	91·87½	89	166·87½	129	241·87½	169	316·87½
10	18·75	**50**	93·75	**90**	168·75	**130**	243·75	**170**	318·75
11	20·62½	51	95·62½	91	170·62½	131	245·62½	171	320·62½
12	22·50	52	97·50	92	172·50	132	247·50	172	322·50
13	24·37½	53	99·37½	93	174·37½	133	249·37½	173	324·37½
14	26·25	54	101·25	94	176·25	134	251·25	174	326·25
15	28·12½	55	103·12½	95	178·12½	135	253·12½	175	328·12½
16	30·00	56	105·00	96	180·00	136	255·00	176	330·00
17	31·87½	57	106·87½	97	181·87½	137	256·87½	177	331·87½
18	33·75	58	108·75	98	183·75	138	258·75	178	333·75
19	35·62½	59	110·62½	99	185·62½	139	260·62½	179	335·62½
20	37·50	**60**	112·50	**100**	187·50	**140**	262·50	**180**	337·50
21	39·37½	61	114·37½	101	189·37½	141	264·37½	181	339·37½
22	41·25	62	116·25	102	191·25	142	266·25	182	341·25
23	43·12½	63	118·12½	103	193·12½	143	268·12½	183	343·12½
24	45·00	64	120·00	104	195·00	144	270·00	184	345·00
25	46·87½	65	121·87½	105	196·87½	145	271·87½	185	346·87½
26	48·75	66	123·75	106	198·75	146	273·75	186	348·75
27	50·62½	67	125·62½	107	200·62½	147	275·62½	187	350·62½
28	52·50	68	127·50	108	202·50	148	277·50	188	352·50
29	54·37½	69	129·37½	109	204·37½	149	279·37½	189	354·37½
30	56·25	**70**	131·25	**110**	206·25	**150**	281·25	**190**	356·25
31	58·12½	71	133·12½	111	208·12½	151	283·12½	191	358·12½
32	60·00	72	135·00	112	210·00	152	285·00	192	360·00
33	61·87½	73	136·87½	113	211·87½	153	286·87½	193	361·87½
34	63·75	74	138·75	114	213·75	154	288·75	194	363·75
35	65·62½	75	140·62½	115	215·62½	155	290·62½	195	365·62½
36	67·50	76	142·50	116	217·50	156	292·50	196	367·50
37	69·37½	77	144·37½	117	219·37½	157	294·37½	197	369·37½
38	71·25	78	146·25	118	221·25	158	296·25	198	371·25
39	73·12½	79	148·12½	119	223·12½	159	298·12½	199	373·12½
40	75·00	**80**	150·00	**120**	225·00	**160**	300·00	**200**	375·00

$\frac{1}{16}$=0·11^{7} $\frac{1}{9}$=0·20^{8} $\frac{1}{8}$=0·23^{4} $\frac{1}{6}$=0·31^{3} $\frac{1}{5}$=0·37^{5} $\frac{1}{4}$=0·46^{9} $\frac{1}{3}$=0·62^{5}

$\frac{3}{8}$=0·70^{3} $\frac{1}{2}$=0·93^{8} $\frac{5}{8}$=1·17^{2} $\frac{3}{4}$=1·40^{6} $\frac{7}{8}$=1·64^{1} $\frac{2}{3}$=1·25^{0} $\frac{5}{6}$=1·56^{3}

$\frac{3}{16}$=0·35^{2} $\frac{5}{16}$=0·58^{6} $\frac{7}{16}$=0·82^{0} $\frac{9}{16}$=1·05^{5} $\frac{11}{16}$=1·28^{9} $\frac{13}{16}$=1·52^{3} $\frac{15}{16}$=1·75^{8}

(0.532=1£) £1.88

1	1·88	41	77·08	81	152·28	121	227·48	161	302·68
2	3·76	42	78·96	82	154·16	122	229·36	162	304·56
3	5·64	43	80·84	83	156·04	123	231·24	163	306·44
4	7·52	44	82·72	84	157·92	124	233·12	164	308·32
5	9·40	45	84·60	85	159·80	125	235·00	165	310·20
6	11·28	46	86·48	86	161·68	126	236·88	166	312·08
7	13·16	47	88·36	87	163·56	127	238·76	167	313·96
8	15·04	48	90·24	88	165·44	128	240·64	168	315·84
9	16·92	49	92·12	89	167·32	129	242·52	169	317·72
10	18·80	**50**	94·00	**90**	169·20	**130**	244·40	**170**	319·60
11	20·68	51	95·88	91	171·08	131	246·28	171	321·48
12	22·56	52	97·76	92	172·96	132	248·16	172	323·36
13	24·44	53	99·64	93	174·84	133	250·04	173	325·24
14	26·32	54	101·52	94	176·72	134	251·92	174	327·12
15	28·20	55	103·40	95	178·60	135	253·80	175	329·00
16	30·08	56	105·28	96	180·48	136	255·68	176	330·88
17	31·96	57	107·16	97	182·36	137	257·56	177	332·76
18	33·84	58	109·04	98	184·24	138	259·44	178	334·64
19	35·72	59	110·92	99	186·12	139	261·32	179	336·52
20	37·60	**60**	112·80	**100**	188·00	**140**	263·20	**180**	338·40
21	39·48	61	114·68	101	189·88	141	265·08	181	340·28
22	41·36	62	116·56	102	191·76	142	266·96	182	342·16
23	43·24	63	118·44	103	193·64	143	268·84	183	344·04
24	45·12	64	120·32	104	195·52	144	270·72	184	345·92
25	47·00	65	122·20	105	197·40	145	272·60	185	347·80
26	48·88	66	124·08	106	199·28	146	274·48	186	349·68
27	50·76	67	125·96	107	201·16	147	276·36	187	351·56
28	52·64	68	127·84	108	203·04	148	278·24	188	353·44
29	54·52	69	129·72	109	204·92	149	280·12	189	355·32
30	56·40	**70**	131·60	**110**	206·80	**150**	282·00	**190**	357·20
31	58·28	71	133·48	111	208·68	151	283·88	191	359·08
32	60·16	72	135·36	112	210·56	152	285·76	192	360·96
33	62·04	73	137·24	113	212·44	153	287·64	193	362·84
34	63·92	74	139·12	114	214·32	154	289·52	194	364·72
35	65·80	75	141·00	115	216·20	155	291·40	195	366·60
36	67·68	76	142·88	116	218·08	156	293·28	196	368·48
37	69·56	77	144·76	117	219·96	157	295·16	197	370·36
38	71·44	78	146·64	118	221·84	158	297·04	198	372·24
39	73·32	79	148·52	119	223·72	159	298·92	199	374·12
40	75·20	**80**	150·40	**120**	225·60	**160**	300·80	**200**	376·00

$\frac{1}{16}=0{\cdot}11^{8}$ $\frac{1}{9}=0{\cdot}20^{9}$ $\frac{1}{8}=0{\cdot}23^{5}$ $\frac{1}{6}=0{\cdot}31^{3}$ $\frac{1}{5}=0{\cdot}37^{6}$ $\frac{1}{4}=0{\cdot}47^{0}$ $\frac{1}{3}=0{\cdot}62^{7}$

$\frac{3}{8}=0{\cdot}70^{5}$ $\frac{1}{2}=0{\cdot}94^{0}$ $\frac{5}{8}=1{\cdot}17^{5}$ $\frac{3}{4}=1{\cdot}41^{0}$ $\frac{7}{8}=1{\cdot}64^{5}$ $\frac{2}{3}=1{\cdot}25^{3}$ $\frac{5}{6}=1{\cdot}56^{7}$

$\frac{3}{16}=0{\cdot}35^{3}$ $\frac{5}{16}=0{\cdot}58^{8}$ $\frac{7}{16}=0{\cdot}82^{3}$ $\frac{9}{16}=1{\cdot}05^{8}$ $\frac{11}{16}=1{\cdot}29^{3}$ $\frac{13}{16}=1{\cdot}52^{8}$ $\frac{15}{16}=1{\cdot}76^{3}$

1	1·88½	41	77·28½	81	152·68½	121	228·08½	161	303·48½
2	3·77	42	79·17	82	154·57	122	229·97	162	305·37
3	5·65½	43	81·05½	83	156·45½	123	231·85½	163	307·25½
4	7·54	44	82·94	84	158·34	124	233·74	164	309·14
5	9·42½	45	84·82½	85	160·22½	125	235·62½	165	311·02½
6	11·31	46	86·71	86	162·11	126	237·51	166	312·91
7	13·19½	47	88·59½	87	163·99½	127	239·39½	167	314·79½
8	15·08	48	90·48	88	165·88	128	241·28	168	316·68
9	16·96½	49	92·36½	89	167·76½	129	243·16½	169	318·56½
10	18·85	**50**	94·25	**90**	169·65	**130**	245·05	**170**	320·45
11	20·73½	51	96·13½	91	171·53½	131	246·93½	171	322·33½
12	22·62	52	98·02	92	173·42	132	248·82	172	324·22
13	24·50½	53	99·90½	93	175·30½	133	250·70½	173	326·10½
14	26·39	54	101·79	94	177·19	134	252·59	174	327·99
15	28·27½	55	103·67½	95	179·07½	135	254·47½	175	329·87½
16	30·16	56	105·56	96	180·96	136	256·36	176	331·76
17	32·04½	57	107·44½	97	182·84½	137	258·24½	177	333·64½
18	33·93	58	109·33	98	184·73	138	260·13	178	335·53
19	35·81½	59	111·21½	99	186·61½	139	262·01½	179	337·41½
20	37·70	**60**	113·10	**100**	188·50	**140**	263·90	**180**	339·30
21	39·58½	61	114·98½	101	190·38½	141	265·78½	181	341·18½
22	41·47	62	116·87	102	192·27	142	267·67	182	343·07
23	43·35½	63	118·75½	103	194·15½	143	269·55½	183	344·95½
24	45·24	64	120·64	104	196·04	144	271·44	184	346·84
25	47·12½	65	122·52½	105	197·92½	145	273·32½	185	348·72½
26	49·01	66	124·41	106	199·81	146	275·21	186	350·61
27	50·89½	67	126·29½	107	201·69½	147	277·09½	187	352·49½
28	52·78	68	128·18	108	203·58	148	278·98	188	354·38
29	54·66½	69	130·06½	109	205·46½	149	280·86½	189	356·26½
30	56·55	**70**	131·95	**110**	207·35	**150**	282·75	**190**	358·15
31	58·43½	71	133·83½	111	209·23½	151	284·63½	191	360·03½
32	60·32	72	135·72	112	211·12	152	286·52	192	361·92
33	62·20½	73	137·60½	113	213·00½	153	288·40½	193	363·80½
34	64·09	74	139·49	114	214·89	154	290·29	194	365·69
35	65·97½	75	141·37½	115	216·77½	155	292·17½	195	367·57½
36	67·86	76	143·26	116	218·66	156	294·06	196	369·46
37	69·74½	77	145·14½	117	220·54½	157	295·94½	197	371·34½
38	71·63	78	147·03	118	222·43	158	297·83	198	373·23
39	73·51½	79	148·91½	119	224·31½	159	299·71½	199	375·11½
40	75·40	**80**	150·80	**120**	226·20	**160**	301·60	**200**	377·00

$\frac{1}{16}$=0·11^{8} $\frac{1}{9}$=0·20^{9} $\frac{1}{8}$=0·23^{6} $\frac{1}{6}$=0·31^{4} $\frac{1}{5}$=0·37^{7} $\frac{1}{4}$=0·47^{1} $\frac{1}{3}$=0·62^{8}

$\frac{3}{8}$=0·70^{7} $\frac{1}{2}$=0·94^{3} $\frac{5}{8}$=1·17^{8} $\frac{3}{4}$=1·41^{4} $\frac{7}{8}$=1·64^{9} $\frac{2}{3}$=1·25^{7} $\frac{5}{6}$=1·57^{1}

$\frac{3}{16}$=0·35^{3} $\frac{5}{16}$=0·58^{9} $\frac{7}{16}$=0·82^{5} $\frac{9}{16}$=1·06^{0} $\frac{11}{16}$=1·29^{6} $\frac{13}{16}$=1·53^{2} $\frac{15}{16}$=1·76^{7}

(0.529=1£) **£1.89**

1	1·89	41	77·49	81	153·09	121	228·69	161	304·29
2	3·78	42	79·38	82	154·98	122	230·58	162	306·18
3	5·67	43	81·27	83	156·87	123	232·47	163	308·07
4	7·56	44	83·16	84	158·76	124	234·36	164	309·96
5	9·45	45	85·05	85	160·65	125	236·25	165	311·85
6	11·34	46	86·94	86	162·54	126	238·14	166	313·74
7	13·23	47	88·83	87	164·43	127	240·03	167	315·63
8	15·12	48	90·72	88	166·32	128	241·92	168	317·52
9	17·01	49	92·61	89	168·21	129	243·81	169	319·41
10	18·90	**50**	94·50	**90**	170·10	**130**	245·70	**170**	321·30
11	20·79	51	96·39	91	171·99	131	247·59	171	323·19
12	22·68	52	98·28	92	173·88	132	249·48	172	325·08
13	24·57	53	100·17	93	175·77	133	251·37	173	326·97
14	26·46	54	102·06	94	177·66	134	253·26	174	328·86
15	28·35	55	103·95	95	179·55	135	255·15	175	330·75
16	30·24	56	105·84	96	181·44	136	257·04	176	332·64
17	32·13	57	107·73	97	183·33	137	258·93	177	334·53
18	34·02	58	109·62	98	185·22	138	260·82	178	336·42
19	35·91	59	111·51	99	187·11	139	262·71	179	338·31
20	37·80	**60**	113·40	**100**	189·00	**140**	264·60	**180**	340·20
21	39·69	61	115·29	101	190·89	141	266·49	181	342·09
22	41·58	62	117·18	102	192·78	142	268·38	182	343·98
23	43·47	63	119·07	103	194·67	143	270·27	183	345·87
24	45·36	64	120·96	104	196·56	144	272·16	184	347·76
25	47·25	65	122·85	105	198·45	145	274·05	185	349·65
26	49·14	66	124·74	106	200·34	146	275·94	186	351·54
27	51·03	67	126·63	107	202·23	147	277·83	187	353·43
28	52·92	68	128·52	108	204·12	148	279·72	188	355·32
29	54·81	69	130·41	109	206·01	149	281·61	189	357·21
30	56·70	**70**	132·30	**110**	207·90	**150**	283·50	**190**	359·10
31	58·59	71	134·19	111	209·79	151	285·39	191	360·99
32	60·48	72	136·08	112	211·68	152	287·28	192	362·88
33	62·37	73	137·97	113	213·57	153	289·17	193	364·77
34	64·26	74	139·86	114	215·46	154	291·06	194	366·66
35	66·15	75	141·75	115	217·35	155	292·95	195	368·55
36	68·04	76	143·64	116	219·24	156	294·84	196	370·44
37	69·93	77	145·53	117	221·13	157	296·73	197	372·33
38	71·82	78	147·42	118	223·02	158	298·62	198	374·22
39	73·71	79	149·31	119	224·91	159	300·51	199	376·11
40	75·60	**80**	151·20	**120**	226·80	**160**	302·40	**200**	378·00

$\frac{1}{16}$=0·11^{8} $\frac{1}{9}$=0·21^{0} $\frac{1}{8}$=0·23^{6} $\frac{1}{6}$=0·31^{5} $\frac{1}{5}$=0·37^{8} $\frac{1}{4}$=0·47^{3} $\frac{1}{3}$=0·63^{0}

$\frac{3}{8}$=0·70^{9} $\frac{1}{2}$=0·94^{5} $\frac{5}{8}$=1·18^{1} $\frac{3}{4}$=1·41^{8} $\frac{7}{8}$=1·65^{4} $\frac{2}{3}$=1·26^{0} $\frac{5}{6}$=1·57^{5}

$\frac{3}{16}$=0·35^{4} $\frac{5}{16}$=0·59^{1} $\frac{7}{16}$=0·82^{7} $\frac{9}{16}$=1·06^{3} $\frac{11}{16}$=1·29^{9} $\frac{13}{16}$=1·53^{6} $\frac{15}{16}$=1·77^{2}

(0.528=1£) **£1.89½**

1	1·89½	41	77·69½	81	153·49½	121	229·29½	161	305·09½
2	3·79	42	79·59	82	155·39	122	231·19	162	306·99
3	5·68½	43	81·48½	83	157·28½	123	233·08½	163	308·88½
4	7·58	44	83·38	84	159·18	124	234·98	164	310·78
5	9·47½	45	85·27½	85	161·07½	125	236·87½	165	312·67½
6	11·37	46	87·17	86	162·97	126	238·77	166	314·57
7	13·26½	47	89·06½	87	164·86½	127	240·66½	167	316·46½
8	15·16	48	90·96	88	166·76	128	242·56	168	318·36
9	17·05½	49	92·85½	89	168·65½	129	244·45½	169	320·25½
10	18·95	**50**	94·75	**90**	170·55	**130**	246·35	**170**	322·15
11	20·84½	51	96·64½	91	172·44½	131	248·24½	171	324·04½
12	22·74	52	98·54	92	174·34	132	250·14	172	325·94
13	24·63½	53	100·43½	93	176·23½	133	252·03½	173	327·83½
14	26·53	54	102·33	94	178·13	134	253·93	174	329·73
15	28·42½	55	104·22½	95	180·02½	135	255·82½	175	331·62½
16	30·32	56	106·12	96	181·92	136	257·72	176	333·52
17	32·21½	57	108·01½	97	183·81½	137	259·61½	177	335·41½
18	34·11	58	109·91	98	185·71	138	261·51	178	337·31
19	36·00½	59	111·80½	99	187·60½	139	263·40½	179	339·20½
20	37·90	**60**	113·70	**100**	189·50	**140**	265·30	**180**	341·10
21	39·79½	61	115·59½	101	191·39½	141	267·19½	181	342·99½
22	41·69	62	117·49	102	193·29	142	269·09	182	344·89
23	43·58½	63	119·38½	103	195·18½	143	270·98½	183	346·78½
24	45·48	64	121·28	104	197·08	144	272·88	184	348·68
25	47·37½	65	123·17½	105	198·97½	145	274·77½	185	350·57½
26	49·27	66	125·07	106	200·87	146	276·67	186	352·47
27	51·16½	67	126·96½	107	202·76½	147	278·56½	187	354·36½
28	53·06	68	128·86	108	204·66	148	280·46	188	356·26
29	54·95½	69	130·75½	109	206·55½	149	282·35½	189	358·15½
30	56·85	**70**	132·65	**110**	208·45	**150**	284·25	**190**	360·05
31	58·74½	71	134·54½	111	210·34½	151	286·14½	191	361·94½
32	60·64	72	136·44	112	212·24	152	288·04	192	363·84
33	62·53½	73	138·33½	113	214·13½	153	289·93½	193	365·73½
34	64·43	74	140·23	114	216·03	154	291·83	194	367·63
35	66·32½	75	142·12½	115	217·92½	155	293·72½	195	369·52½
36	68·22	76	144·02	116	219·82	156	295·62	196	371·42
37	70·11½	77	145·91½	117	221·71½	157	297·51½	197	373·31½
38	72·01	78	147·81	118	223·61	158	299·41	198	375·21
39	73·90½	79	149·70½	119	225·50½	159	301·30½	199	377·10½
40	75·80	**80**	151·60	**120**	227·40	**160**	303·20	**200**	379·00

$\frac{1}{16}$=0·11^{8} $\frac{1}{9}$=0·21^{1} $\frac{1}{8}$=0·23^{7} $\frac{1}{6}$=0·31^{6} $\frac{1}{5}$=0·37^{9} $\frac{1}{4}$=0·47^{4} $\frac{1}{3}$=0·63^{2}

$\frac{3}{8}$=0·71^{1} $\frac{1}{2}$=0·94^{8} $\frac{5}{8}$=1·18^{4} $\frac{3}{4}$=1·42^{1} $\frac{7}{8}$=1·65^{8} $\frac{2}{3}$=1·26^{3} $\frac{5}{6}$=1·57^{9}

$\frac{3}{16}$=0·35^{5} $\frac{5}{16}$=0·59^{2} $\frac{7}{16}$=0·82^{9} $\frac{9}{16}$=1·06^{6} $\frac{11}{16}$=1·30^{3} $\frac{13}{16}$=1·54^{0} $\frac{15}{16}$=1·77^{7}

(0.526=1£) **£1.90**

1	1·90	41	77·90	81	153·90	121	229·90	161	305·90
2	3·80	42	79·80	82	155·80	122	231·80	162	307·80
3	5·70	43	81·70	83	157·70	123	233·70	163	309·70
4	7·60	44	83·60	84	159·60	124	235·60	164	311·60
5	9·50	45	85·50	85	161·50	125	237·50	165	313·50
6	11·40	46	87·40	86	163·40	126	239·40	166	315·40
7	13·30	47	89·30	87	165·30	127	241·30	167	317·30
8	15·20	48	91·20	88	167·20	128	243·20	168	319·20
9	17·10	49	93·10	89	169·10	129	245·10	169	321·10
10	19·00	**50**	95·00	**90**	171·00	**130**	247·00	**170**	323·00
11	20·90	51	96·90	91	172·90	131	248·90	171	324·90
12	22·80	52	98·80	92	174·80	132	250·80	172	326·80
13	24·70	53	100·70	93	176·70	133	252·70	173	328·70
14	26·60	54	102·60	94	178·60	134	254·60	174	330·60
15	28·50	55	104·50	95	180·50	135	256·50	175	332·50
16	30·40	56	106·40	96	182·40	136	258·40	176	334·40
17	32·30	57	108·30	97	184·30	137	260·30	177	336·30
18	34·20	58	110·20	98	186·20	138	262·20	178	338·20
19	36·10	59	112·10	99	188·10	139	264·10	179	340·10
20	38·00	**60**	114·00	**100**	190·00	**140**	266·00	**180**	342·00
21	39·90	61	115·90	101	191·90	141	267·90	181	343·90
22	41·80	62	117·80	102	193·80	142	269·80	182	345·80
23	43·70	63	119·70	103	195·70	143	271·70	183	347·70
24	45·60	64	121·60	104	197·60	144	273·60	184	349·60
25	47·50	65	123·50	105	199·50	145	275·50	185	351·50
26	49·40	66	125·40	106	201·40	146	277·40	186	353·40
27	51·30	67	127·30	107	203·30	147	279·30	187	355·30
28	53·20	68	129·20	108	205·20	148	281·20	188	357·20
29	55·10	69	131·10	109	207·10	149	283·10	189	359·10
30	57·00	**70**	133·00	**110**	209·00	**150**	285·00	**190**	361·00
31	58·90	71	134·90	111	210·90	151	286·90	191	362·90
32	60·80	72	136·80	112	212·80	152	288·80	192	364·80
33	62·70	73	138·70	113	214·70	153	290·70	193	366·70
34	64·60	74	140·60	114	216·60	154	292·60	194	368·60
35	66·50	75	142·50	115	218·50	155	294·50	195	370·50
36	68·40	76	144·40	116	220·40	156	296·40	196	372·40
37	70·30	77	146·30	117	222·30	157	298·30	197	374·30
38	72·20	78	148·20	118	224·20	158	300·20	198	376·20
39	74·10	79	150·10	119	226·10	159	302·10	199	378·10
40	76·00	**80**	152·00	**120**	228·00	**160**	304·00	**200**	380·00

$\frac{1}{16}=0{\cdot}11^{9}$ $\frac{1}{9}=0{\cdot}21^{1}$ $\frac{1}{8}=0{\cdot}23^{8}$ $\frac{1}{6}=0{\cdot}31^{7}$ $\frac{1}{5}=0{\cdot}38^{0}$ $\frac{1}{4}=0{\cdot}47^{5}$ $\frac{1}{3}=0{\cdot}63^{3}$

$\frac{3}{8}=0{\cdot}71^{3}$ $\frac{1}{2}=0{\cdot}95^{0}$ $\frac{5}{8}=1{\cdot}18^{8}$ $\frac{3}{4}=1{\cdot}42^{5}$ $\frac{7}{8}=1{\cdot}66^{3}$ $\frac{2}{3}=1{\cdot}26^{7}$ $\frac{5}{6}=1{\cdot}58^{3}$

$\frac{3}{16}=0{\cdot}35^{6}$ $\frac{5}{16}=0{\cdot}59^{4}$ $\frac{7}{16}=0{\cdot}83^{1}$ $\frac{9}{16}=1{\cdot}06^{9}$ $\frac{11}{16}=1{\cdot}30^{6}$ $\frac{13}{16}=1{\cdot}54^{4}$ $\frac{15}{16}=1{\cdot}78^{1}$

(0.525=1£) **£1.90½**

1	1·90½	41	78·10½	81	154·30½	121	230·50½	161	306·70½
2	3·81	42	80·01	82	156·21	122	232·41	162	308·61
3	5·71½	43	81·91½	83	158·11½	123	234·31½	163	310·51½
4	7·62	44	83·82	84	160·02	124	236·22	164	312·42
5	9·52½	45	85·72½	85	161·92½	125	238·12½	165	314·32½
6	11·43	46	87·63	86	163·83	126	240·03	166	316·23
7	13·33½	47	89·53½	87	165·73½	127	241·93½	167	318·13½
8	15·24	48	91·44	88	167·64	128	243·84	168	320·04
9	17·14½	49	93·34½	89	169·54½	129	245·74½	169	321·94½
10	19·05	**50**	95·25	**90**	171·45	**130**	247·65	**170**	323·85
11	20·95½	51	97·15½	91	173·35½	131	249·55½	171	325·75½
12	22·86	52	99·06	92	175·26	132	251·46	172	327·66
13	24·76½	53	100·96½	93	177·16½	133	253·36½	173	329·56½
14	26·67	54	102·87	94	179·07	134	255·27	174	331·47
15	28·57½	55	104·77½	95	180·97½	135	257·17½	175	333·37½
16	30·48	56	106·68	96	182·88	136	259·08	176	335·28
17	32·38½	57	108·58½	97	184·78½	137	260·98½	177	337·18½
18	34·29	58	110·49	98	186·69	138	262·89	178	339·09
19	36·19½	59	112·39½	99	188·59½	139	264·79½	179	340·99½
20	38·10	**60**	114·30	**100**	190·50	**140**	266·70	**180**	342·90
21	40·00½	61	116·20½	101	192·40½	141	268·60½	181	344·80½
22	41·91	62	118·11	102	194·31	142	270·51	182	346·71
23	43·81½	63	120·01½	103	196·21½	143	272·41½	183	348·61½
24	45·72	64	121·92	104	198·12	144	274·32	184	350·52
25	47·62½	65	123·82½	105	200·02½	145	276·22½	185	352·42½
26	49·53	66	125·73	106	201·93	146	278·13	186	354·33
27	51·43½	67	127·63½	107	203·83½	147	280·03½	187	356·23½
28	53·34	68	129·54	108	205·74	148	281·94	188	358·14
29	55·24½	69	131·44½	109	207·64½	149	283·84½	189	360·04½
30	57·15	**70**	133·35	**110**	209·55	**150**	285·75	**190**	361·95
31	59·05½	71	135·25½	111	211·45½	151	287·65½	191	363·85½
32	60·96	72	137·16	112	213·36	152	289·56	192	365·76
33	62·86½	73	139·06½	113	215·26½	153	291·46½	193	367·66½
34	64·77	74	140·97	114	217·17	154	293·37	194	369·57
35	66·67½	75	142·87½	115	219·07½	155	295·27½	195	371·47½
36	68·58	76	144·78	116	220·98	156	297·18	196	373·38
37	70·48½	77	146·68½	117	222·88½	157	299·08½	197	375·28½
38	72·39	78	148·59	118	224·79	158	300·99	198	377·19
39	74·29½	79	150·49½	119	226·69½	159	302·89½	199	379·09½
40	76·20	**80**	152·40	**120**	228·60	**160**	304·80	**200**	381·00

$\frac{1}{16}=0{\cdot}11^{9}$ $\frac{1}{9}=0{\cdot}21^{2}$ $\frac{1}{8}=0{\cdot}23^{8}$ $\frac{1}{6}=0{\cdot}31^{8}$ $\frac{1}{5}=0{\cdot}38^{1}$ $\frac{1}{4}=0{\cdot}47^{6}$ $\frac{1}{3}=0{\cdot}63^{5}$

$\frac{3}{8}=0{\cdot}71^{4}$ $\frac{1}{2}=0{\cdot}95^{3}$ $\frac{5}{8}=1{\cdot}19^{1}$ $\frac{3}{4}=1{\cdot}42^{9}$ $\frac{7}{8}=1{\cdot}66^{7}$ $\frac{2}{3}=1{\cdot}27^{0}$ $\frac{5}{6}=1{\cdot}58^{8}$

$\frac{3}{16}=0{\cdot}35^{7}$ $\frac{5}{16}=0{\cdot}59^{5}$ $\frac{7}{16}=0{\cdot}83^{3}$ $\frac{9}{16}=1{\cdot}07^{2}$ $\frac{11}{16}=1{\cdot}31^{0}$ $\frac{13}{16}=1{\cdot}54^{8}$ $\frac{15}{16}=1{\cdot}78^{6}$

1	1·91	41	78·31	81	154·71	121	231·11	161	307·51
2	3·82	42	80·22	82	156·62	122	233·02	162	309·42
3	5·73	43	82·13	83	158·53	123	234·93	163	311·33
4	7·64	44	84·04	84	160·44	124	236·84	164	313·24
5	9·55	45	85·95	85	162·35	125	238·75	165	315·15
6	11·46	46	87·86	86	164·26	126	240·66	166	317·06
7	13·37	47	89·77	87	166·17	127	242·57	167	318·97
8	15·28	48	91·68	88	168·08	128	244·48	168	320·88
9	17·19	49	93·59	89	169·99	129	246·39	169	322·79
10	19·10	**50**	95·50	**90**	171·90	**130**	248·30	**170**	324·70
11	21·01	51	97·41	91	173·81	131	250·21	171	326·61
12	22·92	52	99·32	92	175·72	132	252·12	172	328·52
13	24·83	53	101·23	93	177·63	133	254·03	173	330·43
14	26·74	54	103·14	94	179·54	134	255·94	174	332·34
15	28·65	55	105·05	95	181·45	135	257·85	175	334·25
16	30·56	56	106·96	96	183·36	136	259·76	176	336·16
17	32·47	57	108·87	97	185·27	137	261·67	177	338·07
18	34·38	58	110·78	98	187·18	138	263·58	178	339·98
19	36·29	59	112·69	99	189·09	139	265·49	179	341·89
20	38·20	**60**	114·60	**100**	191·00	**140**	267·40	**180**	343·80
21	40·11	61	116·51	101	192·91	141	269·31	181	345·71
22	42·02	62	118·42	102	194·82	142	271·22	182	347·62
23	43·93	63	120·33	103	196·73	143	273·13	183	349·53
24	45·84	64	122·24	104	198·64	144	275·04	184	351·44
25	47·75	65	124·15	105	200·55	145	276·95	185	353·35
26	49·66	66	126·06	106	202·46	146	278·86	186	355·26
27	51·57	67	127·97	107	204·37	147	280·77	187	357·17
28	53·48	68	129·88	108	206·28	148	282·68	188	359·08
29	55·39	69	131·79	109	208·19	149	284·59	189	360·99
30	57·30	**70**	133·70	**110**	210·10	**150**	286·50	**190**	362·90
31	59·21	71	135·61	111	212·01	151	288·41	191	364·81
32	61·12	72	137·52	112	213·92	152	290·32	192	366·72
33	63·03	73	139·43	113	215·83	153	292·23	193	368·63
34	64·94	74	141·34	114	217·74	154	294·14	194	370·54
35	66·85	75	143·25	115	219·65	155	296·05	195	372·45
36	68·76	76	145·16	116	221·56	156	297·96	196	374·36
37	70·67	77	147·07	117	223·47	157	299·87	197	376·27
38	72·58	78	148·98	118	225·38	158	301·78	198	378·18
39	74·49	79	150·89	119	227·29	159	303·69	199	380·09
40	76·40	**80**	152·80	**120**	229·20	**160**	305·60	**200**	382·00

$\frac{1}{16}=0{\cdot}11^{9}$ $\frac{1}{9}=0{\cdot}21^{2}$ $\frac{1}{8}=0{\cdot}23^{9}$ $\frac{1}{6}=0{\cdot}31^{8}$ $\frac{1}{5}=0{\cdot}38^{2}$ $\frac{1}{4}=0{\cdot}47^{8}$ $\frac{1}{3}=0{\cdot}63^{7}$

$\frac{3}{8}=0{\cdot}71^{6}$ $\frac{1}{2}=0{\cdot}95^{5}$ $\frac{5}{8}=1{\cdot}19^{4}$ $\frac{3}{4}=1{\cdot}43^{3}$ $\frac{7}{8}=1{\cdot}67^{1}$ $\frac{2}{3}=1{\cdot}27^{3}$ $\frac{5}{6}=1{\cdot}59^{2}$

$\frac{3}{16}=0{\cdot}35^{8}$ $\frac{5}{16}=0{\cdot}59^{7}$ $\frac{7}{16}=0{\cdot}83^{6}$ $\frac{9}{16}=1{\cdot}07^{4}$ $\frac{11}{16}=1{\cdot}31^{3}$ $\frac{13}{16}=1{\cdot}55^{2}$ $\frac{15}{16}=1{\cdot}79^{1}$

(0.522=1£) **£1.91½**

1	1·91½	41	78·51½	81	155·11½	121	231·71½	161	308·31½
2	3·83	42	80·43	82	157·03	122	233·63	162	310·23
3	5·74½	43	82·34½	83	158·94½	123	235·54½	163	312·14½
4	7·66	44	84·26	84	160·86	124	237·46	164	314·06
5	9·57½	45	86·17½	85	162·77½	125	239·37½	165	315·97½
6	11·49	46	88·09	86	164·69	126	241·29	166	317·89
7	13·40½	47	90·00½	87	166·60½	127	243·20½	167	319·80½
8	15·32	48	91·92	88	168·52	128	245·12	168	321·72
9	17·23½	49	93·83½	89	170·43½	129	247·03½	169	323·63½
10	19·15	**50**	95·75	**90**	172·35	**130**	248·95	**170**	325·55
11	21·06½	51	97·66½	91	174·26½	131	250·86½	171	327·46½
12	22·98	52	99·58	92	176·18	132	252·78	172	329·38
13	24·89½	53	101·49½	93	178·09½	133	254·69½	173	331·29½
14	26·81	54	103·41	94	180·01	134	256·61	174	333·21
15	28·72½	55	105·32½	95	181·92½	135	258·52½	175	335·12½
16	30·64	56	107·24	96	183·84	136	260·44	176	337·04
17	32·55½	57	109·15½	97	185·75½	137	262·35½	177	338·95½
18	34·47	58	111·07	98	187·67	138	264·27	178	340·87
19	36·38½	59	112·98½	99	189·58½	139	266·18½	179	342·78½
20	38·30	**60**	114·90	**100**	191·50	**140**	268·10	**180**	344·70
21	40·21½	61	116·81½	101	193·41½	141	270·01½	181	346·61½
22	42·13	62	118·73	102	195·33	142	271·93	182	348·53
23	44·04½	63	120·64½	103	197·24½	143	273·84½	183	350·44½
24	45·96	64	122·56	104	199·16	144	275·76	184	352·36
25	47·87½	65	124·47½	105	201·07½	145	277·67½	185	354·27½
26	49·79	66	126·39	106	202·99	146	279·59	186	356·19
27	51·70½	67	128·30½	107	204·90½	147	281·50½	187	358·10½
28	53·62	68	130·22	108	206·82	148	283·42	188	360·02
29	55·53½	69	132·13½	109	208·73½	149	285·33½	189	361·93½
30	57·45	**70**	134·05	**110**	210·65	**150**	287·25	**190**	363·85
31	59·36½	71	135·96½	111	212·56½	151	289·16½	191	365·76½
32	61·28	72	137·88	112	214·48	152	291·08	192	367·68
33	63·19½	73	139·79½	113	216·39½	153	292·99½	193	369·59½
34	65·11	74	141·71	114	218·31	154	294·91	194	371·51
35	67·02½	75	143·62½	115	220·22½	155	296·82½	195	373·42½
36	68·94	76	145·54	116	222·14	156	298·74	196	375·34
37	70·85½	77	147·45½	117	224·05½	157	300·65½	197	377·25½
38	72·77	78	149·37	118	225·97	158	302·57	198	379·17
39	74·68½	79	151·28½	119	227·88½	159	304·48½	199	381·08½
40	76·60	**80**	153·20	**120**	229·80	**160**	306·40	**200**	383·00

$\frac{1}{16}=0{\cdot}12^{0}$ $\frac{1}{9}=0{\cdot}21^{3}$ $\frac{1}{8}=0{\cdot}23^{9}$ $\frac{1}{6}=0{\cdot}31^{9}$ $\frac{1}{5}=0{\cdot}38^{3}$ $\frac{1}{4}=0{\cdot}47^{9}$ $\frac{1}{3}=0{\cdot}63^{8}$

$\frac{3}{8}=0{\cdot}71^{8}$ $\frac{1}{2}=0{\cdot}95^{8}$ $\frac{5}{8}=1{\cdot}19^{7}$ $\frac{3}{4}=1{\cdot}43^{6}$ $\frac{7}{8}=1{\cdot}67^{6}$ $\frac{2}{3}=1{\cdot}27^{7}$ $\frac{5}{6}=1{\cdot}59^{6}$

$\frac{3}{16}=0{\cdot}35^{9}$ $\frac{5}{16}=0{\cdot}59^{8}$ $\frac{7}{16}=0{\cdot}83^{8}$ $\frac{9}{16}=1{\cdot}07^{7}$ $\frac{11}{16}=1{\cdot}31^{7}$ $\frac{13}{16}=1{\cdot}55^{6}$ $\frac{15}{16}=1{\cdot}79^{5}$

(0.521=1£) **£1.92**

1	1·92	41	78·72	81	155·52	121	232·32	161	309·12
2	3·84	42	80·64	82	157·44	122	234·24	162	311·04
3	5·76	43	82·56	83	159·36	123	236·16	163	312·96
4	7·68	44	84·48	84	161·28	124	238·08	164	314·88
5	9·60	45	86·40	85	163·20	125	240·00	165	316·80
6	11·52	46	88·32	86	165·12	126	241·92	166	318·72
7	13·44	47	90·24	87	167·04	127	243·84	167	320·64
8	15·36	48	92·16	88	168·96	128	245·76	168	322·56
9	17·28	49	94·08	89	170·88	129	247·68	169	324·48
10	19·20	**50**	96·00	**90**	172·80	**130**	249·60	**170**	326·40
11	21·12	51	97·92	91	174·72	131	251·52	171	328·32
12	23·04	52	99·84	92	176·64	132	253·44	172	330·24
13	24·96	53	101·76	93	178·56	133	255·36	173	332·16
14	26·88	54	103·68	94	180·48	134	257·28	174	334·08
15	28·80	55	105·60	95	182·40	135	259·20	175	336·00
16	30·72	56	107·52	96	184·32	136	261·12	176	337·92
17	32·64	57	109·44	97	186·24	137	263·04	177	339·84
18	34·56	58	111·36	98	188·16	138	264·96	178	341·76
19	36·48	59	113·28	99	190·08	139	266·88	179	343·68
20	38·40	**60**	115·20	**100**	192·00	**140**	268·80	**180**	345·60
21	40·32	61	117·12	101	193·92	141	270·72	181	347·52
22	42·24	62	119·04	102	195·84	142	272·64	182	349·44
23	44·16	63	120·96	103	197·76	143	274·56	183	351·36
24	46·08	64	122·88	104	199·68	144	276·48	184	353·28
25	48·00	65	124·80	105	201·60	145	278·40	185	355·20
26	49·92	66	126·72	106	203·52	146	280·32	186	357·12
27	51·84	67	128·64	107	205·44	147	282·24	187	359·04
28	53·76	68	130·56	108	207·36	148	284·16	188	360·96
29	55·68	69	132·48	109	209·28	149	286·08	189	362·88
30	57·60	**70**	134·40	**110**	211·20	**150**	288·00	**190**	364·80
31	59·52	71	136·32	111	213·12	151	289·92	191	366·72
32	61·44	72	138·24	112	215·04	152	291·84	192	368·64
33	63·36	73	140·16	113	216·96	153	293·76	193	370·56
34	65·28	74	142·08	114	218·88	154	295·68	194	372·48
35	67·20	75	144·00	115	220·80	155	297·60	195	374·40
36	69·12	76	145·92	116	222·72	156	299·52	196	376·32
37	71·04	77	147·84	117	224·64	157	301·44	197	378·24
38	72·96	78	149·76	118	226·56	158	303·36	198	380·16
39	74·88	79	151·68	119	228·48	159	305·28	199	382·08
40	76·80	**80**	153·60	**120**	230·40	**160**	307·20	**200**	384·00

$\frac{1}{16}=0{\cdot}12^{0}$ $\frac{1}{9}=0{\cdot}21^{3}$ $\frac{1}{8}=0{\cdot}24^{0}$ $\frac{1}{6}=0{\cdot}32^{0}$ $\frac{1}{5}=0{\cdot}38^{4}$ $\frac{1}{4}=0{\cdot}48^{0}$ $\frac{1}{3}=0{\cdot}64^{0}$

$\frac{3}{8}=0{\cdot}72^{0}$ $\frac{1}{2}=0{\cdot}96^{0}$ $\frac{5}{8}=1{\cdot}20^{0}$ $\frac{3}{4}=1{\cdot}44^{0}$ $\frac{7}{8}=1{\cdot}68^{0}$ $\frac{2}{3}=1{\cdot}28^{0}$ $\frac{5}{6}=1{\cdot}60^{0}$

$\frac{3}{16}=0{\cdot}36^{0}$ $\frac{5}{16}=0{\cdot}60^{0}$ $\frac{7}{16}=0{\cdot}84^{0}$ $\frac{9}{16}=1{\cdot}08^{0}$ $\frac{11}{16}=1{\cdot}32^{0}$ $\frac{13}{16}=1{\cdot}56^{0}$ $\frac{15}{16}=1{\cdot}80^{0}$

(0.519=1£) £1.92½

1	1·92½	41	78·92½	81	155·92½	121	232·92½	161	309·92½
2	3·85	42	80·85	82	157·85	122	234·85	162	311·85
3	5·77½	43	82·77½	83	159·77½	123	236·77½	163	313·77½
4	7·70	44	84·70	84	161·70	124	238·70	164	315·70
5	9·62½	45	86·62½	85	163·62½	125	240·62½	165	317·62½
6	11·55	46	88·55	86	165·55	126	242·55	166	319·55
7	13·47½	47	90·47½	87	167·47½	127	244·47½	167	321·47½
8	15·40	48	92·40	88	169·40	128	246·40	168	323·40
9	17·32½	49	94·32½	89	171·32½	129	248·32½	169	325·32½
10	19·25	**50**	96·25	**90**	173·25	**130**	250·25	**170**	327·25
11	21·17½	51	98·17½	91	175·17½	131	252·17½	171	329·17½
12	23·10	52	100·10	92	177·10	132	254·10	172	331·10
13	25·02½	53	102·02½	93	179·02½	133	256·02½	173	333·02½
14	26·95	54	103·95	94	180·95	134	257·95	174	334·95
15	28·87½	55	105·87½	95	182·87½	135	259·87½	175	336·87½
16	30·80	56	107·80	96	184·80	136	261·80	176	338·80
17	32·72½	57	109·72½	97	186·72½	137	263·72½	177	340·72½
18	34·65	58	111·65	98	188·65	138	265·65	178	342·65
19	36·57½	59	113·57½	99	190·57½	139	267·57½	179	344·57½
20	38·50	**60**	115·50	**100**	192·50	**140**	269·50	**180**	346·50
21	40·42½	61	117·42½	101	194·42½	141	271·42½	181	348·42½
22	42·35	62	119·35	102	196·35	142	273·35	182	350·35
23	44·27½	63	121·27½	103	198·27½	143	275·27½	183	352·27½
24	46·20	64	123·20	104	200·20	144	277·20	184	354·20
25	48·12½	65	125·12½	105	202·12½	145	279·12½	185	356·12½
26	50·05	66	127·05	106	204·05	146	281·05	186	358·05
27	51·97½	67	128·97½	107	205·97½	147	282·97½	187	359·97½
28	53·90	68	130·90	108	207·90	148	284·90	188	361·90
29	55·82½	69	132·82½	109	209·82½	149	286·82½	189	363·82½
30	57·75	**70**	134·75	**110**	211·75	**150**	288·75	**190**	365·75
31	59·67½	71	136·67½	111	213·67½	151	290·67½	191	367·67½
32	61·60	72	138·60	112	215·60	152	292·60	192	369·60
33	63·52½	73	140·52½	113	217·52½	153	294·52½	193	371·52½
34	65·45	74	142·45	114	219·45	154	296·45	194	373·45
35	67·37½	75	144·37½	115	221·37½	155	298·37½	195	375·37½
36	69·30	76	146·30	116	223·30	156	300·30	196	377·30
37	71·22½	77	148·22½	117	225·22½	157	302·22½	197	379·22½
38	73·15	78	150·15	118	227·15	158	304·15	198	381·15
39	75·07½	79	152·07½	119	229·07½	159	306·07½	199	383·07½
40	77·00	**80**	154·00	**120**	231·00	**160**	308·00	**200**	385·00

$\frac{1}{16}=0{\cdot}12^{0}$ $\frac{1}{9}=0{\cdot}21^{4}$ $\frac{1}{8}=0{\cdot}24^{1}$ $\frac{1}{6}=0{\cdot}32^{1}$ $\frac{1}{5}=0{\cdot}38^{5}$ $\frac{1}{4}=0{\cdot}48^{1}$ $\frac{1}{3}=0{\cdot}64^{2}$

$\frac{3}{8}=0{\cdot}72^{2}$ $\frac{1}{2}=0{\cdot}96^{3}$ $\frac{5}{8}=1{\cdot}20^{3}$ $\frac{3}{4}=1{\cdot}44^{4}$ $\frac{7}{8}=1{\cdot}68^{4}$ $\frac{2}{3}=1{\cdot}28^{3}$ $\frac{5}{6}=1{\cdot}60^{4}$

$\frac{3}{16}=0{\cdot}36^{1}$ $\frac{5}{16}=0{\cdot}60^{2}$ $\frac{7}{16}=0{\cdot}84^{2}$ $\frac{9}{16}=1{\cdot}08^{3}$ $\frac{11}{16}=1{\cdot}32^{3}$ $\frac{13}{16}=1{\cdot}56^{4}$ $\frac{15}{16}=1{\cdot}80^{5}$

(0.518=1£) £1.93

1	1·93	41	79·13	81	156·33	121	233·53	161	310·73
2	3·86	42	81·06	82	158·26	122	235·46	162	312·66
3	5·79	43	82·99	83	160·19	123	237·39	163	314·59
4	7·72	44	84·92	84	162·12	124	239·32	164	316·52
5	9·65	45	86·85	85	164·05	125	241·25	165	318·45
6	11·58	46	88·78	86	165·98	126	243·18	166	320·38
7	13·51	47	90·71	87	167·91	127	245·11	167	322·31
8	15·44	48	92·64	88	169·84	128	247·04	168	324·24
9	17·37	49	94·57	89	171·77	129	248·97	169	326·17
10	19·30	**50**	96·50	**90**	173·70	**130**	250·90	**170**	328·10
11	21·23	51	98·43	91	175·63	131	252·83	171	330·03
12	23·16	52	100·36	92	177·56	132	254·76	172	331·96
13	25·09	53	102·29	93	179·49	133	256·69	173	333·89
14	27·02	54	104·22	94	181·42	134	258·62	174	335·82
15	28·95	55	106·15	95	183·35	135	260·55	175	337·75
16	30·88	56	108·08	96	185·28	136	262·48	176	339·68
17	32·81	57	110·01	97	187·21	137	264·41	177	341·61
18	34·74	58	111·94	98	189·14	138	266·34	178	343·54
19	36·67	59	113·87	99	191·07	139	268·27	179	345·47
20	38·60	**60**	115·80	**100**	193·00	**140**	270·20	**180**	347·40
21	40·53	61	117·73	101	194·93	141	272·13	181	349·33
22	42·46	62	119·66	102	196·86	142	274·06	182	351·26
23	44·39	63	121·59	103	198·79	143	275·99	183	353·19
24	46·32	64	123·52	104	200·72	144	277·92	184	355·12
25	48·25	65	125·45	105	202·65	145	279·85	185	357·05
26	50·18	66	127·38	106	204·58	146	281·78	186	358·98
27	52·11	67	129·31	107	206·51	147	283·71	187	360·91
28	54·04	68	131·24	108	208·44	148	285·64	188	362·84
29	55·97	69	133·17	109	210·37	149	287·57	189	364·77
30	57·90	**70**	135·10	**110**	212·30	**150**	289·50	**190**	366·70
31	59·83	71	137·03	111	214·23	151	291·43	191	368·63
32	61·76	72	138·96	112	216·16	152	293·36	192	370·56
33	63·69	73	140·89	113	218·09	153	295·29	193	372·49
34	65·62	74	142·82	114	220·02	154	297·22	194	374·42
35	67·55	75	144·75	115	221·95	155	299·15	195	376·35
36	69·48	76	146·68	116	223·88	156	301·08	196	378·28
37	71·41	77	148·61	117	225·81	157	303·01	197	380·21
38	73·34	78	150·54	118	227·74	158	304·94	198	382·14
39	75·27	79	152·47	119	229·67	159	306·87	199	384·07
40	77·20	**80**	154·40	**120**	231·60	**160**	308·80	**200**	386·00

$\frac{1}{16}$=0·12^{1} $\frac{1}{9}$=0·21^{4} $\frac{1}{8}$=0·24^{1} $\frac{1}{6}$=0·32^{2} $\frac{1}{5}$=0·38^{6} $\frac{1}{4}$=0·48^{3} $\frac{1}{3}$=0·64^{3}

$\frac{3}{8}$=0·72^{4} $\frac{1}{2}$=0·96^{5} $\frac{5}{8}$=1·20^{6} $\frac{3}{4}$=1·44^{8} $\frac{7}{8}$=1·68^{9} $\frac{2}{3}$=1·28^{7} $\frac{5}{6}$=1·60^{8}

$\frac{3}{16}$=0·36^{2} $\frac{5}{16}$=0·60^{3} $\frac{7}{16}$=0·84^{4} $\frac{9}{16}$=1·08^{6} $\frac{11}{16}$=1·32^{7} $\frac{13}{16}$=1·56^{8} $\frac{15}{16}$=1·80^{9}

1	1·93½	41	79·33½	81	156·73½	121	234·13½	161	311·53½
2	3·87	42	81·27	82	158·67	122	236·07	162	313·47
3	5·80½	43	83·20½	83	160·60½	123	238·00½	163	315·40½
4	7·74	44	85·14	84	162·54	124	239·94	164	317·34
5	9·67½	45	87·07½	85	164·47½	125	241·87½	165	319·27½
6	11·61	46	89·01	86	166·41	126	243·81	166	321·21
7	13·54½	47	90·94½	87	168·34½	127	245·74½	167	323·14½
8	15·48	48	92·88	88	170·28	128	247·68	168	325·08
9	17·41½	49	94·81½	89	172·21½	129	249·61½	169	327·01½
10	19·35	**50**	96·75	**90**	174·15	**130**	251·55	**170**	328·95
11	21·28½	51	98·68½	91	176·08½	131	253·48½	171	330·88½
12	23·22	52	100·62	92	178·02	132	255·42	172	332·82
13	25·15½	53	102·55½	93	179·95½	133	257·35½	173	334·75½
14	27·09	54	104·49	94	181·89	134	259·29	174	336·69
15	29·02½	55	106·42½	95	183·82½	135	261·22½	175	338·62½
16	30·96	56	108·36	96	185·76	136	263·16	176	340·56
17	32·89½	57	110·29½	97	187·69½	137	265·09½	177	342·49½
18	34·83	58	112·23	98	189·63	138	267·03	178	344·43
19	36·76½	59	114·16½	99	191·56½	139	268·96½	179	346·36½
20	38·70	**60**	116·10	**100**	193·50	**140**	270·90	**180**	348·30
21	40·63½	61	118·03½	101	195·43½	141	272·83½	181	350·23½
22	42·57	62	119·97	102	197·37	142	274·77	182	352·17
23	44·50½	63	121·90½	103	199·30½	143	276·70½	183	354·10½
24	46·44	64	123·84	104	201·24	144	278·64	184	356·04
25	48·37½	65	125·77½	105	203·17½	145	280·57½	185	357·97½
26	50·31	66	127·71	106	205·11	146	282·51	186	359·91
27	52·24½	67	129·64½	107	207·04½	147	284·44½	187	361·84½
28	54·18	68	131·58	108	208·98	148	286·38	188	363·78
29	56·11½	69	133·51½	109	210·91½	149	288·31½	189	365·71½
30	58·05	**70**	135·45	**110**	212·85	**150**	290·25	**190**	367·65
31	59·98½	71	137·38½	111	214·78½	151	292·18½	191	369·58½
32	61·92	72	139·32	112	216·72	152	294·12	192	371·52
33	63·85½	73	141·25½	113	218·65½	153	296·05½	193	373·45½
34	65·79	74	143·19	114	220·59	154	297·99	194	375·39
35	67·72½	75	145·12½	115	222·52½	155	299·92½	195	377·32½
36	69·66	76	147·06	116	224·46	156	301·86	196	379·26
37	71·59½	77	148·99½	117	226·39½	157	303·79½	197	381·19½
38	73·53	78	150·93	118	228·33	158	305·73	198	383·13
39	75·46½	79	152·86½	119	230·26½	159	307·66½	199	385·06½
40	77·40	**80**	154·80	**120**	232·20	**160**	309·60	**200**	387·00

$\frac{1}{16}=0{\cdot}12^{1}$ $\frac{1}{9}=0{\cdot}21^{5}$ $\frac{1}{8}=0{\cdot}24^{2}$ $\frac{1}{6}=0{\cdot}32^{3}$ $\frac{1}{5}=0{\cdot}38^{7}$ $\frac{1}{4}=0{\cdot}48^{4}$ $\frac{1}{3}=0{\cdot}64^{5}$

$\frac{3}{8}=0{\cdot}72^{6}$ $\frac{1}{2}=0{\cdot}96^{8}$ $\frac{5}{8}=1{\cdot}20^{9}$ $\frac{3}{4}=1{\cdot}45^{1}$ $\frac{7}{8}=1{\cdot}69^{3}$ $\frac{2}{3}=1{\cdot}29^{0}$ $\frac{5}{6}=1{\cdot}61^{3}$

(0.515=1£) **£1.94**

1	1·94	41	79·54	81	157·14	121	234·74	161	312·34
2	3·88	42	81·48	82	159·08	122	236·68	162	314·28
3	5·82	43	83·42	83	161·02	123	238·62	163	316·22
4	7·76	44	85·36	84	162·96	124	240·56	164	318·16
5	9·70	45	87·30	85	164·90	125	242·50	165	320·10
6	11·64	46	89·24	86	166·84	126	244·44	166	322·04
7	13·58	47	91·18	87	168·78	127	246·38	167	323·98
8	15·52	48	93·12	88	170·72	128	248·32	168	325·92
9	17·46	49	95·06	89	172·66	129	250·26	169	327·86
10	19·40	**50**	97·00	**90**	174·60	**130**	252·20	**170**	329·80
11	21·34	51	98·94	91	176·54	131	254·14	171	331·74
12	23·28	52	100·88	92	178·48	132	256·08	172	333·68
13	25·22	53	102·82	93	180·42	133	258·02	173	335·62
14	27·16	54	104·76	94	182·36	134	259·96	174	337·56
15	29·10	55	106·70	95	184·30	135	261·90	175	339·50
16	31·04	56	108·64	96	186·24	136	263·84	176	341·44
17	32·98	57	110·58	97	188·18	137	265·78	177	343·38
18	34·92	58	112·52	98	190·12	138	267·72	178	345·32
19	36·86	59	114·46	99	192·06	139	269·66	179	347·26
20	38·80	**60**	116·40	**100**	194·00	**140**	271·60	**180**	349·20
21	40·74	61	118·34	101	195·94	141	273·54	181	351·14
22	42·68	62	120·28	102	197·88	142	275·48	182	353·08
23	44·62	63	122·22	103	199·82	143	277·42	183	355·02
24	46·56	64	124·16	104	201·76	144	279·36	184	356·96
25	48·50	65	126·10	105	203·70	145	281·30	185	358·90
26	50·44	66	128·04	106	205·64	146	283·24	186	360·84
27	52·38	67	129·98	107	207·58	147	285·18	187	362·78
28	54·32	68	131·92	108	209·52	148	287·12	188	364·72
29	56·26	69	133·86	109	211·46	149	289·06	189	366·66
30	58·20	**70**	135·80	**110**	213·40	**150**	291·00	**190**	368·60
31	60·14	71	137·74	111	215·34	151	292·94	191	370·54
32	62·08	72	139·68	112	217·28	152	294·88	192	372·48
33	64·02	73	141·62	113	219·22	153	296·82	193	374·42
34	65·96	74	143·56	114	221·16	154	298·76	194	376·36
35	67·90	75	145·50	115	223·10	155	300·70	195	378·30
36	69·84	76	147·44	116	225·04	156	302·64	196	380·24
37	71·78	77	149·38	117	226·98	157	304·58	197	382·18
38	73·72	78	151·32	118	228·92	158	306·52	198	384·12
39	75·66	79	153·26	119	230·86	159	308·46	199	386·06
40	77·60	**80**	155·20	**120**	232·80	**160**	310·40	**200**	388·00

$\frac{1}{16}=0{\cdot}12^{1}$ $\frac{1}{9}=0{\cdot}21^{6}$ $\frac{1}{8}=0{\cdot}24^{3}$ $\frac{1}{6}=0{\cdot}32^{3}$ $\frac{1}{5}=0{\cdot}38^{8}$ $\frac{1}{4}=0{\cdot}48^{5}$ $\frac{1}{3}=0{\cdot}64^{7}$

$\frac{3}{8}=0{\cdot}72^{8}$ $\frac{1}{2}=0{\cdot}97^{0}$ $\frac{5}{8}=1{\cdot}21^{3}$ $\frac{3}{4}=1{\cdot}45^{5}$ $\frac{7}{8}=1{\cdot}69^{8}$ $\frac{2}{3}=1{\cdot}29^{3}$ $\frac{5}{6}=1{\cdot}61^{7}$

$\frac{3}{16}=0{\cdot}36^{4}$ $\frac{5}{16}=0{\cdot}60^{6}$ $\frac{7}{16}=0{\cdot}84^{9}$ $\frac{9}{16}=1{\cdot}09^{1}$ $\frac{11}{16}=1{\cdot}33^{4}$ $\frac{13}{16}=1{\cdot}57^{6}$ $\frac{15}{16}=1{\cdot}81^{9}$

(0.514=1£) **£1.94½**

1	1·94½	41	79·74½	81	157·54½	121	235·34½	161	313·14½
2	3·89	42	81·69	82	159·49	122	237·29	162	315·09
3	5·83½	43	83·63½	83	161·43½	123	239·23½	163	317·03½
4	7·78	44	85·58	84	163·38	124	241·18	164	318·98
5	9·72½	45	87·52½	85	165·32½	125	243·12½	165	320·92½
6	11·67	46	89·47	86	167·27	126	245·07	166	322·87
7	13·61½	47	91·41½	87	169·21½	127	247·01½	167	324·81½
8	15·56	48	93·36	88	171·16	128	248·96	168	326·76
9	17·50½	49	95·30½	89	173·10½	129	250·90½	169	328·70½
10	19·45	**50**	97·25	**90**	175·05	**130**	252·85	**170**	330·65
11	21·39½	51	99·19½	91	176·99½	131	254·79½	171	332·59½
12	23·34	52	101·14	92	178·94	132	256·74	172	334·54
13	25·28½	53	103·08½	93	180·88½	133	258·68½	173	336·48½
14	27·23	54	105·03	94	182·83	134	260·63	174	338·43
15	29·17½	55	106·97½	95	184·77½	135	262·57½	175	340·37½
16	31·12	56	108·92	96	186·72	136	264·52	176	342·32
17	33·06½	57	110·86½	97	188·66½	137	266·46½	177	344·26½
18	35·01	58	112·81	98	190·61	138	268·41	178	346·21
19	36·95½	59	114·75½	99	192·55½	139	270·35½	179	348·15½
20	38·90	**60**	116·70	**100**	194·50	**140**	272·30	**180**	350·10
21	40·84½	61	118·64½	101	196·44½	141	274·24½	181	352·04½
22	42·79	62	120·59	102	198·39	142	276·19	182	353·99
23	44·73½	63	122·53½	103	200·33½	143	278·13½	183	355·93½
24	46·68	64	124·48	104	202·28	144	280·08	184	357·88
25	48·62½	65	126·42½	105	204·22½	145	282·02½	185	359·82½
26	50·57	66	128·37	106	206·17	146	283·97	186	361·77
27	52·51½	67	130·31½	107	208·11½	147	285·91½	187	363·71½
28	54·46	68	132·26	108	210·06	148	287·86	188	365·66
29	56·40½	69	134·20½	109	212·00½	149	289·80½	189	367·60½
30	58·35	**70**	136·15	**110**	213·95	**150**	291·75	**190**	369·55
31	60·29½	71	138·09½	111	215·89½	151	293·69½	191	371·49½
32	62·24	72	140·04	112	217·84	152	295·64	192	373·44
33	64·18½	73	141·98½	113	219·78½	153	297·58½	193	375·38½
34	66·13	74	143·93	114	221·73	154	299·53	194	377·33
35	68·07½	75	145·87½	115	223·67½	155	301·47½	195	379·27½
36	70·02	76	147·82	116	225·62	156	303·42	196	381·22
37	71·96½	77	149·76½	117	227·56½	157	305·36½	197	383·16½
38	73·91	78	151·71	118	229·51	158	307·31	198	385·11
39	75·85½	79	153·65½	119	231·45½	159	309·25½	199	387·05½
40	77·80	**80**	155·60	**120**	233·40	**160**	311·20	**200**	389·00

$\frac{1}{16}=0{\cdot}12^{2}$ $\frac{1}{9}=0{\cdot}21^{6}$ $\frac{1}{8}=0{\cdot}24^{3}$ $\frac{1}{6}=0{\cdot}32^{4}$ $\frac{1}{5}=0{\cdot}38^{9}$ $\frac{1}{4}=0{\cdot}48^{6}$ $\frac{1}{3}=0{\cdot}64^{8}$

$\frac{3}{8}=0{\cdot}72^{9}$ $\frac{1}{2}=0{\cdot}97^{3}$ $\frac{5}{8}=1{\cdot}21^{6}$ $\frac{3}{4}=1{\cdot}45^{9}$ $\frac{7}{8}=1{\cdot}70^{2}$ $\frac{2}{3}=1{\cdot}29^{7}$ $\frac{5}{6}=1{\cdot}62^{1}$

$\frac{3}{16}=0{\cdot}36^{5}$ $\frac{5}{16}=0{\cdot}60^{8}$ $\frac{7}{16}=0{\cdot}85^{1}$ $\frac{9}{16}=1{\cdot}09^{4}$ $\frac{11}{16}=1{\cdot}33^{7}$ $\frac{13}{16}=1{\cdot}58^{0}$ $\frac{15}{16}=1{\cdot}82^{3}$

(0.513=1£) **£1.95**

1	1·95	41	79·95	81	157·95	121	235·95	161	313·95
2	3·90	42	81·90	82	159·90	122	237·90	162	315·90
3	5·85	43	83·85	83	161·85	123	239·85	163	317·85
4	7·80	44	85·80	84	163·80	124	241·80	164	319·80
5	9·75	45	87·75	85	165·75	125	243·75	165	321·75
6	11·70	46	89·70	86	167·70	126	245·70	166	323·70
7	13·65	47	91·65	87	169·65	127	247·65	167	325·65
8	15·60	48	93·60	88	171·60	128	249·60	168	327·60
9	17·55	49	95·55	89	173·55	129	251·55	169	329·55
10	19·50	**50**	97·50	**90**	175·50	**130**	253·50	**170**	331·50
11	21·45	51	99·45	91	177·45	131	255·45	171	333·45
12	23·40	52	101·40	92	179·40	132	257·40	172	335·40
13	25·35	53	103·35	93	181·35	133	259·35	173	337·35
14	27·30	54	105·30	94	183·30	134	261·30	174	339·30
15	29·25	55	107·25	95	185·25	135	263·25	175	341·25
16	31·20	56	109·20	96	187·20	136	265·20	176	343·20
17	33·15	57	111·15	97	189·15	137	267·15	177	345·15
18	35·10	58	113·10	98	191·10	138	269·10	178	347·10
19	37·05	59	115·05	99	193·05	139	271·05	179	349·05
20	39·00	**60**	117·00	**100**	195·00	**140**	273·00	**180**	351·00
21	40·95	61	118·95	101	196·95	141	274·95	181	352·95
22	42·90	62	120·90	102	198·90	142	276·90	182	354·90
23	44·85	63	122·85	103	200·85	143	278·85	183	356·85
24	46·80	64	124·80	104	202·80	144	280·80	184	358·80
25	48·75	65	126·75	105	204·75	145	282·75	185	360·75
26	50·70	66	128·70	106	206·70	146	284·70	186	362·70
27	52·65	67	130·65	107	208·65	147	286·65	187	364·65
28	54·60	68	132·60	108	210·60	148	288·60	188	366·60
29	56·55	69	134·55	109	212·55	149	290·55	189	368·55
30	58·50	**70**	136·50	**110**	214·50	**150**	292·50	**190**	370·50
31	60·45	71	138·45	111	216·45	151	294·45	191	372·45
32	62·40	72	140·40	112	218·40	152	296·40	192	374·40
33	64·35	73	142·35	113	220·35	153	298·35	193	376·35
34	66·30	74	144·30	114	222·30	154	300·30	194	378·30
35	68·25	75	146·25	115	224·25	155	302·25	195	380·25
36	70·20	76	148·20	116	226·20	156	304·20	196	382·20
37	72·15	77	150·15	117	228·15	157	306·15	197	384·15
38	74·10	78	152·10	118	230·10	158	308·10	198	386·10
39	76·05	79	154·05	119	232·05	159	310·05	199	388·05
40	78·00	**80**	156·00	**120**	234·00	**160**	312·00	**200**	390·00

$\frac{1}{16}=0{\cdot}12^{2}$ $\frac{1}{9}=0{\cdot}21^{7}$ $\frac{1}{8}=0{\cdot}24^{4}$ $\frac{1}{6}=0{\cdot}32^{5}$ $\frac{1}{5}=0{\cdot}39^{0}$ $\frac{1}{4}=0{\cdot}48^{8}$ $\frac{1}{3}=0{\cdot}65^{0}$

$\frac{3}{8}=0{\cdot}73^{1}$ $\frac{1}{2}=0{\cdot}97^{5}$ $\frac{5}{8}=1{\cdot}21^{9}$ $\frac{3}{4}=1{\cdot}46^{3}$ $\frac{7}{8}=1{\cdot}70^{6}$ $\frac{2}{3}=1{\cdot}30^{0}$ $\frac{5}{6}=1{\cdot}62^{5}$

$\frac{3}{16}=0{\cdot}36^{6}$ $\frac{5}{16}=0{\cdot}60^{9}$ $\frac{7}{16}=0{\cdot}85^{3}$ $\frac{9}{16}=1{\cdot}09^{7}$ $\frac{11}{16}=1{\cdot}34^{1}$ $\frac{13}{16}=1{\cdot}58^{4}$ $\frac{15}{16}=1{\cdot}82^{8}$

(0.512=1£) £1.95½

1	1·95½	41	80·15½	81	158·35½	121	236·55½	161	314·75½
2	3·91	42	82·11	82	160·31	122	238·51	162	316·71
3	5·86½	43	84·06½	83	162·26½	123	240·46½	163	318·66½
4	7·82	44	86·02	84	164·22	124	242·42	164	320·62
5	9·77½	45	87·97½	85	166·17½	125	244·37½	165	322·57½
6	11·73	46	89·93	86	168·13	126	246·33	166	324·53
7	13·68½	47	91·88½	87	170·08½	127	248·28½	167	326·48½
8	15·64	48	93·84	88	172·04	128	250·24	168	328·44
9	17·59½	49	95·79½	89	173·99½	129	252·19½	169	330·39½
10	19·55	**50**	97·75	**90**	175·95	**130**	254·15	**170**	332·35
11	21·50½	51	99·70½	91	177·90½	131	256·10½	171	334·30½
12	23·46	52	101·66	92	179·86	132	258·06	172	336·26
13	25·41½	53	103·61½	93	181·81½	133	260·01½	173	338·21½
14	27·37	54	105·57	94	183·77	134	261·97	174	340·17
15	29·32½	55	107·52½	95	185·72½	135	263·92½	175	342·12½
16	31·28	56	109·48	96	187·68	136	265·88	176	344·08
17	33·23½	57	111·43½	97	189·63½	137	267·83½	177	346·03½
18	35·19	58	113·39	98	191·59	138	269·79	178	347·99
19	37·14½	59	115·34½	99	193·54½	139	271·74½	179	349·94½
20	39·10	**60**	117·30	**100**	195·50	**140**	273·70	**180**	351·90
21	41·05½	61	119·25½	101	197·45½	141	275·65½	181	353·85½
22	43·01	62	121·21	102	199·41	142	277·61	182	355·81
23	44·96½	63	123·16½	103	201·36½	143	279·56½	183	357·76½
24	46·92	64	125·12	104	203·32	144	281·52	184	359·72
25	48·87½	65	127·07½	105	205·27½	145	283·47½	185	361·67½
26	50·83	66	129·03	106	207·23	146	285·43	186	363·63
27	52·78½	67	130·98½	107	209·18½	147	287·38½	187	365·58½
28	54·74	68	132·94	108	211·14	148	289·34	188	367·54
29	56·69½	69	134·89½	109	213·09½	149	291·29½	189	369·49½
30	58·65	**70**	136·85	**110**	215·05	**150**	293·25	**190**	371·45
31	60·60½	71	138·80½	111	217·00½	151	295·20½	191	373·40½
32	62·56	72	140·76	112	218·96	152	297·16	192	375·36
33	64·51½	73	142·71½	113	220·91½	153	299·11½	193	377·31½
34	66·47	74	144·67	114	222·87	154	301·07	194	379·27
35	68·42½	75	146·62½	115	224·82½	155	303·02½	195	381·22½
36	70·38	76	148·58	116	226·78	156	304·98	196	383·18
37	72·33½	77	150·53½	117	228·73½	157	306·93½	197	385·13½
38	74·29	78	152·49	118	230·69	158	308·89	198	387·09
39	76·24½	79	154·44½	119	232·64½	159	310·84½	199	389·04½
40	78·20	**80**	156·40	**120**	234·60	**160**	312·80	**200**	391·00

$\frac{1}{16}=0{\cdot}12^2$ $\frac{1}{9}=0{\cdot}21^7$ $\frac{1}{8}=0{\cdot}24^4$ $\frac{1}{6}=0{\cdot}32^6$ $\frac{1}{5}=0{\cdot}39^1$ $\frac{1}{4}=0{\cdot}48^9$ $\frac{1}{3}=0{\cdot}65^2$

$\frac{3}{8}=0{\cdot}73^3$ $\frac{1}{2}=0{\cdot}97^8$ $\frac{5}{8}=1{\cdot}22^2$ $\frac{3}{4}=1{\cdot}46^6$ $\frac{7}{8}=1{\cdot}71^1$ $\frac{2}{3}=1{\cdot}30^3$ $\frac{5}{6}=1{\cdot}62^9$

$\frac{3}{16}=0{\cdot}36^7$ $\frac{5}{16}=0{\cdot}61^1$ $\frac{7}{16}=0{\cdot}85^5$ $\frac{9}{16}=1{\cdot}10^0$ $\frac{11}{16}=1{\cdot}34^4$ $\frac{13}{16}=1{\cdot}58^8$ $\frac{15}{16}=1{\cdot}83^3$

1	1·96	41	80·36	81	158·76	121	237·16	161	315·56
2	3·92	42	82·32	82	160·72	122	239·12	162	317·52
3	5·88	43	84·28	83	162·68	123	241·08	163	319·48
4	7·84	44	86·24	84	164·64	124	243·04	164	321·44
5	9·80	45	88·20	85	166·60	125	245·00	165	323·40
6	11·76	46	90·16	86	168·56	126	246·96	166	325·36
7	13·72	47	92·12	87	170·52	127	248·92	167	327·32
8	15·68	48	94·08	88	172·48	128	250·88	168	329·28
9	17·64	49	96·04	89	174·44	129	252·84	169	331·24
10	19·60	**50**	98·00	**90**	176·40	**130**	254·80	**170**	333·20
11	21·56	51	99·96	91	178·36	131	256·76	171	335·16
12	23·52	52	101·92	92	180·32	132	258·72	172	337·12
13	25·48	53	103·88	93	182·28	133	260·68	173	339·08
14	27·44	54	105·84	94	184·24	134	262·64	174	341·04
15	29·40	55	107·80	95	186·20	135	264·60	175	343·00
16	31·36	56	109·76	96	188·16	136	266·56	176	344·96
17	33·32	57	111·72	97	190·12	137	268·52	177	346·92
18	35·28	58	113·68	98	192·08	138	270·48	178	348·88
19	37·24	59	115·64	99	194·04	139	272·44	179	350·84
20	39·20	**60**	117·60	**100**	196·00	**140**	274·40	**180**	352·80
21	41·16	61	119·56	101	197·96	141	276·36	181	354·76
22	43·12	62	121·52	102	199·92	142	278·32	182	356·72
23	45·08	63	123·48	103	201·88	143	280·28	183	358·68
24	47·04	64	125·44	104	203·84	144	282·24	184	360·64
25	49·00	65	127·40	105	205·80	145	284·20	185	362·60
26	50·96	66	129·36	106	207·76	146	286·16	186	364·56
27	52·92	67	131·32	107	209·72	147	288·12	187	366·52
28	54·88	68	133·28	108	211·68	148	290·08	188	368·48
29	56·84	69	135·24	109	213·64	149	292·04	189	370·44
30	58·80	**70**	137·20	**110**	215·60	**150**	294·00	**190**	372·40
31	60·76	71	139·16	111	217·56	151	295·96	191	374·36
32	62·72	72	141·12	112	219·52	152	297·92	192	376·32
33	64·68	73	143·08	113	221·48	153	299·88	193	378·28
34	66·64	74	145·04	114	223·44	154	301·84	194	380·24
35	68·60	75	147·00	115	225·40	155	303·80	195	382·20
36	70·56	76	148·96	116	227·36	156	305·76	196	384·16
37	72·52	77	150·92	117	229·32	157	307·72	197	386·12
38	74·48	78	152·88	118	231·28	158	309·68	198	388·08
39	76·44	79	154·84	119	233·24	159	311·64	199	390·04
40	78·40	**80**	156·80	**120**	235·20	**160**	313·60	**200**	392·00

$\frac{1}{16}=0{\cdot}12^3$ $\frac{1}{9}=0{\cdot}21^8$ $\frac{1}{8}=0{\cdot}24^5$ $\frac{1}{6}=0{\cdot}32^7$ $\frac{1}{5}=0{\cdot}39^2$ $\frac{1}{4}=0{\cdot}49^0$ $\frac{1}{3}=0{\cdot}65^3$

$\frac{3}{8}=0{\cdot}73^5$ $\frac{1}{2}=0{\cdot}98^0$ $\frac{5}{8}=1{\cdot}22^5$ $\frac{3}{4}=1{\cdot}47^0$ $\frac{7}{8}=1{\cdot}71^5$ $\frac{2}{3}=1{\cdot}30^7$ $\frac{5}{6}=1{\cdot}63^3$

$\frac{3}{16}=0{\cdot}36^8$ $\frac{5}{16}=0{\cdot}61^3$ $\frac{7}{16}=0{\cdot}85^8$ $\frac{9}{16}=1{\cdot}10^3$ $\frac{11}{16}=1{\cdot}34^8$ $\frac{13}{16}=1{\cdot}59^3$ $\frac{15}{16}=1{\cdot}83^8$

(0.509=1£) **£1.96½**

1	1·96½	41	80·56½	81	159·16½	121	237·76½	161	316·36½
2	3·93	42	82·53	82	161·13	122	239·73	162	318·33
3	5·89½	43	84·49½	83	163·09½	123	241·69½	163	320·29½
4	7·86	44	86·46	84	165·06	124	243·66	164	322·26
5	9·82½	45	88·42½	85	167·02½	125	245·62½	165	324·22½
6	11·79	46	90·39	86	168·99	126	247·59	166	326·19
7	13·75½	47	92·35½	87	170·95½	127	249·55½	167	328·15½
8	15·72	48	94·32	88	172·92	128	251·52	168	330·12
9	17·68½	49	96·28½	89	174·88½	129	253·48½	169	332·08½
10	19·65	**50**	98·25	**90**	176·85	**130**	255·45	**170**	334·05
11	21·61½	51	100·21½	91	178·81½	131	257·41½	171	336·01½
12	23·58	52	102·18	92	180·78	132	259·38	172	337·98
13	25·54½	53	104·14½	93	182·74½	133	261·34½	173	339·94½
14	27·51	54	106·11	94	184·71	134	263·31	174	341·91
15	29·47½	55	108·07½	95	186·67½	135	265·27½	175	343·87½
16	31·44	56	110·04	96	188·64	136	267·24	176	345·84
17	33·40½	57	112·00½	97	190·60½	137	269·20½	177	347·80½
18	35·37	58	113·97	98	192·57	138	271·17	178	349·77
19	37·33½	59	115·93½	99	194·53½	139	273·13½	179	351·73½
20	39·30	**60**	117·90	**100**	196·50	**140**	275·10	**180**	353·70
21	41·26½	61	119·86½	101	198·46½	141	277·06½	181	355·66½
22	43·23	62	121·83	102	200·43	142	279·03	182	357·63
23	45·19½	63	123·79½	103	202·39½	143	280·99½	183	359·59½
24	47·16	64	125·76	104	204·36	144	282·96	184	361·56
25	49·12½	65	127·72½	105	206·32½	145	284·92½	185	363·52½
26	51·09	66	129·69	106	208·29	146	286·89	186	365·49
27	53·05½	67	131·65½	107	210·25½	147	288·85½	187	367·45½
28	55·02	68	133·62	108	212·22	148	290·82	188	369·42
29	56·98½	69	135·58½	109	214·18½	149	292·78½	189	371·38½
30	58·95	**70**	137·55	**110**	216·15	**150**	294·75	**190**	373·35
31	60·91½	71	139·51½	111	218·11½	151	296·71½	191	375·31½
32	62·88	72	141·48	112	220·08	152	298·68	192	377·28
33	64·84½	73	143·44½	113	222·04½	153	300·64½	193	379·24½
34	66·81	74	145·41	114	224·01	154	302·61	194	381·21
35	68·77½	75	147·37½	115	225·97½	155	304·57½	195	383·17½
36	70·74	76	149·34	116	227·94	156	306·54	196	385·14
37	72·70½	77	151·30½	117	229·90½	157	308·50½	197	387·10½
38	74·67	78	153·27	118	231·87	158	310·47	198	389·07
39	76·63½	79	155·23½	119	233·83½	159	312·43½	199	391·03½
40	78·60	**80**	157·20	**120**	235·80	**160**	314·40	**200**	393·00

$\frac{1}{16}=0{\cdot}12^{3}$	$\frac{1}{9}=0{\cdot}21^{8}$	$\frac{1}{8}=0{\cdot}24^{6}$	$\frac{1}{6}=0{\cdot}32^{8}$	$\frac{1}{5}=0{\cdot}39^{3}$	$\frac{1}{4}=0{\cdot}49^{1}$	$\frac{1}{3}=0{\cdot}65^{5}$
$\frac{3}{8}=0{\cdot}73^{7}$	$\frac{1}{2}=0{\cdot}98^{3}$	$\frac{5}{8}=1{\cdot}22^{8}$	$\frac{3}{4}=1{\cdot}47^{4}$	$\frac{7}{8}=1{\cdot}71^{9}$	$\frac{2}{3}=1{\cdot}31^{0}$	$\frac{5}{6}=1{\cdot}63^{8}$
$\frac{3}{16}=0{\cdot}36^{8}$	$\frac{5}{16}=0{\cdot}61^{4}$	$\frac{7}{16}=0{\cdot}86^{0}$	$\frac{9}{16}=1{\cdot}10^{5}$	$\frac{11}{16}=1{\cdot}35^{1}$	$\frac{13}{16}=1{\cdot}59^{7}$	$\frac{15}{16}=1{\cdot}84^{2}$

(0.508=1£) **£1.97**

1	1·97	41	80·77	81	159·57	121	238·37	161	317·17
2	3·94	42	82·74	82	161·54	122	240·34	162	319·14
3	5·91	43	84·71	83	163·51	123	242·31	163	321·11
4	7·88	44	86·68	84	165·48	124	244·28	164	323·08
5	9·85	45	88·65	85	167·45	125	246·25	165	325·05
6	11·82	46	90·62	86	169·42	126	248·22	166	327·02
7	13·79	47	92·59	87	171·39	127	250·19	167	328·99
8	15·76	48	94·56	88	173·36	128	252·16	168	330·96
9	17·73	49	96·53	89	175·33	129	254·13	169	332·93
10	19·70	**50**	98·50	**90**	177·30	**130**	256·10	**170**	334·90
11	21·67	51	100·47	91	179·27	131	258·07	171	336·87
12	23·64	52	102·44	92	181·24	132	260·04	172	338·84
13	25·61	53	104·41	93	183·21	133	262·01	173	340·81
14	27·58	54	106·38	94	185·18	134	263·98	174	342·78
15	29·55	55	108·35	95	187·15	135	265·95	175	344·75
16	31·52	56	110·32	96	189·12	136	267·92	176	346·72
17	33·49	57	112·29	97	191·09	137	269·89	177	348·69
18	35·46	58	114·26	98	193·06	138	271·86	178	350·66
19	37·43	59	116·23	99	195·03	139	273·83	179	352·63
20	39·40	**60**	118·20	**100**	197·00	**140**	275·80	**180**	354·60
21	41·37	61	120·17	101	198·97	141	277·77	181	356·57
22	43·34	62	122·14	102	200·94	142	279·74	182	358·54
23	45·31	63	124·11	103	202·91	143	281·71	183	360·51
24	47·28	64	126·08	104	204·88	144	283·68	184	362·48
25	49·25	65	128·05	105	206·85	145	285·65	185	364·45
26	51·22	66	130·02	106	208·82	146	287·62	186	366·42
27	53·19	67	131·99	107	210·79	147	289·59	187	368·39
28	55·16	68	133·96	108	212·76	148	291·56	188	370·36
29	57·13	69	135·93	109	214·73	149	293·53	189	372·33
30	59·10	**70**	137·90	**110**	216·70	**150**	295·50	**190**	374·30
31	61·07	71	139·87	111	218·67	151	297·47	191	376·27
32	63·04	72	141·84	112	220·64	152	299·44	192	378·24
33	65·01	73	143·81	113	222·61	153	301·41	193	380·21
34	66·98	74	145·78	114	224·58	154	303·38	194	382·18
35	68·95	75	147·75	115	226·55	155	305·35	195	384·15
36	70·92	76	149·72	116	228·52	156	307·32	196	386·12
37	72·89	77	151·69	117	230·49	157	309·29	197	388·09
38	74·86	78	153·66	118	232·46	158	311·26	198	390·06
39	76·83	79	155·63	119	234·43	159	313·23	199	392·03
40	78·80	**80**	157·60	**120**	236·40	**160**	315·20	**200**	394·00

$\frac{1}{16}=0{\cdot}12^{3}$ $\frac{1}{9}=0{\cdot}21^{9}$ $\frac{1}{8}=0{\cdot}24^{6}$ $\frac{1}{6}=0{\cdot}32^{8}$ $\frac{1}{5}=0{\cdot}39^{4}$ $\frac{1}{4}=0{\cdot}49^{3}$ $\frac{1}{3}=0{\cdot}65^{7}$

$\frac{3}{8}=0{\cdot}73^{9}$ $\frac{1}{2}=0{\cdot}98^{5}$ $\frac{5}{8}=1{\cdot}23^{1}$ $\frac{3}{4}=1{\cdot}47^{8}$ $\frac{7}{8}=1{\cdot}72^{4}$ $\frac{2}{3}=1{\cdot}31^{3}$ $\frac{5}{6}=1{\cdot}64^{2}$

$\frac{3}{16}=0{\cdot}36^{9}$ $\frac{5}{16}=0{\cdot}61^{6}$ $\frac{7}{16}=0{\cdot}86^{2}$ $\frac{9}{16}=1{\cdot}10^{8}$ $\frac{11}{16}=1{\cdot}35^{4}$ $\frac{13}{16}=1{\cdot}60^{1}$ $\frac{15}{16}=1{\cdot}84^{7}$

1	1·97½	41	80·97½	81	159·97½	121	238·97½	161	317·97½
2	3·95	42	82·95	82	161·95	122	240·95	162	319·95
3	5·92½	43	84·92½	83	163·92½	123	242·92½	163	321·92½
4	7·90	44	86·90	84	165·90	124	244·90	164	323·90
5	9·87½	45	88·87½	85	167·87½	125	246·87½	165	325·87½
6	11·85	46	90·85	86	169·85	126	248·85	166	327·85
7	13·82½	47	92·82½	87	171·82½	127	250·82½	167	329·82½
8	15·80	48	94·80	88	173·80	128	252·80	168	331·80
9	17·77½	49	96·77½	89	175·77½	129	254·77½	169	333·77½
10	19·75	**50**	98·75	**90**	177·75	**130**	256·75	**170**	335·75
11	21·72½	51	100·72½	91	179·72½	131	258·72½	171	337·72½
12	23·70	52	102·70	92	181·70	132	260·70	172	339·70
13	25·67½	53	104·67½	93	183·67½	133	262·67½	173	341·67½
14	27·65	54	106·65	94	185·65	134	264·65	174	343·65
15	29·62½	55	108·62½	95	187·62½	135	266·62½	175	345·62½
16	31·60	56	110·60	96	189·60	136	268·60	176	347·60
17	33·57½	57	112·57½	97	191·57½	137	270·57½	177	349·57½
18	35·55	58	114·55	98	193·55	138	272·55	178	351·55
19	37·52½	59	116·52½	99	195·52½	139	274·52½	179	353·52½
20	39·50	**60**	118·50	**100**	197·50	**140**	276·50	**180**	355·50
21	41·47½	61	120·47½	101	199·47½	141	278·47½	181	357·47½
22	43·45	62	122·45	102	201·45	142	280·45	182	359·45
23	45·42½	63	124·42½	103	203·42½	143	282·42½	183	361·42½
24	47·40	64	126·40	104	205·40	144	284·40	184	363·40
25	49·37½	65	128·37½	105	207·37½	145	286·37½	185	365·37½
26	51·35	66	130·35	106	209·35	146	288·35	186	367·35
27	53·32½	67	132·32½	107	211·32½	147	290·32½	187	369·32½
28	55·30	68	134·30	108	213·30	148	292·30	188	371·30
29	57·27½	69	136·27½	109	215·27½	149	294·27½	189	373·27½
30	59·25	**70**	138·25	**110**	217·25	**150**	296·25	**190**	375·25
31	61·22½	71	140·22½	111	219·22½	151	298·22½	191	377·22½
32	63·20	72	142·20	112	221·20	152	300·20	192	379·20
33	65·17½	73	144·17½	113	223·17½	153	302·17½	193	381·17½
34	67·15	74	146·15	114	225·15	154	304·15	194	383·15
35	69·12½	75	148·12½	115	227·12½	155	306·12½	195	385·12½
36	71·10	76	150·10	116	229·10	156	308·10	196	387·10
37	73·07½	77	152·07½	117	231·07½	157	310·07½	197	389·07½
38	75·05	78	154·05	118	233·05	158	312·05	198	391·05
39	77·02½	79	156·02½	119	235·02½	159	314·02½	199	393·02½
40	79·00	**80**	158·00	**120**	237·00	**160**	316·00	**200**	395·00

$\frac{1}{16}=0{\cdot}12^{3}$ $\frac{1}{9}=0{\cdot}21^{9}$ $\frac{1}{8}=0{\cdot}24^{7}$ $\frac{1}{6}=0{\cdot}32^{9}$ $\frac{1}{5}=0{\cdot}39^{5}$ $\frac{1}{4}=0{\cdot}49^{4}$ $\frac{1}{3}=0{\cdot}65^{8}$

$\frac{3}{8}=0{\cdot}74^{1}$ $\frac{1}{2}=0{\cdot}98^{8}$ $\frac{5}{8}=1{\cdot}23^{4}$ $\frac{3}{4}=1{\cdot}48^{1}$ $\frac{7}{8}=1{\cdot}72^{8}$ $\frac{2}{3}=1{\cdot}31^{7}$ $\frac{5}{6}=1{\cdot}64^{6}$

$\frac{3}{16}=0{\cdot}37^{0}$ $\frac{5}{16}=0{\cdot}61^{7}$ $\frac{7}{16}=0{\cdot}86^{4}$ $\frac{9}{16}=1{\cdot}11^{1}$ $\frac{11}{16}=1{\cdot}35^{8}$ $\frac{13}{16}=1{\cdot}60^{5}$ $\frac{15}{16}=1{\cdot}85^{2}$

(0.505=1£) **£1.98**

1	1·98	41	81·18	81	160·38	121	239·58	161	318·78
2	3·96	42	83·16	82	162·36	122	241·56	162	320·76
3	5·94	43	85·14	83	164·34	123	243·54	163	322·74
4	7·92	44	87·12	84	166·32	124	245·52	164	324·72
5	9·90	45	89·10	85	168·30	125	247·50	165	326·70
6	11·88	46	91·08	86	170·28	126	249·48	166	328·68
7	13·86	47	93·06	87	172·26	127	251·46	167	330·66
8	15·84	48	95·04	88	174·24	128	253·44	168	332·64
9	17·82	49	97·02	89	176·22	129	255·42	169	334·62
10	19·80	**50**	99·00	**90**	178·20	**130**	257·40	**170**	336·60
11	21·78	51	100·98	91	180·18	131	259·38	171	338·58
12	23·76	52	102·96	92	182·16	132	261·36	172	340·56
13	25·74	53	104·94	93	184·14	133	263·34	173	342·54
14	27·72	54	106·92	94	186·12	134	265·32	174	344·52
15	29·70	55	108·90	95	188·10	135	267·30	175	346·50
16	31·68	56	110·88	96	190·08	136	269·28	176	348·48
17	33·66	57	112·86	97	192·06	137	271·26	177	350·46
18	35·64	58	114·84	98	194·04	138	273·24	178	352·44
19	37·62	59	116·82	99	196·02	139	275·22	179	354·42
20	39·60	**60**	118·80	**100**	198·00	**140**	277·20	**180**	356·40
21	41·58	61	120·78	101	199·98	141	279·18	181	358·38
22	43·56	62	122·76	102	201·96	142	281·16	182	360·36
23	45·54	63	124·74	103	203·94	143	283·14	183	362·34
24	47·52	64	126·72	104	205·92	144	285·12	184	364·32
25	49·50	65	128·70	105	207·90	145	287·10	185	366·30
26	51·48	66	130·68	106	209·88	146	289·08	186	368·28
27	53·46	67	132·66	107	211·86	147	291·06	187	370·26
28	55·44	68	134·64	108	213·84	148	293·04	188	372·24
29	57·42	69	136·62	109	215·82	149	295·02	189	374·22
30	59·40	**70**	138·60	**110**	217·80	**150**	297·00	**190**	376·20
31	61·38	71	140·58	111	219·78	151	298·98	191	378·18
32	63·36	72	142·56	112	221·76	152	300·96	192	380·16
33	65·34	73	144·54	113	223·74	153	302·94	193	382·14
34	67·32	74	146·52	114	225·72	154	304·92	194	384·12
35	69·30	75	148·50	115	227·70	155	306·90	195	386·10
36	71·28	76	150·48	116	229·68	156	308·88	196	388·08
37	73·26	77	152·46	117	231·66	157	310·86	197	390·06
38	75·24	78	154·44	118	233·64	158	312·84	198	392·04
39	77·22	79	156·42	119	235·62	159	314·82	199	394·02
40	79·20	**80**	158·40	**120**	237·60	**160**	316·80	**200**	396·00

$\frac{1}{16}=0{\cdot}12^{4}$ $\frac{1}{9}=0{\cdot}22^{0}$ $\frac{1}{8}=0{\cdot}24^{8}$ $\frac{1}{6}=0{\cdot}33^{0}$ $\frac{1}{5}=0{\cdot}39^{6}$ $\frac{1}{4}=0{\cdot}49^{5}$ $\frac{1}{3}=0{\cdot}66^{0}$

$\frac{3}{8}=0{\cdot}74^{3}$ $\frac{1}{2}=0{\cdot}99^{0}$ $\frac{5}{8}=1{\cdot}23^{8}$ $\frac{3}{4}=1{\cdot}48^{5}$ $\frac{7}{8}=1{\cdot}73^{3}$ $\frac{2}{3}=1{\cdot}32^{0}$ $\frac{5}{6}=1{\cdot}65^{0}$

$\frac{3}{16}=0{\cdot}37^{1}$ $\frac{5}{16}=0{\cdot}61^{9}$ $\frac{7}{16}=0{\cdot}86^{6}$ $\frac{9}{16}=1{\cdot}11^{4}$ $\frac{11}{16}=1{\cdot}36^{1}$ $\frac{13}{16}=1{\cdot}60^{9}$ $\frac{15}{16}=1{\cdot}85^{6}$

1	1·98½	41	81·38½	81	160·78½	121	240·18½	161	319·58½
2	3·97	42	83·37	82	162·77	122	242·17	162	321·57
3	5·95½	43	85·35½	83	164·75½	123	244·15½	163	323·55½
4	7·94	44	87·34	84	166·74	124	246·14	164	325·54
5	9·92½	45	89·32½	85	168·72½	125	248·12½	165	327·52½
6	11·91	46	91·31	86	170·71	126	250·11	166	329·51
7	13·89½	47	93·29½	87	172·69½	127	252·09½	167	331·49½
8	15·88	48	95·28	88	174·68	128	254·08	168	333·48
9	17·86½	49	97·26½	89	176·66½	129	256·06½	169	335·46½
10	19·85	**50**	99·25	**90**	178·65	**130**	258·05	**170**	337·45
11	21·83½	51	101·23½	91	180·63½	131	260·03½	171	339·43½
12	23·82	52	103·22	92	182·62	132	262·02	172	341·42
13	25·80½	53	105·20½	93	184·60½	133	264·00½	173	343·40½
14	27·79	54	107·19	94	186·59	134	265·99	174	345·39
15	29·77½	55	109·17½	95	188·57½	135	267·97½	175	347·37½
16	31·76	56	111·16	96	190·56	136	269·96	176	349·36
17	33·74½	57	113·14½	97	192·54½	137	271·94½	177	351·34½
18	35·73	58	115·13	98	194·53	138	273·93	178	353·33
19	37·71½	59	117·11½	99	196·51½	139	275·91½	179	355·31½
20	39·70	**60**	119·10	**100**	198·50	**140**	277·90	**180**	357·30
21	41·68½	61	121·08½	101	200·48½	141	279·88½	181	359·28½
22	43·67	62	123·07	102	202·47	142	281·87	182	361·27
23	45·65½	63	125·05½	103	204·45½	143	283·85½	183	363·25½
24	47·64	64	127·04	104	206·44	144	285·84	184	365·24
25	49·62½	65	129·02½	105	208·42½	145	287·82½	185	367·22½
26	51·61	66	131·01	106	210·41	146	289·81	186	369·21
27	53·59½	67	132·99½	107	212·39½	147	291·79½	187	371·19½
28	55·58	68	134·98	108	214·38	148	293·78	188	373·18
29	57·56½	69	136·96½	109	216·36½	149	295·76½	189	375·16½
30	59·55	**70**	138·95	**110**	218·35	**150**	297·75	**190**	377·15
31	61·53½	71	140·93½	111	220·33½	151	299·73½	191	379·13½
32	63·52	72	142·92	112	222·32	152	301·72	192	381·12
33	65·50½	73	144·90½	113	224·30½	153	303·70½	193	383·10½
34	67·49	74	146·89	114	226·29	154	305·69	194	385·09
35	69·47½	75	148·87½	115	228·27½	155	307·67½	195	387·07½
36	71·46	76	150·86	116	230·26	156	309·66	196	389·06
37	73·44½	77	152·84½	117	232·24½	157	311·64½	197	391·04½
38	75·43	78	154·83	118	234·23	158	313·63	198	393·03
39	77·41½	79	156·81½	119	236·21½	159	315·61½	199	395·01½
40	79·40	**80**	158·80	**120**	238·20	**160**	317·60	**200**	397·00

$\frac{1}{16}=0{\cdot}12^{4}$ $\frac{1}{9}=0{\cdot}22^{1}$ $\frac{1}{8}=0{\cdot}24^{8}$ $\frac{1}{6}=0{\cdot}33^{1}$ $\frac{1}{5}=0{\cdot}39^{7}$ $\frac{1}{4}=0{\cdot}49^{6}$ $\frac{1}{3}=0{\cdot}66^{2}$

$\frac{3}{8}=0{\cdot}74^{4}$ $\frac{1}{2}=0{\cdot}99^{3}$ $\frac{5}{8}=1{\cdot}24^{1}$ $\frac{3}{4}=1{\cdot}48^{9}$ $\frac{7}{8}=1{\cdot}73^{7}$ $\frac{2}{3}=1{\cdot}32^{3}$ $\frac{5}{6}=1{\cdot}65^{4}$

$\frac{3}{16}=0{\cdot}37^{2}$ $\frac{5}{16}=0{\cdot}62^{0}$ $\frac{7}{16}=0{\cdot}86^{8}$ $\frac{9}{16}=1{\cdot}11^{7}$ $\frac{11}{16}=1{\cdot}36^{5}$ $\frac{13}{16}=1{\cdot}61^{3}$ $\frac{15}{16}=1{\cdot}86^{1}$

(0.503=1£) **£1.99**

1	1·99	41	81·59	81	161·19	121	240·79	161	320·39
2	3·98	42	83·58	82	163·18	122	242·78	162	322·38
3	5·97	43	85·57	83	165·17	123	244·77	163	324·37
4	7·96	44	87·56	84	167·16	124	246·76	164	326·36
5	9·95	45	89·55	85	169·15	125	248·75	165	328·35
6	11·94	46	91·54	86	171·14	126	250·74	166	330·34
7	13·93	47	93·53	87	173·13	127	252·73	167	332·33
8	15·92	48	95·52	88	175·12	128	254·72	168	334·32
9	17·91	49	97·51	89	177·11	129	256·71	169	336·31
10	19·90	**50**	99·50	**90**	179·10	**130**	258·70	**170**	338·30
11	21·89	51	101·49	91	181·09	131	260·69	171	340·29
12	23·88	52	103·48	92	183·08	132	262·68	172	342·28
13	25·87	53	105·47	93	185·07	133	264·67	173	344·27
14	27·86	54	107·46	94	187·06	134	266·66	174	346·26
15	29·85	55	109·45	95	189·05	135	268·65	175	348·25
16	31·84	56	111·44	96	191·04	136	270·64	176	350·24
17	33·83	57	113·43	97	193·03	137	272·63	177	352·23
18	35·82	58	115·42	98	195·02	138	274·62	178	354·22
19	37·81	59	117·41	99	197·01	139	276·61	179	356·21
20	39·80	**60**	119·40	**100**	199·00	**140**	278·60	**180**	358·20
21	41·79	61	121·39	101	200·99	141	280·59	181	360·19
22	43·78	62	123·38	102	202·98	142	282·58	182	362·18
23	45·77	63	125·37	103	204·97	143	284·57	183	364·17
24	47·76	64	127·36	104	206·96	144	286·56	184	366·16
25	49·75	65	129·35	105	208·95	145	288·55	185	368·15
26	51·74	66	131·34	106	210·94	146	290·54	186	370·14
27	53·73	67	133·33	107	212·93	147	292·53	187	372·13
28	55·72	68	135·32	108	214·92	148	294·52	188	374·12
29	57·71	69	137·31	109	216·91	149	296·51	189	376·11
30	59·70	**70**	139·30	**110**	218·90	**150**	298·50	**190**	378·10
31	61·69	71	141·29	111	220·89	151	300·49	191	380·09
32	63·68	72	143·28	112	222·88	152	302·48	192	382·08
33	65·67	73	145·27	113	224·87	153	304·47	193	384·07
34	67·66	74	147·26	114	226·86	154	306·46	194	386·06
35	69·65	75	149·25	115	228·85	155	308·45	195	388·05
36	71·64	76	151·24	116	230·84	156	310·44	196	390·04
37	73·63	77	153·23	117	232·83	157	312·43	197	392·03
38	75·62	78	155·22	118	234·82	158	314·42	198	394·02
39	77·61	79	157·21	119	236·81	159	316·41	199	396·01
40	79·60	**80**	159·20	**120**	238·80	**160**	318·40	**200**	398·00

$\frac{1}{16}=0{\cdot}12^4$ $\frac{1}{9}=0{\cdot}22^1$ $\frac{1}{8}=0{\cdot}24^9$ $\frac{1}{6}=0{\cdot}33^2$ $\frac{1}{5}=0{\cdot}39^8$ $\frac{1}{4}=0{\cdot}49^8$ $\frac{1}{3}=0{\cdot}66^3$

$\frac{3}{8}=0{\cdot}74^6$ $\frac{1}{2}=0{\cdot}99^5$ $\frac{5}{8}=1{\cdot}24^4$ $\frac{3}{4}=1{\cdot}49^3$ $\frac{7}{8}=1{\cdot}74^1$ $\frac{2}{3}=1{\cdot}32^7$ $\frac{5}{6}=1{\cdot}65^8$

$\frac{3}{16}=0{\cdot}37^3$ $\frac{5}{16}=0{\cdot}62^2$ $\frac{7}{16}=0{\cdot}87^1$ $\frac{9}{16}=1{\cdot}11^9$ $\frac{11}{16}=1{\cdot}36^8$ $\frac{13}{16}=1{\cdot}61^7$ $\frac{15}{16}=1{\cdot}86^6$

(0.501=1£) **£1.99½**

1	1·99½	41	81·79½	81	161·59½	121	241·39½	161	321·19½
2	3·99	42	83·79	82	163·59	122	243·39	162	323·19
3	5·98½	43	85·78½	83	165·58½	123	245·38½	163	325·18½
4	7·98	44	87·78	84	167·58	124	247·38	164	327·18
5	9·97½	45	89·77½	85	169·57½	125	249·37½	165	329·17½
6	11·97	46	91·77	86	171·57	126	251·37	166	331·17
7	13·96½	47	93·76½	87	173·56½	127	253·36½	167	333·16½
8	15·96	48	95·76	88	175·56	128	255·36	168	335·16
9	17·95½	49	97·75½	89	177·55½	129	257·35½	169	337·15½
10	19·95	**50**	99·75	**90**	179·55	**130**	259·35	**170**	339·15
11	21·94½	51	101·74½	91	181·54½	131	261·34½	171	341·14½
12	23·94	52	103·74	92	183·54	132	263·34	172	343·14
13	25·93½	53	105·73½	93	185·53½	133	265·33½	173	345·13½
14	27·93	54	107·73	94	187·53	134	267·33	174	347·13
15	29·92½	55	109·72½	95	189·52½	135	269·32½	175	349·12½
16	31·92	56	111·72	96	191·52	136	271·32	176	351·12
17	33·91½	57	113·71½	97	193·51½	137	273·31½	177	353·11½
18	35·91	58	115·71	98	195·51	138	275·31	178	355·11
19	37·90½	59	117·70½	99	197·50½	139	277·30½	179	357·10½
20	39·90	**60**	119·70	**100**	199·50	**140**	279·30	**180**	359·10
21	41·89½	61	121·69½	101	201·49½	141	281·29½	181	361·09½
22	43·89	62	123·69	102	203·49	142	283·29	182	363·09
23	45·88½	63	125·68½	103	205·48½	143	285·28½	183	365·08½
24	47·88	64	127·68	104	207·48	144	287·28	184	367·08
25	49·87½	65	129·67½	105	209·47½	145	289·27½	185	369·07½
26	51·87	66	131·67	106	211·47	146	291·27	186	371·07
27	53·86½	67	133·66½	107	213·46½	147	293·26½	187	373·06½
28	55·86	68	135·66	108	215·46	148	295·26	188	375·06
29	57·85½	69	137·65½	109	217·45½	149	297·25½	189	377·05½
30	59·85	**70**	139·65	**110**	219·45	**150**	299·25	**190**	379·05
31	61·84½	71	141·64½	111	221·44½	151	301·24½	191	381·04½
32	63·84	72	143·64	112	223·44	152	303·24	192	383·04
33	65·83½	73	145·63½	113	225·43½	153	305·23½	193	385·03½
34	67·83	74	147·63	114	227·43	154	307·23	194	387·03
35	69·82½	75	149·62½	115	229·42½	155	309·22½	195	389·02½
36	71·82	76	151·62	116	231·42	156	311·22	196	391·02
37	73·81½	77	153·61½	117	233·41½	157	313·21½	197	393·01½
38	75·81	78	155·61	118	235·41	158	315·21	198	395·01
39	77·80½	79	157·60½	119	237·40½	159	317·20½	199	397·00½
40	79·80	**80**	159·60	**120**	239·40	**160**	319·20	**200**	399·00

$\frac{1}{16}=0\cdot12^{5}$ $\frac{1}{9}=0\cdot22^{2}$ $\frac{1}{8}=0\cdot24^{9}$ $\frac{1}{6}=0\cdot33^{3}$ $\frac{1}{5}=0\cdot39^{9}$ $\frac{1}{4}=0\cdot49^{9}$ $\frac{1}{3}=0\cdot66^{5}$

$\frac{3}{8}=0\cdot74^{8}$ $\frac{1}{2}=0\cdot99^{8}$ $\frac{5}{8}=1\cdot24^{7}$ $\frac{3}{4}=1\cdot49^{6}$ $\frac{7}{8}=1\cdot74^{6}$ $\frac{2}{3}=1\cdot33^{0}$ $\frac{5}{6}=1\cdot66^{3}$

$\frac{3}{16}=0\cdot37^{4}$ $\frac{5}{16}=0\cdot62^{3}$ $\frac{7}{16}=0\cdot87^{3}$ $\frac{9}{16}=1\cdot12^{2}$ $\frac{11}{16}=1\cdot37^{2}$ $\frac{13}{16}=1\cdot62^{1}$ $\frac{15}{16}=1\cdot87^{0}$

1	2·00	41	82·00	81	162·00	121	242·00	161	322·00
2	4·00	42	84·00	82	164·00	122	244·00	162	324·00
3	6·00	43	86·00	83	166·00	123	246·00	163	326·00
4	8·00	44	88·00	84	168·00	124	248·00	164	328·00
5	10·00	45	90·00	85	170·00	125	250·00	165	330·00
6	12·00	46	92·00	86	172·00	126	252·00	166	332·00
7	14·00	47	94·00	87	174·00	127	254·00	167	334·00
8	16·00	48	96·00	88	176·00	128	256·00	168	336·00
9	18·00	49	98·00	89	178·00	129	258·00	169	338·00
10	20·00	**50**	100·00	**90**	180·00	**130**	260·00	**170**	340·00
11	22·00	51	102·00	91	182·00	131	262·00	171	342·00
12	24·00	52	104·00	92	184·00	132	264·00	172	344·00
13	26·00	53	106·00	93	186·00	133	266·00	173	346·00
14	28·00	54	108·00	94	188·00	134	268·00	174	348·00
15	30·00	55	110·00	95	190·00	135	270·00	175	350·00
16	32·00	56	112·00	96	192·00	136	272·00	176	352·00
17	34·00	57	114·00	97	194·00	137	274·00	177	354·00
18	36·00	58	116·00	98	196·00	138	276·00	178	356·00
19	38·00	59	118·00	99	198·00	139	278·00	179	358·00
20	40·00	**60**	120·00	**100**	200·00	**140**	280·00	**180**	360·00
21	42·00	61	122·00	101	202·00	141	282·00	181	362·00
22	44·00	62	124·00	102	204·00	142	284·00	182	364·00
23	46·00	63	126·00	103	206·00	143	286·00	183	366·00
24	48·00	64	128·00	104	208·00	144	288·00	184	368·00
25	50·00	65	130·00	105	210·00	145	290·00	185	370·00
26	52·00	66	132·00	106	212·00	146	292·00	186	372·00
27	54·00	67	134·00	107	214·00	147	294·00	187	374·00
28	56·00	68	136·00	108	216·00	148	296·00	188	376·00
29	58·00	69	138·00	109	218·00	149	298·00	189	378·00
30	60·00	**70**	140·00	**110**	220·00	**150**	300·00	**190**	380·00
31	62·00	71	142·00	111	222·00	151	302·00	191	382·00
32	64·00	72	144·00	112	224·00	152	304·00	192	384·00
33	66·00	73	146·00	113	226·00	153	306·00	193	386·00
34	68·00	74	148·00	114	228·00	154	308·00	194	388·00
35	70·00	75	150·00	115	230·00	155	310·00	195	390·00
36	72·00	76	152·00	116	232·00	156	312·00	196	392·00
37	74·00	77	154·00	117	234·00	157	314·00	197	394·00
38	76·00	78	156·00	118	236·00	158	316·00	198	396·00
39	78·00	79	158·00	119	238·00	159	318·00	199	398·00
40	80·00	**80**	160·00	**120**	240·00	**160**	320·00	**200**	400·00

$\frac{1}{16}$=0·12^5	$\frac{1}{9}$=0·22^2	$\frac{1}{8}$=0·25^0	$\frac{1}{6}$=0·33^3	$\frac{1}{5}$=0·40^0	$\frac{1}{4}$=0·50^0	$\frac{1}{3}$=0·66^7
$\frac{3}{8}$=0·75^0	$\frac{1}{2}$=1·00^0	$\frac{5}{8}$=1·25^0	$\frac{3}{4}$=1·50^0	$\frac{7}{8}$=1·75^0	$\frac{2}{3}$=1·33^3	$\frac{5}{6}$=1·66^7
$\frac{3}{16}$=0·37^5	$\frac{5}{16}$=0·62^5	$\frac{7}{16}$=0·87^5	$\frac{9}{16}$=1·12^5	$\frac{11}{16}$=1·37^5	$\frac{13}{16}$=1·62^5	$\frac{15}{16}$=1·87^5

(0.446=1£) **£2.24**

1	2·24	41	91·84	81	181·44	121	271·04	161	360·64
2	4·48	42	94·08	82	183·68	122	273·28	162	362·88
3	6·72	43	96·32	83	185·92	123	275·52	163	365·12
4	8·96	44	98·56	84	188·16	124	277·76	164	367·36
5	11·20	45	100·80	85	190·40	125	280·00	165	369·60
6	13·44	46	103·04	86	192·64	126	282·24	166	371·84
7	15·68	47	105·28	87	194·88	127	284·48	167	374·08
8	17·92	48	107·52	88	197·12	128	286·72	168	376·32
9	20·16	49	109·76	89	199·36	129	288·96	169	378·56
10	22·40	**50**	112·00	**90**	201·60	**130**	291·20	**170**	380·80
11	24·64	51	114·24	91	203·84	131	293·44	171	383·04
12	26·88	52	116·48	92	206·08	132	295·68	172	385·28
13	29·12	53	118·72	93	208·32	133	297·92	173	387·52
14	31·36	54	120·96	94	210·56	134	300·16	174	389·76
15	33·60	55	123·20	95	212·80	135	302·40	175	392·00
16	35·84	56	125·44	96	215·04	136	304·64	176	394·24
17	38·08	57	127·68	97	217·28	137	306·88	177	396·48
18	40·32	58	129·92	98	219·52	138	309·12	178	398·72
19	42·56	59	132·16	99	221·76	139	311·36	179	400·96
20	44·80	**60**	134·40	**100**	224·00	**140**	313·60	**180**	403·20
21	47·04	61	136·64	101	226·24	141	315·84	181	405·44
22	49·28	62	138·88	102	228·48	142	318·08	182	407·68
23	51·52	63	141·12	103	230·72	143	320·32	183	409·92
24	53·76	64	143·36	104	232·96	144	322·56	184	412·16
25	56·00	65	145·60	105	235·20	145	324·80	185	414·40
26	58·24	66	147·84	106	237·44	146	327·04	186	416·64
27	60·48	67	150·08	107	239·68	147	329·28	187	418·88
28	62·72	68	152·32	108	241·92	148	331·52	188	421·12
29	64·96	69	154·56	109	244·16	149	333·76	189	423·36
30	67·20	**70**	156·80	**110**	246·40	**150**	336·00	**190**	425·60
31	69·44	71	159·04	111	248·64	151	338·24	191	427·84
32	71·68	72	161·28	112	250·88	152	340·48	192	430·08
33	73·92	73	163·52	113	253·12	153	342·72	193	432·32
34	76·16	74	165·76	114	255·36	154	344·96	194	434·56
35	78·40	75	168·00	115	257·60	155	347·20	195	436·80
36	80·64	76	170·24	116	259·84	156	349·44	196	439·04
37	82·88	77	172·48	117	262·08	157	351·68	197	441·28
38	85·12	78	174·72	118	264·32	158	353·92	198	443·52
39	87·36	79	176·96	119	266·56	159	356·16	199	445·76
40	89·60	**80**	179·20	**120**	268·80	**160**	358·40	**200**	448·00

$\frac{1}{16}=0{\cdot}14^{0}$ $\frac{1}{9}=0{\cdot}24^{9}$ $\frac{1}{8}=0{\cdot}28^{0}$ $\frac{1}{6}=0{\cdot}37^{3}$ $\frac{1}{5}=0{\cdot}44^{8}$ $\frac{1}{4}=0{\cdot}56^{0}$ $\frac{1}{3}=0{\cdot}74^{7}$

$\frac{3}{8}=0{\cdot}84^{0}$ $\frac{1}{2}=1{\cdot}12^{0}$ $\frac{5}{8}=1{\cdot}40^{0}$ $\frac{3}{4}=1{\cdot}68^{0}$ $\frac{7}{8}=1{\cdot}96^{0}$ $\frac{2}{3}=1{\cdot}49^{3}$ $\frac{5}{6}=1{\cdot}86^{7}$

$\frac{3}{16}=0{\cdot}42^{0}$ $\frac{5}{16}=0{\cdot}70^{0}$ $\frac{7}{16}=0{\cdot}98^{0}$ $\frac{9}{16}=1{\cdot}26^{0}$ $\frac{11}{16}=1{\cdot}54^{0}$ $\frac{13}{16}=1{\cdot}82^{0}$ $\frac{15}{16}=2{\cdot}10^{0}$

1 ton=2240 lbs

(0.424=1£) **£2.36**

1	2·36	41	96·76	81	191·16	121	285·56	161	379·96
2	4·72	42	99·12	82	193·52	122	287·92	162	382·32
3	7·08	43	101·48	83	195·88	123	290·28	163	384·68
4	9·44	44	103·84	84	198·24	124	292·64	164	387·04
5	11·80	45	106·20	85	200·60	125	295·00	165	389·40
6	14·16	46	108·56	86	202·96	126	297·36	166	391·76
7	16·52	47	110·92	87	205·32	127	299·72	167	394·12
8	18·88	48	113·28	88	207·68	128	302·08	168	396·48
9	21·24	49	115·64	89	210·04	129	304·44	169	398·84
10	23·60	**50**	118·00	**90**	212·40	**130**	306·80	**170**	401·20
11	25·96	51	120·36	91	214·76	131	309·16	171	403·56
12	28·32	52	122·72	92	217·12	132	311·52	172	405·92
13	30·68	53	125·08	93	219·48	133	313·88	173	408·28
14	33·04	54	127·44	94	221·84	134	316·24	174	410·64
15	35·40	55	129·80	95	224·20	135	318·60	175	413·00
16	37·76	56	132·16	96	226·56	136	320·96	176	415·36
17	40·12	57	134·52	97	228·92	137	323·32	177	417·72
18	42·48	58	136·88	98	231·28	138	325·68	178	420·08
19	44·84	59	139·24	99	233·64	139	328·04	179	422·44
20	47·20	**60**	141·60	**100**	236·00	**140**	330·40	**180**	424·80
21	49·56	61	143·96	101	238·36	141	332·76	181	427·16
22	51·92	62	146·32	102	240·72	142	335·12	182	429·52
23	54·28	63	148·68	103	243·08	143	337·48	183	431·88
24	56·64	64	151·04	104	245·44	144	339·84	184	434·24
25	59·00	65	153·40	105	247·80	145	342·20	185	436·60
26	61·36	66	155·76	106	250·16	146	344·56	186	438·96
27	63·72	67	158·12	107	252·52	147	346·92	187	441·32
28	66·08	68	160·48	108	254·88	148	349·28	188	443·68
29	68·44	69	162·84	109	257·24	149	351·64	189	446·04
30	70·80	**70**	165·20	**110**	259·60	**150**	354·00	**190**	448·40
31	73·16	71	167·56	111	261·96	151	356·36	191	450·76
32	75·52	72	169·92	112	264·32	152	358·72	192	453·12
33	77·88	73	172·28	113	266·68	153	361·08	193	455·48
34	80·24	74	174·64	114	269·04	154	363·44	194	457·84
35	82·60	75	177·00	115	271·40	155	365·80	195	460·20
36	84·96	76	179·36	116	273·76	156	368·16	196	462·56
37	87·32	77	181·72	117	276·12	157	370·52	197	464·92
38	89·68	78	184·08	118	278·48	158	372·88	198	467·28
39	92·04	79	186·44	119	280·84	159	375·24	199	469·64
40	94·40	**80**	188·80	**120**	283·20	**160**	377·60	**200**	472·00

$\frac{1}{16}=0{\cdot}14^{8}$ $\frac{1}{9}=0{\cdot}26^{2}$ $\frac{1}{8}=0{\cdot}29^{5}$ $\frac{1}{6}=0{\cdot}39^{3}$ $\frac{1}{5}=0{\cdot}47^{2}$ $\frac{1}{4}=0{\cdot}59^{0}$ $\frac{1}{3}=0{\cdot}78^{7}$

$\frac{3}{8}=0{\cdot}88^{5}$ $\frac{1}{2}=1{\cdot}18^{0}$ $\frac{5}{8}=1{\cdot}47^{5}$ $\frac{3}{4}=1{\cdot}77^{0}$ $\frac{7}{8}=2{\cdot}06^{5}$ $\frac{2}{3}=1{\cdot}57^{3}$ $\frac{5}{6}=1{\cdot}96^{7}$

$\frac{3}{16}=0{\cdot}44^{3}$ $\frac{5}{16}=0{\cdot}73^{8}$ $\frac{7}{16}=1{\cdot}03^{3}$ $\frac{9}{16}=1{\cdot}32^{8}$ $\frac{11}{16}=1{\cdot}62^{3}$ $\frac{13}{16}=1{\cdot}91^{8}$ $\frac{15}{16}=2{\cdot}21^{3}$

(0.418=1£) **£2.39**

1	2·39	41	97·99	81	193·59	121	289·19	161	384·79
2	4·78	42	100·38	82	195·98	122	291·58	162	387·18
3	7·17	43	102·77	83	198·37	123	293·97	163	389·57
4	9·56	44	105·16	84	200·76	124	296·36	164	391·96
5	11·95	45	107·55	85	203·15	125	298·75	165	394·35
6	14·34	46	109·94	86	205·54	126	301·14	166	396·74
7	16·73	47	112·33	87	207·93	127	303·53	167	399·13
8	19·12	48	114·72	88	210·32	128	305·92	168	401·52
9	21·51	49	117·11	89	212·71	129	308·31	169	403·91
10	23·90	**50**	119·50	**90**	215·10	**130**	310·70	**170**	406·30
11	26·29	51	121·89	91	217·49	131	313·09	171	408·69
12	28·68	52	124·28	92	219·88	132	315·48	172	411·08
13	31·07	53	126·67	93	222·27	133	317·87	173	413·47
14	33·46	54	129·06	94	224·66	134	320·26	174	415·86
15	35·85	55	131·45	95	227·05	135	322·65	175	418·25
16	38·24	56	133·84	96	229·44	136	325·04	176	420·64
17	40·63	57	136·23	97	231·83	137	327·43	177	423·03
18	43·02	58	138·62	98	234·22	138	329·82	178	425·42
19	45·41	59	141·01	99	236·61	139	332·21	179	427·81
20	47·80	**60**	143·40	**100**	239·00	**140**	334·60	**180**	430·20
21	50·19	61	145·79	101	241·39	141	336·99	181	432·59
22	52·58	62	148·18	102	243·78	142	339·38	182	434·98
23	54·97	63	150·57	103	246·17	143	341·77	183	437·37
24	57·36	64	152·96	104	248·56	144	344·16	184	439·76
25	59·75	65	155·35	105	250·95	145	346·55	185	442·15
26	62·14	66	157·74	106	253·34	146	348·94	186	444·54
27	64·53	67	160·13	107	255·73	147	351·33	187	446·93
28	66·92	68	162·52	108	258·12	148	353·72	188	449·32
29	69·31	69	164·91	109	260·51	149	356·11	189	451·71
30	71·70	**70**	167·30	**110**	262·90	**150**	358·50	**190**	454·10
31	74·09	71	169·69	111	265·29	151	360·89	191	456·49
32	76·48	72	172·08	112	267·68	152	363·28	192	458·88
33	78·87	73	174·47	113	270·07	153	365·67	193	461·27
34	81·26	74	176·86	114	272·46	154	368·06	194	463·66
35	83·65	75	179·25	115	274·85	155	370·45	195	466·05
36	86·04	76	181·64	116	277·24	156	372·84	196	468·44
37	88·43	77	184·03	117	279·63	157	375·23	197	470·83
38	90·82	78	186·42	118	282·02	158	377·62	198	473·22
39	93·21	79	188·81	119	284·41	159	380·01	199	475·61
40	95·60	**80**	191·20	**120**	286·80	**160**	382·40	**200**	478·00

$\frac{1}{16}=0{\cdot}14^{9}$	$\frac{1}{9}=0{\cdot}26^{6}$	$\frac{1}{8}=0{\cdot}29^{9}$	$\frac{1}{6}=0{\cdot}39^{8}$	$\frac{1}{5}=0{\cdot}47^{8}$	$\frac{1}{4}=0{\cdot}59^{8}$	$\frac{1}{3}=0{\cdot}79^{7}$
$\frac{3}{8}=0{\cdot}89^{6}$	$\frac{1}{2}=1{\cdot}19^{5}$	$\frac{5}{8}=1{\cdot}49^{4}$	$\frac{3}{4}=1{\cdot}79^{3}$	$\frac{7}{8}=2{\cdot}09^{1}$	$\frac{2}{3}=1{\cdot}59^{3}$	$\frac{5}{6}=1{\cdot}99^{2}$
$\frac{3}{16}=0{\cdot}44^{8}$	$\frac{5}{16}=0{\cdot}74^{7}$	$\frac{7}{16}=1{\cdot}04^{6}$	$\frac{9}{16}=1{\cdot}34^{4}$	$\frac{11}{16}=1{\cdot}64^{3}$	$\frac{13}{16}=1{\cdot}94^{2}$	$\frac{15}{16}=2{\cdot}24^{1}$

1 joule=0·239 calories

(0.405=1£) **£2.47**

1	2·47	41	101·27	81	200·07	121	298·87	161	397·67
2	4·94	42	103·74	82	202·54	122	301·34	162	400·14
3	7·41	43	106·21	83	205·01	123	303·81	163	402·61
4	9·88	44	108·68	84	207·48	124	306·28	164	405·08
5	12·35	45	111·15	85	209·95	125	308·75	165	407·55
6	14·82	46	113·62	86	212·42	126	311·22	166	410·02
7	17·29	47	116·09	87	214·89	127	313·69	167	412·49
8	19·76	48	118·56	88	217·36	128	316·16	168	414·96
9	22·23	49	121·03	89	219·83	129	318·63	169	417·43
10	24·70	**50**	123·50	**90**	222·30	**130**	321·10	**170**	419·90
11	27·17	51	125·97	91	224·77	131	323·57	171	422·37
12	29·64	52	128·44	92	227·24	132	326·04	172	424·84
13	32·11	53	130·91	93	229·71	133	328·51	173	427·31
14	34·58	54	133·38	94	232·18	134	330·98	174	429·78
15	37·05	55	135·85	95	234·65	135	333·45	175	432·25
16	39·52	56	138·32	96	237·12	136	335·92	176	434·72
17	41·99	57	140·79	97	239·59	137	338·39	177	437·19
18	44·46	58	143·26	98	242·06	138	340·86	178	439·66
19	46·93	59	145·73	99	244·53	139	343·33	179	442·13
20	49·40	**60**	148·20	**100**	247·00	**140**	345·80	**180**	444·60
21	51·87	61	150·67	101	249·47	141	348·27	181	447·07
22	54·34	62	153·14	102	251·94	142	350·74	182	449·54
23	56·81	63	155·61	103	254·41	143	353·21	183	452·01
24	59·28	64	158·08	104	256·88	144	355·68	184	454·48
25	61·75	65	160·55	105	259·35	145	358·15	185	456·95
26	64·22	66	163·02	106	261·82	146	360·62	186	459·42
27	66·69	67	165·49	107	264·29	147	363·09	187	461·89
28	69·16	68	167·96	108	266·76	148	365·56	188	464·36
29	71·63	69	170·43	109	269·23	149	368·03	189	466·83
30	74·10	**70**	172·90	**110**	271·70	**150**	370·50	**190**	469·30
31	76·57	71	175·37	111	274·17	151	372·97	191	471·77
32	79·04	72	177·84	112	276·64	152	375·44	192	474·24
33	81·51	73	180·31	113	279·11	153	377·91	193	476·71
34	83·98	74	182·78	114	281·58	154	380·38	194	479·18
35	86·45	75	185·25	115	284·05	155	382·85	195	481·65
36	88·92	76	187·72	116	286·52	156	385·32	196	484·12
37	91·39	77	190·19	117	288·99	157	387·79	197	486·59
38	93·86	78	192·66	118	291·46	158	390·26	198	489·06
39	96·33	79	195·13	119	293·93	159	392·73	199	491·53
40	98·80	**80**	197·60	**120**	296·40	**160**	395·20	**200**	494·00

$\frac{1}{16}=0{\cdot}15^{4}$ $\frac{1}{9}=0{\cdot}27^{4}$ $\frac{1}{8}=0{\cdot}30^{9}$ $\frac{1}{6}=0{\cdot}41^{2}$ $\frac{1}{5}=0{\cdot}49^{4}$ $\frac{1}{4}=0{\cdot}61^{8}$ $\frac{1}{3}=0{\cdot}82^{3}$

$\frac{3}{8}=0{\cdot}92^{6}$ $\frac{1}{2}=1{\cdot}23^{5}$ $\frac{5}{8}=1{\cdot}54^{4}$ $\frac{3}{4}=1{\cdot}85^{3}$ $\frac{7}{8}=2{\cdot}16^{1}$ $\frac{2}{3}=1{\cdot}64^{7}$ $\frac{5}{6}=2{\cdot}05^{8}$

$\frac{3}{16}=0{\cdot}46^{3}$ $\frac{5}{16}=0{\cdot}77^{2}$ $\frac{7}{16}=1{\cdot}08^{1}$ $\frac{9}{16}=1{\cdot}38^{9}$ $\frac{11}{16}=1{\cdot}69^{8}$ $\frac{13}{16}=2{\cdot}00^{7}$ $\frac{15}{16}=2{\cdot}31^{6}$

1 hectare=2·471 acres

(0.397=1£) **£2.52**

1	2·52	41	103·32	81	204·12	121	304·92	161	405·72
2	5·04	42	105·84	82	206·64	122	307·44	162	408·24
3	7·56	43	108·36	83	209·16	123	309·96	163	410·76
4	10·08	44	110·88	84	211·68	124	312·48	164	413·28
5	12·60	45	113·40	85	214·20	125	315·00	165	415·80
6	15·12	46	115·92	86	216·72	126	317·52	166	418·32
7	17·64	47	118·44	87	219·24	127	320·04	167	420·84
8	20·16	48	120·96	88	221·76	128	322·56	168	423·36
9	22·68	49	123·48	89	224·28	129	325·08	169	425·88
10	25·20	**50**	126·00	**90**	226·80	**130**	327·60	**170**	428·40
11	27·72	51	128·52	91	229·32	131	330·12	171	430·92
12	30·24	52	131·04	92	231·84	132	332·64	172	433·44
13	32·76	53	133·56	93	234·36	133	335·16	173	435·96
14	35·28	54	136·08	94	236·88	134	337·68	174	438·48
15	37·80	55	138·60	95	239·40	135	340·20	175	441·00
16	40·32	56	141·12	96	241·92	136	342·72	176	443·52
17	42·84	57	143·64	97	244·44	137	345·24	177	446·04
18	45·36	58	146·16	98	246·96	138	347·76	178	448·56
19	47·88	59	148·68	99	249·48	139	350·28	179	451·08
20	50·40	**60**	151·20	**100**	252·00	**140**	352·80	**180**	453·60
21	52·92	61	153·72	101	254·52	141	355·32	181	456·12
22	55·44	62	156·24	102	257·04	142	357·84	182	458·64
23	57·96	63	158·76	103	259·56	143	360·36	183	461·16
24	60·48	64	161·28	104	262·08	144	362·88	184	463·68
25	63·00	65	163·80	105	264·60	145	365·40	185	466·20
26	65·52	66	166·32	106	267·12	146	367·92	186	468·72
27	68·04	67	168·84	107	269·64	147	370·44	187	471·24
28	70·56	68	171·36	108	272·16	148	372·96	188	473·76
29	73·08	69	173·88	109	274·68	149	375·48	189	476·28
30	75·60	**70**	176·40	**110**	277·20	**150**	378·00	**190**	478·80
31	78·12	71	178·92	111	279·72	151	380·52	191	481·32
32	80·64	72	181·44	112	282·24	152	383·04	192	483·84
33	83·16	73	183·96	113	284·76	153	385·56	193	486·36
34	85·68	74	186·48	114	287·28	154	388·08	194	488·88
35	88·20	75	189·00	115	289·80	155	390·60	195	491·40
36	90·72	76	191·52	116	292·32	156	393·12	196	493·92
37	93·24	77	194·04	117	294·84	157	395·64	197	496·44
38	95·76	78	196·56	118	297·36	158	398·16	198	498·96
39	98·28	79	199·08	119	299·88	159	400·68	199	501·48
40	100·80	**80**	201·60	**120**	302·40	**160**	403·20	**200**	504·00

$\frac{1}{16}=0{\cdot}15^{8}$ $\frac{1}{9}=0{\cdot}28^{0}$ $\frac{1}{8}=0{\cdot}31^{5}$ $\frac{1}{6}=0{\cdot}42^{0}$ $\frac{1}{5}=0{\cdot}50^{4}$ $\frac{1}{4}=0{\cdot}63^{0}$ $\frac{1}{3}=0{\cdot}84^{0}$

$\frac{3}{8}=0{\cdot}94^{5}$ $\frac{1}{2}=1{\cdot}26^{0}$ $\frac{5}{8}=1{\cdot}57^{5}$ $\frac{3}{4}=1{\cdot}89^{0}$ $\frac{7}{8}=2{\cdot}20^{5}$ $\frac{2}{3}=1{\cdot}68^{0}$ $\frac{5}{6}=2{\cdot}10^{0}$

$\frac{3}{16}=0{\cdot}47^{3}$ $\frac{5}{16}=0{\cdot}78^{8}$ $\frac{7}{16}=1{\cdot}10^{3}$ $\frac{9}{16}=1{\cdot}41^{8}$ $\frac{11}{16}=1{\cdot}73^{3}$ $\frac{13}{16}=2{\cdot}04^{8}$ $\frac{15}{16}=2{\cdot}36^{3}$

1 Btu=0·252 calories

(0.394=1£) **£2.54**

1	2·54	41	104·14	81	205·74	121	307·34	161	408·94
2	5·08	42	106·68	82	208·28	122	309·88	162	411·48
3	7·62	43	109·22	83	210·82	123	312·42	163	414·02
4	10·16	44	111·76	84	213·36	124	314·96	164	416·56
5	12·70	45	114·30	85	215·90	125	317·50	165	419·10
6	15·24	46	116·84	86	218·44	126	320·04	166	421·64
7	17·78	47	119·38	87	220·98	127	322·58	167	424·18
8	20·32	48	121·92	88	223·52	128	325·12	168	426·72
9	22·86	49	124·46	89	226·06	129	327·66	169	429·26
10	25·40	**50**	127·00	**90**	228·60	**130**	330·20	**170**	431·80
11	27·94	51	129·54	91	231·14	131	332·74	171	434·34
12	30·48	52	132·08	92	233·68	132	335·28	172	436·88
13	33·02	53	134·62	93	236·22	133	337·82	173	439·42
14	35·56	54	137·16	94	238·76	134	340·36	174	441·96
15	38·10	55	139·70	95	241·30	135	342·90	175	444·50
16	40·64	56	142·24	96	243·84	136	345·44	176	447·04
17	43·18	57	144·78	97	246·38	137	347·98	177	449·58
18	45·72	58	147·32	98	248·92	138	350·52	178	452·12
19	48·26	59	149·86	99	251·46	139	353·06	179	454·66
20	50·80	**60**	152·40	**100**	254·00	**140**	355·60	**180**	457·20
21	53·34	61	154·94	101	256·54	141	358·14	181	459·74
22	55·88	62	157·48	102	259·08	142	360·68	182	462·28
23	58·42	63	160·02	103	261·62	143	363·22	183	464·82
24	60·96	64	162·56	104	264·16	144	365·76	184	467·36
25	63·50	65	165·10	105	266·70	145	368·30	185	469·90
26	66·04	66	167·64	106	269·24	146	370·84	186	472·44
27	68·58	67	170·18	107	271·78	147	373·38	187	474·98
28	71·12	68	172·72	108	274·32	148	375·92	188	477·52
29	73·66	69	175·26	109	276·86	149	378·46	189	480·06
30	76·20	**70**	177·80	**110**	279·40	**150**	381·00	**190**	482·60
31	78·74	71	180·34	111	281·94	151	383·54	191	485·14
32	81·28	72	182·88	112	284·48	152	386·08	192	487·68
33	83·82	73	185·42	113	287·02	153	388·62	193	490·22
34	86·36	74	187·96	114	289·56	154	391·16	194	492·76
35	88·90	75	190·50	115	292·10	155	393·70	195	495·30
36	91·44	76	193·04	116	294·64	156	396·24	196	497·84
37	93·98	77	195·58	117	297·18	157	398·78	197	500·38
38	96·52	78	198·12	118	299·72	158	401·32	198	502·92
39	99·06	79	200·66	119	302·26	159	403·86	199	505·46
40	101·60	**80**	203·20	**120**	304·80	**160**	406·40	**200**	508·00

$\frac{1}{16}=0{\cdot}15^{9}$ $\frac{1}{9}=0{\cdot}28^{2}$ $\frac{1}{8}=0{\cdot}31^{8}$ $\frac{1}{6}=0{\cdot}42^{3}$ $\frac{1}{5}=0{\cdot}50^{8}$ $\frac{1}{4}=0{\cdot}63^{5}$ $\frac{1}{3}=0{\cdot}84^{7}$

$\frac{3}{8}=0{\cdot}95^{3}$ $\frac{1}{2}=1{\cdot}27^{0}$ $\frac{5}{8}=1{\cdot}58^{8}$ $\frac{3}{4}=1{\cdot}90^{5}$ $\frac{7}{8}=2{\cdot}22^{3}$ $\frac{2}{3}=1{\cdot}69^{3}$ $\frac{5}{6}=2{\cdot}11^{7}$

$\frac{3}{16}=0{\cdot}47^{6}$ $\frac{5}{16}=0{\cdot}79^{4}$ $\frac{7}{16}=1{\cdot}11^{1}$ $\frac{9}{16}=1{\cdot}42^{9}$ $\frac{11}{16}=1{\cdot}74^{6}$ $\frac{13}{16}=2{\cdot}06^{4}$ $\frac{15}{16}=2{\cdot}38^{1}$

1 inch=25·4 millimetres

(0.391=1£) **£2.56**

1	2·56	41	104·96	81	207·36	121	309·76	161	412·16
2	5·12	42	107·52	82	209·92	122	312·32	162	414·72
3	7·68	43	110·08	83	212·48	123	314·88	163	417·28
4	10·24	44	112·64	84	215·04	124	317·44	164	419·84
5	12·80	45	115·20	85	217·60	125	320·00	165	422·40
6	15·36	46	117·76	86	220·16	126	322·56	166	424·96
7	17·92	47	120·32	87	222·72	127	325·12	167	427·52
8	20·48	48	122·88	88	225·28	128	327·68	168	430·08
9	23·04	49	125·44	89	227·84	129	330·24	169	432·64
10	25·60	**50**	128·00	**90**	230·40	**130**	332·80	**170**	435·20
11	28·16	51	130·56	91	232·96	131	335·36	171	437·76
12	30·72	52	133·12	92	235·52	132	337·92	172	440·32
13	33·28	53	135·68	93	238·08	133	340·48	173	442·88
14	35·84	54	138·24	94	240·64	134	343·04	174	445·44
15	38·40	55	140·80	95	243·20	135	345·60	175	448·00
16	40·96	56	143·36	96	245·76	136	348·16	176	450·56
17	43·52	57	145·92	97	248·32	137	350·72	177	453·12
18	46·08	58	148·48	98	250·88	138	353·28	178	455·68
19	48·64	59	151·04	99	253·44	139	355·84	179	458·24
20	51·20	**60**	153·60	**100**	256·00	**140**	358·40	**180**	460·80
21	53·76	61	156·16	101	258·56	141	360·96	181	463·36
22	56·32	62	158·72	102	261·12	142	363·52	182	465·92
23	58·88	63	161·28	103	263·68	143	366·08	183	468·48
24	61·44	64	163·84	104	266·24	144	368·64	184	471·04
25	64·00	65	166·40	105	268·80	145	371·20	185	473·60
26	66·56	66	168·96	106	271·36	146	373·76	186	476·16
27	69·12	67	171·52	107	273·92	147	376·32	187	478·72
28	71·68	68	174·08	108	276·48	148	378·88	188	481·28
29	74·24	69	176·64	109	279·04	149	381·44	189	483·84
30	76·80	**70**	179·20	**110**	281·60	**150**	384·00	**190**	486·40
31	79·36	71	181·76	111	284·16	151	386·56	191	488·96
32	81·92	72	184·32	112	286·72	152	389·12	192	491·52
33	84·48	73	186·88	113	289·28	153	391·68	193	494·08
34	87·04	74	189·44	114	291·84	154	394·24	194	496·64
35	89·60	75	192·00	115	294·40	155	396·80	195	499·20
36	92·16	76	194·56	116	296·96	156	399·36	196	501·76
37	94·72	77	197·12	117	299·52	157	401·92	197	504·32
38	97·28	78	199·68	118	302·08	158	404·48	198	506·88
39	99·84	79	202·24	119	304·64	159	407·04	199	509·44
40	102·40	**80**	204·80	**120**	307·20	**160**	409·60	**200**	512·00

$\frac{1}{16}=0{\cdot}16^{0}$	$\frac{1}{9}=0{\cdot}28^{4}$	$\frac{1}{8}=0{\cdot}32^{0}$	$\frac{1}{6}=0{\cdot}42^{7}$	$\frac{1}{5}=0{\cdot}51^{2}$	$\frac{1}{4}=0{\cdot}64^{0}$	$\frac{1}{3}=0{\cdot}85^{3}$
$\frac{3}{8}=0{\cdot}96^{0}$	$\frac{1}{2}=1{\cdot}28^{0}$	$\frac{5}{8}=1{\cdot}60^{0}$	$\frac{3}{4}=1{\cdot}92^{0}$	$\frac{7}{8}=2{\cdot}24^{0}$	$\frac{2}{3}=1{\cdot}70^{7}$	$\frac{5}{6}=2{\cdot}13^{3}$
$\frac{3}{16}=0{\cdot}48^{0}$	$\frac{5}{16}=0{\cdot}80^{0}$	$\frac{7}{16}=1{\cdot}12^{0}$	$\frac{9}{16}=1{\cdot}44^{0}$	$\frac{11}{16}=1{\cdot}76^{0}$	$\frac{13}{16}=2{\cdot}08^{0}$	$\frac{15}{16}=2{\cdot}40^{0}$

(0.383=1£) **£2.61**

1	2·61	41	107·01	81	211·41	121	315·81	161	420·21
2	5·22	42	109·62	82	214·02	122	318·42	162	422·82
3	7·83	43	112·23	83	216·63	123	321·03	163	425·43
4	10·44	44	114·84	84	219·24	124	323·64	164	428·04
5	13·05	45	117·45	85	221·85	125	326·25	165	430·65
6	15·66	46	120·06	86	224·46	126	328·86	166	433·26
7	18·27	47	122·67	87	227·07	127	331·47	167	435·87
8	20·88	48	125·28	88	229·68	128	334·08	168	438·48
9	23·49	49	127·89	89	232·29	129	336·69	169	441·09
10	26·10	**50**	130·50	**90**	234·90	**130**	339·30	**170**	443·70
11	28·71	51	133·11	91	237·51	131	341·91	171	446·31
12	31·32	52	135·72	92	240·12	132	344·52	172	448·92
13	33·93	53	138·33	93	242·73	133	347·13	173	451·53
14	36·54	54	140·94	94	245·34	134	349·74	174	454·14
15	39·15	55	143·55	95	247·95	135	352·35	175	456·75
16	41·76	56	146·16	96	250·56	136	354·96	176	459·36
17	44·37	57	148·77	97	253·17	137	357·57	177	461·97
18	46·98	58	151·38	98	255·78	138	360·18	178	464·58
19	49·59	59	153·99	99	258·39	139	362·79	179	467·19
20	52·20	**60**	156·60	**100**	261·00	**140**	365·40	**180**	469·80
21	54·81	61	159·21	101	263·61	141	368·01	181	472·41
22	57·42	62	161·82	102	266·22	142	370·62	182	475·02
23	60·03	63	164·43	103	268·83	143	373·23	183	477·63
24	62·64	64	167·04	104	271·44	144	375·84	184	480·24
25	65·25	65	169·65	105	274·05	145	378·45	185	482·85
26	67·86	66	172·26	106	276·66	146	381·06	186	485·46
27	70·47	67	174·87	107	279·27	147	383·67	187	488·07
28	73·08	68	177·48	108	281·88	148	386·28	188	490·68
29	75·69	69	180·09	109	284·49	149	388·89	189	493·29
30	78·30	**70**	182·70	**110**	287·10	**150**	391·50	**190**	495·90
31	80·91	71	185·31	111	289·71	151	394·11	191	498·51
32	83·52	72	187·92	112	292·32	152	396·72	192	501·12
33	86·13	73	190·53	113	294·93	153	399·33	193	503·73
34	88·74	74	193·14	114	297·54	154	401·94	194	506·34
35	91·35	75	195·75	115	300·15	155	404·55	195	508·95
36	93·96	76	198·36	116	302·76	156	407·16	196	511·56
37	96·57	77	200·97	117	305·37	157	409·77	197	514·17
38	99·18	78	203·58	118	307·98	158	412·38	198	516·78
39	101·79	79	206·19	119	310·59	159	414·99	199	519·39
40	104·40	**80**	208·80	**120**	313·20	**160**	417·60	**200**	522·00

$\frac{1}{16}=0{\cdot}16^{3}$ $\frac{1}{9}=0{\cdot}29^{0}$ $\frac{1}{8}=0{\cdot}32^{6}$ $\frac{1}{6}=0{\cdot}43^{5}$ $\frac{1}{5}=0{\cdot}52^{2}$ $\frac{1}{4}=0{\cdot}65^{3}$ $\frac{1}{3}=0{\cdot}87^{0}$

$\frac{3}{8}=0{\cdot}97^{9}$ $\frac{1}{2}=1{\cdot}30^{5}$ $\frac{5}{8}=1{\cdot}63^{1}$ $\frac{3}{4}=1{\cdot}95^{8}$ $\frac{7}{8}=2{\cdot}28^{4}$ $\frac{2}{3}=1{\cdot}74^{0}$ $\frac{5}{6}=2{\cdot}17^{5}$

$\frac{3}{16}=0{\cdot}48^{9}$ $\frac{5}{16}=0{\cdot}81^{6}$ $\frac{7}{16}=1{\cdot}14^{2}$ $\frac{9}{16}=1{\cdot}46^{8}$ $\frac{11}{16}=1{\cdot}79^{4}$ $\frac{13}{16}=2{\cdot}12^{1}$ $\frac{15}{16}=2{\cdot}44^{7}$

There are 261 working days in a year of 5 day weeks

(0.369=1£) **£2.71**

1	2·71	41	111·11	81	219·51	121	327·91	161	436·31
2	5·42	42	113·82	82	222·22	122	330·62	162	439·02
3	8·13	43	116·53	83	224·93	123	333·33	163	441·73
4	10·84	44	119·24	84	227·64	124	336·04	164	444·44
5	13·55	45	121·95	85	230·35	125	338·75	165	447·15
6	16·26	46	124·66	86	233·06	126	341·46	166	449·86
7	18·97	47	127·37	87	235·77	127	344·17	167	452·57
8	21·68	48	130·08	88	238·48	128	346·88	168	455·28
9	24·39	49	132·79	89	241·19	129	349·59	169	457·99
10	27·10	**50**	135·50	**90**	243·90	**130**	352·30	**170**	460·70
11	29·81	51	138·21	91	246·61	131	355·01	171	463·41
12	32·52	52	140·92	92	249·32	132	357·72	172	466·12
13	35·23	53	143·63	93	252·03	133	360·43	173	468·83
14	37·94	54	146·34	94	254·74	134	363·14	174	471·54
15	40·65	55	149·05	95	257·45	135	365·85	175	474·25
16	43·36	56	151·76	96	260·16	136	368·56	176	476·96
17	46·07	57	154·47	97	262·87	137	371·27	177	479·67
18	48·78	58	157·18	98	265·58	138	373·98	178	482·38
19	51·49	59	159·89	99	268·29	139	376·69	179	485·09
20	54·20	**60**	162·60	**100**	271·00	**140**	379·40	**180**	487·80
21	56·91	61	165·31	101	273·71	141	382·11	181	490·51
22	59·62	62	168·02	102	276·42	142	384·82	182	493·22
23	62·33	63	170·73	103	279·13	143	387·53	183	495·93
24	65·04	64	173·44	104	281·84	144	390·24	184	498·64
25	67·75	65	176·15	105	284·55	145	392·95	185	501·35
26	70·46	66	178·86	106	287·26	146	395·66	186	504·06
27	73·17	67	181·57	107	289·97	147	398·37	187	506·77
28	75·88	68	184·28	108	292·68	148	401·08	188	509·48
29	78·59	69	186·99	109	295·39	149	403·79	189	512·19
30	81·30	**70**	189·70	**110**	298·10	**150**	406·50	**190**	514·90
31	84·01	71	192·41	111	300·81	151	409·21	191	517·61
32	86·72	72	195·12	112	303·52	152	411·92	192	520·32
33	89·43	73	197·83	113	306·23	153	414·63	193	523·03
34	92·14	74	200·54	114	308·94	154	417·34	194	525·74
35	94·85	75	203·25	115	311·65	155	420·05	195	528·45
36	97·56	76	205·96	116	314·36	156	422·76	196	531·16
37	100·27	77	208·67	117	317·07	157	425·47	197	533·87
38	102·98	78	211·38	118	319·78	158	428·18	198	536·58
39	105·69	79	214·09	119	322·49	159	430·89	199	539·29
40	108·40	**80**	216·80	**120**	325·20	**160**	433·60	**200**	542·00

$\frac{1}{16}=0{\cdot}16^{9}$ $\frac{1}{9}=0{\cdot}30^{1}$ $\frac{1}{8}=0{\cdot}33^{9}$ $\frac{1}{6}=0{\cdot}45^{2}$ $\frac{1}{5}=0{\cdot}54^{2}$ $\frac{1}{4}=0{\cdot}67^{8}$ $\frac{1}{3}=0{\cdot}90^{3}$

$\frac{3}{8}=1{\cdot}01^{6}$ $\frac{1}{2}=1{\cdot}35^{5}$ $\frac{5}{8}=1{\cdot}69^{4}$ $\frac{3}{4}=2{\cdot}03^{3}$ $\frac{7}{8}=2{\cdot}37^{1}$ $\frac{2}{3}=1{\cdot}80^{7}$ $\frac{5}{6}=2{\cdot}25^{8}$

$\frac{3}{16}=0{\cdot}50^{8}$ $\frac{5}{16}=0{\cdot}84^{7}$ $\frac{7}{16}=1{\cdot}18^{6}$ $\frac{9}{16}=1{\cdot}52^{4}$ $\frac{11}{16}=1{\cdot}86^{3}$ $\frac{13}{16}=2{\cdot}20^{2}$ $\frac{15}{16}=2{\cdot}54^{1}$

1 acre foot=271,328 gallons

(0.368=1£) **£2.72**

1	2·72	41	111·52	81	220·32	121	329·12	161	437·92
2	5·44	42	114·24	82	223·04	122	331·84	162	440·64
3	8·16	43	116·96	83	225·76	123	334·56	163	443·36
4	10·88	44	119·68	84	228·48	124	337·28	164	446·08
5	13·60	45	122·40	85	231·20	125	340·00	165	448·80
6	16·32	46	125·12	86	233·92	126	342·72	166	451·52
7	19·04	47	127·84	87	236·64	127	345·44	167	454·24
8	21·76	48	130·56	88	239·36	128	348·16	168	456·96
9	24·48	49	133·28	89	242·08	129	350·88	169	459·68
10	27·20	**50**	136·00	**90**	244·80	**130**	353·60	**170**	462·40
11	29·92	51	138·72	91	247·52	131	356·32	171	465·12
12	32·64	52	141·44	92	250·24	132	359·04	172	467·84
13	35·36	53	144·16	93	252·96	133	361·76	173	470·56
14	38·08	54	146·88	94	255·68	134	364·48	174	473·28
15	40·80	55	149·60	95	258·40	135	367·20	175	476·00
16	43·52	56	152·32	96	261·12	136	369·92	176	478·72
17	46·24	57	155·04	97	263·84	137	372·64	177	481·44
18	48·96	58	157·76	98	266·56	138	375·36	178	484·16
19	51·68	59	160·48	99	269·28	139	378·08	179	486·88
20	54·40	**60**	163·20	**100**	272·00	**140**	380·80	**180**	489·60
21	57·12	61	165·92	101	274·72	141	383·52	181	492·32
22	59·84	62	168·64	102	277·44	142	386·24	182	495·04
23	62·56	63	171·36	103	280·16	143	388·96	183	497·76
24	65·28	64	174·08	104	282·88	144	391·68	184	500·48
25	68·00	65	176·80	105	285·60	145	394·40	185	503·20
26	70·72	66	179·52	106	288·32	146	397·12	186	505·92
27	73·44	67	182·24	107	291·04	147	399·84	187	508·64
28	76·16	68	184·96	108	293·76	148	402·56	188	511·36
29	78·88	69	187·68	109	296·48	149	405·28	189	514·08
30	81·60	**70**	190·40	**110**	299·20	**150**	408·00	**190**	516·80
31	84·32	71	193·12	111	301·92	151	410·72	191	519·52
32	87·04	72	195·84	112	304·64	152	413·44	192	522·24
33	89·76	73	198·56	113	307·36	153	416·16	193	524·96
34	92·48	74	201·28	114	310·08	154	418·88	194	527·68
35	95·20	75	204·00	115	312·80	155	421·60	195	530·40
36	97·92	76	206·72	116	315·52	156	424·32	196	533·12
37	100·64	77	209·44	117	318·24	157	427·04	197	535·84
38	103·36	78	212·16	118	320·96	158	429·76	198	538·56
39	106·08	79	214·88	119	323·68	159	432·48	199	541·28
40	108·80	**80**	217·60	**120**	326·40	**160**	435·20	**200**	544·00

$\frac{1}{16}=0{\cdot}17^{0}$ $\frac{1}{9}=0{\cdot}30^{2}$ $\frac{1}{8}=0{\cdot}34^{0}$ $\frac{1}{6}=0{\cdot}45^{3}$ $\frac{1}{5}=0{\cdot}54^{4}$ $\frac{1}{4}=0{\cdot}68^{0}$ $\frac{1}{3}=0{\cdot}90^{7}$

$\frac{3}{8}=1{\cdot}02^{0}$ $\frac{1}{2}=1{\cdot}36^{0}$ $\frac{5}{8}=1{\cdot}70^{0}$ $\frac{3}{4}=2{\cdot}04^{0}$ $\frac{7}{8}=2{\cdot}38^{0}$ $\frac{2}{3}=1{\cdot}81^{3}$ $\frac{5}{6}=2{\cdot}26^{7}$

$\frac{3}{16}=0{\cdot}51^{0}$ $\frac{5}{16}=0{\cdot}85^{0}$ $\frac{7}{16}=1{\cdot}19^{0}$ $\frac{9}{16}=1{\cdot}53^{0}$ $\frac{11}{16}=1{\cdot}87^{0}$ $\frac{13}{16}=2{\cdot}21^{0}$ $\frac{15}{16}=2{\cdot}55^{0}$

(0.355=1£) **£2.82**

1	2·82	41	115·62	81	228·42	121	341·22	161	454·02
2	5·64	42	118·44	82	231·24	122	344·04	162	456·84
3	8·46	43	121·26	83	234·06	123	346·86	163	459·66
4	11·28	44	124·08	84	236·88	124	349·68	164	462·48
5	14·10	45	126·90	85	239·70	125	352·50	165	465·30
6	16·92	46	129·72	86	242·52	126	355·32	166	468·12
7	19·74	47	132·54	87	245·34	127	358·14	167	470·94
8	22·56	48	135·36	88	248·16	128	360·96	168	473·76
9	25·38	49	138·18	89	250·98	129	363·78	169	476·58
10	28·20	**50**	141·00	**90**	253·80	**130**	366·60	**170**	479·40
11	31·02	51	143·82	91	256·62	131	369·42	171	482·22
12	33·84	52	146·64	92	259·44	132	372·24	172	485·04
13	36·66	53	149·46	93	262·26	133	375·06	173	487·86
14	39·48	54	152·28	94	265·08	134	377·88	174	490·68
15	42·30	55	155·10	95	267·90	135	380·70	175	493·50
16	45·12	56	157·92	96	270·72	136	383·52	176	496·32
17	47·94	57	160·74	97	273·54	137	386·34	177	499·14
18	50·76	58	163·56	98	276·36	138	389·16	178	501·96
19	53·58	59	166·38	99	279·18	139	391·98	179	504·78
20	56·40	**60**	169·20	**100**	282·00	**140**	394·80	**180**	507·60
21	59·22	61	172·02	101	284·82	141	397·62	181	510·42
22	62·04	62	174·84	102	287·64	142	400·44	182	513·24
23	64·86	63	177·66	103	290·46	143	403·26	183	516·06
24	67·68	64	180·48	104	293·28	144	406·08	184	518·88
25	70·50	65	183·30	105	296·10	145	408·90	185	521·70
26	73·32	66	186·12	106	298·92	146	411·72	186	524·52
27	76·14	67	188·94	107	301·74	147	414·54	187	527·34
28	78·96	68	191·76	108	304·56	148	417·36	188	530·16
29	81·78	69	194·58	109	307·38	149	420·18	189	532·98
30	84·60	**70**	197·40	**110**	310·20	**150**	423·00	**190**	535·80
31	87·42	71	200·22	111	313·02	151	425·82	191	538·62
32	90·24	72	203·04	112	315·84	152	428·64	192	541·44
33	93·06	73	205·86	113	318·66	153	431·46	193	544·26
34	95·88	74	208·68	114	321·48	154	434·28	194	547·08
35	98·70	75	211·50	115	324·30	155	437·10	195	549·90
36	101·52	76	214·32	116	327·12	156	439·92	196	552·72
37	104·34	77	217·14	117	329·94	157	442·74	197	555·54
38	107·16	78	219·96	118	332·76	158	445·56	198	558·36
39	109·98	79	222·78	119	335·58	159	448·38	199	561·18
40	112·80	**80**	225·60	**120**	338·40	**160**	451·20	**200**	564·00

$\frac{1}{16}=0{\cdot}17^{6}$ $\frac{1}{9}=0{\cdot}31^{3}$ $\frac{1}{8}=0{\cdot}35^{3}$ $\frac{1}{6}=0{\cdot}47^{0}$ $\frac{1}{5}=0{\cdot}56^{4}$ $\frac{1}{4}=0{\cdot}70^{5}$ $\frac{1}{3}=0{\cdot}94^{0}$

$\frac{3}{8}=1{\cdot}05^{8}$ $\frac{1}{2}=1{\cdot}41^{0}$ $\frac{5}{8}=1{\cdot}76^{3}$ $\frac{3}{4}=2{\cdot}11^{5}$ $\frac{7}{8}=2{\cdot}46^{8}$ $\frac{2}{3}=1{\cdot}88^{0}$ $\frac{5}{6}=2{\cdot}35^{0}$

$\frac{3}{16}=0{\cdot}52^{9}$ $\frac{5}{16}=0{\cdot}88^{1}$ $\frac{7}{16}=1{\cdot}23^{4}$ $\frac{9}{16}=1{\cdot}58^{6}$ $\frac{11}{16}=1{\cdot}93^{9}$ $\frac{13}{16}=2{\cdot}29^{1}$ $\frac{15}{16}=2{\cdot}64^{4}$

1 kilometre per litre=2·825 miles per gallon

1	2·83	41	116·03	81	229·23	121	342·43	161	455·63
2	5·66	42	118·86	82	232·06	122	345·26	162	458·46
3	8·49	43	121·69	83	234·89	123	348·09	163	461·29
4	11·32	44	124·52	84	237·72	124	350·92	164	464·12
5	14·15	45	127·35	85	240·55	125	353·75	165	466·95
6	16·98	46	130·18	86	243·38	126	356·58	166	469·78
7	19·81	47	133·01	87	246·21	127	359·41	167	472·61
8	22·64	48	135·84	88	249·04	128	362·24	168	475·44
9	25·47	49	138·67	89	251·87	129	365·07	169	478·27
10	28·30	**50**	141·50	**90**	254·70	**130**	367·90	**170**	481·10
11	31·13	51	144·33	91	257·53	131	370·73	171	483·93
12	33·96	52	147·16	92	260·36	132	373·56	172	486·76
13	36·79	53	149·99	93	263·19	133	376·39	173	489·59
14	39·62	54	152·82	94	266·02	134	379·22	174	492·42
15	42·45	55	155·65	95	268·85	135	382·05	175	495·25
16	45·28	56	158·48	96	271·68	136	384·88	176	498·08
17	48·11	57	161·31	97	274·51	137	387·71	177	500·91
18	50·94	58	164·14	98	277·34	138	390·54	178	503·74
19	53·77	59	166·97	99	280·17	139	393·37	179	506·57
20	56·60	**60**	169·80	**100**	283·00	**140**	396·20	**180**	509·40
21	59·43	61	172·63	101	285·83	141	399·03	181	512·23
22	62·26	62	175·46	102	288·66	142	401·86	182	515·06
23	65·09	63	178·29	103	291·49	143	404·69	183	517·89
24	67·92	64	181·12	104	294·32	144	407·52	184	520·72
25	70·75	65	183·95	105	297·15	145	410·35	185	523·55
26	73·58	66	186·78	106	299·98	146	413·18	186	526·38
27	76·41	67	189·61	107	302·81	147	416·01	187	529·21
28	79·24	68	192·44	108	305·64	148	418·84	188	532·04
29	82·07	69	195·27	109	308·47	149	421·67	189	534·87
30	84·90	**70**	198·10	**110**	311·30	**150**	424·50	**190**	537·70
31	87·73	71	200·93	111	314·13	151	427·33	191	540·53
32	90·56	72	203·76	112	316·96	152	430·16	192	543·36
33	93·39	73	206·59	113	319·79	153	432·99	193	546·19
34	96·22	74	209·42	114	322·62	154	435·82	194	549·02
35	99·05	75	212·25	115	325·45	155	438·65	195	551·85
36	101·88	76	215·08	116	328·28	156	441·48	196	554·68
37	104·71	77	217·91	117	331·11	157	444·31	197	557·51
38	107·54	78	220·74	118	333·94	158	447·14	198	560·34
39	110·37	79	223·57	119	336·77	159	449·97	199	563·17
40	113·20	**80**	226·40	**120**	339·60	**160**	452·80	**200**	566·00

$\frac{1}{16}=0{\cdot}17^7$ $\frac{1}{9}=0{\cdot}31^4$ $\frac{1}{8}=0{\cdot}35^4$ $\frac{1}{6}=0{\cdot}47^2$ $\frac{1}{5}=0{\cdot}56^6$ $\frac{1}{4}=0{\cdot}70^8$ $\frac{1}{3}=0{\cdot}94^3$

$\frac{3}{8}=1{\cdot}06^1$ $\frac{1}{2}=1{\cdot}41^5$ $\frac{5}{8}=1{\cdot}76^9$ $\frac{3}{4}=2{\cdot}12^3$ $\frac{7}{8}=2{\cdot}47^6$ $\frac{2}{3}=1{\cdot}88^7$ $\frac{5}{6}=2{\cdot}35^8$

$\frac{3}{16}=0{\cdot}53^1$ $\frac{5}{16}=0{\cdot}88^4$ $\frac{7}{16}=1{\cdot}23^8$ $\frac{9}{16}=1{\cdot}59^2$ $\frac{11}{16}=1{\cdot}94^6$ $\frac{13}{16}=2{\cdot}29^9$ $\frac{15}{16}=2{\cdot}65^3$

1 ounce=28·3 grammes

(0.341=1£) **£2.93**

1	2·93	41	120·13	81	237·33	121	354·53	161	471·73
2	5·86	42	123·06	82	240·26	122	357·46	162	474·66
3	8·79	43	125·99	83	243·19	123	360·39	163	477·59
4	11·72	44	128·92	84	246·12	124	363·32	164	480·52
5	14·65	45	131·85	85	249·05	125	366·25	165	483·45
6	17·58	46	134·78	86	251·98	126	369·18	166	486·38
7	20·51	47	137·71	87	254·91	127	372·11	167	489·31
8	23·44	48	140·64	88	257·84	128	375·04	168	492·24
9	26·37	49	143·57	89	260·77	129	377·97	169	495·17
10	29·30	**50**	146·50	**90**	263·70	**130**	380·90	**170**	498·10
11	32·23	51	149·43	91	266·63	131	383·83	171	501·03
12	35·16	52	152·36	92	269·56	132	386·76	172	503·96
13	38·09	53	155·29	93	272·49	133	389·69	173	506·89
14	41·02	54	158·22	94	275·42	134	392·62	174	509·82
15	43·95	55	161·15	95	278·35	135	395·55	175	512·75
16	46·88	56	164·08	96	281·28	136	398·48	176	515·68
17	49·81	57	167·01	97	284·21	137	401·41	177	518·61
18	52·74	58	169·94	98	287·14	138	404·34	178	521·54
19	55·67	59	172·87	99	290·07	139	407·27	179	524·47
20	58·60	**60**	175·80	**100**	293·00	**140**	410·20	**180**	527·40
21	61·53	61	178·73	101	295·93	141	413·13	181	530·33
22	64·46	62	181·66	102	298·86	142	416·06	182	533·26
23	67·39	63	184·59	103	301·79	143	418·99	183	536·19
24	70·32	64	187·52	104	304·72	144	421·92	184	539·12
25	73·25	65	190·45	105	307·65	145	424·85	185	542·05
26	76·18	66	193·38	106	310·58	146	427·78	186	544·98
27	79·11	67	196·31	107	313·51	147	430·71	187	547·91
28	82·04	68	199·24	108	316·44	148	433·64	188	550·84
29	84·97	69	202·17	109	319·37	149	436·57	189	553·77
30	87·90	**70**	205·10	**110**	322·30	**150**	439·50	**190**	556·70
31	90·83	71	208·03	111	325·23	151	442·43	191	559·63
32	93·76	72	210·96	112	328·16	152	445·36	192	562·56
33	96·69	73	213·89	113	331·09	153	448·29	193	565·49
34	99·62	74	216·82	114	334·02	154	451·22	194	568·42
35	102·55	75	219·75	115	336·95	155	454·15	195	571·35
36	105·48	76	222·68	116	339·88	156	457·08	196	574·28
37	108·41	77	225·61	117	342·81	157	460·01	197	577·21
38	111·34	78	228·54	118	345·74	158	462·94	198	580·14
39	114·27	79	231·47	119	348·67	159	465·87	199	583·07
40	117·20	**80**	234·40	**120**	351·60	**160**	468·80	**200**	586·00

$\frac{1}{16}=0{\cdot}18^{3}$ $\frac{1}{9}=0{\cdot}32^{6}$ $\frac{1}{8}=0{\cdot}36^{6}$ $\frac{1}{6}=0{\cdot}48^{8}$ $\frac{1}{5}=0{\cdot}58^{6}$ $\frac{1}{4}=0{\cdot}73^{3}$ $\frac{1}{3}=0{\cdot}97^{7}$

$\frac{3}{8}=1{\cdot}09^{9}$ $\frac{1}{2}=1{\cdot}46^{5}$ $\frac{5}{8}=1{\cdot}83^{1}$ $\frac{3}{4}=2{\cdot}19^{8}$ $\frac{7}{8}=2{\cdot}56^{4}$ $\frac{2}{3}=1{\cdot}95^{3}$ $\frac{5}{6}=2{\cdot}44^{2}$

$\frac{3}{16}=0{\cdot}54^{9}$ $\frac{5}{16}=0{\cdot}91^{6}$ $\frac{7}{16}=1{\cdot}28^{2}$ $\frac{9}{16}=1{\cdot}64^{8}$ $\frac{11}{16}=2{\cdot}01^{4}$ $\frac{13}{16}=2{\cdot}38^{1}$ $\frac{15}{16}=2{\cdot}74^{7}$

1 British Thermal Unit=0·293 watt hours

(0.328=1£) **£3.05**

1	3·05	41	125·05	81	247·05	121	369·05	161	491·05
2	6·10	42	128·10	82	250·10	122	372·10	162	494·10
3	9·15	43	131·15	83	253·15	123	375·15	163	497·15
4	12·20	44	134·20	84	256·20	124	378·20	164	500·20
5	15·25	45	137·25	85	259·25	125	381·25	165	503·25
6	18·30	46	140·30	86	262·30	126	384·30	166	506·30
7	21·35	47	143·35	87	265·35	127	387·35	167	509·35
8	24·40	48	146·40	88	268·40	128	390·40	168	512·40
9	27·45	49	149·45	89	271·45	129	393·45	169	515·45
10	30·50	**50**	152·50	**90**	274·50	**130**	396·50	**170**	518·50
11	33·55	51	155·55	91	277·55	131	399·55	171	521·55
12	36·60	52	158·60	92	280·60	132	402·60	172	524·60
13	39·65	53	161·65	93	283·65	133	405·65	173	527·65
14	42·70	54	164·70	94	286·70	134	408·70	174	530·70
15	45·75	55	167·75	95	289·75	135	411·75	175	533·75
16	48·80	56	170·80	96	292·80	136	414·80	176	536·80
17	51·85	57	173·85	97	295·85	137	417·85	177	539·85
18	54·90	58	176·90	98	298·90	138	420·90	178	542·90
19	57·95	59	179·95	99	301·95	139	423·95	179	545·95
20	61·00	**60**	183·00	**100**	305·00	**140**	427·00	**180**	549·00
21	64·05	61	186·05	101	308·05	141	430·05	181	552·05
22	67·10	62	189·10	102	311·10	142	433·10	182	555·10
23	70·15	63	192·15	103	314·15	143	436·15	183	558·15
24	73·20	64	195·20	104	317·20	144	439·20	184	561·20
25	76·25	65	198·25	105	320·25	145	442·25	185	564·25
26	79·30	66	201·30	106	323·30	146	445·30	186	567·30
27	82·35	67	204·35	107	326·35	147	448·35	187	570·35
28	85·40	68	207·40	108	329·40	148	451·40	188	573·40
29	88·45	69	210·45	109	332·45	149	454·45	189	576·45
30	91·50	**70**	213·50	**110**	335·50	**150**	457·50	**190**	579·50
31	94·55	71	216·55	111	338·55	151	460·55	191	582·55
32	97·60	72	219·60	112	341·60	152	463·60	192	585·60
33	100·65	73	222·65	113	344·65	153	466·65	193	588·65
34	103·70	74	225·70	114	347·70	154	469·70	194	591·70
35	106·75	75	228·75	115	350·75	155	472·75	195	594·75
36	109·80	76	231·80	116	353·80	156	475·80	196	597·80
37	112·85	77	234·85	117	356·85	157	478·85	197	600·85
38	115·90	78	237·90	118	359·90	158	481·90	198	603·90
39	118·95	79	240·95	119	362·95	159	484·95	199	606·95
40	122·00	**80**	244·00	**120**	366·00	**160**	488·00	**200**	610·00

$\frac{1}{16}=0{\cdot}19^{1}$ $\frac{1}{9}=0{\cdot}33^{9}$ $\frac{1}{8}=0{\cdot}38^{1}$ $\frac{1}{6}=0{\cdot}50^{8}$ $\frac{1}{5}=0{\cdot}61^{0}$ $\frac{1}{4}=0{\cdot}76^{3}$ $\frac{1}{3}=1{\cdot}01^{7}$

$\frac{3}{8}=1{\cdot}14^{4}$ $\frac{1}{2}=1{\cdot}52^{5}$ $\frac{5}{8}=1{\cdot}90^{6}$ $\frac{3}{4}=2{\cdot}28^{8}$ $\frac{7}{8}=2{\cdot}66^{9}$ $\frac{2}{3}=2{\cdot}03^{3}$ $\frac{5}{6}=2{\cdot}54^{2}$

$\frac{3}{16}=0{\cdot}57^{2}$ $\frac{5}{16}=0{\cdot}95^{3}$ $\frac{7}{16}=1{\cdot}33^{4}$ $\frac{9}{16}=1{\cdot}71^{6}$ $\frac{11}{16}=2{\cdot}09^{7}$ $\frac{13}{16}=2{\cdot}47^{8}$ $\frac{15}{16}=2{\cdot}85^{9}$

1 foot=0·3048 metres

(0.319=1£) **£3.13**

1	3·13	41	128·33	81	253·53	121	378·73	161	503·93
2	6·26	42	131·46	82	256·66	122	381·86	162	507·06
3	9·39	43	134·59	83	259·79	123	384·99	163	510·19
4	12·52	44	137·72	84	262·92	124	388·12	164	513·32
5	15·65	45	140·85	85	266·05	125	391·25	165	516·45
6	18·78	46	143·98	86	269·18	126	394·38	166	519·58
7	21·91	47	147·11	87	272·31	127	397·51	167	522·71
8	25·04	48	150·24	88	275·44	128	400·64	168	525·84
9	28·17	49	153·37	89	278·57	129	403·77	169	528·97
10	31·30	**50**	156·50	**90**	281·70	**130**	406·90	**170**	532·10
11	34·43	51	159·63	91	284·83	131	410·03	171	535·23
12	37·56	52	162·76	92	287·96	132	413·16	172	538·36
13	40·69	53	165·89	93	291·09	133	416·29	173	541·49
14	43·82	54	169·02	94	294·22	134	419·42	174	544·62
15	46·95	55	172·15	95	297·35	135	422·55	175	547·75
16	50·08	56	175·28	96	300·48	136	425·68	176	550·88
17	53·21	57	178·41	97	303·61	137	428·81	177	554·01
18	56·34	58	181·54	98	306·74	138	431·94	178	557·14
19	59·47	59	184·67	99	309·87	139	435·07	179	560·27
20	62·60	**60**	187·80	**100**	313·00	**140**	438·20	**180**	563·40
21	65·73	61	190·93	101	316·13	141	441·33	181	566·53
22	68·86	62	194·06	102	319·26	142	444·46	182	569·66
23	71·99	63	197·19	103	322·39	143	447·59	183	572·79
24	75·12	64	200·32	104	325·52	144	450·72	184	575·92
25	78·25	65	203·45	105	328·65	145	453·85	185	579·05
26	81·38	66	206·58	106	331·78	146	456·98	186	582·18
27	84·51	67	209·71	107	334·91	147	460·11	187	585·31
28	87·64	68	212·84	108	338·04	148	463·24	188	588·44
29	90·77	69	215·97	109	341·17	149	466·37	189	591·57
30	93·90	**70**	219·10	**110**	344·30	**150**	469·50	**190**	594·70
31	97·03	71	222·23	111	347·43	151	472·63	191	597·83
32	100·16	72	225·36	112	350·56	152	475·76	192	600·96
33	103·29	73	228·49	113	353·69	153	478·89	193	604·09
34	106·42	74	231·62	114	356·82	154	482·02	194	607·22
35	109·55	75	234·75	115	359·95	155	485·15	195	610·35
36	112·68	76	237·88	116	363·08	156	488·28	196	613·48
37	115·81	77	241·01	117	366·21	157	491·41	197	616·61
38	118·94	78	244·14	118	369·34	158	494·54	198	619·74
39	122·07	79	247·27	119	372·47	159	497·67	199	622·87
40	125·20	**80**	250·40	**120**	375·60	**160**	500·80	**200**	626·00

$\frac{1}{16}$=0·19⁶ $\frac{1}{9}$=0·34⁸ $\frac{1}{8}$=0·39¹ $\frac{1}{6}$=0·52² $\frac{1}{5}$=0·62⁶ $\frac{1}{4}$=0·78³ $\frac{1}{3}$=1·04³

$\frac{3}{8}$=1·17⁴ $\frac{1}{2}$=1·56⁵ $\frac{5}{8}$=1·95⁶ $\frac{3}{4}$=2·34⁸ $\frac{7}{8}$=2·73⁹ $\frac{2}{3}$=2·08⁷ $\frac{5}{6}$=2·60⁸

$\frac{3}{16}$=0·58⁷ $\frac{5}{16}$=0·97⁸ $\frac{7}{16}$=1·36⁹ $\frac{9}{16}$=1·76¹ $\frac{11}{16}$=2·15² $\frac{13}{16}$=2·54³ $\frac{15}{16}$=2·93⁴

There are 313 working days in a year (6 day week)

(0.318=1£) **£3.14**

1	3·14	41	128·74	81	254·34	121	379·94	161	505·54
2	6·28	42	131·88	82	257·48	122	383·08	162	508·68
3	9·42	43	135·02	83	260·62	123	386·22	163	511·82
4	12·56	44	138·16	84	263·76	124	389·36	164	514·96
5	15·70	45	141·30	85	266·90	125	392·50	165	518·10
6	18·84	46	144·44	86	270·04	126	395·64	166	521·24
7	21·98	47	147·58	87	273·18	127	398·78	167	524·38
8	25·12	48	150·72	88	276·32	128	401·92	168	527·52
9	28·26	49	153·86	89	279·46	129	405·06	169	530·66
10	31·40	**50**	157·00	**90**	282·60	**130**	408·20	**170**	533·80
11	34·54	51	160·14	91	285·74	131	411·34	171	536·94
12	37·68	52	163·28	92	288·88	132	414·48	172	540·08
13	40·82	53	166·42	93	292·02	133	417·62	173	543·22
14	43·96	54	169·56	94	295·16	134	420·76	174	546·36
15	47·10	55	172·70	95	298·30	135	423·90	175	549·50
16	50·24	56	175·84	96	301·44	136	427·04	176	552·64
17	53·38	57	178·98	97	304·58	137	430·18	177	555·78
18	56·52	58	182·12	98	307·72	138	433·32	178	558·92
19	59·66	59	185·26	99	310·86	139	436·46	179	562·06
20	62·80	**60**	188·40	**100**	314·00	**140**	439·60	**180**	565·20
21	65·94	61	191·54	101	317·14	141	442·74	181	568·34
22	69·08	62	194·68	102	320·28	142	445·88	182	571·48
23	72·22	63	197·82	103	323·42	143	449·02	183	574·62
24	75·36	64	200·96	104	326·56	144	452·16	184	577·76
25	78·50	65	204·10	105	329·70	145	455·30	185	580·90
26	81·64	66	207·24	106	332·84	146	458·44	186	584·04
27	84·78	67	210·38	107	335·98	147	461·58	187	587·18
28	87·92	68	213·52	108	339·12	148	464·72	188	590·32
29	91·06	69	216·66	109	342·26	149	467·86	189	593·46
30	94·20	**70**	219·80	**110**	345·40	**150**	471·00	**190**	596·60
31	97·34	71	222·94	111	348·54	151	474·14	191	599·74
32	100·48	72	226·08	112	351·68	152	477·28	192	602·88
33	103·62	73	229·22	113	354·82	153	480·42	193	606·02
34	106·76	74	232·36	114	357·96	154	483·56	194	609·16
35	109·90	75	235·50	115	361·10	155	486·70	195	612·30
36	113·04	76	238·64	116	364·24	156	489·84	196	615·44
37	116·18	77	241·78	117	367·38	157	492·98	197	618·58
38	119·32	78	244·92	118	370·52	158	496·12	198	621·72
39	122·46	79	248·06	119	373·66	159	499·26	199	624·86
40	125·60	**80**	251·20	**120**	376·80	**160**	502·40	**200**	628·00

$\frac{1}{16}=0{\cdot}19^6$ $\frac{1}{9}=0{\cdot}34^9$ $\frac{1}{8}=0{\cdot}39^3$ $\frac{1}{6}=0{\cdot}52^3$ $\frac{1}{5}=0{\cdot}62^8$ $\frac{1}{4}=0{\cdot}78^5$ $\frac{1}{3}=1{\cdot}04^7$

$\frac{3}{8}=1{\cdot}17^8$ $\frac{1}{2}=1{\cdot}57^0$ $\frac{5}{8}=1{\cdot}96^3$ $\frac{3}{4}=2{\cdot}35^5$ $\frac{7}{8}=2{\cdot}74^8$ $\frac{2}{3}=2{\cdot}09^3$ $\frac{5}{6}=2{\cdot}61^7$

$\frac{3}{16}=0{\cdot}58^9$ $\frac{5}{16}=0{\cdot}98^1$ $\frac{7}{16}=1{\cdot}37^4$ $\frac{9}{16}=1{\cdot}76^6$ $\frac{11}{16}=2{\cdot}15^9$ $\frac{13}{16}=2{\cdot}55^1$ $\frac{15}{16}=2{\cdot}94^4$

$\pi=3{\cdot}142$

(0.311=1£) **£3.22**

1	3·22	41	132·02	81	260·82	121	389·62	161	518·42
2	6·44	42	135·24	82	264·04	122	392·84	162	521·64
3	9·66	43	138·46	83	267·26	123	396·06	163	524·86
4	12·88	44	141·68	84	270·48	124	399·28	164	528·08
5	16·10	45	144·90	85	273·70	125	402·50	165	531·30
6	19·32	46	148·12	86	276·92	126	405·72	166	534·52
7	22·54	47	151·34	87	280·14	127	408·94	167	537·74
8	25·76	48	154·56	88	283·36	128	412·16	168	540·96
9	28·98	49	157·78	89	286·58	129	415·38	169	544·18
10	32·20	**50**	161·00	**90**	289·80	**130**	418·60	**170**	547·40
11	35·42	51	164·22	91	293·02	131	421·82	171	550·62
12	38·64	52	167·44	92	296·24	132	425·04	172	553·84
13	41·86	53	170·66	93	299·46	133	428·26	173	557·06
14	45·08	54	173·88	94	302·68	134	431·48	174	560·28
15	48·30	55	177·10	95	305·90	135	434·70	175	563·50
16	51·52	56	180·32	96	309·12	136	437·92	176	566·72
17	54·74	57	183·54	97	312·34	137	441·14	177	569·94
18	57·96	58	186·76	98	315·56	138	444·36	178	573·16
19	61·18	59	189·98	99	318·78	139	447·58	179	576·38
20	64·40	**60**	193·20	**100**	322·00	**140**	450·80	**180**	579·60
21	67·62	61	196·42	101	325·22	141	454·02	181	582·82
22	70·84	62	199·64	102	328·44	142	457·24	182	586·04
23	74·06	63	202·86	103	331·66	143	460·46	183	589·26
24	77·28	64	206·08	104	334·88	144	463·68	184	592·48
25	80·50	65	209·30	105	338·10	145	466·90	185	595·70
26	83·72	66	212·52	106	341·32	146	470·12	186	598·92
27	86·94	67	215·74	107	344·54	147	473·34	187	602·14
28	90·16	68	218·96	108	347·76	148	476·56	188	605·36
29	93·38	69	222·18	109	350·98	149	479·78	189	608·58
30	96·60	**70**	225·40	**110**	354·20	**150**	483·00	**190**	611·80
31	99·82	71	228·62	111	357·42	151	486·22	191	615·02
32	103·04	72	231·84	112	360·64	152	489·44	192	618·24
33	106·26	73	235·06	113	363·86	153	492·66	193	621·46
34	109·48	74	238·28	114	367·08	154	495·88	194	624·68
35	112·70	75	241·50	115	370·30	155	499·10	195	627·90
36	115·92	76	244·72	116	373·52	156	502·32	196	631·12
37	119·14	77	247·94	117	376·74	157	505·54	197	634·34
38	122·36	78	251·16	118	379·96	158	508·76	198	637·56
39	125·58	79	254·38	119	383·18	159	511·98	199	640·78
40	128·80	**80**	257·60	**120**	386·40	**160**	515·20	**200**	644·00

$\frac{1}{16}=0{\cdot}20^{1}$ $\frac{1}{9}=0{\cdot}35^{8}$ $\frac{1}{8}=0{\cdot}40^{3}$ $\frac{1}{6}=0{\cdot}53^{7}$ $\frac{1}{5}=0{\cdot}64^{4}$ $\frac{1}{4}=0{\cdot}80^{5}$ $\frac{1}{3}=1{\cdot}07^{3}$

$\frac{3}{8}=1{\cdot}20^{8}$ $\frac{1}{2}=1{\cdot}61^{0}$ $\frac{5}{8}=2{\cdot}01^{3}$ $\frac{3}{4}=2{\cdot}41^{5}$ $\frac{7}{8}=2{\cdot}81^{8}$ $\frac{2}{3}=2{\cdot}14^{7}$ $\frac{5}{6}=2{\cdot}68^{3}$

$\frac{3}{16}=0{\cdot}60^{4}$ $\frac{5}{16}=1{\cdot}00^{6}$ $\frac{7}{16}=1{\cdot}40^{9}$ $\frac{9}{16}=1{\cdot}81^{1}$ $\frac{11}{16}=2{\cdot}21^{4}$ $\frac{13}{16}=2{\cdot}61^{6}$ $\frac{15}{16}=3{\cdot}01^{9}$

The acceleration due to gravity is 32·2 feet per second^2

(0.305=1£) **£3.28**

1	3·28	41	134·48	81	265·68	121	396·88	161	528·08
2	6·56	42	137·76	82	268·96	122	400·16	162	531·36
3	9·84	43	141·04	83	272·24	123	403·44	163	534·64
4	13·12	44	144·32	84	275·52	124	406·72	164	537·92
5	16·40	45	147·60	85	278·80	125	410·00	165	541·20
6	19·68	46	150·88	86	282·08	126	413·28	166	544·48
7	22·96	47	154·16	87	285·36	127	416·56	167	547·76
8	26·24	48	157·44	88	288·64	128	419·84	168	551·04
9	29·52	49	160·72	89	291·92	129	423·12	169	554·32
10	32·80	**50**	164·00	**90**	295·20	**130**	426·40	**170**	557·60
11	36·08	51	167·28	91	298·48	131	429·68	171	560·88
12	39·36	52	170·56	92	301·76	132	432·96	172	564·16
13	42·64	53	173·84	93	305·04	133	436·24	173	567·44
14	45·92	54	177·12	94	308·32	134	439·52	174	570·72
15	49·20	55	180·40	95	311·60	135	442·80	175	574·00
16	52·48	56	183·68	96	314·88	136	446·08	176	577·28
17	55·76	57	186·96	97	318·16	137	449·36	177	580·56
18	59·04	58	190·24	98	321·44	138	452·64	178	583·84
19	62·32	59	193·52	99	324·72	139	455·92	179	587·12
20	65·60	**60**	196·80	**100**	328·00	**140**	459·20	**180**	590·40
21	68·88	61	200·08	101	331·28	141	462·48	181	593·68
22	72·16	62	203·36	102	334·56	142	465·76	182	596·96
23	75·44	63	206·64	103	337·84	143	469·04	183	600·24
24	78·72	64	209·92	104	341·12	144	472·32	184	603·52
25	82·00	65	213·20	105	344·40	145	475·60	185	606·80
26	85·28	66	216·48	106	347·68	146	478·88	186	610·08
27	88·56	67	219·76	107	350·96	147	482·16	187	613·36
28	91·84	68	223·04	108	354·24	148	485·44	188	616·64
29	95·12	69	226·32	109	357·52	149	488·72	189	619·92
30	98·40	**70**	229·60	**110**	360·80	**150**	492·00	**190**	623·20
31	101·68	71	232·88	111	364·08	151	495·28	191	626·48
32	104·96	72	236·16	112	367·36	152	498·56	192	629·76
33	108·24	73	239·44	113	370·64	153	501·84	193	633·04
34	111·52	74	242·72	114	373·92	154	505·12	194	636·32
35	114·80	75	246·00	115	377·20	155	508·40	195	639·60
36	118·08	76	249·28	116	380·48	156	511·68	196	642·88
37	121·36	77	252·56	117	383·76	157	514·96	197	646·16
38	124·64	78	255·84	118	387·04	158	518·24	198	649·44
39	127·92	79	259·12	119	390·32	159	521·52	199	652·72
40	131·20	**80**	262·40	**120**	393·60	**160**	524·80	**200**	656·00

$\frac{1}{16}=0{\cdot}20^{5}$ $\frac{1}{9}=0{\cdot}36^{4}$ $\frac{1}{8}=0{\cdot}41^{0}$ $\frac{1}{6}=0{\cdot}54^{7}$ $\frac{1}{5}=0{\cdot}65^{6}$ $\frac{1}{4}=0{\cdot}82^{0}$ $\frac{1}{3}=1{\cdot}09^{3}$

$\frac{3}{8}=1{\cdot}23^{0}$ $\frac{1}{2}=1{\cdot}64^{0}$ $\frac{5}{8}=2{\cdot}05^{0}$ $\frac{3}{4}=2{\cdot}46^{0}$ $\frac{7}{8}=2{\cdot}87^{0}$ $\frac{2}{3}=2{\cdot}18^{7}$ $\frac{5}{6}=2{\cdot}73^{3}$

$\frac{3}{16}=0{\cdot}61^{5}$ $\frac{5}{16}=1{\cdot}02^{5}$ $\frac{7}{16}=1{\cdot}43^{5}$ $\frac{9}{16}=1{\cdot}84^{5}$ $\frac{11}{16}=2{\cdot}25^{5}$ $\frac{13}{16}=2{\cdot}66^{5}$ $\frac{15}{16}=3{\cdot}07^{5}$

1 metre=3·281 feet

(0.293=1£) **£3.41**

1	3·41	41	139·81	81	276·21	121	412·61	161	549·01
2	6·82	42	143·22	82	279·62	122	416·02	162	552·42
3	10·23	43	146·63	83	283·03	123	419·43	163	555·83
4	13·64	44	150·04	84	286·44	124	422·84	164	559·24
5	17·05	45	153·45	85	289·85	125	426·25	165	562·65
6	20·46	46	156·86	86	293·26	126	429·66	166	566·06
7	23·87	47	160·27	87	296·67	127	433·07	167	569·47
8	27·28	48	163·68	88	300·08	128	436·48	168	572·88
9	30·69	49	167·09	89	303·49	129	439·89	169	576·29
10	34·10	**50**	170·50	**90**	306·90	**130**	443·30	**170**	579·70
11	37·51	51	173·91	91	310·31	131	446·71	171	583·11
12	40·92	52	177·32	92	313·72	132	450·12	172	586·52
13	44·33	53	180·73	93	317·13	133	453·53	173	589·93
14	47·74	54	184·14	94	320·54	134	456·94	174	593·34
15	51·15	55	187·55	95	323·95	135	460·35	175	596·75
16	54·56	56	190·96	96	327·36	136	463·76	176	600·16
17	57·97	57	194·37	97	330·77	137	467·17	177	603·57
18	61·38	58	197·78	98	334·18	138	470·58	178	606·98
19	64·79	59	201·19	99	337·59	139	473·99	179	610·39
20	68·20	**60**	204·60	**100**	341·00	**140**	477·40	**180**	613·80
21	71·61	61	208·01	101	344·41	141	480·81	181	617·21
22	75·02	62	211·42	102	347·82	142	484·22	182	620·62
23	78·43	63	214·83	103	351·23	143	487·63	183	624·03
24	81·84	64	218·24	104	354·64	144	491·04	184	627·44
25	85·25	65	221·65	105	358·05	145	494·45	185	630·85
26	88·66	66	225·06	106	361·46	146	497·86	186	634·26
27	92·07	67	228·47	107	364·87	147	501·27	187	637·67
28	95·48	68	231·88	108	368·28	148	504·68	188	641·08
29	98·89	69	235·29	109	371·69	149	508·09	189	644·49
30	102·30	**70**	238·70	**110**	375·10	**150**	511·50	**190**	647·90
31	105·71	71	242·11	111	378·51	151	514·91	191	651·31
32	109·12	72	245·52	112	381·92	152	518·32	192	654·72
33	112·53	73	248·93	113	385·33	153	521·73	193	658·13
34	115·94	74	252·34	114	388·74	154	525·14	194	661·54
35	119·35	75	255·75	115	392·15	155	528·55	195	664·95
36	122·76	76	259·16	116	395·56	156	531·96	196	668·36
37	126·17	77	262·57	117	398·97	157	535·37	197	671·77
38	129·58	78	265·98	118	402·38	158	538·78	198	675·18
39	132·99	79	269·39	119	405·79	159	542·19	199	678·59
40	136·40	**80**	272·80	**120**	409·20	**160**	545·60	**200**	682·00

$\frac{1}{16}=0{\cdot}21^{3}$ $\frac{1}{9}=0{\cdot}37^{9}$ $\frac{1}{8}=0{\cdot}42^{6}$ $\frac{1}{6}=0{\cdot}56^{8}$ $\frac{1}{5}=0{\cdot}68^{2}$ $\frac{1}{4}=0{\cdot}85^{3}$ $\frac{1}{3}=1{\cdot}13^{7}$

$\frac{3}{8}=1{\cdot}27^{9}$ $\frac{1}{2}=1{\cdot}70^{5}$ $\frac{5}{8}=2{\cdot}13^{1}$ $\frac{3}{4}=2{\cdot}55^{8}$ $\frac{7}{8}=2{\cdot}98^{4}$ $\frac{2}{3}=2{\cdot}27^{3}$ $\frac{5}{6}=2{\cdot}84^{2}$

$\frac{3}{16}=0{\cdot}63^{9}$ $\frac{5}{16}=1{\cdot}06^{6}$ $\frac{7}{16}=1{\cdot}49^{2}$ $\frac{9}{16}=1{\cdot}91^{8}$ $\frac{11}{16}=2{\cdot}34^{4}$ $\frac{13}{16}=2{\cdot}77^{1}$ $\frac{15}{16}=3{\cdot}19^{7}$

1 Kilowatthour=341 Btu

(0.285=1£) **£3.51**

1	3·51	41	143·91	81	284·31	121	424·71	161	565·11
2	7·02	42	147·42	82	287·82	122	428·22	162	568·62
3	10·53	43	150·93	83	291·33	123	431·73	163	572·13
4	14·04	44	154·44	84	294·84	124	435·24	164	575·64
5	17·55	45	157·95	85	298·35	125	438·75	165	579·15
6	21·06	46	161·46	86	301·86	126	442·26	166	582·66
7	24·57	47	164·97	87	305·37	127	445·77	167	586·17
8	28·08	48	168·48	88	308·88	128	449·28	168	589·68
9	31·59	49	171·99	89	312·39	129	452·79	169	593·19
10	35·10	**50**	175·50	**90**	315·90	**130**	456·30	**170**	596·70
11	38·61	51	179·01	91	319·41	131	459·81	171	600·21
12	42·12	52	182·52	92	322·92	132	463·32	172	603·72
13	45·63	53	186·03	93	326·43	133	466·83	173	607·23
14	49·14	54	189·54	94	329·94	134	470·34	174	610·74
15	52·65	55	193·05	95	333·45	135	473·85	175	614·25
16	56·16	56	196·56	96	336·96	136	477·36	176	617·76
17	59·67	57	200·07	97	340·47	137	480·87	177	621·27
18	63·18	58	203·58	98	343·98	138	484·38	178	624·78
19	66·69	59	207·09	99	347·49	139	487·89	179	628·29
20	70·20	**60**	210·60	**100**	351·00	**140**	491·40	**180**	631·80
21	73·71	61	214·11	101	354·51	141	494·91	181	635·31
22	77·22	62	217·62	102	358·02	142	498·42	182	638·82
23	80·73	63	221·13	103	361·53	143	501·93	183	642·33
24	84·24	64	224·64	104	365·04	144	505·44	184	645·84
25	87·75	65	228·15	105	368·55	145	508·95	185	649·35
26	91·26	66	231·66	106	372·06	146	512·46	186	652·86
27	94·77	67	235·17	107	375·57	147	515·97	187	656·37
28	98·28	68	238·68	108	379·08	148	519·48	188	659·88
29	101·79	69	242·19	109	382·59	149	522·99	189	663·39
30	105·30	**70**	245·70	**110**	386·10	**150**	526·50	**190**	666·90
31	108·81	71	249·21	111	389·61	151	530·01	191	670·41
32	112·32	72	252·72	112	393·12	152	533·52	192	673·92
33	115·83	73	256·23	113	396·63	153	537·03	193	677·43
34	119·34	74	259·74	114	400·14	154	540·54	194	680·94
35	122·85	75	263·25	115	403·65	155	544·05	195	684·45
36	126·36	76	266·76	116	407·16	156	547·56	196	687·96
37	129·87	77	270·27	117	410·67	157	551·07	197	691·47
38	133·38	78	273·78	118	414·18	158	554·58	198	694·98
39	136·89	79	277·29	119	417·69	159	558·09	199	698·49
40	140·40	**80**	280·80	**120**	421·20	**160**	561·60	**200**	702·00

$\frac{1}{16}$=0·21^{9} $\frac{1}{9}$=0·39^{0} $\frac{1}{8}$=0·43^{9} $\frac{1}{6}$=0·58^{5} $\frac{1}{5}$=0·70^{2} $\frac{1}{4}$=0·87^{8} $\frac{1}{3}$=1·17^{0}

$\frac{3}{8}$=1·31^{6} $\frac{1}{2}$=1·75^{5} $\frac{5}{8}$=2·19^{4} $\frac{3}{4}$=2·63^{3} $\frac{7}{8}$=3·07^{1} $\frac{2}{3}$=2·34^{0} $\frac{5}{6}$=2·92^{5}

$\frac{3}{16}$=0·65^{8} $\frac{5}{16}$=1·09^{7} $\frac{7}{16}$=1·53^{6} $\frac{9}{16}$=1·97^{4} $\frac{11}{16}$=2·41^{3} $\frac{13}{16}$=2·85^{2} $\frac{15}{16}$=3·29^{1}

1 pica point=0·351 mm

(0.283=1£) **£3.53**

1	3·53	41	144·73	81	285·93	121	427·13	161	568·33
2	7·06	42	148·26	82	289·46	122	430·66	162	571·86
3	10·59	43	151·79	83	292·99	123	434·19	163	575·39
4	14·12	44	155·32	84	296·52	124	437·72	164	578·92
5	17·65	45	158·85	85	300·05	125	441·25	165	582·45
6	21·18	46	162·38	86	303·58	126	444·78	166	585·98
7	24·71	47	165·91	87	307·11	127	448·31	167	589·51
8	28·24	48	169·44	88	310·64	128	451·84	168	593·04
9	31·77	49	172·97	89	314·17	129	455·37	169	596·57
10	35·30	**50**	176·50	**90**	317·70	**130**	458·90	**170**	600·10
11	38·83	51	180·03	91	321·23	131	462·43	171	603·63
12	42·36	52	183·56	92	324·76	132	465·96	172	607·16
13	45·89	53	187·09	93	328·29	133	469·49	173	610·69
14	49·42	54	190·62	94	331·82	134	473·02	174	614·22
15	52·95	55	194·15	95	335·35	135	476·55	175	617·75
16	56·48	56	197·68	96	338·88	136	480·08	176	621·28
17	60·01	57	201·21	97	342·41	137	483·61	177	624·81
18	63·54	58	204·74	98	345·94	138	487·14	178	628·34
19	67·07	59	208·27	99	349·47	139	490·67	179	631·87
20	70·60	**60**	211·80	**100**	353·00	**140**	494·20	**180**	635·40
21	74·13	61	215·33	101	356·53	141	497·73	181	638·93
22	77·66	62	218·86	102	360·06	142	501·26	182	642·46
23	81·19	63	222·39	103	363·59	143	504·79	183	645·99
24	84·72	64	225·92	104	367·12	144	508·32	184	649·52
25	88·25	65	229·45	105	370·65	145	511·85	185	653·05
26	91·78	66	232·98	106	374·18	146	515·38	186	656·58
27	95·31	67	236·51	107	377·71	147	518·91	187	660·11
28	98·84	68	240·04	108	381·24	148	522·44	188	663·64
29	102·37	69	243·57	109	384·77	149	525·97	189	667·17
30	105·90	**70**	247·10	**110**	388·30	**150**	529·50	**190**	670·70
31	109·43	71	250·63	111	391·83	151	533·03	191	674·23
32	112·96	72	254·16	112	395·36	152	536·56	192	677·76
33	116·49	73	257·69	113	398·89	153	540·09	193	681·29
34	120·02	74	261·22	114	402·42	154	543·62	194	684·82
35	123·55	75	264·75	115	405·95	155	547·15	195	688·35
36	127·08	76	268·28	116	409·48	156	550·68	196	691·88
37	130·61	77	271·81	117	413·01	157	554·21	197	695·41
38	134·14	78	275·34	118	416·54	158	557·74	198	698·94
39	137·67	79	278·87	119	420·07	159	561·27	199	702·47
40	141·20	**80**	282·40	**120**	423·60	**160**	564·80	**200**	706·00

$\frac{1}{16}=0{\cdot}22^{1}$ $\frac{1}{9}=0{\cdot}39^{2}$ $\frac{1}{8}=0{\cdot}44^{1}$ $\frac{1}{6}=0{\cdot}58^{8}$ $\frac{1}{5}=0{\cdot}70^{6}$ $\frac{1}{4}=0{\cdot}88^{3}$ $\frac{1}{3}=1{\cdot}17^{7}$

$\frac{3}{8}=1{\cdot}32^{4}$ $\frac{1}{2}=1{\cdot}76^{5}$ $\frac{5}{8}=2{\cdot}20^{6}$ $\frac{3}{4}=2{\cdot}64^{8}$ $\frac{7}{8}=3{\cdot}08^{9}$ $\frac{2}{3}=2{\cdot}35^{3}$ $\frac{5}{6}=2{\cdot}94^{2}$

$\frac{3}{16}=0{\cdot}66^{2}$ $\frac{5}{16}=1{\cdot}10^{3}$ $\frac{7}{16}=1{\cdot}54^{4}$ $\frac{9}{16}=1{\cdot}98^{6}$ $\frac{11}{16}=2{\cdot}42^{7}$ $\frac{13}{16}=2{\cdot}86^{8}$ $\frac{15}{16}=3{\cdot}30^{9}$

1 gramme=0·3527 ounces 1 litre=0·353 cubic feet

1	3·54	41	145·14	81	286·74	121	428·34	161	569·94
2	7·08	42	148·68	82	290·28	122	431·88	162	573·48
3	10·62	43	152·22	83	293·82	123	435·42	163	577·02
4	14·16	44	155·76	84	297·36	124	438·96	164	580·56
5	17·70	45	159·30	85	300·90	125	442·50	165	584·10
6	21·24	46	162·84	86	304·44	126	446·04	166	587·64
7	24·78	47	166·38	87	307·98	127	449·58	167	591·18
8	28·32	48	169·92	88	311·52	128	453·12	168	594·72
9	31·86	49	173·46	89	315·06	129	456·66	169	598·26
10	35·40	**50**	177·00	**90**	318·60	**130**	460·20	**170**	601·80
11	38·94	51	180·54	91	322·14	131	463·74	171	605·34
12	42·48	52	184·08	92	325·68	132	467·28	172	608·88
13	46·02	53	187·62	93	329·22	133	470·82	173	612·42
14	49·56	54	191·16	94	332·76	134	474·36	174	615·96
15	53·10	55	194·70	95	336·30	135	477·90	175	619·50
16	56·64	56	198·24	96	339·84	136	481·44	176	623·04
17	60·18	57	201·78	97	343·38	137	484·98	177	626·58
18	63·72	58	205·32	98	346·92	138	488·52	178	630·12
19	67·26	59	208·86	99	350·46	139	492·06	179	633·66
20	70·80	**60**	212·40	**100**	354·00	**140**	495·60	**180**	637·20
21	74·34	61	215·94	101	357·54	141	499·14	181	640·74
22	77·88	62	219·48	102	361·08	142	502·68	182	644·28
23	81·42	63	223·02	103	364·62	143	506·22	183	647·82
24	84·96	64	226·56	104	368·16	144	509·76	184	651·36
25	88·50	65	230·10	105	371·70	145	513·30	185	654·90
26	92·04	66	233·64	106	375·24	146	516·84	186	658·44
27	95·58	67	237·18	107	378·78	147	520·38	187	661·98
28	99·12	68	240·72	108	382·32	148	523·92	188	665·52
29	102·66	69	244·26	109	385·86	149	527·46	189	669·06
30	106·20	**70**	247·80	**110**	389·40	**150**	531·00	**190**	672·60
31	109·74	71	251·34	111	392·94	151	534·54	191	676·14
32	113·28	72	254·88	112	396·48	152	538·08	192	679·68
33	116·82	73	258·42	113	400·02	153	541·62	193	683·22
34	120·36	74	261·96	114	403·56	154	545·16	194	686·76
35	123·90	75	265·50	115	407·10	155	548·70	195	690·30
36	127·44	76	269·04	116	410·64	156	552·24	196	693·84
37	130·98	77	272·58	117	414·18	157	555·78	197	697·38
38	134·52	78	276·12	118	417·72	158	559·32	198	700·92
39	138·06	79	279·66	119	421·26	159	562·86	199	704·46
40	141·60	**80**	283·20	**120**	424·80	**160**	566·40	**200**	708·00

$\frac{1}{16}=0{\cdot}22^{1}$ $\frac{1}{9}=0{\cdot}39^{3}$ $\frac{1}{8}=0{\cdot}44^{3}$ $\frac{1}{6}=0{\cdot}59^{0}$ $\frac{1}{5}=0{\cdot}70^{8}$ $\frac{1}{4}=0{\cdot}88^{5}$ $\frac{1}{3}=1{\cdot}18^{0}$

$\frac{3}{8}=1{\cdot}32^{8}$ $\frac{1}{2}=1{\cdot}77^{0}$ $\frac{5}{8}=2{\cdot}21^{3}$ $\frac{3}{4}=2{\cdot}65^{5}$ $\frac{7}{8}=3{\cdot}09^{8}$ $\frac{2}{3}=2{\cdot}36^{0}$ $\frac{5}{6}=2{\cdot}95^{0}$

$\frac{3}{16}=0{\cdot}66^{4}$ $\frac{5}{16}=1{\cdot}10^{6}$ $\frac{7}{16}=1{\cdot}54^{9}$ $\frac{9}{16}=1{\cdot}99^{1}$ $\frac{11}{16}=2{\cdot}43^{4}$ $\frac{13}{16}=2{\cdot}87^{6}$ $\frac{15}{16}=3{\cdot}31^{9}$

1 mile per gallon=0·354 kilometres per litre

(0.274=1£) **£3.65**

1	3·65	41	149·65	81	295·65	121	441·65	161	587·65
2	7·30	42	153·30	82	299·30	122	445·30	162	591·30
3	10·95	43	156·95	83	302·95	123	448·95	163	594·95
4	14·60	44	160·60	84	306·60	124	452·60	164	598·60
5	18·25	45	164·25	85	310·25	125	456·25	165	602·25
6	21·90	46	167·90	86	313·90	126	459·90	166	605·90
7	25·55	47	171·55	87	317·55	127	463·55	167	609·55
8	29·20	48	175·20	88	321·20	128	467·20	168	613·20
9	32·85	49	178·85	89	324·85	129	470·85	169	616·85
10	36·50	**50**	182·50	**90**	328·50	**130**	474·50	**170**	620·50
11	40·15	51	186·15	91	332·15	131	478·15	171	624·15
12	43·80	52	189·80	92	335·80	132	481·80	172	627·80
13	47·45	53	193·45	93	339·45	133	485·45	173	631·45
14	51·10	54	197·10	94	343·10	134	489·10	174	635·10
15	54·75	55	200·75	95	346·75	135	492·75	175	638·75
16	58·40	56	204·40	96	350·40	136	496·40	176	642·40
17	62·05	57	208·05	97	354·05	137	500·05	177	646·05
18	65·70	58	211·70	98	357·70	138	503·70	178	649·70
19	69·35	59	215·35	99	361·35	139	507·35	179	653·35
20	73·00	**60**	219·00	**100**	365·00	**140**	511·00	**180**	657·00
21	76·65	61	222·65	101	368·65	141	514·65	181	660·65
22	80·30	62	226·30	102	372·30	142	518·30	182	664·30
23	83·95	63	229·95	103	375·95	143	521·95	183	667·95
24	87·60	64	233·60	104	379·60	144	525·60	184	671·60
25	91·25	65	237·25	105	383·25	145	529·25	185	675·25
26	94·90	66	240·90	106	386·90	146	532·90	186	678·90
27	98·55	67	244·55	107	390·55	147	536·55	187	682·55
28	102·20	68	248·20	108	394·20	148	540·20	188	686·20
29	105·85	69	251·85	109	397·85	149	543·85	189	689·85
30	109·50	**70**	255·50	**110**	401·50	**150**	547·50	**190**	693·50
31	113·15	71	259·15	111	405·15	151	551·15	191	697·15
32	116·80	72	262·80	112	408·80	152	554·80	192	700·80
33	120·45	73	266·45	113	412·45	153	558·45	193	704·45
34	124·10	74	270·10	114	416·10	154	562·10	194	708·10
35	127·75	75	273·75	115	419·75	155	565·75	195	711·75
36	131·40	76	277·40	116	423·40	156	569·40	196	715·40
37	135·05	77	281·05	117	427·05	157	573·05	197	719·05
38	138·70	78	284·70	118	430·70	158	576·70	198	722·70
39	142·35	79	288·35	119	434·35	159	580·35	199	726·35
40	146·00	**80**	292·00	**120**	438·00	**160**	584·00	**200**	730·00

$\frac{1}{16}=0{\cdot}22^{8}$ $\frac{1}{9}=0{\cdot}40^{6}$ $\frac{1}{8}=0{\cdot}45^{6}$ $\frac{1}{6}=0{\cdot}60^{8}$ $\frac{1}{5}=0{\cdot}73^{0}$ $\frac{1}{4}=0{\cdot}91^{3}$ $\frac{1}{3}=1{\cdot}21^{7}$

$\frac{3}{8}=1{\cdot}36^{9}$ $\frac{1}{2}=1{\cdot}82^{5}$ $\frac{5}{8}=2{\cdot}28^{1}$ $\frac{3}{4}=2{\cdot}73^{8}$ $\frac{7}{8}=3{\cdot}19^{4}$ $\frac{2}{3}=2{\cdot}43^{3}$ $\frac{5}{6}=3{\cdot}04^{2}$

$\frac{3}{16}=0{\cdot}68^{4}$ $\frac{5}{16}=1{\cdot}14^{1}$ $\frac{7}{16}=1{\cdot}59^{7}$ $\frac{9}{16}=2{\cdot}05^{3}$ $\frac{11}{16}=2{\cdot}50^{9}$ $\frac{13}{16}=2{\cdot}96^{6}$ $\frac{15}{16}=3{\cdot}42^{2}$

365 days=1 year

(0.273=1£) **£3.66**

1	3·66	41	150·06	81	296·46	121	442·86	161	589·26
2	7·32	42	153·72	82	300·12	122	446·52	162	592·92
3	10·98	43	157·38	83	303·78	123	450·18	163	596·58
4	14·64	44	161·04	84	307·44	124	453·84	164	600·24
5	18·30	45	164·70	85	311·10	125	457·50	165	603·90
6	21·96	46	168·36	86	314·76	126	461·16	166	607·56
7	25·62	47	172·02	87	318·42	127	464·82	167	611·22
8	29·28	48	175·68	88	322·08	128	468·48	168	614·88
9	32·94	49	179·34	89	325·74	129	472·14	169	618·54
10	36·60	**50**	183·00	**90**	329·40	**130**	475·80	**170**	622·20
11	40·26	51	186·66	91	333·06	131	479·46	171	625·86
12	43·92	52	190·32	92	336·72	132	483·12	172	629·52
13	47·58	53	193·98	93	340·38	133	486·78	173	633·18
14	51·24	54	197·64	94	344·04	134	490·44	174	636·84
15	54·90	55	201·30	95	347·70	135	494·10	175	640·50
16	58·56	56	204·96	96	351·36	136	497·76	176	644·16
17	62·22	57	208·62	97	355·02	137	501·42	177	647·82
18	65·88	58	212·28	98	358·68	138	505·08	178	651·48
19	69·54	59	215·94	99	362·34	139	508·74	179	655·14
20	73·20	**60**	219·60	**100**	366·00	**140**	512·40	**180**	658·80
21	76·86	61	223·26	101	369·66	141	516·06	181	662·46
22	80·52	62	226·92	102	373·32	142	519·72	182	666·12
23	84·18	63	230·58	103	376·98	143	523·38	183	669·78
24	87·84	64	234·24	104	380·64	144	527·04	184	673·44
25	91·50	65	237·90	105	384·30	145	530·70	185	677·10
26	95·16	66	241·56	106	387·96	146	534·36	186	680·76
27	98·82	67	245·22	107	391·62	147	538·02	187	684·42
28	102·48	68	248·88	108	395·28	148	541·68	188	688·08
29	106·14	69	252·54	109	398·94	149	545·34	189	691·74
30	109·80	**70**	256·20	**110**	402·60	**150**	549·00	**190**	695·40
31	113·46	71	259·86	111	406·26	151	552·66	191	699·06
32	117·12	72	263·52	112	409·92	152	556·32	192	702·72
33	120·78	73	267·18	113	413·58	153	559·98	193	706·38
34	124·44	74	270·84	114	417·24	154	563·64	194	710·04
35	128·10	75	274·50	115	420·90	155	567·30	195	713·70
36	131·76	76	278·16	116	424·56	156	570·96	196	717·36
37	135·42	77	281·82	117	428·22	157	574·62	197	721·02
38	139·08	78	285·48	118	431·88	158	578·28	198	724·68
39	142·74	79	289·14	119	435·54	159	581·94	199	728·34
40	146·40	**80**	292·80	**120**	439·20	**160**	585·60	**200**	732·00

$\frac{1}{16}=0{\cdot}22^{9}$ $\frac{1}{9}=0{\cdot}40^{7}$ $\frac{1}{8}=0{\cdot}45^{8}$ $\frac{1}{6}=0{\cdot}61^{0}$ $\frac{1}{5}=0{\cdot}73^{2}$ $\frac{1}{4}=0{\cdot}91^{5}$ $\frac{1}{3}=1{\cdot}22^{0}$

$\frac{3}{8}=1{\cdot}37^{3}$ $\frac{1}{2}=1{\cdot}83^{0}$ $\frac{5}{8}=2{\cdot}28^{8}$ $\frac{3}{4}=2{\cdot}74^{5}$ $\frac{7}{8}=3{\cdot}20^{3}$ $\frac{2}{3}=2{\cdot}44^{0}$ $\frac{5}{6}=3{\cdot}05^{0}$

$\frac{3}{16}=0{\cdot}68^{6}$ $\frac{5}{16}=1{\cdot}14^{4}$ $\frac{7}{16}=1{\cdot}60^{1}$ $\frac{9}{16}=2{\cdot}05^{9}$ $\frac{11}{16}=2{\cdot}51^{6}$ $\frac{13}{16}=2{\cdot}97^{4}$ $\frac{15}{16}=3{\cdot}43^{1}$

366 days=1 leap year

(0.254=1£) **£3.94**

1	3·94	41	161·54	81	319·14	121	476·74	161	634·34
2	7·88	42	165·48	82	323·08	122	480·68	162	638·28
3	11·82	43	169·42	83	327·02	123	484·62	163	642·22
4	15·76	44	173·36	84	330·96	124	488·56	164	646·16
5	19·70	45	177·30	85	334·90	125	492·50	165	650·10
6	23·64	46	181·24	86	338·84	126	496·44	166	654·04
7	27·58	47	185·18	87	342·78	127	500·38	167	657·98
8	31·52	48	189·12	88	346·72	128	504·32	168	661·92
9	35·46	49	193·06	89	350·66	129	508·26	169	665·86
10	39·40	**50**	197·00	**90**	354·60	**130**	512·20	**170**	669·80
11	43·34	51	200·94	91	358·54	131	516·14	171	673·74
12	47·28	52	204·88	92	362·48	132	520·08	172	677·68
13	51·22	53	208·82	93	366·42	133	524·02	173	681·62
14	55·16	54	212·76	94	370·36	134	527·96	174	685·56
15	59·10	55	216·70	95	374·30	135	531·90	175	689·50
16	63·04	56	220·64	96	378·24	136	535·84	176	693·44
17	66·98	57	224·58	97	382·18	137	539·78	177	697·38
18	70·92	58	228·52	98	386·12	138	543·72	178	701·32
19	74·86	59	232·46	99	390·06	139	547·66	179	705·26
20	78·80	**60**	236·40	**100**	394·00	**140**	551·60	**180**	709·20
21	82·74	61	240·34	101	397·94	141	555·54	181	713·14
22	86·68	62	244·28	102	401·88	142	559·48	182	717·08
23	90·62	63	248·22	103	405·82	143	563·42	183	721·02
24	94·56	64	252·16	104	409·76	144	567·36	184	724·96
25	98·50	65	256·10	105	413·70	145	571·30	185	728·90
26	102·44	66	260·04	106	417·64	146	575·24	186	732·84
27	106·38	67	263·98	107	421·58	147	579·18	187	736·78
28	110·32	68	267·92	108	425·52	148	583·12	188	740·72
29	114·26	69	271·86	109	429·46	149	587·06	189	744·66
30	118·20	**70**	275·80	**110**	433·40	**150**	591·00	**190**	748·60
31	122·14	71	279·74	111	437·34	151	594·94	191	752·54
32	126·08	72	283·68	112	441·28	152	598·88	192	756·48
33	130·02	73	287·62	113	445·22	153	602·82	193	760·42
34	133·96	74	291·56	114	449·16	154	606·76	194	764·36
35	137·90	75	295·50	115	453·10	155	610·70	195	768·30
36	141·84	76	299·44	116	457·04	156	614·64	196	772·24
37	145·78	77	303·38	117	460·98	157	618·58	197	776·18
38	149·72	78	307·32	118	464·92	158	622·52	198	780·12
39	153·66	79	311·26	119	468·86	159	626·46	199	784·06
40	157·60	**80**	315·20	**120**	472·80	**160**	630·40	**200**	788·00

$\frac{1}{16}=0{\cdot}24^{6}$ $\frac{1}{9}=0{\cdot}43^{8}$ $\frac{1}{8}=0{\cdot}49^{3}$ $\frac{1}{6}=0{\cdot}65^{7}$ $\frac{1}{5}=0{\cdot}78^{8}$ $\frac{1}{4}=0{\cdot}98^{5}$ $\frac{1}{3}=1{\cdot}31^{3}$

$\frac{3}{8}=1{\cdot}47^{8}$ $\frac{1}{2}=1{\cdot}97^{0}$ $\frac{5}{8}=2{\cdot}46^{3}$ $\frac{3}{4}=2{\cdot}95^{5}$ $\frac{7}{8}=3{\cdot}44^{8}$ $\frac{2}{3}=2{\cdot}62^{7}$ $\frac{5}{6}=3{\cdot}28^{3}$

$\frac{3}{16}=0{\cdot}73^{9}$ $\frac{5}{16}=1{\cdot}23^{1}$ $\frac{7}{16}=1{\cdot}72^{4}$ $\frac{9}{16}=2{\cdot}21^{6}$ $\frac{11}{16}=2{\cdot}70^{9}$ $\frac{13}{16}=3{\cdot}20^{1}$ $\frac{15}{16}=3{\cdot}69^{4}$

1 centimetre=0·3937 inches

(0.252=1£) **£3.97**

1	3·97	41	162·77	81	321·57	121	480·37	161	639·17
2	7·94	42	166·74	82	325·54	122	484·34	162	643·14
3	11·91	43	170·71	83	329·51	123	488·31	163	647·11
4	15·88	44	174·68	84	333·48	124	492·28	164	651·08
5	19·85	45	178·65	85	337·45	125	496·25	165	655·05
6	23·82	46	182·62	86	341·42	126	500·22	166	659·02
7	27·79	47	186·59	87	345·39	127	504·19	167	662·99
8	31·76	48	190·56	88	349·36	128	508·16	168	666·96
9	35·73	49	194·53	89	353·33	129	512·13	169	670·93
10	39·70	**50**	198·50	**90**	357·30	**130**	516·10	**170**	674·90
11	43·67	51	202·47	91	361·27	131	520·07	171	678·87
12	47·64	52	206·44	92	365·24	132	524·04	172	682·84
13	51·61	53	210·41	93	369·21	133	528·01	173	686·81
14	55·58	54	214·38	94	373·18	134	531·98	174	690·78
15	59·55	55	218·35	95	377·15	135	535·95	175	694·75
16	63·52	56	222·32	96	381·12	136	539·92	176	698·72
17	67·49	57	226·29	97	385·09	137	543·89	177	702·69
18	71·46	58	230·26	98	389·06	138	547·86	178	706·66
19	75·43	59	234·23	99	393·03	139	551·83	179	710·63
20	79·40	**60**	238·20	**100**	397·00	**140**	555·80	**180**	714·60
21	83·37	61	242·17	101	400·97	141	559·77	181	718·57
22	87·34	62	246·14	102	404·94	142	563·74	182	722·54
23	91·31	63	250·11	103	408·91	143	567·71	183	726·51
24	95·28	64	254·08	104	412·88	144	571·68	184	730·48
25	99·25	65	258·05	105	416·85	145	575·65	185	734·45
26	103·22	66	262·02	106	420·82	146	579·62	186	738·42
27	107·19	67	265·99	107	424·79	147	583·59	187	742·39
28	111·16	68	269·96	108	428·76	148	587·56	188	746·36
29	115·13	69	273·93	109	432·73	149	591·53	189	750·33
30	119·10	**70**	277·90	**110**	436·70	**150**	595·50	**190**	754·30
31	123·07	71	281·87	111	440·67	151	599·47	191	758·27
32	127·04	72	285·84	112	444·64	152	603·44	192	762·24
33	131·01	73	289·81	113	448·61	153	607·41	193	766·21
34	134·98	74	293·78	114	452·58	154	611·38	194	770·18
35	138·95	75	297·75	115	456·55	155	615·35	195	774·15
36	142·92	76	301·72	116	460·52	156	619·32	196	778·12
37	146·89	77	305·69	117	464·49	157	623·29	197	782·09
38	150·86	78	309·66	118	468·46	158	627·26	198	786·06
39	154·83	79	313·63	119	472·43	159	631·23	199	790·03
40	158·80	**80**	317·60	**120**	476·40	**160**	635·20	**200**	794·00

$\frac{1}{16}=0{\cdot}24^{8}$ $\frac{1}{9}=0{\cdot}44^{1}$ $\frac{1}{8}=0{\cdot}49^{6}$ $\frac{1}{6}=0{\cdot}66^{2}$ $\frac{1}{5}=0{\cdot}79^{4}$ $\frac{1}{4}=0{\cdot}99^{3}$ $\frac{1}{3}=1{\cdot}32^{3}$

$\frac{3}{8}=1{\cdot}48^{9}$ $\frac{1}{2}=1{\cdot}98^{5}$ $\frac{5}{8}=2{\cdot}48^{1}$ $\frac{3}{4}=2{\cdot}97^{8}$ $\frac{7}{8}=3{\cdot}47^{4}$ $\frac{2}{3}=2{\cdot}64^{7}$ $\frac{5}{6}=3{\cdot}30^{8}$

$\frac{3}{16}=0{\cdot}74^{4}$ $\frac{5}{16}=1{\cdot}24^{1}$ $\frac{7}{16}=1{\cdot}73^{7}$ $\frac{9}{16}=2{\cdot}23^{3}$ $\frac{11}{16}=2{\cdot}72^{9}$ $\frac{13}{16}=3{\cdot}22^{6}$ $\frac{15}{16}=3{\cdot}72^{2}$

1	4·05	41	166·05	81	328·05	121	490·05	161	652·05
2	8·10	42	170·10	82	332·10	122	494·10	162	656·10
3	12·15	43	174·15	83	336·15	123	498·15	163	660·15
4	16·20	44	178·20	84	340·20	124	502·20	164	664·20
5	20·25	45	182·25	85	344·25	125	506·25	165	668·25
6	24·30	46	186·30	86	348·30	126	510·30	166	672·30
7	28·35	47	190·35	87	352·35	127	514·35	167	676·35
8	32·40	48	194·40	88	356·40	128	518·40	168	680·40
9	36·45	49	198·45	89	360·45	129	522·45	169	684·45
10	40·50	**50**	202·50	**90**	364·50	**130**	526·50	**170**	688·50
11	44·55	51	206·55	91	368·55	131	530·55	171	692·55
12	48·60	52	210·60	92	372·60	132	534·60	172	696·60
13	52·65	53	214·65	93	376·65	133	538·65	173	700·65
14	56·70	54	218·70	94	380·70	134	542·70	174	704·70
15	60·75	55	222·75	95	384·75	135	546·75	175	708·75
16	64·80	56	226·80	96	388·80	136	550·80	176	712·80
17	68·85	57	230·85	97	392·85	137	554·85	177	716·85
18	72·90	58	234·90	98	396·90	138	558·90	178	720·90
19	76·95	59	238·95	99	400·95	139	562·95	179	724·95
20	81·00	**60**	243·00	**100**	405·00	**140**	567·00	**180**	729·00
21	85·05	61	247·05	101	409·05	141	571·05	181	733·05
22	89·10	62	251·10	102	413·10	142	575·10	182	737·10
23	93·15	63	255·15	103	417·15	143	579·15	183	741·15
24	97·20	64	259·20	104	421·20	144	583·20	184	745·20
25	101·25	65	263·25	105	425·25	145	587·25	185	749·25
26	105·30	66	267·30	106	429·30	146	591·30	186	753·30
27	109·35	67	271·35	107	433·35	147	595·35	187	757·35
28	113·40	68	275·40	108	437·40	148	599·40	188	761·40
29	117·45	69	279·45	109	441·45	149	603·45	189	765·45
30	121·50	**70**	283·50	**110**	445·50	**150**	607·50	**190**	769·50
31	125·55	71	287·55	111	449·55	151	611·55	191	773·55
32	129·60	72	291·60	112	453·60	152	615·60	192	777·60
33	133·65	73	295·65	113	457·65	153	619·65	193	781·65
34	137·70	74	299·70	114	461·70	154	623·70	194	785·70
35	141·75	75	303·75	115	465·75	155	627·75	195	789·75
36	145·80	76	307·80	116	469·80	156	631·80	196	793·80
37	149·85	77	311·85	117	473·85	157	635·85	197	797·85
38	153·90	78	315·90	118	477·90	158	639·90	198	801·90
39	157·95	79	319·95	119	481·95	159	643·95	199	805·95
40	162·00	**80**	324·00	**120**	486·00	**160**	648·00	**200**	810·00

$\frac{1}{16}=0{\cdot}25^{3}$ $\frac{1}{9}=0{\cdot}45^{0}$ $\frac{1}{8}=0{\cdot}50^{6}$ $\frac{1}{6}=0{\cdot}67^{5}$ $\frac{1}{5}=0{\cdot}81^{0}$ $\frac{1}{4}=1{\cdot}01^{3}$ $\frac{1}{3}=1{\cdot}35^{0}$

$\frac{3}{8}=1{\cdot}51^{9}$ $\frac{1}{2}=2{\cdot}02^{5}$ $\frac{5}{8}=2{\cdot}53^{1}$ $\frac{3}{4}=3{\cdot}03^{8}$ $\frac{7}{8}=3{\cdot}54^{4}$ $\frac{2}{3}=2{\cdot}70^{0}$ $\frac{5}{6}=3{\cdot}37^{5}$

$\frac{3}{16}=0{\cdot}75^{9}$ $\frac{5}{16}=1{\cdot}26^{6}$ $\frac{7}{16}=1{\cdot}77^{2}$ $\frac{9}{16}=2{\cdot}27^{8}$ $\frac{11}{16}=2{\cdot}78^{4}$ $\frac{13}{16}=3{\cdot}29^{1}$ $\frac{15}{16}=3{\cdot}79^{7}$

1 acre=0·4047 hectares

(0.176=1£) **£5.68**

1	5·68	41	232·88	81	460·08	121	687·28	161	914·48
2	11·36	42	238·56	82	465·76	122	692·96	162	920·16
3	17·04	43	244·24	83	471·44	123	698·64	163	925·84
4	22·72	44	249·92	84	477·12	124	704·32	164	931·52
5	28·40	45	255·60	85	482·80	125	710·00	165	937·20
6	34·08	46	261·28	86	488·48	126	715·68	166	942·88
7	39·76	47	266·96	87	494·16	127	721·36	167	948·56
8	45·44	48	272·64	88	499·84	128	727·04	168	954·24
9	51·12	49	278·32	89	505·52	129	732·72	169	959·92
10	56·80	**50**	284·00	**90**	511·20	**130**	738·40	**170**	965·60
11	62·48	51	289·68	91	516·88	131	744·08	171	971·28
12	68·16	52	295·36	92	522·56	132	749·76	172	976·96
13	73·84	53	301·04	93	528·24	133	755·44	173	982·64
14	79·52	54	306·72	94	533·92	134	761·12	174	988·32
15	85·20	55	312·40	95	539·60	135	766·80	175	994·00
16	90·88	56	318·08	96	545·28	136	772·48	176	999·68
17	96·56	57	323·76	97	550·96	137	778·16	177	1005·36
18	102·24	58	329·44	98	556·64	138	783·84	178	1011·04
19	107·92	59	335·12	99	562·32	139	789·52	179	1016·72
20	113·60	**60**	340·80	**100**	568·00	**140**	795·20	**180**	1022·40
21	119·28	61	346·48	101	573·68	141	800·88	181	1028·08
22	124·96	62	352·16	102	579·36	142	806·56	182	1033·76
23	130·64	63	357·84	103	585·04	143	812·24	183	1039·44
24	136·32	64	363·52	104	590·72	144	817·92	184	1045·12
25	142·00	65	369·20	105	596·40	145	823·60	185	1050·80
26	147·68	66	374·88	106	602·08	146	829·28	186	1056·48
27	153·36	67	380·56	107	607·76	147	834·96	187	1062·16
28	159·04	68	386·24	108	613·44	148	840·64	188	1067·84
29	164·72	69	391·92	109	619·12	149	846·32	189	1073·52
30	170·40	**70**	397·60	**110**	624·80	**150**	852·00	**190**	1079·20
31	176·08	71	403·28	111	630·48	151	857·68	191	1084·88
32	181·76	72	408·96	112	636·16	152	863·36	192	1090·56
33	187·44	73	414·64	113	641·84	153	869·04	193	1096·24
34	193·12	74	420·32	114	647·52	154	874·72	194	1101·92
35	198·80	75	426·00	115	653·20	155	880·40	195	1107·60
36	204·48	76	431·68	116	658·88	156	886·08	196	1113·28
37	210·16	77	437·36	117	664·56	157	891·76	197	1118·96
38	215·84	78	443·04	118	670·24	158	897·44	198	1124·64
39	221·52	79	448·72	119	675·92	159	903·12	199	1130·32
40	227·20	**80**	454·40	**120**	681·60	**160**	908·80	**200**	1136·00

$\frac{1}{16}=0{\cdot}35^5$ $\frac{1}{9}=0{\cdot}63^1$ $\frac{1}{8}=0{\cdot}71^0$ $\frac{1}{6}=0{\cdot}94^7$ $\frac{1}{5}=1{\cdot}13^6$ $\frac{1}{4}=1{\cdot}42^0$ $\frac{1}{3}=1{\cdot}89^3$

$\frac{3}{8}=2{\cdot}13^0$ $\frac{1}{2}=2{\cdot}84^0$ $\frac{5}{8}=3{\cdot}55^0$ $\frac{3}{4}=4{\cdot}26^0$ $\frac{7}{8}=4{\cdot}97^0$ $\frac{2}{3}=3{\cdot}78^7$ $\frac{5}{6}=4{\cdot}73^3$

$\frac{3}{16}=1{\cdot}06^5$ $\frac{5}{16}=1{\cdot}77^5$ $\frac{7}{16}=2{\cdot}48^5$ $\frac{9}{16}=3{\cdot}19^5$ $\frac{11}{16}=3{\cdot}90^5$ $\frac{13}{16}=4{\cdot}61^5$ $\frac{15}{16}=5{\cdot}32^5$

1 pint=0·568 litres

(0.164=1£) **£6.10**

1	6·10	41	250·10	81	494·10	121	738·10	161	982·10
2	12·20	42	256·20	82	500·20	122	744·20	162	988·20
3	18·30	43	262·30	83	506·30	123	750·30	163	994·30
4	24·40	44	268·40	84	512·40	124	756·40	164	1000·40
5	30·50	45	274·50	85	518·50	125	762·50	165	1006·50
6	36·60	46	280·60	86	524·60	126	768·60	166	1012·60
7	42·70	47	286·70	87	530·70	127	774·70	167	1018·70
8	48·80	48	292·80	88	536·80	128	780·80	168	1024·80
9	54·90	49	298·90	89	542·90	129	786·90	169	1030·90
10	61·00	**50**	305·00	**90**	549·00	**130**	793·00	**170**	1037·00
11	67·10	51	311·10	91	555·10	131	799·10	171	1043·10
12	73·20	52	317·20	92	561·20	132	805·20	172	1049·20
13	79·30	53	323·30	93	567·30	133	811·30	173	1055·30
14	85·40	54	329·40	94	573·40	134	817·40	174	1061·40
15	91·50	55	335·50	95	579·50	135	823·50	175	1067·50
16	97·60	56	341·60	96	585·60	136	829·60	176	1073·60
17	103·70	57	347·70	97	591·70	137	835·70	177	1079·70
18	109·80	58	353·80	98	597·80	138	841·80	178	1085·80
19	115·90	59	359·90	99	603·90	139	847·90	179	1091·90
20	122·00	**60**	366·00	**100**	610·00	**140**	854·00	**180**	1098·00
21	128·10	61	372·10	101	616·10	141	860·10	181	1104·10
22	134·20	62	378·20	102	622·20	142	866·20	182	1110·20
23	140·30	63	384·30	103	628·30	143	872·30	183	1116·30
24	146·40	64	390·40	104	634·40	144	878·40	184	1122·40
25	152·50	65	396·50	105	640·50	145	884·50	185	1128·50
26	158·60	66	402·60	106	646·60	146	890·60	186	1134·60
27	164·70	67	408·70	107	652·70	147	896·70	187	1140·70
28	170·80	68	414·80	108	658·80	148	902·80	188	1146·80
29	176·90	69	420·90	109	664·90	149	908·90	189	1152·90
30	183·00	**70**	427·00	**110**	671·00	**150**	915·00	**190**	1159·00
31	189·10	71	433·10	111	677·10	151	921·10	191	1165·10
32	195·20	72	439·20	112	683·20	152	927·20	192	1171·20
33	201·30	73	445·30	113	689·30	153	933·30	193	1177·30
34	207·40	74	451·40	114	695·40	154	939·40	194	1183·40
35	213·50	75	457·50	115	701·50	155	945·50	195	1189·50
36	219·60	76	463·60	116	707·60	156	951·60	196	1195·60
37	225·70	77	469·70	117	713·70	157	957·70	197	1201·70
38	231·80	78	475·80	118	719·80	158	963·80	198	1207·80
39	237·90	79	481·90	119	725·90	159	969·90	199	1213·90
40	244·00	**80**	488·00	**120**	732·00	**160**	976·00	**200**	1220·00

$\frac{1}{16}=0{\cdot}38^{1}$ $\frac{1}{9}=0{\cdot}67^{8}$ $\frac{1}{8}=0{\cdot}76^{3}$ $\frac{1}{6}=1{\cdot}01^{7}$ $\frac{1}{5}=1{\cdot}22^{0}$ $\frac{1}{4}=1{\cdot}52^{5}$ $\frac{1}{3}=2{\cdot}03^{3}$

$\frac{3}{8}=2{\cdot}28^{8}$ $\frac{1}{2}=3{\cdot}05^{0}$ $\frac{5}{8}=3{\cdot}81^{3}$ $\frac{3}{4}=4{\cdot}57^{5}$ $\frac{7}{8}=5{\cdot}33^{8}$ $\frac{2}{3}=4{\cdot}06^{7}$ $\frac{5}{6}=5{\cdot}08^{3}$

$\frac{3}{16}=1{\cdot}14^{4}$ $\frac{5}{16}=1{\cdot}90^{6}$ $\frac{7}{16}=2{\cdot}66^{9}$ $\frac{9}{16}=3{\cdot}43^{1}$ $\frac{11}{16}=4{\cdot}19^{4}$ $\frac{13}{16}=4{\cdot}95^{6}$ $\frac{15}{16}=5{\cdot}71^{9}$

1 cubic centimetre=0·0610 cubic inches

(0.161=1£) **£6.21**

1	6·21	41	254·61	81	503·01	121	751·41	161	999·81
2	12·42	42	260·82	82	509·22	122	757·62	162	1006·02
3	18·63	43	267·03	83	515·43	123	763·83	163	1012·23
4	24·84	44	273·24	84	521·64	124	770·04	164	1018·44
5	31·05	45	279·45	85	527·85	125	776·25	165	1024·65
6	37·26	46	285·66	86	534·06	126	782·46	166	1030·86
7	43·47	47	291·87	87	540·27	127	788·67	167	1037·07
8	49·68	48	298·08	88	546·48	128	794·88	168	1043·28
9	55·89	49	304·29	89	552·69	129	801·09	169	1049·49
10	62·10	**50**	310·50	**90**	558·90	**130**	807·30	**170**	1055·70
11	68·31	51	316·71	91	565·11	131	813·51	171	1061·91
12	74·52	52	322·92	92	571·32	132	819·72	172	1068·12
13	80·73	53	329·13	93	577·53	133	825·93	173	1074·33
14	86·94	54	335·34	94	583·74	134	832·14	174	1080·54
15	93·15	55	341·55	95	589·95	135	838·35	175	1086·75
16	99·36	56	347·76	96	596·16	136	844·56	176	1092·96
17	105·57	57	353·97	97	602·37	137	850·77	177	1099·17
18	111·78	58	360·18	98	608·58	138	856·98	178	1105·38
19	117·99	59	366·39	99	614·79	139	863·19	179	1111·59
20	124·20	**60**	372·60	**100**	621·00	**140**	869·40	**180**	1117·80
21	130·41	61	378·81	101	627·21	141	875·61	181	1124·01
22	136·62	62	385·02	102	633·42	142	881·82	182	1130·22
23	142·83	63	391·23	103	639·63	143	888·03	183	1136·43
24	149·04	64	397·44	104	645·84	144	894·24	184	1142·64
25	155·25	65	403·65	105	652·05	145	900·45	185	1148·85
26	161·46	66	409·86	106	658·26	146	906·66	186	1155·06
27	167·67	67	416·07	107	664·47	147	912·87	187	1161·27
28	173·88	68	422·28	108	670·68	148	919·08	188	1167·48
29	180·09	69	428·49	109	676·89	149	925·29	189	1173·69
30	186·30	**70**	434·70	**110**	683·10	**150**	931·50	**190**	1179·90
31	192·51	71	440·91	111	689·31	151	937·71	191	1186·11
32	198·72	72	447·12	112	695·52	152	943·92	192	1192·32
33	204·93	73	453·33	113	701·73	153	950·13	193	1198·53
34	211·14	74	459·54	114	707·94	154	956·34	194	1204·74
35	217·35	75	465·75	115	714·15	155	962·55	195	1210·95
36	223·56	76	471·96	116	720·36	156	968·76	196	1217·16
37	229·77	77	478·17	117	726·57	157	974·97	197	1223·37
38	235·98	78	484·38	118	732·78	158	981·18	198	1229·58
39	242·19	79	490·59	119	738·99	159	987·39	199	1235·79
40	248·40	**80**	496·80	**120**	745·20	**160**	993·60	**200**	1242·00

$\frac{1}{16}=0{\cdot}38^{8}$ $\frac{1}{9}=0{\cdot}69^{0}$ $\frac{1}{8}=0{\cdot}77^{6}$ $\frac{1}{6}=1{\cdot}03^{5}$ $\frac{1}{5}=1{\cdot}24^{2}$ $\frac{1}{4}=1{\cdot}55^{3}$ $\frac{1}{3}=2{\cdot}07^{0}$

$\frac{3}{8}=2{\cdot}32^{9}$ $\frac{1}{2}=3{\cdot}10^{5}$ $\frac{5}{8}=3{\cdot}88^{1}$ $\frac{3}{4}=4{\cdot}65^{8}$ $\frac{7}{8}=5{\cdot}43^{4}$ $\frac{2}{3}=4{\cdot}14^{0}$ $\frac{5}{6}=5{\cdot}17^{5}$

$\frac{3}{16}=1{\cdot}16^{4}$ $\frac{5}{16}=1{\cdot}94^{1}$ $\frac{7}{16}=2{\cdot}71^{7}$ $\frac{9}{16}=3{\cdot}49^{3}$ $\frac{11}{16}=4{\cdot}26^{9}$ $\frac{13}{16}=5{\cdot}04^{6}$ $\frac{15}{16}=5{\cdot}82^{2}$

1 kilometre=0·6213 miles

(0.161=1£) **£6.23**

1	6·23	41	255·43	81	504·63	121	753·83	161	1003·03
2	12·46	42	261·66	82	510·86	122	760·06	162	1009·26
3	18·69	43	267·89	83	517·09	123	766·29	163	1015·49
4	24·92	44	274·12	84	523·32	124	772·52	164	1021·72
5	31·15	45	280·35	85	529·55	125	778·75	165	1027·95
6	37·38	46	286·58	86	535·78	126	784·98	166	1034·18
7	43·61	47	292·81	87	542·01	127	791·21	167	1040·41
8	49·84	48	299·04	88	548·24	128	797·44	168	1046·64
9	56·07	49	305·27	89	554·47	129	803·67	169	1052·87
10	62·30	**50**	311·50	**90**	560·70	**130**	809·90	**170**	1059·10
11	68·53	51	317·73	91	566·93	131	816·13	171	1065·33
12	74·76	52	323·96	92	573·16	132	822·36	172	1071·56
13	80·99	53	330·19	93	579·39	133	828·59	173	1077·79
14	87·22	54	336·42	94	585·62	134	834·82	174	1084·02
15	93·45	55	342·65	95	591·85	135	841·05	175	1090·25
16	99·68	56	348·88	96	598·08	136	847·28	176	1096·48
17	105·91	57	355·11	97	604·31	137	853·51	177	1102·71
18	112·14	58	361·34	98	610·54	138	859·74	178	1108·94
19	118·37	59	367·57	99	616·77	139	865·97	179	1115·17
20	124·60	**60**	373·80	**100**	623·00	**140**	872·20	**180**	1121·40
21	130·83	61	380·03	101	629·23	141	878·43	181	1127·63
22	137·06	62	386·26	102	635·46	142	884·66	182	1133·86
23	143·29	63	392·49	103	641·69	143	890·89	183	1140·09
24	149·52	64	398·72	104	647·92	144	897·12	184	1146·32
25	155·75	65	404·95	105	654·15	145	903·35	185	1152·55
26	161·98	66	411·18	106	660·38	146	909·58	186	1158·78
27	168·21	67	417·41	107	666·61	147	915·81	187	1165·01
28	174·44	68	423·64	108	672·84	148	922·04	188	1171·24
29	180·67	69	429·87	109	679·07	149	928·27	189	1177·47
30	186·90	**70**	436·10	**110**	685·30	**150**	934·50	**190**	1183·70
31	193·13	71	442·33	111	691·53	151	940·73	191	1189·93
32	199·36	72	448·56	112	697·76	152	946·96	192	1196·16
33	205·59	73	454·79	113	703·99	153	953·19	193	1202·39
34	211·82	74	461·02	114	710·22	154	959·42	194	1208·62
35	218·05	75	467·25	115	716·45	155	965·65	195	1214·85
36	224·28	76	473·48	116	722·68	156	971·88	196	1221·08
37	230·51	77	479·71	117	728·91	157	978·11	197	1227·31
38	236·74	78	485·94	118	735·14	158	984·34	198	1233·54
39	242·97	79	492·17	119	741·37	159	990·57	199	1239·77
40	249·20	**80**	498·40	**120**	747·60	**160**	996·80	**200**	1246·00

$\frac{1}{16}=0{\cdot}38^{9}$ $\frac{1}{9}=0{\cdot}69^{2}$ $\frac{1}{8}=0{\cdot}77^{9}$ $\frac{1}{6}=1{\cdot}03^{8}$ $\frac{1}{5}=1{\cdot}24^{6}$ $\frac{1}{4}=1{\cdot}55^{8}$ $\frac{1}{3}=2{\cdot}07^{7}$

$\frac{3}{8}=2{\cdot}33^{6}$ $\frac{1}{2}=3{\cdot}11^{5}$ $\frac{5}{8}=3{\cdot}89^{4}$ $\frac{3}{4}=4{\cdot}67^{3}$ $\frac{7}{8}=5{\cdot}45^{1}$ $\frac{2}{3}=4{\cdot}15^{3}$ $\frac{5}{6}=5{\cdot}19^{2}$

$\frac{3}{16}=1{\cdot}16^{8}$ $\frac{5}{16}=1{\cdot}94^{7}$ $\frac{7}{16}=2{\cdot}72^{6}$ $\frac{9}{16}=3{\cdot}50^{4}$ $\frac{11}{16}=4{\cdot}28^{3}$ $\frac{13}{16}=5{\cdot}06^{2}$ $\frac{15}{16}=5{\cdot}84^{1}$

1 cubic foot=6·23 gallons 1 ounce per gallon=6·236 grammes per litre

(0.160=1£) **£6.25**

1	6·25	41	256·25	81	506·25	121	756·25	161	1006·25
2	12·50	42	262·50	82	512·50	122	762·50	162	1012·50
3	18·75	43	268·75	83	518·75	123	768·75	163	1018·75
4	25·00	44	275·00	84	525·00	124	775·00	164	1025·00
5	31·25	45	281·25	85	531·25	125	781·25	165	1031·25
6	37·50	46	287·50	86	537·50	126	787·50	166	1037·50
7	43·75	47	293·75	87	543·75	127	793·75	167	1043·75
8	50·00	48	300·00	88	550·00	128	800·00	168	1050·00
9	56·25	49	306·25	89	556·25	129	806·25	169	1056·25
10	62·50	**50**	312·50	**90**	562·50	**130**	812·50	**170**	1062·50
11	68·75	51	318·75	91	568·75	131	818·75	171	1068·75
12	75·00	52	325·00	92	575·00	132	825·00	172	1075·00
13	81·25	53	331·25	93	581·25	133	831·25	173	1081·25
14	87·50	54	337·50	94	587·50	134	837·50	174	1087·50
15	93·75	55	343·75	95	593·75	135	843·75	175	1093·75
16	100·00	56	350·00	96	600·00	136	850·00	176	1100·00
17	106·25	57	356·25	97	606·25	137	856·25	177	1106·25
18	112·50	58	362·50	98	612·50	138	862·50	178	1112·50
19	118·75	59	368·75	99	618·75	139	868·75	179	1118·75
20	125·00	**60**	375·00	**100**	625·00	**140**	875·00	**180**	1125·00
21	131·25	61	381·25	101	631·25	141	881·25	181	1131·25
22	137·50	62	387·50	102	637·50	142	887·50	182	1137·50
23	143·75	63	393·75	103	643·75	143	893·75	183	1143·75
24	150·00	64	400·00	104	650·00	144	900·00	184	1150·00
25	156·25	65	406·25	105	656·25	145	906·25	185	1156·25
26	162·50	66	412·50	106	662·50	146	912·50	186	1162·50
27	168·75	67	418·75	107	668·75	147	918·75	187	1168·75
28	175·00	68	425·00	108	675·00	148	925·00	188	1175·00
29	181·25	69	431·25	109	681·25	149	931·25	189	1181·25
30	187·50	**70**	437·50	**110**	687·50	**150**	937·50	**190**	1187·50
31	193·75	71	443·75	111	693·75	151	943·75	191	1193·75
32	200·00	72	450·00	112	700·00	152	950·00	192	1200·00
33	206·25	73	456·25	113	706·25	153	956·25	193	1206·25
34	212·50	74	462·50	114	712·50	154	962·50	194	1212·50
35	218·75	75	468·75	115	718·75	155	968·75	195	1218·75
36	225·00	76	475·00	116	725·00	156	975·00	196	1225·00
37	231·25	77	481·25	117	731·25	157	981·25	197	1231·25
38	237·50	78	487·50	118	737·50	158	987·50	198	1237·50
39	243·75	79	493·75	119	743·75	159	993·75	199	1243·75
40	250·00	**80**	500·00	**120**	750·00	**160**	1000·00	**200**	1250·00

$\frac{1}{16}=0{\cdot}39^{1}$ $\frac{1}{9}=0{\cdot}69^{4}$ $\frac{1}{8}=0{\cdot}78^{1}$ $\frac{1}{6}=1{\cdot}04^{2}$ $\frac{1}{5}=1{\cdot}25^{0}$ $\frac{1}{4}=1{\cdot}56^{3}$ $\frac{1}{3}=2{\cdot}08^{3}$

$\frac{3}{8}=2{\cdot}34^{4}$ $\frac{1}{2}=3{\cdot}12^{5}$ $\frac{5}{8}=3{\cdot}90^{6}$ $\frac{3}{4}=4{\cdot}68^{8}$ $\frac{7}{8}=5{\cdot}46^{9}$ $\frac{2}{3}=4{\cdot}16^{7}$ $\frac{5}{6}=5{\cdot}20^{8}$

$\frac{3}{16}=1{\cdot}17^{2}$ $\frac{5}{16}=1{\cdot}95^{3}$ $\frac{7}{16}=2{\cdot}73^{4}$ $\frac{9}{16}=3{\cdot}51^{6}$ $\frac{11}{16}=4{\cdot}29^{7}$ $\frac{13}{16}=5{\cdot}07^{8}$ $\frac{15}{16}=5{\cdot}85^{9}$

(0.142=1£) **£7.03**

1	7·03	41	288·23	81	569·43	121	850·63	161	1131·83
2	14·06	42	295·26	82	576·46	122	857·66	162	1138·86
3	21·09	43	302·29	83	583·49	123	864·69	163	1145·89
4	28·12	44	309·32	84	590·52	124	871·72	164	1152·92
5	35·15	45	316·35	85	597·55	125	878·75	165	1159·95
6	42·18	46	323·38	86	604·58	126	885·78	166	1166·98
7	49·21	47	330·41	87	611·61	127	892·81	167	1174·01
8	56·24	48	337·44	88	618·64	128	899·84	168	1181·04
9	63·27	49	344·47	89	625·67	129	906·87	169	1188·07
10	70·30	**50**	351·50	**90**	632·70	**130**	913·90	**170**	1195·10
11	77·33	51	358·53	91	639·73	131	920·93	171	1202·13
12	84·36	52	365·56	92	646·76	132	927·96	172	1209·16
13	91·39	53	372·59	93	653·79	133	934·99	173	1216·19
14	98·42	54	379·62	94	660·82	134	942·02	174	1223·22
15	105·45	55	386·65	95	667·85	135	949·05	175	1230·25
16	112·48	56	393·68	96	674·88	136	956·08	176	1237·28
17	119·51	57	400·71	97	681·91	137	963·11	177	1244·31
18	126·54	58	407·74	98	688·94	138	970·14	178	1251·34
19	133·57	59	414·77	99	695·97	139	977·17	179	1258·37
20	140·60	**60**	421·80	**100**	703·00	**140**	984·20	**180**	1265·40
21	147·63	61	428·83	101	710·03	141	991·23	181	1272·43
22	154·66	62	435·86	102	717·06	142	998·26	182	1279·46
23	161·69	63	442·89	103	724·09	143	1005·29	183	1286·49
24	168·72	64	449·92	104	731·12	144	1012·32	184	1293·52
25	175·75	65	456·95	105	738·15	145	1019·35	185	1300·55
26	182·78	66	463·98	106	745·18	146	1026·38	186	1307·58
27	189·81	67	471·01	107	752·21	147	1033·41	187	1314·61
28	196·84	68	478·04	108	759·24	148	1040·44	188	1321·64
29	203·87	69	485·07	109	766·27	149	1047·47	189	1328·67
30	210·90	**70**	492·10	**110**	773·30	**150**	1054·50	**190**	1335·70
31	217·93	71	499·13	111	780·33	151	1061·53	191	1342·73
32	224·96	72	506·16	112	787·36	152	1068·56	192	1349·76
33	231·99	73	513·19	113	794·39	153	1075·59	193	1356·79
34	239·02	74	520·22	114	801·42	154	1082·62	194	1363·82
35	246·05	75	527·25	115	808·45	155	1089·65	195	1370·85
36	253·08	76	534·28	116	815·48	156	1096·68	196	1377·88
37	260·11	77	541·31	117	822·51	157	1103·71	197	1384·91
38	267·14	78	548·34	118	829·54	158	1110·74	198	1391·94
39	274·17	79	555·37	119	836·57	159	1117·77	199	1398·97
40	281·20	**80**	562·40	**120**	843·60	**160**	1124·80	**200**	1406·00

$\frac{1}{16}=0{\cdot}43^{9}$ $\frac{1}{9}=0{\cdot}78^{1}$ $\frac{1}{8}=0{\cdot}87^{9}$ $\frac{1}{6}=1{\cdot}17^{2}$ $\frac{1}{5}=1{\cdot}40^{6}$ $\frac{1}{4}=1{\cdot}75^{8}$ $\frac{1}{3}=2{\cdot}34^{3}$

$\frac{3}{8}=2{\cdot}63^{6}$ $\frac{1}{2}=3{\cdot}51^{5}$ $\frac{5}{8}=4{\cdot}39^{4}$ $\frac{3}{4}=5{\cdot}27^{3}$ $\frac{7}{8}=6{\cdot}15^{1}$ $\frac{2}{3}=4{\cdot}68^{7}$ $\frac{5}{6}=5{\cdot}85^{8}$

$\frac{3}{16}=1{\cdot}31^{8}$ $\frac{5}{16}=2{\cdot}19^{7}$ $\frac{7}{16}=3{\cdot}07^{6}$ $\frac{9}{16}=3{\cdot}95^{4}$ $\frac{11}{16}=4{\cdot}83^{3}$ $\frac{13}{16}=5{\cdot}71^{2}$ $\frac{15}{16}=6{\cdot}59^{1}$

(0.134=1£) **£7.46**

1	7·46	41	305·86	81	604·26	121	902·66	161	1201·06
2	14·92	42	313·32	82	611·72	122	910·12	162	1208·52
3	22·38	43	320·78	83	619·18	123	917·58	163	1215·98
4	29·84	44	328·24	84	626·64	124	925·04	164	1223·44
5	37·30	45	335·70	85	634·10	125	932·50	165	1230·90
6	44·76	46	343·16	86	641·56	126	939·96	166	1238·36
7	52·22	47	350·62	87	649·02	127	947·42	167	1245·82
8	59·68	48	358·08	88	656·48	128	954·88	168	1253·28
9	67·14	49	365·54	89	663·94	129	962·34	169	1260·74
10	74·60	**50**	373·00	**90**	671·40	**130**	969·80	**170**	1268·20
11	82·06	51	380·46	91	678·86	131	977·26	171	1275·66
12	89·52	52	387·92	92	686·32	132	984·72	172	1283·12
13	96·98	53	395·38	93	693·78	133	992·18	173	1290·58
14	104·44	54	402·84	94	701·24	134	999·64	174	1298·04
15	111·90	55	410·30	95	708·70	135	1007·10	175	1305·50
16	119·36	56	417·76	96	716·16	136	1014·56	176	1312·96
17	126·82	57	425·22	97	723·62	137	1022·02	177	1320·42
18	134·28	58	432·68	98	731·08	138	1029·48	178	1327·88
19	141·74	59	440·14	99	738·54	139	1036·94	179	1335·34
20	149·20	**60**	447·60	**100**	746·00	**140**	1044·40	**180**	1342·80
21	156·66	61	455·06	101	753·46	141	1051·86	181	1350·26
22	164·12	62	462·52	102	760·92	142	1059·32	182	1357·72
23	171·58	63	469·98	103	768·38	143	1066·78	183	1365·18
24	179·04	64	477·44	104	775·84	144	1074·24	184	1372·64
25	186·50	65	484·90	105	783·30	145	1081·70	185	1380·10
26	193·96	66	492·36	106	790·76	146	1089·16	186	1387·56
27	201·42	67	499·82	107	798·22	147	1096·62	187	1395·02
28	208·88	68	507·28	108	805·68	148	1104·08	188	1402·48
29	216·34	69	514·74	109	813·14	149	1111·54	189	1409·94
30	223·80	**70**	522·20	**110**	820·60	**150**	1119·00	**190**	1417·40
31	231·26	71	529·66	111	828·06	151	1126·46	191	1424·86
32	238·72	72	537·12	112	835·52	152	1133·92	192	1432·32
33	246·18	73	544·58	113	842·98	153	1141·38	193	1439·78
34	253·64	74	552·04	114	850·44	154	1148·84	194	1447·24
35	261·10	75	559·50	115	857·90	155	1156·30	195	1454·70
36	268·56	76	566·96	116	865·36	156	1163·76	196	1462·16
37	276·02	77	574·42	117	872·82	157	1171·22	197	1469·62
38	283·48	78	581·88	118	880·28	158	1178·68	198	1477·08
39	290·94	79	589·34	119	887·74	159	1186·14	199	1484·54
40	298·40	**80**	596·80	**120**	895·20	**160**	1193·60	**200**	1492·00

$\frac{1}{16}=0{\cdot}46^{6}$ $\frac{1}{9}=0{\cdot}82^{9}$ $\frac{1}{8}=0{\cdot}93^{3}$ $\frac{1}{6}=1{\cdot}24^{3}$ $\frac{1}{5}=1{\cdot}49^{2}$ $\frac{1}{4}=1{\cdot}86^{5}$ $\frac{1}{3}=2{\cdot}48^{7}$

$\frac{3}{8}=2{\cdot}79^{8}$ $\frac{1}{2}=3{\cdot}73^{0}$ $\frac{5}{8}=4{\cdot}66^{3}$ $\frac{3}{4}=5{\cdot}59^{5}$ $\frac{7}{8}=6{\cdot}52^{8}$ $\frac{2}{3}=4{\cdot}97^{3}$ $\frac{5}{6}=6{\cdot}21^{7}$

$\frac{3}{16}=1{\cdot}39^{9}$ $\frac{5}{16}=2{\cdot}33^{1}$ $\frac{7}{16}=3{\cdot}26^{4}$ $\frac{9}{16}=4{\cdot}19^{6}$ $\frac{11}{16}=5{\cdot}12^{9}$ $\frac{13}{16}=6{\cdot}06^{1}$ $\frac{15}{16}=6{\cdot}99^{4}$

1 British Horse power=746 Watts

(0.131=1£) **£7.65**

1	7·65	41	313·65	81	619·65	121	925·65	161	1231·65
2	15·30	42	321·30	82	627·30	122	933·30	162	1239·30
3	22·95	43	328·95	83	634·95	123	940·95	163	1246·95
4	30·60	44	336·60	84	642·60	124	948·60	164	1254·60
5	38·25	45	344·25	85	650·25	125	956·25	165	1262·25
6	45·90	46	351·90	86	657·90	126	963·90	166	1269·90
7	53·55	47	359·55	87	665·55	127	971·55	167	1277·55
8	61·20	48	367·20	88	673·20	128	979·20	168	1285·20
9	68·85	49	374·85	89	680·85	129	986·85	169	1292·85
10	76·50	**50**	382·50	**90**	688·50	**130**	994·50	**170**	1300·50
11	84·15	51	390·15	91	696·15	131	1002·15	171	1308·15
12	91·80	52	397·80	92	703·80	132	1009·80	172	1315·80
13	99·45	53	405·45	93	711·45	133	1017·45	173	1323·45
14	107·10	54	413·10	94	719·10	134	1025·10	174	1331·10
15	114·75	55	420·75	95	726·75	135	1032·75	175	1338·75
16	122·40	56	428·40	96	734·40	136	1040·40	176	1346·40
17	130·05	57	436·05	97	742·05	137	1048·05	177	1354·05
18	137·70	58	443·70	98	749·70	138	1055·70	178	1361·70
19	145·35	59	451·35	99	757·35	139	1063·35	179	1369·35
20	153·00	**60**	459·00	**100**	765·00	**140**	1071·00	**180**	1377·00
21	160·65	61	466·65	101	772·65	141	1078·65	181	1384·65
22	168·30	62	474·30	102	780·30	142	1086·30	182	1392·30
23	175·95	63	481·95	103	787·95	143	1093·95	183	1399·95
24	183·60	64	489·60	104	795·60	144	1101·60	184	1407·60
25	191·25	65	497·25	105	803·25	145	1109·25	185	1415·25
26	198·90	66	504·90	106	810·90	146	1116·90	186	1422·90
27	206·55	67	512·55	107	818·55	147	1124·55	187	1430·55
28	214·20	68	520·20	108	826·20	148	1132·20	188	1438·20
29	221·85	69	527·85	109	833·85	149	1139·85	189	1445·85
30	229·50	**70**	535·50	**110**	841·50	**150**	1147·50	**190**	1453·50
31	237·15	71	543·15	111	849·15	151	1155·15	191	1461·15
32	244·80	72	550·80	112	856·80	152	1162·80	192	1468·80
33	252·45	73	558·45	113	864·45	153	1170·45	193	1476·45
34	260·10	74	566·10	114	872·10	154	1178·10	194	1484·10
35	267·75	75	573·75	115	879·75	155	1185·75	195	1491·75
36	275·40	76	581·40	116	887·40	156	1193·40	196	1499·40
37	283·05	77	589·05	117	895·05	157	1201·05	197	1507·05
38	290·70	78	596·70	118	902·70	158	1208·70	198	1514·70
39	298·35	79	604·35	119	910·35	159	1216·35	199	1522·35
40	306·00	**80**	612·00	**120**	918·00	**160**	1224·00	**200**	1530·00

$\frac{1}{16}=0\cdot47^{8}$ $\frac{1}{9}=0\cdot85^{0}$ $\frac{1}{8}=0\cdot95^{6}$ $\frac{1}{6}=1\cdot27^{5}$ $\frac{1}{5}=1\cdot53^{0}$ $\frac{1}{4}=1\cdot91^{3}$ $\frac{1}{3}=2\cdot55^{0}$

$\frac{3}{8}=2\cdot86^{9}$ $\frac{1}{2}=3\cdot82^{5}$ $\frac{5}{8}=4\cdot78^{1}$ $\frac{3}{4}=5\cdot73^{8}$ $\frac{7}{8}=6\cdot69^{4}$ $\frac{2}{3}=5\cdot10^{0}$ $\frac{5}{6}=6\cdot37^{5}$

$\frac{3}{16}=1\cdot43^{4}$ $\frac{5}{16}=2\cdot39^{1}$ $\frac{7}{16}=3\cdot34^{7}$ $\frac{9}{16}=4\cdot30^{3}$ $\frac{11}{16}=5\cdot25^{9}$ $\frac{13}{16}=6\cdot21^{6}$ $\frac{15}{16}=7\cdot17^{2}$

1 cubic yard=0·765 cubic metres

(0.120=1£) **£8.36**

1	8·36	41	342·76	81	677·16	121	1011·56	161	1345·96
2	16·72	42	351·12	82	685·52	122	1019·92	162	1354·32
3	25·08	43	359·48	83	693·88	123	1028·28	163	1362·68
4	33·44	44	367·84	84	702·24	124	1036·64	164	1371·04
5	41·80	45	376·20	85	710·60	125	1045·00	165	1379·40
6	50·16	46	384·56	86	718·96	126	1053·36	166	1387·76
7	58·52	47	392·92	87	727·32	127	1061·72	167	1396·12
8	66·88	48	401·28	88	735·68	128	1070·08	168	1404·48
9	75·24	49	409·64	89	744·04	129	1078·44	169	1412·84
10	83·60	**50**	418·00	**90**	752·40	**130**	1086·80	**170**	1421·20
11	91·96	51	426·36	91	760·76	131	1095·16	171	1429·56
12	100·32	52	434·72	92	769·12	132	1103·52	172	1437·92
13	108·68	53	443·08	93	777·48	133	1111·88	173	1446·28
14	117·04	54	451·44	94	785·84	134	1120·24	174	1454·64
15	125·40	55	459·80	95	794·20	135	1128·60	175	1463·00
16	133·76	56	468·16	96	802·56	136	1136·96	176	1471·36
17	142·12	57	476·52	97	810·92	137	1145·32	177	1479·72
18	150·48	58	484·88	98	819·28	138	1153·68	178	1488·08
19	158·84	59	493·24	99	827·64	139	1162·04	179	1496·44
20	167·20	**60**	501·60	**100**	836·00	**140**	1170·40	**180**	1504·80
21	175·56	61	509·96	101	844·36	141	1178·76	181	1513·16
22	183·92	62	518·32	102	852·72	142	1187·12	182	1521·52
23	192·28	63	526·68	103	861·08	143	1195·48	183	1529·88
24	200·64	64	535·04	104	869·44	144	1203·84	184	1538·24
25	209·00	65	543·40	105	877·80	145	1212·20	185	1546·60
26	217·36	66	551·76	106	886·16	146	1220·56	186	1554·96
27	225·72	67	560·12	107	894·52	147	1228·92	187	1563·32
28	234·08	68	568·48	108	902·88	148	1237·28	188	1571·68
29	242·44	69	576·84	109	911·24	149	1245·64	189	1580·04
30	250·80	**70**	585·20	**110**	919·60	**150**	1254·00	**190**	1588·40
31	259·16	71	593·56	111	927·96	151	1262·36	191	1596·76
32	267·52	72	601·92	112	936·32	152	1270·72	192	1605·12
33	275·88	73	610·28	113	944·68	153	1279·08	193	1613·48
34	284·24	74	618·64	114	953·04	154	1287·44	194	1621·84
35	292·60	75	627·00	115	961·40	155	1295·80	195	1630·20
36	300·96	76	635·36	116	969·76	156	1304·16	196	1638·56
37	309·32	77	643·72	117	978·12	157	1312·52	197	1646·92
38	317·68	78	652·08	118	986·48	158	1320·88	198	1655·28
39	326·04	79	660·44	119	994·84	159	1329·24	199	1663·64
40	334·40	**80**	668·80	**120**	1003·20	**160**	1337·60	**200**	1672·00

$\frac{1}{16}=0{\cdot}52^{3}$ $\frac{1}{9}=0{\cdot}92^{9}$ $\frac{1}{8}=1{\cdot}04^{5}$ $\frac{1}{6}=1{\cdot}39^{3}$ $\frac{1}{5}=1{\cdot}67^{2}$ $\frac{1}{4}=2{\cdot}09^{0}$ $\frac{1}{3}=2{\cdot}78^{7}$

$\frac{3}{8}=3{\cdot}13^{5}$ $\frac{1}{2}=4{\cdot}18^{0}$ $\frac{5}{8}=5{\cdot}22^{5}$ $\frac{3}{4}=6{\cdot}27^{0}$ $\frac{7}{8}=7{\cdot}31^{5}$ $\frac{2}{3}=5{\cdot}57^{3}$ $\frac{5}{6}=6{\cdot}96^{7}$

$\frac{3}{16}=1{\cdot}56^{8}$ $\frac{5}{16}=2{\cdot}61^{3}$ $\frac{7}{16}=3{\cdot}65^{8}$ $\frac{9}{16}=4{\cdot}70^{3}$ $\frac{11}{16}=5{\cdot}74^{8}$ $\frac{13}{16}=6{\cdot}79^{3}$ $\frac{15}{16}=7{\cdot}83^{8}$

1 square yard=0·836 square metres

(0.109=1£) £9.14

1	9·14	41	374·74	81	740·34	121	1105·94	161	1471·54
2	18·28	42	383·88	82	749·48	122	1115·08	162	1480·68
3	27·42	43	393·02	83	758·62	123	1124·22	163	1489·82
4	36·56	44	402·16	84	767·76	124	1133·36	164	1498·96
5	45·70	45	411·30	85	776·90	125	1142·50	165	1508·10
6	54·84	46	420·44	86	786·04	126	1151·64	166	1517·24
7	63·98	47	429·58	87	795·18	127	1160·78	167	1526·38
8	73·12	48	438·72	88	804·32	128	1169·92	168	1535·52
9	82·26	49	447·86	89	813·46	129	1179·06	169	1544·66
10	91·40	**50**	457·00	**90**	822·60	**130**	1188·20	**170**	1553·80
11	100·54	51	466·14	91	831·74	131	1197·34	171	1562·94
12	109·68	52	475·28	92	840·88	132	1206·48	172	1572·08
13	118·82	53	484·42	93	850·02	133	1215·62	173	1581·22
14	127·96	54	493·56	94	859·16	134	1224·76	174	1590·36
15	137·10	55	502·70	95	868·30	135	1233·90	175	1599·50
16	146·24	56	511·84	96	877·44	136	1243·04	176	1608·64
17	155·38	57	520·98	97	886·58	137	1252·18	177	1617·78
18	164·52	58	530·12	98	895·72	138	1261·32	178	1626·92
19	173·66	59	539·26	99	904·86	139	1270·46	179	1636·06
20	182·80	**60**	548·40	**100**	914·00	**140**	1279·60	**180**	1645·20
21	191·94	61	557·54	101	923·14	141	1288·74	181	1654·34
22	201·08	62	566·68	102	932·28	142	1297·88	182	1663·48
23	210·22	63	575·82	103	941·42	143	1307·02	183	1672·62
24	219·36	64	584·96	104	950·56	144	1316·16	184	1681·76
25	228·50	65	594·10	105	959·70	145	1325·30	185	1690·90
26	237·64	66	603·24	106	968·84	146	1334·44	186	1700·04
27	246·78	67	612·38	107	977·98	147	1343·58	187	1709·18
28	255·92	68	621·52	108	987·12	148	1352·72	188	1718·32
29	265·06	69	630·66	109	996·26	149	1361·86	189	1727·46
30	274·20	**70**	639·80	**110**	1005·40	**150**	1371·00	**190**	1736·60
31	283·34	71	648·94	111	1014·54	151	1380·14	191	1745·74
32	292·48	72	658·08	112	1023·68	152	1389·28	192	1754·88
33	301·62	73	667·22	113	1032·82	153	1398·42	193	1764·02
34	310·76	74	676·36	114	1041·96	154	1407·56	194	1773·16
35	319·90	75	685·50	115	1051·10	155	1416·70	195	1782·30
36	329·04	76	694·64	116	1060·24	156	1425·84	196	1791·44
37	338·18	77	703·78	117	1069·38	157	1434·98	197	1800·58
38	347·32	78	712·92	118	1078·52	158	1444·12	198	1809·72
39	356·46	79	722·06	119	1087·66	159	1453·26	199	1818·86
40	365·60	**80**	731·20	**120**	1096·80	**160**	1462·40	**200**	1828·00

$\frac{1}{16}$=0·57^{1} $\frac{1}{9}$=1·01^{6} $\frac{1}{8}$=1·14^{3} $\frac{1}{6}$=1·52^{3} $\frac{1}{5}$=1·82^{8} $\frac{1}{4}$=2·28^{5} $\frac{1}{3}$=3·04^{7}

$\frac{3}{8}$=3·42^{8} $\frac{1}{2}$=4·57^{0} $\frac{5}{8}$=5·71^{3} $\frac{3}{4}$=6·85^{5} $\frac{7}{8}$=7·99^{8} $\frac{2}{3}$=6·09^{3} $\frac{5}{6}$=7·61^{7}

$\frac{3}{16}$=1·71^{4} $\frac{5}{16}$=2·85^{6} $\frac{7}{16}$=3·99^{9} $\frac{9}{16}$=5·14^{1} $\frac{11}{16}$=6·28^{4} $\frac{13}{16}$=7·42^{6} $\frac{15}{16}$=8·56^{9}

1 yard=0·914 metres

(0.108=1£) **£9.29**

1	9·29	41	380·89	81	752·49	121	1124·09	161	1495·69
2	18·58	42	390·18	82	761·78	122	1133·38	162	1504·98
3	27·87	43	399·47	83	771·07	123	1142·67	163	1514·27
4	37·16	44	408·76	84	780·36	124	1151·96	164	1523·56
5	46·45	45	418·05	85	789·65	125	1161·25	165	1532·85
6	55·74	46	427·34	86	798·94	126	1170·54	166	1542·14
7	65·03	47	436·63	87	808·23	127	1179·83	167	1551·43
8	74·32	48	445·92	88	817·52	128	1189·12	168	1560·72
9	83·61	49	455·21	89	826·81	129	1198·41	169	1570·01
10	92·90	**50**	464·50	**90**	836·10	**130**	1207·70	**170**	1579·30
11	102·19	51	473·79	91	845·39	131	1216·99	171	1588·59
12	111·48	52	483·08	92	854·68	132	1226·28	172	1597·88
13	120·77	53	492·37	93	863·97	133	1235·57	173	1607·17
14	130·06	54	501·66	94	873·26	134	1244·86	174	1616·46
15	139·35	55	510·95	95	882·55	135	1254·15	175	1625·75
16	148·64	56	520·24	96	891·84	136	1263·44	176	1635·04
17	157·93	57	529·53	97	901·13	137	1272·73	177	1644·33
18	167·22	58	538·82	98	910·42	138	1282·02	178	1653·62
19	176·51	59	548·11	99	919·71	139	1291·31	179	1662·91
20	185·80	**60**	557·40	**100**	929·00	**140**	1300·60	**180**	1672·20
21	195·09	61	566·69	101	938·29	141	1309·89	181	1681·49
22	204·38	62	575·98	102	947·58	142	1319·18	182	1690·78
23	213·67	63	585·27	103	956·87	143	1328·47	183	1700·07
24	222·96	64	594·56	104	966·16	144	1337·76	184	1709·36
25	232·25	65	603·85	105	975·45	145	1347·05	185	1718·65
26	241·54	66	613·14	106	984·74	146	1356·34	186	1727·94
27	250·83	67	622·43	107	994·03	147	1365·63	187	1737·23
28	260·12	68	631·72	108	1003·32	148	1374·92	188	1746·52
29	269·41	69	641·01	109	1012·61	149	1384·21	189	1755·81
30	278·70	**70**	650·30	**110**	1021·90	**150**	1393·50	**190**	1765·10
31	287·99	71	659·59	111	1031·19	151	1402·79	191	1774·39
32	297·28	72	668·88	112	1040·48	152	1412·08	192	1783·68
33	306·57	73	678·17	113	1049·77	153	1421·37	193	1792·97
34	315·86	74	687·46	114	1059·06	154	1430·66	194	1802·26
35	325·15	75	696·75	115	1068·35	155	1439·95	195	1811·55
36	334·44	76	706·04	116	1077·64	156	1449·24	196	1820·84
37	343·73	77	715·33	117	1086·93	157	1458·53	197	1830·13
38	353·02	78	724·62	118	1096·22	158	1467·82	198	1839·42
39	362·31	79	733·91	119	1105·51	159	1477·11	199	1848·71
40	371·60	**80**	743·20	**120**	1114·80	**160**	1486·40	**200**	1858·00

$\frac{1}{16}=0{\cdot}58^{1}$ $\frac{1}{9}=1{\cdot}03^{2}$ $\frac{1}{8}=1{\cdot}16^{1}$ $\frac{1}{6}=1{\cdot}54^{8}$ $\frac{1}{5}=1{\cdot}85^{8}$ $\frac{1}{4}=2{\cdot}32^{3}$ $\frac{1}{3}=3{\cdot}09^{7}$

$\frac{3}{8}=3{\cdot}48^{4}$ $\frac{1}{2}=4{\cdot}64^{5}$ $\frac{5}{8}=5{\cdot}80^{6}$ $\frac{3}{4}=6{\cdot}96^{8}$ $\frac{7}{8}=8{\cdot}12^{9}$ $\frac{2}{3}=6{\cdot}19^{3}$ $\frac{5}{6}=7{\cdot}74^{2}$

$\frac{3}{16}=1{\cdot}74^{2}$ $\frac{5}{16}=2{\cdot}90^{3}$ $\frac{7}{16}=4{\cdot}06^{4}$ $\frac{9}{16}=5{\cdot}22^{6}$ $\frac{11}{16}=6{\cdot}38^{7}$ $\frac{13}{16}=7{\cdot}54^{8}$ $\frac{15}{16}=8{\cdot}70^{9}$

1 square foot=0·0929 square metres

(0.102=1£) **£9.81**

1	9·81	41	402·21	81	794·61	121	1187·01	161	1579·41
2	19·62	42	412·02	82	804·42	122	1196·82	162	1589·22
3	29·43	43	421·83	83	814·23	123	1206·63	163	1599·03
4	39·24	44	431·64	84	824·04	124	1216·44	164	1608·84
5	49·05	45	441·45	85	833·85	125	1226·25	165	1618·65
6	58·86	46	451·26	86	843·66	126	1236·06	166	1628·46
7	68·67	47	461·07	87	853·47	127	1245·87	167	1638·27
8	78·48	48	470·88	88	863·28	128	1255·68	168	1648·08
9	88·29	49	480·69	89	873·09	129	1265·49	169	1657·89
10	98·10	**50**	490·50	**90**	882·90	**130**	1275·30	**170**	1667·70
11	107·91	51	500·31	91	892·71	131	1285·11	171	1677·51
12	117·72	52	510·12	92	902·52	132	1294·92	172	1687·32
13	127·53	53	519·93	93	912·33	133	1304·73	173	1697·13
14	137·34	54	529·74	94	922·14	134	1314·54	174	1706·94
15	147·15	55	539·55	95	931·95	135	1324·35	175	1716·75
16	156·96	56	549·36	96	941·76	136	1334·16	176	1726·56
17	166·77	57	559·17	97	951·57	137	1343·97	177	1736·37
18	176·58	58	568·98	98	961·38	138	1353·78	178	1746·18
19	186·39	59	578·79	99	971·19	139	1363·59	179	1755·99
20	196·20	**60**	588·60	**100**	981·00	**140**	1373·40	**180**	1765·80
21	206·01	61	598·41	101	990·81	141	1383·21	181	1775·61
22	215·82	62	608·22	102	1000·62	142	1393·02	182	1785·42
23	225·63	63	618·03	103	1010·43	143	1402·83	183	1795·23
24	235·44	64	627·84	104	1020·24	144	1412·64	184	1805·04
25	245·25	65	637·65	105	1030·05	145	1422·45	185	1814·85
26	255·06	66	647·46	106	1039·86	146	1432·26	186	1824·66
27	264·87	67	657·27	107	1049·67	147	1442·07	187	1834·47
28	274·68	68	667·08	108	1059·48	148	1451·88	188	1844·28
29	284·49	69	676·89	109	1069·29	149	1461·69	189	1854·09
30	294·30	**70**	686·70	**110**	1079·10	**150**	1471·50	**190**	1863·90
31	304·11	71	696·51	111	1088·91	151	1481·31	191	1873·71
32	313·92	72	706·32	112	1098·72	152	1491·12	192	1883·52
33	323·73	73	716·13	113	1108·53	153	1500·93	193	1893·33
34	333·54	74	725·94	114	1118·34	154	1510·74	194	1903·14
35	343·35	75	735·75	115	1128·15	155	1520·55	195	1912·95
36	353·16	76	745·56	116	1137·96	156	1530·36	196	1922·76
37	362·97	77	755·37	117	1147·77	157	1540·17	197	1932·57
38	372·78	78	765·18	118	1157·58	158	1549·98	198	1942·38
39	382·59	79	774·99	119	1167·39	159	1559·79	199	1952·19
40	392·40	**80**	784·80	**120**	1177·20	**160**	1569·60	**200**	1962·00

$\frac{1}{16}=0{\cdot}61^{3}$ $\frac{1}{9}=1{\cdot}09^{0}$ $\frac{1}{8}=1{\cdot}22^{6}$ $\frac{1}{6}=1{\cdot}63^{5}$ $\frac{1}{5}=1{\cdot}96^{2}$ $\frac{1}{4}=2{\cdot}45^{3}$ $\frac{1}{3}=3{\cdot}27^{0}$

$\frac{3}{8}=3{\cdot}67^{9}$ $\frac{1}{2}=4{\cdot}90^{5}$ $\frac{5}{8}=6{\cdot}13^{1}$ $\frac{3}{4}=7{\cdot}35^{8}$ $\frac{7}{8}=8{\cdot}58^{4}$ $\frac{2}{3}=6{\cdot}54^{0}$ $\frac{5}{6}=8{\cdot}17^{5}$

$\frac{3}{16}=1{\cdot}83^{9}$ $\frac{5}{16}=3{\cdot}06^{6}$ $\frac{7}{16}=4{\cdot}29^{2}$ $\frac{9}{16}=5{\cdot}51^{8}$ $\frac{11}{16}=6{\cdot}74^{4}$ $\frac{13}{16}=7{\cdot}97^{1}$ $\frac{15}{16}=9{\cdot}19^{7}$

The acceleration due to gravity=9·81 metres per second^{2}

(0.102=1£) **£9.84**

1	9·84	41	403·44	81	797·04	121	1190·64	161	1584·24
2	19·68	42	413·28	82	806·88	122	1200·48	162	1594·08
3	29·52	43	423·12	83	816·72	123	1210·32	163	1603·92
4	39·36	44	432·96	84	826·56	124	1220·16	164	1613·76
5	49·20	45	442·80	85	836·40	125	1230·00	165	1623·60
6	59·04	46	452·64	86	846·24	126	1239·84	166	1633·44
7	68·88	47	462·48	87	856·08	127	1249·68	167	1643·28
8	78·72	48	472·32	88	865·92	128	1259·52	168	1653·12
9	88·56	49	482·16	89	875·76	129	1269·36	169	1662·96
10	98·40	**50**	492·00	**90**	885·60	**130**	1279·20	**170**	1672·80
11	108·24	51	501·84	91	895·44	131	1289·04	171	1682·64
12	118·08	52	511·68	92	905·28	132	1298·88	172	1692·48
13	127·92	53	521·52	93	915·12	133	1308·72	173	1702·32
14	137·76	54	531·36	94	924·96	134	1318·56	174	1712·16
15	147·60	55	541·20	95	934·80	135	1328·40	175	1722·00
16	157·44	56	551·04	96	944·64	136	1338·24	176	1731·84
17	167·28	57	560·88	97	954·48	137	1348·08	177	1741·68
18	177·12	58	570·72	98	964·32	138	1357·92	178	1751·52
19	186·96	59	580·56	99	974·16	139	1367·76	179	1761·36
20	196·80	**60**	590·40	**100**	984·00	**140**	1377·60	**180**	1771·20
21	206·64	61	600·24	101	993·84	141	1387·44	181	1781·04
22	216·48	62	610·08	102	1003·68	142	1397·28	182	1790·88
23	226·32	63	619·92	103	1013·52	143	1407·12	183	1800·72
24	236·16	64	629·76	104	1023·36	144	1416·96	184	1810·56
25	246·00	65	639·60	105	1033·20	145	1426·80	185	1820·40
26	255·84	66	649·44	106	1043·04	146	1436·64	186	1830·24
27	265·68	67	659·28	107	1052·88	147	1446·48	187	1840·08
28	275·52	68	669·12	108	1062·72	148	1456·32	188	1849·92
29	285·36	69	678·96	109	1072·56	149	1466·16	189	1859·76
30	295·20	**70**	688·80	**110**	1082·40	**150**	1476·00	**190**	1869·60
31	305·04	71	698·64	111	1092·24	151	1485·84	191	1879·44
32	314·88	72	708·48	112	1102·08	152	1495·68	192	1889·28
33	324·72	73	718·32	113	1111·92	153	1505·52	193	1899·12
34	334·56	74	728·16	114	1121·76	154	1515·36	194	1908·96
35	344·40	75	738·00	115	1131·60	155	1525·20	195	1918·80
36	354·24	76	747·84	116	1141·44	156	1535·04	196	1928·64
37	364·08	77	757·68	117	1151·28	157	1544·88	197	1938·48
38	373·92	78	767·52	118	1161·12	158	1554·72	198	1948·32
39	383·76	79	777·36	119	1170·96	159	1564·56	199	1958·16
40	393·60	**80**	787·20	**120**	1180·80	**160**	1574·40	**200**	1968·00

$\frac{1}{16}=0{\cdot}61^{5}$ $\frac{1}{9}=1{\cdot}09^{3}$ $\frac{1}{8}=1{\cdot}23^{0}$ $\frac{1}{6}=1{\cdot}64^{0}$ $\frac{1}{5}=1{\cdot}96^{8}$ $\frac{1}{4}=2{\cdot}46^{0}$ $\frac{1}{3}=3{\cdot}28^{0}$

$\frac{3}{8}=3{\cdot}69^{0}$ $\frac{1}{2}=4{\cdot}92^{0}$ $\frac{5}{8}=6{\cdot}15^{0}$ $\frac{3}{4}=7{\cdot}38^{0}$ $\frac{7}{8}=8{\cdot}61^{0}$ $\frac{2}{3}=6{\cdot}56^{0}$ $\frac{5}{6}=8{\cdot}20^{0}$

$\frac{3}{16}=1{\cdot}84^{5}$ $\frac{5}{16}=3{\cdot}07^{5}$ $\frac{7}{16}=4{\cdot}30^{5}$ $\frac{9}{16}=5{\cdot}53^{5}$ $\frac{11}{16}=6{\cdot}76^{5}$ $\frac{13}{16}=7{\cdot}99^{5}$ $\frac{15}{16}=9{\cdot}22^{5}$

1 tonne=0·984 tons

(0.024=1£) **£41.25**

1	41·25	41	1691·25	81	3341·25	121	4991·25	161	6641·25
2	82·50	42	1732·50	82	3382·50	122	5032·50	162	6682·50
3	123·75	43	1773·75	83	3423·75	123	5073·75	163	6723·75
4	165·00	44	1815·00	84	3465·00	124	5115·00	164	6765·00
5	206·25	45	1856·25	85	3506·25	125	5156·25	165	6806·25
6	247·50	46	1897·50	86	3547·50	126	5197·50	166	6847·50
7	288·75	47	1938·75	87	3588·75	127	5238·75	167	6888·75
8	330·00	48	1980·00	88	3630·00	128	5280·00	168	6930·00
9	371·25	49	2021·25	89	3671·25	129	5321·25	169	6971·25
10	412·50	**50**	2062·50	**90**	3712·50	**130**	5362·50	**170**	7012·50
11	453·75	51	2103·75	91	3753·75	131	5403·75	171	7053·75
12	495·00	52	2145·00	92	3795·00	132	5445·00	172	7095·00
13	536·25	53	2186·25	93	3836·25	133	5486·25	173	7136·25
14	577·50	54	2227·50	94	3877·50	134	5527·50	174	7177·50
15	618·75	55	2268·75	95	3918·75	135	5568·75	175	7218·75
16	660·00	56	2310·00	96	3960·00	136	5610·00	176	7260·00
17	701·25	57	2351·25	97	4001·25	137	5651·25	177	7301·25
18	742·50	58	2392·50	98	4042·50	138	5692·50	178	7342·50
19	783·75	59	2433·75	99	4083·75	139	5733·75	179	7383·75
20	825·00	**60**	2475·00	**100**	4125·00	**140**	5775·00	**180**	7425·00
21	866·25	61	2516·25	101	4166·25	141	5816·25	181	7466·25
22	907·50	62	2557·50	102	4207·50	142	5857·50	182	7507·50
23	948·75	63	2598·75	103	4248·75	143	5898·75	183	7548·75
24	990·00	64	2640·00	104	4290·00	144	5940·00	184	7590·00
25	1031·25	65	2681·25	105	4331·25	145	5981·25	185	7631·25
26	1072·50	66	2722·50	106	4372·50	146	6022·50	186	7672·50
27	1113·75	67	2763·75	107	4413·75	147	6063·75	187	7713·75
28	1155·00	68	2805·00	108	4455·00	148	6105·00	188	7755·00
29	1196·25	69	2846·25	109	4496·25	149	6146·25	189	7796·25
30	1237·50	**70**	2887·50	**110**	4537·50	**150**	6187·50	**190**	7837·50
31	1278·75	71	2928·75	111	4578·75	151	6228·75	191	7878·75
32	1320·00	72	2970·00	112	4620·00	152	6270·00	192	7920·00
33	1361·25	73	3011·25	113	4661·25	153	6311·25	193	7961·25
34	1402·50	74	3052·50	114	4702·50	154	6352·50	194	8002·50
35	1443·75	75	3093·75	115	4743·75	155	6393·75	195	8043·75
36	1485·00	76	3135·00	116	4785·00	156	6435·00	196	8085·00
37	1526·25	77	3176·25	117	4826·25	157	6476·25	197	8126·25
38	1567·50	78	3217·50	118	4867·50	158	6517·50	198	8167·50
39	1608·75	79	3258·75	119	4908·75	159	6558·75	199	8208·75
40	1650·00	**80**	3300·00	**120**	4950·00	**160**	6600·00	**200**	8250·00

$\frac{1}{16}=2{\cdot}57^{8}$ $\frac{1}{9}=4{\cdot}58^{3}$ $\frac{1}{8}=5{\cdot}15^{6}$ $\frac{1}{6}=6{\cdot}87^{5}$ $\frac{1}{5}=8{\cdot}25^{0}$ $\frac{1}{4}=10{\cdot}31^{2}$ $\frac{1}{3}=13{\cdot}75^{0}$

$\frac{3}{8}=15{\cdot}46^{8}$ $\frac{1}{2}=20{\cdot}62^{5}$ $\frac{5}{8}=25{\cdot}78^{1}$ $\frac{3}{4}=30{\cdot}93^{7}$ $\frac{7}{8}=36{\cdot}09^{3}$ $\frac{2}{3}=27{\cdot}50^{0}$ $\frac{5}{6}=34{\cdot}37^{5}$

$\frac{3}{16}=7{\cdot}73^{4}$ $\frac{5}{16}=12{\cdot}89^{0}$ $\frac{7}{16}=18{\cdot}04^{7}$ $\frac{9}{16}=23{\cdot}20^{3}$ $\frac{11}{16}=28{\cdot}35^{9}$ $\frac{13}{16}=33{\cdot}51^{6}$ $\frac{15}{16}=38{\cdot}67^{2}$

Standard rate of income tax on units of £100, 1967–1970